Motor Proteins

Front cover: double immunofluorescence staining of PtK_1 cells, showing that anti-kinesin stains membrane-bound vesicles associated with the microtubular network. See article by Brady and Pfister.

Half title: myosin I and myosin II differential immunolocalisation by confocal microscopy of avian intestinal epithelial brush border cells. Myosin I is more closely apposed to the membrane. See article by Collins and Matsudaira.

ISBN: 0 948601 29 9

Motor Proteins

A volume based on the EMBO Workshop
Cambridge, September 1990

Organized and Edited by

R. A. Cross and **J. Kendrick-Jones**
Medical Research Council
Laboratory of Molecular Biology
Cambridge

SUPPLEMENT 14 1991
JOURNAL OF CELL SCIENCE
Published by THE COMPANY OF BIOLOGISTS LIMITED, Cambridge

Typeset, Printed and Published by
THE COMPANY OF BIOLOGISTS LIMITED
Department of Zoology, University of Cambridge, Downing Street,
Cambridge CB2 3EJ

ISBN: 0 948601 29 9

JOURNAL OF CELL SCIENCE SUPPLEMENTS

This volume is the latest in a continuing series on important topics in cell and molecular biology. All supplements are available free to subscribers to *Journal of Cell Science* or may be purchased separately from Portland Press Ltd, PO Box 32, Commerce Way, Colchester CO2 8HP, UK.

1 (1984) Higher Order Structure in the Nucleus (US$23.00)
2 (1985) The Cell Surface in Plant Growth and Development (US$30.00)
3 (1985) Growth Factors: Structure and Function (US$30.00)
4 (1986) Prospects in Cell Biology (US$30.00) **SOLD OUT**
5 (1986) The Cytoskeleton: Cell Function and Organization (US$30.00) **SOLD OUT**
6 (1987) Molecular Biology of DNA Repair (US$70.00)
7 (1987) Virus Replication and Genome Interactions (US$70.00)
8 (1987) Cell Behaviour: Shape, Adhesion and Motility (US$60.00)
9 (1988) Macrophage Plasma Membrane Receptors: Structure and Function (US$50.00)
10 (1988) Stem Cells (US$65.00)
11 (1989) Protein Targeting (US$65.00)
12 (1989) The Cell Cycle (US$60.00)
13 (1990) Growth Factors in Cell and Developmental Biology (US$55.00)
14 (1991) Motor Proteins (US$60.00)

Preface

Molecular motors are miniscule force-generating machines that move along tracks inside the cell, doing work against a load. Evidence is accumulating that cells recruit appropriate motors whenever they need to change their shape, move, phagocytose or divide, as well as for the intracellular translocation of organelles and membranous vesicles. This EMBO workshop brought together people working on many different aspects of motors, principally in order to encourage the spread of emerging techniques. This technical emphasis continues into the pages of this book, for which we have asked contributors to provide a critical commentary on their methods, as well as clear instructions for their use.

Recent technical advances have made it possible to measure the performance of single molecules, and thereby to narrow considerably the range of thermodynamically possible mechanisms for their action. Contributions to the book deal with measurements of force production, the speed of motion and the directionality of force, for both single molecules and polymeric assemblies of motors. Several contributors describe the effects on performance of other track-binding proteins (including other motors) and/or of experimentally altering the geometry of interaction (myosin at least can be twisted through 180 degrees and still work). Other contributions concern mechanisms for the regulated self-assembly of tracks (which on occasion may itself work as a motor) and for the regulated self-assembly of myosin polymers. Lastly, a crucial problem is to work out the mechanism by which motors are deployed appropriately by the cell to produce the right amount of force at the right time and in the right place. Contributions concerning cells and cell extracts address these questions.

One motor molecule, exerting about 1 pN, will move a vesicle with strict polarity along its track, an actin filament or a microtubule. In a different context, the same molecules are very highly collaborative: for a muscle to lift 1 kg, about 10^{13} myosin motors must share the effort. Both environments require molecular motors to be able efficiently to transduce energy whilst tolerating large elastic deformations. There is no obviously analogous man-made device and, perhaps in consequence, the precise mechanism by which a motor exerts force on its track is still unknown. What is needed is nothing less than a complete molecular explanation of energy transduction through the motor molecule. A few years ago there was doubt that this was possible. This is no longer the case.

R. A. Cross
J. Kendrick Jones
Cambridge, September 1990

CONTENTS

Kinetic analysis of regulated myosin ATPase activity using single and limited turnover assays

R. J. ANKRETT, A. R. WALMSLEY and C. R. BAGSHAW

Department of Biochemistry, University of Leicester, Leicester LE1 7RH, UK

Summary

Myosin from molluscan adductor muscle is regulated directly by Ca^{2+} binding. In the absence of Ca^{2+} the ATPase activity is greatly inhibited. We review the application of transient kinetic methods to this system and show how they can be simple to perform and less ambiguous than steady-state methods.

Key words: transient kinetics, calcium, molluscan myosin.

Introduction

A characteristic feature of motor proteins is the need for the efficient inhibition of their ATPase activity during periods of inactivity. In turn, this property provides a way of characterising regulatory systems *in vitro*; for example, the Ca^{2+} sensitivity of the steady-state ATPase activity provided the initial clue in the discovery of the diverse mechanisms involved in the regulation of actomyosin-based motors. The degree of control based on steady-state measurement can, however, be ambiguous. Thus a five-fold inhibition of the ATPase on removal of Ca^{2+} might indicate that all the molecules are subject to limited control, or that 20 % of the molecules are unregulated and display a permanently high ATPase activity while the remaining 80 % are inhibited by an unknown factor of $\geqslant$5-fold. This problem can be overcome by investigating a single or limited turnover of the ATPase where the putative regulated and unregulated populations are resolved in time.

Provided the enzyme active site concentration exceeds the K_m of the reaction for ATP, then on addition of stoichiometric amounts of ATP, the binding phase will be nearly complete before the enzyme–nucleotide intermediates begin to decay. The latter process can then be analysed in terms of a multiexponential decay, the amplitudes and rate constants of which reflect the relative concentrations and ATPase activities of each population. Alternatively, a 2- to 20-fold molar excess of ATP may be added to the enzyme, so that the unregulated population contributes a short steady-state phase while the regulated population undergoes a single turnover in the final phase of the reaction (Fig. 1). Variations of this method include chasing a fluorescent ATP analogue with a non-fluorescent one to follow the fate of enzyme-bound nucleotide, or addition of Ca^{2+} during the final phase to determine the degree of activation directly. All that is required for the application of these methodologies is a means for detecting the enzyme–nucleotide intermediate. In many cases this need not entail specialised equipment and indeed the measurement can be simpler and quicker than performing steady-state kinetic analysis.

In this paper we summarise the methods that we have used to study the regulatory system of the myosin ATPase of molluscan striated adductor muscle. Steady-state ATPase assays were instrumental in demonstrating the role of the myosin regulatory light chain subunits in the suppression of the ATPase activity in the absence of Ca^{2+} (Szent-Györgyi *et al.* 1973), but transient methods have provided new insight into the coupling between the Ca^{2+} and ATPase sites on the myosin head and the degree of regulation achieved by this mechanism.

Methods and Results

The application of transient-state kinetic methods requires the ability to distinguish between ligand-bound and ligand-free states of the enzyme. While the following methods refer specifically to the myosin ATPase, many of the principles may well apply to other motor proteins (cf. kinesin, Hackney *et al.* 1989).

Physical separation

The ATPase activity of the inhibited states of the myosin ATPase are sufficiently slow that the kinetics of product release may be followed by physical separation of the protein from the ligands using rapid gel permeation chromatography. For example, column centrifugation was used to follow a single turnover of scallop striated adductor muscle heavy meromyosin (HMM). Stoichiometric [γ-^{32}P]-ATP was added to 20 μM HMM and after incubation times varying from 1 to 30 min, a 0.1 ml sample was applied to a 1 ml G-50 Sephadex column in a tuberculin syringe. By spinning for 1 min in a benchtop centrifuge, the HMM together with bound P_i was eluted, while the free P_i remained in the column. The rate constant for P_i release from the regulated population was determined as 0.002 s^{-1} in the absence of Ca^{2+}, but >0.02 s^{-1} in its presence (Wells and Bagshaw, 1985). A comparable time resolution of separation (30 s) was also achieved using a small (5×40 mm) Superose 6 column (Pharmacia) attached to an HPLC pump operating at a flow rate of 1 ml min^{-1}. Nevertheless both of these separation methods are limited to processes with rate constants of <0.02 s^{-1}. The optical methods described below provide a much better time resolution but they are indirect.

Turbidity and light scattering assays

A motor protein is almost invariably involved in the transient association with another protein during the course of its mechanochemical cycle. In solution this can

Journal of Cell Science, Supplement 14, 1–5 (1991)

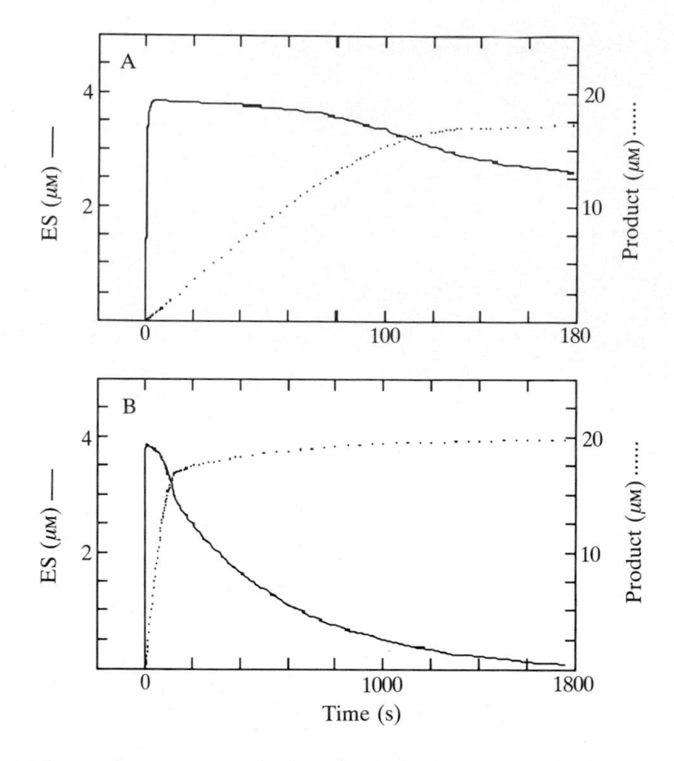

Fig. 1. Computer simulation of a limited turnover of substrate by two populations of enzyme. 20 μM substrate (S) was added to 4 μM enzyme (E) which comprised (A) a 25 % active (unregulated) population (k_{cat}=0.2 s^{-1}) and (B) a 75 % inhibited (regulated) population (k_{cat}=0.002 s^{-1}). Both populations were assigned a second order association rate constant of 10^5 M^{-1} s^{-1}. All reverse reactions were ignored. These parameters were assigned to simulate the turnover of FTP by scallop heavy meromyosin in the absence of Ca^{2+}. The plots show the time course of formation and decay of the total ES intermediate and the formation of product (P). Note that the bulk of the substrate (about 17 μM) is turned over by the unregulated population during the initial steady-state phase (first 90 s) (A), while the regulated population undergoes an effective single turnover and yields an exponential profile with a rate constant=0.002 s^{-1} (i.e. k_{cat}) (B). The amplitudes of the fast and slow phases of the ES decay reflect the 1:3 ratio of the relative concentrations of the unregulated and regulated populations.

lead to a change in light scattering or turbidity of the sample. These signals may therefore be used to follow the kinetics of association or dissociation of the proteins, which may in turn be limited by some other chemical step.

In the absence of ATP, scallop HMM formed a rigor complex with actin which was turbid. Addition of a molar excess of ATP to the complex resulted in its rapid dissociation, and the solution remained clear during the steady-state hydrolysis phase (since the predominant intermediate is the dissociated M.ADP.P$_i$ species). When the ATP was exhausted the actomyosin rigor complex quickly reformed to give a turbidity similar to its initial value. In the absence of Ca^{2+} the final phase was more complicated. Only 20 % of the turbidity was rapidly regained, the remaining 80 % recovered slowly with an exponential profile (k=0.002 s^{-1}). It was concluded that the HMM sample comprised a mixture of 20 % unregulated molecules with a permanently high actin-activated ATPase activity (1.2 s^{-1}) and 80 % regulated molecules whose ATPase was suppressed some 650-fold in the absence of Ca^{2+} (Wells and Bagshaw, 1984). In the absence of Ca^{2+}, the steady-state ATPase was dominated by the unregulated fraction which contributed a turnover of 0.24 s^{-1} (0.2×1.2 s^{-1}) while the regulated fraction made a negligible contribution. Thus the degree of Ca^{2+} activation of the steady-state ATPase rate (5×) reflected the proportion of regulated molecules rather than the true degree of activation of the turnover rate itself. In conjunction with the rapid column separation experiments described above, it was concluded that Ca^{2+} was controlling the rate of P$_i$ release, rather than the actin association steps which remained practically Ca^{2+} independent.

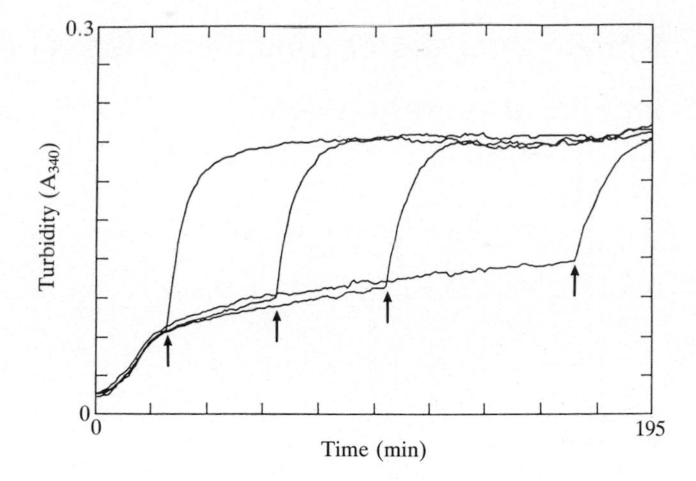

Fig. 2. Filament formation by 10S scallop myosin on addition of Ca^{2+}. Scallop striated adductor myosin, initially in high [NaCl], was diluted down to conditions which favour the 10S conformer (175 mM NaCl, 5 mM $MgCl_2$, 75 μM ATP, 0.1 mM EGTA, 25 mM imidazole at pH 7.0 and 20° C). Filments were removed by brief centrifugation and the supernatant was portioned into four cuvettes in a Perkin Elmer Lambda 5 spectrophotometer. Turbidity was monitored at 340 nm for all four samples using an autochanger accessory. In the absence of Ca^{2+} the majority of the molecules were trapped in the 10S monomeric conformation and filament formation was very slow. Addition of 0.1 mM free Ca^{2+} after 25, 62, 102 or 169 min (indicated by arrows) caused an immediate rise in turbidity with about the same rate constant (0.002 s^{-1}) and end point but with a progressively reduced amplitude.

Turbidity changes are also associated with the formation of myosin filaments from monomeric myosin. Myosins from muscles which display thick-filament regulation share the characteristic of forming a folded, 10S conformation, which traps nucleotide at the active site and is inert with respect to actin activation and filament assembly (Cross *et al.* 1988). In order to form filaments, the 10S conformer must first release its nucleotide and straighten to the elongated 6S conformer. Filament formation, as monitored by a turbidity change, therefore provides a measure of the kinetics of product release from the 10S conformer. Using this approach, we have recently shown that scallop striated adductor muscle myosin is capable of forming a 10S conformer. This conformer is destabilised by the addition of Ca^{2+}, the removal of the regulatory light chains or the labelling of the heavy chain reactive thiol (Ankrett *et al.* 1990). Fig. 2 shows the turbidity profiles of scallop myosin which had been solubilised by the addition of a 20 fold molar excess of ATP in the absence of Ca^{2+}. As the ATP was exhausted there was an initial small increase in turbidity corresponding to filament formation by a population of extended 6S

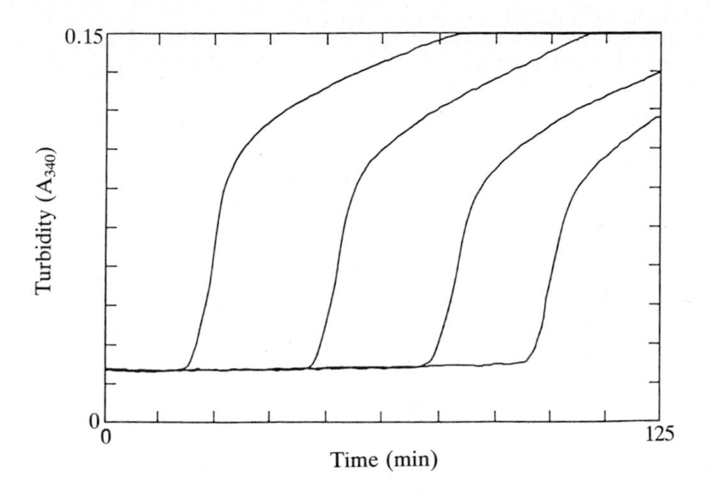

Fig. 3. Effect of ATP on the formation of filaments by scallop myosin. 10S scallop myosin was generated and filament formation was monitored as described in Fig. 2, but using an initial [ATP] of 100 μM. The profiles show the effect of adding additional ATP (0, 20, 40 and 60 μM, left to right respectively) to the assay following the centrifugation step. The additional ATP prolongs the steady-state phase, but has little effect on the amplitude of the rapid recovery (attributed to the active 6S population which dominates the turnover) or the final slow decay (attributed to unfolding and filament formation by the 10S population).

molecules, followed by a much slower rise ($t_{\frac{1}{2}}$ of several hours) as the 10S monomers unfolded and participated in filament assembly. When Ca^{2+} was added to the assay at any point in time there was a rapid increase in the rate of filament formation owing to the activation of the unfolding transition. The pre-existing unfolded molecules dominated the steady-state ATPase and accounted for the initial phase of the reaction which led to the exhaustion of the ATP. Thus, if additional ATP was added to the assay in the absence of Ca^{2+}, filament assembly was delayed by a time proportional to the added [ATP], but the final slow turnover of ATP by the 10S conformer was unaffected (Fig. 3). In this case it is clear that conventional steady-state assays of the preparation would not yield any information about the kinetics of product release from the 10S conformer. From these studies it is concluded that the 6S–10S transition is too slow to be involved in regulation of contraction, but may be involved in control of filament assembly during growth and development.

The choice of measuring turbidity or light scattering depends on several experimental factors. In both cases a wavelength (typically 320 to 340 nm) is selected just beyond the protein absorption band and preferably in the range where the tungsten lamp can be used. Light scattering can be measured in commercial fluorimeters with good sensitivity, but it is more prone to artifacts from aggregates and arc light sources tend to be unstable. Furthermore the signal is recorded on an arbitrary scale which makes comparison between experiments performed on different instruments more difficult. Turbidity can be recorded using a conventional spectrophotometer, although the reading will depend to a limited extent on the degree to which the optics reject forward low-angle scattering. For the kinds of measurement described here, the amplitudes are interpreted in relative terms and therefore these instrumental characteristics are rarely critical.

Fluorescence assays

Myosin is not particularly specific with regard to its substrate, ATP, and a number of fluorescent analogues have been developed which show large enhancements in the emission signal on binding to the active site. We have used formycin triphosphate (FTP) extensively, which is commercially available from Sigma Chemical Co., although it is considerably cheaper to purchase formycin monophosphate and convert it to the triphosphate enzymically (Jackson and Bagshaw, 1988*a*). By selecting a λ_{ex}=313 nm, the background contribution from protein fluorescence is reduced while a convenient line may be selected when using a Hg arc lamp. For extension into *in vivo* measurements, fluorophores responding to longer wavelengths may be required, such as mant-ATP (λ_{ex}=350 nm; Hiratsuka, 1985).

In a typical experiment, a 5-fold excess of FTP was added to scallop HMM and the fluorescence emission was monitored at 350 nm. This ratio is a compromise. When less was used (or a true single turnover measured) the FTP binding phase was not complete before the product decay started (particularly when manual addition was employed) and hence the ratio of the regulated and unregulated HMM fractions could not be determined. When more FTP was added, the background fluorescence from the free formycin nucleotide was higher and hence the change in signal during binding and turnover was reduced. In the absence of Ca^{2+} a limited FTP turnover assay has a similar profile to that observed using turbidity measurements described above. Following the binding phase there is a short steady-state phase of enhanced fluorescence during which time the unregulated HMM turns over the bulk of the FTP. A partial drop in fluorescence then occurs when the FTP is exhausted, leaving a final exponential phase of fluorescence decay as the products are released from the regulated fraction. The profile, however, is not easy to analyse since an undetermined amount of product FDP remains bound to the active site. For this reason, and also to obtain an estimate for the rate of product release from the unregulated fraction, we normally 'chase' the FTP turnover with a large excess of ATP during the brief steady-state phase (Jackson and Bagshaw, 1988*a*). This method is particularly successful with myosin because the back dissociation rate constant for bound triphosphate is extremely small (Bagshaw and Trentham, 1973) leaving turnover the only route available for the bound FTP. More recently we have used pyrophosphate (PP_i) as the non-fluorescent chasing agent since it is not hydrolysed and also causes less problems with solubilisation when assaying filamentous myosin (Walmsley *et al.* 1990; Ankrett *et al.* 1990). In the chase assay, the product release steps are revealed as an exponential with at least two phases corresponding to the unregulated and regulated myosin populations. The FTP turnover assays provide a convenient way of following product release steps and confirm the kinetic parameters deduced by the other assays described above. The initial association rate of FTP to myosin, and many other ATP requiring enzymes, is reduced by about 20-fold, but once bound the kinetics appear to mimic those of ATP closely. FTP also becomes trapped at the active site of 10S myosin and provides a sensitive assay for this state (Cross *et al.* 1988; Citi *et al.* 1989; Ankrett *et al.* 1990).

Fluorescence spectroscopy also provides a convenient means of following the kinetics of Ca^{2+} binding to proteins. In practice, many Ca^{2+}-binding proteins are near saturated at contaminant levels of Ca^{2+} and it is more

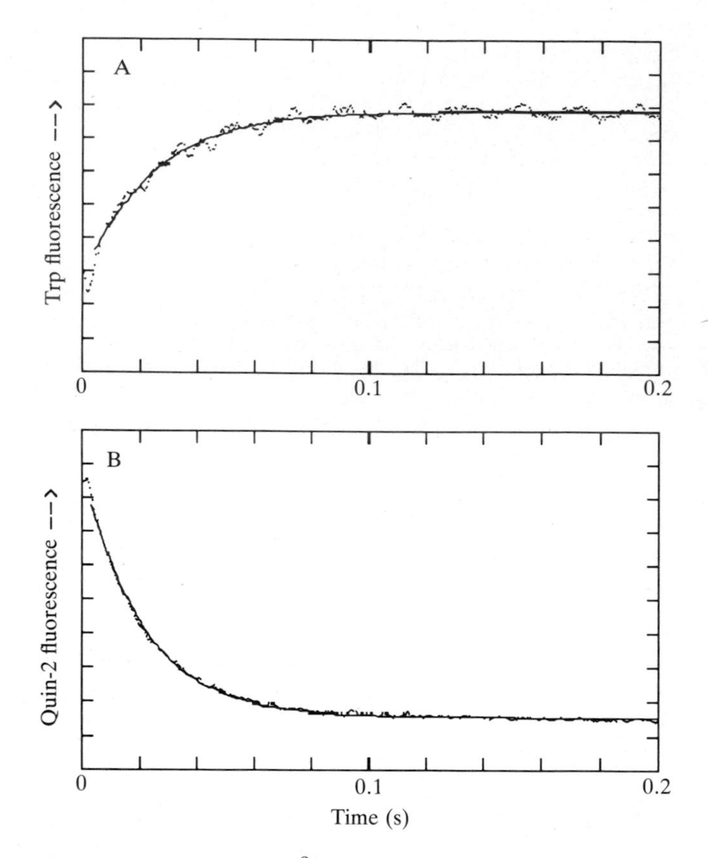

Fig. 4. The kinetics of Ca^{2+} release from scallop heavy meromyosin. (A) 10 μM heavy meromyosin plus 10 μM Ca^{2+} was mixed with 400 μM EGTA in an Applied Photophysics SF.17 MV Stopped Flow apparatus and protein fluorescence was monitored at λ_{ex}=295 nm and λ_{em}=350 nm. The observed rate constant for the release of Ca^{2+} was 40 s^{-1}. (B) Substituting 200 μM quin-2 for the EGTA provides better signal-to-noise (λ_{ex}=336 nm, λ_{em}=510 nm) and yielded an observed rate constant of 49 s^{-1}. Both experiments were carried out in 20 mM NaCl, 10 mM Tes at pH 7.5 and 20 °C. The experiments were repeated at different chelator concentrations and the true rate constant for the dissociation of Ca^{2+} from the protein was determined by extrapolation to infinite [chelalor] (cf. Jackson *et al.* 1987).

convenient to follow the kinetics of dissociation. In any event such measurements, in conjunction with the equilibrium constant, indicate that the association reaction would be too fast to measure by rapid mixing methods at the concentrations required to obtain an adequate signal. Intrinsic protein fluorescence (tryptophan or tyrosine) often provides a convenient probe. Fig. 4A shows the release of Ca^{2+} from the specific, high affinity sites of scallop heavy meromyosin on mixing with EGTA. Better signal-to-noise ratios can be obtained by using the fluorescent indicator, quin-2 as a combined chelator and probe (Fig. 4B). These procedures can potentially monitor different steps in the Ca^{2+} release mechanism, but in this case they are probably both limited by the same conformational change which occurs at about 40 s^{-1}.

Instrumental considerations

As mentioned above many of the key features of the regulatory mechanism of the molluscan myosin ATPase activity were evaluated using a standard spectrophotometer and fluorimeter. The recent introduction of ultramicrocuvettes taking 0.1 ml samples (Hellma Co.) has increased the scope of these measurements for cases where material is limiting. There are, however, a number of assays where more specialised approaches are warranted. The ATPase rates when activated by Ca^{2+} are generally too fast to be evaluated by manual mixing methods and require stopped-flow instruments for transient kinetic analysis. Such apparatus is available commercially (e.g. Applied Photophysics Ltd, Hi-Tech Ltd) or can be constructed in the laboratory from commercial components (e.g. Ealing Electro-optics Ltd, Oriel Corporation). In either case the light source and detector are modular and therefore can be arranged to suit the experiment at hand. For example, we have made extensive use of these components in conjunction with a custom-built fluorescence housing which accommodates a standard cuvette (Jackson and Bagshaw 1988*b*). The instrument comprises a light guide input for excitation, two 90° photomultipliers for fluorescence and/or light scattering detection and a photodiode for transmittance detection. The latter has proved of great value for detecting artifacts in the fluorescence signal due to inner filter effects, arc lamp jumps or changes in the assembly state of the myosin. Components may be added to the cuvette through a syringe needle and the contents stirred by an overhead paddle, allowing readings to be obtained within about 2 s.

The spectroscopic signals are digitised and stored using a microcomputer for subsequent analysis. The main problem facing the experimentalist these days is not so much finding an appropriate system for data capture but achieving compatability between different instruments so that data may be moved freely between systems for analysis and presentation. The latter can be achieved by hardwire connections through the serial ports and a communications package for ASCII files or by using emulation software.

Transient kinetic profiles usually comprise one or more exponential phases and non-linear, least squares analysis can be used to extract the amplitudes and rate constants. The efficiency of the control of the scallop myosin ATPase leads to a wide dynamic range in the observed transients so that selecting an appropriate timebase for the record can cause difficulties. In these critical situations we have digitised records using a logarithmic timebase to ensure that the fast and slow phases are defined by a sufficient number of points and are subject to even weighting in the analysis (Walmsley and Bagshaw, 1989; Walmsley *et al.* 1990).

Computers also find extensive use in simulation of transients in mechanisms for which the analytical solution is difficult or impossible to derive (e.g. Fig. 1). The principle here is to calculate the change in concentration of each species over a small time interval from the differential rate equations, assuming a linear relationship, for example:

$$\Delta[\mathrm{X}] = -k.[\mathrm{X}].\Delta t, \qquad (1)$$

and to construct the entire profile by repeated calculation,

$$[\mathrm{X}]_{\Delta t} = [\mathrm{X}] - \Delta[\mathrm{X}]. \qquad (2)$$

As in all hypothesis testing, a match between the observed and simulated records does not prove that a mechanism is correct, but an inappropriate mechanism may be ruled out.

We are currently testing whether the trapped nucleotide states of scallop myosin (M) observed under relaxing conditions (i.e. $-Ca^{2+}$) represent a kinetic dead-end (Eqn. 3), rather than an in-line intermediate of the pathway.

$$\begin{array}{c} M.^{*}ADP.P_i \\ \updownarrow \\ M+ATP \leftrightarrow M.ATP \leftrightarrow M.ADP.P_i \leftrightarrow M+ADP+P_i \end{array} \qquad (3)$$

These mechanisms only become distinguishable if the rate constant for trapping is comparable or slower than the rate constant for turnover, in which case the pathway will take several turnovers to build up the trapped state ($M.^{*}ADP.P_i$) to its steady-state value. Thus the extent of trapping, as estimated from the amplitude of the slow phase of product release, will be dependent on the time of chasing the system during a limited turnover experiment.

Conclusion

Motor proteins require rigorous control so as not to waste chemical energy during periods of inactivity. *In vitro* studies of myosin have now revealed a number of preparations where long-lived intermediates exist under relaxing conditions, but which are capable of rapid activation on addition of Ca^{2+}. The true dynamic range of these preparations has only been revealed by transient state methods. However, because the time scale of these transients extends over several minutes, or even hours, specialised rapid reaction equipment is not required for their detection. This methodology should find application in other motor systems.

We are grateful to the SERC for financial support.

References

Ankrett, R. J., Rowe, A. J. and Bagshaw, C. R. (1990). The characterisation of a 10S conformer of myosin from scallop striated muscle. *J. Mus. Res. Cell Motil.* **11**, 60–61.

Bagshaw, C. R. and Trentham, D. R. (1973). The reversibility of adenosine tripohosphate cleavage by myosin. *Biochem. J.* **133**, 323–328.

Citi, S., Cross, R. A., Bagshaw, C. R. and Kendrick-Jones, J. (1989). Parallel modulation of brush border myosin conformation and enzyme activity induced by monoclonal antibodies. *J. Cell Biol.* **109**, 549–556.

Cross, R. A., Jackson, A. P., Citi, S., Kendrick-Jones, J. and Bagshaw, C. R. (1988). Active site trapping of nucleotides by smooth and non-muscle myosins. *J. molec. Biol.* **203**, 173–181.

Hackney, D. D., Malik, A. and Wright, K. W. (1989). Nucleotide-free kinesin hydrolyses ATP with burst kinetics. *J. biol. Chem.* **264**, 15 943–15 948.

Hiratsuka, T. (1983). New ribose-modified fluorescent analogs of adenine and guanine nucleotides available as substrates for various enzymes. *Biochim. biophys. Acta* **742**, 496–508.

Jackson, A. P. and Bagshaw, C. R. (1988*a*). Kinetic trapping of intermediates of the scallop heavy meromyosin adenosine triphosphatase reaction revealed by formycin nucleotides. *Biochem. J.* **251**, 527–540.

Jackson, A. P. and Bagshaw, C. R. (1988*b*). Transient kinetic studies of the adenosine triphosphatase activity of scallop heavy meromyosin. *Biochem. J.* **251**, 515–526.

Jackson, A. P., Timmerman, M. P., Bagshaw, C. R. and Ashley, C. C. (1987). The kinetics of calcium binding to fura-2 and indo-1. *FEBS Lett.* **216**, 35–39.

Szent-Györgyi, A. G., Szentkiralyi, E. M. and Kendrick-Jones, J. (1973). The light chains of scallop myosin as regulatory subunits. *J. molec. Biol.* **74**, 179–203.

Walmsley, A. R. and Bagshaw, C. R. (1989). A logarithmic timebase for stopped-flow data aquisition and analysis. *Analyt. Biochem.* **176**, 313–318.

Walmsley, A. R., Evans, G. E. and Bagshaw, C. R. (1990). The calcium ion dependence of scallop myosin ATPase activity. *J. Musc. Res. Cell Motil.* **11**, 512–521.

Wells, C. and Bagshaw, C. R. (1984). The Ca^{2+} sensitivity of the actin activated ATPase of scallop heavy meromyosin. *FEBS Lett.* **168**, 260–264.

Wells, C. and Bagshaw, C. R. (1985). Calcium regulation of molluscan myosin ATPase in the absence of actin. *Nature* **313**, 696–697.

Molecular basis of myosin assembly: coiled-coil interactions and the role of charge periodicities

SIMON J. ATKINSON and MURRAY STEWART

MRC Laboratory of Molecular Biology, Hills Rd., Cambridge CB2 2QH, England

Summary

Complementation of alternating zones of positive and negative charge in the myosin rod enables molecules to interact in a number of ways. This accounts for the complexity of the molecular organisation of thick filaments. However, directed mutagenesis of expressed LMM cDNA indicated that charge zone complementation is not a major driving force in myosin polymerisation. Instead, it probably serves to prevent unfavourable interaction geometries.

Key words: myosin, assembly, structure, mutagenesis, expression.

Introduction

The function of myosin II as a motor protein in muscle and non-muscle cells depends on its ability to polymerize into macromolecular assemblies (thick filaments) that anchor the molecules and enable them to produce force concertedly (see, for example, Squire, 1981). Skeletal muscle thick filaments are stable structures that reflect the rigid architecture of a permanent force-producing machine. In non-muscle cells, however, the organization of myosin is more dynamic, adapting to the needs of cell motility, the maintenance of cell shape, and cell division. Bipolar filamentous molecular arrays similar to native thick filaments can be produced using purified myosin alone. Therefore, although other factors are almost certainly important for the precise arrangement of molecules found in thick filaments *in vivo*, the assembly properties of myosin are largely determined by the interacting surfaces of myosin molecules themselves. Here we combine structural and protein engineering data to explore the molecular basis of the interactions between myosin molecules.

Structure of myosin supramolecular assemblies

Muscle thick filaments are the archetypical myosin assembly. Myosin molecules have rod-like tails and two globular heads and pack into bipolar molecular arrays in which the tails form the thick filament shaft, with the heads arranged on the filament surface in an approximately helical manner. Arthropod thick filaments are 4-stranded helical structures and 3-D reconstructions show a regular arrangement of overlapping myosin heads on their surface (Crowther *et al.* 1985; Stewart *et al.* 1985). Vertebrate thick filaments are 3-stranded and contain 294 myosin molecules. These filaments have a more complicated structure in which the myosin heads are displaced from the positions expected in a perfectly regular helix (Stewart and Kensler, 1986). It is likely that perturbations in the positions of the heads on the filament surface may reflect the fact that the underlying tails are not all arranged equivalently.

Although a good deal is known about the positions of the myosin heads in filaments, there is only scant information about the positions of myosin tails. Transverse sections suggest that there are subunits spaced about 3–4 nm apart, which probably represent molecular dimers (Stewart *et al.* 1981), but it has not been possible to establish the position of individual myosin molecules in filaments. It is important to note that there are a number of key differences between *in vivo* and *in vitro* myosin assemblies and so artificial myosin aggregates may not always reflect the *in vivo* packing precisely. Also, the assembly properties of smooth muscle and non-muscle myosins appear to differ in a number of ways from vertebrate skeletal myosin. In particular, their assembly properties may be modulated *in vivo* by phosphorylation.

Myosin coiled-coil structure

The tails of myosin molecules are 2-stranded alpha-helical coiled-coils. Coiled-coils are a common structural motif in fibrous proteins (e.g. intermediate filament proteins and tropomyosin) and are also present in a range of other proteins where they may form important interaction sites involving such motifs as 'leucine zippers'. In skeletal muscle, the myosin coiled-coil tails interact in the core of thick filaments, with the globular myosin heads arranged in a roughly helical manner on the surface. Coiled-coils are formed by alpha helices winding around one another so that the side chains from one helix fit into the gaps between the side chains of the other. This 'knobs into holes' packing is favoured if the side chains at the interface are hydrophobic. Because the alpha helix has roughly seven residues in two turns, this produces a periodic heptad repeat, *a-b-c-d-e-f-g*, with residues *a* and *d* hydrophobic. Because small changes in the dimensions of the alpha helix produce large variations in the coiled-coil pitch, it is difficult to determine this important parameter from theoretical considerations alone (Fraser and MacRae, 1973).

Direct determination of the coiled-coil pitch was achieved by using electron microscopy and image processing (Quinlan and Stewart, 1987) to solve the structure of crystalline sheets of a proteolytic fragment of myosin rod

Journal of Cell Science, Supplement 14, 7–10 (1991)

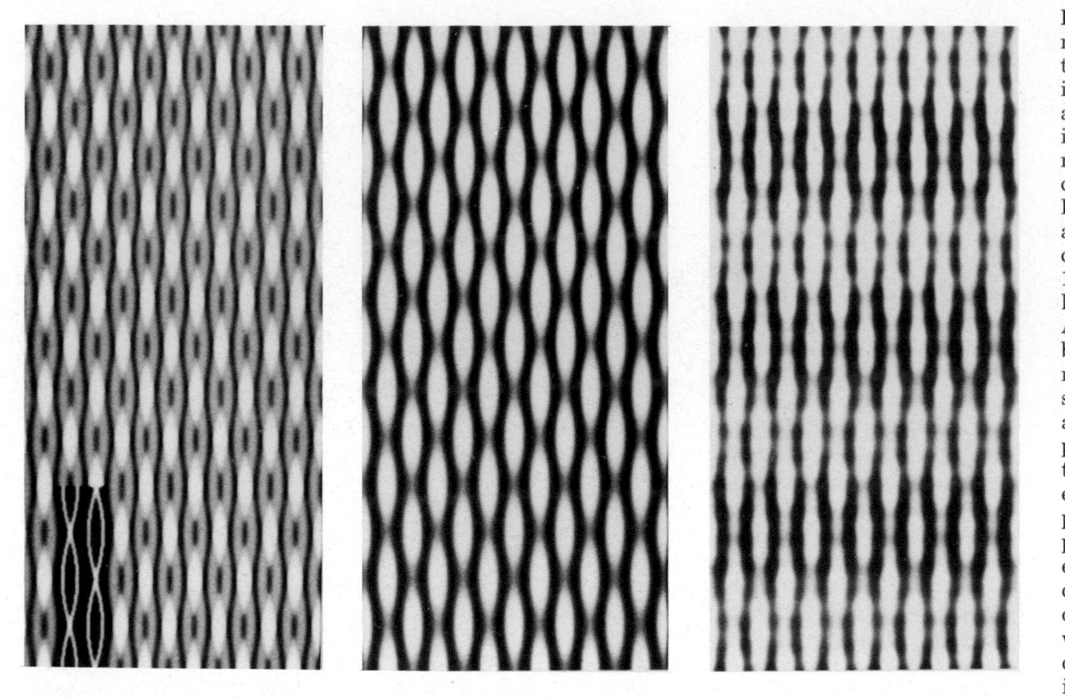

Fig. 1. Crystalline sheets of myosin long S-2 that show the coiled-coil. Schematic illustration (*left*) of the arrangement of coiled-coils in the projection of a negatively stained crystalline sheet of long S-2. Protein is shown in white and stain in black. The coiled-coils have a pitch of 14.1 nm and are spaced laterally at 2 nm intervals. Adjacent coils are staggered by a quarter pitch. A number of molecules are superimposed on one another perpendicular to the plane of the sheet, with their coils staggered by an even number of quarter pitches. The inset shows the path taken by the centre of each alpha-helix. The density is almost uniform over the array, but doubles where the coils cross over one another. The same image at a lower resolution of 1.9 nm (*centre*) shows only a pattern of light areas at the crossovers, and closely resembles the pattern seen in the image reconstructed from the electron micrographs (*right*). Reproduced from The Journal of Cell Biology, Volume 105, pp. 403–415, by copyright permission of the Rockefeller University Press.

(long S-2) to a resolution of 2 nm (Fig. 1). The sheets give diffraction patterns close to those seen by X-ray diffraction of whole muscle, and so probably give a reliable picture of both molecular structure and interaction geometry. In these sheets the pitch of the myosin coiled-coil was found to be close to 14 nm.

Because they interact over long distances, there are often long-range periodicities in coiled-coil protein sequences (Stewart *et al.* 1989). The 1000-residue myosin rod sequence contains a strong 28-residue repeat in the charged residues (Fig. 2) which produces a series of alternating positive and negative zones along the sequence. Complementation of these zones by an appropriate molecular stagger (an odd multiple of 28/2 residues) is thought to be a powerful determinant of the detailed geometry of their interaction (McLachlan, 1984). For example, the axial stagger that generates the 14.3 nm period seen in muscle and paracrystals of myosin rod fragments corresponds to a stagger of 7×28/2 residues.

In the long S-2 sheets, the coiled-coils interact in two different geometries: in the plane of the sheet, the coils are staggered by an odd number of quarter pitches, whereas perpendicular to the plane of the sheet, coils are staggered by an even number of quarter pitches. Only the second arrangement is consistent with the complementation of the alternating charged zones; the interaction in the plane of the sheets may involve the periodicities at 28/3 residues (see Fig. 2).

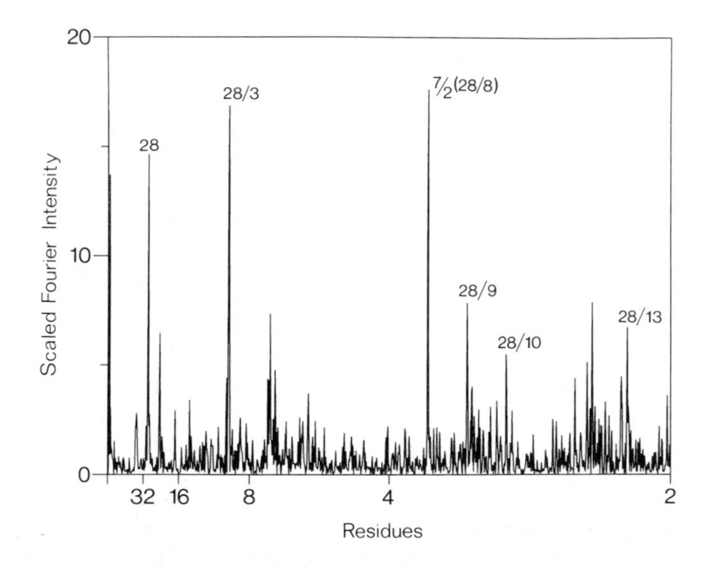

Fig. 2. Fourier transform of the distribution of the acidic residues in the nematode myosin rod sequence. The transform is calculated by first converting the sequence into mathematical form by setting acidic residues to one and all other residues to zero. The transform shows, in addition to the 7/2 and 7/3 peaks derived from the coiled-coil heptad repeat, peaks at 28 residues and overtones thereof. The peaks at 28/3 and 28/9 are particularly strong. These peaks derive from the bands of alternating charge that are thought to play an important role in the molecular interactions between myosin molecules in thick filaments.

The role of myosin rod charge periodicities

We investigated the role of the 28-residue charge periodicity in myosin assembly using a number of modified recombinant proteins produced by expressing in *E. coli* a cDNA clone corresponding broadly to the light meromyosin (LMM) region of rabbit myosin rod (Atkinson, 1990). Full-length recombinant LMM (rLMM) had solubility properties similar to proteolytic rabbit LMM and formed paracrystals with characteristic 43 nm and

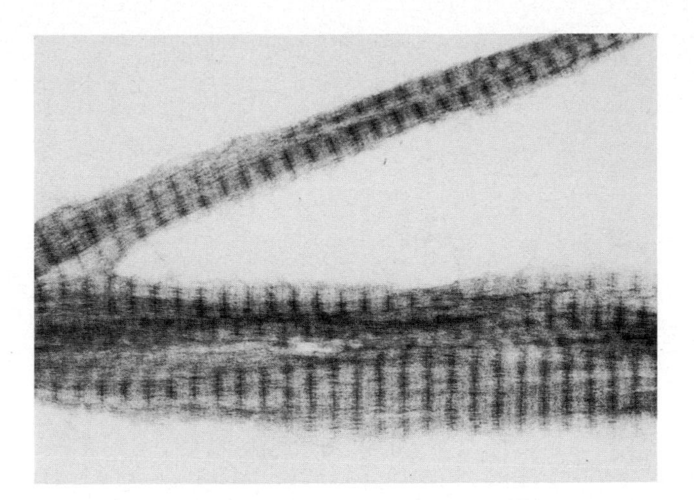

Fig. 3. Paracrystals of LMM made from recombinant material expressed in *E. coli*, formed in low ionic strength buffers and negatively stained with uranyl acetate. The paracrystals show a prominent axial banding pattern with a repeat of 43 nm. Images of metal-shadowed paracrystals confirm that the dark stripes correspond to gaps between the ends of molecules filled with stain, whereas the light bands correspond to the region in which the molecules overlap.

14.3 nm axial repeats (Fig. 3), and so clearly retained the assembly properties of native material. Several polymorphic paracrystal forms, in which the molecules were related by different staggers, were also observed to coexist in the paracrystal preparation (compare Katsura and Noda, 1973). Deletions from the N terminus of up to half the length of rLMM did not produce a marked change in its solubility properties, but deletions of as little as 90 residues from its C terminus greatly increased its solubility at low ionic strength, indicating that an important determinant of solubility is located near the rod C terminus. A similar effect has been observed with proteolytic fragments (Nyitray *et al.* 1983), but studies of deletion mutants of expressed *Dictyostelium* myosin rod (O'Halloran *et al.* 1990) indicated that in this case the solubility determining region lay about $34\times10^3\,M_r$ from the C terminus. Since the 28-residue periodicity was present throughout the rod sequence, this result indicated that complementation of the charged zones could not be the only sort of interaction important in the interaction between myosin molecules in aggregates.

The contribution of the 28-residue charged zones was evaluated more precisely using site-specific mutagenesis to construct two series of recombinant molecules containing internal deletions or insertions of 14 residues (Atkinson, 1990). These mutations change the phase of the 28-residue repeat and so prevent complementation of the charged zones over part of the region in which the molecules overlap. Remarkably, the solubility properties of these mutants were indistinguishable from those of the native material, indicating that the complementation of the zones of alternating charge made a negligible contribution to the free energy of association of the myosin coiled-coils in these aggregates. However, electron microscopy showed that the mutant proteins formed distinctly different aggregates. The wild-type protein formed paracrystals with a 43 nm axial repeat in which molecules overlapped over most of their length. The mutants formed paracrystals with a 65 nm axial repeat, in which molecules overlapped only at their ends, excluding the mutated region from the overlap, thereby avoiding unfavourable interactions involving apposition of zones of like charge. Deletions of 28 residues restored the phase of the charge repeats and yielded some paracrystals in which the 43 nm repeat and complete molecular overlap were restored.

Molecular interactions in thick filaments

The structural results obtained on vertebrate thick filaments indicate that it is unlikely that all the myosin tails in these aggregates interact in an equivalent manner. This contrasts with many other regular biological assemblies (such as virus particles, F-actin and microtubules) in which the interactions between subunits are equivalent or nearly so. However, a range of different molecular interactions is easily accommodated by the 28-residue charge repeat in the sequence, since the charged zones can be complemented by a whole range of molecular staggers. This is confirmed by the presence of polymorphic paracrystal forms. The precise molecular interaction geometries employed must await determination of the positions of the tails in the thick filament shaft. The driving force behind myosin assembly is still not completely clear, since the directed mutagenesis studies indicate that the role of the charged zones is in preventing unfavourable interactions rather than facilitating favourable ones. A number of different axial staggers appear to have very similar free energies of association. The C-terminal region of the molecule clearly has a vital role. The sequence of this part of the molecule is not remarkably different from the remainder of the rod, so it is difficult to conceive that there is a distinct myosin binding site. A concentration of positive charge in this region may be significant. Clearly further work will be required to define precisely the driving force behind myosin assembly and to identify the key interactions that determine axial and azimuthal packing of molecules. It is likely that there will be several structural levels in assembly and so different interactions may be involved in, for example, the dimerisation of molecules, the assembly of dimers into 3–4 nm protofilaments, the assembly of protofilaments into subfilaments, and the final aggregation of subfilaments into filaments. In addition, the interactions between antiparallel molecules in the bare zone will be different to those between parallel molecules in the bulk of the filaments and probably different again to the interactions at the filament tips. Mutagenised myosin molecules may be useful in disecting this hierachy of molecular interactions.

The mechanism by which *in vivo* vertebrate skeletal muscle thick filaments are precisely regulated in both length and diameter is also incompletely understood. It is difficult to account for their precise structure if these filaments arise by simple self-assembly alone, and it has not been possible to produce *in vitro* aggregates that accurately mirror the *in vivo* structure. It seems likely that some accessory proteins, perhaps analogous to those shown to be involved in virus assembly, will be required to act as a template on which the thick filament is assembled. These accessory proteins may not only regulate length and diameter, but may also specify particular interaction geometries for the incorporating monomers. Moreover, once the filaments are formed, these accessory molecules may well dissociate from the filaments and so be difficult to detect in filaments isolated from whole muscle. The study of the molecular basis of mutant muscle phenotypes may provide the key to their identification.

References

Atkinson, S. J. (1990). Molecular Determinants of Myosin Assembly. PhD Thesis, University of Cambridge.

Crowther, R. A., Padron, R. and Craig, R. (1985). Arrangement of the heads of myosin in relaxed thick filaments from tarantula muscle. *J. molec. Biol.* **184**, 429–439.

Fraser, R. D. B. and MacRae, T. P. (1973). *Conformation in Fibrous Proteins*. Academic Press: New York.

Katsura, I. and Noda, H. (1973). Structure and polymorphisms of light meromyosin aggregates. *J. Biochem.* (Tokyo). **73**, 257–268.

McLachlan, A. D. (1984). Structural implications of the myosin rod amino acid sequence. *A. Rev. Biophys. Bioeng.* **13**, 167–189.

Nyitray, L., Mocz, G., Szilagyi, L., Balint, M., Lu, R. C., Wong, A. and Gergely, J. (1983). The proteolytic substructure of light meromyosin. Location of a region responsible for the low ionic strength insolubility of myosin. *J. biol. Chem.* **258**, 13 213–13 220.

O'Halloran, T. J., Ravid, S. and Spudich, J. A. (1990). Expression of *Dictyostelium* myosin tail fragments in *E. coli*: domains required for assembly and phosphorylation. *J. Cell Biol.* **110**, 63–70.

Quinlan, R. A. and Stewart, M. (1987). Crystalline tubes of myosin long subfragment-2 showing the coiled-coil pitch and molecular interaction geometry. *J. Cell Biol.* **105**, 403–415.

Squire, J. M. (1981). *The Structural Basis of Muscle Contraction.* Plenum Press: New York.

Stewart, M., Ashton, F. T., Lieberson, R. and Pepe, F. A. (1981). The myosin filament IX. Determination of subunit positions by computer image processing. *J. molec. Biol.* **153**, 381–392.

Stewart, M. and Kensler, R. W. (1986). The arrangement of myosin crossbridges in relaxed frog muscle thick filaments. *J. molec. Biol.* **192**, 831–851.

Stewart, M., Kensler, R. W. and Levine, R. J. C. (1985). Three-dimensional reconstruction of thick filaments from *Limulus* and scorpion muscle. *J. Cell Biol.* **101**, 402–411.

Differential regulation of vertebrate myosins I and II

KATHLEEN COLLINS and PAUL MATSUDAIRA

M. I. T. and Whitehead Institute for Biomedical Research, Cambridge, MA 02142, USA

Summary

Cell motility events require movement of the cytoskeleton. Actin-based movement is catalyzed by the mechanoenzyme myosin, which translocates toward the barbed end of actin filaments in an ATP-dependent fashion. There are two subclasses of myosin with different structures and functions: conventional filamentous myosin (myosin II) and monomeric myosin I. Vertebrate non-muscle myosins I and II function as similar actin motors *in vitro*, catalyzing virtually identical actin-activated MgATP hydrolysis and motility. The functional diversification of these two enzymes results from their differential regulation. Calcium and tropomyosin, which activate the MgATP hydrolysis and motility of vertebrate non-skeletal muscle myosin II proteins, inhibit vertebrate (brush border) myosin I. The activities and regulation of brush border myosin I provide insight into conserved and unique features of the myosin mechanoenzymes and suggest how the functions of myosins I and II are divided in vertebrate cells. Brush border myosin I as an enzyme also contributes to our understanding of the molecular mechanism of motility.

Key words: cell motility, myosin regulation, tropomyosin, calmodulin.

Introduction

Myosin, isolated as a force-generating component of muscle, was the first cellular enzyme studied for its ability to produce mechanical force from the energy of ATP hydrolysis. In the presence of ATP and calcium, myosin in muscle bipolar thick filaments induces sliding of actin-containing thin filaments toward each other. This results in a net contraction, or shortening, of muscle. The discovery of molecules similar to skeletal muscle myosin in non-muscle cells suggested that myosin-mediated contraction might be involved in cell shape change, locomotion, cytokinesis, phagocytosis and intracellular transport. The requirement for conventional myosin, termed myosin II, for non-muscle cell contraction and cytokinesis has been shown by selective depletion experiments using extraction (Keller *et al.* 1985), antibodies (Mabuchi and Okuno, 1977), antisense RNA (Knecht and Loomis, 1987) and gene disruption (De Lozanne and Spudich, 1987). These same experiments, however, demonstrate that other actin-based motility events do not require myosin II. Cells lacking myosin II still extend pseudopods, phagocytose and chemotax.

Conventional myosins, or myosin II proteins, consist of a hexamer of two associated heavy chains (of about $200\times10^3 M_r$), each of which binds two light chains ($16-20\times10^3 M_r$). The C-terminal 'tail' half of the heavy chain dimerizes to form a parallel α-helical coiled coil. This tail domain also directs the staggered assembly of heavy-chain dimers into the large, bipolar filaments characteristic of skeletal muscle. The globular, N-terminal 'head' domain of myosin binds actin, ATP and light chains, and by itself retains myosin's two defining activities: actin-activated hydrolysis of MgATP and actin filament translocation (see Warrick and Spudich, 1987, for a review of myosin II). These two activities are thought to be coupled, with efficient ATP hydrolysis linked to productive mechanical work.

The first actin-activated ATP-hydrolyzing enzyme not conforming to the structure of myosin II was discovered in the single-celled eukaryote *Acanthamoeba* (Pollard and Korn, 1973). This new myosin, because it lacks a tail capable of dimerization, has been termed myosin I. The generality of this myosin variant to other cells has only recently become clear. Myosin I proteins from *Acanthamoeba*, the slime mold *Dictyostelium* and vertebrate cells all share several defining characteristics. Structurally, all myosin I proteins exist as monomers and have relatively shorter, non-helical tail domains with some unique and some shared features (Fig. 1). The tails of myosin I proteins from *Acanthamoeba* and *Dictyostelium* contain a glycine/proline/alanine or threonine-rich region (Jung *et al.* 1987; Jung *et al.* 1989) with a second (ATP-independent) actin-binding site (Lynch *et al.* 1986). The vertebrate (brush border) myosin I tail contains unique regions with homology to concensus calmodulin binding sites (Hoshimaru *et al.* 1989; unpublished observation). *Acanthamoeba*, *Dictyostelium*, and brush border myosin I tails all share a region which binds phospholipids *in vitro* (Adams and Pollard, 1989).

Myosin I enzymes also share similar cellular localizations. Whereas myosin II is distributed in cytoplasm, actin bundles or ordered actin arrays, myosin I is associated predominantly with cellular membranes. *Acanthamoeba* myosin I localizes to and cofractionates with the plasma membrane and membrane vesicles (Gadasi and Korn, 1980; Adams and Pollard, 1986). Motile *Dictyostelium* cells concentrate myosin I specifically at the leading edge, with myosin II distributed in the posterior cortex (Fukui *et al.* 1989). In vertebrate intestinal epithelial cells, myosin I is localized to apical and basolateral plasma membranes and to the vesicle-containing terminal web region (Coudrier *et al.* 1981; Fig. 2). Because of its monomeric nature and membrane localization, myosin I is likely to be involved in events such as membrane transport, cortical flow and phagocytosis, events shown not to depend on myosin II.

Journal of Cell Science, Supplement 14, 11–16 (1991)
Printed in Great Britain © The Company of Biologists Limited 1991

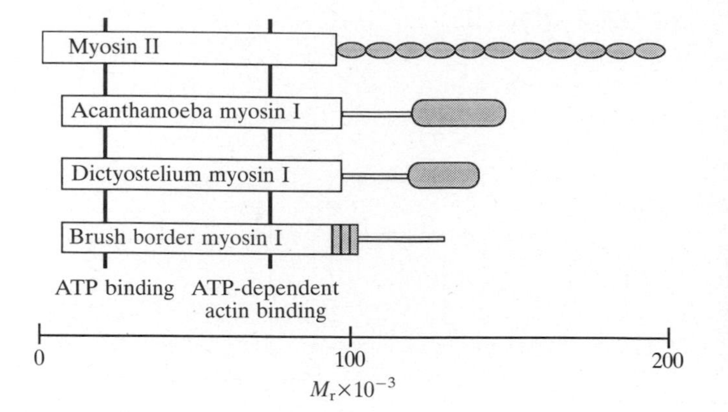

Fig. 1. Comparison of common and distinguishing structural features of myosins. All myosins share a conserved N-terminal domain which binds ATP and actin. The C-terminal domains of the myosins, however, have unique features (shaded). Myosin II proteins possess a tail domain which dimerizes to form α-helical coiled coil. *Acanthamoeba* (Jung *et al.* 1987) and *Dictyostelium* (Jung *et al.* 1989) myosin I have a second, ATP-independent actin-binding site. The brush border myosin I tail (Hoshimaru and Nakanishi, 1987; Garcia *et al.* 1989) contains three sites which by sequence homology are likely to bind the three calmodulins of the complex. All myosin I proteins share a region which binds phospholipids *in vitro* (thin rod).

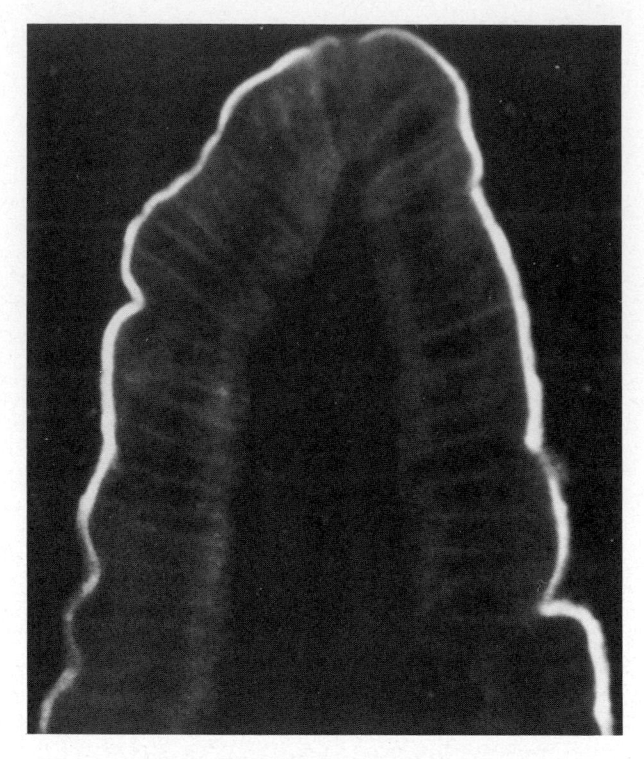

Fig. 2. Localization of brush border myosin I in intestinal epithelial cells. Sections of fixed, frozen adult mouse intestine were stained with affinity-purified chicken brush border myosin I antiserum and fluorescein-conjugated secondary antibody. A villus cross-section is shown, with columnar epithelial cells (a single, polarized sheet) exposed to the lumen (lumen is on the top, left, and right of the figure) at their apical surface. Most myosin I is localized apically in epithelial cells, to microvilli and the terminal web. Lighter staining of the cells' basolateral membrane is also observed (appears as thin, parallel stripes).

The activities of brush border myosin I

To study the properties of myosin I and the division of cellular labor between myosins I and II in a vertebrate system, we have characterized myosin I from intestinal epithelial cells. These cells have an apical domain specialized to absorb essential nutrients from the lumen of the gut, separated from the basolateral cell domain by membrane tight junctions and a circumferential belt of actin filaments. The apical membrane is highly increased in surface area by packed, finger-like protrusions known as microvilli, each stabilized by a central core bundle of 15–20 actin filaments extending into the cytosol. Microvilli and the actin-rich region beneath them, the terminal web, together constitute the brush border, which can be isolated intact after cell homogenization (see Louvard, 1989, and Mooseker, 1985, for reviews of brush border). Myosin I was discovered in brush border as the lateral linkage of the microvillus actin core to the plasma membrane by virtue of observations that ATP extracted the linkage from actin (Matsudaira and Burgess, 1979) and that the purified linker complex catalyzed a weakly actin-activated MgATPase activity (Collins and Borysenko, 1984). Brush border also contains myosin II (Mooseker *et al.* 1978).

We purify myosin I by MgATP extraction of isolated chicken brush borders (Collins *et al.* 1990). The purified complex consists of a $110\times10^3 M_r$ heavy chain associated with $18\times10^3 M_r$ subunits. The $18\times10^3 M_r$ protein has been identified as calmodulin by a variety of criteria, including calcium-dependent apparent molecular weight shift on SDS gels, heat stability, and phosphodiesterase activation. We have confirmed the exact identity of the $18\times10^3 M_r$ polypeptide to be calmodulin by amino acid sequencing. Calmodulin, which serves as a regulatory subunit for several cellular enzymes, changes conformation with the cooperative binding of four calcium ions. The function of calmodulin as a subunit of brush border myosin I is not known, but the association of the myosin I heavy chain with a common regulatory protein is in contrast to the association of specialized light chains with myosin II.

The localization of myosin I in intestinal epithelial cells is shown by immunofluorescence microscopy in Fig. 2. Myosin I antiserum was raised and affinity-purified against chicken brush border myosin I and detects only a single $110\times10^3 M_r$ protein in total epithelial cell homogenates (Collins *et al.* 1990). The apical region of the epithelial cells, containing microvilli and the terminal web, is labeled most strongly, but the basolateral membrane also contains myosin I. This localization is similar to a previous report (Coudrier *et al.* 1981). The distribution of 'brush border' myosin I in several regions of the cell suggests that although a subset of myosin I laterally links the microvillar actin core to the plasma membrane, it probably performs other functions as well.

Purified brush border myosin I demonstrates the simple-hyperbolic actin-activation of MgATPase activity characteristic of myosin II (Collins *et al.* 1990). As for non-muscle myosin II, V_{max} increases about 40-fold in the presence of actin. At 37 °C and a saturating actin concentration, brush border myosin I hydrolyzes MgATP at approximately $1\,s^{-1}$, with a K_m of actin-activation of 20–40 μM. Brush border myosin I also demonstrates properties of actin filament translocation similar to myosin II. Using the *in vitro* sliding filament assay, in which fluorescently labeled actin filaments are translocated over a nitrocellulose-coated, myosin-bound coverslip (Kron and Spudich, 1986), we determine a myosin I-mediated rate of motility of 0.08 $\mu m\,s^{-1}$ at 37 °C. This rate of motility is identical to that we measure for brush border myosin II in the same assay, but it differs from a previous report of 10-fold slower brush border myosin I motility (Mooseker and Coleman, 1989). In addition, brush border myosin I and rabbit skeletal myosin II (Sheetz *et al.* 1984) both have the same ATP-concentration dependence of motility, and as for all myosins, brush border myosin I translocates actin independent of actin filament length or myosin concentration above a minimum threshhold. Myosin I motility is more stable than myosin II motility *in vitro*: the rate of translocation by brush border myosin I on a single coverslip is constant for at least two hours at room temperature, with virtually all filaments which attach to the coverslip in constant motion. Preparations of myosin I stored at 4 °C show full activity for at least 1–2 months, in contrast with the very rapid loss, in days, of observable myosin II-mediated motility.

As described above, the actin-activated MgATPase activity and motility of vertebrate non-muscle cell myosins I and II are very similar. Therefore, the myosins probably do not serve unique functions by virtue of unique enzymatic activities. This similarity of myosin I and II enzymes appears to be specific to the vertebrate system. Myosin I from *Acanthamoeba*, but not myosin II, demonstrates complex kinetics of actin-activated MgATPase activity due to the additional, ATP-independent actin-binding site (Korn *et al.* 1988). Even without the extra binding site, however, *Acanthamoeba* myosin I retains unique activities including an actin-activated MgATPase activity 10-fold greater than myosin II (Lynch *et al.* 1986). The reason for the similarity of myosin I and II enzymes in vertebrate cells in particular is not clear. It is possible that vertebrate myosin I isoforms discovered in the future may possess activities unlike myosin II, and that *Acanthamoeba* myosin I isoforms not yet purified may more closely resemble myosin II.

Calcium regulation

The striking similarity of the enzymatic activities of myosins I and II was unexpected, but is consistent with their structural homology. This identity does, however, leave unexplained the basis for division of cell functions. If myosins I and II are similar enzymes, what suits the two proteins to different motility functions in a cell? As one answer to this question, we find that myosins I and II are regulated distinctly: conditions which activate myosin I inhibit smooth and non-muscle myosin II, and *vice versa*.

For smooth and non-muscle myosin II, calcium induces the phosphorylation of regulatory light chain by myosin light chain kinase. This light chain phosphorylation greatly stimulates myosin II actin-activated MgATPase activity and motility. Brush border myosin I is not phosphorylated by myosin light chain kinase (Keller and Mooseker, 1982; unpublished observation). In contrast with its activation of myosin II, we find that calcium inhibits brush border myosin I, and does so directly rather than through phosphorylation (Collins *et al.* 1990). One to 100 μM calcium causes a gradual inhibition of myosin I actin-activated MgATPase activity, resulting from the gradual dissociation of a subset of calmodulin *in vitro.*

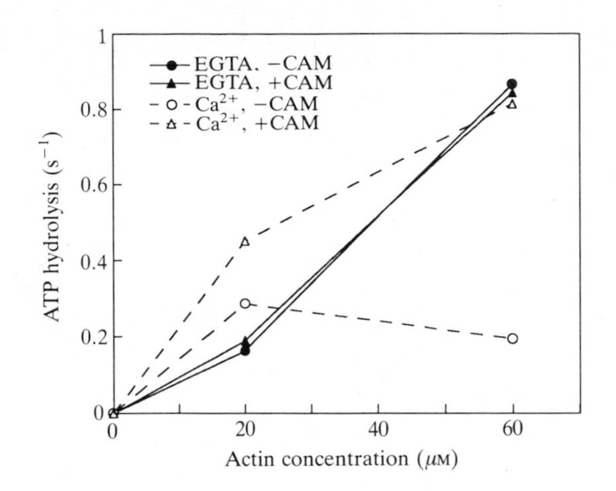

Fig. 3. Calcium regulation of brush border myosin I (data from Collins *et al.* 1990). Myosin I-mediated actin-activated MgATPase activity is measured as a function of increasing actin concentration in 1 mM EGTA (solid lines, filled points) or 1 mM calcium (dashed lines, open points). Actin-independent ATPase activity is subtracted from all points. Circles indicate myosin I alone; triangles indicate the presence of 5–10 μM calmodulin (CAM). The actin-activation of myosin I MgATPase activity is inhibited in calcium, but inhibition is reversible with addition of exogenous calmodulin. Calmodulin has no effect on MgATPase activity in EGTA. The K_m of actin-activation decreases in calcium, an effect separate from calmodulin dissociation. This second effect of calcium is responsible for the greater activity of myosin I at 20 μM actin in calcium *versus* EGTA.

Readdition of calmodulin (5–10 μM) completely restores MgATPase activity even in buffers of high calcium (1 mM), demonstrating that inhibition results from calmodulin dissociation rather than calcium-binding (Fig. 3). By determining the stoichiometry of $110\times10^3 M_r$ and $18\times10^3 M_r$ myosin I subunits in calcium and EGTA, we find that three calmodulin molecules are bound per heavy chain in EGTA while only two remain associated after repurification of the complex in calcium. These molar ratios of $110\times10^3 M_r$ to $18\times10^3 M_r$ polypeptides were derived from comparison of integrated peak areas of subunits fractionated by gel filtration in guanidine, monitoring peptide bond absorbance (unpublished observations). The affinity of the dissociable calmodulin for the heavy chain can be determined from the calmodulin-concentration dependence of reactivation of myosin I MgATPase activity. The regulatory calmodulin binds to the myosin I heavy chain with an affinity in the nanomolar range in EGTA, but in calcium this affinity decreases to micromolar (unpublished observations).

High calcium also completely inhibits brush border myosin I motility (Collins *et al.* 1990). In the coverslip motility assay, only a small number of active myosin heads are necessary to promote the maximum rate of motility if inactive heads do not act as a brake. Brush border myosin I in high calcium still binds actin but releases it in the presence of ATP. Therefore, myosin I-mediated motility is not inhibited *in vitro* until complete dissociation of the regulatory calmodulin subunit has occurred, at 100 μM calcium. Movement ceases abruptly rather than gradually with increasing calcium. Inhibition of motility, as for MgATPase activity, is reversible with an excess of calmodulin.

The direct calcium inhibition of brush border myosin I is significant in its contrast with the regulation of myosin II and also in its mechanism. The reversible dissociation of a calmodulin from the complex provides a model system for studying the coupling of MgATP hydrolysis and translocation in the myosin kinetic cycle. Brush border myosin I without regulatory calmodulin still binds actin in an ATP-sensitive manner, but does not translocate filaments. It still binds and hydrolyzes MgATP, dependent on the association of actin, but at a 10-fold lower rate and independent of actin concentration between 0.5 and 60 μM. Therefore, although the actin- and ATP-binding sites are still functional and active, the cycle of MgATP hydrolysis coupled to force production has been eliminated. Removal of scallop myosin II light chain has a similar effect: motility is reversibly inactivated and MgATPase activity becomes deregulated, in this case constitutively high (Vale *et al.* 1984). These induced deregulations of the myosin I and II mechanoenzymes suggest that even for myosins with divergent heavy- and light-chain structures, the association of regulatory light chain or its equivalent may coordinate heavy chain activities. Because the mechanism of energy coupling appears to be conserved between myosins, the more easily reversible inhibition of myosin I makes it an excellent model system to study this basic feature of motor proteins.

Tropomyosin regulation

The mechanisms of regulation described above rely on modification of myosin structure and thus are termed myosin-linked forms of regulation. A second type of myosin regulation is mediated by actin structure, in particular by the actin-binding protein tropomyosin. Actin-linked regulation is most thoroughly characterized in skeletal muscle, where tropomyosin and the tropomyosin-associated troponin complex saturate the actin of thin filaments. Troponin–tropomyosin inhibits myosin cross-bridge cycling on actin in low calcium, maintaining the relaxed state. An increase in calcium concentration induces a conformational change in the troponin–tropomyosin complex which activates myosin-mediated ATP hydrolysis and muscle contraction. Isoforms of tropomyosin also exist in smooth and non-muscle cells, but the significance of actin-linked regulation in these cells is less clear. These tropomyosin isoforms stimulate rather than inhibit smooth and non-muscle myosin II motility and MgATPase activity by 2- to 3-fold (Umemoto *et al.* 1989).

To determine if actin-linked as well as myosin-linked regulation differentiates vertebrate myosins I and II, we tested the effect of smooth (turkey gizzard) and non-muscle (chicken fibroblast, chicken brush border) isoforms of tropomyosin on the activities of chicken brush border myosin I (K. Collins, J. Sellers, P. Matsudaira, manuscript in preparation). Although these isoforms activate smooth and non-muscle myosin II actin-activated MgATPase activity and motility, they inhibit brush border myosin I. For smooth muscle tropomyosin (a high molecular weight, high actin-affinity tropomyosin isoform), the actin-activated MgATPase activity of myosin I decreases with increasing tropomyosin concentration until a saturating ratio of 1 tropomyosin dimer per 7 actin monomers is reached. Brush border tropomyosin, a low molecular weight tropomyosin with a lower affinity for actin under the assay conditions, also inhibits myosin I MgATPase activity, but a higher concentration of tropomyosin is required for the same effect. The rate of MgATP hydrolysis at maximal inhibition is approximately 15–20% of the uninhibited rate. This residual MgATPase activity, as with the MgATPase activity remaining after calcium inhibition, is not coupled to motility. In the motility assay, actin filaments which are bound to the myosin-coated coverslip under conditions of tropomyosin saturation do not move at all, to within the error of our measurements (0.004 $\mu m\,s^{-1}$, 20-fold less than normal velocity).

We also assayed the dependence of brush border myosin I actin-binding on tropomyosin. Saturating concentrations of smooth and non-muscle tropomyosin inhibit the affinity of myosin I for actin by 25–50% in the presence of ATP. This binding inhibition is not sufficient to explain the inhibition of MgATPase activity and motility. Therefore, as is the case with inhibition of skeletal myosin by skeletal tropomyosin, smooth and non-muscle tropomyosins most likely inhibit myosin I MgATPase activity and motility by inhibiting a kinetic step of the cross-bridge cycle. Although the inhibition of myosin I actin-binding by tropomyosin is not responsible for the inhibition of myosin I MgATPase activity and motility, it may be relevant to myosin I localization *in vivo*. We find that the localization of myosin I in intestinal epithelial cells is consistent with a competition of myosin I and at least a low molecular weight isoform of tropomyosin for binding to actin (unpublished observations). In contrast, myosin II colocalizes with tropomyosin, perhaps with a high molecular weight tropomyosin isoform(s) in particular. Although myosin and tropomyosin localizations are consistent with an effect of tropomyosin on myosin distribution *in vivo*, myosin I also has membrane-binding activities not shared with myosin II which are equally as likely to contribute to differential localization.

Structural revelations

Although vertebrate myosin I and smooth or non-muscle myosin IIs bind actin with similar affinity or display similar kinetics of MgATPase activity actin-activation, they are differentially regulated by tropomyosin. This suggests a divergence of fine structure in the otherwise conserved actin-binding site. The structure of the myosin II subfragment-1 (S-1)–tropomyosin–actin complex in the absence of ATP has been resolved to 2–3 nm by reconstructions from electron micrographs (Milligan and Flicker, 1987). Actin saturated with S-1 in this fashion has a distinctive arrowhead appearance, due to the helical repeat of the complex. Brush border myosin I similarly decorates actin filaments (Coluccio and Bretscher, 1987; Fig. 4A). Reconstruction of the myosin I-actin complex will allow comparison of the myosin I and II actin-binding sites in the absence of ATP. In calcium, myosin I arrowheads have a different structure (Fig. 4B), half as wide and angled more sharply from the filament axis. This calcium-induced change could represent the loss of a calmodulin subunit or it could also reflect a different state of actin–myosin association, induced by the inhibition of force production. Reconstruction of the calcium and EGTA complexes should provide information about the location of the dissociable calmodulin and the extent of conformational change in the rigor complex with inhibition of translocation.

Implications

The regulation of brush border myosin I, as well as myosin I localization and structure, suggest that myosins I and II fulfill complementary functions in the cell. Due to the differential regulation of myosins I and II by calcium,

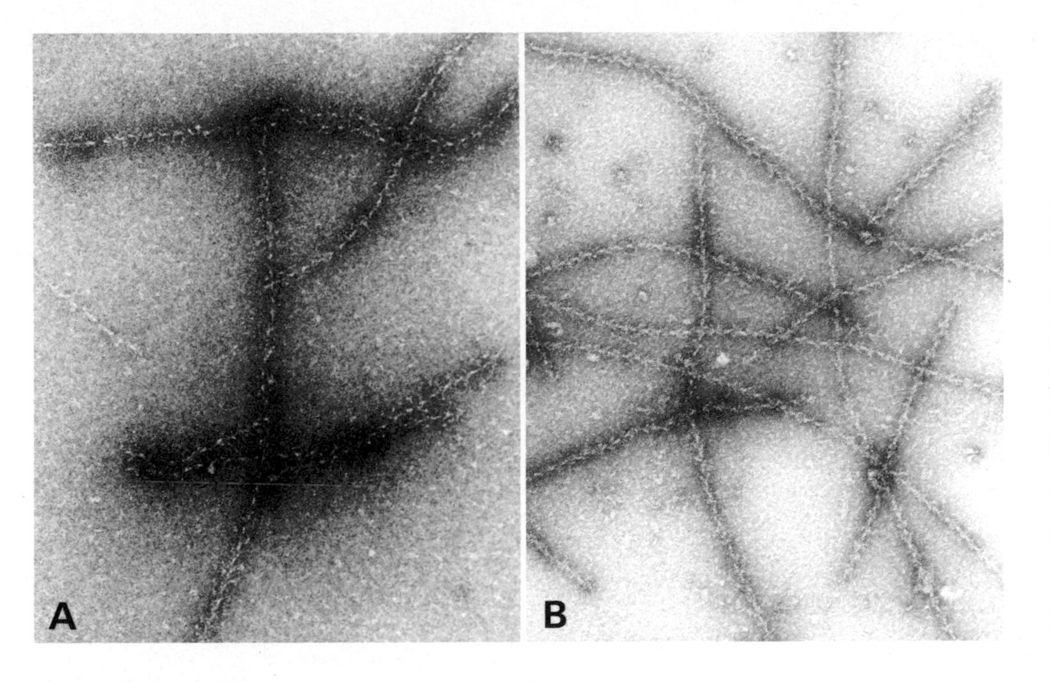

Fig. 4. Actin filaments decorated with brush border myosin I in EGTA (A) or calcium (B). An excess of myosin I was incubated with 0.5–1.0 μM actin in the standard myosin I purification buffer (Collins *et al.* 1990) to form decorated filaments. Calcium was added in excess of EGTA to the myosin–actin mixture in (B). Samples were stained with 2 % uranyl acetate on carbon-coated 400-mesh grids and examined at 80 kV with a Phillips 410 electron microscope. Final magnification in (A) and (B) is approximately ×50 000.

myosin I would remain constitutively active in the low calcium concentrations predominant in cells *in vivo*, whereas myosin II, lacking light chain phosphorylation, would be predominantly inactive. Constitutive cellular motility processes include vesicle transport, membrane movement, cortical flow, and phagocytosis, events also implicated as myosin I functions on the basis of myosin I localization and the lack of requirement for myosin II. The increase in calcium concentration necessary to activate myosin II, occurring when a cell is stimulated to divide or contract, would inhibit myosin I. This inhibition of myosin I functions with myosin II activation would reduce the use of cell energy for housekeeping purposes during the period the cell is responding to a contractile stimulus. Calcium regulation partitions actin motility events between cell states; any cell state or compartment with functioning myosin I would not have active myosin II and *vice versa*.

Differential regulation extends to actin-linked as well as myosin-linked mechanisms. The reciprocal effects of tropomyosin on myosins I and II suggests that even if both myosins could associate with the same actin population *in vivo*, the enzymes could not both be maximally active. Myosin I, however, does not demonstrate the apparent *in vivo* preference for actin structures containing tropomyosin that myosin II does. As a consequence, myosin I would be associated with a less stable subset of actin filaments in the cell. Stabilized actin arrays may not be as necessary for myosin I function as for the contraction mediated by myosin II.

Myosin I enzymes may serve a multiplicity of functions. The localization of brush border myosin I in the terminal web and at the apical and basolateral membranes of intestinal epithelial cells suggests that it could perform several different membrane-related tasks. Lower eukaryotic myosin I also seems to be a diverse cellular workhorse: *Dictyostelium* and *Acanthamoeba* express a large number of myosin I isoforms but only one myosin II. The existence of multiple myosin I isoforms in vertebrates is hinted by cloning (Montell and Rubin, 1988) but deserves future experimental attention. Although the localization and *in vitro* activities of vertebrate myosin I suggest possible functions for this protein, definitive demonstration of the role of myosin I in intact cells is difficult. Gene disruption and cell assays are more complex with vertebrate cells than with simpler organisms, but microinjection of antibodies which specifically inhibit myosin I-mediated motility *in vitro* (unpublished observations) might halt myosin I-requiring activities in the cell. A better understanding of the structure and composition of 'motile' cell regions will hopefully complement the investigation of purified motor proteins to determine the operating principles of cell motility.

Conclusions

Vertebrate actin-based cell motility has diversified by the evolution of novel regulatory mechanisms for a single family of enzymes, the myosins. There is not yet evidence for a multiplicity of different classes of motor enzymes capable of translocating on actin, in contrast with the several different categories of microtubule-translocating enzymes that have been discovered (Gelfand, 1989). Myosin I motors may perform a diversity of actin-based motility events previously ascribed to myosin II, particularly functions involving membranes or membrane-associated actin. The differential regulation of myosins I and II provides a mechanism for diversification of the two enzymes, and suggests a balance of myosin I-mediated constitutive motor activities with myosin II-mediated contractile events induced by specific stimuli or cell states. Future study of myosin I will provide new insights into both the cellular and molecular bases of motility.

This work was funded by grants from the American Heart Association, The Pew Foundation, and the Markey Foundation.

References

ADAMS, R. J. AND POLLARD, T. D. (1986). Propulsion of organelles isolated from *Acanthamoeba* along actin filaments by myosin-I. *Nature* **322**, 754–756.

ADAMS, R. J. AND POLLARD, T. D. (1989). Binding of myosin I to membrane lipids. *Nature* **340**, 565–568.

Collins, J. H. and Borysenko, C. W. (1984). The 110,000-dalton actin- and calmodulin-binding protein from intestinal brush border is a myosin-like ATPase. *J. biol. Chem.* **259**, 14 128–14 135.

Collins, K., Sellers, J. R. and Matsudaira, P. (1990). Calmodulin dissociation regulates brush border myosin I (110-kD-calmodulin) mechanochemical activity *in vitro*. *J. Cell Biol.* **110**, 1137–1147.

Coluccio, L. M. and Bretscher, A. (1987). Calcium-regulated cooperative binding of the microvillar 110K-calmodulin complex to F-actin: formation of decorated filaments. *J. Cell Biol.* **105**, 325–333.

Coudrier, E., Reggio, H. and Louvard, D. (1981). Immunolocalization of the 110,000 molecular weight cytoskeletal protein of intestinal microvilli. *J. molec. Biol.* **152**, 49–66.

De Lozanne, A. and Spudich, J. A. (1987). Disruption of the *Dictyostelium* myosin heavy chain gene by homologous recombination. *Science* **236**, 1086–1091.

Fukui, Y., Lynch, T. J., Brzeska, H. and Korn, E. D. (1989). Myosin I is located at the leading edges of locomoting *Dictyostelium* amoebae. *Nature* **341**, 328–331.

Gadasi, H. and Korn, E. D. (1980). Evidence for differential intracellular localization of the *Acanthamoeba* myosin isoenzymes. *Nature* **286**, 452–456.

Garcia, A., Coudrier, E., Carboni, J., Anderson, J., Vanderkhove, J., Mooseker, M., Louvard, D. and Arpin, M. (1989). Partial deduced sequence of the 110-kD-calmodulin complex of the avian intestinal microvillus shows that this mechanoenzyme is a member of the myosin I family. *J. Cell Biol.* **109**, 2895–2903.

Gelfand, V. I. (1989). Cytoplasmic microtubular motors. *Current Opinion in Cell Biology* **1**, 63–66.

Hoshimaru, M., Fujio, Y., Sobue, K., Sugimoto, T. and Nakanishi, S. (1989). Immunochemical evidence that myosin I heavy chain-like protein is identical to the 110-kilodalton brush-border protein. *J. Biochem.* **106**, 455–459.

Hoshimaru, M. and Nakanishi, S. (1987). Identification of a new type of mammalian myosin heavy chain by molecular cloning: overlap of its mRNA with preprotachykinin B mRNA. *J. biol. Chem.* **262**, 14 625–14 632.

Jung, G., Korn, E. D. and Hammer, J. A. III. (1987). The heavy chain of *Acanthamoeba* myosin IB is a fusion of myosin-like and non-myosin-like sequences. *Proc. natn. Acad. Sci. U.S.A.* **84**, 6720–6724.

Jung, G., Saxe, C. L. III, Kimmel, A. R. and Hammer, J. A. III. (1989). *Dictyostelium discoideum* contains a gene encoding a myosin I heavy chain. *Proc. natn. Acad. Sci. U.S.A.* **86**, 6186–6190.

Keller, T. C. S. III, Conzelman, K. A., Chasan, R. and Mooseker. M. S. (1985). Role of myosin in terminal web contraction in isolated intestinal epithelial brush borders. *J. Cell Biol.* **100**, 1647–1655.

Keller, T. C. S. III and Mooseker, M. S. (1982). Ca^{++}-calmodulin-dependent phosphorylation of myosin, and its role in brush border contraction *in vitro*. *J. Cell Biol.* **95**, 943–959.

Knecht, D. A. and Loomis, W. F. (1987). Antisense RNA inactivation of myosin heavy chain gene expression in *Dictyostelium discoideum*. *Science* **236**, 1081–1086.

Korn, E. D., Atkinson, M. A. L., Brzeska, H., Hammer, J. A. III, Jung, G. and Lynch, T. J. (1988). Structure–function studies on *Acanthamoeba* myosins IA, IB, and II. *J. Cell Biochem.* **36**, 37–50.

Kron, S. J. and J. A. Spudich. (1986). Fluorescent actin filaments move on myosin fixed to a glass surface. *Proc. natn. Acad. Sci. U.S.A.* **83**, 6272–6276.

Louvard, D. (1989). The function of the major cytoskeletal components of the brush border. *Current Opinion in Cell Biology* **1**, 51–57.

Lynch, T. J., Albanesi, J. P., Korn, E. D., Robinson, E. A., Bowers, B. and Fujisaki, H. (1986). ATPase activities and actin-binding properties of subfragments of *Acanthamoeba* myosin IA. *J. biol. Chem.* **261**, 17 156–17 162.

Mabuchi, I. and Okuno, M. (1977). The effect of myosin antibody on the division of starfish blastomeres. *J. Cell Biol.* **74**, 251–263.

Matsudaira, P. T. and Burgess, D. R. (1979). Identification and organization of the components in the isolated microvillus cytoskeleton. *J. Cell Biol.* **83**, 667–673.

Milligan, R. A. and Flicker, P. F. (1987). Structural relationships of actin, myosin, and tropomyosin revealed by cryo-electron microscopy. *J. Cell Biol.* **105**, 29–39.

Montell, C. and Rubin, G. M. (1988). The *Drosophila ninaC* locus encodes two photoreceptor cell specific proteins with domains homologous to protein kinases and the myosin heavy chain head. *Cell* **52**, 757–772.

Mooseker, M. S. (1985). Organization, chemistry, and assembly of the cytoskeletal apparatus of the intestinal brush border. *A. Rev. Cell Biol.* **1**, 209–241.

Mooseker, M. S. and Coleman, T. R. (1989). The 110-kD protein-calmodulin complex of the intestinal microvillus (brush border myosin I) is a mechanoenzyme. *J. Cell Biol.* **108**, 2395–2400.

Mooseker, M. S., Pollard, T. D. and Fujiwara, K. (1978). Characterization and localization of myosin in the brush border of intestinal epithelial cells. *J. Cell Biol.* **79**, 444–453.

Pollard, T. D. and Korn, E. D. (1973). *Acanthamoeba* myosin: isolation from *Acanthamoeba castellanii* of an enzyme similar to muscle myosin. *J. biol. Chem.* **248**, 4682–4690.

Sheetz, M. P., Chasan, R. and Spudich, J. A. (1984). ATP-dependent movement of myosin *in vitro*: characterization of a quantitative assay. *J. Cell Biol.* **99**, 1867–1871.

Umemoto, S., Bengur, A. R. and Sellers, J. R. (1989). Effect of multiple phosphorylations of smooth muscle and cytoplasmic myosins on movement in an *in vitro* motility assay. *J. biol. Chem.* **264**, 1431–1436.

Vale, R. D., Szent-Gyorgyi, A. G. and Sheetz, M. P. (1984). Movement of scallop myosin on *Nitella* actin filaments: regulation by calcium. *Proc. natn. Acad. Sci. U.S.A.* **81**, 6775–6778.

Warrick, H. M. and Spudich, J. A. (1987). Myosin structure and function in cell motility. *A. Rev. Cell Biol.* **3**, 379–421.

Self-assembly pathway of nonsarcomeric myosin II

R. A. CROSS, T. P. HODGE and J. KENDRICK-JONES

MRC Laboratory of Molecular Biology, Hills Rd, Cambridge CB2 2QH, UK

Summary

Cells need to control the location and timing of actomyosin-dependent force generation, and appear to do so in the first instance by regulating myosin filament self-assembly (Yumura and Fukui, 1985). The mechanism of the self-assembly is little understood. *In vitro* it is a true self-assembly, which requires a short domain at the C terminus of the myosin molecule. The availability of this domain appears suppressed by the folding of the molecule into a compact, looped state. *In vitro*, the rate at which these looped molecules unfold turns out to be a key determinant of filament number and filament length.

Key words: myosin assembly, side-polar, 6 S–10 S transition, nucleation, elongation.

What are the interesting questions?

Long-tail, double-headed myosins self-assemble spontaneously in buffers of physiological ionic strength. The self assembly tethers the myosin head-pairs into ordered arrays, constraining sets of tens to several hundreds of them to join forces and pull along the same axis. It is not known if the proximity of other heads causes a tethered head-pair to function more efficiently. Certainly the time spent searching for a binding site on actin will be different for a head in a filament compared to one in solution, and it is possible that the heads in an array may restrict or guide each others' diffusion in other ways. We would like to know more about the way self-assembly affects motor function, and to do this we need to understand the design principles of the myosin filament, particularly which of them are to do with the dynamics of assembly, and which are to do with optimising force generation. Specifically, we would like to know:

1. Building units. Are filaments assembled from subfilaments or oligomers, or from monomers?
2. Molecular packing. Is the relationship of one head to its partner optimised in some way? How much space does a myosin head-pair have, and how much freedom of movement radially and azimuthally? Are there cooperative structural effects?
3. The choosing of partners. What are the dynamics of the assembly pathway? How fast can a filament be put together, and how fast can it be taken apart? What is the rate limiting step in both cases?
4. Template-proteins. Which steps on the assembly pathway are perturbed by myosin-binding proteins?
5. Molecular exchange. How fast do molecules exchange from the interior of the filament, relative to the ends? Can the rate of these processes also be affected by myosin-binding proteins? Does the presence of actin, for example, affect the dynamics of myosin self-assembly? How do other forces in the cell (mechanical, electrostatic, hydrostatic) affect exchange?
6. Domain function. Which regions of the myosin molecule are important for assembly? Why does the myosin molecule have a long tail?

Here we discuss our own recent attempts to answer some of these questions, using smooth muscle and brush border myosins as model systems. On the basis of the data, we draw some tentative inferences about how *ad hoc* assembly and disassembly of myosin filaments might be achieved in motile cells.

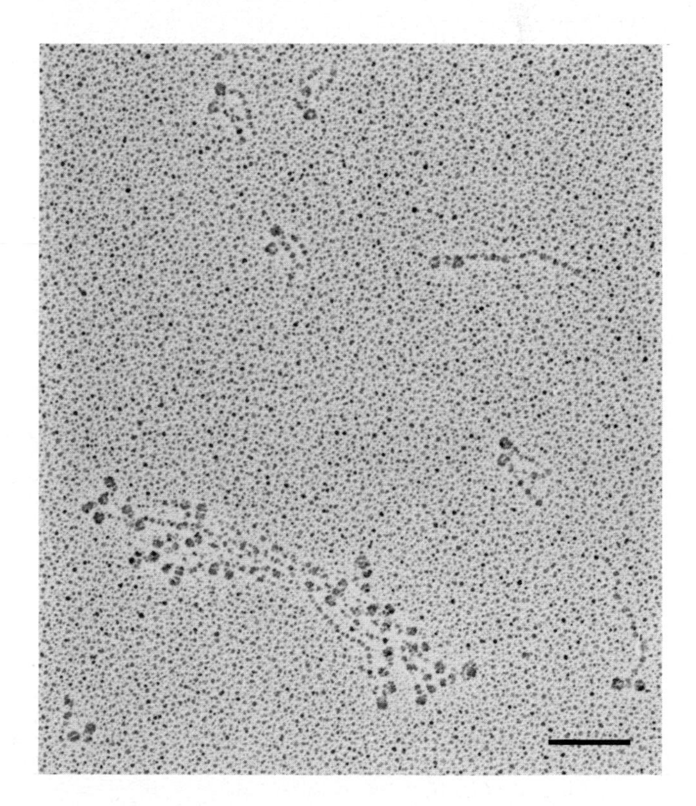

Fig. 1. Conformational species involved in myosin self-assembly. Straight-tail monomers equilibrate with filaments, and folded-tail monomers with straight-tail monomers. Myosin in solution in high salt was rapidly diluted into a large volume of 50 mM sodium formate, pH 7.0, to produce a final protein concentration of about 40 μg ml^{-1}. This sample was diluted with one volume of glycerol, vortexed, sprayed as a fine mist on to 5 mm squares of freshly-cleaved mica, dried under a vacuum of 2.10^{-5} Torr for 30 min, and rotary shadowed at an angle of 6–8 degrees by slowly (30 s) evaporating 2 cm of 0.25 mm o.d. platinum wire wrapped around a tungsten carrier filament suspended about 10 cm from the specimen. Bar, 100 nm.

Journal of Cell Science, Supplement 14, 17–21 (1991)

Monomer pool

Long-tail myosins equilibrate between two global conformations: a straight conformation, which self-assembles into filaments, and a looped or bent conformation which is incapable of self-assembly (Fig. 1). The assembly-incompetent folded conformation is populated to a different extent for different myosin isoforms, and according to the ionic conditions. We know that for the myosins we have studied, it is greatly stabilised by the binding of MgATP into the myosin active sites. The nucleotide is trapped in the active sites by the folding reaction (Cross *et al.* 1986, 1988).

Switching molecules from the looped into the straight-tail state causes filament assembly. This can be effected (for smooth muscle and brush border myosins) by phosphorylation of the regulatory light chain, or (for molluscan myosins) by Ca^{2+}-binding to the regulatory light chain, or (in both cases) by depletion of the MgATP in the medium. Self-assembly following depletion of the MgATP is rate-limited by the very slow ($0.0002\,s^{-1}$) loss of nucleotide from the folded molecules, thought to be due to transient excursions which they make into the straight-tail state. On the basis of this assumption, it can be calculated that the straight-tail molecules comprise rather less than 1% of the monomer population in equilibrium with filaments. Readdition of nucleotide to assembled filaments causes disassembly, as the folded monomer state is repopulated (Cross *et al.* 1986, 1988).

Filament packing

Smooth muscle and nonmuscle myosin filaments are

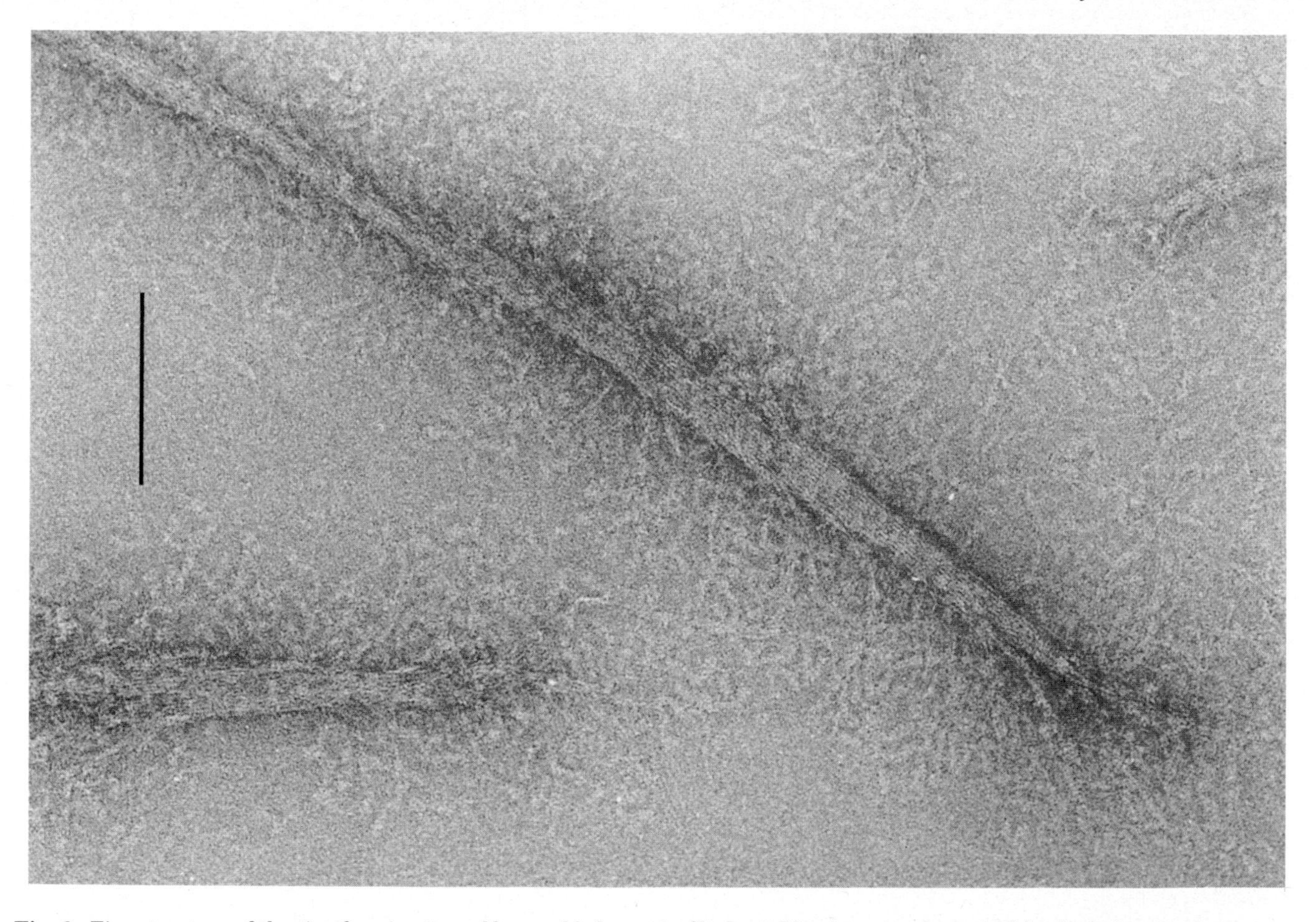

Fig. 2. Fine structure of the tip of an *in vitro* self-assembled myosin filament. Myosin was polymerised by depleting the free MgATP in a $200\,\mu g\,ml^{-1}$ solution of folded-tail monomers in 150 mM NaCl, 2 mM sodium phosphate, 10 mM imidazole, 1 mM $MgCl_2$, 0.5 mM DTT, 0.5 mM EGTA, pH 7.3 (20°C). After the attainment of steady state, a 2 μl drop of filaments was applied to a glow discharged carbon grid. After 5 s, the grid was drained by touching Whatman no. 1 filter paper to its edge, then rinsed with 6 drops of 1% uranyl acetate, drained again and dried in air at room temperature. The grids were photographed at a nominal magnification of 43 000 under low dose conditions in a Philips CM12 electron microscope with a LaB_6 filament operated at 80 kV. Bar, 100 nm.

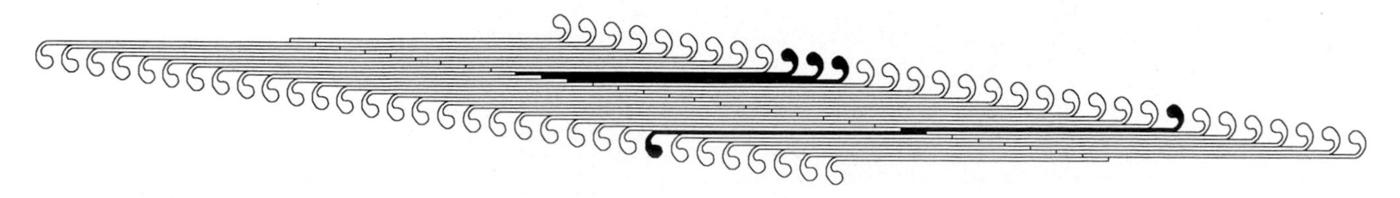

Fig. 3. Scale diagram of filament packing. For clarity, only one head of each molecule is shown.

thought to be composed only of myosin, no core or accessory proteins having been described. For smooth muscle native myosin thick filaments and for *in vitro*-assembled smooth muscle myosin, micrographs show filaments up to 5 μm long, composed of many hundreds of myosin molecules, with a 14.3 nm axial repeat in the head array, and/or a side–polar type of appearance (Sobieszek, 1972; Craig and Megerman, 1977; Hinssen *et al.* 1978; Cooke *et al.* 1989). In recent work, we have duplicated these findings and also obtained high resolution micrographs of the backbone structure of self-assembled smooth muscle myosin filaments, using a combination of improved specimen preparation and low dose electron microscopy. The images (Fig. 2) reveal that these filaments consist of sheets of straight, close-packed myosin molecules. The observation of close-packed, straight molecules assembled in a side polar arrangement with a 14.3 nm repeat defines the packing model shown in Fig. 3.

Disassembly occurs by progressive loss of molecules from filament ends

We have found that the rate of MgATP-induced disassembly depends on the filament geometry. Addition of ATP to very short myosin filaments (formed by rapid dilution of monomers into assembly conditions) causes them to disassemble completely within a few seconds. Longer filaments (formed by a slower reduction of the ionic strength at the same total myosin concentration (Sobieszek, 1972)), disassemble much more slowly, with a half time typically of about 30 s (Cross *et al.* 1986). This suggests that disassembly proceeds from the filament ends in both cases, since the only difference between the two populations is the number of filament ends. Electron microscopy of the disassembly reaction confirmed that this was so (Fig. 4). We concluded that the filament ends exchange molecules significantly faster than the interior of the filament, since otherwise the addition of MgATP would dissociate molecules from the interior of the filament and sever it.

Existing filaments as templates for reassembly

The recovery phase which follows MgATP-induced disassembly yields further information about the assembly process. Titration of short myosin filaments with incremental amounts of MgATP produces incremental amounts of disassembly, and, as discussed above, subsequent reassembly. Fig. 5 shows that for substoichiometric amounts of MgATP, recovery produces a regain of essentially the original length distribution of the filaments. Recovery from superstoichiometric MgATP produces a dramatically different steady state, in which the filaments are much longer and fewer. We think the explanation for this is that reassembling molecules prefer

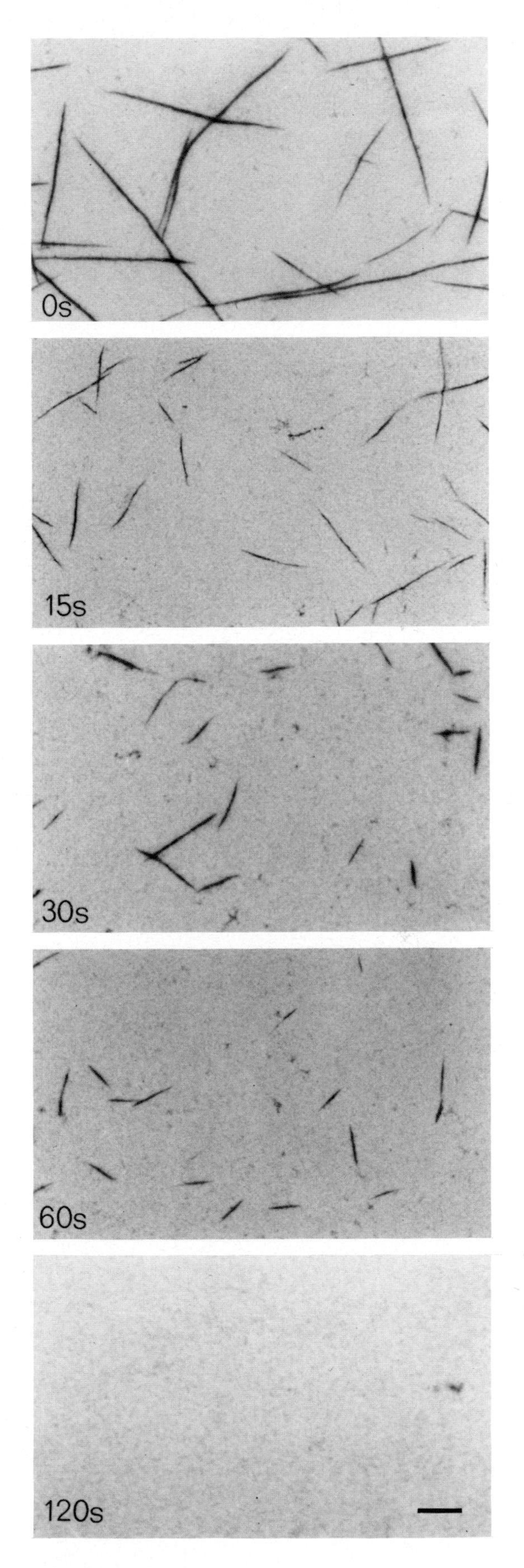

Fig. 4. Endwise disassembly of filaments, visualised by negative stain microscopy. Disassembly of a population of long filaments (top frame) was initiated by addition of an excess of MgATP. Samples were withdrawn from the reaction mixture at the intervals shown and applied to glow-discharged carbon grids. After 2–3 s the grid was rinsed with 1 % uranyl acetate, drained and the next sample taken. Conventional dose micrographs were recorded at a nominal magnification of 2500, which allowed entire grid squares to be observed. The areas shown were judged to be representative of the populations. Bar, 1 μm.

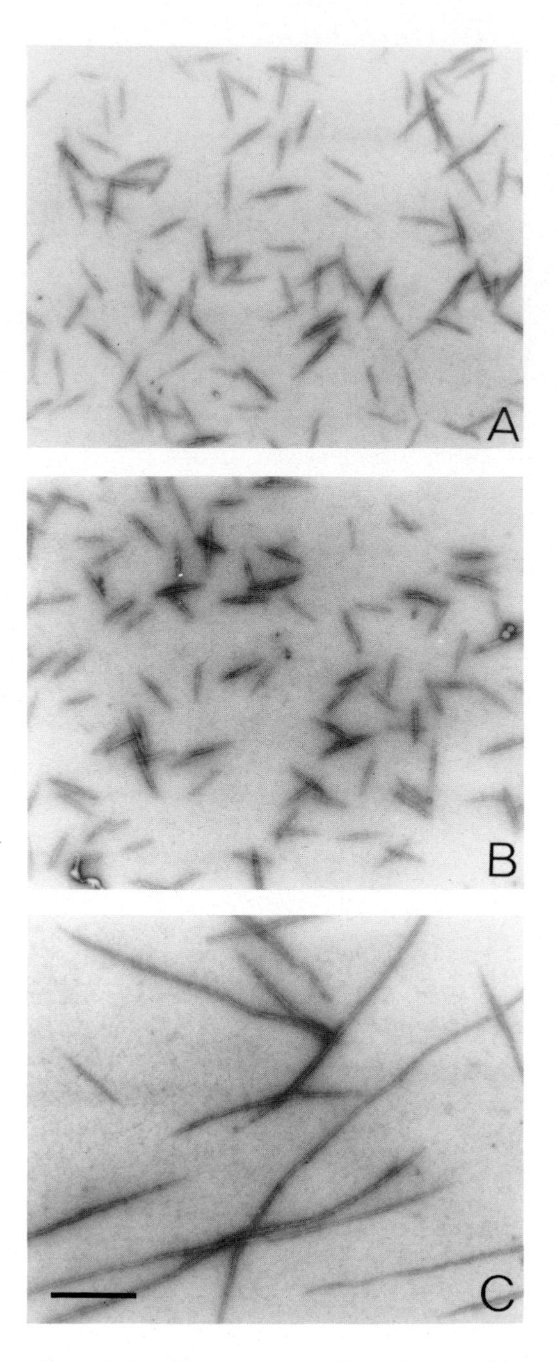

Fig. 5. Pre-existing filaments as templates for assembly. Different amounts of MgATP were added to samples of the filament preparation shown in (A), causing various degrees of disassembly (not shown). As the MgATP was hydrolysed, reassembly occurred (Cross *et al.* 1986). The final length distribution of the filaments varied according to how much MgATP was added. For substoichiometric MgATP, disassembly was limited, and reassembly occurred on to the ends of existing filaments, producing a regrowth of the original population (B). For superstoichiometric MgATP, disassembly was complete, and reassembly required fresh nuclei to be formed. Nucleation is clearly unfavourable compared to elongation: rather few nuclei are formed, and much growth occurs, producing the population of long filaments shown in (C). Note that (B) and (C) are both stable situations. There was no detectable drift of the length distribution in either case. Bar, 1 μm.

to join on to the ends of existing filaments, rather than initiate new filaments. Substoichiometric MgATP shortens filaments but does not reduce their number, and reassembling molecules simply rejoin their ends. Superstoichiometric MgATP produces complete disassembly, so that reassembly must involve initiation of new filaments. Since this is unfavourable compared to growth, few filaments are initiated, and much growth occurs, resulting in a few, long filaments.

Role of subdomains of the myosin tail in self-assembly

The repeating pattern of charge along the myosin tail has been suggested to specify the molecular overlaps for thick filament assembly (see the accompanying article by Atkinson and Stewart). Proteolysis data have gone some way towards testing this proposal, and have shown, contrary to expectation, that regions at or close to the C terminus of the myosin rod are crucial for self-assembly (Warrick and Spudich, 1987; Castellani *et al.* 1988). The precise location of the subdomain involved appears to vary according to the myosin isoform. In many myosins (but not mammalian skeletal muscle myosins) the C terminus of the tail carries an extension which is predicted to be nonhelical. This region typically contains phosphorylation sites, whose phosphorylation acts to prevent assembly.

Deletion mutagenesis in *E. coli* provides a powerful tool with which to dissect the domain function of the myosin tail at high resolution. Fig. 6 summarises recent results from our own laboratory in which progressively shorter

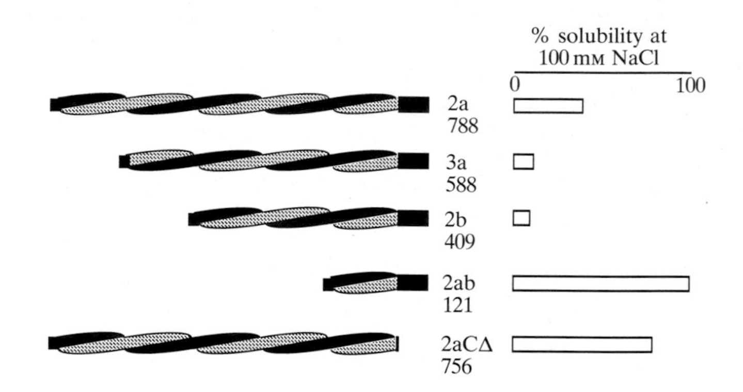

Fig. 6. The solubility of brush border myosin II rods, truncated by deletion mutagenesis. Mutant proteins were expressed in *E. coli* using a PIN expression vector. Fragments 2a, 3a and 2b were produced from cDNA clones isolated from a *lgt*11 library, 2ab and 2aCΔ were constructed from the 2a DNA clone using restriction endonuclease sites within the clone. The sizes of the expressed fragments are shown as the number of peptide residues including those derived from the fusion between the myosin coding region and the *E. coli lpp* gene in the expression vector.The solubility of expressed myosin rod fragments was tested by dialysing 100 μl (0.5 mg ml^{-1}) into 100 mM NaCl, 25 mM sodium phosphate, pH 7.5, 0.5 mM DTT and 2 mM $MgCl_2$ for 5 h. The recovered samples were weighed to determine final volume, then centrifuged at 100 000 ***g*** for 20 min in a Beckman airfuge. A supernatant sample was taken for SDS–PAGE and the remainder of the supernatant removed. The pellet was re-suspended in 100 μl 0.6 M NaCl, 25 mM sodium phosphate, pH 7.0, 10 mM Tris, pH 7.5, 0.5 mM sodium azide and 2 mM DTT overnight. Pellet and supernatant samples were run on SDS–PAGE. The gel was stained with Coomassie Blue, destained and the percentage of myosin fragment in the supernatant fraction determined by densitometry of the gel bands.

versions of the brush border myosin II tail were expressed, purified, and assayed for self-assembly by airfuge ultracentrifugation and by electron microscopy. The results were striking: progressive deletion from the N terminus of the tail produced a progressive, but slight, increase in the solubility of the myosin tail. In contrast, deletion of the C terminus completely prevented self-assembly, even for otherwise full-length transcripts (Fig. 6).

How does this C terminal domain confer the ability to self-assemble? The evidence suggests that interactions between the myosin tails in this region are stronger than in others. In Fig. 2 the N terminal part of the tail can be seen to flare away from the thick filament backbone a considerable distance, leaving the C terminal parts of the tails interacting and buried. In the absence of other evidence, we imagine that the C termini interact directly with one another, and that this strong interaction is likely to be important for an initial recognition between myosin molecules which serves to orient them at an early stage of their binding to a growing filament.

Discussion

The loose coupling of assembly and contractile activity

In several systems, myosin self-assembly occurs coincidently with an increase in the myosin MgATPase activity. It has accordingly been unclear if filament self-assembly is required for high MgATPase activity and, by implication, for contraction. Recent experiments using monoclonal antibodies have enabled the two functions of assembly and the MgATPase to be dissected away from one another: thus molecules prevented by monoclonal antibodies from folding became locked into filaments, yet were still regulated by phosphorylation. Furthermore, molecules locked into the straight-tail monomer conformation had essentially the same MgATPase activity as those in filaments. It can be concluded that in these cases, *i.e.* brush border (Citi *et al.* 1989) and smooth muscle (Trybus *et al.* 1989), assembly is not necessary for enzyme activity, but is loosely coupled to it by virtue of having the same switch.

Concluding comments

The experiments discussed have revealed some of the component reactions of myosins II self-assembly, and some unexpected features of their dynamics. Key findings are that assembly depends critically on a narrow domain at the C terminus of the myosin tail, and that molecular exchange is at the filament ends, is rapid, and occurs in preference to the initiation of fresh filaments. A predicted consequence for *ad hoc* self-assembly is that monomers will be have to be supplied to the assembly reaction faster than they can add to the ends of existing filaments, if new filaments are to be formed. A second predicted property of the system is that it should be possible to stabilise myosin filaments by capping their ends, in much the same way that F-actin can be capped.

Regarding the myosin molecule itself, the packing diagram obtained offers a rationale for the observed importance of the C-terminal region in self assembly. This domain terminates in a nonhelical region whose function is unknown, but which can be suspected of an involvement in self assembly, perhaps in the initial binding of an attacking monomer. The subsequent stages by which such a molecule samples alternative orientations and settles into position are obscure. It may be possible to throw light on them using further deletion mutagenesis of the myosin tail. The greater part of the myosin tail appears simply to act as a spacer arm. The binding of this section of the tail to its neighbours is weaker than that of the C-terminal part to its neighbours, but this may be significant in providing for slippage at some stage of monomer addition, or in achieving the delicate balance between the mechanical integrity of the filament and the exchangeability of its component molecules. Again, expression mutagenesis should tell us the answer. The distribution of myosin heads produced by the self-assembly of wild-type myosin tails can reasonably be expected to have been functionally optimised by the forces of evolution. On the other hand recent work has shown that kinesin is functional when conjugated to a tail made of part of spectrin (Yang *et al.* 1990). Is the myosin tail really such an exquisite piece of engineering? Can we conceive of an artificially improved mounting for the myosin head?

References

Castellani, L., Elliott, B. W. and Cohen, C. (1988). Phosphorylatable serine residues are located in a non-helical tailpiece of a catch muscle myosin. *J. Mus. Res. Cell Mot.* **9**, 533–540.

Citi, S., Cross, R. A., Bagshaw, C. R. and Kendrick-Jones, J. (1989). Parallel modulation of brush border myosin conformation and enzyme activity induced by monoclonal antibodies. *J. Cell Biol.* **109**, 549–556.

Cooke, P. H., Fay, F. S. and Craig, R. (1989). Myosin filaments isolated from skinned amphibian smooth muscle cells are side polar. *J. Mus. Res. Cell Mot.* **10**, 206–220.

Craig, R. and Megerman, J. (1977). Assembly of smooth muscle myosin into side-polar filaments. *J. Cell Biol.* **75**, 990–996.

Cross, R. A. (1988). What is 10S myosin for? *J. Mus. Res. Cell Mot.* **9**, 108–110.

Cross, R. A., Cross, K. E. and Sobieszek, A. (1986). ATP-linked monomer–polymer equilibrium of smooth muscle myosin: the free folded monomer traps ADP.P$_i$. *EMBO J.* **5**, 2637–2641.

Hinnsen, H., DiHaese, J., Small, J. V. and Sobieszek, A. (1978). Mode of filament assembly of myosins from muscle and nonmuscle cells. *J. Ultrastruc. Res.* **64**, 282–302.

Sobieszek, A. (1972). Crossbridges on synthetic smooth muscle myosin filaments. *J. molec. Biol.* **70**, 741–744.

Trybus, K. M. (1989). Filamentous smooth muscle myosin is regulated by phoshorylation. *J. Cell Biol.* **109**, 2887–2994.

Warrick, H. M. and Spudich, J. A. (1987). Myosin structure and function in cell motility. *A. Rev. Cell Biol.* **3**. 379–421.

Yang, J. T., Saxton, W. M., Stewart, R. J., Raff, E. C. and Goldstein, L. S. B. (1990). Evidence that the head of kinesin is sufficient for force generation and motility *in vitro*. *Science* **249**, 42–47.

Yumura, S. and Fukui, Y. (1985). Reversible cyclic AMP-dependent change in distribution of myosin filaments in *Dictyostelium*. *Nature* **314**, 195–196.

Small angle X-ray scattering studies on myosin

A. R. FARUQI, R. A. CROSS and J. KENDRICK-JONES

MRC Laboratory of Molecular Biology, Hills Road, Cambridge CB2 2QH, UK

Summary

Small angle X-ray scattering (SAXS) is a potentially powerful method for obtaining structural information from biological molecules in solution. The use of this technique in the laboratory has hitherto been limited by the long exposures necessary to obtain patterns on photographic film. Multi-wire area detectors, due to their high efficiency and absence of noise, enable patterns to be collected much more rapidly, typically in 1–2 h for a typical protein using laboratory sources. This opens up the possibility of using the technique on a semi-routine basis for a wide variety of problems. We outline the use of SAXS to characterise a large conformational change of myosin.

Key words: myosin, X-ray scattering, 6 S–10 S transition.

Small angle X-ray scattering – an introduction

X-ray scattering from a dilute, mono-disperse solution of particles is the sum of the intensities scattered by all the individual particles, i.e. the shape of the scattering curve relates to a single particle (a comprehensive treatment of the theory and practice of SAXS can be found in Glatter and Kratky, 1982). An important qualification in making such measurements is that the molecules should not have significant amounts of aggregation as the scattering shapes would then be 'contaminated' by a scattering contribution from the larger, i.e. wrong, shape. Many biological molecules in solution have been studied with this method to obtain a variety of structural parameters, discussed in greater detail below. The information is at low resolution compared to what can be obtained with protein crystallography, but it is possible to work in solution at near physiological pH and the tedious process of growing single crystals is not necessary. Scattering of X-rays takes place from the electron clouds in the atoms, and proteins have a higher electron density compared to the surrounding medium giving rise to an 'excess' of X-ray scattering which constitutes the signal. Electrons from different parts of the molecule act as secondary scatterers of the incident radiation; the interference between the different wavelets results in the characteristic scattering pattern.

Radius of gyration

A very useful parameter which characterises the shape of proteins is the radius of gyration, R_g, which is obtained from the slope of the inner part of the scattering curve when it is plotted as the logarithm of intensity, ln(I), as a function of a parameter, h^2, where h is proportional to the scattering angle:

$$h = 4\pi \sin\theta/\lambda, \qquad (1)$$

where 2θ is the scattering angle, and λ is the wavelength of radiation used. Some authors use Q instead of h, especially in neutron scattering. Another useful parameter, s is defined as:

$$s = 2\sin\theta/\lambda. \qquad (2)$$

Both h and s are independent of the wavelength used and have the dimensions of reciprocal angstrom (or nanometers). The radius of gyration is the root mean square distance of the electrons from the centre of gravity of the molecule and is a quantity which is very similar to the moment of inertia of the molecule. Two different shapes can, however, lead to the same radius of gyration, something one has to be aware of! Katoaka *et al.* (1989) have measured the conformational changes in calmodulin due to melittin binding in the presence and absence of calcium by measuring the radius of gyration of calmodulin and the calmodulin–mellitin complex. The surprising finding was that the radius of gyration decreased from 20.9Å for calmodulin to 18.0Å for the complex, but only in the presence of calcium. It is suggested that this reduction is due to a change from a 'dumbbell' configuration to a more globular form which is spatially less extended. In a similar study on another calcium binding protein, Troponin C, Hubbard *et al.* (1988) have found that the conformational change in Troponin C on binding calcium is too small to be measured. In terms of accuracy of measurement, one can say that, in general, if the protein is not too sensitive to radiation (and denatures) it is possible to measure radius of gyration of, say 30Å to an accuracy of better than 0.5Å in a reasonable time in the laboratory (a few hours, or a few minutes at a synchrotron). If the intensity of the direct beam is known, one can also obtain the molecular weight of the protein; alternatively, one can callibrate the intensity by using a protein of similar size and known molecular weight.

Radius of gyration of cross-section

For molecules which are much bigger in one direction, i.e. are rod-shaped, one can get another useful parameter, the radius of gyration of cross-section which, as the name implies, is a measure of the spatial extent of the cross-sectional area and is independent of the length. In order to measure the radius of gyration of cross-section, one needs to measure the slope of the inner portion of the curve showing the relation between ln(Ih) and h^2. As in the case discussed above, if the intensity of the direct beam can be measured then it is possible to calculate the mass per unit length of the 'rod-like' molecule. Sund *et al.* (1969) have used this method to study the aggregation of beef liver

Journal of Cell Science, Supplement 14, 23–26 (1991)

glutamate dedydrogenase between concentrations of 1–33 mg ml^{-1} when the molecular weight increases from 0.5 to 2.0×10^{6}. They found that though the molecular weight increased by a large amount, the radius of gyration of cross-section stayed remarkably constant, suggesting linear aggregation along the long axis without altering the cross-section. In another example (Sperling and Tardieu, 1976), the mass per unit length of the '100Å' chromatin fibre was measured to find out the density of the nucleosome core particle along the fibre, in order to decide between two models. The transition of the 100Å chromatin fibre to a 300Å filament in response to salt changes has also been studied with this technique by Widom (1986) and by Greulich *et al.* (1987).

Time resolved studies

It is possible to obtain a greatly increased flux of X-rays from synchrotron radiation making it possible to make time resolved measurements. In ordered systems like muscle, time resolutions of 1 millisecond were achieved some years back (reviewed by Huxley and Faruqi, 1983), but even in solution scattering, where the scattering signal is much weaker, about 100 millisecond time resolutions have been achieved and two examples are discussed below. The assembly and disassembly of microtubule protein caused by temperature changes was studied by Bordas *et al.* (1983) at the EMBL Outstation in Hamburg. Modelling the intermediate and final states, they were able to establish that microtubule nucleation started at 22 °C and elongation took place between 25–30 °C. The elongation was due mainly to the addition of tubulin subunits and not whole rings. Conversely, disassembly was induced by lowering the temperature and proceeded by removal of subunits. In a separate study on the kinetics of actin polymerisation, Matsudaira *et al.* (1987) demonstrated the absence of a rate-limiting nucleation step in the salt-induced assembly of actin filaments. Actin filaments were found to be approximately 300Å long within about 10 s of initiation.

X-ray studies on myosin

Myosin from smooth muscle, certain vertebrate muscle and some non-muscle cells exists in a 'rolled-up' configuration where the tail seems to be attached to the neck region and has a sedimentation coefficient of 10 S, compared to 6 S for the unfolded form. In the 10 S state the products of ATP cleavage are trapped by the folding and the enzymatic activity of myosin is greatly reduced; a transition to the 6 S state is probably essential before the monomers can be incorporated into filaments (Cross, 1988). Phosphorylation of the regulatory light chains allows actin activation and inhibits the transition to the 10 S form, biasing the equilibrium towards the unfolded form which can then be assembled into filaments.

We have measured the small angle scattering from both the 6 S and the 10 S forms of myosin in the range 350–25Å^{-1} with the aim of defining the scattering signatures and establishing significant differences, if such exist. As this work is still in progress, we are presenting only a small part of the data which has been analysed; data has been compared with modelling studies based on predicted scattering curves derived from 'expected' shapes of the two conformations of myosin. Previous work by Mendelsohn and Wagner (1984) has centred on the shape of the S-1 (myosin head) subunit and on single-headed myosin to find if there was a change in the structure of the S-1 when not attached to the rod; no significant differences could be detected. Gorrigos and Vachette (1989) have more recently measured the radius of gyration of S-1 as 37Å. However, to our knowledge, there has been no study on whole myosin monomers in either the unfolded or folded forms.

The small angle camera

The small angle camera, consisting of a double focusing mirror arrangement, is set on a fine focus, rotating anode, X-ray generator operating at 2.4 kW; apart from the entrance slits before the first mirror, there are two sets of carefully machined and tantalum-lined slits which collimate the beam accurately and cut out stray radiation (discussed by Widom, 1986). The camera background is extremely low and in our measurements is not corrected for; however, when subtracting buffer scattering from the protein, it is automatically eliminated. The beam path subsequent to the first slit up to the detector is in vacuum except for the specimen which is in air. The 'specimen to detector' distance was set to 750 mm for the present set of experiments, which permits data to be recorded from about 350Å^{-1} to about 25Å^{-1}. Due to the longer 'source to mirror' distance in the horizontal plane, the resolution is in fact better, but as we are circularly averaging the data the poorer resolution is applicable. The scattering pattern is recorded on a multiwire area detector (Faruqi, 1988), 100 mm by 100 mm, filled with a mixture of 90% xenon and 10% methane. The efficiency of the detector for 8 keV photons is about 80% including absorption effects in the 1 mm beryllium entrance window. The detector has two cathode planes with delay lines sandwiching an anode plane which only provides the event trigger pulse and pulse height information. Start and stop pulses from the cathode plane are converted to digital form in a time-to-digital converter and combined to form a 16 bit address which is used for incrementing a histogramming memory. The digital electronics, which is in CAMAC, is controlled by a HYTEC Crate Controller and connected to a DEC LSI 11/23 Plus computer which also has a fast data link to a VAX 8600 computer. During an experiment, data is accumulated in the histogramming memory and displayed on a colour monitor using a MATROX driver board along with a linear slice of data selected from the area. Data are transferred to the VAX 8600 on completion of the experiment, where the data is further analysed using a range of programs before being archived. Measurements on detector properties and demonstration of test patterns are given in Faruqi and Andrews (1989).

Data analysis and modelling studies

As solution scattering data from molecules with random orientations is isotropic the area detector output data was circularly averaged and the buffer scattering subtracted to leave just the protein contribution. The 'linear' data was analysed using a modified version of OTOKO, a general analysis program developed at the EMBL, Hamburg (Koch and Bendall, 1981), to obtain the plots of intensity as a function of scattering angle and measurements of radius of gyration, radius of gyration of cross-section, etc. Comparison of myosin scattering data was made with 'expected' scattering curves calculated from model structures for the molecule using the program ATOMIN, a program developed for generating scattering curves from objects using the Debye equations (Bordas *et al.* 1983, with additions at Daresbury). For the purposes of modelling, myosin was represented by three basic units: a 20Å diameter, 1500Å long rod with two heads at one end whose

A

Fig. 1. Model structures composed of small spheres used in the construction of 6 S (A) and 10 S (B) conformations of the myosin molecule.

shape was taken from recent X-ray and electron microscopic data (Winkelman *et al.* 1985) and shown schematically in Fig. 1A. The maximum length of the head was just over 160Å, with a distinct curved region near the thinner end of the head. The radius of gyration of the (model) head is approximately 36Å, which is close to the experimentally measured values of 37Å for rabbit S-1 (Garrigos and Vachette, 1989). The rod was represented by spheres of 20Å diameter and the head by 10Å diameter spheres, which provides adequate resolution in comparison with the data. The 'looped' rod structure and the heads, representing a possible 10 S configuration, are shown in Fig. 1B. The diameter of the rod is again 20Å and it is 1500Å long but, about 1000Å from the tail end it is attached to the neck region, as observed by electron microscopy (Cross *et al.* 1986). In order to quantitate the 'goodness of fit' between calculated and measured scattering curves, the calculated curves were first normalised to the measured curves using the total scattered intensity in the region (taken between 250Å^{-1} and 40Å^{-1}), followed by a calculation of a simple residual defined as:

$$F\chi^2 = \text{weight}^*((I_{\text{obs}} - I_{\text{calc}})^2)/N^*I_{\text{obs}}\,, \qquad (3)$$

where weight$=1/I_{\text{obs}}$.

Results

The scattering from myosin in both the 6 S and the 10 S forms is dominated by the two 'heads'; the 'rolled-up' rod does increase the 'compactness' of the molecule however, as evidenced by plots of the radius of gyration of cross-section in Fig. 2, which also shows the most striking difference in the shape of scattering from the two forms. The 'apparent' radius of gyration of cross-section is about 17Å for the 6 S and 32Å for the 10 S myosin in these preparations. The scattering from the 6 S was measured with concentrations ranging from 2 mg ml^{-1} to 25 mg ml^{-1} and the 10 S form betwen 2 mg ml^{-1} to 8 mg ml^{-1}, the latter being the highest concentration feasible for the 10 S form for these experiments. The source of 6 S was smooth muscle from chicken gizzard and scallop muscle; all the 10 S was from smooth muscle. We present an example from 6 S and one from 10 S scattering to illustrate the general features of the scattering pattern and to obtain a 'goodness-of-fit' estimate with the proposed model. The 6 S scattering curve shown in Fig. 3A is from scallop muscle (see article by Ankrett *et al.* in this volume) from a 6 h exposure and it is compared with the model curve in the same plot. A similar plot for the 10 S form is shown in Fig. 3B from a shorter exposure of 2 h. Again, the model curve is shown in the same plot. The model is derived from the structure described in the previous section. Clearly the 6 S model is better than the 10 S model and we are investigating with different models and improved specimen preparations possible ways to improve the fit. It is possible to vary model parameters and compare with experimental data to investigate how the fits vary with the parameter. We have varied the angle between the two heads and calculated the scattering from the new shapes. Somewhat surprisingly, as shown in Table 1, the fit appears to improve when the heads are close together and gets worse as the inter-head angle is increased, becoming worst when the angle is 180 degrees; models in which the heads are close together but the 'couplet' is at different angles to the rod are difficult to distinguish at the resolution at which we are working.

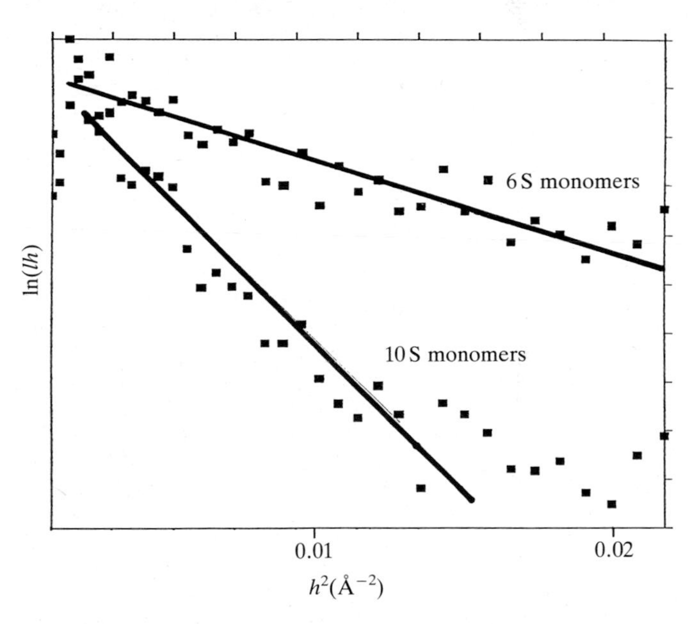

Fig. 2. Plots of the scattering curves from the 6 S and 10 S monomers showing the change in the radius of gyration of cross-section between the two forms when plotted against the logarithm of intensity. The 6 S data were collected over 4.5 h and the 10 S data over 2 h.

Table 1. *The 'goodness-of-fit' factor as a function of inter-head angle*

Angle between heads (degrees)	$F\chi^2$
35	13.1
45	9.6
90	24.2
180	29.6
270	34.2
315	15.8
325	8.1

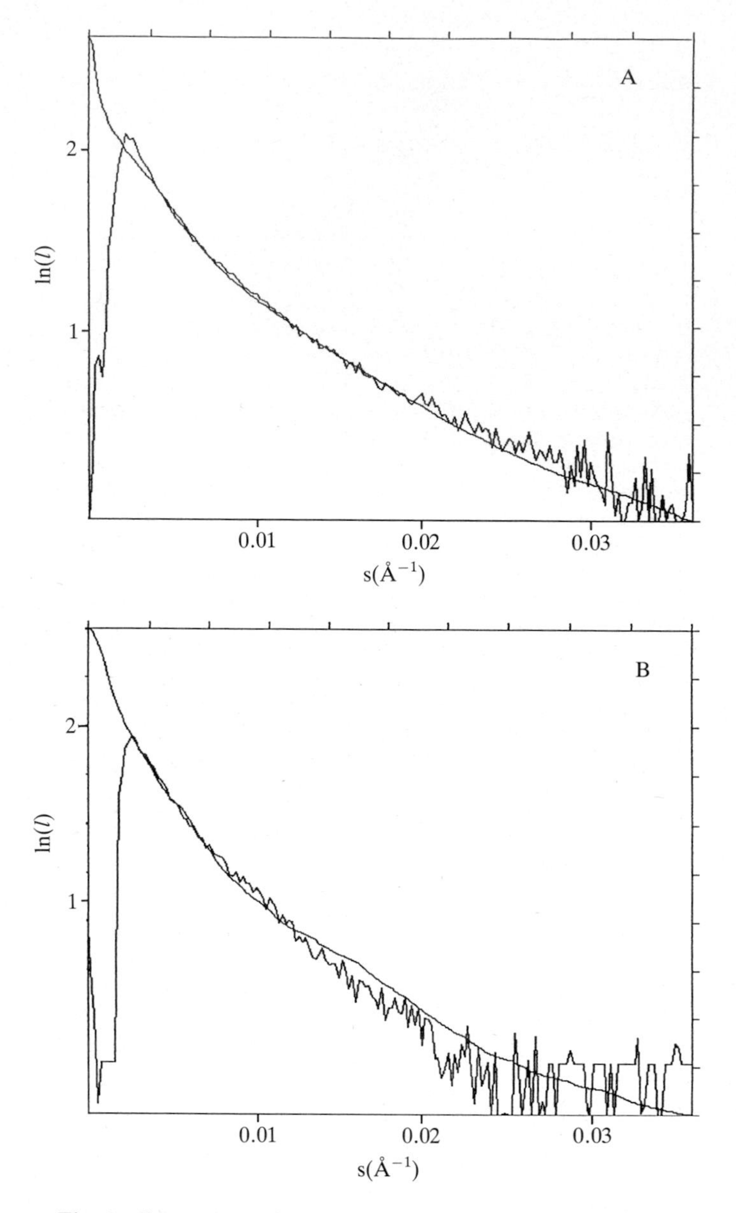

Fig. 3. Comparison of scattering from 6 S monomers and from the 'model' structure in (A) and the same for the 10 S in (B). More details are given in the text.

Conclusions

The information obtained from X-ray scattering diagrams is at relatively low resolution, and accordingly the method or self-association. Electronic area detectors render it possible to obtain such X-ray scattering data from small volumes (100 μl) of protein solutions at concentrations of a few mg ml^{-1} in about an hour. Preliminary experiments suggest it will be possible dynamically to study the 6 S to 10 S transition and other early events of myosin assembly at the Daresbury synchrotron.

References

Bordas, J., Mandelkow, E. M. and Mandelkow, E. (1983). Stages of tubulin assembly and disassembly studied by time-resolved synchrotron X-ray scattering. *J. molec. Biol.* **164**, 89–135.

Cross, R. A. (1988). What is 10S myosin for? *J. Musc. Res. Cell Mot.* **9**, 108–109.

Cross, R. A., Cross, K. E. and Sobieszek, A. (1986). ATP-linked monomer–polymer equilibrium of smooth muscle myosin: the free folded monomer traps ADP.P_i. *EMBO J.* **5**, 2637–2641.

Faruqi, A. R. (1988). A high-resolution multiwire area detector for X-ray scattering. *Nucl. Instrum. Meth.* A**273**, 754–763.

Faruqi, A. R. and Andrews, H. (1989). Development and application of multiwire detectors in biological X-ray studies. *Nucl. Instrum. Meth.* A**283**, 445–447.

Glatter, O. and Kratky, O. (1982). *Small Angle X-Ray Scattering.* Academic Press: New York and London.

Gorrigos, M. and Vachette, P. (1989). Solution X-ray scattering studies of rabbit myosin subfragment-1. *Biophys. J.* **55**, 80a.

Greulich, K. O., Wachtel, E., Ausio, J., Seger, D. and Eisenberg, H. (1987). Transition of chromatin from the 10 nm lower order structure to the 30 nm higher order structure as followed by small angle X-ray scattering. *J. molec. Biol.* **193**, 709–721.

Hubbard, S. R., Hodgson, K. O. and Doniach, S. (1988). Small angle X-ray scattering investigation of the solution structure of troponin C. *J. biol. Chem.* **263**, 4151–4158.

Huxley, H. E. and Faruqi, A. R. (1983). Time resolved X-ray diffraction studies on vertebrate striated muscle. *A. Rev. Biophys. Bioeng.* **12**, 381–417.

Kataoka, M., Head, J., Seaton, B. S. and Engelman, D. M. (1989). Mellitin binding causes a large calcium ion dependent conformational change in calmodulin. *Proc. natn. Acad. Sci. U.S.A.* **86**, 6944–6948.

Koch, M. H. J. and Bendall, P. (1981). *Proc. Dig. Equip. Users Soc.*, 13–16.

Matsudaira, P., Bordas, J. and Koch, M. H. J. (1987). Synchrotron X-ray diffraction studies of actin structure during polymerisation. *Proc. natn. Acad. Sci.* **84**, 3151–3155.

Mendelsohn, R. A. and Wagner, P. (1984). X-ray scattering by single-headed heavy meromyosin cleavage of the myosin head from the rod does not change its shape. *J. molec. Biol.* **177**, 153–171.

Sperling, L. and Tardieu, A. (1976). The mass per unit length of chromatin by low angle X-ray scattering. *FEBS Lett.* **64**, 89–91.

Sund, H., Pilz, I. and Herbert, M. (1969). Studies of glutamate dehydrogenase. 5. The X-ray small angle investigation of beef liver glutamate dehydrogenase. *Eur. J. Biochem.* **7**, 517–525.

Widom, J. (1986). Physicochemical studies of the folding of the 100Å nucleosome filament into the 300Å filament. *J. molec. Biol.* **190**, 411–424.

Winkelman, D. A., Mckeel, H. and Rayment, I. (1985). Packing analysis of crystalline myosin subfragment-1: implications for the size and shape of the myosin head. *J. molec. Biol.* **181**, 487–501.

From genes to tensile forces: genetic dissection of contractile protein assembly and function in *Drosophila melanogaster*

ERIC FYRBERG*, CLIFFORD BEALL and CHRISTINE C. FYRBERG

Department of Biology, The Johns Hopkins University, Charles and 34th Streets, Baltimore, Maryland 21218, USA

* Corresponding Author

Summary

Myofibrils, the contractile organelles of skeletal muscle, are highly ordered and precisely regulated actomyosin networks. Investigations of myofibril assembly are revealing the cellular mechanisms by which contractile components are arranged and regulated. In order to facilitate this research we have developed formal molecular genetics for myofibrillar proteins of *Drosophila* flight muscle. Presently, mutations can be used systematically to perturb or eliminate any of the classical myofibrillar proteins within these fibers, and the *in vivo* consequences can be conveniently evaluated using protein electrophoresis, electron microscopy, or by assaying flight performance. Here we review some recent progress.

Key words: muscle myofibrils, contractile proteins, *Drosophila*.

Introduction

Our work aims to further elucidate the assembly and functioning of myofibrils, the contractile organelles of skeletal muscle, and to extend this knowledge to actomyosin networks of the cellular cytoskeleton. Our approach combines molecular genetics with cell biology and physiology. We use mutations to eliminate or perturb particular contractile and cytoskeletal components in the fruit fly, *Drosophila melanogaster*, then observe the effects on muscle structure and function. The genetic approach allows us to manipulate muscle proteins *in vivo*, and thus to bridge the gulf that continues to exist between biochemical studies of purified proteins and ultrastructural, biophysical, and immunohistochemical studies of cells, tissues, and organisms. *Drosophila*, one of the most completely understood metazoans, is an excellent subject for such work because its complex physiology and simple genome are amenable both to genetic and cell biological approaches (refer to Fyrberg and Goldstein, 1990).

Of several types of adult fibers, *Drosophila* indirect flight muscles are best suited for experimental work. Within the laboratory these muscles are expendable, hence mutations that perturb their structure and function are in many instances not lethal, merely causing the loss of flight. This factor considerably simplifies mutant analyses. Also noteworthy is the fact that myofibrils of these fibers are highly ordered, and thus facilitate discernment of the structural perturbations caused by particular mutations using transmission electron microscopy. Finally, although much smaller than vertebrate muscles, the fibers are large enough to provide material for gel electrophoretic, biochemical, and physiological analyses.

Before describing our strategies and observations we should clarify how it is possible to perturb or eliminate contractile proteins of flight muscles without affecting those of essential muscle groups. The most straightforward approach is to mutate a contractile protein gene or exon that is expressed only within flight muscles. However, one is by no means limited to that strategy, as it is clear that particular missense or regulatory mutations which impede contractile protein accumulation and function are in many instances manifested chiefly in flight muscles (refer to Fyrberg and Beall, 1990, for further discussion).

Summing the parts: dissecting myofibril assembly using null alleles

Perhaps the most direct method for elaborating the flow of information during myofibril assembly is to eliminate particular components and assess how much of the remaining sarcomeric structure is able to form. These experiments are made possible by a growing number of gene mutations that eliminate the synthesis or accumulation of major myofibrillar components (refer to Fyrberg and Beall, 1990; Fyrberg *et al.* 1990*a*). For example, *Act88F*^{KM88}, an actin gene allele wherein the codon for tryptophan 79 (TGG) is converted to an opal terminator (TGA), eliminates all actin and thin filaments in flight muscles (Okamoto *et al.* 1986; Beall *et al.* 1989). In the absence of actin no thin filaments or Z-discs form, but reasonably well ordered arrays of thick filaments and M-lines nevertheless assemble. In the converse experiment we have used a flight muscle-specific myosin heavy chain allele, *Ifm(2)2* to eliminate thick filaments. In this case arrays of thin filaments and Z-lines form. These two experiments demonstrate that thick filament/M-line and thin filament/Z-disc arrays assemble independently, that thick filaments are probably not integral Z-disc components, and further suggest that M-lines and Z-discs are organizing centers for myofibril formation. Analyses of actin and myosin heavy chain, null allele, heterozygotes (flies having one normal allele and one null allele) have demonstrated the importance of filament stoichiometry for normal myofibrillar assembly (Beall *et al.* 1989).

Analyses of muscles having tropomyosin, troponin-T, and alpha-actinin null alleles are similarly defining the roles of the corresponding proteins in myofibril assembly and maintenance. In the absence of troponin-T or tropomyosin, thin filaments are largely absent from sarcomeres (Fyrberg *et al.* 1990*b*; our unpublished obser-

Journal of Cell Science, Supplement 14, 27–29 (1991)

vations), demonstrating the importance of these proteins in the stabilization of F-actin. Absence of the principal actin filament crosslinking protein, alpha-actinin, leads to progressive muscle degeneration and paralysis in the *Drosophila* larva, consistent with the notion that these crosslinks are essential for fixing thin filaments in place during contraction, a role similar to that proposed for spectrin in the membrane-associated cytoskeleton.

The crux of myofibril assembly: aligning and integrating thick and thin filaments

A central question of myofibril assembly is how the thick and thin filament networks are aligned and integrated. Several hypotheses have been invoked to explain this facet of the assembly process, and genetic experiments have begun to test critically some of these ideas. One notion, based upon electron microscopy of differentiating muscles, is that the initial alignment and integration of thick and thin filaments is mediated in some as yet undetermined fashion by transient arrays of microtubules. Microtubule bundles or arrays can typically be seen in differentiating muscle (Fischman, 1967). Our recent survey of *Drosophila* myofibrillogenesis (done in collaboration with Mary Reedy of Duke University Medical Center) has revealed that in developing insect muscle microtubules are found closely apposed to, or surrounding, nascent myofibrils (M. C. Reedy, personal communication; refer also to work of Auber and Couteaux, 1963). Antin *et al.* (1981) previously showed that microtubules can integrate into the myofibrillar lattice in taxol-treated, postmitotic, chicken myoblasts. However, since no specific interaction between a microtubule-associated protein and a myofibrillar component has been demonstrated, a causal relationship between microtubules and myofibrillogenesis has yet to be established. The recent demonstration by Leiss *et al.* (1988) and Kimble *et al.* (1989) that a particular isoform of *Drosophila* tubulin (β3) is transiently expressed in all developing muscle cells has rekindled interest in the issue, and provided a system in which to investigate rigorously the role of transient microtubule arrays in myofibrillogenesis. Preliminary results from Raff and collaborators suggest that certain mutations in this tubulin gene reduce muscle mass and perturb contractile function. Examination of these muscles in the electron microscope should establish whether the corresponding microtubules serve as a nonspecific scaffold or as a specific template for sarcomere formation.

A second hypothesis is that the long elastic protein referred to as titin connects M-lines with Z-discs, and thus facilitates integration of thick and thin filament arrays, also possibly serving as a molecular ruler for establishing thick filament length (Whiting *et al.* 1989). A related molecule named twitchin has been characterized in nematodes, and the effects of a number of mutant alleles on muscle formation have been evaluated in this organism. On the basis of these results, Benian *et al.* (1989) concluded that twitchin is involved in regulating the contraction–relaxation cycle and has little, if any, role in myofibril assembly. In order to use genetics to evaluate titin/twitchin function in a second system we have isolated and partially characterized the *Drosophila* analogue. Using antibodies against insect titin, Belinda Bullard and collaborators (EMBL, Heidelberg) isolated the corresponding cDNA of the giant waterbug, *Lethocerus*, from expression libraries (Lakey *et al.* 1990). By hybridizing this cDNA to *Drosophila* genomic and cDNA libraries at low stringency we isolated the corresponding genes. Partial sequencing of several cDNAs has confirmed the identity of the *Drosophila* gene. *In situ* hybridization of the gene to polytene chromosomes has revealed that the gene is within subdivision 102 of the fourth chromosome, the site of a number of interesting candidate mutations. Presently, we are analyzing a variety of these putative mutants in order to establish which lack titin, and how that deficit affects *Drosophila* myofibril formation.

Disrupting contractility: analyses of point mutations that preclude normal myofibrillar function

It is possible, in *Drosophila*, to mutate any residue of a contractile protein, then to test functioning of the mutant derivatives by introducing them back into the corresponding null mutant. Alternatively, one can characterize a variety of randomly induced point mutations within a contractile protein gene, correlating the molecular changes to the most interesting and informative phenotypes engendered. We are conducting both approaches successfully. Our purposes in these studies are to define better how actin and myosin generate force, and to delineate how force production is regulated by various regulatory proteins.

To understand force generation we are focusing our mutagenesis efforts on actin. Of the several actin mutations analyzed in our laboratory, probably the most informative involved conversion of glycine six and alanine seven to alanine and threonine, respectively. Thick and thin filaments within the perpipheral regions of myofibrils containing this actin are out of register, and one occasionally finds thick filaments flanked by actin filaments having opposite polarities. Electron microscopy of these arrays revealed that crossbridges can form with both of the thin filaments, revealing the extreme torsional flexibility resident within the myosin head or head–tail junction (Reedy *et al.* 1989). Presently, we are manufacturing a number of additional actin mutations in order to test various structure/function relationships, a project that will benefit considerably from the recent solution of the structure of actin at atomic resolution (Kabsch *et al.* 1990). One obvious experiment involves evaluating the function of the single methylhistidine residue of actin. We are presently converting it to both arginine and tyrosine. We are also using alanine scanning mutagenesis (Cunningham and Wells, 1989) to perturb regions of actin known to interact with myosin, the hope being that functional testing of such derivatives will ultimately define the nature of the conformational change proposed to accompany the development of tension.

In order to understand better how the contraction–relaxation cycle is regulated, we have developed molecular genetics for tropomyosin and the three subunits of troponin. The most interesting observation we have made to date is that certain missense mutations in troponin-T and -I assemble into normal myofibrils but subsequently degenerate. These abnormalities appear to be due to aberrant interactions of actin and myosin, as elimination of myosin using a heavy chain null allele prevents the degeneration (Fyrberg *et al.* 1990*a*). The most straightforward interpretation of these results is that the mutant troponin subunits foster abnormal interactions of actin and myosin that destabilize the myofibrillar lattices. More detailed functional observations should reveal the precise cause of the myofibrillar degeneration and may ultimately elucidate mechanisms driving the regulated crossbridge cycle.

References

ANTIN, P. B., FORRY-SCHAUDIES, S., FRIEDMAN, T. M., TAPSCOTT, S. J. AND HOLTZER, H. (1981). Taxol induces postmitotic myoblasts to assemble interdigitating microtubule–myosin arrays that exclude actin filaments. *J. Cell Biol.* **90**, 300–8.

AUBER, J. AND COUTEAUX, R. (1963). Ultrastructure de la strié Z dans des muscles du *Dipteres*. *J. Microsc.* **2**, 309–24.

BEALL, C. J., SEPANSKI, M. A. AND FYRBERG, E. A. (1989). Genetic dissection of *Drosophila* myofibril formation: effects of actin and myosin heavy chain null alleles. *Genes Dev.* **3**, 131–40.

BENIAN, G. M., KIFF, J. E., NECKELMANN, N., MOERMAN, D. G. AND WATERSTON, R. H. (1989). Sequence of an unusually large protein implicated in regulation of myosin activity in *C. elegans*. *Nature* **342**, 45–50.

CUNNINGHAM, B. C. AND WELLS, J. A. (1989). High-resolution epitope mapping of hGH-receptor interactions by alanine-scanning mutagenesis. *Science* **244**, 1081–85.

FISCHMAN, D. A. (1967). An electron microscopy study of myofibril formation in embryonic chick skeletal muscle. *J. Cell Biol.* **32**, 557–75.

FYRBERG, E. AND BEALL, C. J. (1990). Genetic approaches to myofibril form and function in *Drosophila*. *Trends Genet.* **6**, 126–31.

FYRBERG, E., FYRBERG, C. C., BEALL, C. AND SAVILLE, D. L. (1990*a*). *Drosophila melanogaster* troponin-T mutations engender three distinct syndromes of myofibrillar abnormalities. *J. molec. Biol.* In press.

FYRBERG, E. AND GOLDSTEIN, L. S. B. (1990). The *Drosophila* cytoskeleton. *A. Rev. Cell Biol.* **6**: In press.

FYRBERG, E., KELLY, M., BALL, E., FYRBERG, C. AND REEDY, M. C. (1990*b*). Molecular genetics of *Drosophila* alpha-actinin: mutant alleles disrupt Z discs integrity and muscle insertions. *J. Cell Biol.* **110**, 1999–2012.

KABSCH, W., MANNHERZ, H. G., SUCK, D., PAI, E. F. AND HOLMES, K. C. (1990). Atomic structure of the actin:DNase I complex. *Nature* **347**, 37–44.

KIMBLE, M., INCARDONA, J. P. AND RAFF, E. C. (1989). A variant β-tubulin isoform of *Drosophila melanogaster* (β3) is expressed primarily in tissues of mesodermal origin in embryos and pupae, and is utilized in populations of transient microtubules. *Devl Biol.* **131**, 415–29.

LAKEY, A., FERGUSON, C., LABEIT, S., REEDY, M., BUTCHER, G., LEONARD, K. AND BULLARD, B. (1990). Identification and localization of high molecular weight proteins in insect flight and leg muscle. *EMBO J.* In press.

LEISS, D., HINZ, U., GASCH, A., MERTZ, R. AND RENKAWITZ-POHL, R. (1988). β3 tubulin expression characterizes the differentiating mesodermal germ layer during *Drosophila* embryogenesis. *Development* **104**, 525–31.

OKAMOTO, H., HIROMI, Y., ISHIKAWA, E., YAMADA, T., ISODA, K., MAEKAWA, H. AND HOTTA, Y. (1986). Molecular characterization of mutant actin genes which induce heat-shock proteins in *Drosophila* flight muscles. *EMBO J.* **5**, 589–96.

REEDY, M. C., BEALL, C. J. AND FYRBERG, E. A. (1989). Myosin heads can form reverse rigor chevrons. *Nature* **339**, 481–83.

WHITING, A., WARDALE, J. AND TRINICK, J. (1989). Does titin regulate the length of thick filaments? *J. molec. Biol.* **205**, 263–68.

The influence of pressure on actin and myosin interactions in solution and in single muscle fibres

M. A. GEEVES

Department of Biochemistry, School of Medical Sciences, University of Bristol, Bristol BS8 1TD, UK

Summary

Studies of the molecular mechanism of motile activity require the capacity to examine the properties of individual, isolated molecular components and the properties of these same molecular components in the organised system. Pressure perturbation is one method which can be applied to motile systems at different levels of organisation. We show here that pressure perturbs a specific interaction between actin and myosin in solution and also perturbs the cycling crossbridge in a contracting muscle.

Key words: actin, myosin, muscle, pressure.

Introduction

Muscle contraction is the result of a dynamic interaction between several proteins within the muscle fibre. A complete molecular description of this process requires information on both the behaviour of individual isolated proteins and the behaviour of these same proteins when assembled into the contracting muscle. Few methods can be readily applied to proteins both in solution and in the cellular environment. Over the past 5 years we have been exploring the use of pressure perturbation methods for probing protein–protein interactions at different levels of organisation in muscle.

Before detailing the principal results we have obtained, a few words are appropriate about the background and methods of this novel approach.

The use of pressure to perturb a system at equilibrium (or in a steady state) depends upon there being a difference in the volume of the components of the equilibrium mixture.

$$A \overset{K_{eq}}{\rightleftharpoons} B, \qquad (1)$$

where K_{eq} is the equilibrium constant.

If the molar volume of A differs from that of B by ΔV then for a small pressure change ΔP:

$$\Delta K/K = -\Delta V \Delta P/\boldsymbol{RT}, \qquad (2)$$

where ΔK is the pressure-induced change in equilibrium constant, $\boldsymbol{R}$ is the gas constant and T is the absolute temperature. In solution, volume changes normally result from changes in the local water structure when the number of ionic and/or hydrophobic groups exposed to the solvent is changed. Protein–protein and protein–ligand interactions or the resulting protein conformational changes frequently involve changes in the side chains that are exposed to solvent and so can have significant volume changes.

The magnitudes of the perturbation induced in the concentrations of A and B depend not only on ΔV but also on the magnitude of K. A 10 % change in K will result in a much greater change in [B]/([A]+[B]) when K=1 than when K is much greater or much less than 1.

Materials and methods

We have been using a pressure chamber for both solution and muscle fibre studies that is based on the equipment built by Davis and Gutfreund (1976). They originally used this apparatus to study the assembly of myosin into filaments (Davis, 1981*a*,*b*). The apparatus shown in Fig. 1A applies hydrostatic pressure *via* an oil line to a solution in the centre of a 3×3×3 in stainless-steel block. The oil is separated from the experimental solution by a deformable Teflon membrane. When the solution is stabilised at the elevated pressure, the pressure can be released *via* a specially constructed valve that returns the pressure to 101.3 kPa (=1 atm) within 200 μs.

For optical studies of protein solutions the cell has three sapphire windows, which allow both transmission and fluorescence measurements to be made. For mechanical studies on muscle fibres one of the windows is removed and is replaced by a tension transducer mounted on a pressure sealing plug (Fig. 1B). The muscle fibre is glued between a hook attached to the transducer and a second hook fixed to the transducer mounting before the whole assembly is inserted into the pressure chamber.

Results

The effect of pressure on actin–myosin subfragment 1 in solution

In 1982 we published experiments which showed that the interaction between actin and myosin subfragment 1 (actin.S1) was pressure sensitive (Geeves and Gutfreund, 1982). Exposure of a solution of actin.S1 (or actin.S1.ADP) to hydrostatic pressures of up to 20 MPa. (200 atm) resulted in the dissociation of a small fraction of the complex as judged by the change in light scattering. Rapid release of pressure back to 101 kPa resulted in an exponential increase in the concentration of the actin.S1 complex. A series of such relaxations induced by different pressure jumps is shown in Fig. 2. For a simple one step binding reaction the relaxation time is defined by:

$$\tau^{-1} = k_{+1}([\bar{A}] + [\bar{M}]) + k_{-1}, \qquad (3)$$

where the bars over concentrations define equilibrium concentrations. The concentration dependence of τ^{-1}

Journal of Cell Science, Supplement 14, 31–35 (1991)

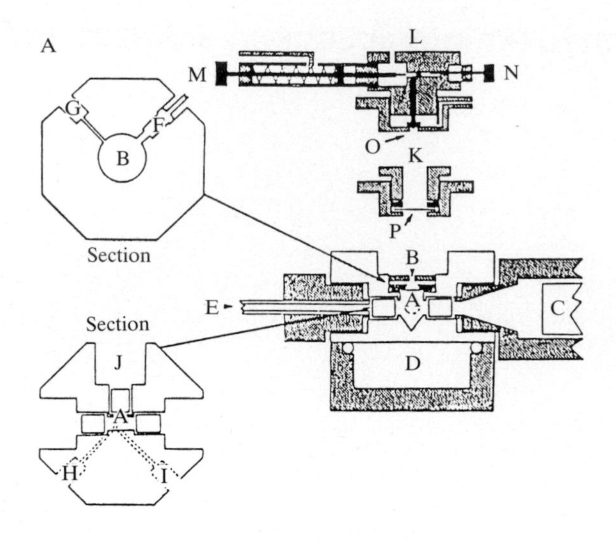

B

Fig. 1. Diagram of the pressure equipment. (A) A cross section of the pressure cell. The chamber is milled from a single piece of stainless steel (3×3×3 in). A, observation cell; B, hydraulic chamber; C, absorbancy photomultiplier; D, thermostated base; E, quartz fibre optic from light source; F, pressure transducer; G, pressure line; H and I, ports for filling sample chamber; J, fluorescence window; K, bursting disc release valve; M, trigger mechanism; N, reset mechanism; O, valve seat. (B) Schematic drawing of the tension transducer assembly. a, strain gauge with hook enclosed in glass tube (d); b, stainless steel tube housing the insulated terminals of the transducer element; c, epoxy resin seal at the junction of the glass and steel tubes; e and f, Main body of the transducer plug comprising a stainless steel cap (e) and brass body (f); g, rubber 'O' rings to maintain pressure seal; h, transducer terminals.

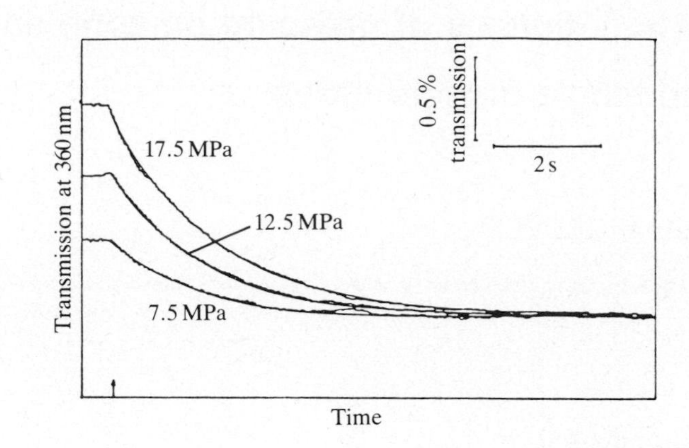

Fig. 2. The pressure-induced changes in light scattering of actin.S1.ADP. The three traces represent the observed changes in light scattering (monitored as transmission changes at 360 nm) for pressure jumps of 17.5, 12.5, and 7.5 MPa. The arrow indicates the time at which pressure was released. Each trace is the average of five successive relaxations and the single exponential fit is superimposed. The reciprocal relaxation times were 0.56, 0.65, and 0.62 s^{-1} respectively. Conditions: pH 8, 21 °C and ionic strength of 0.135 M. Vertical bar, 0.5 % transmission. Horizontal bar, 2 s.

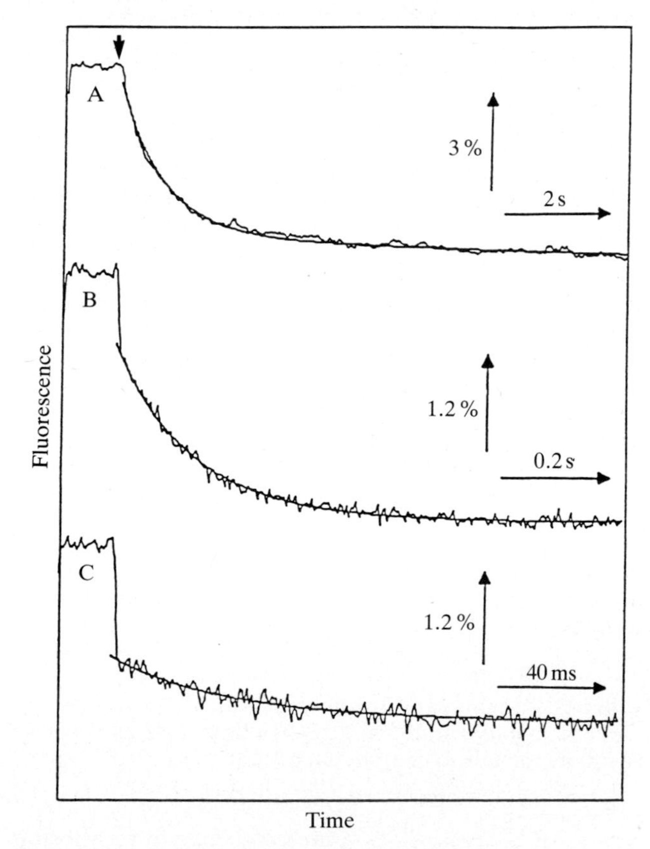

Fig. 3. The pressure-induced changes in fluorescence of pyrene–actin.S1. The arrow indicates the time of pressure release from 10 MPa to 0.1 MPa. The traces are an average of five successive relaxations on the same solution with the best-fit single exponential to the slow phase superimposed. The [S1] and the reciprocal relaxation times in each case were: (A) 2.8 μM, 1.24 s^{-1}; (B) 8.4 μM, 7.32 s^{-1}; (C) 19.5 μM, 24.3 s^{-1}. The fluorescence is expressed relative to the signal at high pressure. Conditions: actin 2.5 μM, pH 7.5, 20 °C and ionic strength 0.13 M.

defines k_{+1} and k_{-1} and the values obtained, over a range of reaction conditions, agreed with the results of earlier stopped-flow studies (White and Taylor, 1976; Trybus and Taylor, 1976; Marston, 1982).

The fluorescence of a pyrene group covalently attached to Cys 374 of actin is quenched by 70 % when S1 binds to actin (Kouyama and Mihashi, 1981; Criddle *et al.* 1985) and provides a more sensitive monitor of the actin.S1 interaction. Pressure jumps were performed on a solution of actin.S1 in the absence of nucleotide and show two clear relaxations (Fig. 3). No relaxation is observed in the absence of S1 and, therefore, the two relaxations must represent two relaxations of the acto.S1 complex. Monitoring the same reaction by light scattering shows only a single relaxation and the concentration dependences of the relaxation time and amplitude are identical to those of the slower relaxation observed by pyrene fluorescence. The fast relaxation was complete in the pressure release time (200 μs) and the amplitude increased with protein concentrations. The amplitude of the slow relaxation decreased at higher protein concentration. The experimental data were interpreted in terms of the following model in which step 2 is fast, pressure-sensitive and results in the change in pyrene fluorescence. Step 1 is a second-order reaction and results in a change in light scattering (Coates *et al.* 1985):

$$\mathrm{A+M} \overset{K_1}{\rightleftharpoons} \mathrm{A\text{–}M} \overset{K_2}{\rightleftharpoons} \mathrm{A.M}. \qquad (4)$$

Thus, an increase in pressure reduces K_2 and will lead to an overall decrease in affinity of actin for S1. Only at low concentrations of protein will this result in dissociation of a significant fraction of the complex. Under these conditions, rapid release of pressure will result in rapid re-equilibration of step 2, observed as a change in fluorescence. This will be followed by a slower rebinding of S1 to actin which can be observed by both light scattering and fluorescence. However, at very high protein concentration, increases in pressure will cause no dissociation of the two proteins but will still perturb the first-order isomerization of step 2.

Geeves *et al.* (1984) had earlier proposed that such a two-step binding of actin to myosin and myosin–nucleotide complexes was a general property of the proteins and that the isomerization step may be associated with the force generating event of the crossbridge cycle. This view has been supported by a range of biochemical studies in solution, including both measurements of K_2 for a range of nucleotides and nucleotide analogues and evidence that the rate of product release from actin.S1 remains slow unless the isomerization takes place (Geeves *et al.* 1986; Geeves, 1989; Geeves and Jeffries, 1988; Woodward *et al.* 1991).

In order for the isomerization to be responsible for force generation it must involve a substantial structural change in the actin.S1 complex. The pressure sensitivity of the isomerization step allows an estimation of a volume change of 100 $\mathrm{cm^3\,mole^{-1}}$ which suggests a significant rearrangement of the protein (Coates *et al.* 1985). The isomerization has also been shown to perturb the environment around Cys374 of actin and the 2′/3′ hydroxyl groups of the ADP ribose simultaneously, even though these groups are believed to be some distance apart within the acto.S1.nucleotide complex (Woodward *et al.* 1991).

If the isomerization is coupled to the force-generating event then factors that affect the isomerization should also effect the generation of force. Both increases in ethylene glycol concentration and increases in ionic strength reduce the equilibrium constant for the isomerization step and reduce the level of isometric force generated in a muscle fibre (Coates *et al.* 1985; Clarke *et al.* 1980; Gulati and Podolsky, 1981). Increased pressure would also be expected to decrease isometric force. To test this prediction, we set out to measure the effect of hydrostatic pressure on isometric force in single muscle fibres.

The effect of pressure on muscle fibres

In the 1930s a series of studies reported the effects of changes in hydrostatic pressure on whole muscle from the frog and turtle (Brown, 1934*a*,*b*; Cattel and Edwards, 1928). Increases in pressure caused reversible changes in both tetanic and twitch tension; the size and magnitude of the effects depended upon the exact conditions of the experiment. To distinguish the effect of pressure on the crossbridge cycle from effects on excitation–contraction coupling we used chemically skinned rabbit psoas fibres. Fibres were fully calcium activated by exchanging the bathing solution around the fibre whilst it was mounted inside the pressure chamber. The results of a series of stepped increases in pressure on the measured tension of a single muscle fibre are shown in Fig. 4. The results show that in the relaxed state, changes in pressure cause no measurable change in the output of the tension trans-

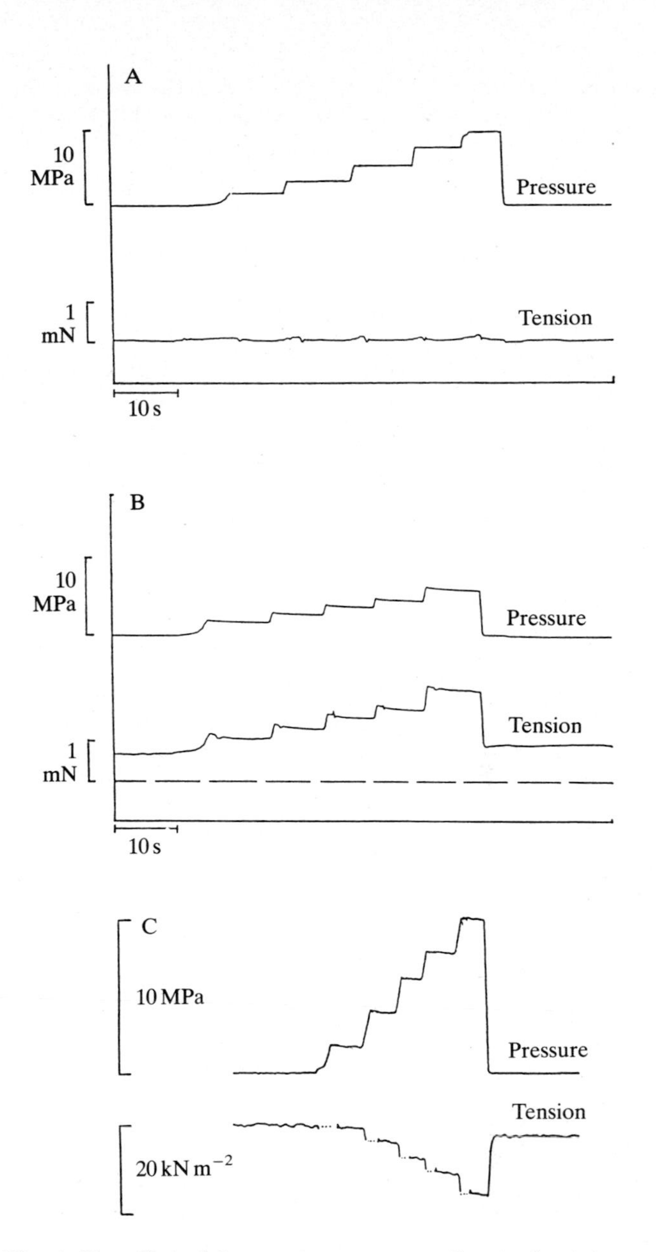

Fig. 4. The effect of changes in pressure on the steady-state tension of glycerinated muscle fibres from rabbit psoas. (A) A small fibre bundle under relaxing conditions; 10 mM ATP, low Ca^{2+}, 20°C, pH 7, ionic strength 200 mM. (B) The same fibre bundle holding a rigor tension of 520 μN. Conditions as in (A) but with no ATP. (C) A single fibre holding a steady fully active isometric tension of 1980 $\mathrm{kN\,m^{-2}}$. Conditions as in (A) but with 30 μM free calcium.

ducer. The same result was observed if the relaxed fibre was stretched to hold a large passive tension before the pressure changes were applied. A similar negligible change in measured tension was observed for a stretched rubber filament (Geeves and Ranatunga, 1987; Fortune *et al.* 1989*a*,*b*).

A fibre holding rigor tension showed an increase in measured tension with increases in pressure. The increases in tension were approximately linearly related to pressure over the range 0.1–8 MPa, with some irreversible loss of tension occurring at higher pressures. Each increase in tension corresponded to that observed following a 0.3 % length increase. Increases in tension were also observed for stretched filaments of glass, copper, silk, collagen and human hair (Ranatunga *et al.* 1990). The magnitude of the increase was specific to each material and is believed to reflect the volume changes in the material in response to a length change (defined by the Poissons ratio for a pure material). The relaxed fibre is rubber-like in its response to a pressure change, i.e. there is no volume change of the system on stretching. The fibre in rigor has a different response and this must represent some additional load-bearing element present in the fibre in rigor. The only known difference is the presence of attached crossbridges and the pressure sensitivity is therefore most likely to reflect the mechanical response of the non-cycling crossbridge or a structural element in series with the crossbridge.

The steady isometric tension of a muscle fibre in a fully activated contraction decreases as the pressure is increased (by approximately 8 % per 10 MPa pressure increase) and the tension is fully recovered on pressure release over the pressure range 0.1–15 MPa (Fig. 4C) (Geeves and Ranatunga, 1987; Fortune *et al.* 1989*a*). This decrease is specific to the active muscle fibre and must represent perturbation of some component which is not present in the rigor fibre, i.e it is likely to represent a perturbation of the dynamic crossbridge.

To investigate this possibility further we examined the rate of tension change following a rapid release of 10 MPa pressure. The results for rigor and active muscle fibres are shown in Fig. 5; the relaxed fibre showed no tension transient. The fibre in rigor showed a stepped decrease in pressure, occurring in phase with the pressure change (complete in approximately 2 ms) compatible with a mechanical relaxation of an elastic element. The active fibre, in contrast, showed a complex response. Initially, a rapid decrease in tension was seen in phase with the pressure release. This is similar to the response observed in the rigor fibre and may represent a perturbation of the same elastic element. If this rigor-like response were the only perturbation present in the active fibre, then a very rapid recovery of tension to the initial level might be expected, as is seen in response to a small length change. However, we observed a recovery of tension back to that originally observed at 0.1 MPa. The time course of the recovery can be fitted by two exponentials with $1/\tau_1$=60 s^{-1} and $1/\tau_2$=5.5 s^{-1} (Fortune *et al.* 1989*b*; Geeves *et al.* 1990). These two additional relaxations are only observed in the active muscle and suggest that a crossbridge event is being perturbed. The solution studies show that the isomerization step of actin.S1 association is pressure sensitive and this isomerization remains the most likely candidate for the pressure response seen in the muscle fibre.

Two additional observations support this view. Increases in pressure in the range used here caused no measurable change in the equatorial X-ray scattering from relaxed, rigor or active muscle fibres, suggesting no change in lattice spacing or in the number of attached crossbridges (Knight *et al.* 1990). The presence of phosphate in the active solution increases the observed $1/\tau_2$ in the active fibre, consistent with this process involving a reversible binding of phosphate (N. S. Fortune, M. A. Geeves and K. W. Ranatunga, unpublished data).

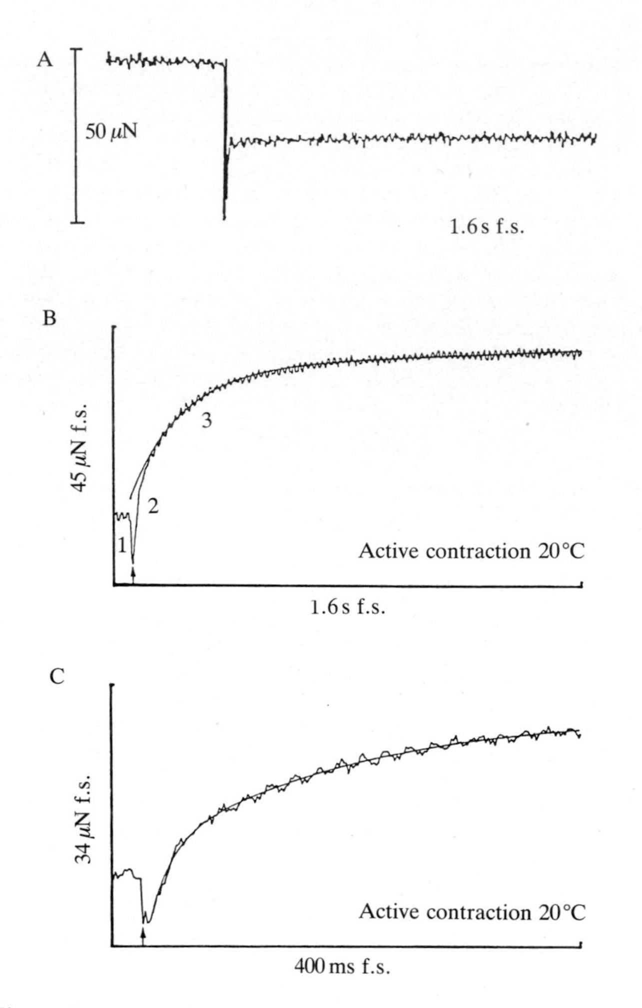

Fig. 5. Pressure induced tension transients in single glycerinated rabbit psoas fibres. Pressure was released from 10 MPa to 0.1 MPa in about 1 ms. Solution conditions were the same as in Fig. 4. (A) A single fibre holding a rigor tension of 180 μN. (B) and (C) A single fibre holding a steady active tension of 520 μN. The numbers in (B) identify the three principal phases observed. The same relaxation is shown on two time scales with a single exponential fit to the slowest phase (5.5 s^{-1}) in (B) and a double exponential fit (5.6 and 60 s^{-1}) in (C). The arrows indicate the time of pressure release. f.s., full scale.

The series of studies outlined here has resulted in the identification of a pressure-sensitive transition occurring both in solutions of purified actin and myosin and in contracting muscle fibres. The results so far are consistent with a direct correlation between the specific isomerization of actin.S1 in solution, the phosphate release step and a force-generating transition in the crossbridge cycle.

This work has been supported by the Wellcome Trust and the European Economic Community. The author is a Royal Society University Research Fellow.

References

Brown, D. E. S. (1934). Effect of rapid changes in pressure upon the contraction of skeletal muscle. *J. cell. comp. Physiol.* **4**, 257–281.

Brown, D. E. S. (1936). Effect of rapid compression on events in the isometric contraction of skeletal muscle. *J. cell. comp. Physiol.* **8**, 141–157.

Cattel, McK. and Edwards, D. J. (1928). The energy changes of skeletal muscle accompanying contraction under high pressure. *Am. J. Physiol.* **86**, 370–382.

Clarke, M. L., Rodger, C. D., Tregear, R. T., Bordas, J. and Koch, M. (1980). Effect of ethylene glycol and low temperature on the structure and function of insect flight muscle. *J. Mus. Res. Cell Motil.* **1**, 195–196.

Coates, J. H., Criddle, A. H. and Geeves, M. A. (1985). Pressure relaxation studies of pyrene labelled actin and myosin subfragment 1 from rabbit muscle. Evidence for two states of acto-subfragment 1. *Biochem. J.* **232**, 351–356.

Criddle, A. H., Geeves, M. A. and Jeffries, T. (1985). The use of actin labelled with *N*-(1-pyrenyl)iodoacetamide to study the interaction of actin with myosin subfragments and troponin/tropomyosin. *Biochem. J.* **232**, 343–349.

Davis, J. S. (1981*a*). The influence of pressure on the self assembly of the thick filament from myosin of vertebrate skeletal muscle. *Biochem. J.* **197**, 301–308.

Davis, J. S. (1981*b*). Pressure jump studies of the length regulation kinetics of the self assembly of myosin from vertebrate skeletal muscle into thick filaments. *Biochem. J.* **197**, 309–314.

Davis, J. S. and Gutfreund, H. (1976). The scope of moderate pressure changes for kinetic and equilibrium studies of biochemical systems. *FEBS Lett.* **72**, 199–207.

Fortune, N. S., Geeves, M. A. and Ranatunga, K. W. (1989*a*). Pressure sensitivity of active tension in glycerinated rabbit psoas muscle fibres; effects of ADP and phosphate. *J. Muscle Res. Cell Motil.* **10**, 113–123.

Fortune, N. S., Geeves, M. A. and Ranatunga, K. W. (1989*b*). Tension transients initiated by pressure perturbations in skinned rabbit muscle fibres. *J. Physiol.* **418**, 158P.

Geeves, M. A. (1989). Dynamic interaction between actin and myosin subfragment 1 in the presence of ADP. *Biochemistry* **28**, 5864–5871.

Geeves, M. A., Fortune, N. S. and Ranatunga, K. W. (1990). Temperature dependence of pressure induced tension transients in skinned rabbit psoas fibres. *Biophys. J.* **57**, 409a.

Geeves, M. A., Goody, R. S. and Gutfreund, H. (1984). The kinetics of acto-S1 interaction as a guide to a model for the crossbridge cycle. *J. Muscle Res. Cell. Motil.* **5**, 351–361.

Geeves, M. A. and Gutfreund, H. (1982). The use of pressure perturbations to investigate the interaction of rabbit muscle myosin subfragment 1 with actin in the presence of ADP. *FEBS Lett.* **140**, 11–15.

Geeves, M. A. and Jeffries, T. E. (1988). Effect of nucleotide upon a specific isomerisation of acto.S1. *Biochem. J.* **256**, 41–46.

Geeves, M. A., Jeffries, T. E. and Millar, N. C. (1986). The ATP induced dissociation of rabbit skeletal actomyosin subfragment 1. Characterization of an isomerization of the ternary acto.S1.ATP complex. *Biochemistry* **25**, 8454–8458.

Geeves, M. A. and Ranatunga, K. W. (1987). Tension responses to increased hydrostatic pressure in glycerinated rabbit psoas muscle fibres. *Proc. R. Soc.* B **232**, 217–226.

Gulati, J. and Podolsky, R. J. (1981). Isotonic contraction of skinned muscle fibres on a slow time base. *J. gen. Physiol.* **78**, 233–257.

Knight, P., Fortune, N. S. and Geeves, M. A. (1990). Structural basis of reduction of active muscle tension under pressure. *J. Muscle Res. Cell Motil.* **11**, 73.

Kouyama, T. and Mihashi, K. (1981). Fluorimetry study of *N*-(1-pyrenyl)iodoacetamide labelled F-actin. *Eur. J. Biochem.* **114**, 33–38.

Marston, S. B. (1982). Formation and dissociation of actomyosin complexes. *Biochem. J.* **203**, 453–460.

Ranatunga, K. W., Fortune, N. S. and Geeves, M. A. (1990). Hydrostatic compression in glycerinated rabbit muscle fibres. *Biophys. J.* (in press).

Trybus, K. M. and Taylor, E. W. (1980). Kinetic studies of the cooperative binding of S1 to regulated actin. *Proc. natn. Acad. Sci. U.S.A.* **77**, 7209–7213.

White, H. and Taylor, E. W. (1976). Energetics and mechanism of action of actomyosin ATPase. *Biochemistry* **15**, 5818–5826.

Woodward, S. K. A., Eccleston, J. F. and Geeves, M. A. (1990). Kinetics of the interaction of mant.ATP with myosin S1 and acto.S1. Characterization of two acto.S1 ADP complexes. *Biochemistry* (in press).

Molecular cloning of protozoan myosin I heavy chain genes

JOHN A. HAMMER III and GOEH JUNG

Laboratory of Cell Biology, National Heart, Lung, and Blood Institute, National Institutes of Health, Bethesda, MD 20892, USA

Summary

Myosins I are ubiquitous, nonfilamentous, actin-based mechanoenzymes originally discovered in protozoa. The extensive *in vitro* biochemical studies of purified protozoan myosins I are now being complemented with *in vivo* studies using cloned myosin I heavy chain genes and gene targeting techniques. Here we review briefly the systems and methods being used in these efforts to dissect protozoan myosin I structure and function using molecular genetic approaches.

Key words: myosin I, molecular cloning, structure–function, gene targeting, polymerase chain reaction.

Introduction

The myosins I are low molecular weight, nonfilamentous, actin-activated MgATPases, originally isolated from the soil amoeba *Acanthamoeba castellanii* almost 20 years ago by Tom Pollard and Ed Korn (1973) (reviewed by Korn and Hammer, 1988, 1989; Pollard *et al.* 1990). Recently, interest in these proteins has increased since it has been shown (1) that they are capable of supporting motile and contractile-like activities *in vitro*, (2) that *Dictyostelium* cells which contain myosin I but lack conventional type myosin (myosin II) retain the ability to locomote, chemotax and phagocytose (reviewed by Spudich, 1989), (3) that myosins I and II are differentially localized in both migrating and dividing cells and the pattern of localization is consistent with myosin I playing a crucial role in supporting myosin II-independent motile functions (Fukui *et al.* 1989), and (4) that myosins I exist in higher eukaryotes (Montell and Rubin, 1988; Hoshimaru and Nakanishi, 1987; Garcia *et al.* 1989). In this article we describe briefly how the myosin I heavy chain (MIHC) genes of *Acanthamoeba* and *Dictyostelium* have been identified, what is known currently concerning the complexity of the MIHC gene families in these two protozoans, and what might be the best approaches for cloning MIHC genes in other organisms.

Cloning of MIHC genes from *Acanthamoeba castellanii*

In 1984 we set out to clone the heavy chain gene for *Acanthamoeba* myosin I for several reasons. While a great deal was already known about the biochemistry of the amoeba myosins I, and while there was immunological evidence that they were true gene products, there remained a suspicion in the field that these nonfilamentous myosins might in fact be simply proteolytic breakdown products of a conventional-type amoeba myosin (in point of fact, type II myosins can be proteolyzed to a soluble active fragment, subfragment 1). Cloning the MIHC gene would resolve this question unambiguously, would indicate to what extent the primary structure of the myosin I heavy chain resembled that of conventional myosins and to what extent it differed, and would hopefully allow the use of molecular genetic techniques to examine myosin I function in living amoebae. In 1987 we published the sequence of *Acanthamoeba* myosin IC* (Jung *et al.* 1987), one of the three known isoforms of *Acanthamoeba* myosin I (IA, IB and IC), and the first MIHC sequence described. Since then we have completed the sequence of myosin IB (Jung *et al.* 1989) and the partial sequence of a third isoform (J. Gu, G. Jung and J. A. Hammer III, unpublished observations).

All three of these MIHC genes, as well as an *Acanthamoeba* myosin II heavy chain gene (Hammer *et al.* 1987) and the heavy chain for an apparent high molecular weight form of myosin I (HMWMI; see below), were isolated as genomic clones from an *Acanthamoeba Sau*-3A partial-digest genomic DNA library constructed in phage λ2001 (Hammer *et al.* 1986). In every case, the genes were identified using a 2.7 kb *Bam*H1 fragment from the nematode *unc 54* muscle myosin heavy chain gene (Karn *et al.* 1983) as a heterologous probe. This probe spans residues 35–760 of the nematode myosin heavy chain, or approximately 90 % of the globular head domain coding sequence. This portion of the nematode heavy chain contains several regions of sequence that are highly conserved among all myosins, e.g. the ATP binding site sequence and the sequence surrounding the reactive thiols. Unlike *Dictyostelium* (see below), nematode and *Acanthamoeba* share a similar codon bias (strong preference for G and C in the third position), which was crucial to the success of this approach. In addition, before actually screening the library, we found that this nematode probe would detect at moderate stringency both discrete bands in Southern blots of restricted *Acanthamoeba* DNA and RNA transcripts of the size expected for the MIHC in Northern blots of amoeba poly(A)$^+$ RNA (Hammer *et al.*

*The sequencing of MIHC peptides (Brzeska *et al.* (1989). *J. biol. Chem.* **264**, 19 340–19 348) has revealed that the first amoeba MIHC sequenced (Jung *et al.* 1987), which was originally called 1B, is in fact 1C, while the second MIHC sequenced (Jung *et al.* 1989), originally called 1L, is in fact 1B.

Journal of Cell Science, Supplement 14, 37–40 (1991)

1986, 1987). These preliminary studies gave us confidence that this approach would work.

All of the *Acanthamoeba* MIHC genes cloned to date are highly interrupted by introns (myosin IB, 17 introns; myosin IC, 23 introns; HMWMI, 17 introns; the partially-sequenced MIHC isoform referred to above is a cDNA clone). For the portion of these genes that encodes the globular head-like sequence, intron positions and reading frame were determined by homology with other sequenced myosins. Determining the intron positions and reading frame for the tail domains, which show no homology to any portion of conventional myosin sequences, was particularly difficult for the first myosin I sequenced (IC) and required very careful sequencing as well as mapping of all intron splice sites by S1-nuclease protection analyses (Jung *et al.* 1987). The sequencing of *Acanthamoeba* myosin IB, as well as several of the *Dictyostelium* MIHC isoforms (see below), was simplified by the fact that they demonstrate striking homology throughout their tail domains to *Acanthamoeba* myosin IC (see below).

The heavy chain of *Acanthamoeba* myosin IC revealed a polypeptide of approximately $127\times10^3\,M_r$ composed of a $77\times10^3\,M_r$ S1-like domain fused to a $51\times10^3\,M_r$ C-terminal domain whose sequence shows no significant similarity to any portion of conventional myosin sequences and lacks completely the characteristic sequence repeats found throughout the coiled-coil rod-like tail of type II myosins. These results, in addition to proving unequivocally that myosin I is a true gene product, provided a way to integrate further the biochemical properties of the purified protein. Specifically, the presence of a slightly truncated* S1-like domain was consistent with the expression of actin-activated Mg^{2+}-ATPase by myosin I and its ability to support actin-based contractile activities *in vitro* (indeed, S1 alone has been shown to be sufficient to generate both movement and force *in vitro* (Toyoshima *et al.* 1987; Kishino and Yanagida, 1988)). In addition, the lack of a coiled-coil tail sequence was consistent with the monomeric nature of the protein (one heavy chain and one head per molecule) and its inability to self-associate into filaments at low ionic strength like myosin II. This result further supported the conclusion, drawn from studies of the purified protein (reviewed by Korn and Hammer, 1988; Korn *et al.* 1988), that nature has devised ways to power actin-based motility other than by bipolar filaments of myosin II. The results obtained from the subsequent cloning and sequencing of *Acanthamoeba* myosin IB (Jung *et al.* 1988), as well as the partial sequence of a third isoform, are completely consistent with all of these conclusions.

One of the important observations to come from determining the sequences of several *Acanthamoeba* MIHCs is that, in addition to the high degree of similarity between head sequences, tail domain sequences are conserved throughout (Jung *et al.* 1988). Specifically, the tail domains contain in every case three distinct and conserved regions of sequence, which we refer to as tail homology regions 1, 2 and 3 (numbered from the N terminus). Tail homology region 1 (TH-1) spans approximately 220 residues, is about 60 %† conserved between the isoforms, and is characterized by clusters of basic residues. Tail homology region 2 (TH-2) spans approximately 180 residues and is characterized by being extremely rich in glycine, proline and alanine residues (GPA-rich; these residues appear in part as irregular short repeats) and by having a strong net positive charge. While the GPA-rich regions from different isoforms cannot be aligned in any unique way, they are conserved in their unusual amino acid composition and net positive charge. Tail homology region 3 (TH-3) spans approximately 50 residues, is about 75 % conserved between the isoforms, and is also found in a wide range of other proteins that associate with the submembraneous cytoskeleton, including all members of the nonreceptor tyrosine kinase family (such as pp60–src, in which this sequence motif is referred to as src homology region 3, or SH3), phospholipase cγ, brain spectrin, and a yeast actin binding protein (Rodaway *et al.* 1989; Drubin *et al.* 1990).

What gives these conserved tail sequences particular significance is that their positions correlate very well with the locations of the two apparent functional domains within the tail, as defined *in vitro* using proteolytic fragments of the heavy chain and/or heavy chain fragments expressed in *E. coli*. These studies have shown that the myosin I tail domain contains two potential anchoring sites for the S1-motor domain: a membrane binding site and a second actin binding site. The interaction site for anionic phospholipid membranes has been mapped to the N-terminal half of the tail, i.e. within TH-1 (reviewed by Adams and Pollard, 1989), while the second actin binding site has been mapped to the C-terminal half of the tail, i.e. within the sequences corresponding to TH-2 plus TH-3 (reviewed by Korn *et al.* 1988). While mapping studies may further refine the positions of these two interaction sites (for example, whether both TH-2 and TH-3, or just one of these regions, contribute to formation of the second actin binding site), the current data provide a valuable correlation between tail domain sequence and function. This correlation should prove to be of benefit in classifying additional MIHCs as they are sequenced (see below), and in exploring further the relationships between structure and function in myosin I.

One striking exception to the results described above is a high molecular weight form of *Acanthamoeba* myosin I (HMWMI) (Horowitz and Hammer, 1990). The deduced sequence of this protein reveals a polypeptide of approximately $177\times10^3\,M_r$ composed of an S1-like domain fused to an approximately 800 residue tail domain, which shows essentially no homology to the tail domains of either myosins I or II. The only exception to this is the last ~50 residues of the HMWMI, which is 50 % similar to the TH-3 region of the smaller myosins I. Because the tail domain of HMWMI clearly cannot form a coiled-coil structure, we predict that the protein will be single headed and nonfilamentous. For this reason we have tentatively classified it as a high molecular weight form of myosin I (high molecular weight because its heavy chain is ~50 % bigger than that of *Acanthamoeba* myosins IA, IB and IC; this difference is due to an additional 350 or so residues in the HMWMI tail domain relative to the smaller amoeba myosins I). While immunological evidence that the protein exists in cells has been obtained, and the protein has been partially purified, biochemical studies directed at correlating function and sequence of the HMWMI tail domain remain to be done.

Are there additional MIHC genes in *Acanthamoeba*?

* All MIHCs sequenced to date lack approximately 80 amino acids at their N terminus, relative to type II myosins, and approximately 50 residues at the base of the head, again relative to type II myosins.

† Per cent similarities given are the sum of exact matches and conservative substitutions.

The answer appears to be yes, as Southern blots probed at moderate stringency using the TH-3 region of *Acanthamoeba* myosin IB are consistent with there being up to six MIHC genes. Furthermore, the partially sequenced MIHC gene (J. Gu, G. Jung and J. A. Hammer III, unpublished data) does not contain the peptide sequence obtained from myosin IA, indicating that it represents a fourth MIHC isoform (fifth, counting HMWMI).

Cloning of MIHC genes from *Dictyostelium discoideum*

One of our principle goals has been to use MIHC genes as tools to define the functions of the myosins I *in vivo*. Because our efforts to do these experiments in *Acanthamoeba* have been unsuccessful, we have recently switched our emphasis to the protozoan *Dicytostelium discoideum*. Like *Acanthamoeba*, *Dictyostelium* is a highly motile amoeboid cell. Unlike *Acanthamoeba*, *Dictyostelium* is also haploid, easily transformed and demonstrates high frequencies of homologous recombination (reviewed by DeLozanne, 1989). This latter characteristic allows one to target genes for disruption and, based on analyses of the phenotype of the resultant cells, to infer the *in vivo* functions of the protein. Furthermore, *Dictyostelium* has already been shown to contain myosin I as well as myosin II (Coté *et al.* 1985) and to retain many motile functions, including the abilities to locomote, chemotax and phagocytose, when only myosin I is expressed (reviewed by Spudich, 1989). We set out, therefore, to clone the gene or genes for MIHC in *Dictyostelium* and to determine their physiological roles in the cell using gene targeting techniques. Pilot experiments like those described above indicated that cloning *Dictyostelium* MIHC genes using heterologous probes (e.g. *Acanthamoeba* MIHC gene fragments) would not work, almost certainly because of the very unusual codon bias in *Dictyostelium* (very strong preference for A and T over G and C). Instead, we screened a λgt11 cDNA expression library using an antibody prepared against the whole heavy chain of purified *Dictyostelium* myosin I (Fukui *et al.* 1989). The partial cDNA clone obtained was used to identify two genomic clones which together spanned the entire $124\times10^3 M_r$ *Dictyostelium* MIHC (Jung *et al.* 1989). This gene turned out to encode a bonafide MIHC, as evidenced by the fact that its entire amino acid sequence was similar to that of the *Acanthamoeba* MIHCs.

When we probed Southern blots of *Dictyostelium* DNA at moderate stringency using a fragment that encodes the ATP-binding site region from this MIHC gene, we found evidence for up to three additional MIHC genes (Jung and Hammer, 1990). Using this probe, we were able to clone from a *Dictyostelium* genomic library three additional MIHC genes which corresponded to the additional bands seen in the Southern blot (Jung and Hammer, 1990). In parallel, Titus *et al.* (1989) cloned three MIHC genes using as a probe a fragment encoding the ATP-binding site of *Dictyostelium* myosin II. Comparisons of our results with theirs indicated that there are at least five MIHC genes in *Dictyostelium*, which we have named myosins IA, IB, IC, ID and IE (with IB corresponding to the gene described by Jung *et al.* 1989). The complete sequences of IA (Titus *et al.* 1989) and IB (Jung *et al.* 1989) have been published.

As mentioned above, the deduced $124\times10^3 M_r$ (approx.) heavy chain sequence of *Dictyostelium* myosin IB is very similar throughout to the *Acanthamoeba* MIHCs and contains all three tail homology regions described above (TH-1, TH-2 and TH-3). Interestingly, the sequence of *Dictyostelium* myosin IA reveals a somewhat truncated heavy chain (approximately $113\times10^3 M_r$) which terminates just after tail homology region I, i.e. it lacks the sequences that correlate with the second actin binding site (TH-2 plus TH-3) (Titus *et al.* 1989). These observations suggest that *Dictyostelium* myosin IA would be limited to driving movements of membranes relative to actin, while myosin IB could power the movement of one actin filament relative to another as well as the movement of membranes relative to actin. Preliminary results from our laboratory indicate that *Dictyostelium* myosin ID (G. Jung, R. Urrutia and J. A. Hammer III, unpublished data) is very similar to IB, while IE (R. Urrutia, G. Jung and J. A. Hammer III, unpublished data) is very similar to IA, suggesting that these multiple isoforms may indeed fall into at least two groups based on tail domain sequence. While we can only speculate for now, it seems possible that these two groups of isoforms might be largely, if not solely, responsible for supporting different motile functions in the cell. If so, then it would probably be best when creating cells carrying more than one disrupted gene to target members of the same group first. This approach of prioritizing gene targeting experiments will be necessary if efforts to block the expression of all myosin I isoforms simultaneously are unsuccessful.

Are there more than five MIHC genes in *Dictyostelium*? The answer is probably yes, since low stringency Southern blots using a battery of ATP binding site probes from the various cloned MIHC genes indicate that there may be as many as four more MIHC genes (G. Jung, R. Urrutia and J. A. Hammer III, unpublished data). Interestingly, none of these additional putative MIHC genes (nor myosin IC) cross hybridize at low stringency with high specific activity probes containing the tail homology 3 regions of IB and ID. This finding suggests that IB and ID may be the only isoforms in their group. For this reason, our current efforts have been directed at disrupting the IB and ID genes in *Dictyostelium*. Cells lacking the IB isoform, which is known to localize to the leading edge of migrating cells, have been created (Jung and Hammer, 1990). These cells show delayed and somewhat inefficient chemotactic aggregation, impaired uptake of bacteria, three-fold slower growth rate on bacteria and abnormal fruit morphology (Jung and Hammer, 1990). Efforts are now underway to create a IB^-/ID^- double mutant, with the expectation that this mutant will show even greater impairment of chemotaxis and phagocytosis.

Cloning of MIHC genes from other organisms

One approach for identifying MIHC genes in other organisms would be to use portions of the tail domains of the *Acanthamoeba* and/or *Dictyostelium* MIHCs as heterologous probes. If such probes were to work, they would have detected only myosins of the type I class. While this approach might work well when searching for additional members of a family of MIHC genes within one organism, it unfortunately may not work well when crossing species because of (1) differences in codon bias, (2) the large number of conservative substitutions that occur between isoforms within tail homology region 1, and (3) the lack of direct sequence homology within the GPA-rich TH-2 region. Furthermore, the two known examples of MIHC genes in higher eukaryotes, *Drosophila* nina C (Montell and Rubin, 1988) and brush border myosin I (Hoshimaru and Nakanishi, 1987), share very little tail sequence homology with the protozoan myosins I (the best match is between the TH-1 region of *Acanthamoeba* myosin IB and the last approximately 20×10^3 M_r of bovine brush border myosin I, which are only about 45% similar; Jung *et al.* 1989). Nevertheless, this approach might represent a reasonable starting point, especially since pilot experiments in which heterologous probes are checked against Southern and Northern blots for the organism of interest are simple to do. Of all the regions within the protozoan MIHC tail domains to try, probes encoding tail homology region 3 may be the best.

An approach which is in essence the opposite of the approach described above is again to use heterologous probes, but ones encoding the most highly conserved regions within the myosin globular head domain, i.e. regions that are conserved between type I and type II myosins. This approach worked quite well for *Acanthamoeba* (in which a nematode muscle myosin head probe identified both amoeba myosins I and myosin II), although a similar codon bias is again a prerequisite for the success of this approach. Once pure clones are obtained, Northern blots can be used to quickly sort myosin I clones from myosin II clones, based on the large difference in their transcript sizes. As a variation on this approach, oligonucleotides made against these highly conserved head sequences could be used to clone MIHC genes by polymerase chain reaction (PCR) from genomic DNA or first-strand cDNAs. Owing to the large degree of divergence seen between the tail domain sequences of MIHCs from protozoans and metazoans, this approach of targeting highly conserved head sequences (e.g. ATP binding site, reactive thiol region), whether *via* heterologous probes or PCR, may be the best approach available.

Antibodies can of course be used in combination with cDNA expression libraries to clone MIHC genes. In many cases, however, sufficient myosin I cannot be purified from the organism of interest in order to generate specific antibodies (lack of the pure protein also precludes determination of protein sequence, from which specific oligonucleotides could be made for use as probes or in PCR reactions). Antibodies (both monoclonal and polyclonal) against the protozoan MIHCs are available, however, and their efficacy can be checked rapidly by screening a Western blot of a whole cell extract from the organism of interest. One should also be aware that the number of recombinant MIHC clones within a cDNA expression library may vary considerably, depending on the state of the cells or their developmental stage at the time the RNA was extracted for library construction.

Once a MIHC clone has been obtained, an ATP-binding-site probe from the gene may work well for finding other members of the family within that organism. This approach has worked very well for *Dictyostelium* (Jung and Hammer, 1990; Titus *et al.* 1989). Again, myosin I clones can be distinguished from myosin II clones early on by back screening first-round positives with a myosin II-specific probe or by probing Northerns with clone inserts.

We thank Edward D. Korn for advice on the manuscript and for his support of this work.

References

Adams, R. J. and Pollard, T. D. (1989). Membrane bound myosin I provides new mechanisms in cell motility. *Cell Motil. Cytoskel.* **14**, 178–182.

Coté, G. P., Albanesi, J. P., Ueno, T., Hammer, J. A. III. and Korn, E. D. (1985). Purification from *Dictyostelium discoideum* of a low molecular weight myosin that resembles myosin I from *Acanthamoeba castellanii*. *J. biol. Chem.* **260**, 4543–4546.

DeLozanne, A. (1989). Gene targeting and cell motility. *Cell Motil. Cytoskel.* **14**, 62–68.

Drubin, D. G., Mulholland, J., Zhu, Z. and Botstein, D. (1990). Homology of a yeast actin-binding protein to signal transduction proteins and myosin I. *Nature* **343**, 288–290.

Fukui, Y., Lynch, T. J., Brzeska, H. and Korn, E. D. (1989). Myosin I is located at the leading edges of locomoting *Dictyostelium* amoebae. *Nature* **341**, 328–331.

Garcia, A. E., Coudrier, E., Carboni, J., Anderson, J., Vandekerchhove, J., Mooseker, M. S., Louvard, D. and Arpin, M. (1989). Partial deduced sequence of the 110 kD calmodulin complex of the avian intestinal microvillus shows that this mechanoenzyme is a member of the myosin I family. *J. Cell Biol.* **109**, 2895–2903.

Hammer, J. A. III, Bowers, B., Paterson, B. M. and Korn, E. D. (1987). Complete nucleotide sequence and deduced polypeptide sequence of a nonmuscle myosin heavy chain gene from *Acanthamoeba*: evidence of a hinge in the rod-like tail. *J. Cell Biol.* **105**, 913–925.

Hammer, J. A. III, Jung, G. and Korn, E. D. (1986). Genetic evidence that *Acanthamoeba* myosin I is a true myosin. *Proc. natn. Acad. Sci. U.S.A.* **83**, 4655–4659.

Hammer, J. A., III, Korn, E. D. and Paterson, B. M. (1986). Isolation of a nonmuscle myosin heavy chain gene from *Acanthamoeba*. *J. biol. Chem.* **261**, 1949–1959.

Horowitz, J. and Hammer, J. A. III (1990). *J. biol. Chem.* (in press).

Hoshimaru, M. and Nakanishi, J. (1987). Identification of a new type of mammalian myosin heavy chain by molecular cloning: overlap of its mRNA with preprotachykinin B mRNA. *J. biol. Chem.* **262**, 14 525–14 532.

Jung, G. and Hammer, J. A. III (1990). Generation and characterization of *Dictyostelium* cells deficient in a myosin I heavy chain isoform. *J. Cell Biol.* **110**, 1955–1964.

Jung, G., Korn, E. D. and Hammer, J. A. III (1987). The heavy chain of *Acanthamoeba* myosin IB is a fusion of myosin-like and non-myosin-like sequences. *Proc. natn. Acad. Sci. U.S.A.* **84**, 6720–6724.

Jung, G., Saxe, C. L., Kimmel, A. R. and Hammer, J. A. III (1989). *Dictyostelium discoideum* contains a gene encoding a myosin I heavy chain. *Proc. natn. Acad. Sci. U.S.A.* **86**, 6186–6190.

Jung, G., Schmidt, C. J. and Hammer, J. A. III (1989). Myosin I heavy chain genes of *Acanthamoeba*: cloning of a second gene and evidence for the existence of a third isoform. *Gene* **82**, 269–280.

Karn, J., Brenner, S. and Barnett, L. (1983). Protein structural domains in *Caenorhabditis elegans* unc 54 myosin heavy chain gene are not separated by introns. *Proc. natn. Acad. Sci. U.S.A.* **80**, 4253–4257.

Kishino, A. and Yanagida, T. (1988). Force measurements by micromanipulation of a single actin filament by glass needles. *Nature* **334**, 74–76.

Korn, E. D., Atkinson, M. A., Brzeska, H., Hammer, J. A. III, Jung, G. and Lynch, T. J. (1988). Structure-function studies on *Acanthamoeba* myosin IA, myosin IB and myosin II. *J. Cell Biochem.* **36**, 37–50.

Korn, E. D. and Hammer, J. A. III (1988). Myosins of nonmuscle cells. *A. Rev. Biophys. Biophys. Chem.* **17**, 23–45.

Korn, E. D. and Hammer, J. A. III (1989). Myosin I. *Curr. Opin. cell Biol.* **2**, 57–62.

Montell, C. and Rubin, G. (1988). The *Drosophila* ninaC locus encodes two photoreceptor cell specific proteins with domains homologous to protein kinases and the myosin heavy chain head. *Cell* **52**, 757–772.

Pollard, T. D., Doberstein, S. K. and Zot, H. G. (1990). Myosin I. *A. Rev. Physiol.* (in press).

Pollard, T. D. and Korn, E. D. (1973). *Acanthamoeba* myosin. Isolation from *Acanthamoeba castellanii* of an enzyme similar to muscle myosin. *J. biol. Chem.* **248**, 4682–4690.

Rodaway, A. R. F., Steinberg, M. J. E. and Bentley, D. L. (1989). Similarity in membrane proteins. *Nature* **342**, 624.

Spudich, J. A. (1989). In pursuit of myosin function. *Cell Reg.* **1**, 1–11.

Titus, M. A., Warrick, H. and Spudich, J. A. (1989). Multiple actin-based motor genes in *Dictyostelium*. *Cell Reg.* **1**, 55–63.

Toyoshima, Y. Y., Kron, S. J., McNally, E. M., Niebling, K. R., Toyoshima, C. and Spudich, J. A. (1987). Myosin subfragment-1 is sufficient to move actin filaments *in vitro*. *Nature* **328**, 536–539.

Antibodies against vertebrate microfilament proteins in the analysis of cellular motility and adhesion

BRIGITTE M. JOCKUSCH, BARBARA ZUREK, RALPH ZAHN, ANNETTE WESTMEYER and ANNETTE FÜCHTBAUER

Cell Biology Group, University of Bielefeld, POB 8640, D-4800 Bielefeld, FRG

Summary

Microinjection of specific antibodies can be an alternative and a supplement to genetic engineering in dissecting the function of individual cytoskeletal components. In this report, we describe some of the requirements for using this technique, its potential application in conjunction with morphological and biochemical analyses, and its limitations. Examples are given for the injection of antibodies to α-actinin, vinculin and myosin, and the effects of such treatment on adhesion, motility and cytokinesis of the recipient cells.

Key words: anti-α-actinin, anti-myosin, anti-vinculin, microinjection.

Why would one like to microinject antibodies?

The complexity of cytoskeletal components and their interactions (Jockusch, 1983; Stossel *et al.* 1986) are serious obstacles on the way to understanding cell motility at the molecular level. While an ever increasing number of proteins are discovered and subsequently characterised with respect to their sequence, genomic arrangement and biochemical properties, our knowledge of their roles in motile processes in the living cell is restricted to a few selected examples.

In primitive eukaryotes like yeast or cellular slime moulds, molecular genetics has recently offered an elegant way to gain information on the function of specific cytoskeletal proteins by gene disruption. In these haploid organisms, homologous recombination between mutant DNA introduced by suitable plasmids offers a straightforward approach to understanding the role of a given microfilament protein by interfering with its expression in the recombinant. This strategy has for example allowed conclusions to be drawn about the role of myosin II in *Dictyostelium* (De Lozanne and Spudich, 1987; Manstein *et al.* 1989; Fukui *et al.* 1990) and of actin (Novick and Botstein, 1985) and profilin (Haarer *et al.* 1990) in yeast. In vertebrate cells, however, the situation is more complicated. The cells are diploid and homologous recombination is not or not yet a reliable event. Rather than elimination of the endogenous protein by gene disruption, overexpression of mutated sequences or of the wildtype sequence in a foreign environment may be chosen as a promising approach towards learning the function of individual microfilament proteins. An elegant example in this field is the expression of villin in transfected fibroblasts (Friederich *et al.* 1989). However, so far, the number of proteins that can be analysed in this manner is rather limited. Moreover, some of the plasmids used as vectors themselves affect cell morphology and microfilament organisation.

As an alternative and supplement to genetic engineering, we have used microinjection of antibodies by glass capillaries as a means of interfering selectively with individual cytoskeletal components directly in the cytoplasm of tissue culture cells. In this article, we describe some of the parameters that should be considered for a successful application of this technique, and show the results obtained for a few selected examples.

Which vertebrate cells can be used?

A number of primary cells of avian or mammalian origin, as well as established lines, have been used successfully. These include fibroblastic, endothelial and epithelial cells that express a well organised cytoskeleton and are firmly attached to the culture dish, like chicken, rat or human fibroblasts, pig and rat kangaroo epitheloid lines. Highly motile cells, virus-transformed cells or cells with a transformed morphology (like macrophages, BHK, CHO, HeLa or Cos cells) can also be injected, although they are not as flat and firmly attached as the cells mentioned before. However, it is difficult to make predictions as to which cells tolerate microinjection. For example, we have not been successful with injecting amphibian epithelial cells. Such cells, which can be easily obtained from tadpoles, show a beautifully organised cytoskeleton, spread very well and locomote at a surprising speed on glass surfaces (Bereiter-Hahn *et al.* 1981), but they do not survive even buffer injections. In general, it should be kept in mind that only a limited number of tissue culture cells have been successfully used so far.

Which antibodies can be injected?

We have injected polyclonal as well as monoclonal antibodies against a variety of actin-binding proteins. All our antibodies had been tested for their specificity and reactivity with the cellular antigen in immunoblots and by indirect immunofluorescence. In general, we inject them in at least threefold molar excess over the antigen. This implies that they have to be concentrated to values between 5 and 25 mg ml^{-1}. These concentrations can be easily obtained with purified rabbit or mouse antibodies of the IgG class. In our hands, IgM tends to aggregate at high concentrations. Also, since IgM does not spread easily

Journal of Cell Science, Supplement 14, 41–47 (1991)

throughout the cytoplasm and since many of them are unstable, we do not inject antibodies of this class.

What happens after microinjection of antibodies?

In general, microinjection of physiological buffers by glass capillaries into tolerant cells increases the cell's volume by approximately 10 %. This by itself leads to a variety of distinct changes in the microscopic appearance of the living cell: within seconds, ruffles collapse, vesicles appear in the perinuclear area and a few focal contacts disappear. These initial 'wounding reactions' are rapidly overcome by the cell types suitable for microinjection, usually within 10–15 min. Immunoglobulin injection *per se* induces no more severe nor longer lasting reactions, even at concentrations of more than 25 mg ml^{-1}. Control antibodies are distributed throughout the cytoplasm, as are other injected proteins. These, as well as antibodies binding to cytoskeletal components, can be detected there for surprisingly long periods: at concentrations above 20 mg ml^{-1} in the injection solution, antibodies were detected by fluorescence microscopy in the cytoplasm of the progeny of the injected cells up to 3 days after injection (Zurek *et al.* 1990).

Obviously, the effects of antibodies in the injected cell must be transient, since growth and proliferation are not inhibited.

What types of analysis are feasable?

Microinjection by glass capillaries into single cells is of course ideally suited to following the effects of interfering with individual proteins in living cells. For this purpose, microinjection may be combined with phase, Nomarski or reflection contrast microscopy. Video microscopy and image processing allow for a detailed study of individual cells over longer periods of time. Examples involving antibodies against microfilament proteins include studies on granule movement (Hegman *et al.* 1989) and on locomotory behaviour and cytokinesis (Höner *et al.* 1988; Zurek *et al.* 1990). However, it is also possible to obtain data for a large number of cells: morphological criteria (changes in microtubule and microfilament patterns or focal contacts as a result of antibody injections) can be evaluated by injecting more than 100 cells within a few minutes and processing the recipients for indirect immunofluorescence at discrete time points after injection (Füchtbauer *et al.* 1985; Höner and Jockusch, 1988; Höner *et al.* 1988; Zurek *et al.* 1990; Westmeyer *et al.* 1990). Microinjected cells can also be subjected to radioactive labeling and analysed for synthesis of individual cytoskeletal proteins in 2D autoradiograms, as has already been shown for cells injected with α-actinin and riboprobes specifically suppressing α-actinin synthesis (Schulze *et al.* 1989).

What can be concluded from antibody injections?

In general, antibodies against microfilament proteins might be expected to bind to and immobilise immediately those molecules that are not structure-bound but present in a soluble pool. In this case, any effect on microfilament organisation would be indirect: microfilaments might disintegrate because of a shift in the equilibrium between soluble and structure-bound components. This would be especially expected for polyclonal, high affinity antibodies against proteins that are only occasionally or loosely structure-bound, like gelsolin, profilin or even α-actinin. In contrast to the situation in muscle, more than 50 % of the nonmuscle α-actinin does not sediment with the cytoskeleton in cell lysates (Schulze *et al.* 1989). When we injected polyclonal, affinity purified antibodies against α-actinin into PtK$_2$ cells, large precipitates of antibody–antigen complexes were found concentrated in the perinuclear area. In addition, the antibody labeled α-actinin in the expected periodic pattern along microfilament bundles in these epitheloid cells, i.e. peripheral belts and radial stress fibres (Fig. 1). However, there was no effect on microfilament organisation or function: the injected cells within the epitheloid sheet appeared normal, and cells at the periphery of the sheet displayed a ruffling activity of marginal veils quite similar to that of their noninjected neighbours. When the injected cells approached mitosis, their microfilament bundles were dissolved normally. After cytokinesis, antibody molecules were still detectable at the newly formed stress fibres and peripheral bundles (Fig. 1). In this system, antibody-tagging of a large proportion of the α-actinin molecules does obviously not interfere with organisation, *de novo* formation or function of microfilaments.

In contrast, microfilament organisation proved very sensitive to the injection of antibodies against vinculin. Several polyclonal, affinity purified rabbit antibodies against chicken gizzard vinculin were found to alter drastically the vinculin distribution in chicken embryo and rat fibroblasts: instead of the normal, arrowhead-like localisation at the terminal portions of stress fibres, abnormally wide, vinculin-positive patches were formed (Fig. 2 E–H) that were only partially coincident with adhesion sites. Focal contacts and stress fibres were seen to disintegrate in parallel with the formation of these abnormalities (Jockusch and Füchtbauer, 1983). Monoclonal antibodies against epitopes located in the region of the talin-binding domain of vinculin (residues 167–207; Jones *et al.* 1989) show the same effect (Fig. 2 A–D), while antibodies directed against the actin-binding domain of vinculin (residues 587–850) altered neither microfilament organisation nor cell adhesion after microinjection (Westmeyer *et al.* 1990). These data suggest that antibodies binding close to or within the talin-binding domain of

Fig. 1. PtK$_2$ cells injected with a polyclonal, affinity-purified rabbit antibody against chicken gizzard α-actinin (5 mg ml^{-1}). (A, C, E, G) Actin filament organisation as revealed by Rh–phalloidin. (B, D, F) The distribution of the injected antibody, as revealed by FITC-sheep-anti-rabbit IgG. (H) The distribution of PtK$_2$ α-actinin in control cells, as revealed by indirect immunofluorescence with the same antibody used for injections in (A)–(F). (A, B) Single cell injected with anti-α-actinin, 10 min past injection. The antibody has formed large precipitates in the center of the cell, but is also bound to α-actinin in stress fibres and at the periphery of the cell (inset). Actin organisation appears normal (A). (C, D) A single cell had been injected 12 h before and has obviously divided since then (D). Both daughter cells contain large antibody clusters in the center and additional antibody in microfilament bundles (inset in D) that appear completely normal (C). (E, F) Similar to C, D, but cells were fixed 24 h after antibody injection. Antibody is still found in all structures where α-actinin is expected, with the exception of marginal ruffling veils (compare arrowheads in E, F, with the control, G, H). (G, H) Noninjected control cells. Bar, 10 μm (valid for A–H). The insets are enlargements by 2×.

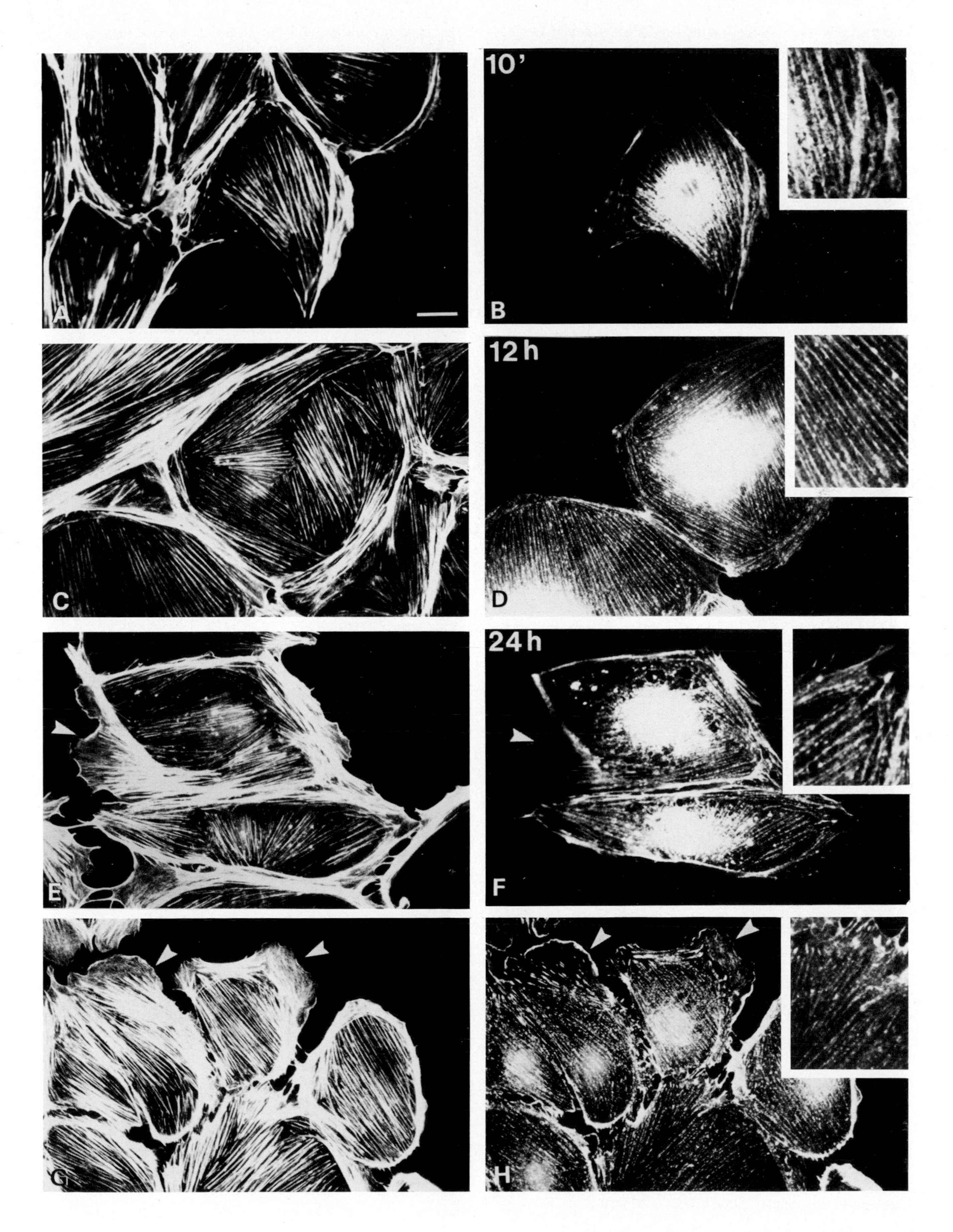
10'
12 h
24 h
A
B
C
D
E
F
G
H

vinculin directly interfere with the talin–vinculin interaction that is mediating the attachment between actin filaments and the plasma membrane in cellular adhesion sites. Cells injected with such antibodies lost their firm anchorage to the substratum, but did not grossly alter their morphology. Their locomotory activity was not different from that of controls.

Quite different observations were made when epitheloid (rat, kangaroo and pig) cells, rat or chicken fibroblasts were injected with antibodies against pig brain myosin.

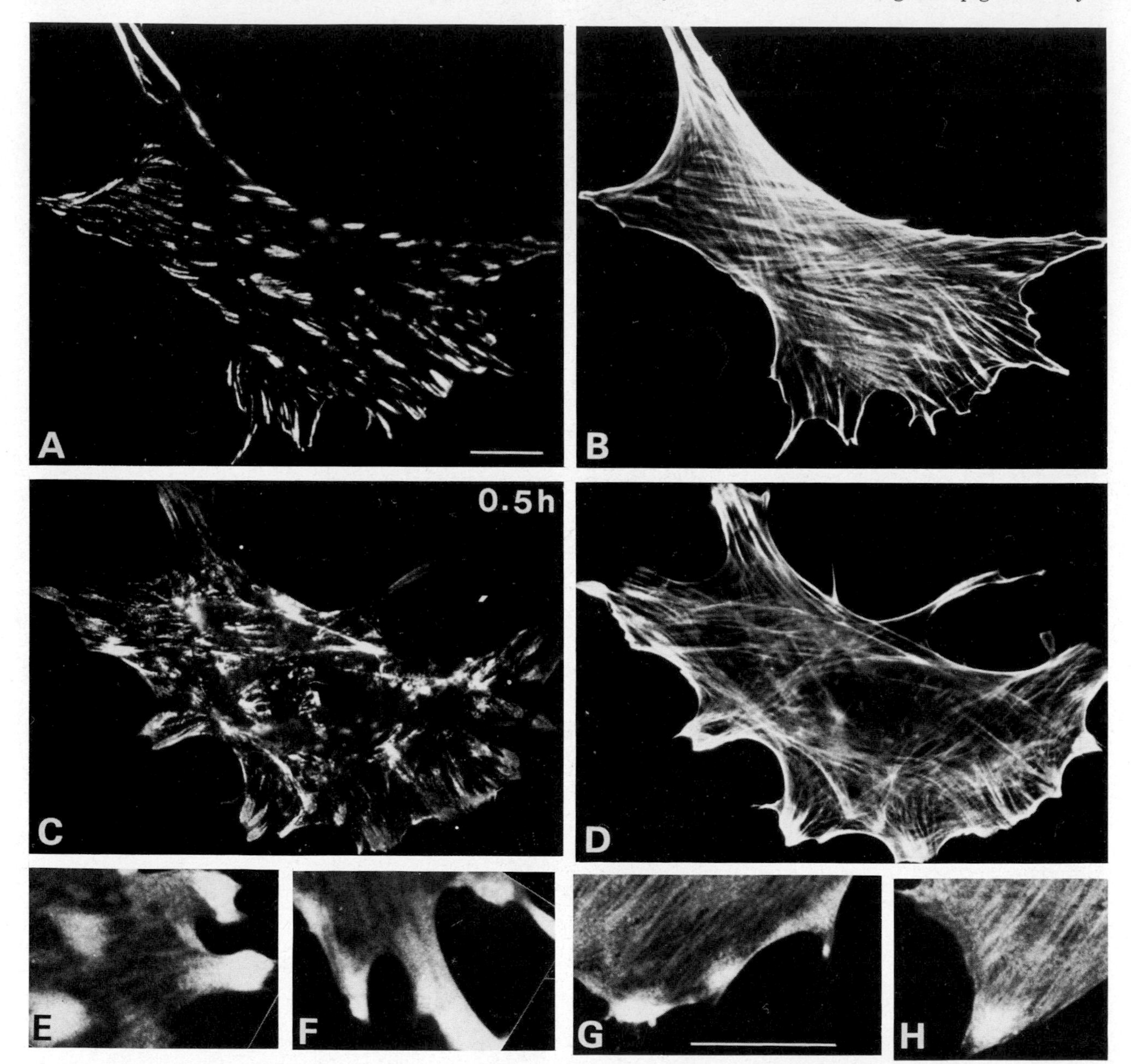

Fig. 2. Chicken embryo fibroblasts injected with antibodies against vinculin. (A) Distribution of vinculin, and (B) F-actin in a control cell, as revealed by a monoclonal mouse anti-vinculin and FITC-rabbit-anti-mouse IgG (A) and Rh–phalloidin (B). (C) Distribution of vinculin, and (D) F-actin in a cell injected 30 min earlier with 5 mg ml^{-1} of the same antibody used in (A). (E–H) Enlargements of vinculin patches induced by the injection of a polyclonal affinity purified antibody (2 mg ml^{-1}) against chicken gizzard vinculin, as seen 60 min past injection by FITC-sheep-anti-rabbit IgG. Bars: (A) 10 μm (valid for A–D), (G) 10 μm (valid for E–H).

Fig. 3. Rat fibroblasts injected with monoclonal antimyosin (Antibody 5 in Fig. 4, 12 mg ml^{-1}). (A, C, E, G) Myosin distribution as revealed by the monoclonal antibody and FITC-rabbit-anti-mouse IgG. (B, D, F, H) F-actin distribution as seen by Rh–phalloidin. (A, B) Part of an uninjected control cell. (C, D) Two fibroblasts injected 30 min earlier. Myosin–antibody complexes cluster mainly in the central part of the cells (C), stress fibres have disintegrated (D), and the cells appear much smaller. (E, F) Similar to (C, D), but 4 h after injection. The upper cell shows part of its myosin (E) and F-actin (F) in stress fibres, while the lower one is more severely affected at this time point. (G, H) At 24 h after injection, this cell is apparently on its way to recovery: myosin (G) and actin (H) are largely organised in stress fibres, the cell is flattening out again. Bar, 20 μm (valid for A–H).

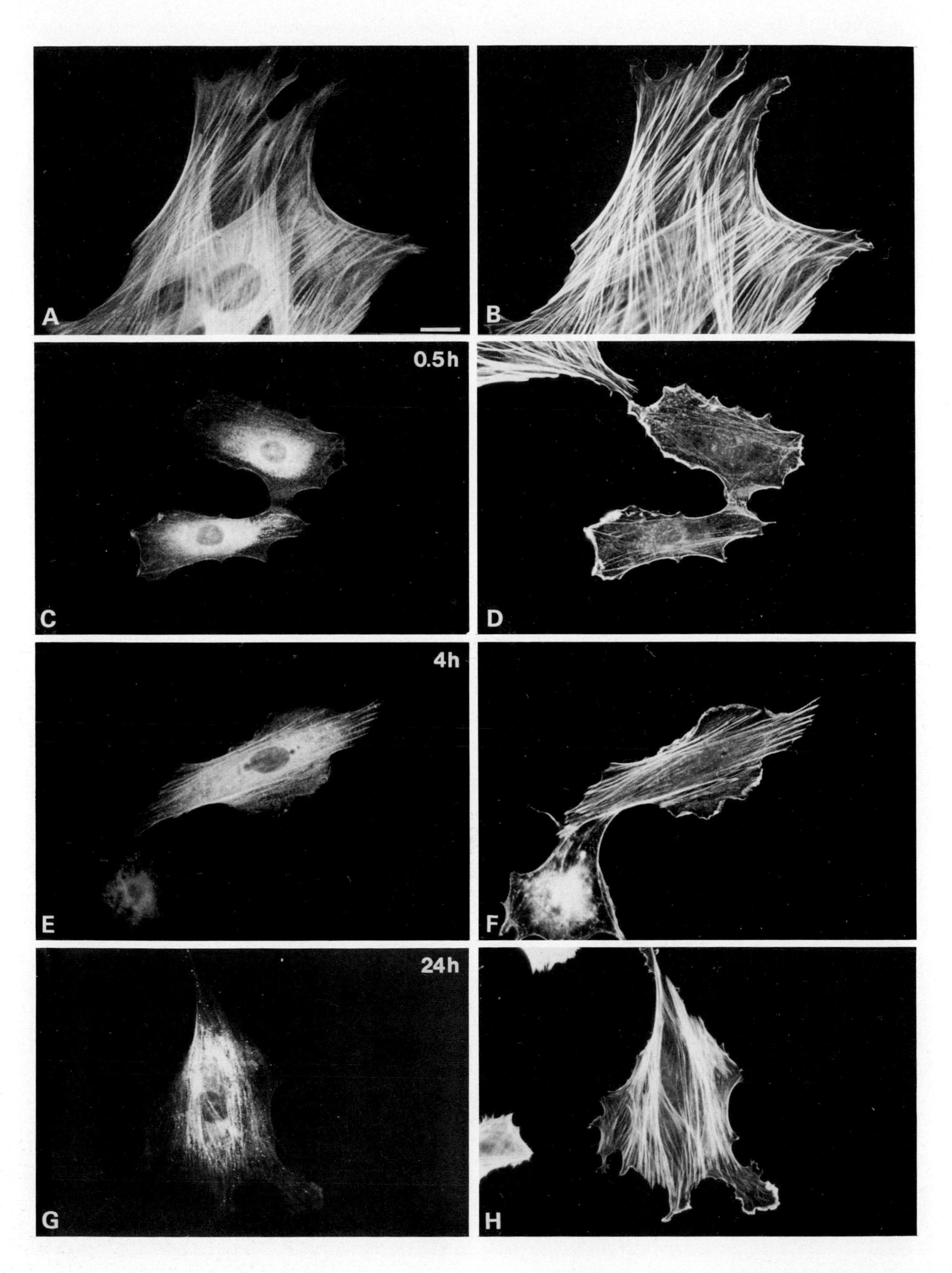
A
B
0.5h
C
D
4h
E
F
24h
G
H

The monoclonal antibodies used in this case all reacted with epitopes at the rod portion of the myosin heavy chain (Fig. 4; Barylko *et al.* 1989; Zurek *et al.* 1989, 1990). When injected into rat fibroblasts, they caused a rapid disintegration of stress fibres, concomitant with a change of cellular morphology: the cells became less flat and showed an increase in peripheral ruffling veils (Fig. 3). Their locomotory activity was markedly increased as compared with buffer-injected or noninjected controls (Fig. 4). Similar observations had previously been made with mono-

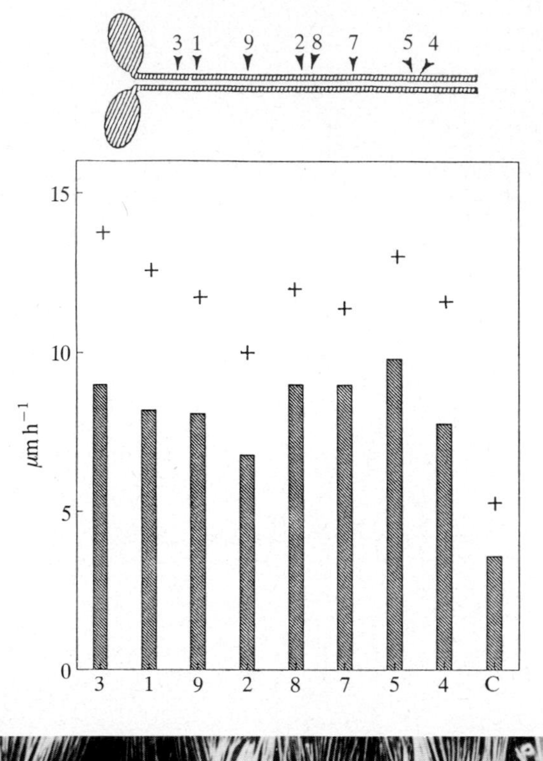

Fig. 4. Locomotory activity of rat fibroblasts injected with monoclonal antibodies against pig brain myosin. The cartoon on top of the graph shows the location of the epitopes determined for the various antibodies. Cells were injected with an at least threefold molar excess of antibody. C, control fibroblasts, not injected. The bars give the mean value of locomotory velocity, as determined by locating the cells' nuclei at various time points within 24 h after injection. (+), standard deviations.

Fig. 5. PtK_2 cells injected with the monoclonal antibodies against pig brain myosin. (A, B) Fluorescence images of uninjected controls, to reveal the normal distribution of myosin with the same antibody used for injection followed by FITC–rabbit-anti-mouse IgG (B), and of actin with Rh–phalloidin (A). (C, D) Daughters of a single cell injected 24 h earlier with 20 mg ml^{-1} of antibody 5 (Fig. 4). The antibody is revealed by FITC-rabbit-anti-mouse IgG. Bar, 20 μm (valid for A–D).

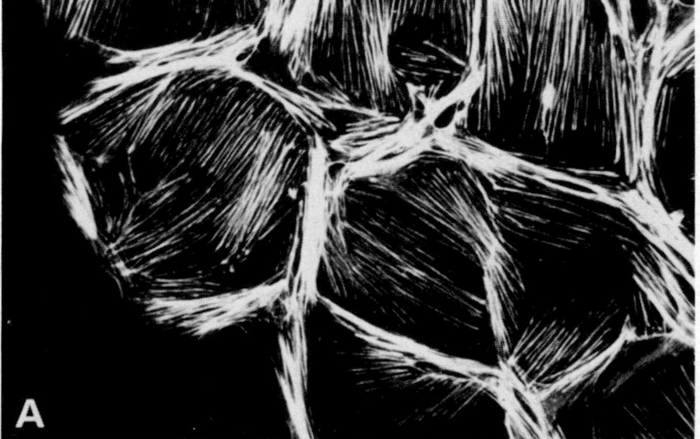

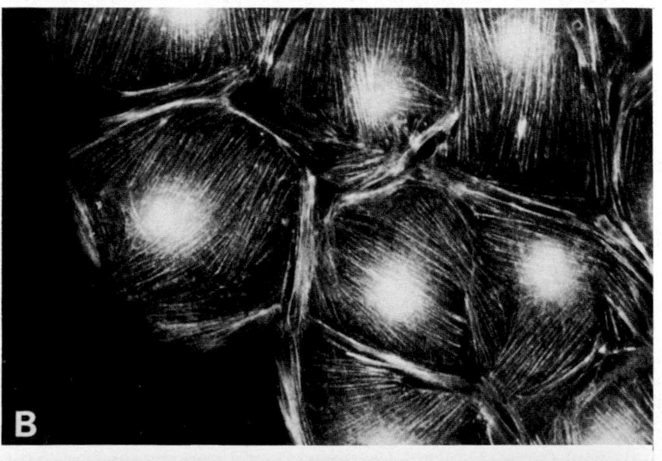

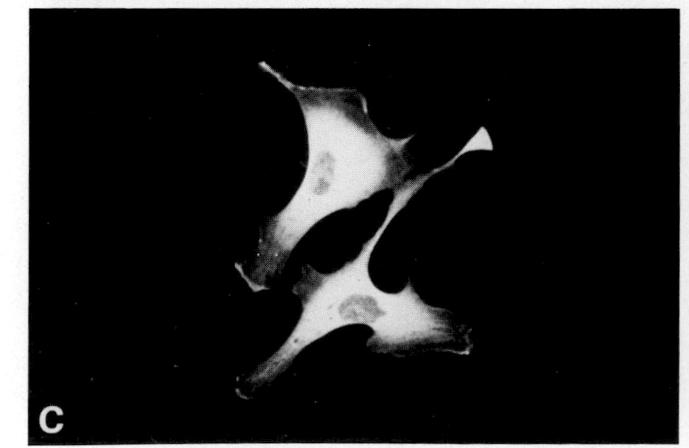

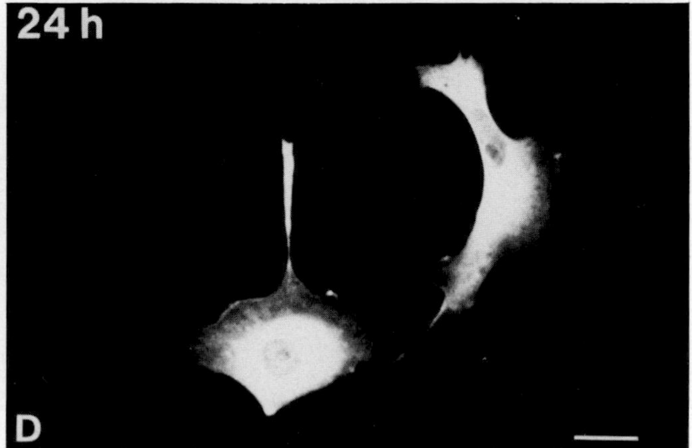

clonal antibodies against the tail portion of chicken brush border myosin (Höner *et al.* 1988). In this case, the recipient chicken embryo fibroblasts exhibited very abnormal, ribbon-like shapes. Changes in cellular shape, adhesion and locomotory activity were even more striking in epitheloid cells (Zurek *et al.* 1989, 1990). Here, the injection of antimyosin especially disturbed the assembly of the peripheral actomyosin belts in postmitotic cells that are essential for the establishment and maintenance of epitheloid sheets (Fig. 5). Bizarre cell shapes and an increase in locomotion were frequent among the progeny of injected cells. The percentage of such extreme shapes, as depicted in Fig. 5, was clearly dependent on the amount of antibody injected: at an approximately twelvefold molar excess over myosin, roughly 30% of the cells showed this morphology. After a second or third division they still looked abnormal, and the injected antibody could be detected in the cytoplasm by fluorescence microscopy. However, even in these cases, the cells subsequently returned to their normal morphology and formed epitheloid sheets.

Cleavage was not inhibited by the injection of these antimyosins. However, while anaphase movement of chromosomes was normal, both assembly and constriction of the contractile ring were markedly prolonged and very variable between individual cells. The contractile rings in the injected cells contained less myosin, but not less actin than controls as seen by indirect immunofluorescence (Zurek *et al.* 1990).

No differences were seen in all these effects between the various antibodies. From these observations, one can conclude that antibodies binding to the myosin tail interfere with cell morphology and epitheloid tissue formation, and delay cytokinesis. The observed increase in locomotory activity was unexpected and surprising, in particular since injection of antimyosin into *Acanthamoeba* cells had decreased their migratory velocity (Sinard and Pollard, 1989). It seems unlikely that the antimyosin-induced stimulation of locomotion observed in vertebrate cells is simply a consequence of the release of the recipients from their focal adhesion sites, since these contacts were more rapidly and more severely attacked by antivinculin than by antimyosin. Yet, only in the latter case did we find an increase in locomotory activity. A possible explanation for these results may be that the antimyosin-triggered disintegration of actomyosin complexes allows actin filaments to interact with other motor proteins, for example minimyosins. Such interactions in the vertebrate cell's cortex may be essential for membrane ruffling and locomotion.

References

Barylko, B., Zurek, B. and Jockusch, B. M. (1989). Brain myosin assembly: characterization of aggregation-competent fragments by antibodies. *Eur. J. Cell Biol.* **50**, 41–47.

Bereiter-Hahn, J., Strohmeier, R., Kunzenbacher, I., Beck, K. and Vöth, M. (1981). Locomotion of *Xenopus* epidermis cells in primary culture. *J. Cell Sci.* **52**, 289–311.

De Lozanne, A. and Spudich, J. A. (1987). Disruption of the *Dictyostelium* myosin heavy chain gene by homologous recombination. *Science* **236**, 1086–1091.

Friederich, E., Huet, C., Arpin, M. and Louvard, D. (1989). Villin induces microvilli growth and actin redistribution in transfected fibroblasts. *Cell* **59**, 461–475.

Füchtbauer, A., Herrmann, M., Mandelkow, E. M. and Jockusch, B. M. (1985). Disruption of microtubules in living cells and cell models by high affinity antibodies to beta-tubulin. *EMBO J.* **4**, 2807–2814.

Haarer, B. K., Lillie, S. H., Adams, A. E. M., Magdolen, V., Bandlow, W. and Brown, S. S. (1990). Purification of profilin from *Saccharomyces cerevisiae* and analysis of profilin-deficient cells. *J. Cell Biol.* **110**, 105–114.

Hegman, T. E., Lin, J. L. and Lin, J. J. (1989). Probing the role of nonmuscle tropomyosin isoforms in intracellular granule movement by miroinjection of monoclonal antibodies. *J. Cell Biol.* **109**, 1141–1152.

Höner, B., Citi, S., Kendrick-Jones, J. and Jockusch, B. M. (1988). Modulation of cellular morphology and locomotory activity by antibodies against myosin. *J. Cell Biol.* **107**, 2181–2189.

Höner, B. and Jockusch, B. M. (1988). Stress fiber dynamics as probed by antibodies against myosin. *Eur. J. Cell Biol.* **47**, 14–21.

Jockusch, B. M. (1983). Patterns of microfilament organization in animal cells. *Molec. Cell Endocrinol.* **29**, 1–19.

Jockusch, B. M. and Füchtbauer, A. (1983). Organization and function of structural elements in focal contacts of tissue culture cells. *Cell Motil.* **3**, 391–397.

Jones, P., Jackson, P., Price, G. J., Patel, B., Ohanion, V., Lear, A. L. and Critchley, D. R. (1989). Identification of a talin binding site in the cytoskeletal protein vinculin. *J. Cell Biol.* **109**, 2917–2927.

Manstein, D. J., Titus, M. A., De lozanne, A. and Spudich, J. A. (1989). Gene replacement in *Dictyostelium*: generation of myosin null mutants. *EMBO J.* **8**, 923–932.

Novick, P. and Botstein, D. (1985). Phenotypic analysis of temperature-sensitive yeast actin mutants. *Cell* **40**, 405–416.

Schulze, H., Huckriede, A., Noegel, A. A., Schleicher, M. and Jockusch, B. M. (1989). α-actinin synthesis can be modulated by antisense probes and is autoregulated in non-muscle cells. *EMBO J.* **8**, 3587–3593.

Sinard, J. H. and Pollard, T. D. (1989). Microinjection into *Acanthamoeba castellanii* of monoclonal antibodies to myosin-II slows but does not stop cell locomotion. *Cell Motil. Cytoskel.* **12**, 42–52.

Stossel, T. P., Chaponnier, C., Ezzell, R. M., Hartwig, J. H., Janmey, P. A., Kwiatkowski, D. J., Lind, S. E., Smith, D. B., Southwick, F. S., Yin, H. L. and Zaner, K. S. (1985). Nonmuscle actin-binding proteins. *A. Rev. Cell Biol.* **1**, 353–402.

Westmeyer, A., Ruhnau, K., Wegner, A. and Jockusch, B. M. (1990). Antibody mapping of functional domains in vinculin. *EMBO J.* **9**, 2071–2078.

Zurek, B., Höner, B. and Jockusch, B. M. (1989). Scientific Program, I. Structural bases of cell functions: the role of myosin filaments in nonmuscle cells. *Verh. Dtsch. Zool. Ges.* **82**, 5–16.

Zurek, B., Sanger, J. M., Sanger, J. W. and Jockusch, B. M. (1990). Differential effects of myosin–antibody complexes on contractile rings and circumferential belts in epitheloid cells. *J. Cell Sci.* **97**, 297–306.

Phosphorylation of vertebrate smooth muscle and nonmuscle myosin heavy chains *in vitro* and in intact cells

CHRISTINE A. KELLEY, SACHIYO KAWAMOTO, MARY ANNE CONTI and ROBERT S. ADELSTEIN*

Laboratory of Molecular Cardiology, National Heart, Lung, and Blood Institute, National Institutes of Health, Bethesda, MD 20892, USA

* Author for correspondence

Summary

In this article we summarize our recent experiments studying the phosphorylation of vertebrate myosin heavy chains by protein kinase C and casein kinase II. Protein kinase C phosphorylates vertebrate nonmuscle myosin heavy chains both *in vitro* and in intact cells. A single serine residue near the end of the helical portion of the myosin rod is the only site phosphorylated in a variety of vertebrate nonmuscle myosin heavy chains. There does not appear to be a site for protein kinase C phosphorylation in vertebrate smooth muscle myosin heavy chains. Casein kinase II phosphorylates a single serine residue located near the carboxyl terminus of the 204×10^3 M_r smooth muscle myosin heavy chain *in vitro* as well as in cultured smooth muscle cells. It does not phosphorylate the 200×10^3 M_r smooth muscle myosin heavy chain. However, the site is present in vertebrate nonmuscle myosin heavy chains. The 204×10^3 M_r myosin heavy chain of embryonic chicken gizzard smooth muscle is exceptional in not containing a site for casein kinase II phosphorylation.

Key words: myosin, phosphorylation, protein kinase C, casein kinase II.

Introduction

Myosin is a major contractile protein that is present in all eukaryotic cells (for reviews, see Citi and Kendrick-Jones, 1987; Sellers and Adelstein, 1987; Hartshorne, 1987; Korn and Hammer, 1988). In vertebrate cells, conventional myosin molecules consist of a pair of heavy chains ($\sim200\times10^3$ M_r) and two pairs of light chains (see Fig. 1) and are referred to as myosin II. In addition to the 480×10^3 M_r myosin isoforms, *Acanthamoeba*, *Dictyostelium* and the epithelial cells of vertebrate intestinal brush border contain a smaller (110×10^3 M_r) myosin molecule, myosin I, that preserves many of the structural and biological properties of the globular (head) domain, but lacks the tail portion (Korn and Hammer, 1988).

The purpose of this paper is to describe what is known about the phosphorylation of myosin II heavy chains (MHCs) in vertebrate nonmuscle and smooth muscle cells. To date, most of the experiments examining vertebrate myosin phosphorylation have focused on the 20×10^3 M_r myosin light chain (MLC) (Citi and Kendrick-Jones, 1987; Sellers and Adelstein, 1987; Hartshorne, 1987; Korn and Hammer, 1988). Vertebrate smooth muscle and nonmuscle myosin can be phosphorylated on at least five different sites present in the 20×10^3 M_r MLC. Ser-1, Ser-2 and Thr-9 (using the sequence based on the avian gizzard 20×10^3 M_r MLC, but also applicable to other smooth muscle and nonmuscle myosins) can be phosphorylated *in vitro* by protein kinase C (Bengur *et al.* 1987; Ikebe *et al.* 1987). In intact cells, only the serine residues have been shown to be phosphorylated (Kawamoto *et al.* 1989; Ikebe and Reardon, 1990; Sutton and Haeberle, 1990; Kamm *et al.* 1989). However, the possibility that the Thr-9 residue may first undergo phosphorylation, but then serve as a substrate for phosphatase(s), cannot be ruled out. The exact function of protein kinase C phosphorylation is uncertain in smooth muscle and nonmuscle cells. In the case of nonmuscle myosin, understanding the function of MLC phosphorylation has been complicated by the relatively recent finding that there is a single site on the MHC that can also be phosphorylated by protein kinase C in intact cells and *in vitro*.

In vitro studies have demonstrated that phosphorylation of the 20×10^3 M_r MLC on smooth muscle heavy meromyosin (HMM) makes the HMM a poorer substrate for myosin light chain kinase by decreasing the apparent affinity of myosin light chain kinase for HMM (Nishikawa *et al.* 1984). Studies using an *in vitro* motility assay system have not shown any difference between the movement of beads coated with myosin phosphorylated by both protein kinase C and myosin light chain kinase and beads coated with myosin that has been phosphorylated by myosin light chain kinase alone. Vertebrate nonmuscle and smooth muscle myosin that has only been phosphorylated by protein kinase C behaves like unphosphorylated myosin and fails to support bead movement (Umemoto *et al.* 1989).

Myosin light chain kinase phosphorylates Ser-19 and Thr-18 in the 20×10^3 M_r MLC (reviewed by Citi and Kendrick-Jones, 1987; Sellers and Adelstein, 1987; Hartshorne, 1987). This phosphorylation appears to play a role in the initiation of contractile activity in smooth muscle and nonmuscle cells as well as in the initiation of filament formation (Scholey *et al.* 1980). Studies using an *in vitro* motility assay system suggest that myosin light chain kinase phosphorylation is required for the movement of myosin coated beads (Sellers *et al.* 1985).

In addition to serving as a substrate for protein kinase C and myosin light chain kinase, the 20×10^3 M_r MLC can also be phosphorylated by calcium/calmodulin-dependent protein kinase II. The same serine residue that undergoes phosphorylation by myosin light chain kinase is phos-

Journal of Cell Science, Supplement 14, 49–54 (1991)

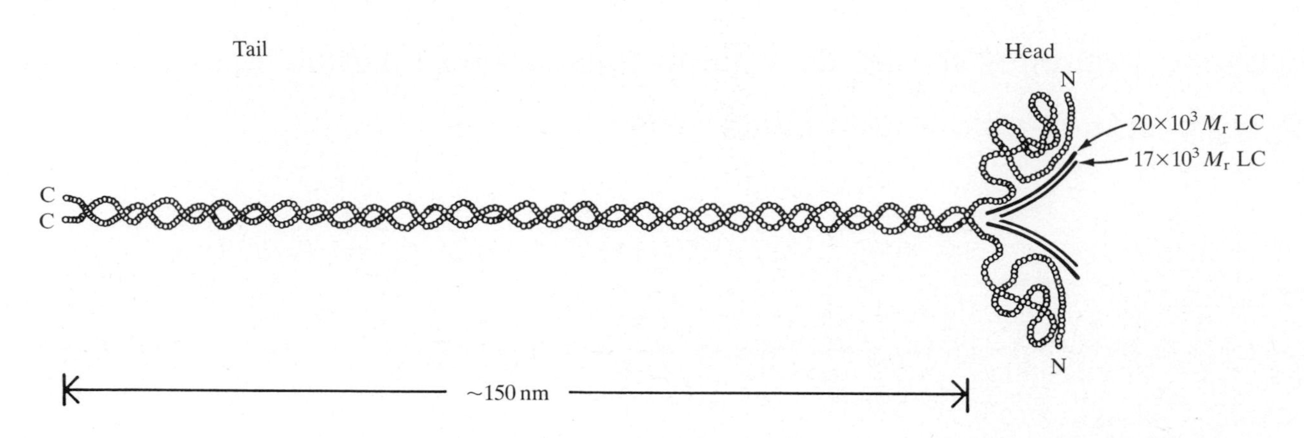

Fig. 1. Diagrammatic representation of the myosin heavy chains and light chains. The diagram illustrates the six polypeptide chains and the globular head (right) and helical (tail) domains. The 20×10^3 M_r light chain (LC) and the carboxyl-terminal region (C) of the heavy chain are phosphorylated in vertebrate nonmuscle and smooth muscle myosins (reproduced from Sellers and Adelstein, 1987).

phorylated by calcium/calmodulin-dependent protein kinase II (C. A. Kelley, unpublished observation). The significance of this phosphorylation is unclear at present.

Myosin heavy chain phosphorylation

Although the importance of myosin heavy chain phosphorylation in *Acanthamoeba* and *Dictyostelium* has been appreciated for a number of years (Korn and Hammer, 1988; Warwick and Spudich, 1987), the phosphorylation of vertebrate nonmuscle and smooth muscle myosin heavy chains has not been studied as intensively. A calcium/calmodulin-dependent kinase has been shown to catalyze phosphorylation of chicken intestinal epithelial cell myosin II (Rieker *et al.* 1987) as well as the heavy chain of brain myosin (Tanaka *et al.* 1986). Casein kinase II has also been shown to catalyze the phosphorylation of brain MHCs, and the serine residue phosphorylated by casein kinase II has been identified and shown to be close to the carboxyl-terminal end of the molecule (Murakami *et al.* 1990).

Our laboratory has been studying phosphorylation of vertebrate nonmuscle MHCs by protein kinase C and smooth muscle MHCs by casein kinase II. Below, we summarize some of our results, both in intact cells and *in vitro*, and discuss them in the context of the findings from other laboratories.

Protein kinase C

When human platelet myosin is incubated with protein kinase C, phosphate is covalently incorporated into the 20×10^3 M_r MLC (see above) as well as into the 200×10^3 M_r MHC (Kawamoto *et al.* 1989). Two-dimensional tryptic peptide maps of the 200×10^3 M_r MHC, yielded a single phosphopeptide (Fig. 2). When human platelets were incubated with phorbol ester, 0.7 moles of phosphate were incorporated into the 200×10^3 M_r MHC and 1.2 moles of phosphate were incorporated into the 20×10^3 M_r MLC (Kawamoto *et al.* 1989). Two-dimensional tryptic peptide mapping of the MHC revealed the same phosphopeptide that was found following *in vitro* phosphorylation of platelet myosin with protein kinase C (see Fig. 2). Further evidence that a unique site for protein kinase C is present on the 200×10^3 M_r MHC was obtained using RBL-2H3 (rat basophilic leukemia) cells. These cells contain receptors for IgE on their surface which aggregate following treatment with IgE and an antigen for the IgE. Subsequently, histamine and serotonin are released. It has also been shown that receptor aggregation results in activation of protein kinase C and the stoichiometric phosphorylation of the 200×10^3 M_r MHC at the same site that was phosphorylated by protein kinase C *in vitro* using human platelet myosin (Ludowyke *et al.* 1989).

Recently, the tryptic hexapeptide containing the serine phosphorylated by protein kinase C from human platelets was purified from human platelet MHCs and sequenced (Conti *et al.* 1990). A single serine residue, located ten amino acids amino-terminal to the proline that disrupts the coiled coil alpha-helix of the myosin rod, and approximately 50 residues from the carboxyl terminus of the molecule, was found to be the site phosphorylated by protein kinase C. The location of the phosphorylated serine was identified by comparing the amino acid sequence of the hexapeptide (-Glu-Val-Ser-*Ser**-Leu-Lys-) with the known sequences for the vertebrate nonmuscle MHCs (Shohet *et al.* 1989; Saez *et al.* 1990).

This site for protein kinase C phosphorylation appears to be present in all vertebrate nonmuscle MHCs examined to date. Of note is its absence from all vertebrate smooth muscle MHCs, although the amino acid sequence on both sides of the phosphorylated serine is conserved. This striking difference between vertebrate smooth muscle and nonmuscle MHCs is one of the first to be reported and points to putative differences in the regulation of these two molecules.

The biological consequences of protein kinase C phosphorylation of the MHCs are unknown. One possibility is that activation of kinase C, an enzyme that is known to be present near or at the cell membrane, results in disassembly of the myosin filaments in the cell cortex. Evidence supporting this idea comes from the work of A. Spudich (Stanford University). She showed, by immunofluorescence microscopy using antibodies to nonmuscle myosin, that the concentration of myosin in the cell cortex decreased dramatically following antigenic activation of RBL-2H3 cells (Spudich *et al.* 1990). One possibility is that protein kinase C phosphorylation of both the MHC and MLC is necessary for filament disassembly, so that the histamine and serotonin containing granules can reach the cell membrane for release. Whereas phosphorylation of the 20×10^3 M_r MLC by protein kinase C might act to

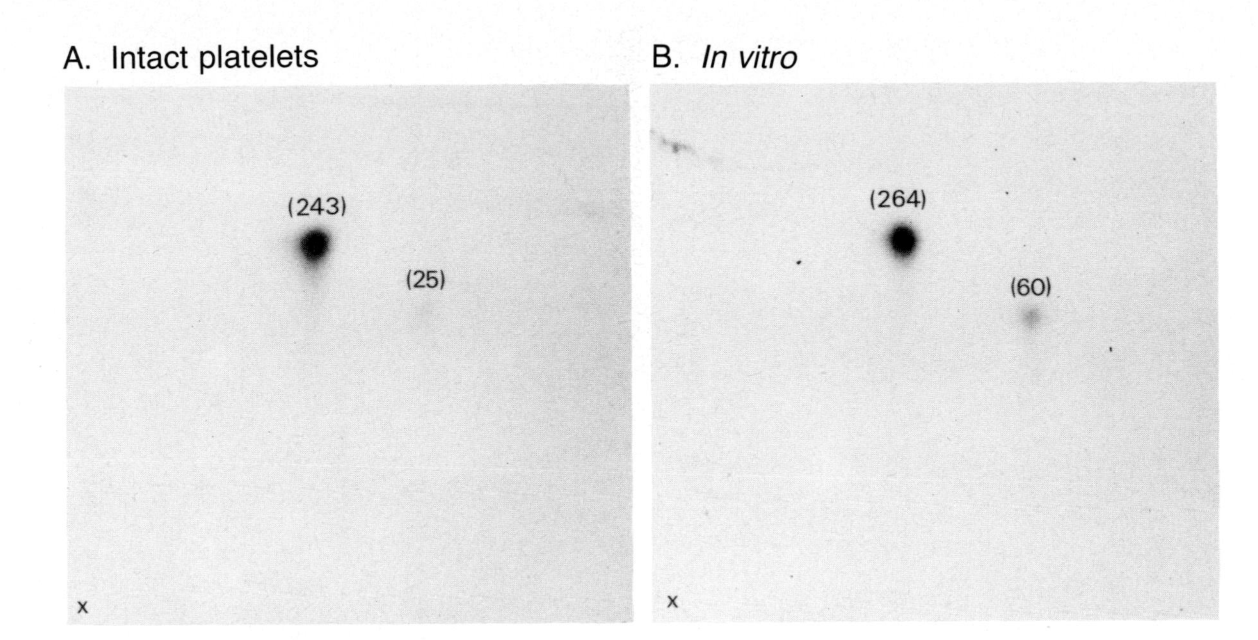

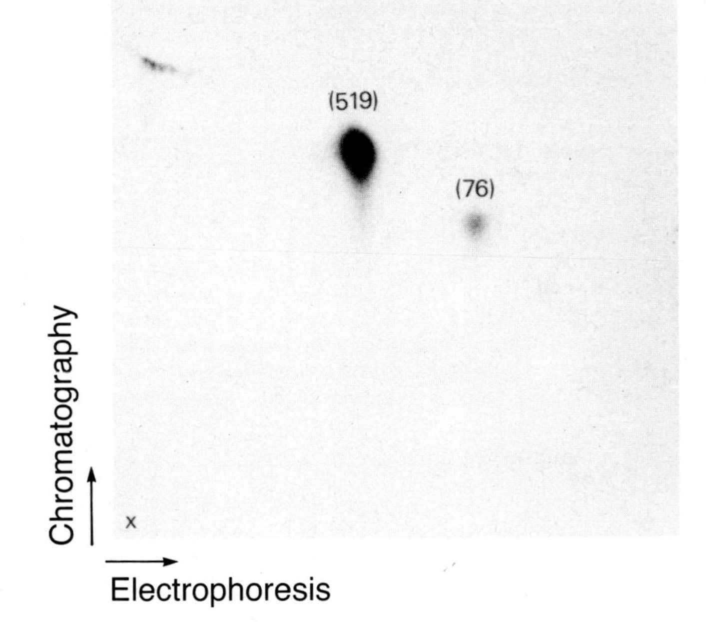

Fig. 2. Identification of phosphorylated myosin heavy chain tryptic peptides in TPA-treated platelets. MHCs were isolated by SDS–PAGE using 5% polyacrylamide gels. Tryptic digests of MHC eluted from the polyacrylamide gels were analyzed by two-dimensional peptide mapping. (A) Tryptic phosphopeptide map from intact platelets treated with TPA (162 nM, 15 min); (B) peptide map from purified human platelet myosin phosphorylated *in vitro* with protein kinase C. In (C) a 1:1 mixture of the samples mapped in A and B was spotted. Origin is marked with X. The numbers in parentheses indicate counts min^{-1} of ^{32}P radioactivity scraped from the thin-layer plates at the spots indicated on the autoradiogram. This figure is from Kawamoto *et al.* (1989).

inhibit filament formation, by rendering the MLC a poorer substrate for myosin light chain kinase (which favors filament formation), phosphorylation of the MHC by protein kinase C might act to dissociate the myosin filaments. Thus, MHC and MLC phosphorylation by protein kinase C would appear to complement each other.

Casein kinase II

Unlike protein kinase C, which phosphorylates a serine residue in a consensus sequence present in the vertebrate nonmuscle MHCs, but not in the 204 or $200\times10^3 M_r$ smooth muscle MHC, casein kinase II can phosphorylate both vertebrate nonmuscle (Murakami *et al.* 1990) and smooth muscle MHCs (Kelley and Adelstein, 1990). It is of note, however, that the $204\times10^3 M_r$ smooth muscle MHC, and not the $200\times10^3 M_r$ smooth muscle MHC isoform, contains a site for casein kinase II phosphorylation. Murakami *et al.* (1990) showed that bovine brain MHC contains a serine residue, which can be phosphorylated by casein kinase II, within the sequence -Leu-Glu-Leu-*Ser**-Asp-Asp-Asp-Asp-Glu- (see Fig. 5). They demonstrated

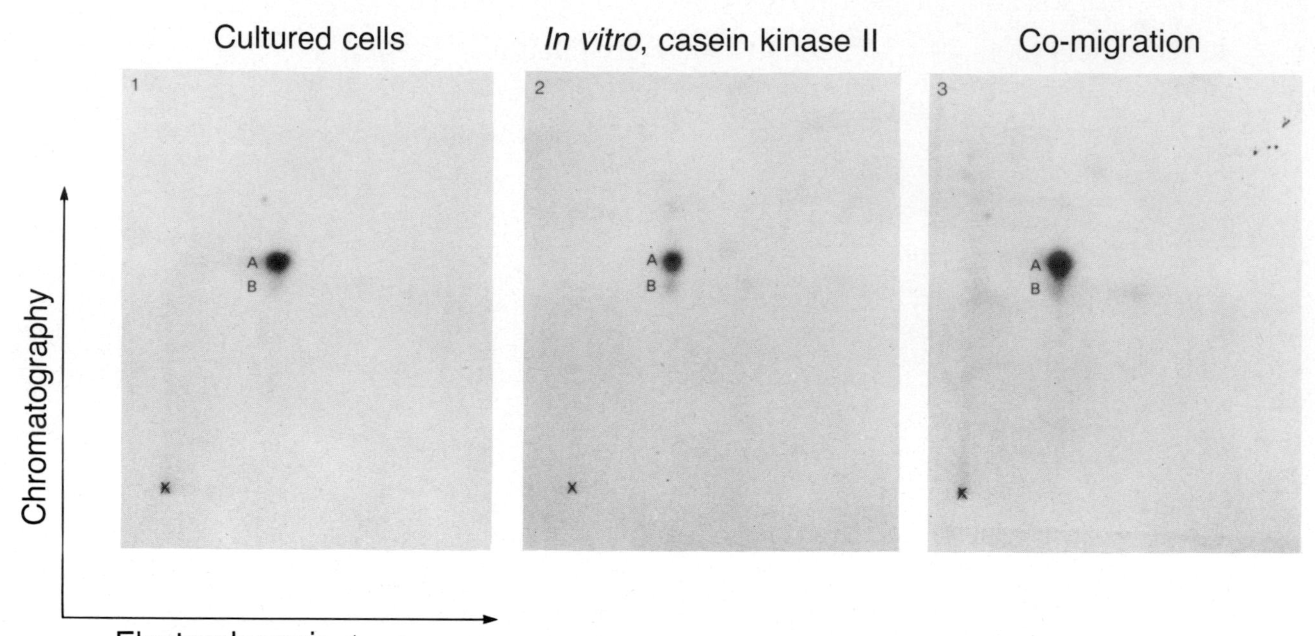

Fig. 3. Identification of tryptic phosphopeptides of smooth muscle myosin heavy chains phosphorylated in cultured aortic smooth muscle cells or *in vitro* with casein kinase II. The $204\times10^3\,M_r$ heavy chains of smooth muscle myosin immunoprecipitated from cultured cells labeled with [^{32}P]orthophosphate or purified myosin phosphorylated with casein kinase II *in vitro* were separated by SDS–PAGE (5% gel) and digested with trypsin. The resulting tryptic peptides were separated by two-dimensional thin layer electrophoresis and chromatography. The phosphopeptides were located by autoradiography. Panel 1, phosphopeptides of heavy chains from intact cells; panel 2, phosphopeptides of heavy chains phosphorylated *in vitro* with casein kinase II; panel 3, co-mapping of 1 and 2 above. The origin is marked with an ×. The letters indicate the common phosphopeptides. The directions of electrophoresis and chromatography are marked by the arrows. This figure is from Kelley and Adelstein (1990).

that this peptide was derived from the carboxyl-terminal region of the bovine brain MHC.

Kelley and Adelstein (1990) demonstrated that the $204\times10^3\,M_r$ MHC of bovine aortic smooth muscle cells could undergo phosphorylation both *in vitro* and in cultured aortic cells. When purified bovine aortic smooth muscle myosin, which consists of equal amounts of the two heavy chain isoforms, was incubated with casein kinase II, only the $204\times10^3\,M_r$ MHC, and not the $200\times10^3\,M_r$ MHC, was phosphorylated. Two-dimensional tryptic phosphopeptide mapping of the $204\times10^3\,M_r$ MHC shows 1 major and 1 minor peptide (Fig. 3, panel 2). When first passage cultures of bovine aortic cells were incubated with ^{32}P-labeled orthophosphate and the MHC subsequently purified and subjected to tryptic peptide mapping, the same two phosphopeptides were detected (Fig. 3, panels 1 and 3). These cultured cells only contain the $204\times10^3\,M_r$ MHC. This suggests that casein kinase II is the enzyme responsible for phosphorylation of the $204\times10^3\,M_r$ smooth muscle MHC in cultured aortic cells.

Calcium/calmodulin-dependent protein kinase II catalyzed phosphorylation of the smooth muscle MHC *in vitro*, but tryptic phosphopeptide mapping of the MHC revealed a different peptide map from that seen in intact cells. Phosphorylation of smooth muscle MHCs by protein kinase C, cAMP-dependent protein kinase and cGMP-dependent protein kinase did not result in a significant amount of phosphate incorporation into the smooth muscle MHCs. Prior to incubation with protein kinase C, the smooth muscle MHCs were also subjected to dephosphorylation using a variety of phosphatases. However, there was no change in the low amount of phosphate incorporated by this enzyme (Kelley and Adelstein, 1990). This was important because the stoichiometry of phosphate incorporation increased significantly in the case of casein kinase II, when the myosin was dephosphorylated prior to addition of the enzyme (from 0.2 to 0.6 moles of phosphate per mole of MHC).

Purification of the tryptic phosphopeptides. Fig. 4 shows the elution profile of tryptic peptides derived from the $204\times10^3\,M_r$ smooth muscle MHC phosphorylated by

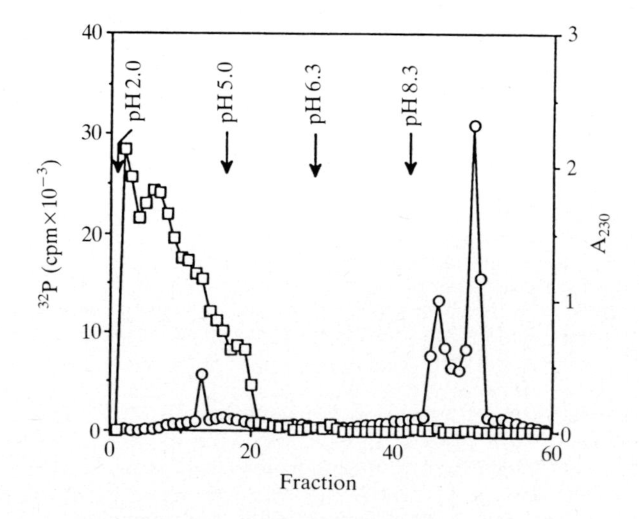

Fig. 4. Fe^{3+}-affinity column chromatography of tryptic digests of myosin heavy chains phosphorylated by casein kinase II *in vitro*. Arrows represent the changes in pH for the column washes and elution buffer. Radioactivity was detected by Cerenkov counting of the fractions (circles). Peptides were detected by A_{230} (squares). This figure is from Kelley and Adelstein (1990).

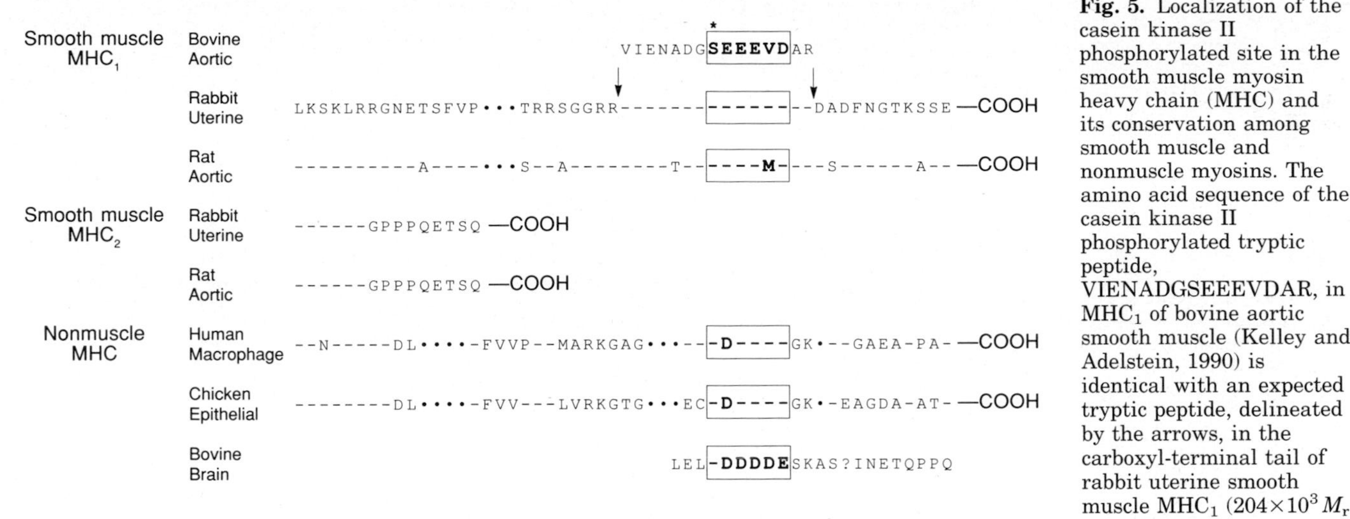

Fig. 5. Localization of the casein kinase II phosphorylated site in the smooth muscle myosin heavy chain (MHC) and its conservation among smooth muscle and nonmuscle myosins. The amino acid sequence of the casein kinase II phosphorylated tryptic peptide, VIENADGSEEEVDAR, in MHC_1 of bovine aortic smooth muscle (Kelley and Adelstein, 1990) is identical with an expected tryptic peptide, delineated by the arrows, in the carboxyl-terminal tail of rabbit uterine smooth muscle MHC_1 ($204\times10^3 M_r$ MHC; Nagai *et al.* 1989). This region is absent in rabbit uterine MHC_2 ($200\times10^3 M_r$ MHC). Casein kinase II phosphorylated the serine marked with an asterisk in the bovine aortic sequence. The amino acid sequence and casein kinase II phosphorylation site of this tryptic peptide are also highly conserved in rat aortic MHC_1 and are absent from rat aortic MHC_2 (Babij and Periasamy, 1989). The casein kinase II phosphorylation site and surrounding consensus sequence required for phosphorylation (boxed bold letters) are also conserved among nonmuscle myosins. The human macrophage myosin sequence is from Saez *et al.* (1990). The chicken epithelial myosin sequence is from Shohet *et al.* 1989. The bovine brain myosin sequence is from Murakami *et al.* (1990). This figure is from Kelley and Adelstein (1990).

casein kinase II. The column utilized was a Fe^{3+}–iminodiacetic acid–Sepharose affinity column (1×5 cm, Pharmacia) that could selectively bind phosphopeptides, which are eluted at pH 8.3 (see Fig. 4; Kelley and Adelstein, 1990). Tryptic digests of the phosphorylated smooth muscle MHC were applied to the column and the doublet peak (which appears to be a single phosphopeptide) eluted at pH 8.3 was pooled and purified on reverse phase HPLC chromatography. The amino acid sequence derived from the purified phosphopeptide is shown in Fig. 5, where it is compared to both vertebrate smooth muscle and nonmuscle isoforms. The box denotes the casein kinase II phosphorylation site and surrounding consensus sequence for phosphorylation. The figure shows that both the rabbit uterine and rat aortic $204\times10^3 M_r$ smooth muscle MHCs, but not the $200\times10^3 M_r$ MHCs, contain the casein kinase II phosphorylation site. Previous work has demonstrated that the rat aortic $204\times10^3 M_r$ MHC can be phosphorylated (Kawamoto and Adelstein, 1988), although the site of phosphorylation was not identified. Two-dimensional mapping of the tryptic phosphopeptide resulted in a map similar to that shown in Fig. 3.

With respect to vertebrate nonmuscle MHCs, previous work has demonstrated that there are at least two major $200\times10^3 M_r$ isoforms, which we refer to as MHC-A and B. Human macrophage (Saez *et al.* 1990) and chicken intestinal epithelial cell (Shohet *et al.* 1989) MHCs are examples of the MHC-A isoform. Both, as shown in Fig. 5, contain a serine residue that is followed by five acidic residues. We would, therefore, predict that this serine would be a substrate for casein kinase II. The bovine brain sequence is most likely an example of MHC-B, as suggested by Northern analysis of avian brain tissues (Katsuragawa *et al.* 1989; Kawamoto and Adelstein, 1990). It also preserves the same consensus sequence and phosphorylatable site. One important exception can be found in the sequence of embryonic chicken gizzard smooth muscle myosin which contains the necessary consensus sequence, but has a glycine residue in place of the phosphorylatable serine (Yanagisawa *et al.* 1987).

What is the effect of casein kinase II phosphorylation? To date, we cannot answer this question, but a number of studies are suggested by the above data. These include studies on *in vitro* filament formation, *in vitro* motility assays and possible effects on actin activated MgATPase activity. In addition, the effects of casein kinase II phosphorylation on MHC turnover should be considered. All of these experiments must also take into account a possible important contribution by the phosphorylation of the $20\times10^3 M_r$ MLC. The recent cloning of cDNA for vertebrate smooth muscle (Yanagisawa *et al.* 1987; Nagai *et al.* 1989; Babij and Periasamy, 1989) and nonmuscle (Shohet *et al.* 1989; Saez *et al.* 1990) MHCs will also permit the application of techniques utilizing molecular genetics to studies on the role of vertebrate smooth muscle and nonmuscle MHC phosphorylation in cellular function.

The authors are grateful for the expert editorial assistance of Catherine S. Magruder.

References

Babij, P. and Periasamy, M. (1989). Myosin heavy chain isoform diversity in smooth muscle is produced by differential RNA processing. *J. molec. Biol.* **210**, 673–679.

Bengur, A. R., Robinson, E. A., Appella, E. and Sellers, J. R. (1987). Sequence of the sites phosphorylated by protein kinase C in the smooth muscle myosin light chain. *J. biol. Chem.* **262**, 7613–7617.

Citi, S. and Kendrick-Jones, J. (1987). Regulation of non-muscle myosin structure and function. *Bioessays* **7**, 155–159.

Conti, M. A., Sellers, J. R., Adelstein, R. S. and Elzinga, M. (1991). Identification of the serine residue phosphorylated by protein kinase C in vertebrate nonmuscle myosin heavy chains. *Biochemistry* (in press).

Hartshorne, D. J. (1987). Biochemistry of the contractile process in smooth muscle. In *Physiology of the Gastrointestinal Tract* (ed. L. R. Johnson), vol. 2, pp. 423–482. Raven Press, New York.

Ikebe, M., Hartshorne, D. J. and Elzinga, M. (1987). Phosphorylation of the 20,000 dalton light chain of smooth muscle myosin by the calcium-activated, phospholipid-dependent protein kinase. *J. biol. Chem.* **262**, 9569–9573.

Ikebe, M. and Reardon, S. (1990). Phosphorylation of bovine platelet myosin by protein kinase C. *Biochemistry* **28**, 2713–2720.
Kamm, K. E., Hsu, L.-C., Kubota, Y. and Stull, J. T. (1989). Phosphorylation of smooth muscle myosin heavy and light chains: effects of phorbol dibutyrate and agonists. *J. biol. Chem.* **264**, 21 223–21 229.
Katsuragawa, Y., Yanagisawa, M., Inoue, A. and Masaki, T. (1989). Two distinct nonmuscle myosin-heavy-chain mRNAs are differentially expressed in various chicken tissues: Identification of a novel gene family of vertebrate non-sarcomeric myosin heavy chains. *Eur. J. Biochem.* **184**, 611–616.
Kawamoto, S. and Adelstein, R. S. (1988). The heavy chain of smooth muscle myosin is phosphorylated in aorta cells. *J. biol. Chem.* **263**, 1099–1102.
Kawamoto, S. and Adelstein, R. S. (1990). Chicken nonmuscle myosin heavy chains: Differential expression of two mRNAs and evidence for two different polypeptides. *J. Cell Biol.* (in press).
Kawamoto, S., Bengur, A. R., Sellers, J. R. and Adelstein, R. S. (1989). *In situ* phosphorylation of human platelet myosin heavy and light chains by protein kinase C. *J. biol. Chem.* **264**, 2258–2265.
Kelley, C. A. and Adelstein, R. S. (1991). The 204-kDa smooth muscle myosin heavy chain is phosphorylated in intact cells by casein kinase II on a serine near the carboxyl terminus. *J. biol. Chem.* **265**, 17 876–17 882.
Korn, E. D. and Hammer, J. A. (1988). Myosins of nonmuscle cells. *A. Rev. Biophys. biophys. Chem.* **17**, 23–45.
Ludowyke, R. I., Peleg, I., Beaven, M. A. and Adelstein, R. S. (1989). Antigen-induced secretion of histamine and the phosphorylation of myosin by protein kinase C in rat basophilic leukemia cells. *J. biol. Chem.* **264**, 12 492–12 501.
Murakami, N., Healy-Louie, G. and Elzinga, M. (1990). Amino acid sequence around the serine phosphorylated by casein kinase II in brain myosin heavy chain. *J. biol. Chem.* **265**, 1041–1047.
Nagai, R., Kuro-o, M., Babij, P. and Periasamy, M. (1989). Identification of two types of smooth muscle myosin heavy chain isoforms by cDNA cloning and immunoblot analysis. *J. biol. Chem.* **264**, 9734–9737.
Nishikawa, M., Sellers, J. R., Adelstein, R. S. and Hidaka, H. (1984). Protein kinase C modulates *in vitro* phosphorylation of the smooth muscle heavy meromyosin by myosin light chain kinase. *J. biol. Chem.* **259**, 8808–8814.
Rieker, J. P., Swanljung-Collins, H. and Collins, J. H. (1987). Purification and characterization of a calmodulin-dependent myosin heavy chain kinase from intestinal brush border. *J. biol. Chem.* **262**, 15 262–15 268.
Saez, C. G., Meyers, J. C., Shows, T. B. and Leinwand, L. A. (1990). Human nonmuscle myosin heavy chain mRNA: Generation of diversity through alternative polyadenylation. *Proc. natn. Acad. Sci. U.S.A.* **87**, 1164–1168.
Scholey, J. M., Taylor, K. A. and Kendrick-Jones, J. (1980). Regulation of non-muscle myosin assembly by calmodulin-dependent light chain kinase. *Nature* **287**, 233–235.
Sellers, J. R. and Adelstein, R. S. (1987). Regulation of contractile activity. In *The Enzymes* (ed. P. D. Boyer and E. G. Krebs), vol. 18, pp. 381–418. Academic Press: Orlando, FL.
Sellers, J. R., Spudich, J. A. and Sheetz, M. P. (1985). Light chain phosphorylation regulates the movement of smooth muscle myosin on actin filaments. *J. Cell Biol.* **101**, 1897–1902.
Shohet, R. V., Conti, M. A., Kawamoto, S., Preston, Y. A., Brill, D. A. and Adelstein, R. S. (1989). Cloning of the cDNA encoding the myosin heavy chain of a vertebrate cellular myosin. *Proc. natn. Acad. Sci. U.S.A.* **86**, 7726–7730.
Spudich, A., Wrenn, J. T. and Meyer, T. (1990). Dynamic changes in organization of myosin II in RBL cells undergoing antigen-mediated secretion. *J. Cell Biol.* **111**, 425a.
Sutton, T. A. and Haeberle, J. R. (1990). Phosphorylation by protein kinase C of the 20,000 dalton light chain of myosin in intact and chemically skinned vascular smooth muscle. *J. biol. Chem.* **265**, 2749–2754.
Tanaka, E., Fukunaga, K., Yamamoto, H., Iwasa, T. and Miyamoto, E. (1986). Regulation of the actin-activated Mg-ATPase of brain myosin via phosphorylation by the brain Ca^{2+}, calmodulin-dependent protein kinases. *J. Neurochem.* **47**, 254–262.
Umemoto, S., Bengur, A. R. and Sellers, J. R. (1989). Effect of multiple phosphorylations of smooth muscle and cytoplasmic myosins on movement in an *in vitro* motility assay. *J. biol. Chem.* **246**, 1431–1436.
Warrick, H. M. and Spudich, J. A. (1987). Myosin structure and function in cell motility. *A. Rev. Cell Biol.* **3**, 379–421.
Yanagisawa, M., Hamada, Y., Katsuragawa, Y., Imamura, M., Mikawa, T. and Masaki, T. (1987). Complete primary structure of vertebrate smooth muscle myosin heavy chain deduced from its complementary DNA sequence: Implications on topography and function of myosin. *J. molec. Biol.* **198**, 143–157.

Recombinant DNA approaches to study the role of the regulatory light chains (RLC) using scallop myosin as a test system

JOHN KENDRICK-JONES, ANA CLAUDIA RASERA DA SILVA*, FERNANDO C. REINACH*, NEIL MESSER, TONY ROWE and PAUL MCLAUGHLIN

MRC Laboratory of Molecular Biology, Hills Road, Cambridge CB2 2QH, UK

*Departamento de Bioquimica, Instituto de Quimica, Universidade de Sao Paulo, CP 20.780, CEP 01398 Sao Paulo SP, Brazil

Summary

The ability to exchange reversibly the regulatory light chains (RLCs) from scallop myosin has provided us with a test system to probe the mechanisms of regulation mediated by the RLCs from vertebrate skeletal, vertebrate smooth and molluscan myosins. The cloning and expression of these RLCs, together with domain-swapping and site-directed mutagenesis approaches, has allowed us to explore further the mechanisms involved and identify the functional importance of specific regions of the RLC molecule; for example, the presence of a high affinity metal binding site in the N-terminal domain and its interaction with the intact C-terminal domains are required for regulation.

Key words: calcium regulation, myosin II, regulatory light chains, metal binding domains, protein engineering.

Regulation of contraction in scallop muscle is accomplished by direct calcium binding to the myosin filament (Kendrick-Jones *et al.* 1970; Lehman and Szent-Györgyi, 1975). As in all conventional myosins, the myosin isolated from the scallop adductor muscle contains two heavy chains and four light chains. Associated with each myosin head are two light chains, one essential light chain (ELC) and one regulatory light chain (RLC), and they are responsible for regulating the interaction of the myosin head with actin (Kendrick-Jones and Scholey, 1981). Both types of light chains belong to the EF-hand family of calcium binding proteins, which includes calmodulin and troponin C. Structural predictions based on their amino-acid sequences indicate that each light chain is composed of four EF-hands or domains composed of a helix–loop–helix motif (Weeds and McLachlan, 1974; Kretsinger, 1980). Three of the domains however are incapable of binding metal, due to deletions and substitutions in the calcium coordinating positions (Collins, 1976; Kendrick-Jones and Jakes, 1976). Calcium binding experiments show that scallop myosin contains four metal binding sites, two in each head (Bagshaw and Kendrick-Jones, 1979). One of the sites, located in the RLC, has a high affinity but is not calcium specific since it binds both calcium and magnesium (Bagshaw and Kendrick-Jones, 1980; Bagshaw and Kendrick-Jones, 1979). Due to the relative affinities of calcium and magnesium for this RLC site it is believed that it is occupied by magnesium in the resting muscle, and during contraction the magnesium off rate is slow, so that calcium only slowly replaces it (Bagshaw and Reed, 1977). This site is present in all RLCs and is therefore considered to be a nonspecific 'structural' metal binding site. The second site, present in each scallop myosin head, is the regulatory site and calcium binding to this site triggers muscle contraction (Szent-Györgyi *et al.* 1973). This calcium specific site in scallop myosin is located in the ELC but can only be detected when this light chain is associated with the RLC and part of the heavy chain (Kwon *et al.* 1990).

The feasibility of exchanging the RLCs in scallop myosin with RLCs from different myosins (Szent-Györgyi *et al.* 1973; Kendrick-Jones, 1974; Chantler and Szent-Györgyi, 1980) has allowed us to study the role of the RLC in the regulation of myosin–actin interaction. It is still the only 'regulated' myosin where the RLCs can be exchanged reversibly with ease. Initial experiments using this approach indicated that the RLC has an inhibitory function since a myosin depleted of RLCs was no longer calcium regulated, i.e. myosin–actin interaction as measured *in vitro* by the actin-activated myosin MgATPase activity remained high in the presence or absence of calcium. Calcium regulation was restored when the scallop RLCs were added back. The exchange of scallop RLCs with foreign RLCs permitted a direct comparison of the functional capabilities of the different RLCs (Sellers *et al.* 1980; Kendrick-Jones *et al.* 1983). The behaviour of these foreign RLCs when incorporated into scallop myosin depends on the kind of regulation present in the myosin from which the RLC originated. For example, the RLCs from vertebrate skeletal muscle, where regulation is mediated by calcium binding to the troponin complex on the thin filament, conferred a calcium independent inhibitory behaviour on scallop myosin–actin interaction, whereas RLCs from vertebrate smooth muscle and non-muscle myosins, which are regulated by phosphorylation of the RLCs, when incorporated into scallop myosin restored calcium regulation to the myosin. The regulatory switch, in this case, could be achieved either by direct calcium binding if the smooth RLCs are unphosphorylated, or by phosphorylation of RLCs using a calcium dependent or independent myosin light chain kinase. Reconstitution with RLCs from other molluscan muscles restored the regulation modulated by calcium binding to the myosin head (Kendrick-Jones and Scholey, 1981; Simmons and Szent-Gyorgyi, 1980). These results indicated that although all these RLCs are very similar, slight

Journal of Cell Science, Supplement 14, 55–58 (1991)

differences in their sequences are responsible for their different regulatory behaviours.

With the cloning and expression in bacteria of a series of RLCs from vertebrate skeletal (Reinach and Fischman, 1985), scallop (Goodwin *et al.* 1987) and vertebrate smooth myosins (Messer and Kendrick-Jones, 1988), it is now possible to use new approaches, such as site-directed mutagenesis and domain-exchange experiments, to analyze RLC function and assign to specific regions in the molecule the specific properties observed when RLCs of different origins are used to reconstitute the scallop myosin.

Using site-directed mutagenesis, the role of the metal binding site in the N-terminal domain of the RLC has been studied (see Fig. 1). Two groups (Reinach *et al.* 1986; Goodwin *et al.* 1990) have destroyed this metal binding site in vertebrate skeletal and scallop RLCs by making alterations in the residues responsible for coordinating the metal ion in the metal binding loop. When the aspartic acid residues present in either the first or the last coordinating positions (X and −Y in Fig. 2) were replaced with alanine, calcium binding was disrupted. Upon hybridisation with scallop myosin, it was shown that these mutations result in a loss in the inhibitory activity of the RLCs. This led to the conclusion that the integrity of the metal binding site in the RLC is crucial for inhibition and

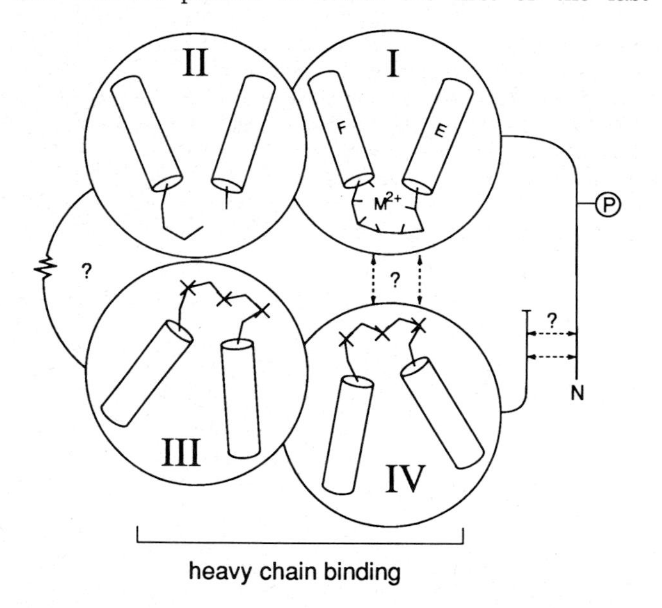

Fig. 1. Cartoon showing a possible arrangement of the domains in the regulatory light chains. Sequence comparisons among the Ca^{2+} binding protein family have established that the RLCs contain four EF-hand domains each composed of a helix E–loop–helix F motif. Domain I is a high affinity Ca^{2+}/Mg^{2+} binding site (structural, non regulatory site) whereas domains II, III and IV have lost metal binding due to deletions and substitutions in their binding loops. (P) denotes the position of the phosphorylatable serine residue in the smooth muscle RLC. Whether the oppositely charged N and C termini interact is speculative and is currently being tested. Evidence suggests that the C-terminal domains are involved in binding to the heavy chain and in regulation (Kendrick-Jones and Jakes, 1976; Kendrick-Jones *et al.* 1982; Goodwin *et al.* 1990). The pairs of domains I+II and III+IV may be linked by a long central helix as seen in the X-ray structures of troponin C (Herzberg and James, 1988) and calmodulin (Babu *et al.* 1987), which may be bent to allow interactions between the domains in the RLCs. Evidence for interactions between domains is provided by the observations (Szent-Györgyi *et al.* 1973; Reinach *et al.* 1986) that metal binding to the N-terminal domain is required for tight binding of the C-terminal domain to the heavy chain and for regulation. N.B. Shown is a highly speculative working model of the RLC which is currently being tested by experiments using crosslinking between domains, domain exchange/rearrangements between light chains, RLC mutants with deletions in selected regions etc.

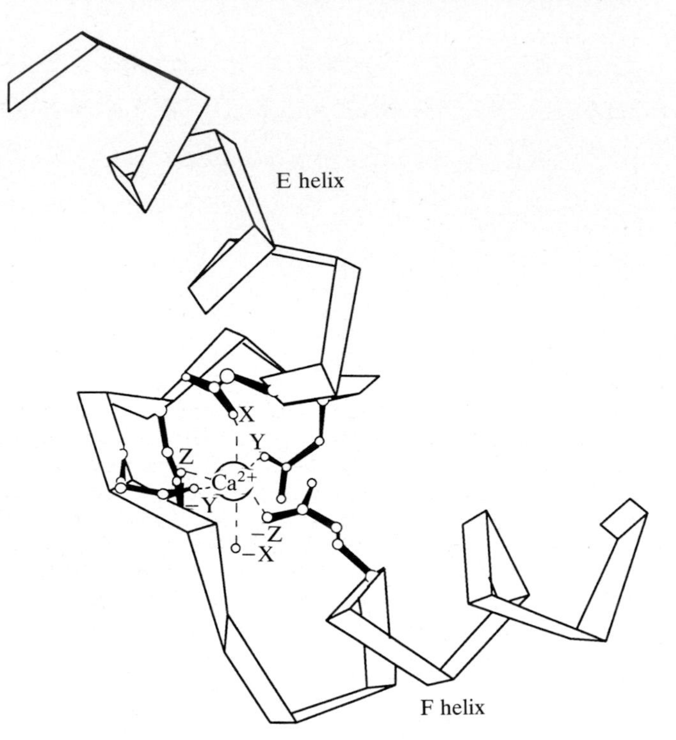

Co-ordinating Positions

	x		y		z		−y		−x			−z
Skeletal RLC	D	Q	N	R	D	G	I	I	D	K	D	D
Smooth RLC	D	Q	N	R	D	G	F	I	D	K	E	D
Skeletal TnC	D	K	N	A	D	G	F	I	D	I	E	E

Fig. 2. A Ca^{2+}/Mg^{2+} binding site (loop III) taken from the refined crystal structure of troponin C from Turkey skeletal muscle at 2.0Å resolution (Herzberg and James, 1988). The co-ordinates are from the Protein Data Bank entry 5TNC (Herzberg and James, 1988). The protein backbone is shown as a 'ribbon' where each rectangle represents the plane of the peptide bond for each amino acid. The side chains are only shown for the co-ordinating residues (except for the −Y position, where the carbonyl oxygen is the ligand and only the backbone is shown, and for the −X position where the ligand is a water molecule). This diagram was produced by a computer programme written by Lesk and Hardman (1982). As described by Strynadka and James (1989) one can visualise a metal binding domain as a cup with its inside walls lined with relatively exposed hydrophobic residues and its bottom containing the metal binding loop. The amino acid sequences shown are those of the Ca^{2+}/Mg^{2+} binding loops in domain I of both RLCs and in domain III of turkey skeletal muscle TnC (Collins, 1976; Kretsinger, 1980). The boxed residues indicate: (1) the aspartate residues at the X and −Z co-ordinating positions which were changed to alanine resulting in a loss of Ca^{2+} binding and regulation (Reinach *et al.* 1986; Goodwin *et al.* 1990) and (2) the invariant glycine residue between the Z and −Y co-ordinating positions.

regulation. However, an experiment in which the last eleven amino acids of the scallop RLC were substituted with a tail of 15 unrelated amino acids indicated that the C-terminal domain is also important for inhibition of myosin–actin interaction (Goodwin *et al.* 1990). When this particular mutant was hybridised with scallop myosin, the calcium regulatory site was regained and functional, but the actin-activated myosin ATPase was not modulated by calcium, the myosin remaining in the 'ON' state irrespective of the calcium concentration. Previous spin label, chemical and protease fragmentation studies on the RLCs have also indicated the importance of the C-terminal domains for regulatory function and heavy chain binding (Kendrick-Jones and Jakes, 1976; Bagshaw and Kendrick-Jones, 1980). These results demonstrate that the inhibitory/regulatory capacity of RLCs depends on the integrity of both the N-terminal metal binding site and the C-terminal domain. Further support for this conclusion was obtained in the following domain swapping experiments.

Chimaeric proteins containing domains derived from vertebrate skeletal RLC, smooth RLC and troponin C were constructed to try to understand the roles and interactions between the different domains of the RLC. One of these hybrids containing a calcium specific EF-hand from TnC replacing the calcium/magnesium site present in the chicken skeletal RLC was constructed, in an attempt to obtain a RLC with a calcium specific site, i.e. it contained domains $TnC_ISK_{II,III,IV}$ (A. da Silva, J. Kendrick-Jones and F. Reinach, personal communication). It was expected that with such a hybrid it would be possible to generate a site that would be empty in the relaxed muscle and occupied by calcium in the active muscle (high calcium). Unfortunately, calcium binding experiments using this chimaera indicated that after transplantation, the TnC site had lost part of the affinity and specificity it possessed while in the TnC background. Thus, the calcium binding properties of this site are determined partly by interactions within the EF-hand domain and partly by interactions of the EF-hand with the remainder of the protein. When tested in the scallop myosin system this TnC/RLC chimaera behaved like those mutant RLCs that do not have a functional calcium binding site, i.e. it had lost the inhibitory capacity and the scallop myosin remained unregulated. This result clearly shows that intra- and inter-domain interactions are important in maintaining the inhibitory capability of the RLC.

These results also pose the fundamental question which has been raised many times before (Kretsinger, 1980): what are the primary, secondary and tertiary structural elements that determine affinity and specificity in a calcium binding site? (see Fig. 2). The RLC model for the study of metal specificity and affinity has been largely unexplored. By altering the sequence of the calcium/magnesium binding site of RLC using site directed mutagenesis, it should be possible to study the determinants of affinity and specificity in a metal binding site. The results obtained with this system should be easier to analyze when compared with other systems such as troponin-C and calmodulin with their multi-binding sites, since in the RLC we have just one site to analyze and the complications due to interactions between sites can be overcome. It should also be interesting to analyze the functional implications of changes in specificity and affinity by reconstituting the mutated RLCs back into scallop myosin.

Using the domain swapping approach two hybrids between skeletal and smooth RLCs were constructed (Messer and Kendrick-Jones, 1990). The hybrid Sk/Sm RLC contained the N-terminal and metal binding site (N-terminal domain) from skeletal RLC and the remaining domains from smooth RLC i.e. domains $Sk_ISm_{II,III,IV}$. This hybrid was tested in the scallop myosin system and it behaved like the smooth RLC, i.e. it conferred calcium regulation to the actin-activated scallop myosin ATPase. The complementary hybrid, Sm/Sk, containing the N-terminal domain from smooth RLC and the C-terminal domains from skeletal RLC, i.e. domains $Sm_ISk_{II,III,IV}$, behaved like the skeletal RLC in the scallop myosin system, i.e. it inhibited the ATPase irrespective of the calcium concentration (Messer and Kendrick-Jones, 1990). Thus the type of regulation which these chimaeric RLCs confer on the myosin is determined by the origins of domains II to IV rather than that of domain I.

The results obtained with the chimaera indicate that the C-terminal domains play the major role in determining the characteristic properties of the RLC. However, it seems that a fine tuned interaction between the N-terminal and C-terminal domains is necessary to maintain the functional activity of the RLC. The data support the idea that the inhibitory element in the scallop myosin system is the RLC and that the inhibitory capacity of this subunit depends on the presence of a functional metal binding site and its interaction with the C-terminal domain of the protein.

We still know very little about how the RLC works. At a basic level we can draw an analogy between myosin essential and regulatory light chain (ELC/RLC) regulation in scallop muscle and troponin-C and troponin-I (TnC/TnI) regulation on the actin filaments in vertebrate skeletal muscles. Functionally the RLC works like TnI, inhibiting muscle contraction in the absence of calcium. This inhibition is relieved when calcium binds to the ELC site (in scallop myosin) or to the TnC component in skeletal muscle (repression/derepression control). In vertebrate smooth muscle and nonmuscle myosins, calcium activates a specific kinase which phosphorylates the RLC and this modification in the N-terminal region of the RLC overcomes its inhibitory effect. The observation that when the smooth muscle RLC is incorporated into scallop myosin it can function under phosphorylation or direct calcium control suggests that both regulatory mechanisms are similar. Furthermore it has recently been demonstrated that in the absence of calcium, scallop myosin can be induced to fold into a 10S monomer conformation (a filament assembly and ATPase blocked state) and addition of calcium causes the myosin tail to unfold, allowing it to assemble into filaments (Ankrett *et al.* 1991). This folded 10S myosin conformation was previously believed to be restricted to vertebrate smooth muscle and nonmuscle myosins (Trybus and Lowey, 1984; Craig *et al.* 1983; Cross *et al.* 1988). Thus the scallop myosin 'test' system can be used to study not only RLC regulation of actin–myosin interaction but also the role of the RLC in the folded 10S monomer/filament assembly process under direct calcium or phosphorylation control.

In conclusion, as we have seen in the examples discussed, the ability to incorporate and test RLCs with different functional properties in scallop myosin, combined with detailed protein engineering on recombinant RLCs, should provide us with interesting insights into the mechanisms of regulation of myosin–actin interaction and monomer–filament assembly controlled by the RLCs. However, the fine details of the mechanisms involved await the elucidation of the atomic structures of the light chains on the myosin head in the ON and OFF regulatory states.

References

Ankrett, R. J., Rowe, A. J., Cross, R. A., Kendrick-Jones, J. and Bagshaw, C. R. (1991). A folded (10S) conformer of myosin from a striated muscle and its implications for regulation of ATPase activity. *J. molec. Biol.*, (in press).

Babu, Y. S., Bugg, C. E. and Cook, W. J. (1987). X-ray diffraction studies of Calmodulin. *Meth. Enzym.* **139**, 632–642.

Bagshaw, C. R. and Kendrick-Jones, J. (1979). Characterization of homologous divalent metal ion binding sites of vertebrate and molluscan myosins using electron paramagnetic resonance spectroscopy. *J. molec. Biol.* **130**, 317–336.

Bagshaw, C. R. and Kendrick-Jones, J. (1980). Indentificaton of the divalent metal ion binding domain of myosin regulatory light chains using spin-labelling techniques. *J. molec. Biol.* **140**, 411–433.

Bagshaw, C. R. and Reed, G. H. (1977). The significance of the slow dissociation of divalent metal ions from myosin regulatory light chains. *FEBS Lett.* **81**, 386–390.

Chantler, P. D. and Szent-Györgyi, A. G. (1980). Regulatory light-chains and scallop myosin: full dissociation, reversibility and co-operative effects. *J. molec. Biol.* **138**, 473–492.

Collins, J. H. (1976). Homology of myosin DTNB light chain with alkali light chains, troponin C and parvalbumin. *Nature* **259**, 699–700.

Craig, R., Smith, R. and Kendrick-Jones, J. (1983). Light chain phosphorylation controls the conformation of vertebrate non-muscle and smooth muscle myosin molecules. *Nature* **302**, 436–439.

Cross, R. A., Jackson, A. P., Citi, S., Kendrick-Jones, J. and Bagshaw, C. R. (1988). Active site trapping of nucleotide by smooth and non-muscle myosin. *J. molec. Biol.* **203**, 173–181.

Goodwin, E. B., Leinwand, L. A. and Szent-Györgyi, A. G. (1990). Regulation of scallop myosin by mutant regulatory light chain. *J. molec. Biol.* **216**, 85–93.

Goodwin, E. B., Szent-Györgyi, A. G. and Leinwand, L. A. (1987). Cloning and characterization of the scallop essential and regulatory myosin light chain cDNAs. *J. biol. Chem.* **262**, 11 052–11 056.

Herzberg, O. and James, M. N. G. (1988). Refined crystal structure of troponin C from Turkey skeletal muscle at 2.0Å resolution. *J. molec. Biol.* **203**, 761–779.

Kendrick-Jones, J. (1974). Role of myosin light chains in calcium regulation. *Nature* **249**, 631–634.

Kendrick-Jones, J., Cande, W. Z., Tooth, P. J., Smith, R. C. and Scholey, J. M. (1983). Studies on the effect of phosphorylation of the 20,000 M_r light chain of vertebrate smooth muscle myosin. *J. molec. Biol.* **165**, 139–154.

Kendrick-Jones, J. and Jakes, R. (1976). Myosin-linked regulation: a chemical aproach. In *International Symposium on Myocardial Failure* (ed. G. Rieker, A. Weber, and J. Goodwin), pp. 28–40. Springer-Verlag Berlin: Heidelberg-New York.

Kendrick-Jones, J., Jakes, R., Tooth, P., Craig, R. and Scholey, J. M. (1982). Role of the myosin light chains in the regulation of contractile activity. In *Basic Biology of Muscles: A Comparative Approach* (ed. Twarog, B. M., Levine, R. J. C. and Dewey, M. M.), pp. 255–272. Raven Press, New York.

Kendrick-Jones, J., Lehman, W. and Szent-Györgyi, A. G. (1970). Regulation in molluscan muscles. *J. molec. Biol.* **54**, 313–326.

Kendrick-Jones, J. and Scholey, J. M. (1981). Myosin-linked regulatory systems. *J. Mus. Res. Cell Mot.* **2**, 347–372.

Kretsinger, R. H. (1980). Structure and evolution of calcium-modulated proteins. *CRC Crit. Rev. Biochem.* **8**, 119–174.

Kwon, H., Goodwin, E. B., Nyitray, L., Berliner, E., O'Neall-Hennessey, E., Melandri, F. D. and Szent-Györgyi, A. G. (1990). Isolation of the regulatory domain of scallop myosin: role of the essential light chain in calcium binding. *Proc. natn. Acad. Sci. U.S.A.* **87**, 4771–4775.

Lehman, W. and Szent-Györgyi, A. G. (1975). Regulation of muscular contraction. *J. gen. Physiol.* **66**, 1–30.

Lesk, A. and Hardman, K. (1982). Computer generated schematic diagrams of protein structures. *Science* **216**, 539–540.

Messer, N. G. and Kendrick-Jones, J. (1988). Molecular cloning and sequencing of the chicken smooth muscle myosin regulatory light chain. *FEBS Lett.* **234**, 49–52.

Messer, N. G. and Kendrick-Jones, J. (1991). Chimaeric myosin regulatory light chains. Domain-switching experiments to analyse the function of the N-terminal domain. *J. molec. Biol.* (in press).

Reinach, F. C. and Fischman, D. A. (1985). Recombinant DNA approach for defining the primary structure of monoclonal antibody epitopes. The analysis of a conformational-specific antibody to myosin light chain two. *J. molec. Biol.* **181**, 411–422.

Reinach, F. C., Nagai, K. and Kendrick-Jones, J. (1986). Site directed mutagenesis of the regulatory light chain Ca^{2+}/Mg^{2+} binding site and its role in hybrid myosins. *Nature* **322**, 80–83.

Sellers, J. R., Chantler, P. D. and Szent-Györgyi, A. G. (1980). Hybrid formation between scallop myofibrils and foreign regulatory light chains. *J. molec. Biol.* **144**, 223–245.

Simmons, R. M. and Szent-Györgyi, A. G. (1980). Control of tension development in scallop muscle fibres with foreign regulatory light chains. *Nature* **286**, 626–628.

Strynadka, N. C. J. and James, M. N. G. (1989). Crystal structures of the helix–loop–helix calcium-binding proteins. *A. Rev. Biochem.* **58**, 951–998.

Szent-Györgyi, A. G., Szentkiralyi, E. M. and Kendrick- Jones, J. (1973). The light chains of scallop myosin as regulatory subunits. *J. molec. Biol.* **74**, 179–203.

Trybus, K. M. and Lowey, S. (1984). Conformational states of smooth muscle myosin. Effects of light chain phosphorylation and ionic strength. *J. biol. Chem.* **259**, 8564–8571.

Weeds, A. G. and McLachlan, A. D. (1974). Structural homology of myosin alkali light chains, troponin C and carp calcium binding protein. *Nature* **859**, 646–649.

A myosin-like protein from smooth muscle

KAZUHIRO KOHAMA, YUAN LIN, HIROMI TAKANO-OHMURO and RYOKI ISHIKAWA

Department of Pharmacology, Gunma University School of Medicine, Maebashi, Gunma 371, Japan

Summary

A protein was purified from chicken gizzard smooth muscle. It bound ATP and actin. Actin activated the Mg^{2+}-ATPase activity of this protein. The Ca^{2+}-ATPase activity was lower than K^+-EDTA ATPase activity. Thus, it appears that this protein is akin to myosin I rather than to conventional myosin. However, ATPase activities of the protein were much lower than those of myosin I. A protein cofactor, such as protein kinase, which would enhance these activities remains to be purified from the smooth muscle.

Key words: smooth muscle, myosin, actin.

Introduction

The contraction of smooth muscle is modified by various pharmacological agents. The sites of action of such agents are mostly associated with their respective receptors on the cell membrane. However, several agents have recently been shown to modify the contraction by directly affecting cytosolic proteins related to motor proteins (Hidaka *et al.* 1985). They affect the activities of kinase(s) or phosphatase(s), thereby modifying the extent of phosphorylation levels of myosin (Yabu *et al.* 1990), with resultant changes in myosin ATPase activities.

The analysis of contractions induced by phorbol ester (Park and Rasmussen, 1985) has not yet yielded such explicit results. The phorbol ester stimulated the activity of protein kinase C (PKC). However, smooth muscle myosin, the light chain of which was phosphorylated by PKC, was not in an active form (Nishikawa *et al.* 1983; Takano-Ohmuro and Kohama, 1987). Thus, it is tempting to speculate that a myosin-like protein other than conventional myosin may be phosphorylated by PKC to produce the contraction.

In two lower eukaryotes, *Acanthamoeba* and *Dictyostelium*, two classes of myosin have been described: myosin I, which is composed of a single heavy chain of about 120×10^3 M_r and a single light chain, and conventional myosin, which is composed of two heavy chains and two pairs of light chains (Warrick and Spudich, 1987; Korn and Hammer, 1988). In vertebrates, there is accumulating evidence to indicate that the complex between the 110×10^3 M_r protein and calmodulin (110K–calmodulin complex) is a protein that resembles myosin I (Hoshimaru and Nakanishi, 1987; Collins *et al.* 1990). The complex is detectable in intestinal brush-border and some nervous tissues.

Although there are no reports that myosin I is subject to regulation by phorbol ester and/or PKC, we have been engaged in a search for a myosin-like protein in smooth muscle.

Materials and methods

The method is based on that described by Lin *et al.* (1989). Minced smooth muscle from chicken gizzard was homogenized with a Virtis homogenizer at 45 000 rpm for 30 sec in 6 volumes of 0.5 M NaCl, 0.1 M sucrose, 20 mM Tris–HCl (pH 7.5), 10 mM $Na_4P_2O_7$, 2 mM EGTA, 14 mM 2-mercaptoethanol (ME), 1 mM NaN_3 and protease inhibitors, i.e. 0.2 mM PMSF, 0.1 mM E64C. The homogenate was centrifuged at 250 000 ***g*** for 60 min. The supernatant, i.e. the crude extract, was applied to a column of hydroxylapatite (HAP; Bio-Rad) equilibrated with buffer that contained 0.5 M NaCl, 20 mM Tris–HCl (pH 7.5), 1 mM NaN_3, 14 mM ME, 2 mM ATP and protease inhibitors. The column was washed with the same buffer, and then a 200 ml gradient of 0 M to 1 M KH_2PO_4 (pH 7.0) was applied. The fractions eluted from the column were subjected to the assay for K^+-EDTA ATPase activity and to SDS–PAGE. Fractions containing the ATPase activity but without the 200×10^3 M_r heavy chain of conventional myosin were pooled and dialyzed against a dialysis buffer that contained 50 mM NaCl, 0.2 mM NaN_3, 20 mM Tris–HCl (pH 7.5), 14 mM ME and protease inhibitors. The dialyzate was made 2 mM in ATP and applied to a column of DEAE cellulose (DE-52 Whatman) equilibrated with buffer that contained 50 mM NaCl, 20 mM Tris–HCl (pH 7.5), 2 mM ATP, 0.2 mM NaN_3 and protease inhibitors. After the column was washed with the same buffer, a 160 ml gradient of 0.05 M to 0.8 M NaCl in the same buffer was applied to the column. Fractions eluted from the DE52 column were subjected to SDS–PAGE, and those containing only a band at 130×10^3 M_r were pooled and concentrated by dialysis against polyethylene glycol (PEG, molecular mass $15–20\times10^3$; Sigma). The resulting concentrate was dialyzed against the dialysis buffer mentioned above and used as the 130×10^3 M_r protein.

Results and discussion

Search for myosin-like protein in smooth muscle

Both myosin I and conventional myosin exhibit high K^+-EDTA ATPase activity, which is quite distinct from that of other ATPases (Warrick and Spudich, 1987). Therefore a crude extract of chicken gizzard smooth muscle was fractionated by molecular size, and the resultant fractions were subjected to measurement of K^+-EDTA ATPase activity. As shown in Fig. 1, the ATPase activity was detected not only in the fraction that contained conventional myosin but also in the fractions devoid of conventional myosin. We speculated that the latter fractions should contain some form of myosin other than conventional myosin.

Journal of Cell Science, Supplement 14, 59–61 (1991)

Characterization of the 130×10^3 M_r protein

ATP-binding activity of the 130×10^3 M_r protein was examined by the photoaffinity-labelling technique (Fig. 2). The protein bound ATP to a level comparable with that bound by conventional myosin. Its ability to hydrolyze ATP is shown in Table 1. As in the case of conventional myosin, actin activated the Mg^{2+}-ATPase activity. A remarkable feature of the protein was that K^+-EDTA ATPase activity was higher than Ca^{2+}-ATPase activity. Since Ca^{2+}-ATPase activity of conventional myosin is relatively high (Korn and Hammer, 1988), the 130×10^3 M_r protein has similar properties in terms of its ATPase activities to those of myosin I. However, unlike myosin I, the protein has no light chain (Fig. 3A).

As expected from the activating effect of actin on the ATPase activity, actin bound to the 130×10^3 M_r protein up to a level of 0.2 mol mol^{-1} of actin (Fig. 4). The dissociation constant was in the micromolar range.

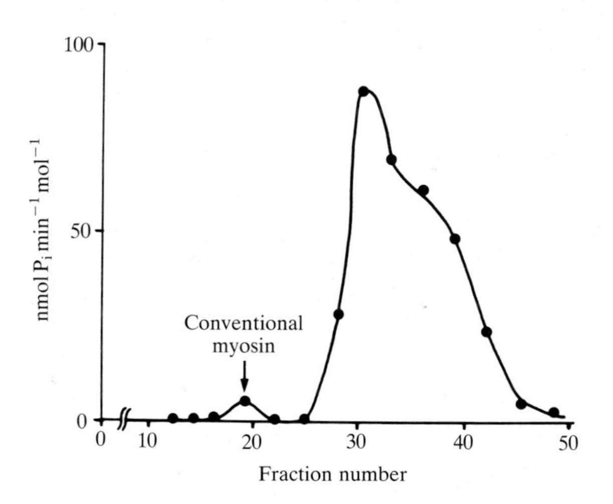

Fig. 1. K^+-EDTA ATPase activities of an extract of chicken gizzard after gel filtration. Crude extract, prepared as described in the text, was fractionated by HPLC on a column of TSKG3000SW in 0.5 M NaCl, 20 mM Tris–HCl (pH 7.5) and 14 mM ME. K^+-EDTA ATPase activities of each fractions were determined as described in the legend to Table 1. An arrow shows the fraction that contained conventional myosin. P_i, phosphate released.

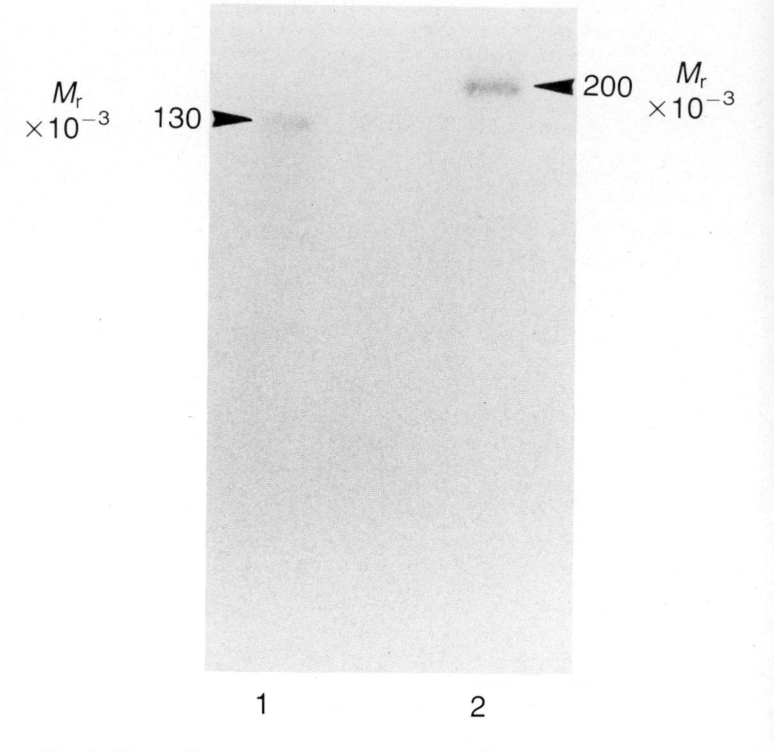

Fig. 2. Photoaffinity labelling of the 130×10^3 M_r protein and conventional myosin. As described by Kohama *et al.* (1988), ATP alpha^{32}P was mixed with the protein (130×10^3 M_r, lane 1) or conventional myosin from gizzard (200×10^3 M_r, lane 2). The mixture was illuminated with a UV lamp to label at the ATPase active site of the ATPase. Then it was subjected to SDS–PAGE and subsequent autoradiography.

A

Lc

1 2 3

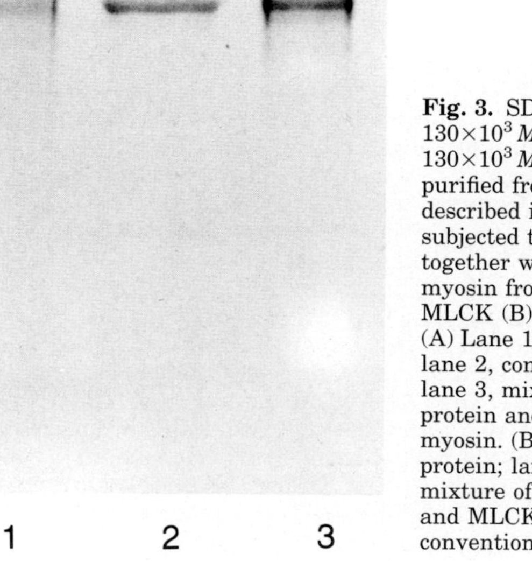

Fig. 3. SDS–PAGE of the 130×10^3 M_r protein. A 130×10^3 M_r protein was purified from gizzard as described in the text and was subjected to SDS–PAGE together with conventional myosin from gizzard (A) or MLCK (B) from gizzard. (A) Lane 1, 130×10^3 M_r protein; lane 2, conventional myosin; lane 3, mixture of 130×10^3 M_r protein and conventional myosin. (B) Lane 1, 130×10^3 M_r protein; lane 2, MLCK; lane 3, mixture of 130×10^3 M_r protein and MLCK. Lc, light chains of conventional myosin.

Table 1. *ATPase activities of the 130×10³ M_r protein**

Activity measured	P_i released†
Ca^{2+}-ATPase	4.0
K^+-EDTA ATPase	13.8
Mg^{2+}-ATPase	3.1
Actin-activated ATPase	9.5

ATPase activity was measured by the colorimetric determination of inorganic phosphate released from ATP (Youngberg and Youngberg, 1930). The conditions for assessment of Ca^{2+}-ATPase activities were: 0.6 M KCl, 20 mM Tris–HCl (pH 7.5), 3.8 mM $CaCl_2$ and 2.5 mM ATP; those for K^+-EDTA ATPase activities were: 0.6 M KCl, 1 mM EDTA, 20 mM Tris–HCl (pH 7.5) and 2.5 mM ATP; those for Mg^{2+}-ATPase and actin-activated ATPase activities were 25 mM KCl, 2.8 mM $MgCl_2$, 20 mM Tris–HCl (pH 7.5), 0.1 mM $CaCl_2$ and 2.5 mM ATP. The concentrations of F-actin and 130×10^3 M_r protein were 0.61 mg ml^{-1} and 0.13 mg ml^{-1}, respectively.

* Lin *et al.* (1989).

† P_i, nmol P_i min^{-1} mg^{-1} at 25 °C.

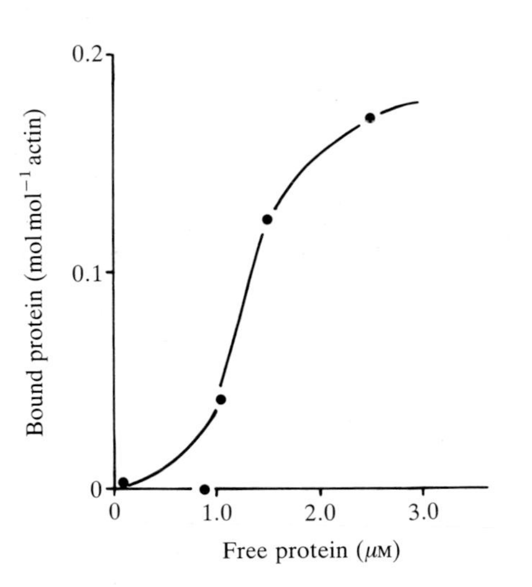

Fig. 4. Binding of the 130×10^3 M_r protein to actin. F-actin (0.5 mg ml^{-1}) was mixed with various amounts of the 130×10^3 M_r protein in 250 mM KCl, 3.8 mM $MgCl_2$, 0.1 mM $CaCl_2$ and 20 mM Tris–HCl (pH 7.5). The mixture was centrifuged in a Beckman Airfuge at 140 000 ***g*** for 30 min. The amounts of 130×10^3 M_r protein and actin in both supernatants and precipitates were determined quantitatively by densitometry, as described previously (Ishikawa *et al.* 1989).

Evaluation of the 130×10³ M_r protein as a myosin-like protein

The 130×10^3 M_r protein shares the ability to bind both ATP and actin with conventional myosin and myosin I, as mentioned above. Furthermore, the ATPase activities show that it is more like myosin I than like conventional myosin. However, it is not yet appropriate to call the protein the myosin I of smooth muscle, because its ATPase activities (Table 1) are lower than those of myosin I. We are now examining cofactor(s) that enhance the ATPase activities of the 130×10^3 M_r protein.

It must be noted that myosin light chain kinase (MLCK) of smooth muscle is known to interact with ATP and actin, as does the 130×10^3 M_r protein (Yamazaki *et al.* 1987). However, the 130×10^3 M_r protein did not co-migrate with MLCK during SDS–PAGE (Fig. 3B), indicating that the proteins differ from one another.

The 110K–calmodulin complex of brush-border has Mg-ATPase activity which is activated by actin. The 130×10^3 M_r protein gave, however, no indication of any ability to bind calmodulin, as judged by a gel-overlay experiment with ^{125}I-calmodulin (data not shown). Thus, the 130×10^3 M_r protein may not be related to such a complex.

The purification of the 130×10^3 M_r protein was carried out at 0–4 °C in the presence of sufficient levels of protease inhibitors to minimize its proteolysis. The polyclonal antibodies raised against the 130×10^3 M_r protein did not react with heavy chain of conventional myosin from gizzard smooth muscle (not shown). Thus, the 130×10^3 M_r protein may not be a proteolytic product of conventional myosin.

References

Collins, K., Sellers, R. and Matsudaira, P. (1990). *J. Cell Biol.* **110**, 1137–1147.

Hidaka, H. and Tanaka, T. (1985). In *Calmodulin Antagonists and Cellular Physiology* (ed. Hidaka, H. and Hartshorne, D. T.), pp. 13–22. Academic Press: New York.

Hoshimaru, M. and Nakanishi, S. (1987). *J. biol. Chem.* **262**, 14 625–14 632.

Ishikawa, R., Yamashiro, S. and Matsumura, F. (1989). *J. biol. Chem.* **264**, 7490–7497.

Kohama, K., Sohda, M., Murayama, K. and Okamoto, Y. (1988). *Protoplasma* (Suppl. 2), 37–47.

Korn, E. D. and Hammer, J. A. III. (1988). *A. Rev. Biophys. Biophys. Chem.* **17**, 23–45.

Lin, Y., Takano-Ohmuro, H. and Kohama, K. (1989). *Proc. Japan Acad.* **65**B, 203–206.

Nishikawa, M., Hidaka, H. and Adelstein, R. S. (1983). *J. biol. Chem.* **258**, 14 069–14 072.

Park, S. and Rasmussen, H. (1985). *Proc. natn. Acad. Sci. U.S.A.* **82**, 8835–8839.

Takano-Ohmuro, H. and Kohama, K. (1987). *J. Biochem.* **102**, 971–974.

Warrick, H. M. and Spudich, J. A. (1987). *A. Rev. Cell Biol.* **3**, 379–421.

Yabu, H., Yoshino, M., Usuki, T., Someya, T., Obara, K., Ozaki, H. and Karaki, H. (1990). In *Frontiers in Smooth Muscle Research* (ed. Sperelakis, N. and Wood, J. D.), pp. 623–626. Wiley-Liss: New York.

Yamazaki, K., Ito, K., Sobue, K., Mori, T. and Shibata, N. (1987). *J. Biochem.* **101**, 1–9.

Youngberg, G. E. and Youngberg, M. V. (1930). *J. Lab. Clin. Med.* **16**, 158–166.

Manipulation and expression of molecular motors in *Dictyostelium discoideum*

D. J. MANSTEIN

National Institute for Medical Research, London NW7 1AA, UK

K. M. RUPPEL, L. KUBALEK and J. A. SPUDICH

Departments of Cell Biology and Developmental Biology, Stanford University, Stanford CA 94305, USA

Summary

The eukaryote *Dictyostelium discoideum* is an attractive model organism for the study of cytoskeletal proteins and cell motility. The appearance and behavior of this cell closely resembles that of mammalian cells, but unlike mammalian cells, *Dictyostelium* offers the opportunity specifically to alter the cell physiology by molecular genetic approaches.

Key words: *Dictyostelium discoideum*, cytoskeleton, myosin, gene targeting, expression.

We are currently witnessing remarkable advances in our knowledge of the cytoskeleton and its motile elements. This progress can in part be attributed to the fact that the majority of cytoskeletal proteins are highly conserved throughout the eukaryotic kingdom, and to the recent progress made in the application of molecular genetic techniques for the manipulation of lower eukaryotic organisms such as *Saccharomyces cerevisiae*, *Aspergillus nidulans* and *Dictyostelium discoideum*. The simplicity of the genome of these organisms, and the result that transfections lead to a high frequency of homologous recombination events relative to random integration events, allows one to modify the structure and expression of the gene encoding a particular protein in a specific manner. As a subject for study of the cytoskeleton, *Dictyostelium* stands out among these model organisms in that it displays many forms of cell motility. It extends filopodia and pseudopodia, undergoes chemotactic movements and changes shape in response to signals, such as those operating during mitosis. In doing so, the appearance and behavior of this cell closely resembles that of mammalian cells.

The use of *Dictyostelium* to study the cellular function of myosin has been particularly rewarding. Two distinct classes of myosins are known to occur in nonmuscle cells. The first class of nonmuscle myosins are those which, like their muscle-located relatives, consist of two identical heavy chains (about $200\times10^3\,M_r$) and two pairs of light chains ($14–24\times10^3\,M_r$) and can form bipolar filament assemblies. These are generally referred to as myosin and sometimes as conventional myosin or myosin II. The second class, called myosin I, consist of a single, smaller heavy chain ($110–140\times10^3\,M_r$) and one or more light chains and cannot form bipolar filaments. The amino terminal part (approx. $90\times10^3\,M_r$) of the heavy chains of both myosin and myosin I form a globular head and it is this region which shows the highest degree of sequence conservation among all classes of myosins. It is also this part of the molecule, often referred to as the myosin head or subfragment-1 (S1), that contains the binding sites for the light chains and that has the catalytic and force generating properties of the myosin molecule (Toyoshima *et al.* 1987).

The high frequency of homologous recombination events in *Dictyostelium* was initially exploited in order to create cells which express a heavy meromyosin (HMM) fragment and only <0.1 % of wild type levels of intact myosin (De Lozanne and Spudich, 1987) and more recently cells in which the entire coding region for the single myosin heavy chain gene (*mhc*A) was deleted (Manstein *et al.* 1989). In complementary experiments antisense RNA was used to reduce the expression of myosin to <1 % of wild type levels (Knecht and Loomis, 1987). It became clear from these studies that in *Dictyostelium*, myosin is required for the furrowing event of cytokinesis, is essential for cell-surface receptor capping but not patching, contributes to the establishment of cell polarity and is necessary for morphogenetic changes associated with development. However, even in the absence of myosin, cells display many forms of movement, including intracellular particle movement, formation of cell-surface extensions, karyokinesis and cell migration, albeit in a less polarized form (Spudich, 1989). Definitive genetic proof that the phenotype attributed to the myosin null mutant is in fact due solely to the elimination of the *mhc*A gene and not to secondary mutations, which might have been induced during their generation, was achieved by the reintroduction of the cloned gene (Egelhoff *et al.* 1990), resulting in complementation of all features of the mutant phenotype.

Another important feature of *Dictyostelium* is that it can be used as an expression system. *Dictyostelium* is biochemically well characterized, can be grown readily in the laboratory on a large scale (200 g of cells are obtained from 30 to 40 liters of shaken culture), and carries out mammalian like post-translational modifications. However, *Dictyostelium* is a phagocytic cell, abundant in lysosomal proteases, and special precautions should be taken to protect proteins against their activity (Spudich, 1987). An effective way to eliminate most of the protease activity is to starve cells for approximately four hours. This prevents further synthesis of proteases and stimu-

Journal of Cell Science, Supplement 14, 63–65 (1991)

lates their secretion. Secretion can be further stimulated by adding sucrose or other disaccharides to the starvation medium (Dimond *et al.* 1981; North *et al.* 1990). Additionally, cell lysis and protein purification should always be performed in the presence of protective agents such as phenylmethylsulfonyl fluoride (0.5 mM), benzamidine (5 mM), tosyl lysine chloromethyl ketone (0.25 mM) and leupeptin (0.01 mM). The addition of 40 mM sodium pyrophosphate and 30 % (w/v) sucrose to the lysis buffer is also recommended.

The expression and purification of functional myosin head fragments represents a potentially powerful tool for the study of chemo-mechanical coupling. Understanding the mechanism by which myosin catalyzes the transduction of energy stored in chemical bonds into mechanical work requires high resolution structural information. However, obtaining crystals and structural data of S1 has proved to be a difficult task. This is undoubtedly due to heterogeneities in the starting material when S1 is prepared from muscle myosin. Muscle tissue typically contains multiple isoforms of the myosin heavy and light chains and further heterogeneities are introduced during the proteolytic digestion of myosin. Therefore, it was a most promising breakthrough when Rayment and Winkelmann (1984) obtained crystals of S1 from chicken pectoralis muscle. Although these crystals might be sufficiently ordered to diffract X-rays to 0.4 nm resolution, the heterogenious nature of the starting material might still present a problem. One way to avoid all these problems is to express an S1-like fragment in a host like *Dictyostelium*. An additional benefit of working with a recombinant protein is that it is accessible to molecular genetic manipulation. For example, the availability of site-directed mutagenesis might potentially facilitate the analysis of the S1-structure on several levels. One major obstacle in protein crystallography, the trial and error search for suitable heavy atom derivatives, can be overcome by generating S1 molecules with engineered heavy atom binding sites. The ability to alter the protein by site-directed mutagenesis will also provide ways for testing working models for the mechanism of chemo-mechanical coupling.

So far, two functional myosin head fragments have been expressed successfully in *Dictyostelium*. These fragments correspond to the proteolytically defined muscle meromyosin heavy chain (HMM–140), which forms a dimer consisting of two globular heads and 50 nm of alpha-helical, coiled-coil rod, and to a monomeric S1-like myosin head fragment (MHF) of the myosin heavy chain that extends 46 amino acids beyond the proline which marks the region of proteolytic cleavage in muscle myosin. Their expression in a functional form relies in part on the fact that *Dictyostelium* can supply the expressed fragments with its endogenous myosin light chains. Indeed, both recombinant proteins, HMM–140 and MHF, copurify with the essential and regulatory myosin light chain, decorate actin filaments, display actin-activated adenosine triphosphatase activity, and support sliding movement of actin filaments in the *in vitro* motility assay developed by Kron and Spudich (1986). Recombinant fragments are expressed in a wild-type background and not in myosin null cells. The disadvantage of having to purify the myosin fragments away from endogenous myosin is fully compensated for by the ability of the wild-type cells to grow in suspension. The differences in size and and solubility properties make it very easy to separate myosin and myosin head fragments. Typically, one obtains 1–2 mg of purified HMM–140 or MHF from 100 g of cells (Ruppel *et al.* 1990; Manstein *et al.* 1989). Higher expression levels can in principle be achieved. However, cells that produce larger quantities of myosin fragments have phenotypic characteristics similar to myosin null cells and are difficult to grow in suspension.

Crystallization trials using MHF are now in progress. Needle shaped microcrystals can be readily obtained at neutral pH, in the presence of the magnesium salts of either adenosine-5′-diphosphate, adenylyl-imidodiphosphate, or pyrophosphate, and with 8 to 10 % (w/v) polyethylene glycol 4000 as precipitating agent. Electron micrographs of these needles, which are 0.1 to 0.5 μm in width and often exceed 100 μm in length, show extensive order even after preservation in heavy metal stain (uranyl acetate) (Fig. 1). Electron microscopic studies of these needles in the frozen hydrated state should yield more structural information, possibly at higher resolution (Stokes and Green, 1990). However, our efforts are still focused on the generation of crystals suitable for high resolution X-ray analysis. Dramatic differences in the crystallization behavior of a protein are often caused by minor differences in amino acid sequence, as often found among homologous enzymes isolated from different species. Therefore, we are currently exploiting the possibility of expressing different types of myosin head fragments in *Dictyostelium*.

The production of force and movement by S1 is thought to be linked to large conformational changes within the protein, occurring upon binding and interaction with nucleotide and actin. Determining the structure of S1 in these different structural states might therefore prove crucial for the understanding of the mechanism of chemo-mechanical coupling. The information necessary for the modelling of these associated states could be obtained by the crystallization of the myosin head complexed to actin. One way to achieve this is by crystallizing an actin:DNAaseI:S1 complex. Even if the sheer size of this complex creates a formidable crystallographic problem this route should not be ignored, especially since Kabsch, Holmes and coworkers have recently solved the structure

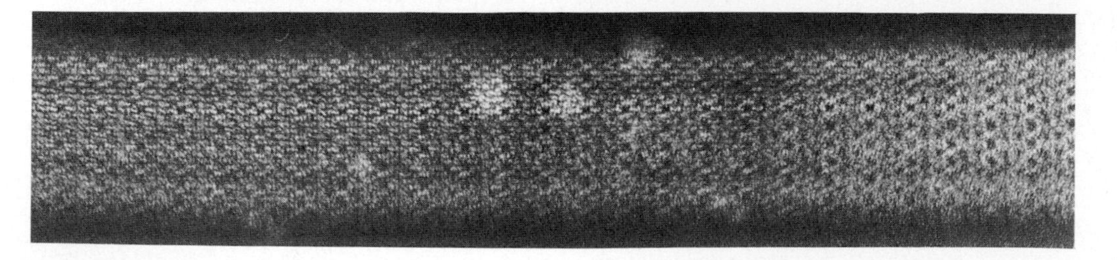

Fig. 1. Electron micrograph of a microcrystal of a *Dictyostelium* myosin head fragment (MHF) negatively stained with uranyl acetate. The microcrystal is about 200 nm in width and shows extensive order even after preservation in 1 % (w/v) uranyl acetate.

of the actin:DNAaseI complex in the ATP and ADP forms at an effective resolution of 0.28 nm and 0.30 nm (Kabsch *et al.* 1990). A second complementary way would be to crystallize a myosin head complexed to a nonpolymerizable monomeric actin. Such a complex may be generated from myosin head fragments and G-actin treated with m-maleimidobenzoyl-*N*-hydroxysuccinimide ester (Bettache *et al.* 1989).

Dictyostelium proved to be an ideal system for the elucidation of the cellular function of the conventional myosin. The application of molecular genetics in this system, directed towards structural studies of the myosin motor, in combination with the recent progress made in the *in vitro* study of this molecular motor (Kron *et al.* 1990) should soon provide us with a better understanding of chemo-mechanical coupling.

References

Bettache, N., Bertrand, R. and Kassab, R. (1989). Coupling of nonpolymerizable monomeric actin to the F-actin binding region of the myosin head. *Proc. natn. Acad. Sci. U.S.A.* **86**, 6028–6032.

De Lozanne, A. and Spudich, J. A. (1987). Disruption of the *Dictyostelium* myosin heavy chain by homologous recombination. *Science* **238**, 1086–1091.

Dimond, R. L., Burns, R. A. and Jordan, K. B. (1981). Secretion of lysosomal enzymes in the cellular slime mold *Dictyostelium discoideum*. *J. biol. Chem.* **256**, 6565–6572.

Egelhoff, T. T., Manstein, D. J. and Spudich, J. A. (1990). Complementation of myosin null mutants in *Dictyostelium discoideum* by direct functional selection. *Devl Biol.* **137**, 359–367.

Kabsch, W., Mannherz, H. G., Suck, D., Pai, E. F. and Holmes, K. C. (1990). Atomic structure of the Actin:DNase complex. *Nature* **347**, 37–44.

Knecht, D. and Loomis, W. (1987). Antisense RNA inactivation of myosin heavy chain gene expression in *Dictyostelium discoideum*. *Science* **236**, 1081–1086.

Kron, S. J. and Spudich, J. A. (1986). Fluorescent actin filaments move on myosin fixed to a glass surface. *Proc. natn. Acad. Sci. U.S.A.* **83**, 6272–6276.

Kron, S. J., Toyoshima, Y. Y., Uyeda, T. Q. P. and Spudich, J. A. (1990). Assays for actin sliding over myosin coated surfaces. *Methods in Enzymology*, in press.

Manstein, D. J., Ruppel, K. M. and Spudich, J. A. (1989). Expression of a functional myosin head fragment in *Dictyostelium discoideum*. *Science* **246**, 656–658.

Manstein, D. J., Titus, M. A., De Lozanne, A. and Spudich, J. A. (1989). Gene replacement in *Dictyostelium*: generation of myosin null mutants. *EMBO J.* **8**, 923–932.

North, M. J., Franek, K. J. and Cotter, D. A. (1990). Differential secretion of *Dictyostelium discoideum* proteinases. *J. gen. Microbiol.* **136**, 827–833.

Rayment, I. and Winkelmann, D. A. (1984). Crystallization of myosin subfragment 1. *Proc. natn. Acad. Sci. U.S.A.* **81**, 4378–4380.

Ruppel, K. M., Egelhoff, T. T. and Spudich, J. A. (1990). Purification of a functional recombinant myosin fragment from *Dictyostelium discoideum*. *Ann. N.Y. Acad. Sci.* **582**, 147–155.

Spudich, J. A. (1987). *Dictyostelium discoideum*: molecular approaches to cell biology. In *Methods in Cell Biology* (ed. J. A. Spudich), pp. 3–8, vol. 28. Academic Press: Orlando, FL.

Spudich, J. A. (1989). In pursuit of myosin function. *Cell Regulation* **1**, 1–11.

Stokes, D. L. and Green, N. M. (1990). Structure of CaATPase: electron microscopy of frozen-hydrated crystals at 6Å resolution in projection. *J. mol. Biol.* **213**, 529–538.

Toyoshima, Y. Y., Kron, S. J., McNally, E. M., Niebling, K. R., Toyoshima, C. and Spudich, J. A. (1987). Myosin Subfragment-1 is sufficient to move actin filaments *in vitro*. *Nature* **328**, 536–539.

The use of native thick filaments in *in vitro* motility assays

JAMES R. SELLERS*, YUNG JIN HAN

Laboratory of Molecular Cardiology, National Heart, Lung and Blood Institute, National Institutes of Health, Bethesda, MD 20892, USA

and BECHARA KACHAR

Laboratory of Molecular Otology, National Institute on Deafness and Other Communication Disorders, National Institutes of Health, Bethesda, MD 20892, USA

* Author for correspondence

Summary

Native thick filaments from the clam, *Mercinaria mercinaria* translocate actin filaments both toward and away from the center of the thick filament in an *in vitro* motility assay. The thick filaments from the adductor muscle are about 10 μm long whereas those from the catch muscle are 30–50 μm long. These thick filaments should prove useful in understanding the mechanism of myosin-dependent movement of actin filaments.

Key words: motility, myosin, actin, mollusc, filaments.

Introduction

The interaction of actin and myosin has been studied for many years using a wide variety of methods. These could, in general, be grouped into studies of isolated proteins using various biochemical assays or of studies using intact or permeabilized muscle fibers. A great deal has been learnt about actomyosin interactions using these combined approaches, but in some cases there have been difficulties in relating the *in vitro* biochemical studies with the *in vivo* physiological studies. The biochemical studies often provide simple interpretations of the results and are free from many of the complications of muscle fibers. Although one could easily measure the actin-activation of MgATPase activity of myosin *in vitro*, fundamental processes of muscle such as force production and active sliding of the two filament systems were not measured. Similarly in fiber systems, one could measure the latter two parameters but these measurements were complicated by a complex structural system and diffusion barriers.

Recently, two different types of *in vitro* motility assays have been developed that measure the relative movement of actin and myosin (Sheetz *et al.* 1984; Kohama and Shimmen, 1985; Kron and Spudich, 1986; Harada *et al.* 1987). These systems provide a bridge between the schools of biochemistry and physiology. The first system involves visualization of the movement of myosin-coated beads over organized arrays of actin filaments derived from microdissection of giant alga cells such as *Nitella axillaris* (Sheetz *et al.* 1984; Kohama and Shimmen, 1985). It has been used to characterize the movement of myosin filaments from a number of sources including vertebrate striated and smooth muscle (Sheetz *et al.* 1984; Sellers *et al.* 1985; Umemoto *et al.* 1989; Umemoto and Sellers, 1990), invertebrate muscle (Yamada *et al.* 1989; Vale *et al.* 1984) and nonmuscle myosins from both vertebrate (Umemoto *et al.* 1989; Umemoto and Sellers, 1990) and lower eukaryotic sources (Kohama and Shimmen, 1985; Albanesi *et al.* 1985; Flicker *et al.* 1985). The assay has been shown to be reproducible and quantitative, despite the fact that the actin substratum is derived from microdissection of a plant.

The second system uses only purified proteins. It involves observation of the movement of fluorescently labelled actin filaments over a myosin-coated surface (Kron and Spudich, 1986; Harada *et al.* 1987). Individual actin filaments, which are too small to be seen in the light microscope, are labelled with rhodamine phalloidin which binds stoichiometrically to the actin molecules within a filament. Upon excitation with light, this labelled actin filament emits a bright image which can easily be detected using fluorescence microscopy. These filaments can be seen to move in a serpentine manner over the myosin-coated surface in the presence of MgATP. Both polymerized myosin filaments and myosin monomers can be used to coat the surface (Umemoto and Sellers, 1990; Toyoshima *et al.* 1987). In addition, the proteolytic fragments of myosin, heavy meromyosin and subfragment-1 can also be used in the sliding actin assay (Toyoshima *et al.* 1987; Kishimo and Yanagida, 1988; Takiguchi *et al.* 1990).

Despite the dissimilarity in the geometry of the two assay systems there is surprisingly good agreement between the relative rates of movement of actin and myosin when the same myosin is used (Umemoto and Sellers, 1990). It is noted in both systems that the source of myosin is the basic determinant of the speed of movement, whereas the polarity of actin is thought to determine the direction of movement. This is particularly important in the sliding actin filament assay system where the myosin molecules or filaments are generally bound to the glass surface in a random manner.

The range of rates produced by different myosin molecules is astounding given the degree of sequence conservation that is seen in the head region of myosin molecules from different sources. In our laboratory we have measured rates of movement of purified myosins that extend from a pedestrian 0.04 $\mu m\,s^{-1}$ for human platelet myosin to about 6 $\mu m\,s^{-1}$ for clam adductor myosin under the same ionic conditions. In preliminary experiments, rates of 30–60 $\mu m\,s^{-1}$ have been measured for sliding of purified actin filaments over coverslips coated with high

Journal of Cell Science, Supplement 14, 67–71 (1991)

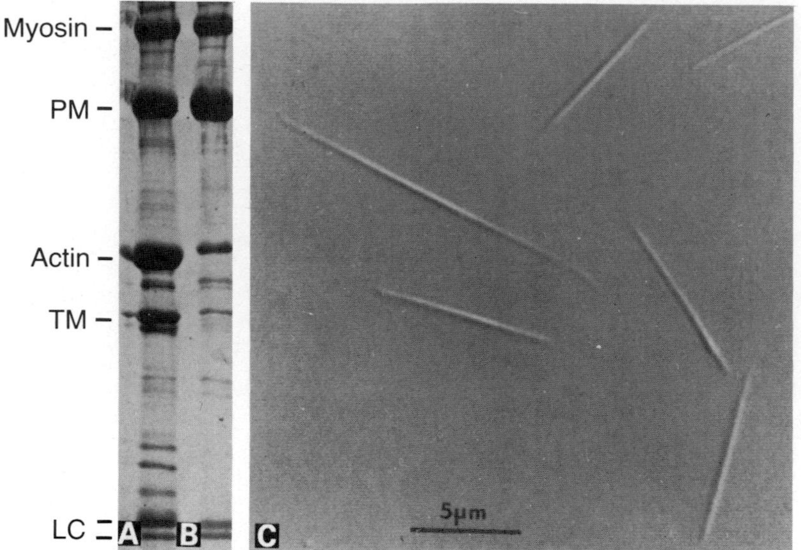

Fig. 1. SDS polyacrylamide gel of (A) whole clam adductor muscle homogenate, (B) isolated native thick filaments from clam adductor muscle. (C) Video enhanced light micrograph of isolated native myosin filaments from clam muscle. The mean myosin filament length was 9.5±3.8 μm.

speed supernatants of soluble extracts of *Nitella axillaris* protoplasm, giving a total range of movements of about 1500 fold for the rate of sliding of purified actin filaments under the same ionic conditions (Rivolta *et al.* 1990).

In the sliding actin assay, the myosin bound to the surface is not usually visualized because of the small size of monomeric myosin and synthetic thick filaments. Recently, we (Sellers and Kachar, 1990) have been observing the movement of fluorescently labelled actin filaments over large native thick filaments isolated from molluscan muscle, using a gentle technique developed by Yamada *et al.* (1989). These thick filaments consist of myosin molecules packed in a bipolar manner around a core of paramyosin (Fig. 1B). It is the paramyosin core which allows these thick filaments to obtain lengths up to 50 μm depending on the source of the muscle. Using video-enhanced differential interference contrast (DIC) microscopy, individual thick filaments can be directly imaged in the light microscope (Fig. 1C). The quality of the image is markedly improved by the use of computer image processing such as background subtraction, frame averaging and contrast enhancement.

The ability to image both actin and myosin filaments has allowed us to correlate directly the movement of actin filaments imaged by fluorescence with their position on the myosin filament imaged by DIC. Direct comparison between the images revealed that actin filaments could bind at any position on the myosin filament and commence directed movement (Fig. 2). Actin filaments moved both towards and away from the center of the myosin filaments. The movement of actin filaments away from the center of the myosin filament is opposite that which normally occurs in contracting muscle, where it is thought that myosin pulls actin filaments only toward the center of the thick filament. Analysis of the movement demonstrated that those actin filaments traveling toward the center of the thick filament moved at a rate of $8.8 \pm 1.4\ \mu m\ s^{-1}$, whereas those traveling away from the center of the thick filament moved at a much slower rate of $1.0 \pm 0.3\ \mu m\ s^{-1}$ (Fig. 3). Some actin filaments traveled the entire length of the thick filament, first at the fast rate and then changing abruptly to the slower rate as it encountered myosin heads of the opposite polarity. It was observed that the direction of travel was determined by the polarity of the actin

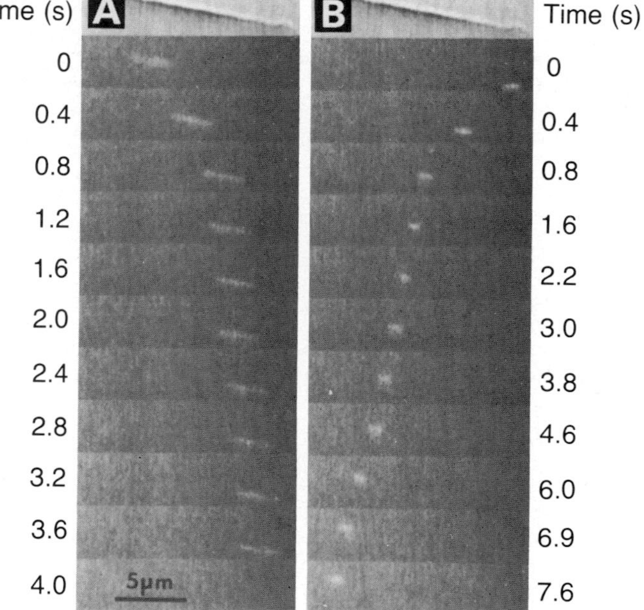

Fig. 2. Video analysis of the bidirectional sliding rates of fluorescently labeled actin filaments along single myosin filaments. Sequential fluorescence images of actin filaments sliding along a single myosin filament (upper panel). Note that in each case (both left and right sequences) when an actin filament was moving toward the center of the myosin filament the velocity was fast and when it was moving away from the center the velocity was slow.

filament. A single actin filament could reverse direction of travel by detaching and rapidly reattaching after undergoing a 180° flip (Fig. 4A). This occurred most frequently with short actin filaments which undergo rapid Brownian movement upon detachment. In some cases the leading end of a longer actin filament sliding off the end of a thick filament will reattach and commence moving back toward the center. Occasionally very small actin filaments can be observed to move back and forth along a thick filament reversing its direction up to 7 times with no apparent dissociation. We believe that in these cases the actin

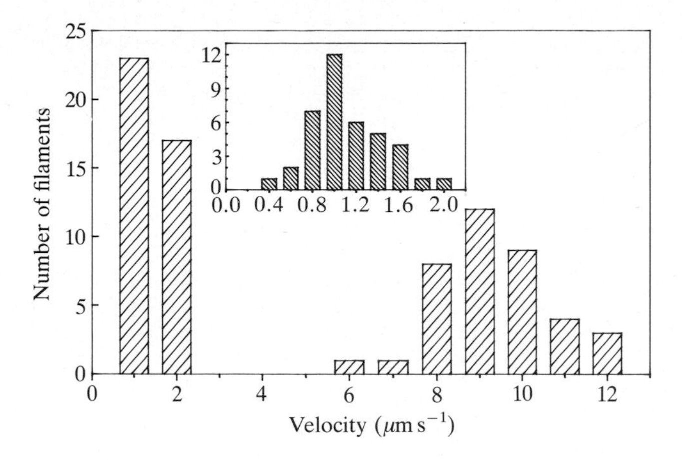

Fig. 3. Distribution of velocities. The mean and standard deviation of the slow rates was $1.0\pm0.3\,\mu m\,s^{-1}$ while that of the fast rates was $8.8\pm1.4\,\mu m\,s^{-1}$. The inset shows an expansion of the data from 0 to $2.0\,\mu m\,s^{-1}$. All filaments moving $2.0\,\mu m\,s^{-1}$ or less were used for the computation of the standard deviation of the velocity of the slow rates, whereas all filaments moving at $6.0\,\mu m\,s^{-1}$ or greater were used for the fast rates.

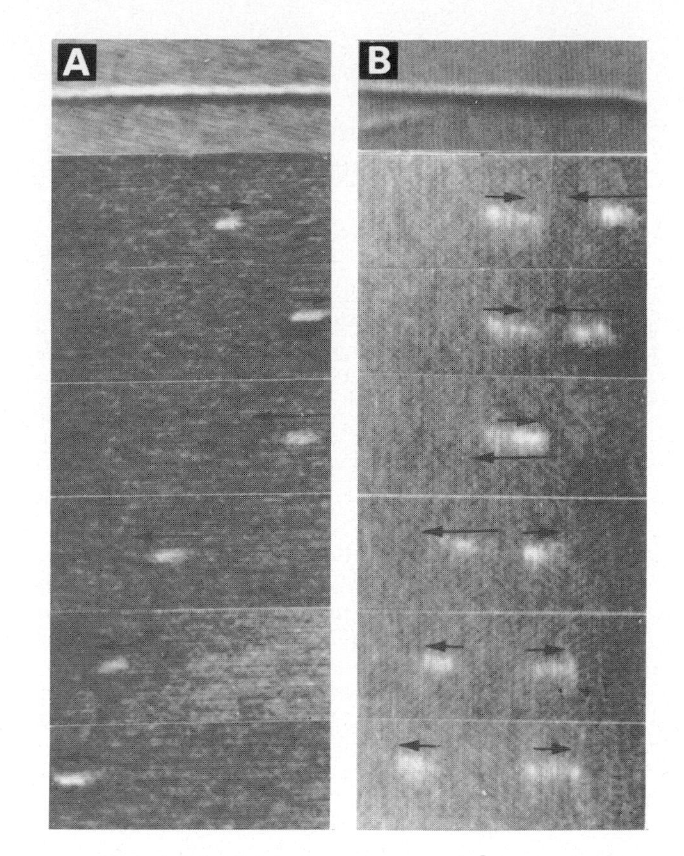

Fig. 4. Sequential video images showing different trajectories of fluorescently labeled actin filaments translocating along individual myosin filaments. The DIC image of the myosin filament is shown at the beginning of each sequence of fluorescence images. (A) An actin filament sliding away from the center of the myosin filament momentarily detached and then reattached to the same myosin filament and started sliding in the reverse direction. Arrows show the direction of movement and the length of the arrow indicates whether the filament was moving at the slow (short arrow) or the fast velocity (long arrow). (B) Two filaments passed each other unimpeded and changed velocities on a single thick filament. This figure is taken from Sellers and Kachar (1990).

filament may be held instantaneously by only a single myosin head allowing for Brownian movement in two dimensions which can result in a 180° rotation and reversal of direction of travel.

The native thick filaments have a large diameter and can support the movement of more than one actin filament at the same time. Actin filaments can pass one another unimpeded on the same thick filament (Fig. 4B). When this occurred one filament would be moving at the fast rate and the other at the slow rate. Actin filaments can also interact simultaneously with two adjacent myosin filaments after undergoing large changes in angle when switching from one filament to another.

Several reports have shown that the heads of isolated myosin molecules are capable of large rotational movements about the subfragment 1–subfragment 2 junction (Winklemann and Lowey, 1986; Craig *et al.* 1980). Similar observations have been made on myosin heads projecting from myosin filaments (Walker and Trinick, 1986, 1988). Recently it has been demonstrated that in mutated *Drosophila* flight muscle, in which peripheral thick and thin filaments are misregistered, the heads of myosin can bind with opposite rigor cross-bridge angles to flanking actin filaments that are of the opposite polarity (Reedy *et al.* 1989). Toyoshima *et al.* (1989) described bidirectional movement of actin filaments along tracks of myosin heads created by binding actin filaments that had been decorated with skeletal muscle heavy meromyosin to the surface of the coverslip. It was also shown that the myofibrillar Mg^{2+}ATPase activity does not decrease at short sarcomere length, suggesting that actin of the opposite polarity can activate the Mg^{2+}ATPase activity of myosin (Stephenson *et al.* 1989). We now provide definitive data that the myosin within a native thick filament can also generate the force to translocate actin both toward and away from the central bare zone.

In summary, we propose a model whereby the heads of myosin have considerable rotational flexibility. The direction of movement is determined by the polarity of actin and myosin heads can rotate perhaps 180° in order to interact with actin (Fig. 5). The reason for the slower rate of motility when actin is sliding over myosin heads of the opposite polarity is not known, but may be related to altered cross-bridge kinetics of strained heads.

Most of our experiments have been performed with native thick filaments isolated from the adductor (pink) or catch (white) muscle of the clam, *Mercinaria mercinaria*. The thick filaments from the adductor muscle average about 9 μm in length whereas those of the catch muscle are considerably longer (30–50 μm). We concentrated most of our effort on the smaller adductor muscle thick filaments in order to avoid complications that may arise from the potentially more complex regulation of the catch muscle thick filaments. We have also prepared thick filaments from the mussel, *Mytillus edulis*. All of these thick filaments, while exhibiting their own characteristic rates of movement, nonetheless show a 8–10 fold slower rate when the actin filaments are moving away from the center of the thick filaments.

The structures of several invertebrate native thick filaments have been examined by high resolution electron microscopy coupled with optical diffraction techniques (Kensler and Levine, 1982; Castellani *et al.* 1983; Vibert and Craig, 1983; Crowther *et al.* 1985). Clam thick

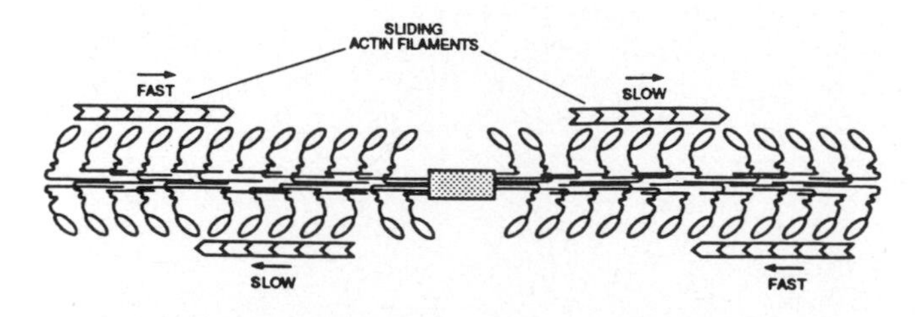

Fig. 5. Schematic diagram showing the allowed sliding interactions of the polar actin filaments with the bipolar myosin filament. The arrows indicate the direction of movement. The myosin heads are schematically shown at the ends of their power strokes. The 'reverse chevrons' concept for the myosin heads contacting actin moving away from the center of the myosin filament is taken from Reedy *et al.* (1989). The cross hatched area represents the bare zone. This diagram is taken from Sellers and Kachar (1990).

filaments have not been studied in such detail at present. For this reason, thick filaments from other species may prove to be more interesting. In particular, *Limulus* muscle has thick filaments of about 10 μm in length which can be isolated using gentle means. The structure of this thick filament has been extensively studied (Kensler and Levine, 1982). To date, we have not been able to use successfully *Limulus* thick filaments in the motility assay, probably due to the presence of excess actin in the preparation.

The ability to image both the moving actin filament and the thick filament which is generating the motile force represents a significant advance in the utility of *in vitro* motility assays. It should allow more direct observation of actin–myosin interactions and may be of use in systems designed to measure force *in vitro* using such techniques as actin-coated microneedles (Kishino and Yanagida, 1988) and optical tweezers (Block, 1990). In addition, these molluscan myosin filaments are regulated by calcium binding to myosin (Kendrick-Jones *et al.* 1970). This affords the possibility of studying the regulation of movement *in vitro*. Preliminary experiments show that movement of actin filaments along clam thick filaments is dependent upon micromolar levels of calcium.

In order to aid others who wish to use the system of native thick filaments in *in vitro* motility assays, we will conclude by giving the details of the preparation and the assay. The thick filament preparation was modified from that described by Yamada *et al.* (1989). Muscle from either the pink adductor muscle or the white catch muscle of the clam, *Mercinaria mercinaria* was removed from the shell, diced into small pieces and placed in buffer A (10 mM ATP, 10 mM $MgCl_2$, 1 mM EGTA, 20 mM MOPS (pH 7.0), 3 mM NaN_3, 1 mM dithiothreitol, 0.1 mM PMSF). The muscle was rinsed twice in buffer A and then homogenized twice in 5 ml of buffer A in a Sorvall Omnimixer (small cup) at a setting of 7.5, for 7 s on ice. The homogenized muscle was mixed with an equal vol of buffer A containing 0.1 % Triton X-100. After sitting for 5 min on ice, the homogenate was sedimented at 500 ***g*** for 5 min. The supernatant was carefully removed and sedimented for 30 min (adductor muscle) or 20 min (catch muscle) at 5000 ***g***. The pellet was gently resuspended in 30 ml of buffer A and sedimented at 500 ***g*** for 5 min. The supernatant was removed and sedimented at 5000 ***g*** for 20 or 30 min (see above). The two centrifugation steps were repeated once more and the resulting pellet was gently resuspended in 1–2 ml of buffer A, incubated for 20 min on ice and sedimented at 500 ***g*** for 5 min. The supernatant was recovered and stored on ice until used. An SDS polyacrylamide gel of the sample is shown in Fig. 1B. A band migrating near the position of actin is observed, but lack of staining of the preparation with rhodamine phalloidin suggests that it is not actin.

We have modified the published procedures for sliding actin motility assays for use with the native thick filaments. Glass coverslips can be used instead of nitrocellulose-coated coverslips. A 40 μl aliquot of thick filaments (30–200 μg ml^{-1} diluted with buffer A) was applied to the slide and a flow cell consisting of two slivers of coverslip flanked by a ribbon of Apiezon grease topped by a coverslip was constructed around the droplet. This was done in order to avoid orientation of the thick filaments under flow. After 30–60 s the flow cell was washed with 20 mM KCl, 20 mM MOPS (pH 7.2), 5 mM $MgCl_2$, 0.1 mM EGTA, 2 mM DTT, 0.5 mg ml^{-1} bovine serum albumin. After 1 min incubation, the flow cell was washed with motility buffer (20 mM KCl, 10 mM MOPS (pH 7.2), 5 mM $MgCl_2$, 1 mM ATP, 0.2 mM $CaCl_2$, 0.1 mM EGTA, 2 mM DTT, 2.5 mg ml^{-1} glucose, 0.1 mg ml^{-1} glucose oxidase, 0.02 mg ml^{-1} catalase) containing 2 nM rhodamine phalloidin actin. In some cases, care was taken to minimize the contribution of rigor-like heads in the preparation which bind to actin in an ATP-independent manner and tether actin filaments. To accomplish this, after the bovine serum albumin blocking step, the flow cell was washed with motility buffer containing 1 μM phalloidin-labelled actin (note: not rhodamine-labelled phalloidin). After 1–5 min incubation this solution was washed out with motility buffer and replaced with the motility buffer containing 2 nM rhodamine phalloidin-labelled actin. Images were recorded on either VHS or U-matic videotapes or in some cases on optical memory disks. In order to obtain nonoverlap of the thick filaments, concentrations between 30–50 μg ml^{-1} total protein must be applied.

Bright rhodamine–phalloidin labelled actin is prepared by drying 60 μl of a 3.3 μM rhodamine phalloidin stock (Molecular Probes, Oregon, USA) under vacuum. The sample is resuspended in 90 μl of 4 mM imidazole (pH 7.0), 2 mM $MgCl_2$, 1 mM DTT, 0.1 mM EGTA, 3 mM NaN_3 and sonicated in an ultrasonic cleaner for 10 min. F-actin (10 μl) is added to a final concentration of 2 μM from a freshly diluted 20 μM solution of F-actin in the same buffer as above. This is allowed to sit on ice for at least 1–2 h before use. The rhodamine–phalloidin labelled actin solution is usually good for about 1 week.

To obtain video-enhanced DIC images of the thick filaments a Zeiss axiomat microscope was used in the critical illumination mode (Kachar *et al.* 1987) equipped with a ×100, 1.4 NA objective. The video image was background subtracted, averaged and contrast enhanced using the Image 1 image processor (Universal Imaging, Pennsylvania, USA). Higher resolution images were obtained using a Newvicon video camera (Dage-MTI), but for direct comparison between the fluorescent image and the DIC image a SIT camera (Dage-MTI) was used. In some cases the fluorescent image was captured with a intensified Newvicon (KS 1380, Videoscope International). The rate of movement was quantified using the computer assisted tracking program of Steven Block (1990).

References

Albanesi, J. P., Fujisaki, H., Hammer, J. A., III, Korn, E. D., Jones,

R. and Sheetz, M. P. (1985). Monomeric acanthamoeba myosins I support movement *in vitro*. *J. biol. Chem.* **260**, 8649–8652.
Block, S. M. (1990). Optical tweezers: A new tool for biophysics. In *Noninvasive Techniques in Cell Biology*, pp. 375–401, Wiley-Liss: New York.
Castellani, L., Vibert, P. and Cohen, C. (1983). Structure of myosin/paramyosin filaments from a molluscan smooth muscle. *J. molec. Biol.* **167**, 853–872.
Craig, R., Szent-Györgyi, A. G., Beese, L., Flicker, P., Vibert, P. and Cohen, C. (1980). Electron microscopy of thin filaments decorated with a Ca^{2+}-regulated myosin. *J. molec. Biol.* **140**, 35–55.
Crowther, R. A., Padron, R. and Craig, R. (1985). Arrangement of the heads of myosin in relaxed thick filaments from tarantula muscle. *J. molec. Biol.* **184**, 429–439.
Flicker, P. F., Peltz, G., Sheetz, M. P., Parham, P. and Spudich, J. A. (1985). Site-specific inhibition of myosin-mediated motility *in vitro* by monoclonal antibodies. *J. Cell Biol.* **100**, 1024–1030.
Harada, Y., Noguchi, A., Kishino, A. and Yanagida, T. (1987). Sliding movement of single actin filaments on one-headed myosin filaments. *Nature* **326**, 8005–808.
Kachar, B., Bridgman, P. C. and Reese, T. S. (1987). Dynamic shape changes of cytoplasmic organelles translocating along microtubules. *J. Cell Biol.* **105**, 1267–1271.
Kendrick-Jones, J., Lehman, W. and Szent-Györgyi, A. G. (1970). Regulation in molluscan muscles. *J. mol. Biol.* **54**, 313–326.
Kensler, R. W. and Levine, R. J. C. (1982). Determination of the handedness of the crossbridge helix of *Limulus* thick filaments. *J. Muscle Res. Cell Motil.* **3**, 349–361.
Kishino, A. and Yanagida, T. (1988). Force measurements by micromanipulation of a single actin filament by glass needles. *Nature* **334**, 74–76.
Kohama, K. and Shimmen, T. (1985). Inhibitory Ca^{2+}-control of movement of beads coated with Physarum myosin along actin-cables in chara internodal cells. *Protoplasma* **129**, 88–91.
Kron, S. J. and Spudich, J. A. (1986). Fluorescent actin filaments move on myosin fixed to a glass surface. *Proc. natn. Acad. Sci. U.S.A.* **83**, 6272–6276.
Reedy, M. C., Beall, C. and Fyrberg, E. (1989). Formation of reverse rigor chevrons by myosin heads. *Nature* **339**, 481–483.
Rivolta, M. N., Urrutia, R., Sellers, J. R. and Kachar, B. (1990). Preliminary characterization of an actin based organelle translocator from *Nitella*. *Biophys. J.* **57**, 535a.
Sellers, J. R. and Kachar, B. (1990). Polarity and velocity of sliding filaments: Control of direction by actin and of speed by myosin. *Science* **249**, 406–408.
Sellers, J. R., Spudich, J. A. and Sheetz, M. P. (1985). Light chain phosphorylation regulates the movement of smooth muscle myosin on actin filaments. *J. Cell Biol.* **101**, 1897–1902.
Sheetz, M. P., Chasan, R. and Spudich, J. A. (1984). ATP-dependent movement of myosin *in vitro*: characterization of a quantitative assay. *J. Cell Biol.* **99**, 867–871.
Sheetz, M. P. and Spudich, J. A. (1983). ATP-dependent movement of myosin in vitro: characterization of a quantitative assay. *Nature* **303**, 31–35.
Stephenson, D. G., Stewart, A. W. and Wilson, G. J. (1989). Dissociation of force from myofibrillar MgATPase and stiffness at short sarcomere length in rat and toad skeletal muscle. *J. Physiol. (Lond.)* **410**, 351–366.
Takiguchi, K., Hayashi, H., Kurimoto, E. and Higashi-Fujime, S. (1990). *In vitro* motility of skeletal muscle myosin and its proteolytic fragments. *J. Biochem. (Tokyo)* **107**, 671–679.
Toyoshima, Y. Y., Kron, S. J., McNally, E. M., Niebling, K. R., Toyoshima, C. and Spudich, J. A. Myosin subfragment-1 is sufficient to move actin filaments *in vitro*. *Nature* **328**, 536–539.
Toyoshima, Y. Y., Toyoshima, C. and Spudich, J. A. (1989). Bidirectional movement of actin filaments along tracks of myosin heads. *Nature* **341**, 154–156.
Umemoto, S., Bengur, A. R. and Sellers, J. R. (1989). Effect of multiple phosphorylation of smooth muscle and cytoplasmic myosins on movement in an *in vitro* motility assay. *J. biol. Chem.* **264**, 1431–1436.
Umemoto, S. and Sellers, J. R. (1990). Characterization of *in vitro* motility assays using smooth muscle and cytoplasmic myosins. *J. biol. Chem.* **265**, 14864–14869.
Vale, R. D., Szent-Gyorgyi, A. G. and Sheetz, M. P. (1984). Movement of scallop myosin on *Nitella* actin filaments: regulation by calcium. *Proc. natn. Acad. Sci. U.S.A.* **81**, 6775–6778.
Vibert, P. and Craig, R. (1983). Electron microscopy and image analysis of myosin filaments from scallop striated muscle. *J. molec. Biol.* **165**, 303–320.
Walker, M. and Trinick, J. (1986). Electron microscope study of the effect of temperature on the length of the tail of the myosin molecule. *J. molec. Biol.* **192**, 661–667.
Walker, M. and Trinick, J. (1988). Visualization of domains in native and nucleotide-trapped myosin heads by negative staining. *J. Muscle Res. Cell Motil.* **9**, 359–366.
Winklemann, D. A. and Lowey, S. (1986). Probing myosin head structure with monoclonal antibodies. *J. molec. Biol.* **188**, 595–612.
Yamada, A., Ishii, N., Shimmen, T. and Takahashi, K. (1989). MgATPase activity and motility of native thick filaments isolated from the anterior byssus retractor muscle of *Mytilus edulis*. *J. Muscle. Res. Cell. Motility* **10**, 124–134.

Note added in proof

Similar conclusions to those reached in this article have been reported recently by A. Yamada, N. Ishii and K. Takahashi, who used thick filaments isolated from *Mytilus edulis* (see *J. Biochem.* **108**, 341–343 (1990)).

Protein engineering and the study of muscle contraction in *Drosophila* flight muscles

JOHN SPARROW, DOUGLAS DRUMMOND, MICHELLE PECKHAM, EMMA HENNESSEY and DAVID WHITE

Department of Biology, University of York, York Y01 5DD, UK

Summary

We describe an experimental approach to the use of genetics to study muscle contraction in *Drosophila melanogaster*. Mutations induced by *in vitro* mutagenesis are inserted into the genome of flies using P-element mediated transformation, permitting the effects of the mutant genes to be studied *in vivo* in the indirect flight muscles (IFMs). Details of how mechanical experiments can be performed on skinned IFMs, despite their small size, are provided. The effects of two *in vitro* actin mutations, *G368E* and *E316K*, are described. The problems of performing biochemical and biophysical experiments on the IFMs and their myofibrillar proteins are described, together with indications as to how these may be overcome.

Key words: *Drosophila*, flight muscle, mutants, actin, muscle mechanics.

Introduction

The application of protein engineering techniques to the study of the relationship between the amino acid sequence and function of proteins involved in muscle contraction is already producing useful insights (Hitchcock-DeGregori and Heald, 1987; Hitchcock-DeGregori and Varnell, 1990; Grabarek *et al.* 1990; Fujimori *et al.* 1990). In these studies mutant muscle proteins, such as tropomyosin or troponin C, have been expressed in *E. coli* and studied by various techniques *in vitro*. Particularly important for the study of muscle contraction *per se* is the exchange of these proteins *in vitro* to fibres denuded of the wild-type proteins, followed by mechanical measurements on the fibres. The potential of this approach has been demonstrated in the recent study of troponin C mutants by Fujimori *et al.* (1990). This approach is not feasible with the myosin heavy chain or actin which, as major components of the thick and thin filaments, cannot be removed and exchanged into fibres. For these proteins, and many others, the effects of mutants on muscle contraction can only be studied using systems where the mutant proteins assemble into fibres *in vivo*.

The '*in vivo*' approach requires a species whose genetics are well known, whose genes are easily manipulated, and whose muscles can be analysed by conventional structural and functional techniques, including mechanical recording. We have developed a model system for the genetical analysis of muscle contraction using the indirect flight muscles (IFMs) of the fruitfly *Drosophila melanogaster* (Fig. 1) which, to date, uniquely fulfils these requirements.

Background

Genetics

Many *D. melanogaster* muscle protein genes have been cloned and sequenced, including the myosin heavy chain (see reviews by Emerson and Bernstein, 1987; George *et al.* 1989), both myosin light chains (Falkenthal *et al.* 1984; Parker *et al.* 1985; Toffenetti *et al.* 1987), tropomyosin (Karlik *et al.* 1984), troponin-I (Beall and Fyrberg, 1990), troponin-T (Bullard *et al.* 1988) α-actinin (Fyrberg *et al.* 1990), troponin-H (an IFM-specific heavy tropomyosin with additional properties) (Hanke and Storti, 1988) and the IFM-specific actin gene, *Act88F* (Sanchez *et al.* 1983). Mutations have been recovered in some of these genes by selecting for flightlessness following chemical mutagenesis of flies (e.g. Mogami and Hotta, 1981). Most of these mutations lead to structurally aberrant flight muscles which are useful for analysing myofibril assembly but are not appropriate for the study of muscle contraction. In *D. melanogaster*, cloned genes can be re-introduced into the germline using the now commonplace technique of P-element mediated transformation (Rubin and Spradling, 1982; Spradling and Rubin, 1982; see e.g. Roberts, 1986 for practical details). This permits the effects of mutations induced by *in vitro* mutagenesis techniques to be tested *in vivo*.

The P-element is a *Drosophila* transposable genetic element able to move, under certain conditions, from one part of the *Drosophila* genome to another. This property is used for re-introducing a cloned gene into the germline by placing the gene in a plasmid between the ends of a P-element (Fig. 2). The DNA construct also carries a 'selectable' marker, the wild-type gene for *rosy*, which permits the identification, by eye colour, of those flies which have been successfully transformed. Transformation is accomplished by micro-injecting the DNA into the posterior end of pre-blastoderm stage embryos. Transformed copies of cloned genes insert almost at random into the genome. However, they show normal gene expression, being correctly expressed in both a tissue- and stage-specific manner. Among a number of transformant lines set up from different transformant flies, it is usual to find some with wild-type levels of gene expression. This is an important point in the use of this technique with muscle protein genes, since it appears that normal muscle

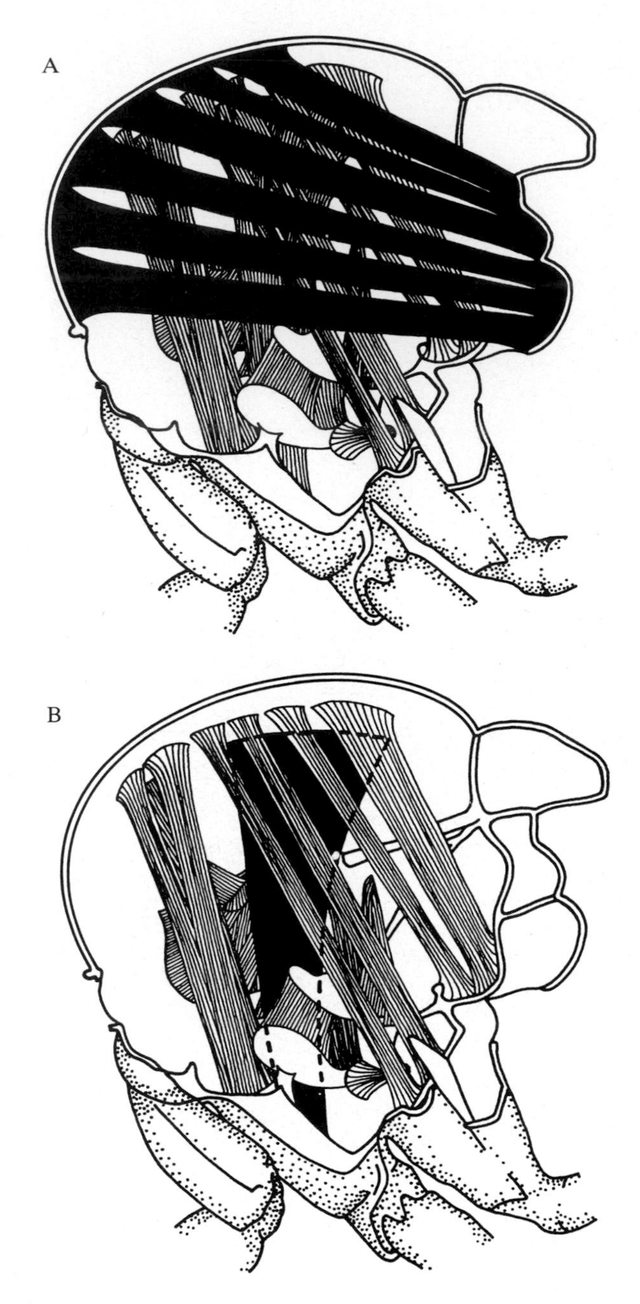

Fig. 1. Positions of (A) the dorsolongitudinal (DLM) indirect flight muscles and (B) the tergal depressor of trochanter muscle (TDT) in the thorax of *Drosophila melanogaster*.

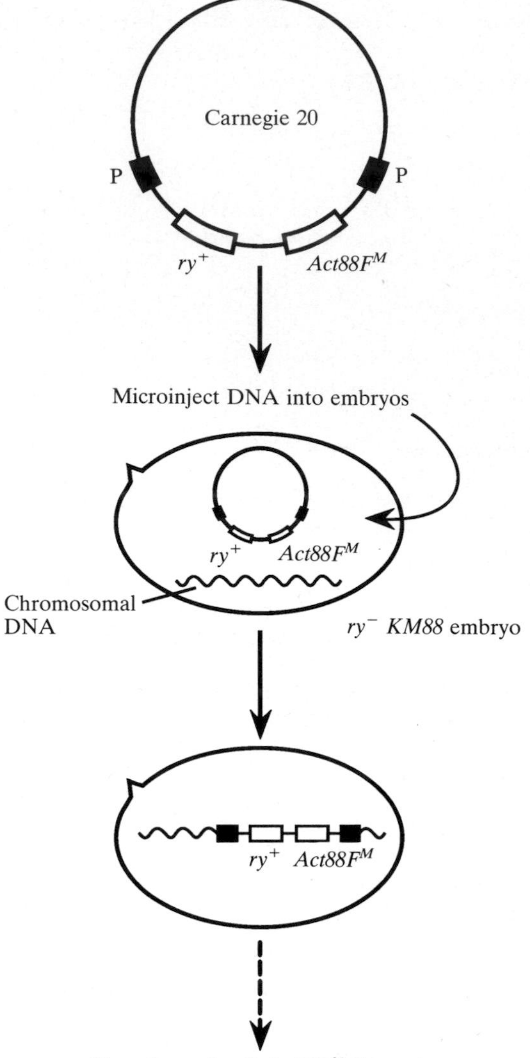

Fig. 2. P-element mediated transformation of ry^- $Act88F^{KM88}$ embryos with DNA contained between the two P-element ends in the Carnegie 20 plasmid vector. The transforming DNA contains a wild-type copy of the *rosy* gene and a copy of the *Act88F* gene which has undergone *in vitro* mutagenesis, ^{M}.

structure is only achieved when there are wild-type levels of gene expression.

So far only two muscle protein genes have been successfully re-introduced into the fly genome. These are the tropomyosin gene, *Tm1*, (Fyrberg and Karlik, 1987) and the IFM-specific actin gene, *Act88F* (Mahaffey *et al.* 1985; Hiromi *et al.* 1986). There are two reasons why so few muscle protein genes have been transformed. First, if one wishes to study the expression of a transformed wild-type gene copy, or the effect of a mutant gene, it is useful, if not necessary, for the wild-type chromosomal copy to be functionally inactive. Null mutations exist for the *Act88F* gene (*KM88*, Hiromi and Hotta, 1985), the MLC-2 myosin light chain gene (Warmke *et al.* 1989) and one of the products of the tropomyosin gene *Tm1* (*Ifm(3)3*, Fyrberg and Karlik, 1987). Second, the efficiency of P-element transformation decreases with the size of the cloned gene that is being transformed. So, for instance, the whole muscle myosin heavy chain gene, *Mhc*, which is approximately 10 times larger than *Act88F*, has yet to be re-inserted in the genome using P-elements, although DNA fragments this long can be transformed. IFM-specific null mutations are available for this gene and Bernstein (personal communication) has obtained significant levels of expression of a construct based on a cDNA copy of an IFM-specific myosin heavy chain transcript.

Muscles

IFMs are especially suitable for the genetical study of muscle structure and function for two reasons. First, because their function is not necessary for viability or fertility and second, because these muscles contain specific isoforms of some myofibrillar proteins, including the myosin heavy chain (O'Donnell *et al.* 1989), actin (Ball *et al.* 1987) and tropomyosin as well as novel, IFM-specific proteins such as Tn-H (heavy troponin, Bullard *et al.* 1988) and arthrin, a conjugate of actin and ubiquitin (Ball *et al.* 1987). Mutations affecting these isoforms either do not, or will not, affect the function of the other muscle types necessary for life functions.

The IFMs of *Drosophila* are typical of insect fibrillar flight muscle (Pringle, 1978). They are asynchronous; their nerve input occurs at a much lower frequency than the wingbeat frequency of 200 Hz (Levine, 1973; Levine and Wyman, 1973). They are fibrillar, implying a low sarcoplasmic reticulum content (Pak and Grabowski, 1978) and striated, with a sarcomere length of about 3.4 μm (Shafiq, 1963*a*,*b*; 1964). They have very short I-bands.

These *Drosophila* muscles are not ideal for mechanical experiments on two counts. First, the IFMs are short (about 1 mm long) and second, they have fast contraction rates. This was predicted from the high wingbeat frequency (200 Hz) of the flies (Pringle, 1957) and demonstrated to be an inherent property of the isolated, demembranated fibres (Molloy *et al.* 1987). We have been able to obtain consistent and reliable mechanical data from sinusoidal and step-length changes applied to chemically demembranated ('skinned') fibres from the largest pair of dorsal longitudinal fibres (IFMs) attached to a mechanical apparatus (Molloy *et al.* 1987; Peckham *et al.* 1990). The use of aluminium T-clips (Goldman and Simmons, 1984) to mount the fibres (Fig. 3) and a mechanical apparatus able to respond to fast changes in the tension transients were crucial to this achievement. Briefly, the thoraces of adult flies are bisected along their midline and the fibres 'skinned' using either a combination of glycerol and Triton X-100 or saponin (see Peckham *et al.* 1990 for details). After skinning, the fibres of the two longest dorsolongitudinal 'muscles' are carefully cut from the cuticle, pared down to about 70 μm in diameter and the ends crimped in T-clips leaving about 0.5 mm exposed between the T-clips. All these manipulations are performed in a bathing solution and once crimped the fibres are mounted on the apparatus without passing through an air/liquid interface as this very easily damages the fibres. Fibres are kept immersed in solution on the apparatus by changing solutions using flow-through system. Further details of experimental protocols and solutions used are given elsewhere (Peckham *et al.* 1990).

We have shown that the *D. melanogaster* IFMs share many of the physiological properties of those from the giant water-bug, *Lethocerus*, although there are significant differences in the rate constants which correlate with the very different wingbeat frequencies of these insects (Peckham *et al.* 1990). In addition, we have also successfully performed mechanical experiments on a *Drosophila* synchronous muscle, the tergal depressor of trochanter muscle (Figs 1 and 3), the so-called 'jump' muscle (Peckham *et al.* 1990). This muscle has physiological properties which are more similar to vertebrate striated muscle than those of the indirect flight muscles. The ability to perform mechanical experiments on the two muscle types will enable a molecular genetic comparison to be made of the protein and physiological differences between these muscle types.

The Act88F *gene*

We have chosen to begin our protein engineering analysis of muscle contraction by concentrating on the *Act88F* gene. This gene, one of the six actin genes in the *Drosophila* genome (Tobin *et al.* 1980; Fyrberg *et al.* 1980), is expressed only in the indirect flight muscles and encodes the only actin isoform found in these muscles (Ball *et al.* 1987). This has two important consequences: (1) *Act88F* mutations will not affect the function of other muscles (and will not affect fly viability) and (2) only the mutant actin will be present in these muscles. The wild-type gene, which has been cloned and sequenced, can be re-inserted into the *Drosophila* genome using P-element transformation (Mahaffey *et al.* 1985; Hiromi *et al.* 1986) and 'rescues' the flight ability of flies homozygous for *KM88*, an *Act88F* null mutation (Hiromi and Hotta, 1985) which is a useful transformation host for *in vitro* mutants.

Progress

In the absence of information about the importance of specific amino acids in actin, we chose to make a number of mutations in that part of the gene which corresponds to amino acids 314–375 (C terminus), a region known to contain a contact site for one of the myosin light chains (see Hambly *et al.* 1986). Mutations were induced in the corresponding gene fragment by random bisulphite mutagenesis and six gene fragments encoding single amino acid changes were recovered. The mutants are *E316K* (where glutamate (E) at amino acid position 316 is replaced by the lysine (K)), *E334K*, *V339I*, *E364K*, *G366D* and *G368E*. Each one was reconstructed into complete gene copies in P-element vectors and transformed into hosts homozygous for the *KM88* mutation.

Phase microscopy of IFM myofibrils from each transformed *KM88* strain revealed that only the *E316K* and *G368E* mutants permit the assembly of cross-striated myofibrils. The others produce aberrant myofibrils (*E334K*) or no fibrillar material at all (*V339I*, *E364K*, *G366D*). Electron microscopy of *E316K* and *G368E* showed that these mutants produce essentially normal myofibrillar structure, although in both some minor lattice defects are seen (Fig. 4). These occur near the periphery of the myofibrils in *G368E* and internally in *E316K* myofibrils. In addition more severe abnormalities, which are restricted to the ends of the fibres, close to the muscle attachment sites, have been seen in *E316K* (Mary Reedy, personal communication). Both mutants produced normal actin/myosin ratios (from scans of 1-D SDS–PAGE gels) and only the expected charge-changed actins on 2-D gels. Subsequent analysis of muscle contraction has concentrated on these two actin mutations (Drummond *et al.* 1990).

To test the effects of the mutations on flight muscle function, we measured flight ability. *E316K* flies were as flightless as *KM88* homozygotes, probably due to the aberrant fibre ends described above; *G368E* flies flew but less well than wild-type or a *KM88* strain homozygous for transformed copies of a wild-type *Act88F* gene. Clearly the reduced flight ability was due to the *G368E* mutation. The wingbeat frequencies of wild-type and *G368E* flies were measured using an optical tachometer (Unwin and Ellington, 1979). *G368E* flies had a wingbeat frequency of 178±3 Hz (mean±S.E.M.) which was reduced by about 20 % compared to wild-type at 227±4 Hz under the same conditions.

Both the *E316K* and *G368E* mutations altered the mechanics of demembranated IFM fibres, compared to those of wild-type (Fig. 5). In *G368E*, the rate constant for the delayed increase in tension which followed a quick

Table 1. *Delayed tension rate constants from the mechanical transients obtained from skinned fibres from the IFMs of wild-type,* Act88F–G368E *and* Act88F–E316K *flies*

	Genotype		
	Wild-type	*G368E*	*E316K*
Rate constant (s^{-1})	222	162*	342**
S.D.	49	61	86
n	11	13	6

The rate constants are estimated using exponential curve-fitting methods (Provencher, 1976*a*,*b*).

S.D., standard deviation of the mean.

n, number of fibres studied.

*, significant difference from wild-type at 5% levels, and **, at 10% levels (Student's *t*-test).

stretch in activating solution was reduced by 30% (Table 1). The rate constants for tension relaxation during phases 2 and 4 were also reduced, although our data were not significant, perhaps because determination of these rate constants is subject to greater experimental error. Steady state active tension and stiffness were increased, while those in rigor were not. Relaxed stiffness was unchanged. In *E316K* the rate constant for the delayed increase in tension following a quick stretch was increased by about 30% (Table 1) as was the relaxation of tension in phase 2. Steady state stiffness and tension were unchanged in the different solutions.

These changes cannot be explained by the rather minor disturbances of myofibrillar structure, as similar mechanical experiments on *KM88/+* fibres, which have considerably more myofibrillar disruption, show a pattern of changes in the mechanical parameters which are unlike those found in either *G368E* or *E316K* fibres. Both these mutations occur within conserved regions of the actin amino acid sequence. In all actins, amino acid 368 is either glycine or serine while amino acid 316 is always glutamic acid, except for a single case where it is aspartic acid. Actin is a highly conserved protein and it has been speculated that this results either from the constraints of folding a protein which undergoes large conformational changes (Schutt and Lindberg, 1989) or the requirement for actin to interact with a large number of other proteins (Moir and Levine, 1986). Our results demonstrate that in at least some cases these constraints are not so severe that even non-conservative changes cannot be tolerated. They also demonstrate that amino acid changes in actin can have significant effects on the mechanical properties of the muscle. Amino acid 368 is in a region involved in the binding of tropomyosin and the alkaline myosin light chain so it is not surprising that a change here affects muscle function. In contrast, amino acid 316 is in the large domain, distant from known myosin contacts, suggesting that *E316K* may be affecting the interaction with myosin by a long range conformational change in the actin. It is something of a relief to find that even in a highly conserved protein like actin, amino acid substitutions in conserved residues can result in proteins with subtle enough changes in function to permit an analysis of the relationship between actin amino acid sequence and function.

Problems and limitations

The major problem with the use of *Drosophila* for muscle protein engineering studies is the very small size of the organism and its tissue heterogeneity. Although the IFMs are the major muscles, flies contain many muscles with different myofibrillar proteins and physiological properties (Peckham *et al.* 1990). It is therefore difficult to obtain muscle, fibres, myofibrils or myofibrillar proteins from a single muscle type for the more traditional approaches of muscle investigators.

We have solved the problems of scale for mechanical experiments on fibres from IFMs and the TDT, but mechanical data on their own are insufficient to exploit the potential of the mutants for the study of muscle contraction. Recently, assays have been developed to measure IFM fibre ATPase activities (J. Molloy and D. Maughan, personal communication) and to study the effects of calcium concentration on mechanical parameters in mutants with myosin light chain defects (Yamakawa *et al.* 1990). Mechanical experiments on fibres in which caged-ATP is used have been performed (M. Peckham and M. Ferenczi, unpublished data) and techniques such as X-ray diffraction and different forms of spectroscopy are probably feasible although *Drosophila* fibres, like *Lethocerus*, contain a lot of mitochondria which are not conducive to good spectroscopic measurements.

Biochemical experiments with isolated proteins are essential for the interpretation of the effects of the mutations on protein structure and function. Major problems of scale arise when purifying the *Drosophila* IFM proteins. However, some progress has been made in resolving these problems.

We have used an *in vitro* transcription/translation system to study the *Act88F* mutations we have recovered. The six mutant fragments have been re-assembled into a gene copy which lacks introns and is under the control of the T7 RNA polymerase promoter. This DNA construct is used to synthesise RNA, which is then translated in the presence of radioactive amino acids in a rabbit reticulocyte lysate system. By this means, small (nanogram) quantities of actins with a high specific activity of incorporated radioisotope are recovered. We have used these to test the ability of the mutant actins to polymerise, using a co-polymerisation assay (Solomon and Rubenstein, 1987) with either rabbit or *Lethocerus* bulk actin added to achieve the critical concentration. The ability of radioactive mutant actins to bind actin binding proteins (DNase I, profilin, or heavy meromyosin) in affinity chromatography columns has been used to test the properties of the mutant actin molecules. The *in vitro* approach has particular advantages. We have found that all of our mutants, even those which do not accumulate *in vivo*, form stable actins *in vitro* which have altered, but rarely absent, abilities to function in these assays. We have also used this approach to examine the effects of inhibiting N-terminal processing on actin function.

We have expressed the *Act88F* gene in *E. coli* either directly under the control of a prokaryotic promotor or as a fusion protein. As others have found with different actins, the products are very insoluble and we have been unable to make even reproducible cyanogen bromide fragments. Recently, native actin has been successfully recovered from an *E. coli* expression system (Frankel *et al.* 1990). In general this problem should not affect the study of many other *Drosophila* muscle proteins since their homologues from other species can be expressed in *E. coli*. Eukaryotic expression systems such as yeast and baculovirus are available and may be useful for many muscle proteins, particularly since some post-translational

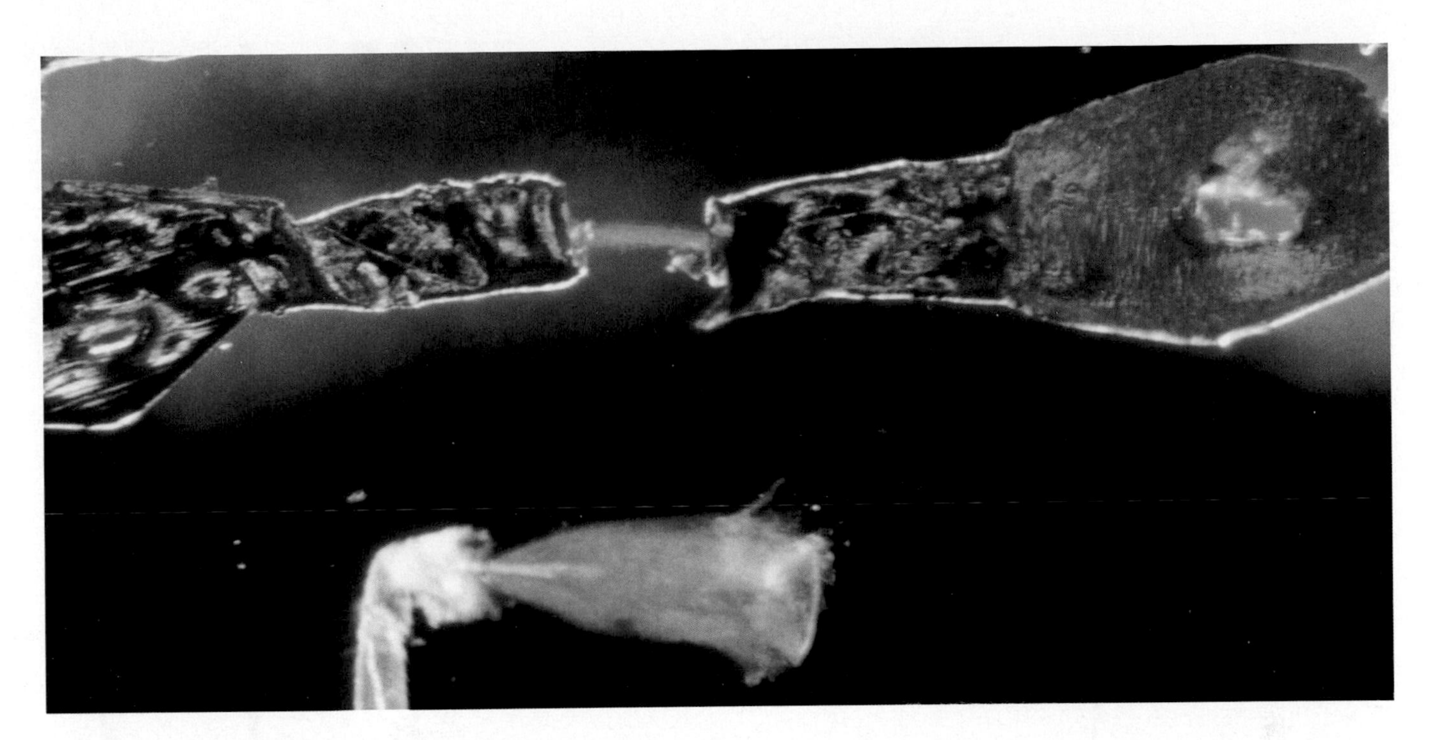

Fig. 3. Photomicrograph of a dorsolongitudinal muscle fibre crimped between two aluminium T-clips prior to mounting on the mechanical testing apparatus (above) and (below) a dissected tergal depressor of the trochanter muscle. The length of DLM muscle exposed between the clips is about 0.5 mm.

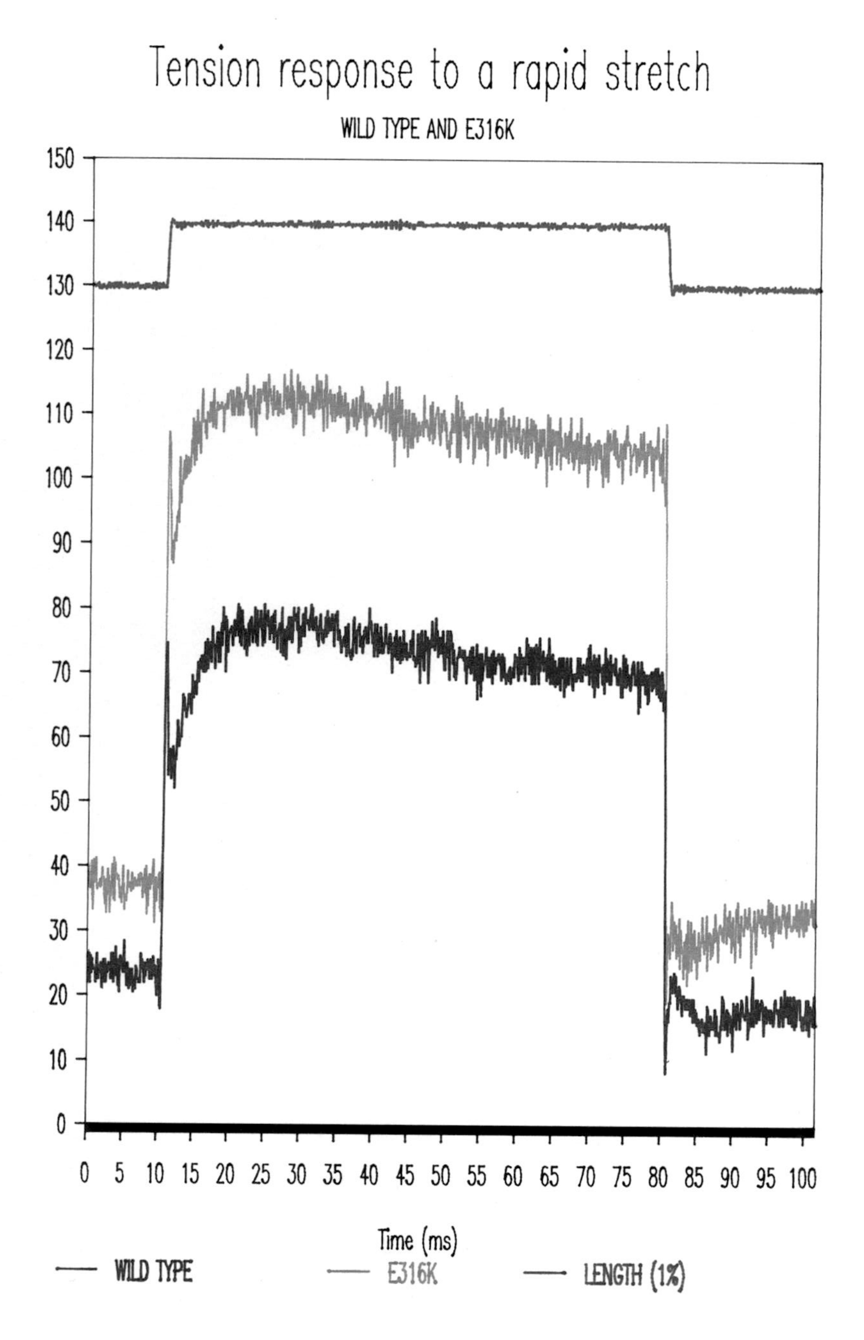

Tension response to a rapid stretch
WILD TYPE AND E316K
Tension (pN/A-filament)
Time (ms)
WILD TYPE
E316K
LENGTH (1%)

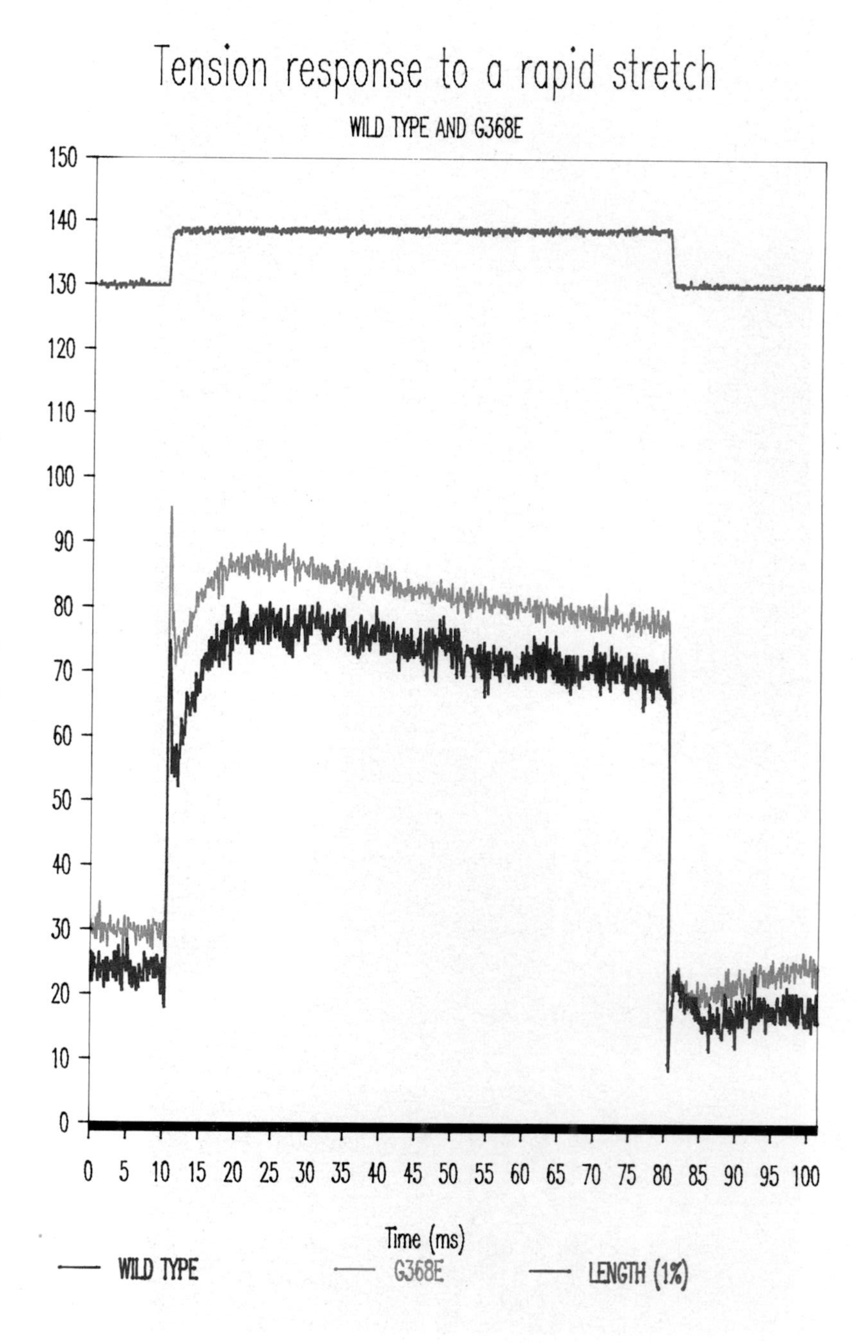

Tension response to a rapid stretch
WILD TYPE AND G368E
Tension (pN/A-filament)
Time (ms)
WILD TYPE
G368E
LENGTH (1%)

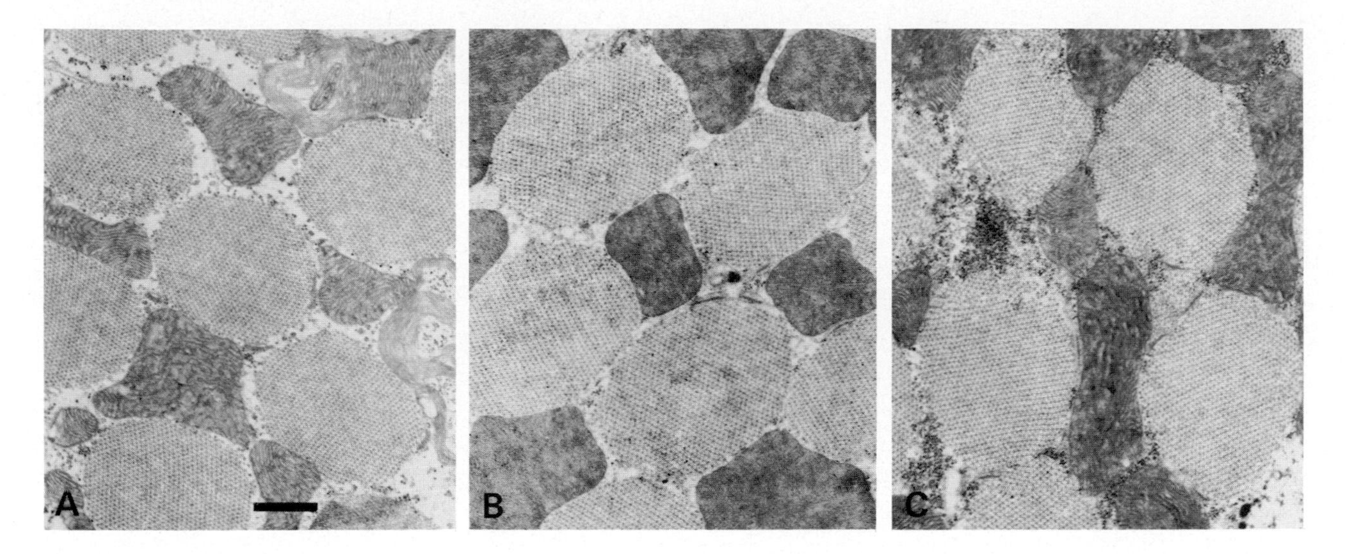

Fig. 4. Electronmicrographs of transverse sections of dorsolongitudinal muscles from (A) *Oregon* (wild-type), (B) *E316K* and (C) *G368E*. Bar, 500 nm.

modifications occur in these systems. However, for actin, which is a ubiquitous eukaryotic cellular protein, these systems may not be the answer. Mutant actins will probably disrupt the cytoskeleton and overexpression, even of wild-type actins, is likely to be deleterious to the host cells. This has already been seen with yeast actins (K. Wertman, personal communication). A further problem is to purify the expressed mutant actins from the endogenous ones in the cell, though this may be possible where a charge change is introduced into the mutant actin.

Recently we have been using the mass isolation technique developed for the IFMs by Saide *et al.* (1989). From 5 g of flies (about 5000 flies) we can obtain 20 mg of IFM myofibrillar protein. This is a relatively small number of flies and larger quantities are easily grown. Procedures for the isolation of insect muscle proteins have been developed (Bullard, 1983; Bullard *et al.* 1973, 1988) and small but adequate quantities of major myofibrillar proteins should now be obtainable for ATPase measurements, *in vitro* binding assays, filament assembly assays, and a variety of spectroscopic analyses including NMR. A limitation in using this approach with mutants is that it will only be applicable to those mutations which are stable and accumulate *in vivo*.

The future

The goal is to understand the molecular biology of muscle contraction by having a description of the protein conformational changes and interactions which are an integral part of converting ATP into work. Protein engineering of muscle proteins and an ability to study the effects of amino acid changes, both *in vitro* and '*in vivo*', is a powerful way to study this. A major technical challenge with the *Drosophila* system is the development of ways to apply the many techniques which are currently used to study muscle contraction and muscle proteins in larger, biochemically more amenable muscles to the study of the mutant proteins.

Fig. 5. Tension responses of chemically demembranated indirect flight muscle fibres to a rapid 1 % stretch in activating conditions. The phases of the tension response are (see text) 2, for the descending phase following the initial tension 'spike', 3, the delayed increase in tension and 4, the slow decrease of tension which follows. For each tension transient the rate constants (r_2, r_3, r_4) are determined for each of these phases by curve-fitting these phases by three exponential rate constants using the procedure of Provencher (1976*a*,*b*).

No single mutation will solve the mechanism of how muscle contraction works, and protein engineering approaches are only capable of answering those questions about muscle contraction that can be formulated at the level of the amino acid. Such questions will concern protein structure, ligand binding properties, conformational changes and protein–protein interactions. Large numbers of mutations in the different muscle proteins will be needed to study not only the role of single proteins, but also the effects of combinations of mutants in different proteins. As the size and numbers of proteins being studied increase, the number of possible amino acid changes assumes astronomical proportions.

The *in vivo* system offers a powerful solution to this. For instance, we now have actin mutations which subtly alter the myofibrillar kinetics and reduce the flight ability of flies. By performing whole organism mutagenesis of the flightless flies and selecting for increased flight ability, it should be possible to obtain further mutants which compensate for the effects of the original mutation in one of two ways. First, they may occur at other sites within the actin gene itself or second, in other genes, such as the myosin heavy or light chain genes, and compensate for the original mutation at the level of protein interaction. In either case they will be informative about the role of specific amino acids in muscle protein structure and function. The advantage of this approach is that it lets the organism tell us what is important.

References

Ball, E., Karlik, C. C., Beall, C. J., Saville, D. L., Sparrow, J. C., Bullard, B. and Fyrberg, E. A. (1987). Arthrin, a myofibrillar protein of insect flight muscle, is an actin-ubiquitin conjugate. *Cell* **51**, 221–228.

Beall, C. J. and Fyrberg, E. A. (1990). Mutations in the *Drosophila* troponin I (*heldup*) gene cause a variety of defects in sarcomeric structure. *J. Cell Biochem.* **14a**, 11.

BULLARD, B. (1983). Contractile proteins of insect flight muscle. *Trends biochem. Sci.* **8**, 68–70.

BULLARD, B., DABROWSKA, R. AND WINKELMAN, L. (1973). The contractile and regulatory proteins of insect flight muscle. *Biochem. J.* **135**, 277–286.

BULLARD, B., LEONARD, K., LARKINS, A., BUTCHER, G., KARLIK, C. C. AND FYRBERG, E. A. (1988). The troponin of asynchronous flight muscle. *J. molec. Biol.* **204**, 621–637.

DRUMMOND, D. R., PECKHAM, M., SPARROW, J. C. AND WHITE, D. C. S. (1990). Alteration in crossbridge kinetics caused by mutations in actin. *Nature* **348**, 440–442.

EMERSON, C. P., JR AND BERNSTEIN, S. I. (1987). Molecular genetics of myosin. *A. Rev. Biochem.* **56**, 695–726.

FALKENTHAL, S., PARKER, V. P., MATTOX, W. W. AND DAVIDSON, N. (1984). *Drosophilia melanogaster* has only one myosin alkali light-chain gene which encodes a protein with considerable amino acid sequence homology to chicken myosin alkali light chains. *Molec. cell. Biol.* **4**, 956–965.

FRANKEL, S., CONDEELIS, J. AND LEINWAND, L. (1990). Expression of actin in *Escherichia coli*. *J. biol. Chem.* **265**, 17 980–17 987.

FUJIMORI, K., SORENSON, M., HERZBERG, O., MOULT, J. AND REINACH, F. C. (1990). Probing the calcium-induced conformational transition of troponin C with site-directed mutants. *Nature* **345**, 182–184.

FYRBERG, E. A. AND KARLIK, C. C. (1987). Genetic rescue of muscle defects associated with a mutant *Drosophila melanogaster* tropomyosin allele. *Molec. cell. Biol.* **7**, 2977–2980.

FYRBERG, E. A., KELLY, M., BALL, E., FYRBERG, C. AND REEDY, M. C. (1990). Molecular genetics of *Drosophila* alpha-actinin; mutant alleles disrupt Z-disc integrity and muscle insertions. *J. Cell Biol.* **110**, 1999–2011.

FYRBERG, E. A., KINDLE, K. L., DAVIDSON, N. AND SODJA, A. (1980). The actin genes of *Drosophila*: a dispersed multigene family. *Cell* **19**, 365–378.

GEORGE, E., OBER, M. AND EMERSON, C. P., JR (1989). Functional domains of the *Drosophila melanogaster* muscle myosin heavy chain gene are encoded by alternatively spliced axons. *Molec. cell. Biol.* **9**, 2957–2974.

GOLDMAN, Y. E. AND SIMMONS, R. M. (1984). Control of sarcomere length in skinned muscle fibres of *Rana temporaria* during mechanical transients. *J. Physiol.* **350**, 497–518.

GRABAREK, Z., TAN, R-Y., WANG, J., TAO, T. AND GERGELY, J. (1990). Inhibition of mutant troponin C activity by an intra-domain disulphide bond. *Nature* **345**, 132–135.

HAMBLY, B. D., BARDEN, A., MIKI, M. AND DOS REMEDIOS, C. G. (1986). Structural and functional domains on actin. *BioEssays* **4**, 124–128.

HANKE, P. D. AND STORTI, R. V. (1988). The *Drosophila melanogaster* tropomyosin II gene produces multiple proteins by use of alternative tissue-specific promoters and alternative splicing. *Molec. cell. Biol.* **8**, 3591–3602.

HIROMI, Y. AND HOTTA, Y. (1985). Actin gene mutations in *Drosophila*: heat-shock activation in the indirect flight muscles. *EMBO J.* **4**, 1681–1687.

HIROMI, Y., OKAMOTO, H., GEHRING, W. J. AND HOTTA, Y. (1986). Germline transformation with Drosophila mutant actin genes induces constitutive expression of heat shock genes. *Cell* **44**, 293–301.

HITCHCOCK-DEGREGORI, S. E. AND HEALD, R. W. (1987). Altered actin and troponin binding of amino-terminal variants of the chicken striated muscle-tropomyosin expressed in *Escherichia coli*. *J. Biol. Chem.* **262**, 9730–9735.

HITCHCOCK-DEGREGORI, S. E. AND VARNELL, T. A. (1990). Tropomyosin has discrete actin-binding sites with sevenfold and fourteenfold periodicities. *J. molec. Biol.* **214**, 885–896.

KARLIK, C. C., MAHAFFEY, J. W., COUTU, M. D. AND FYRBERG, E. A. (1984). Organization of contractile protein genes within the 88F subdivision of the *D. melanogaster* third chromosome. *Cell* **37**, 469–481.

LEVINE, J. D. (1973). Properties of the nervous system controlling flight in *Drosophila melanogaster*. *J. Comp. Physiol.* **84**, 129–166.

LEVINE, J. D. AND WYMAN, R. J. (1973). Neurophysiology of flight in wild-type and a mutant *Drosophila*. *Proc. natn. Acad. Sci. U.S.A.* **70**, 1050–1054.

MAHAFFEY, J. W., COUTU, M. D., FYRBERG, E. A. AND INWOOD, W. (1985). The flightless Drosophila mutant *raised* has two distinct genetic lesions affecting accumulation of myofibrillar proteins in flight muscles. *Cell* **40**, 101–110.

MOGAMI, K. AND HOTTA, Y. (1981). Isolation of *Drosophila* flightless mutations which affect myofibrillar proteins of indirect flight muscles. *Mol. gen. Genet.* **183**, 409–417.

MOIR, A. J. G. AND LEVINE, B. (1986). Protein cognitive sites on the surface of actin. A proton NMR study. *J. Inorg. Chem.* **28**, 271–278.

MOLLOY, J. E., KYRTATAS, V., SPARROW, J. C. AND WHITE, D. C. S. (1987). Kinetics of flight muscles from insects with different wingbeat frequencies. *Nature* **328**, 449–451.

O'DONNELL, P. T., COLLIER, V. L., MOGAMI, K. AND BERNSTEIN, S. I. (1989). Ultrastructural and molecular analyses of homozygous-viable *Drosophila melanogaster* muscle mutants indicates there is a complex pattern of myosin heavy chain isoform distribution. *Genes Dev.* **3**, 1233–1246.

PAK, W. L. AND GRABOWSKI, S. R. (1978). Physiology of the visual and flight systems. In *The Genetics and Biology of Drosophila* (ed. M. Ashburner and T. R. F. Wright) **2a**, 553–604, Academic Press: New York and London.

PARKER, V. P., FALKENTHAL, S. AND DAVIDSON, N. (1985). Characterisation of the myosin light-chain-2 gene of *Drosophila melanogaster*. *Molec. cell. Biol.* **5**, 3058–3068.

PECKHAM, M., MOLLOY, J. E., SPARROW, J. C. AND WHITE, D. C. S. (1990). Physiological properties of the dorsal longitudinal flight muscle and the tergal depressor of the trochanter muscle of *Drosophila melanogaster*. *J. Muscle Res. Cell Motil.* **11**, 203–215.

PRINGLE, J. W. S. (1957). Insect Flight. Cambridge University Press: Cambridge.

PRINGLE, J. W. S. (1978). Stretch activation of muscle: function and mechanism. *Proc. R. Soc. Lond. B* **201**, 107–130.

PROVENCHER, S. W. (1976*a*). A Fourier method for the analysis of exponential decay curves. *Biophys. J.* **16**, 27–41.

PROVENCHER, S. W. (1976*b*). An eigenfunction expansion method of exponential decay curves. *J. chem. Phys.* **64**, 2772–2777.

ROBERTS, D. B. (1986). Drosophila*: A Practical Approach.* IRL Press, Oxford.

RUBIN, G. M. AND SPRADLING, A. C. (1982). Genetic transformation of *Drosophila* with transposable element vectors. *Science* **218**, 348–353.

SAIDE J. D., CHIN-BOW, S., HOGAN-SHELDON, J., BUSQUETS-TURNER, L., VIGOREAUX, J., VALGEIRSDOTTIR, K. AND PARDUE, M. L. (1989). Characterisation of components of Z-bands in the fibrillar flight muscle of *Drosophila melanogaster*. *J. Cell Biol.* **109**, 2157–2167.

SANCHEZ, F., TOBIN, S. L., RDEST, U., ZULAUF, E. AND MCCARTHY, B. J. (1983). Two *Drosophila* actin genes in detail. Gene structure, protein structure and transcription during development. *J. molec. Biol.* **163**, 533–551.

SCHUTT, C. E. AND LINDBERG, U. (1989). The nature of the actin molecule. In *Molecular Mechanisms in Muscular Contraction* (ed. J. M. Squire). Topics in Molecular and Structural Biology Vol. **10**, pp. 49–63. Macmillan: London.

SHAFIQ, S. A. (1963*a*). Electron microscope studies of the indirect flight muscles of *Drosophilia melangaster*. I. Structure of the myofibrils. *J. Cell Biol.* **17**, 351–362.

SHAFIQ, S. A. (1963*b*). Electron microscope studies of the indirect flight muscles of *Drosophila melanogaster*. II. Differentiation of myofibrils. *J. Cell Biol.* **17**, 363–374.

SHAFIQ, S. A. (1964). An electron microscopical study of the innervation and sarcoplasmic reticulum of the fibrillar flight muscle of *Drosophila melanogaster*. *Q. Jl. micros. Sci.* **105**, 1–6.

SOLOMON, L. R. AND RUBENSTEIN, P. A. (1987). Studies on the role of actin's N-methylhistidine using oligodeoxynucleotide-directed site-specific mutagenesis. *J. biol. Chem.* **262**, 11 382–11 388.

SPRADLING, A. C. AND RUBIN, G. M. (1982). Transposition of cloned P element into *Drosophila* germline chromosomes. *Science* **218**, 341–347.

TOBIN, S. L., ZULAUF, E., SANCHEZ, F., CRAIG, E. A. AND MCCARTHY, B. J. (1980). Multiple actin-related sequences in the *Drosophila melanogaster* genome. *Cell* **19**, 121–131.

TOFFENETTI, J., MISCHKE, D. AND PARDUE, M. L. (1987). Isolation and characterisation of the gene for myosin light chain two of *Drosophila melanogaster*. *J. Cell Biol.* **104**, 19–28.

UNWIN, D. M. AND ELLINGTON, C. P. (1979). An optical tachometer for measurement of the wing-beat frequency of free-flying insects. *J. exp. Biol.* **82**, 377–378.

WARMKE, J. W., KREUZ, A. J. AND FALKENTHAL, S. (1989). Co-localisation to chromosome bands 99E1-3 of the *Drosophila melanogaster* myosin light chain-2 gene and a haploinsufficient locus that affects flight behaviour. *Genetics* **122**, 139–151.

YAMAKAWA, M., WARMKE, J., FALKENTHAL, S. AND MAUGHAN, D. (1990). pCa-tension curves and kinetics of stretch activation in skinned single muscle fibers from *Drosophila melanogaster*. *Biophys. J.* **57**, 411a.

Nuclear envelope dynamics and nucleocytoplasmic transport

MURRAY STEWART, SUE WHYTOCK and ROBERT D. MOIR

MRC Laboratory of Molecular Biology, Hills Rd., Cambridge CB2 2QH, England

Summary

We have combined structural, biochemical and recombinant DNA methods to explore molecular interactions involved in nuclear envelope assembly dynamics and nucleocytoplasmic transport. Electron microscopy has established the overall architecture of the envelope and the relationship between nuclear pores, lamina fibres and pore-connecting fibrils. The lamin proteins that constitute the lamina resemble intermediate filament proteins, and assemble and disassemble during mitosis in response to phosphorylation. Lamins have been expressed in *E. coli* to facilitate structural investigations and the exploration of interaction sites with other envelope components. Disruption of envelopes has shown that nuclear pores are constructed from a central cylinder with cytoplasmic and nucleoplasmic rings. Examination of envelopes transporting gold-labelled nucleoplasmin has indicated that the transport pathway is complex and probably involves ring components in addition to the central cylinder. Molecular motors may be involved in changes in pore shape to enable transport and in the translocation mechanism.

Key words: lamins, nuclear pores, interactions, transport, structure.

Nuclear envelope structure

The nuclear envelope separates the chromosomes from the cytoplasm of eucaryotic cells and controls transport into and out of the nucleus. It is constructed from a double membrane perforated by nuclear pore complexes thought to be responsible for selective transport (reviewed by Newport and Forbes, 1987). A fibrous lamina, composed of lamins, lies under the nucleoplasmic face of the envelope, below the level of the pores (see Aebi *et al.* 1986; Stewart and Whytock, 1988). The lamins closely resemble intermediate filaments in sequence and structure (McKeon *et al.* 1986), forming fibres about 10 nm in diameter that are sometimes found in a remarkably regular basketweave pattern (Aebi *et al.* 1986). The lamina is thought to be important in organising the nuclear pore complexes and possibly also chromosomes within the nucleus (Gerace, 1986; Newport and Forbes, 1987). In addition to the lamina, there are also pore-connecting fibrils that link pore complexes directly to each other (Maul, 1977; Stewart and Whytock, 1988; Reichelt *et al.* 1990).

Shadowed nuclear envelopes from which membranes have been removed show pore complexes and lamina clearly and enable cytoplasmic and nucleoplasmic faces to be recognised (Aebi *et al.* 1986; Stewart and Whytock, 1988). Cytoplasmic faces are dominated by nuclear pores which are often linked by connecting fibrils (PCFs). Stereo pairs show that the PCFs attach to the pores near their cytoplasmic end and micrographs of disrupted pore complexes indicate that the PCFs attach to the cytoplasmic ring (Stewart *et al.* 1990). The nucleoplasmic face has a distinctively different appearance and the fibrous lamina is much more prominent. In areas where pore complexes are close together, the lamina fibres dominate the image and often form a fibrous felt-like mat that, in many places, is elevated above the support film (Fig. 1). In areas where the pores are more separated they have a distinctive star-like outline due to the attachment of lamina fibres. Although clearly present, the pore complexes are not a striking feature of nucleoplasmic faces. Fig. 2 shows a schematic representation of the structure of the *Xenopus* oocyte nuclear envelope.

Proteolysis has provided direct evidence that the 10 nm fibres seen in shadowed preparations are composed from lamins (Whytock *et al.* 1990). Increasing protease concentrations progressively removed the lamina fibres with a corresponding decrease in lamin signal using fluorescence microscopy and 'dot-blots' of whole nuclei. PCFs were more resistant to proteolysis and remained in areas of digested samples where the lamina had been removed. Similarly, the morphology of the nuclear pore complexes was not markedly altered by the degree of proteolysis needed to remove the lamina. Some digestion of the nuclear pore complexes and PCFs was only seen when high concentrations of trypsin were used. When nuclear envelopes were first digested with high concentrations of trypsin and then extracted with Triton X-100, the nuclear envelopes disintegrated completely. A similar effect of protease concentration was observed with the proteolysis of material that had been first extracted with Triton. These observations provide direct evidence that the lamina has a role in maintaining the structural integrity of the envelope (Whytock *et al.* 1990).

Lamin molecular structure and interaction

Lamins have been identified in a wide range of organisms (see Krohne and Benevente, 1986), with mammals expressing three lamin isoforms (A, B, and C), although lamins A and C are very similar and arise by alternative splicing (McKeon *et al.* 1986). When the nuclear envelope disassembles during mitosis, B-type lamins remain associated with membranous vesicles, whereas A-type lamins become freely soluble in the cytoplasm (Gerace and Blobel, 1980). The precise pattern of lamin expression often varies in response to cell differentiation and development. Although B-type lamins appear to be ubiquitous, lamins A and C are usually expressed at low levels during early development but increase with cell differentiation (Lehner

Fig. 1. Shadowed *Xenopus* oocyte nuclear envelope, with its nucleoplasmic face uppermost, showing the dense canopy of lamina fibres supported by underlying nuclear pore complexes. Reproduced from Stewart and Whytock (1988). Bar, 0.5 μm.

et al. 1987; Stewart and Burke, 1987). Lamin sequences (McKeon *et al.* 1986) show strong homologies to intermediate filaments (IFs) and indicate a three-domain model for lamins, with a central fibrous rod having principally a coiled-coil conformation, with non-helical N- and C-terminal domains (reviewed by Stewart, 1990). The lamin N-terminal domain is small compared with most other IF proteins, whereas the lamin C-terminal domain is comparatively large. Electron microscopy of shadowed single molecules and paracrystals (Aebi *et al.* 1986) support this model and show a rod-shaped molecule, about 52 nm long, with a globular domain at one end. Although there is substantial promiscuity in the assembly of IF proteins (see Stewart, 1990), they appear not to co-assemble with lamins. A 42-residue insert in the lamin rod domain may be associated with this behaviour. The assembly properties of both lamins and IFs appear to be altered by phosphorylation which may have a role in the dynamics of both filament systems during cell division (Peter *et al.* 1990).

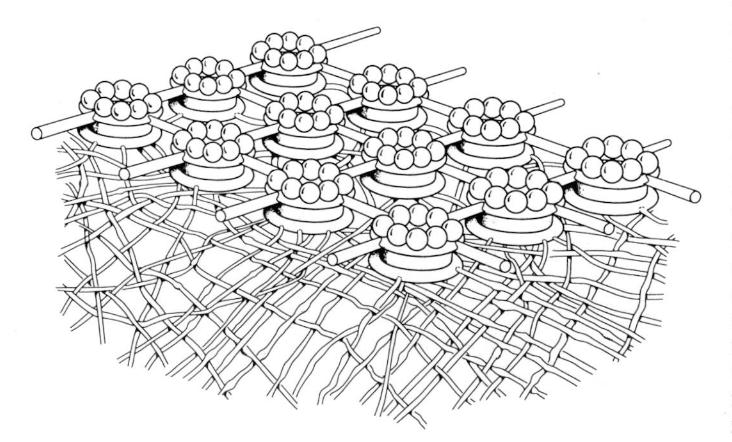

Fig. 2. Schematic illustration of the arrangement of the principal non-membrane components of *Xenopus* oocyte nuclear envelopes. The cylindrical nuclear pore complexes are constructed from nucleoplasmic and cytoplasmic rings and are linked by pore-connecting fibrils. There is a coronet of eight granules on the cytoplasmic face of the pore complex, whereas the 10 nm lamina fibres attach to the nucleoplasmic face. In some areas the lamina fibres are arranged in a basket-weave pattern.

Lamin expression

Because the volume fraction of cells represented by nuclear envelope is small, difficulties can be experienced in preparing large quantities of lamins. Moreover, native material usually is modified post-transcriptionally. Expression in *E. coli* circumvents these difficulties and also facilitates the production of fragments and modified material, which can be used to explore the molecular basis of assembly and domain structure, and complement expression studies in cultured cells. We initially expressed human lamin C cDNA using the pLcII vector (Moir *et al.* 1990), but this produced a fusion with part of the lambda cII protein which we were unable to remove. To circumvent these difficulties, we instead used the T7 expression system to produce substantial quantities of recombinant human lamins A and C. In addition to expressing full length lamins A and C, we used site-directed mutagenesis to modify the lamin cDNA and so express a number of modified protein molecules that lacked the N- or C-terminal non-helical domains (or both).

All the expressed proteins had SDS–PAGE mobilities that corresponded to those predicted from their sequence. Shadowed preparations showed distinctive rod-like molecular profiles which generally resembled native material (Aebi *et al.* 1986). When the C-terminal domain was retained, shadowed particles had a prominent globular domain at one end that frequently could be observed to bifurcate to give the typical two-headed appearance observed on native lamins (Aebi *et al.* 1986). In material that lacked the C-terminal, non-helical domain, these globular domains were absent. Suberimidate crosslinking increased the M_r fourfold, indicating that all the expressed fragments aggregated into four-chain 'tetramer' units analogous to those observed with other intermediate filament proteins.

The solubility of lamins A and C decreased by about an order of magnitude when the salt concentration was decreased from 0.4 to 0.1 M and solubility also decreased when the pH was lowered. Fragments in which either the N- or C-terminal non-helical domain had been deleted were much more soluble. Lamins A and C both formed well-ordered paracrystals with an axial repeat of 22 nm, similar to those observed by Aebi *et al.* (1986) for a mixture of rat lamins A and C. Lamin C with its N-terminal domain deleted formed similar paracrystals. We were unable to form paracrystals from lamins lacking the C-terminal domain, which always yielded 10-nm filaments that closely resembled cytoplasmic IFs. The lamin rod domain formed well ordered paracrystals with a 45 nm axial repeat in which there were two 10 nm wide light bands separated by about 12 and 13 nm alternately.

Nuclear pores and nucleocytoplasmic transport

Nuclear pore complexes mediate both active and passive exchange of material between nucleus and cytoplasm and are roughly cylindrical structures with eight internal

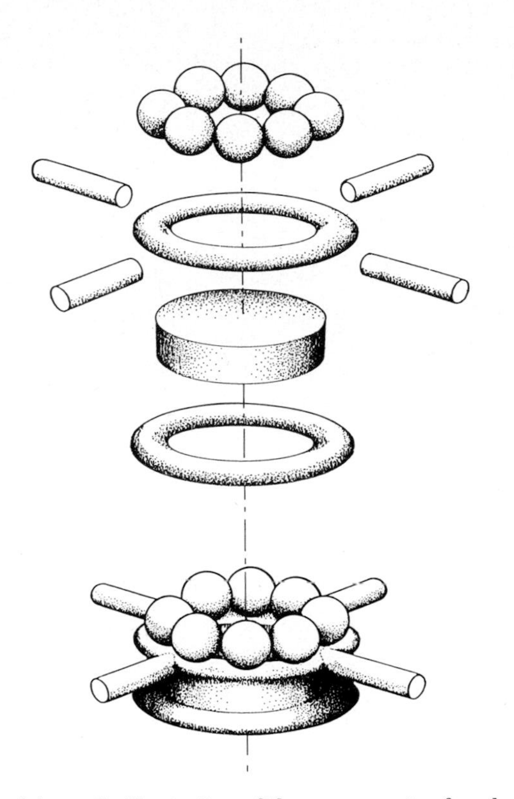

Fig. 3. Schematic illustration of the components of nuclear pores. Nucleoplasmic and cytoplasmic rings are linked to a central cylinder. These components can be seen most easily in disrupted nuclear envelope preparations such as that in Fig. 4. Reproduced from Stewart *et al.* (1990).

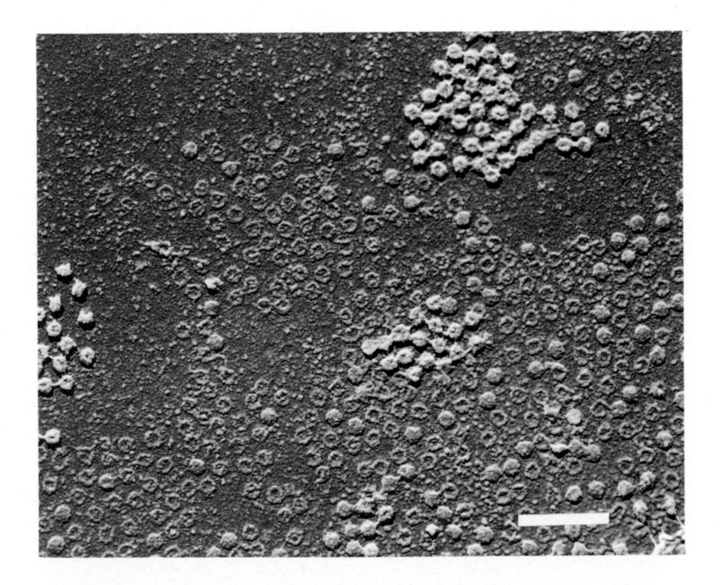

Fig. 4. Shadowed preparation of a disrupted *Xenopus* nuclear envelope showing the rings and other pore components, together with some islands of pores shed from the envelope. Bar, 0.5 μm.

spoke-like units arranged between cytoplasmic and nucleoplasmic rings (see Akey, 1989; Reichelt *et al.* 1990; Unwin and Milligan, 1982). Fig. 3 illustrates the major structural components of nuclear pore complexes, which were most easily seen in partially disrupted envelopes produced by low ionic strength (Fig. 4). Here islands of pores, often still connected by pore-connecting fibrils, were seen frequently, together with pores in various degrees of disintegration.

We used nucleoplasmin to investigate transport. Colloidal gold conjugated with nucleoplasmin (Au-NP) microinjected into *Xenopus* oocytes accumulates in the nucleus over a period of hours, and sections of embedded material show many pores with colloidal gold markers located on their axis (Felhderr *et al.* 1984). When envelopes from nucleoplasmin-microinjected oocytes were disrupted, a substantial number of rings had centrally-located gold particles. When Au-NP was applied to envelopes after isolation, a clear pattern of decoration was not seen. This was consistent with studies that have shown that additional factors from the cell cytoplasm are required for nuclear transport to be effected *in vitro* (Newport and Forbes, 1987). When BSA-conjugated gold was microinjected instead of nucleoplasmin, the density of gold particles attached to isolated nuclear envelopes was greatly reduced.

The observation of nucleoplasmin labelling of rings indicated that at least part of the active transport mechanism resided in this morphological component. However, we think it is likely that additional components located in the central body of the pores are also involved in active transport. The mechanism of transport through nuclear pores is not understood in detail and probably involves a number of steps and interactions with different pore components. Substantial structural changes appear to accompany transport (Akey and Goldfarb, 1989) and molecular motors may well be involved in this process. Molecular motors may also participate in the translocation of material, which has to move about 50 nm to traverse the entire nuclear envelope. Although clearly more work will be required to establish in detail the steps involved in nucleocytoplasmic transport, the labelling of rings we have observed here provides evidence for the location of some components of the transport machinery in the ring components at the pore periphery.

We thank Patrick Sadler for artwork. RDM was supported by the Alberta Heritage Foundation for Medical Research.

References

Aebi, U., Cohn, J., Buhle, L. and Gerace, L. (1986). The nuclear lamina is a meshwork of intermediate-type filaments. *Nature* **323**, 560–564.

Akey, C. W. (1989). Interactions and structure of the nuclear pore complex revealed by cryo-electron microscopy. *J. Cell Biol.* **109**, 955–970.

Akey, C. W. and Goldfarb, D. S. (1989). Protein import through the nuclear pore complex is a multistage process. *J. Cell Biol.* **109**, 971–982.

Feldherr, C. M., Kallenback, E. and Schuyltz, N. (1984). Movement of a karyophilic protein through the nuclear pores of oocytes. *J Cell Biol.* **99**, 2216–2222.

Gerace, L. (1986). Nuclear lamina and organisation of nuclear architecture. *Trends Biochem. Sci.* **11**, 443–446.

Gerace, L. and Blobel, G. (1980). The nuclear envelope lamina is reversibly depolymerised during mitosis. *Cell* **19**, 277–287.

Krohne, G. and Benevente, R. (1986). The nuclear lamins. A multigene family of proteins in evolution and differentiation. *Expl Cell Res.* **162**, 1–10.

Lehner, C. F., Stick, R., Eppenberger, H. M. and Nigg, E. A. (1987). Differential expression of nuclear lamin proteins during chick development. *J. Cell Biol.* **105**, 577–587.

Maul, G. G. (1977). The nuclear and cytoplasmic pore complexes: structure, dynamics, distribution and evolution. *Int. Rev. Cytol. Suppl.* **6**, 75–186.

McKeon, F. D., Kirschner, M. W. and Caput, D. (1986). Homologies in both primary and secondary structure between nuclear envelope and intermediate filament proteins. *Nature* **319**, 463–468.

Moir, R. D., Quinlan, R. A. and Stewart, M. (1990). Expression and characterization of human lamin C. *FEBS Lett.* **268**, 301–305.
Newport, J. W. and Forbes, D. J. (1987). The nucleus: structure, function and dynamics. *A. Rev. Biochem.* **56**, 535–565.
Peter, M., Nakagawa, J., Doree, M., Labbe, J. C. and Nigg, E. A. (1990). *In vitro* disassembly of the nuclear lamina and M phase-specific phosphorylation of lamins by cdc2 kinase. *Cell* **61**, 591–602.
Reichelt, R., Holzenburg, A., Buhle, E. L., Jarnik, M., Engel, A. and Aebi, U. (1990). Correlation between structure and mass distribution of the nuclear pore complex and of distinct pore components. *J. Cell Biol.* **110**, 883–894.
Stewart, M. (1990). Intermediate filaments: structure, assembly and molecular interactions. *Curr. Opinion Cell Biol.* **2**, 91–100.
Stewart, C. and Burke, B. (1987). Teratocarcinoma stem cells and early mouse embryos contain only a single major lamin polypeptide closely resembling lamin B. *Cell* **51**, 383–392.
Stewart, M. and Whytock, S. (1988). The structure and interactions of components of nuclear envelopes of *Xenopus* oocyte germinal vesicles observed by heavy metal shadowing. *J. Cell Sci.* **90**, 409–423.
Stewart, M., Whytock, S. and Mills, A. D. (1990). Association of gold-labelled nucleoplasmin with the centres of ring components of *Xenopus* oocyte nuclear pore complexes. *J. molec. Biol.* **213**, 575–582.
Unwin, P. N. T. and Milligan, R. A. (1982). A large particle associated with the perimeter of the nuclear pore complex. *J. Cell Biol.* **93**, 63–75.
Whytock, S., Moir, R. D. and Stewart, M. (1990). Selective digestion of nuclear envelopes from *Xenopus* oocyte germinal vesicles: possible structural role for the nuclear lamina. *J. Cell Sci.* **97**, 571–580.

Bidirectional movement of actin filaments along tracks of heavy meromyosin and native thick filaments

YOKO YANO TOYOSHIMA

Department of Biology, Ochanomizu University, Tokyo 112, Japan

Summary

Flexibility of the myosin molecule was studied by an *in vitro* motility assay in terms of the direction of actin movement. Actin filaments can move in both directions on tracks of heavy meromyosin made on a nitrocellulose surface, and, furthermore, along the native thick filaments passing over their central bare zone. These observations indicate that the myosin molecule has a considerable flexibility in interacting with actin filaments.

Key words: motility assay, actin, myosin, flexibility, polarity.

The recent development of *in vitro* motility assay systems has revealed that the polarity of actin filaments is important in determining the direction of actin–myosin sliding movement. We have developed an *in vitro* motility assay system for actin and myosin fragments (Kron *et al.* 1990). This assay system uses a flow cell consisting of a nitrocellulose film cast on a coverslip. Myosin fragments are put on the nitrocellulose film and the movement of rhodamine–phalloidin labelled actin filaments are observed by fluorescence microscopy. In the first series of experiments, we demonstrated that S1 attached to nitrocellulose film can support the movement of actin filaments (Toyoshima *et al.* 1987).

In this assay system, there may be a carpet of myosin fragments or a lawn of them on the nitrocellulose film. Actin filaments move smoothly and nearly always continuously at a constant speed. Actin filaments follow winding paths and frequently change direction, but never move backward reversing the polarity of the movement. Apparently, the direction of movement is determined by the polarity of the actin filament. Since the movement is really smooth even with S1, there must be considerable flexibility in myosin heads.

To see the rigor binding of heavy meromyosin (HMM) to actin filaments, the surface was examined by electron microscopy. Fig. 1A shows a negatively stained image of HMM-bound surface. This field corresponds to about 0.25 μm^2, and from the density estimate about 250 HMM molecules are in it. We can see arrowhead-like patterns along actin filaments. Thus, HMM can form rigor complexes even though they were scattered and tethered to the surface.

A similar surface was examined by freeze-etching and rotary-shadowing (Fig. 1B). HMM heads appeared to bind to actin filaments from the top as well as from the bottom. Nothing special was apparently caused by tethering HMM to the surface. These observations suggest that HMM heads on the nitrocellulose surface can move to bind actin filaments with considerable freedom.

The next set of experiments was to investigate the movement of decorated actin filaments (Toyoshima *et al.* 1989). First, the decorated filaments were formed in a test tube by mixing stoichiometric amounts (head to actin monomer) of HMM and fluorescent actin. By placing decorated filaments on the nitrocellulose film we were able to make tracks of HMM. Since the HMM binds to actin filaments in a specific configuration, polarity of myosin heads must be the same. Then BSA solution was applied to block the free surface from nonspecific binding of actin filaments. Finally, ATP was added to start the movement.

On adding ATP, actin filaments moved continuously following the same path with one filament length, then dissociated from the surface when the trailing end of the filament passed the opposite end of the track. This was expected because the viscosity of the solution was very high and thus the inertia could be neglected.

After initial actin filaments have gone, HMMs are left on the nitrocellulose surface forming tracks. We have confirmed that the tracks are well preserved by examining the same field with a fluorescence microscope (to see the initial location) and with an electron microscope using immuno-gold labelling (to see the tracks).

Patterns of actin movement observed on the tracks are summarized in Fig. 2. After initial actin filaments have moved and dissociated, other filaments released from HMM tracks elsewhere in the flow cell came to the tracks

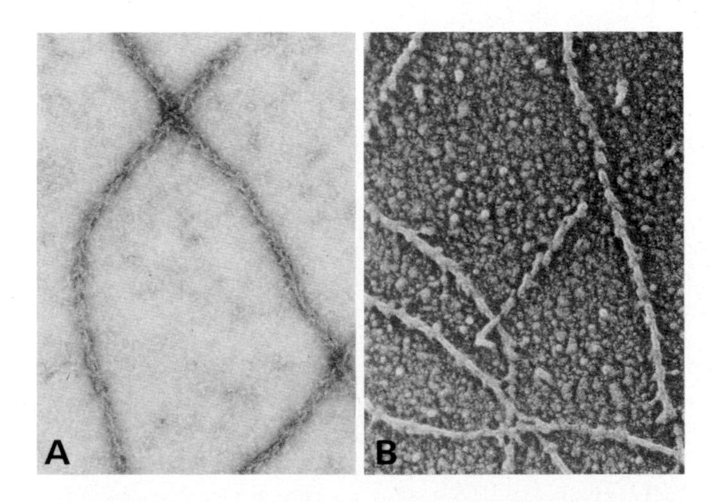

Fig. 1. Electron micrographs of actin filaments on HMM lawn. (A) Negatively stained with uranyl acetate. (B) Freeze-etched and rotary shadowed.

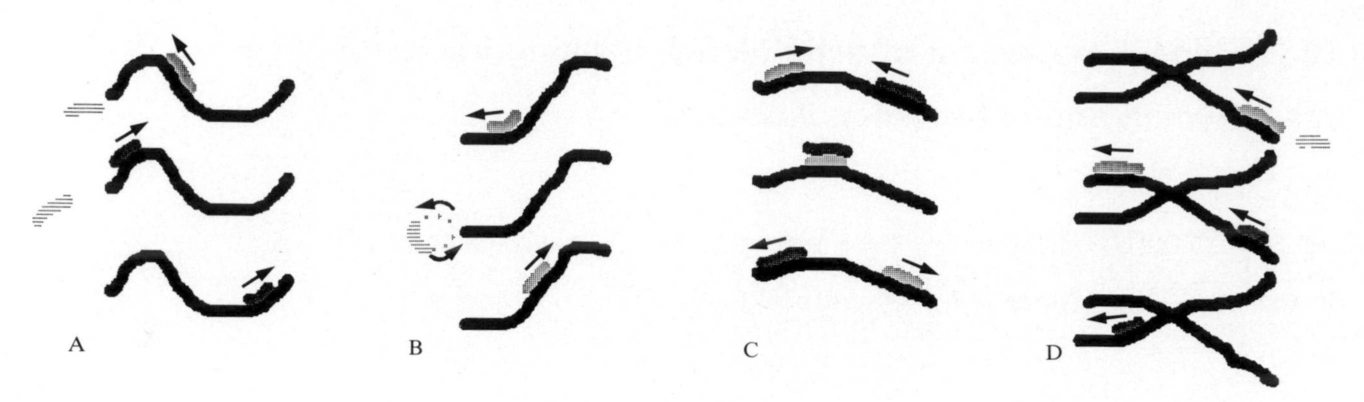

Fig. 2. Various patterns of actin movement observed on the tracks of HMM.

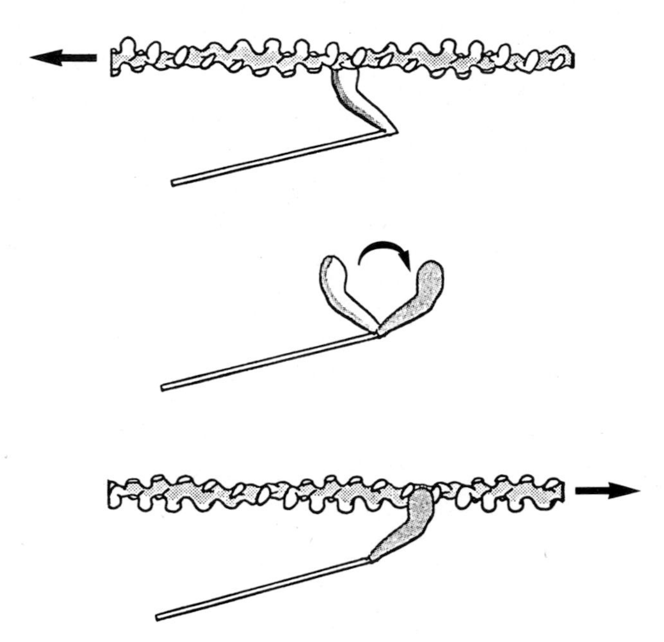

Fig. 3. A cartoon showing how HMM can support the movement of actin filaments in both directions.

and followed the same tracks very accurately (Fig. 2). Other filaments came to the track from different directions and followed the same path toward the other end (Fig. 2A). Sometimes, after having moved and dissociated at the end of the track, actin filaments flipped in solution and returned back to the other end (Fig. 2B). Further, two or more filaments came to the same track from different directions and passed over each other (Fig. 2C).

These observations suggest that HMMs are tethered to the nitrocellulose with their S2 portion, and their heads (S1 portion) can rotate freely to support the movement of actin in both directions (Fig. 3).

There are several pieces of evidence that the myosin head can rotate freely around its axis. For example, two heads of HMM can bind to consecutive actin monomers located on the same strand (Craig *et al.* 1980), showing that the two heads are translationally related in this bound configuration. Further, a single IgG molecule cross-links two heads of myosin (Winkelmann and Lowey, 1986; Miyanishi *et al.* 1988), showing that the two heads can be arranged so that the same epitopes on the two heads face each other. All this means that the myosin head can rotate around its axis more than 180 degrees, and suggests that myosin heads rotate very rapidly to find the correct binding site on the actin molecule.

From experiments to measure tension development at various sarcomere lengths (Gordon *et al.* 1966), it has been suggested that myosin heads produce force so that thin filaments slide to the center of the sarcomere, and that thin filaments cannot interact productively with another half of myosin heads over the central zone. The question arises whether actin filaments can move along isolated thick filaments up to the center or over the center to the other end.

To answer this question, we isolated long native thick filaments; long enough to be observed in a light microscope. Crustacean skeletal muscles are known to have long sarcomeres and long thick filaments. As material, we used crayfish claw muscles which have sarcomere lengths of about 10 μm. To isolate thick filaments, claw muscle is homogenized gently and centrifuged at a low speed. The supernatant contains isolated thick filaments.

Electron microscopy showed many narrow thick filaments of about 5 to 6 μm in length. They have the central bare zone of 0.2 μm. The thickness of the isolated thick filaments is about 25 nm and comparable to that of microtubules. Thus, the isolated native thick filaments can be visualized under dark field illumination as has been shown for microtubules (Miki-Noumura and Kamiya, 1976).

When applied to glass slides, fluorescent actin filaments as well as thick filaments are visible under dark field illumination. As the rhodamine–phalloidin labelled actin filaments emit red light, we can distinguish actin filaments from thick filaments by eye. Unfortunately, the SIT camera is color blind, everything becomes black and white. Nevertheless, the intensity of the portion where an actin filament is moving along the filament becomes higher; thus it is possible to extract the image of actin filaments by subtracting the image when no actin filament is moving as the reference. After storing sequences of images using a frame grabber, the computer painted actin and myosin filaments with different colors.

On the native thick filaments, actin filaments moved from one end to the other end passing over the central bare zone. No difference in velocity was found between the movements towards the center (regular orientation) and from the center (so-called wrong orientation). Although the velocity is the same in both orientations, we do not know if myosin can produce the same amount of force in both directions.

In conclusion, the direction of the movement is determined by the polarity of actin filaments and myosin has a large flexibility. It is marvelous that myosin can produce this large force in spite of its large flexibility. This flexibility must be taken into account when exploring the cross-bridge mechanism.

References

Craig, R., Szent-Györgyi, A. G., Beese, L., Flicker, P., Vibert, P. and Cohen, C. (1980). Electron microscopy of thin filaments decorated with a Ca^{2+}-regulated myosin. *J. molec. Biol.* **140**, 35–55.

Gordon, A. M., Huxley, A. F. and Julian, F. J. (1966). The variation in isometric tension with sarcomere length in vertebrate muscle fibers. *J. Physiol.* **184**, 170–192.

Kron, S. J., Toyoshima, Y. Y., Uyeda, T. Q. P. and Spudich, J. A. (1990). Assays for actin sliding movement over myosin coated surfaces. *Meth. Enzymol.* (in press).

Miki-Noumura, T. and Kamiya, R. (1976). Shape of microtubules in solutions. *Expl Cell Res.* **97**, 451–453.

Miyanishi, T., Toyoshima, C., Wakabayashi, T. and Matsuda, G. (1988). Electron microscopic study on the location of 23K and 50K fragments in skeletal myosin head. *J. Biochem.* **103**, 458–462.

Toyoshima, Y. Y., Kron, S. J., McNally, E. M., Niebling, K. R., Toyoshima, C. and Spudich, J. A. (1987). Myosin subfragment-1 is sufficient to move actin filaments *in vitro*. *Nature* **328**, 536–539.

Toyoshima, Y. Y., Toyoshima, C. and Spudich, J. A. (1989). Bidirectional movement of actin filaments along tracks of myosin heads. *Nature* **341**, 154–156.

Winkelmann, D. A. and Lowey, S. J. (1986). Probing myosin head structure with monoclonal antibodies. *J. molec. Biol.* **188**, 595–612.

Regulation of the interaction between smooth muscle myosin and actin

KATHLEEN M. TRYBUS

Rosenstiel Research Center, Brandeis University, Waltham, MA 02254, USA

and DAVID M. WARSHAW

Department of Physiology and Biophysics, University of Vermont, College of Medicine, Burlington, VT 05405, USA

Summary

Phosphorylation of the regulatory light chain of smooth muscle myosin efficiently regulates the actin-activated ATPase activity of myosin filaments in solution and actin movement in an *in vitro* motility assay, independently of thin-filament regulatory proteins. Filaments containing both phosphorylated and dephosphorylated heads move actin at intermediate rates, depending on the relative proportions of the two myosin species. The decrease in velocity can be accounted for by mechanical interactions between phosphorylated heads and 'weak-binding' dephosphorylated crossbridges. These results imply that shortening velocity could be modulated in any muscle by varying the relative proportions of two populations of crossbridges with different cycling rates.

Key words: smooth muscle myosin, light chain phosphorylation, motility assay.

Introduction

Phosphorylation of serine 19 of the $20\times10^3 M_r$ regulatory light chain by myosin light chain kinase (MLCK) is a key event regulating smooth muscle actomyosin interactions and force development. The calcium sensitivity of smooth muscle contraction arises because MLCK alone is inactive, while the calcium–calmodulin–MLCK complex is active. Although regulation of smooth muscle myosin is considered to be primarily myosin-based, thin-filament associated proteins such as caldesmon and calponin have recently been given considerable attention as putative regulatory proteins (Fig. 1). Caldesmon, an $87\times10^3 M_r$ 74 nm long protein that binds actin, tropomyosin, calmodulin and myosin, inhibits actin-activated ATPase activity. It has been suggested that caldesmon sterically blocks binding of the weak-binding M.ATP and M.ADP.P_i complexes to actin (Hemric and Chalovich, 1988). Calponin, a $35\times10^3 M_r$ calcium-binding, calmodulin-binding, troponin-T like protein also inhibits actin-activated ATPase activity (Takahashi *et al.* 1988). Recent evidence suggests that phosphorylation of calponin reverses the inhibition of activity (Winder and Walsh, 1990).

Phosphorylation regulates the activity of filamentous myosin

The presence of inhibitory actin-binding proteins raises the question of whether phosphorylation–dephosphorylation alone can regulate the activity of myosin filaments to a high degree, or whether a second regulatory system is required to depress the activity of dephosphorylated filaments. In solution, it has been difficult to determine the activity of dephosphorylated filaments because of the lability of this structure in the presence of MgATP. Dephosphorylated filaments readily dissociate to a monomer in which the tail is folded into thirds and the heads are bent toward the rod. Single-turnover measurements, in which each active site is given only one ATP molecule, established that this conformation traps the products of ATP hydrolysis, and releases them at the very low rate of $0.0005\,s^{-1}$ (Cross *et al.* 1986). Phosphorylation of the regulatory light chain causes reassembly into filaments and a several hundred-fold increase in actin-activated activity to $0.3\,s^{-1}$ (Fig. 2). It was not clear from these observations, however, whether conformation or phosphorylation was the major determinant of activity. By analogy with other myosins, the regulatory light chain is located near the head/rod junction, at least 10 nm from the active site; thus even a 'direct' effect of phosphorylation implies communication over distances equal to half the length of the myosin head. In smooth muscle, the $17\times10^3 M_r$ light chain has been shown by photoaffinity labelling to be near the active site (Okamoto *et al.* 1986), and this subunit might be the link between the two regions.

The extent to which phosphorylation dependent assembly–disassembly occurs *in vivo* has not been well established, although in at least one type of smooth muscle a change in filament density between rest and contraction has been observed (Gillis *et al.* 1988). Nevertheless, it has been established that relaxed smooth muscle cells contain some dephosphorylated filaments (Somlyo *et al.* 1981), and a mechanism to inhibit their activity must exist. If dephosphorylation is not sufficient, then thin-filament proteins may be required to inhibit their activity.

In order to block disassembly of dephosphorylated filaments in the presence of MgATP, monoclonal antibodies with epitopes along the central portion of the myosin were used to stabilize the polymer, without affecting the activity of phosphorylated filaments. The actin-activated ATPase of antibody-stabilized, dephosphorylated filaments, obtained by steady-state and single-turnover measurements, was approximately $0.002\,s^{-1}$, well below the rate obtained with phosphorylated filaments, and only 4–5 fold higher than the value obtained with folded myosin (Trybus, 1989). These results suggest that three levels of activity can be distinguished: (1) enzymatically-incompetent folded myosin, (2) inactive dephosphorylated filaments and (3) active phosphorylated filaments. Interaction of the myosin rod with the neck in the folded conformation appears to stabilize ADP and phosphate at the active site to a degree that cannot quite be achieved in the filament.

Journal of Cell Science, Supplement 14, 87–89 (1991)

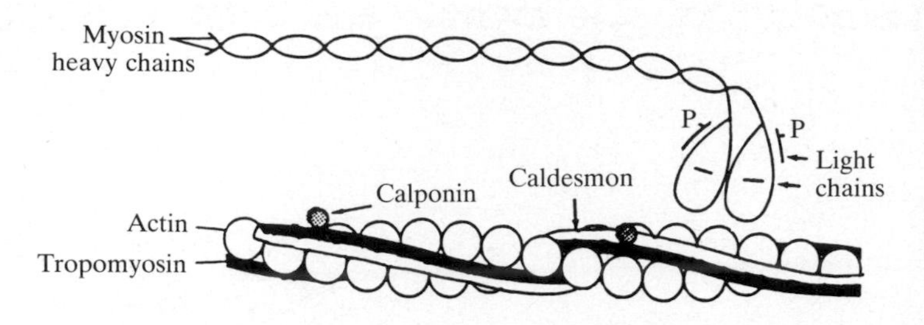

Fig. 1. Schematic diagram of smooth muscle myosin regulatory components.

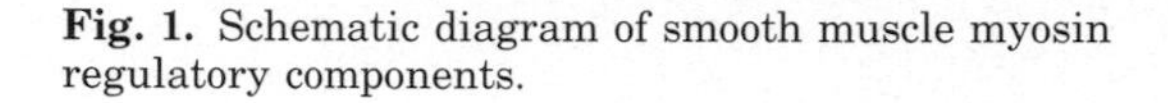
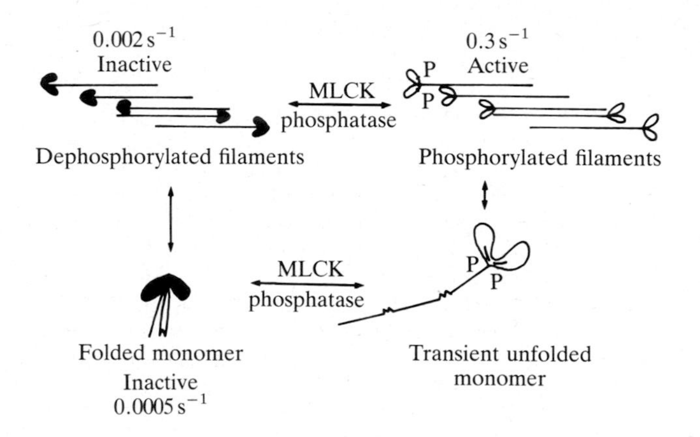

Fig. 2. Conformational states adopted by smooth muscle myosin, and their actin-activated ATPase rates.

Based on these experiments, a second regulatory system is not required to inhibit the enzymatic activity of dephosphorylated filaments. Phosphorylation, however, acts predominantly on a kinetic step affecting product release, and has only minor effects on binding of M.ATP and M.ADP.P_i to actin (Sellers, 1985). The role of caldesmon might then simply be to inhibit binding of myosin to actin in a relaxed muscle, although cycling between weak and strong binding states could be adequately regulated by phosphorylation. If filament disassembly to the folded conformation occurred, it would only enhance an already efficient regulatory system. It would also produce a soluble species that does not bind actin, and which in principle could be recruited to different areas of the cell more readily than filamentous myosin.

Dephosphorylated myosin slows movement of actin by phosphorylated myosin

Actin-activated ATPases in solution showed a large difference in ATP turnover by phosphorylated and dephosphorylated myosin, but is this factor enough to control movement of actin by myosin? The degree to which phosphorylation regulates the motion of single actin filaments by myosin was observed by an *in vitro* motility assay. Phosphorylated filaments moved actin at 0.2–0.4 $\mu m\ s^{-1}$ (at 22°C), but dephosphorylated filaments held actin in an immobile, rigor-like conformation even in the presence of MgATP (Warshaw *et al.* 1990). The surprising aspect of this observation was that dephosphorylated myosin, which from solution studies is predominantly in the 'weak-binding' AM.ADP.P_i state, maintained strong enough interactions with actin that the polymer did not diffuse away into solution (Warshaw *et al.* 1990).

During a contraction there are many times when myosin is not fully dephosphorylated or phosphorylated, but only partially phosphorylated. As calcium levels increase and myosin gets phosphorylated, the muscle begins to rapidly shorten and produce force (Fig. 3). With time, although force levels remain high, the muscle shortens more slowly, and levels of light chain phosphorylation decrease as phosphatase activity predominates over kinase activity. This apparent modulation of shortening velocity by phosphorylation implies that phosphorylation–dephosphorylation may be more than a simple 'on–off' switch. Shortening velocity is independent of the number of phosphorylated heads that move actin, so changes in velocity cannot be simply explained by changes in the number of active crossbridges.

Two ways that intermediate shortening velocities could be obtained are: (1) dephosphorylated bridges act as a load against which the phosphorylated bridges must work, thus slowing their velocity, or (2) the intrinsic rate of ATP

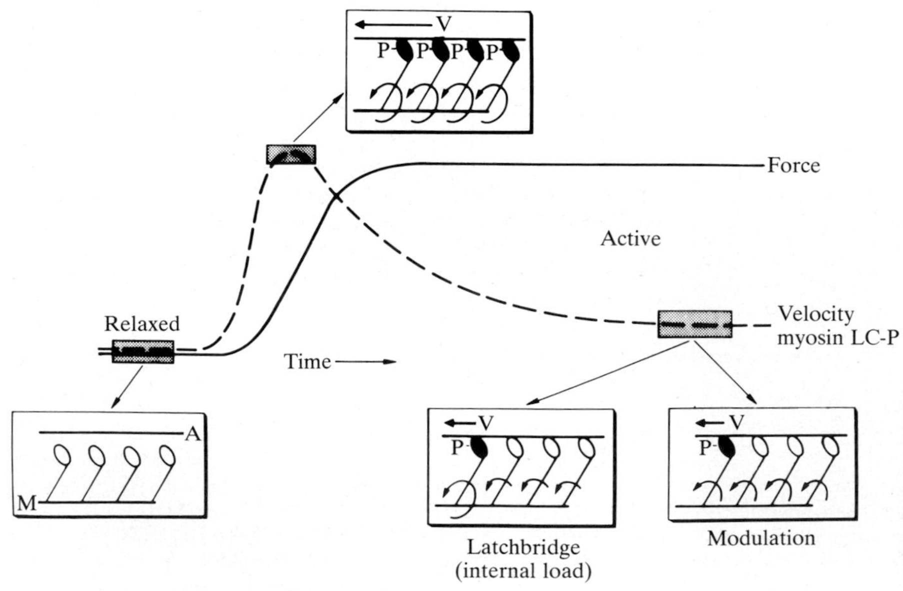

Fig. 3. Schematic diagram of the response of a smooth muscle to calcium. Dashed line indicates calcium concentration, degree of light chain phosphorylation and shortening velocity, while the solid line shows the level of force production. Curved arrows represent the crossbridge cycling rate.

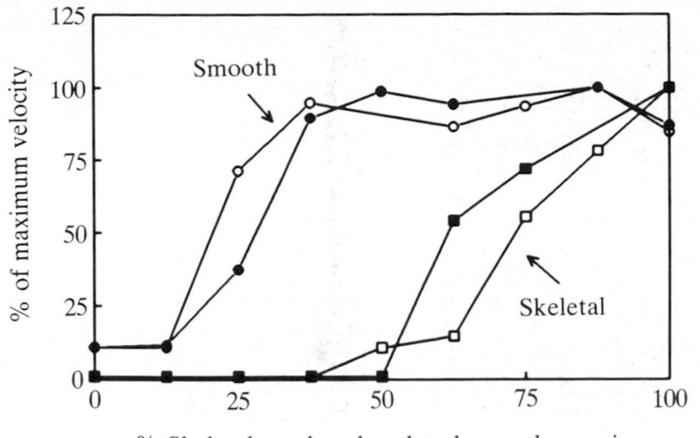

Fig. 4. Effect of 'weak-binding' bridges on actin movement. Dephosphorylated smooth muscle myosin slows actin movement by phosphorylated smooth muscle myosin (filled circles) and skeletal muscle myosin (filled boxes). The effect of pPDM–skeletal myosin, a non-cycling analog of the 'weak-binding' conformation, on phosphorylated smooth (open circles) or skeletal myosin (open boxes) is very similar to that obtained with dephosphorylated smooth muscle myosin. Conditions: 25 mM KCl, 4 mM $MgCl_2$, 1 mM EGTA, 1 mM DTT, pH 7.4, 22°C.

turnover by phosphorylated myosin is decreased as the number of neighboring dephosphorylated molecules in the filament increases. The motility assay is considered to be the *in vitro* correlate of unloaded shortening velocity, thus this technique in combination with solution studies should allow one to distinguish between the above possibilities.

The rate at which actin was moved by phosphorylated smooth muscle myosin slowed markedly as the proportion of dephosphorylated myosin in the filament increased to greater than 60% of the total (Fig. 4, 25 mM KCl). Intrinsic changes in ATPase activity were ruled out both by solution actin-activated ATPases, and by the observation that mixtures of monomers resulted in changes in velocity similar to that observed for copolymers (unpublished data, D. M. Warshaw and K. M. Trybus). Surprisingly, dephosphorylated myosin had an even more profound effect on the movement of actin by skeletal muscle myosin (Fig. 4), suggesting that under these conditions skeletal myosin has relatively fewer crossbridges in a 'strong-binding' state compared to smooth muscle myosin.

Another way of observing the effect of 'weak-binding' bridges on movement is to use the well-characterized chemically modified analog, pPDM-skeletal myosin, which is thought to resemble a non-cycling M.ADP.P_i conformation. This analog behaved mechanically in the same way as dephosphorylated myosin, based on its effect on actin movement by phosphorylated smooth and skeletal muscle myosin (Fig. 4). In contrast, skeletal myosin extensively modified with NEM, which is an analog of a 'strong-binding' conformation, completely abolished movement when present as only 1% of the total myosin.

Modulation of actin velocity by active crossbridges with different cycling rates

Actin velocity was not only modulated when copolymers contained 'weak-binding' bridges, but could also be observed when filaments were formed from two populations of active bridges that cycle at different rates, such as phosphorylated smooth muscle myosin and skeletal myosin. The observed velocity of actin movement by smooth–skeletal copolymers was not a simple linear relationship that depended only on the proportion of fast and slow myosin. A nonlinear relationship, reflecting a mechanical interaction between the two myosin species, could arise if the crossbridge force–velocity relationships for the two species are assumed to be hyperbolic and different in curvature (Warshaw *et al.* 1990). Mixtures of smooth and skeletal myosin have no counterpart in the cell, but these results could be generally applied to muscles where multiple myosin isoforms with different turnover times are expressed, such as in developing muscles or in the heart.

Future prospects

The observations described here could explain why shortening velocity correlates with the degree of light chain phosphorylation, but they do not offer any information as to why a smooth muscle can maintain force at low phosphorylation levels (Fig. 3). The latchbridge hypothesis of Murphy and co-workers (Hai and Murphy, 1989) suggests that phosphorylation is required for the myosin head to attach to actin, but once attached, it can be dephosphorylated by phosphatases and become a 'latchbridge'. This slowly detaching, force-producing, dephosphorylated latchbridge allows force to remain high while phosphorylation levels decrease. The experiments described here analyze the properties of dephosphorylated myosin that has not gone through the pathway required of a latchbridge (i.e phosphorylation, attachment to actin, and subsequent dephosphorylation), yet these dephosphorylated bridges are having profound effects on the observed movement of actin by phosphorylated myosin. Direct measurement of the force produced by dephosphorylated and phosphorylated crossbridges interacting with a single actin filament (Kishino and Yanagida, 1988) should help resolve some of these questions.

References

Cross, R. A., Cross, K. E. and Sobieszek, A. (1986). ATP-linked monomer–polymer equilibrium of smooth muscle myosin: the free folded monomer traps ADP.P_i. *EMBO J.* **5**, 2637–2641.

Gillis, J. M., Cao, M. L. and Godfraind-DeBecker, A. (1988). Density of myosin filaments in the rat anococcygeus muscle, at rest and in contraction. II. *J. Muscle Res. Cell Motil.* **9**, 18–28.

Hai, C. and Murphy, R. A. (1989). Ca^{2+}, crossbridge phosphorylation and contraction. *A. Rev. Physiol.* **51**, 285–298.

Hemric, M. E. and Chalovich, J. M. (1988). Effect of caldesmon on the ATPase activity and the binding of smooth and skeletal myosin subfragments to actin. *J. biol. Chem.* **263**, 1878–1885.

Kishino, A. and Yanagida, T. (1988). Force measurements by micromanipulation of a single actin filament by glass needles. *Nature* **334**, 74–76.

Okamoto, Y., Sekine, T., Grammer, J. and Yount, R. G. (1986). The essential light chains constitute part of the active site of smooth muscle myosin. *Nature* **324**, 78–80.

Sellers, J. R. (1985). Mechanism of the phosphorylation dependent regulation of smooth muscle heavy meromyosin. *J. biol. Chem.* **260**, 15 815–15819.

Somlyo, A. V., Butler, T. M., Bond, M. and Somlyo, A. P. (1981). Myosin filaments have non-phosphorylated light chains in relaxed smooth muscle. *Nature* **294**, 567–569.

Takahashi, K., Hiwada, K. and Kokubu, T. (1988). Vascular smooth muscle calponin. A novel troponin T-like protein. *Hypertension* **11**, 620–626.

Trybus, K. M. (1989). Filamentous smooth muscle myosin is regulated by phosphorylation. *J. Cell Biol.* **109**, 2887–2894.

Warshaw, D. M., Desrosiers, J. M., Work, S. S. and Trybus, K. M. (1990). Smooth muscle myosin cross-bridge interactions modulate actin filament sliding velocity *in vitro*. *J. Cell Biol.* **111**, 453–463.

Winder, S. J. and Walsh, M. P. (1990). Smooth muscle calponin. Inhibition of actomyosin MgATPase and regulation by phosphorylation. *J. biol. Chem.* **265**, 10 148–10 155.

Molecular biology of actin binding proteins: evidence for a common structural domain in the F-actin binding sites of gelsolin and α-actinin

MICHAEL WAY, BRIAN POPE and ALAN WEEDS

MRC Laboratory of Molecular Biology, Hills Road, Cambridge CB2 2QH, UK

Summary

We review the impact of molecular biology on actin binding proteins, in particular on sequence relationships and expression of clones to dissect properties *in vitro*. Significant homologies exist between proteins in each class, but we propose, in addition, that common structural features exist between the F-actin binding sites of severing and cross-linking proteins.

Key words: actin binding proteins, α-actinin, gelsolin, expression, homologies.

Introduction

An ever increasing number of actin binding proteins have been purified from a wide variety of eukaryotic cells and tissues. They have been classified according to their effects on actin *in vitro* (Bennett and Weeds, 1986). However, particular actin binding properties, e.g. cross-linking, are not unique to single proteins. Different proteins co-existing in the same cell are capable of showing a similar phenotype. Futhermore, functional analogues of individual proteins appear to be universally distributed in eukaryotes.

Understanding the structural relationship between polymorphic forms of actin binding proteins of very different sizes has progressed at a furious rate due to the impact of molecular biology. This reflects improvements in techniques, including protein sequencing, cDNA library construction and screening by PCR, as well as more general cloning methods; all of these facilitate easier identification and analysis. Often the hardest part is finding the right cDNA library in the first place! It is impossible in this review to describe the almost endless list of proteins that have now been cloned. Here we restrict our discussion to the actin severing and capping proteins typified by gelsolin and the cross-linking proteins typified by α-actinin, which we regard as paradigms for others.

Primary sequence relationships

Actin severing and capping proteins

The properties and limited information obtained from protein sequencing of actin severing and capping proteins suggested that they would all be related. The cDNA sequence of gelsolin indicated that the N- and C-terminal halves evolved by gene duplication (Kwiatkowski *et al.* 1986). Furthermore the sequences of severin and fragmin are more closely related to the N-terminal half of gelsolin than its C terminus (Ampe and Vandekerckhove, 1987; André *et al.* 1988). This, together with their similar properties, suggests that fragmin and severin may be ancestral prototypes (Yin *et al.* 1990). Villin has an almost identical sequence to gelsolin but contains an additional actin binding 'head-piece' at its C terminus, which accounts for its unique cross-linking activity (Glenney *et al.* 1981). More detailed sequence analysis suggests that all these proteins have a more complex evolution from an ancestral actin binding protein of about 15 000 M_r, based on a weaker six-fold segmental repeat in gelsolin and a corresponding three-fold repeat in the smaller homologues (Way and Weeds, 1988; Bazari *et al.* 1988).

The three actin binding domains in gelsolin identified by limited proteolysis (Bryan, 1988) are distributed unevenly within the six segmental repeat (S1–6). Segment 1 contains a strong calcium-independent monomer binding site. A second monomer binding site exists in S4–6, which differs from S1 in showing much weaker affinity and strict calcium dependence. The third binding site is located in S2–3 and it is unique in being specific for filamentous actin. We currently do not understand the relationship between the actin binding domains and the six-fold repeats.

Multiple alignments of all the sequences show that the greatest homology is found in segment 1, the smallest actin binding domain. Two other classes of monomer binding proteins of similar size have been described: the profilins, which sequester monomers and show weak capping activity (Pollard and Cooper, 1984) and a class of severing proteins including actin depolymerizing factor, cofilin and destrin (reviewed by Vandekerckhove, 1990). Interestingly, there are regions of sequence homology between these two classes and segment 1 (reviewed by Vandekerckhove, 1989). While it may be tempting to speculate that all these proteins have evolved from a more primitive prototype, functional similarity of proteins may arise by convergent rather than divergent evolution. It is important therefore to assess the significance of these sequences in relation to actin binding activity (see later).

Actin cross-linking proteins

Proteolysis of α-actinin has identified an N-terminal, 27 000 M_r, F-actin binding domain (Mimura and Asano, 1987). This shows no obvious sequence relationships to S23 of gelsolin, but it is highly homologous to a domain in a number of other actin cross-linking proteins, including fimbrin, filamin, spectrin, the 120×10^3 M_r gelation factor of *Dictyostelium discoideum* and dystrophin (Blanchard *et al.* 1989; de Arruda *et al.* 1990), though actin binding by

Journal of Cell Science, Supplement 14, 91–94 (1991)

dystrophin has yet to be demonstrated. Although these proteins contain a related actin binding site, they are of very different sizes and form very different supramolecular structures with F-actin. Fimbrin is the smallest: it contains two actin binding domains and bundles filaments (de Arruda *et al.* 1990). All the others contain only a single actin site; hence cross-linking requires self-association. Furthermore, the nature and rigidity of the networks formed depends on the flexibility of the molecules and their mode of self-association. Filamin and gelation factor form flexible networks, while spectrin appears to generate a more rigid and organised lattice structure, probably largely due to the small size of the actin protofilaments and the additional membrane interactions *via* ankyrin.

Within this common structure, four regions of 30–40 residues show an even greater conservation of sequence than the whole domain (de Arruda *et al.* 1990). Furthermore, comparison of these sequences suggests that the domain evolved by gene duplication of two 125-residue repeats. At least the first of these repeats contains an actin binding site based on co-sedimentation of a trypic peptide of β-spectrin (Ala47 and Lys186) with F-actin (Karinch *et al.* 1990). An even shorter 27-amino acid sequence essential for actin binding has been identified in the $120\times10^3\,M_r$ gelation factor of *Dictyostelium*. This is located near the C terminus of the first 125-residue repeat, but there is no corresponding sequence in the second (Bresnick *et al.* 1990). These relationships are shown in Fig. 1.

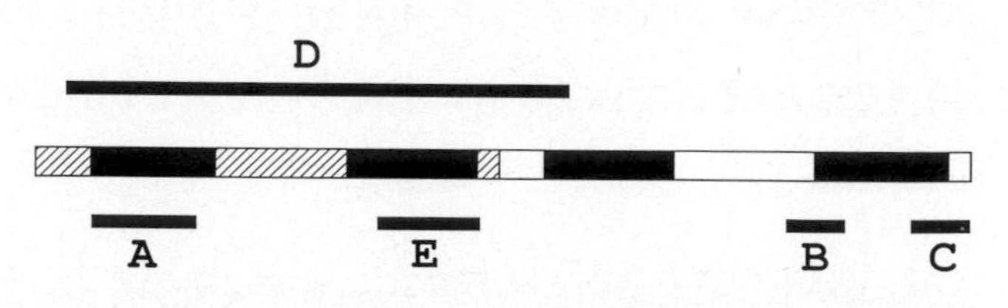

Fig. 1. Schematic representation of the actin binding site of actin cross-linking proteins. The 125-residue tandem repeat is indicated along with the four regions of 30–40 residues which identify this duplication (shaded). A,B and C correspond to the sequences common to gelsolin and α-actinin shown in Fig. 2; D is the tryptic peptide of β-spectrin which binds F-actin and E the 27-amino acid sequence essential for binding in *Dictyostelium* $120\times10^3\,M_r$ gelation factor. The first 125-residue repeat is cross-hatched.

Chemical cross-linking with zero length cross-linkers has been used to identify sites on actin for a number of actin binding proteins. In the case of α-actinin, these experiments suggested two contact sites on actin, one within the N-terminal residues 1–12 and the other between residues 86 and 119. The 3-dimensional structure of actin shows that these sequences are spatially close together in subdomain 1 (Kabsch *et al.* 1990). This domain lies at high radius in the filament and also contains the binding site for myosin as well as a number of other actin binding proteins (Holmes *et al.* 1990). It will be interesting to see whether S23 can be cross-linked to the same region of actin.

As already stated, alignments of the actin binding site of α-actinin and S23 of gelsolin show no obvious overall relationship. However, detailed comparison of the most conserved regions in the actin cross-linking domain with S23 shows evidence of limited homology between these two different classes of actin binding proteins (Fig. 2). In addition there is an overall similarity of these F-actin binding sites. Both are contained in proteolytically stable domains of about 250 residues, which can be subdivided into two repeats of 120–125 amino acids. Both bind at a 1:1 ratio to F-actin subunits and in a calcium-insensitive manner. The limited nature of the homologies might indicate evolution from a common ancestor. An alternative explanation, which we favour, is that the highly conserved nature of the interaction site in actin necessitates conserved structural features in proteins associating with the same monomer subdomain in the filament. This is analogous to the common motifs used to identify nucleotide binding sites in myosins, kinases and related proteins. This hypothesis can be tested by using chemical cross-linking to identify contact sites. It will also be interesting to find out whether the domains can be functionally exchanged between the cross-linking and severing classes of actin binding proteins. We are currently attempting to test this, using chimeras in which the F-actin binding domains between the two proteins are exchanged.

A

PGS	194	W C G S N S N R Y E R L K A - T Q V S K G I R D N E R S G R A H V H V	227
HGS	200	W C G S N S N R Y E R L K A - T Q V S K G I R D N E R S G R A H V H V	233
CSM	41	W C - - N S - H L - R - K A G T Q I E N - I E E D F R D G - L K L M L	68
CSK	50	W C - - N S - H L - R - K A G T Q I E N - I E E D E R D G - L K L M L	77

B

PGS	316	L K T A S D F I - S K - M N Y P K	330
HGS	333	L K T A S D N I - T K - M D Y P K	347
CSM	204	L N T A F D - V A E K Y L D I P K	218
CSK	215	I N L A M E - I A E K H L D I P K	229

C

PGS	354	R D P D Q V D G P G L S Y L S S	369
HGS	370	R D P D Q V D G L G L S Y L S S	385
CSM	229	R - P D E K - A I - M T Y V S S	242
CSK	241	K - P D E R - A I - M T Y V S C	255

Fig. 2. Sequence homologies in three regions (A), (B) and (C) of pig plasma gelsolin (PGS), human plasma gelsolin (HGS), chick smooth muscle α-actinin (CSM) and chick skeletal muscle α-actinin (CSK). Sequences were aligned with a minimum number of padding characters and the numbers refer to the positions in the sequences from their N-terminal ends. Vertical boxes with shading denote identities and without shading, conservative substitutions.

Expression of clones to dissect function

Cloning techniques have provided a means of producing large amounts of proteins that may exist in minute quantities in tissues or cells. The simplest and cheapest methods rely on expression in *E. coli*, but expressed proteins are frequently recovered in an insoluble form and may not easily be renatured. While we have had no problems in obtaining native gelsolin, the difficulties encountered with other proteins, including actin itself, testify to the limitations of *E. coli*. In many such cases, this difficulty can be circumvented by expression in yeast or tissue culture cells. Although gelsolin has been studied in greatest detail to date, a wide variety of actin binding proteins covering all classes have now been expressed either in their entirety or as functional domains. In each case it is important to compare the properties of an expressed protein with its natural counterpart, since renaturation *in vitro* does not necessarily recover full cellular activity.

The properties of the expressed gelsolin from both cos cells and *E. coli* are identical to those of native plasma gelsolin. Two questions have been addressed: (1) what are the functions of the three different domains in the activities of gelsolin and (2) what consititutes a minimal actin binding site? Experiments with segmentally deleted and C-terminally truncated mutants have addressed the first question (Kwiatkowski *et al.* 1989; Way *et al.* 1989). More recently we have attempted to answer the second question in respect of the smallest actin binding domain, S1, using site-directed mutagenesis (Way *et al.* 1990).

We have found that truncation of the N terminus by 11 residues had no effect on actin binding, but removal of a further 8 residues eliminated binding altogether. We cannot rule out that these 8 residues are critical for actin binding, but it is equally possible that loss of phenotype may reflect incorrect folding of this mutant. Deletions at the C terminus have a more progressive effect on actin binding and provide new information about the involvement of calcium in interaction by this domain. Removal of 19 residues gives identical actin affinity in calcium, but compared to intact S1 this mutant has a 100-fold lower affinity in EGTA. Deletion of an additional 5 residues reduces the affinity for actin in calcium by over 50-fold and there is no detectable binding in EGTA. Further truncation results in progressively weaker binding until it is eliminated when 29 residues have been removed.

The appearance of calcium-sensitive binding was unexpected, since earlier results had suggested that this domain was calcium insensitive (Kwiatkowski *et al.* 1985). However, indirect evidence based on nucleotide exchange rates suggested that actin binding may be different in calcium and EGTA (Bryan, 1988). Our calcium-binding experiments showed that the interaction between S1 and actin creates an additional calcium binding site not seen in S1 or actin individually. This calcium site probably corresponds to the calcium found 'trapped' in EGTA-stable binary complexes of gelsolin and actin (Bryan and Kurth, 1984). Thus the appearance of calcium-sensitive binding in our experiments is simply due to reduction of actin binding affinity by the mutant into a range where calcium dependence can be observed in equilibrium experiments.

It is noteworthy that recovery of calcium sensitivity was also observed by Kwiatkowski *et al.* (1989) in an analysis of severing activity by C-terminally truncated mutants of gelsolin. However, we do not believe that the severing activity observed in these heavily truncated mutants occurs by the same mechanism as that in intact gelsolin. Our experiments have shown that while our expressed mutant N160 (corresponding to PG160 (Kwiatkowski *et al.* 1989)) increases the rate of depolymerization of F-actin in a manner consistent with severing, there is no significant capping by this mutant, because the F-actin viscosity quickly recovers to near its original value (B. Pope *et al.* unpublished work). These properties are similar to the behaviour of actin depolymerizing factor and suggest a common mechanism.

The identification of a region at the C terminus of S1 has enabled us to focus on the role of invidual amino acid residues in actin binding within this area. Although over 20 such mutants have been analyzed, we have been unable to identify any residue critical for actin binding, as their properties are almost all indistinguishable from native S1. One outcome of this work has been to show that particular residues within sequences showing high homology between different classes of monomer binding proteins are not essential for actin binding (Way *et al.*, unpublished).

Probing cellular functions

Identifying binding sites and understanding the activities of actin binding proteins *in vitro* does not reveal their cellular functions. However, cDNA clones can be used to probe mRNA and, in conjunction with antibodies, protein levels in cells. In this way it may be possible to monitor the cellular distribution of individual proteins and correlate this with morphological or motile changes. It is important to assess changes in the levels of different cytoskeletal proteins simultaneously, as phenotypic changes are likely to depend on the balance of the various actin binding proteins.

Two other approaches have been used to explore the function of particular proteins which rely (1) on eliminating expression of the endogenous gene or (2) overexpression of a gene which may or may not be found in that cell type. (1) Gene disruption has been achieved either by using chemical mutagenesis (e.g. by André *et al.* 1989 for producing mutants of *Dictyostelium* deficient in severin) or by homologous recombination (e.g. by Manstein *et al.* 1989 for generating myosin II null mutants, also in *Dictyostelium*). (2) The best example of this approach to date is that of Friederich *et al.* (1989), who monitored the effects of expressing villin in fibroblasts (which do not normally contain this protein). We have recently reviewed these and other experiments on the functions of actin binding proteins *in vivo* (Way and Weeds, 1990) and it is clear that our current understanding is rudimentary.

References

AMPE, C. AND VANDEKERCKHOVE, J. (1987). The F-actin capping proteins of *Physarum polycephalum*: cap42(a) is very similar, if not identical, to fragmin and is structually and functionally very homologous to gelsolin; cap42(b) is *Physarum* actin. *EMBO J.* **6**, 4149–4157.

ANDRÉ, E., BRINK, M., GERISCH, G., ISENBERG, G., NOEGEL, A., SCHLEICHER, M., SEGALL, J. E. AND WALLRAFF, E. (1989). A *Dictyostelium* mutant deficient in severin, an F-actin fragmenting protein, shows normal motility and chemotaxis. *J. Cell Biol.* **108**, 985–995.

ANDRÉ, E., LOTTSPEICH, F., SCHLEICHER, M. AND NOEGEL, A. (1988). Severin, gelsolin, and villin share a homologous sequence in regions presumed to contain F-actin severing domains. *J. biol. Chem.* **263**, 722–728.

DEARRUDA, M. V., WATSON, S., LIN, C.-S., LEVITT, J. AND MATSUDAIRA, P. (1990). Fimbrin is a homologue of the cytoplasmic phosphoprotein

plastin and has domains homologous with calmodulin and actin gelation proteins. *J. Cell Biol.* **111**, 1069–1079.
BAZARI, W. L., MATSUDAIRA, P., WALLEK, M., SMEAL, T., JAKES, R. AND AHMED, Y. (1988). Villin sequence and peptide map identify six homologous domains. *Proc. natn. Acad. Sci. U.S.A.* **85**, 4986–4990.
BENNETT, J. AND WEEDS, A. G. (1986). Calcium and the cytoskeleton. *Br. med. Bull.* **42**, 385–390.
BLANCHARD, A., OHANIAN, V. AND CRITCHLEY, D. (1989). The structure and function of α-actinin. *J. Mus. Res. and Cell Mot.* **10**, 280–289.
BRESNICK, A., WARREN, V. AND CONDEELIS, J. (1990). Identification of a short sequence essential for actin binding by *Dictyostelium* ABP-120. *J. biol. Chem.* **265**, 9236–9240.
BRYAN, J. (1988). Gelsolin has three actin-binding sites. *J. Cell Biol.* **106**, 1553–1562.
BRYAN, J. AND KURTH, M. C. (1984). Actin–gelsolin interactions: evidence for two actin-binding domains. *J. biol. Chem.* **259**, 7480–7487.
FRIEDERICH, E., HUET, C., ARPIN, M. AND LOUVARD, D. (1989). Villin induces microvilli growth and actin redistribution in transfected fibroblasts. *Cell* **59**, 461–475.
GLENNEY, J. R., GEISLER, N., KAULFUS, P. AND WEBER, K. (1981). Demonstration of at least two different actin-binding sites in villin, a calcium-regulated modulator of F-actin organization. *J. biol. Chem.* **256**, 8156–8161.
HOLMES, K. C., POP, D., GEBHARD, W. AND KABSCH, W. (1990). Atomic model of the actin filament. *Nature* **347**, 44–49.
KABSCH, W., MANNHERZ, H. G., SUCK, D., PAI, E. F. AND HOLMES, K. C. (1990). Atomic structure of the actin: DNase I complex. *Nature* **347**, 37–44.
KARINCH, A. M., ZIMMER, W. E. AND GOODMAN, S. R. (1990). The identification and sequence of the actin-binding domain of human red blood cell β-spectrin. *J. biol. Chem.* **265**, 11 833–11 840.
KWIATKOWSKI, D. J., JANMEY, P. A., MOLE, J. E. AND YIN, H. L. (1985). Isolation and properties of two actin-binding domains in gelsolin. *J. biol. Chem.* **260**, 15 232–15 238.
KWIATKOWSKI, D. J., JANMEY, P. A. AND YIN, H. L. (1989). Identification of critical functional and regulatory domains in gelsolin. *J. Cell Biol.* **108**, 1717–1726.
KWIATKOWSKI, D. J., STOSSEL, T. P., ORKIN, S. H., MOLE, J. E., COLTEN, H. R. AND YIN, H. L. (1986). Plasma and cytoplasmic gelsolins are encoded by a single gene and contain a duplicated actin-binding domain. *Nature* **323**, 455–458.
MANSTEIN, D. J., TITUS, M. A., DE LOZANNE, A. AND SPUDICH, J. A. (1989). Gene replacement in *Dictyostelium*: generation of myosin null mutants. *EMBO J.* **8**, 923–932.
MIMURA, N. AND ASANO, A. (1987). Further characterization of a conserved actin-binding 27-kDa fragment of actinogelin and α-actinins and mapping of their binding sites on the actin molecule by chemical crosslinking. *J. biol. Chem.* **262**, 4717–4723.
POLLARD, T. D. AND COOPER, J. A. (1984). Quantitative analysis of the effect of *Acanthamoeba* profilin on actin filament nucleation and elongation. *Biochemistry* **23**, 6631–6641.
VANDEKERCKHOVE, J. (1989). Structural principles of actin-binding proteins. *Current Opinion in Cell Biology* **1**, 15–22.
VANDEKERCKHOVE, J. (1990). Actin-binding proteins. *Current Opinion in Cell Biology* **2**, 41–50.
WAY, M., GOOCH, J., POPE, B. AND WEEDS, A. G. (1989). Expression of human plasma gelsolin in *E. coli* and dissection of actin binding sites by segmental deletion mutagenesis. *J. Cell Biol.* **109**, 593–605.
WAY, M., POPE, B., GOOCH, J., HAWKINS, M. AND WEEDS, A. G. (1990). Identification of a region of segment 1 of gelsolin critical for actin binding. *EMBO J.* **9**, 4103–4109.
WAY, M. AND WEEDS, A. G. (1988). Nucleotide sequence of pig plasma gelsolin. Comparison of protein sequence with human gelsolin and other actin-severing proteins shows strong homologies and evidence for large internal repeats. *J. molec. Biol.* **203**, 1127–1133.
WAY, M. AND WEEDS, A. G. (1990). Cytoskeletal ups and downs. *Nature* **344**, 292–294.
YIN, H. L., JANMEY, P. A. AND SCHLEICHER, M. (1990). Severin is a gelsolin prototype. *FEBS Lett.* **264**, 78–80.

The bending of sliding microtubules imaged by confocal light microscopy and negative stain electron microscopy

L. A. AMOS and W. B. AMOS

MRC Laboratory of Molecular Biology, Cambridge, UK

Summary

Individual microtubules can be visualised by confocal microscopy in reflection mode; when associated with a glass surface, they show up as black lines against the bright reflection from the surface. The high contrast imaging allows details of the behaviour of sliding microtubules to be studied easily.

Taxol-stabilised microtubules sliding over kinesin-coated surfaces are normally straight, but can bend into tight loops if the leading end sticks to the surface. Some remain curved after release and move in circles. In such cases, the microtubule lattice must have become stably deformed. Electron microscopy of microtubules fixed during sliding shows no gross rearrangement of the subunit lattice and indicates that microtubule bending is mainly achieved by increased twisting of the longitudinal protofilaments around the microtubule.

Key words: tubulin, microtubule lattice, kinesin, confocal microscopy.

Introduction

The sliding of microtubules over glass surfaces was discovered using video-enhanced differential interference contrast (DIC) light microscopy (Allen *et al.* 1985) and most studies of individual microtubule behaviour by light microscopy since then have employed the same technique. Dark-field microscopy, with or without video-enhancement, has also been used to obtain important new results in this field (e.g. Summers and Gibbons, 1971; Horio and Hotani, 1986). Recent developments in laser-scanning confocal microscopy (White *et al.* 1987; Amos *et al.* 1987) have improved the resolution obtainable by reflection interference microscopy to the extent that we have yet another means of observing movement of individual microtubules. Confocal reflection interference microscopy shares the advantage of DIC light microscopy that objects not close to the relevant glass surface are excluded from the image, but it produces a higher contrast than DIC; it offers the additional possibility of simultaneously imaging a fluorescence image in perfect registration with the reflection image. This report demonstrates some results obtained with just reflection imaging, followed by observation of some specimens by electron microscopy.

Materials and methods

(a) Preparation of microtubules and kinesin

Reassembled brain microtubules were stabilised with 20 μM taxol after two cycles of assembly and disassembly and MAPs were removed in medium containing high salt (Vallee, 1982). The pelletted MAP-depleted microtubules were resuspended in Pipes buffer (0.1 M Pipes–KOH, pH 6.9, 1 mM EGTA) and stored in small samples in liquid nitrogen. Samples were thawed when required and diluted in media of varying pH (0.1 M Mes, pH 6.6; 0.1 M Pipes, pH 6.9; 0.1 M Hepes, pH 7.4 or 7.8; plus 1 mM EGTA, 10 μM taxol).

Kinesin was purified from pig brain using the protocol described by Amos (1989) and also stored in liquid nitrogen. For experiments on microtubule sliding, 5–10 μl of kinesin, at protein concentrations of 0.1–1.0 mg ml^{-1}, were spread over the central area of a glass coverslip and incubated in a moist chamber for 1–5 min, as described by Vale *et al.* (1985). The coverslips were normally used untreated but for some experiments the surface was coated with a film of nitrocellulose, as described by Kron *et al.* (1990). The latter treatment had no obvious effect on the attachment of kinesin to the surface or to its interaction with microtubules. For observation of some samples by electron microscopy, copper grids were sandwiched between coverslip and nitrocellulose film.

5 μl of diluted microtubule solution was added to the sample on the coverslip which was then inverted on to a slide where it was supported with strips of coverslip, also as described in detail by Kron *et al.* (1990). Additional buffer was added from either side of the coverslip, to fill the space, together with 5 mM MgATP.

(b) Confocal microscopy

A MRC500 manufactured by BioRad was used to scan the specimens through a Nikon Optiphot with a X60 Planapo objective. A 15 mW argon ion laser was used, with a narrow-band filter to isolate the 488 nm line. The standard reflection filter block was used, without polarising anti-reflection components. Laser power was varied from 1% to 100%; surprisingly, the strongest illumination used seemed to have no effect on kinesin-induced motility. However, the gain in terms of noise reduction, in increasing the power from 10% to 100%, was small. The full field that can be scanned using the ×60 objective has a diameter of 200 μm but it is possible to zoom in on any part of the field using a restricted range of scanning angles; at maximum zoom, this lens gives a field width of 25 μm. Microtubules can be seen most clearly with a field width of 50 μm or less. A framing rate of 1 per second was judged to be optimum both for direct viewing and for recording on video; no averaging of frames was carried out for moving specimens as this blurred the ends of microtubules.

Image contrast could be enhanced on-line at the analogue stage and after conversion to digital form. Some series of images were filed directly from the frame store on to a Winchester or optical disc. Others were fed back through the system from video-tape recordings; this gave a slight reduction in image quality but made the choice of interesting frames easier.

Journal of Cell Science, Supplement 14, 95–101 (1991)

(c) Electron microscopy

EM grids trapped under plastic on sample-covered coverslips were lifted off, dipped briefly in wash buffer (1 mM Hepes or Pipes, 10 mM potassium acetate) and negatively stained with 1% aqueous uranyl acetate.

Results

(a) Confocal microscopy

Fig. 1 shows the appearance of microtubules in a confocal reflection interference image averaged over many frames. Microtubules just below the coverslip are visible as lines approximately 0.2 μm thick; those very close to the glass surface appear black, others roughly half a wavelength further away appear brighter than the surface. Microtubules between these extremes would be expected to contrast less with the surface and, occasionally, grey segments can be identified.

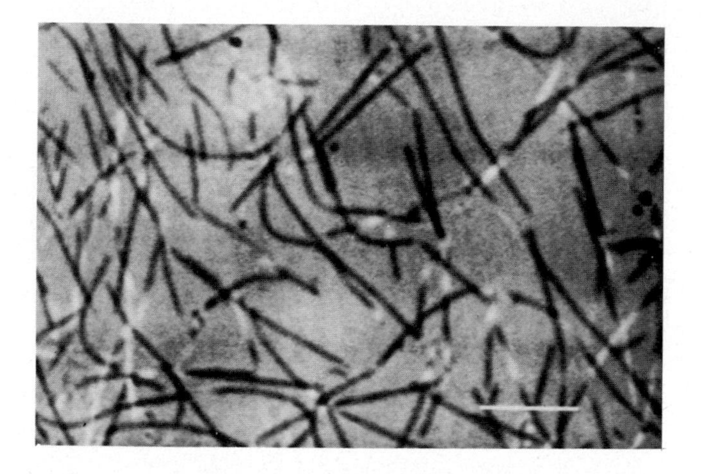

Fig. 1. Confocal laser scanning image of microtubules, stabilized with taxol and free of associated proteins, adhering to the surface of a glass coverslip. Individual microtubules, interfering with reflection from the surface, produce black or white lines in the image. Averaged over 50 scanning frames and computer enhanced. Specimens in Pipes buffer, pH 6.9. Scale bar, 4 μm.

Fig. 2 consists of a series of images extracted from single frames to illustrate the sliding of straight microtubules. Kinesin-induced movement was investigated in buffers with a range of pH values. Sliding occurred at a rate of 0.3 $\mu m\,s^{-1}$ in 0.1 M Mes, pH 6.6; at 0.4 $\mu m\,s^{-1}$ in 0.1 M Pipes, pH 6.9; at 0.6 $\mu m\,s^{-1}$ in 0.1 M Hepes, pH 7.4 or 7.8.

Microtubules were observed to bend in all these buffers; it appeared that the leading end would become stuck to the surface and pushing from the rear would produce bends near the front (e.g. Fig. 3B–D). This behaviour is similar to that described by Allen *et al.* (1985) as *fishtailing*. If the front end remained stuck, the rest of the microtubule would often rotate around it in a spiral. If the end of such a microtubule became detached again, it would sometimes re-straighten, especially in the Mes and Pipes buffers. Under conditions used in these experiments, other microtubules remained in a curved configuration and continued to move in circles (Fig. 4). The bends formed at pH values above 7.0 tended to be tighter and were more often retained after detachment. Breakage during bending (see Fig. 3) was also much commoner at higher pH. An estimate, from the confocal images, of the radii of the smallest circles seen is around 0.5 μm (i.e. a maximum curvature, before breakage, of 2 radians per micron). Figs 3 and 5 illustrate the point that initially straight microtubules can become stably bent but that stably bent microtubules can subsequently straighten again.

(b) Electron Microscopy

To discover how the microtubule lattice was accommodating to such pronounced bending, EM grids were attached under a thin film of nitrocellulose to the surface of a coverslip, on which kinesin and microtubules were combined as before. The reflection interference image was unfortunately degraded within the squares of the EM grid

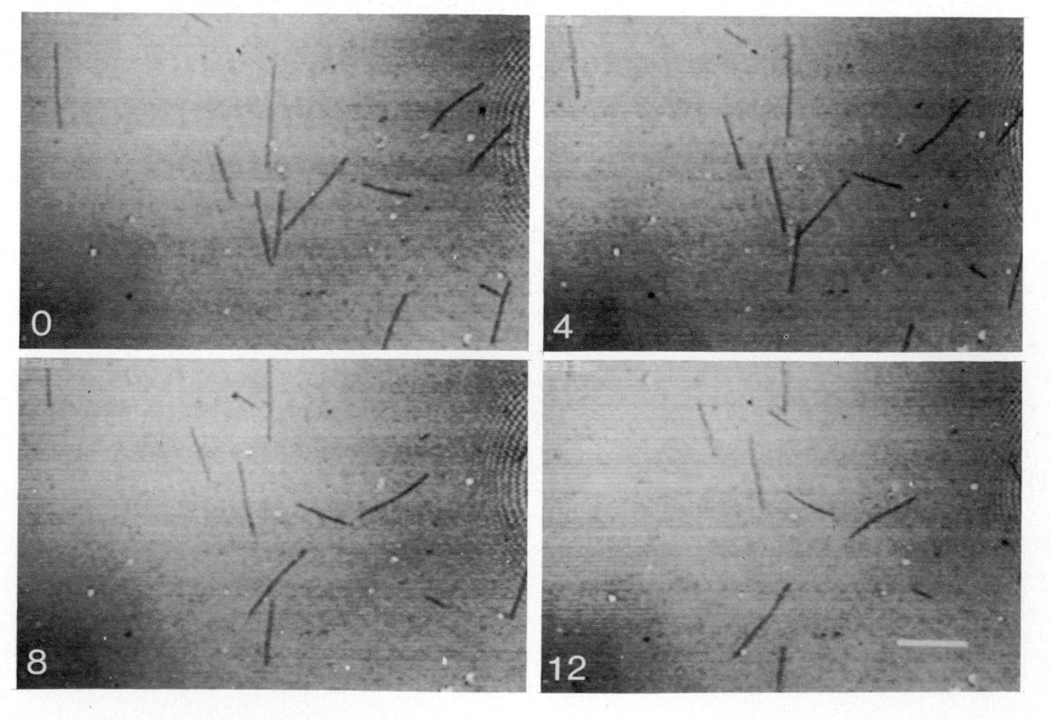

Fig. 2. Series of individual scanning frames recorded on video and then played back through the confocal frame board; frames at 4 s intervals have been selected. Specimen in Hepes buffer, pH 7.4. Diffraction ripples on the right of each image come from a small central spot of light reflected from the eye-piece. White scale bar, 6 μm.

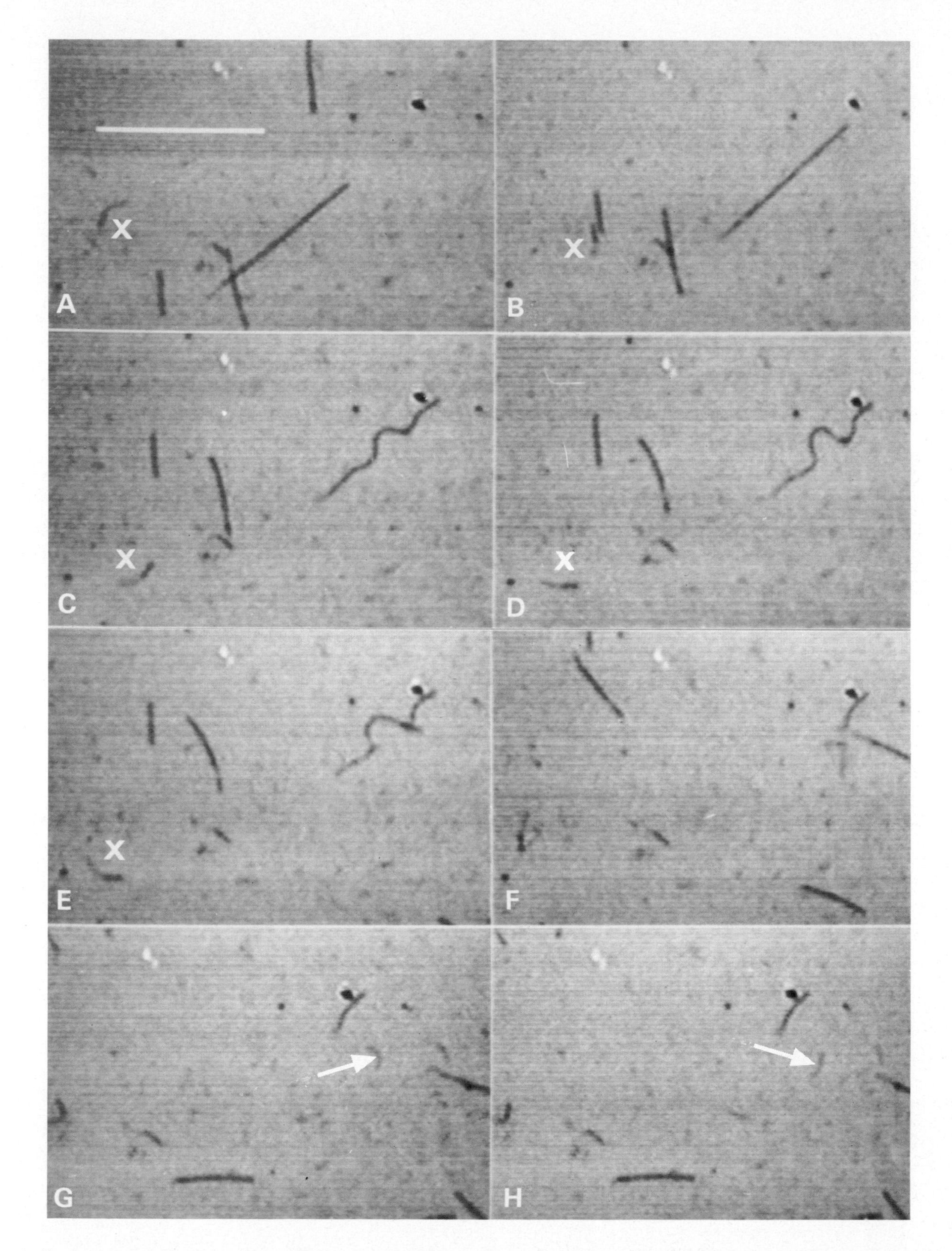

Fig. 3. Confocal series, similar to Fig. 2. The long microtubule in the centre of (A) was moving in a straight line until its leading end became attached to the glass (B–C). After pronounced bending (D), the tubule broke (E). A major portion re-straightened and slid off in a straight line (F). A smaller fragment remained curved and moved in a circle (arrows in G–H). Meanwhile, a similar fragment (x) produced by an earlier event rotated until its path changed (F). Hepes buffer, pH 7.4. Time intervals shown are unequal. White scale bar, 6 μm.

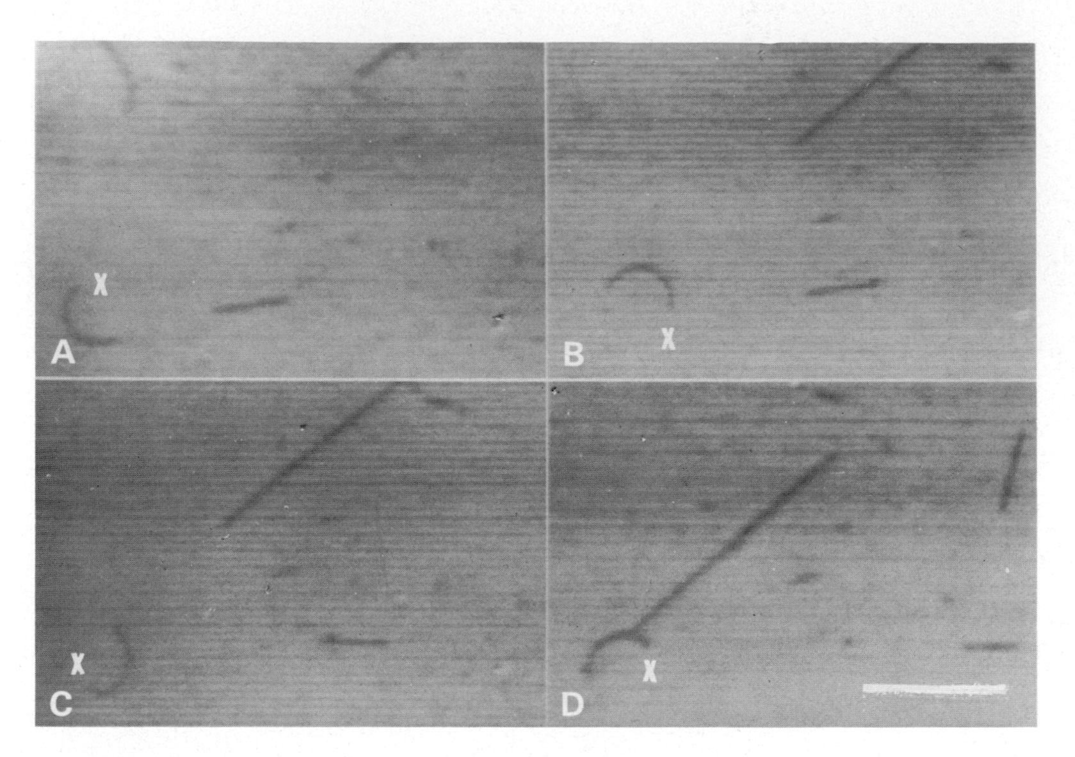

Fig. 4. Confocal image series showing an arc-shaped microtubule moving around a circle. ×marks the leading end. Hepes buffer, pH 7.4. 3–4 s between images. Scale bar, 3 μm.

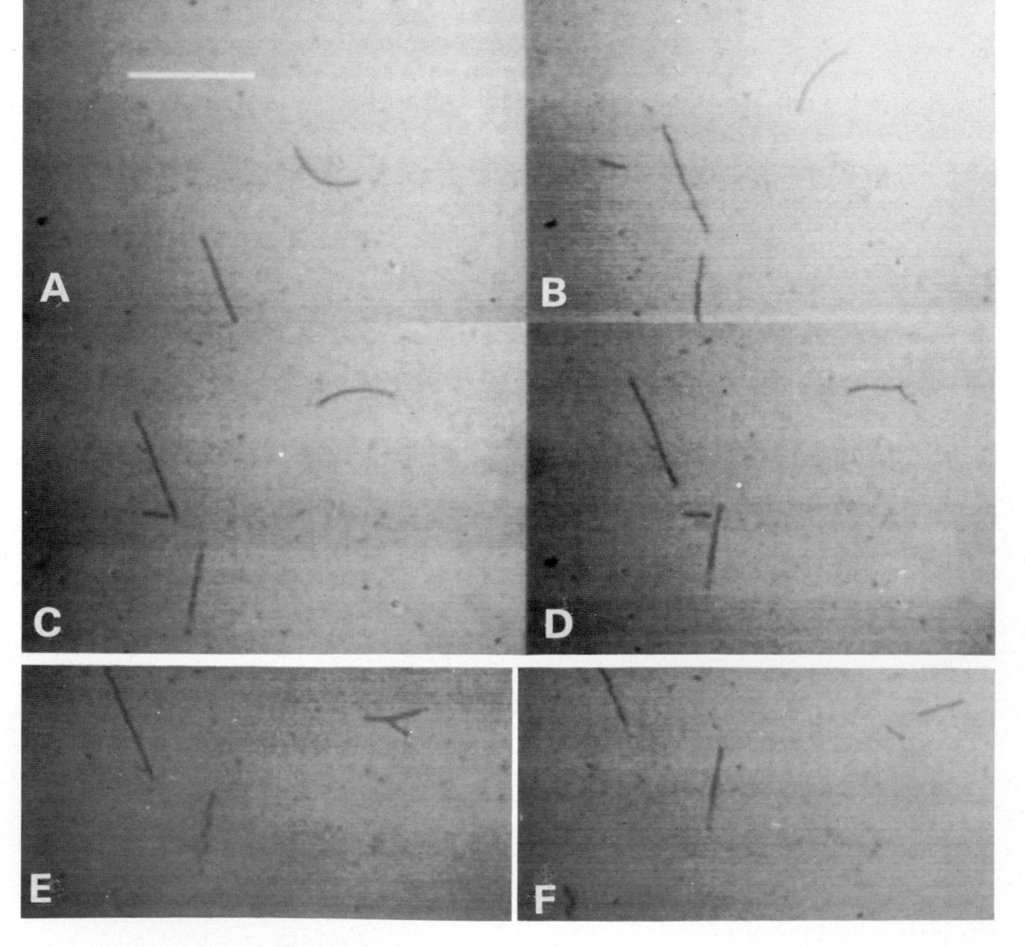

Fig. 5. Confocal series in which an arc-shaped microtubule was seen moving initially in a circle (A–C). The original leading end stuck to the surface, producing a break (D). The rear portion became straight and slid off in a straight line (F). Hepes buffer, pH 7.4. Time intervals unequal, chosen to show particular events. Scale bar, 6 μm.

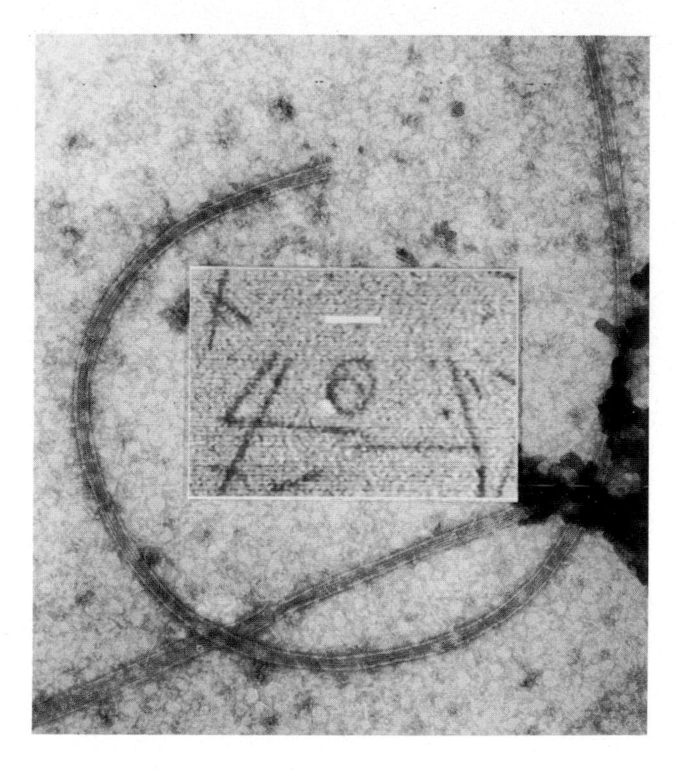

Fig. 6. Electron micrograph of a microtubule bent almost into a circle with a radius of 500 nm. Specimen in Hepes buffer, pH 7.4, before staining. Inset shows a confocal light microscope image including a similarly bent microtubule that was observed rotating. White scale bar, 100 nm for EM image, 1 μm for confocal image.

(presumably because the contact here between glass surface and nitrocellulose film was not close enough) but movement observed on the surface adjacent to each grid appeared to be as usual.

Normally, in the absence of kinesin, microtubule lattice angles and spacings appear to be strongly maintained. Although the longitudinal protofilaments (pfs) run straight in a 13-pf tubule, they twist slowly around the axes of the 14-pf or 15-pf tubules commonly found in reassembled microtubule samples; in each case, the twist is sufficient to preserve the lengths and angles of the standard 13-pf surface lattice (Wade *et al.*. 1990). EM images such as Figs 6 and 7 show that curved microtubules remain tubular and apparently retain their subunits in basically the same helical lattice as in straight microtubules; there is no radical rearrangement of whole subunits.

Bending without damage to the lattice could be achieved by contraction of the lengths of all the subunits along the inner circumference of a curved tubule or extension of subunit spacings on the outer circumference; in other words, by contraction or extension along the protofilaments. The latter could, however, remain roughly constant in length if they were to twist around the microtubule at a suitable rate, different from that in a straight 14- or 15-pf microtubule; this would mean the lattice angles being distorted from normal, however.

The lattice parameters of curved segments of microtubule are not easily determined, since optical diffraction requires straight subjects. Analysis of such images by computer will be reported elsewhere. Meanwhile, direct inspection of the images indicates that the protofilaments

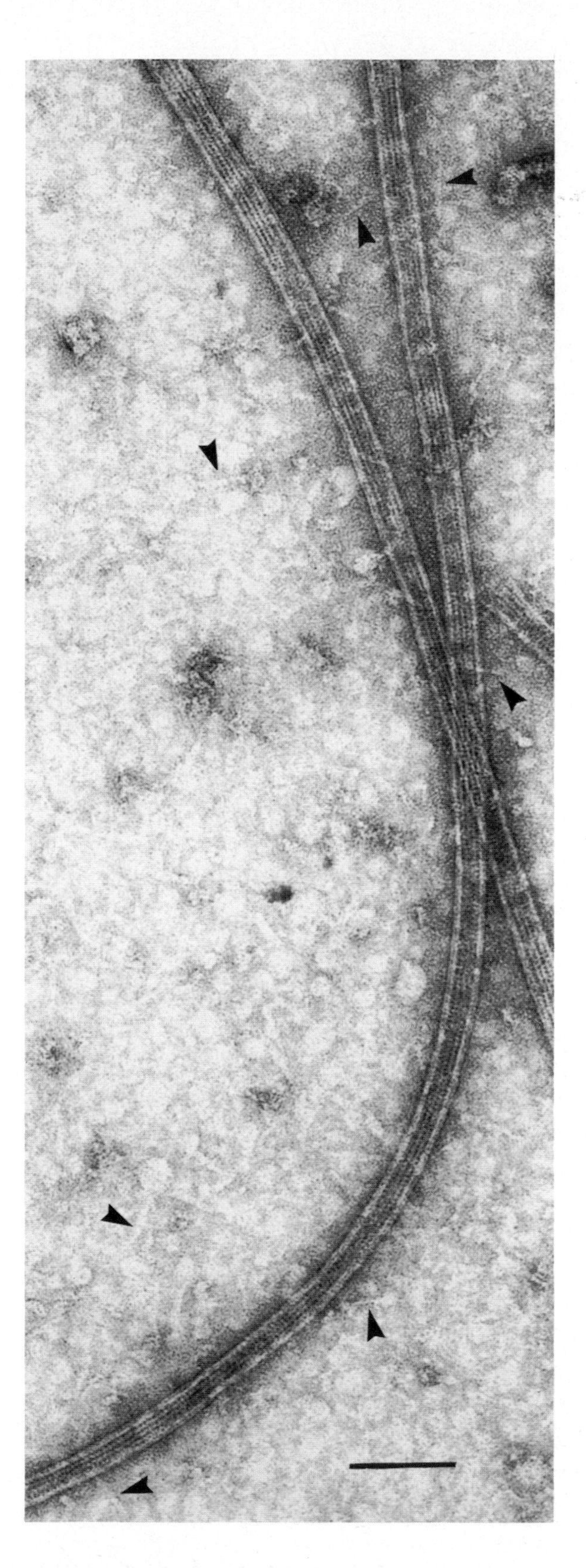

Fig. 7. Electron micrograph of parts of a curved and a straight microtubule on a kinesin coated plastic surface. Kinesin molecules, appearing as fine rods (indicated by arrow-heads), are not easily distinguished against the irregular background of negative stain. Hepes buffer, pH 7.4. Scale bar, 100 nm.

do twist around curved 14-pf microtubules more rapidly than around straight 14-pf microtubules. Straight microtubules on the same grids had presumably also been sliding over the surface not long before being fixed by negative stain. Optical diffraction patterns obtained from many of these images showed complex variations in the lattice parameters along individual microtubules; these are also in the process of being analyzed. There appear to be variations in the protofilament twist angle and even small amounts of contraction (or extension) of subunit spacings along the protofilaments.

Discussion

Tubulin is an allosteric protein whose subunits appear to change conformation substantially between the disassembled and polymerized states (Ventilla *et al.* 1972). The results reported here suggest that even an intact tubular polymer is capable of holding more than one conformational state without the help of an external force; otherwise, a curved microtubule with a freed leading end would always revert to a straight form as it slid forwards. Instead, the lattice presumably drops into a local energy minimum; the number and depths of such minima may depend on conditions in the media, such as the buffer pH and the temperature. The possibility that transitions can be induced from one structural state to another in the assembled microtubule lattice has implications for the precise role of microtubules in motility and may also be relevant to their behaviour during assembly and disassembly.

The microtubules studied here were stabilized using taxol, so it is necessary to consider the question of how different they might be from 'normal' microtubules. First, there is no obvious difference in gross structure, as seen by electron microscopy, from microtubules reassembled without taxol. Second, purified kinesin appears to interact with them in the same manner as with axonemal microtubules (see Porter *et al.* 1988), which are naturally stable. The main effect of taxol seems to be enhancement of assembly at low temperature or low tubulin concentration (Schiff *et al.* 1979) Exactly how taxol exerts its stabilising effect is not known; however, it seems likely that it inhibits disassembly by binding to individual subunits and maintaining them in a conformation or state that strongly favours their polymerization. Observations such as those reported here show that it does not suppress all changes in tubulin conformation. The components that stabilise axonemal microtubules also seem to allow them to bend in any direction, so it seems reasonable to suppose that taxol-stabilized tubules are not significantly different from naturally-occurring cold-stable microtubules.

Although the two-dimensional lattice formed by tubulin during assembly is apparently quite strictly defined, it seems that the lattice may subsequently be altered in a non-elastic fashion. The lack of evidence for an 8 nm axial periodicity in the optical diffraction patterns indicates that structural changes in these sliding microtubules affect alpha- and beta-tubulin subunits equally.

Analysis of protofilament structure in three dimensions by electron microscopy and X-ray diffraction (see Beese *et al.* 1987; Amos and Baker, 1979) shows each tubulin monomer to consist of two or more structural domains. Most likely, conformational changes in tubulin, as in other proteins, depend largely on the movement of whole domains relative to one another. An important special feature is that there must be coordinated changes to whole groups of neighbouring subunits in order to distort the tubulin lattice.

It has previously been observed that flagellar microtubules can occur in stable curly and helical conformations (e.g. Costello *et al.* 1973; Amos, 1978; Miki-Noumura and Kamiya, 1976, 1979). The results reported here indicate that the tubulin lattice itself, rather than any associated protein, is primarily responsible for maintaining such configurations, though the doublet structure and presence of specialized associated proteins probably influence the precise helical conformations (Miki-Noumura and Kamiya, 1979). Bending is obviously an important part of microtubule function in flagellar motility. The ability of the tubulin lattice to assume conformations such that bent microtubules do not produce an elastic restoring force seems likely to be an important feature of flagellar beating. A cooperative change in the tubulin lattice produced by interaction with dynein might even form part of the mechanism by which the relative sliding of adjacent doublet microtubules is converted synchronously into bending.

The possible relevance of different conformational states to the function of cytoplasmic microtubules is less obvious. There is no evidence of any type of beating motion for single microtubules *in vivo*. However, preliminary observations on straight sliding microtubules suggest that variations in the twist of the lattice are not necessarily associated with obvious bending. It is not clear whether lattice distortion is a normal feature of the interaction of a microtubule with motor molecules, or simply due, in the present situation, to the attachment of kinesin to an immovable surface; but it is possible that each hydrolysis event involves structural changes in a series of neighbouring subunits. The rotation of microtubules sliding over surfaces covered with single-headed dynein (Vale and Toyoshima, 1989) may perhaps be a related phenomenon.

Finally, it may be significant that bending and breakage are more pronounced and occur more frequently at pH values above 7, a condition that is generally unfavourable for microtubule assembly *in vitro*. Possibly, in the absence of taxol or of a class of associated proteins that confer a high level of stability, some lattice changes might lead to depolymerization. Since a lattice distortion affects many individual subunits, lattice changes associated with disassembly would be highly cooperative and might possibly explain the rapid disassembly phases of dynamically unstable microtubules.

We thank John White and other colleagues for help with the confocal equipment and software. Taxol was kindly provided by Dr Matthew Suffness of the National Cancer Institute, Bethesda, Maryland, USA.

References

Allen, R. D., Weiss, D. G., Hayden, J. H., Brown, D. T., Fujiwake, H. and Simpson, M. (1985). Gliding movement of and bidirectional transport along single native microtubules from squid axoplasm: evidence for an active role of microtubules in cytoplasmic transport. *J. Cell Biol.* **100**, 1736–1752.

Amos, L. A. (1978). Changes in tubulin conformation during bending of flagellar microtubules. *6th Int. Biophys. Congr., Tokyo*, 1978.

Amos, L. A. (1989). Brain dynein crossbridges microtubules into bundles. *J. Cell Sci.* **93**, 19–28.

Amos, L. A. and Baker, T. S. (1979). Three-dimensional image of tubulin in zinc-induced sheets, reconstructed from electron micrographs. *Int. J. Biol. Macromolec.* **1**, 146–156.

Amos, W. B., White, J. G. and Fordham, M. (1987). Use of confocal

imaging in the study of biological structures. *Appl. Optics* **26**, 3239–3243.
BEESE, L., STUBBS, G. AND COHEN, C. (1987). Microtubule structure at 18Å resolution. *J. molec. Biol.* **194**, 257–264.
COSTELLO, D. P. (1973). A new theory of the mechanism of ciliary and flagellar motility: I. Supporting observations. *Biol. Bull. (Woods Hole)* **145**, 279–308.
HORIO, T. AND HOTANI, H. (1986). Visualization of the dynamic instability of individual microtubules by dark-field microscopy. *Nature (Lond.)* **321**, 605–607.
KRON, S. J., TOYOSHIMA, Y. Y., UYEDA, T. Q. P. AND SPUDICH, J. A. (1990). Assays for actin sliding movement over myosin coated surfaces. *Meth. Enzym.* (in press).
MIKI-NOUMURA, T. AND KAMIYA, R. (1976). Shape of microtubules in solutions. *Expl Cell Res.* **97**, 451–453.
MIKI-NOUMURA, T. AND KAMIYA, R. (1979). Conformational change in the outer doublet microtubules from sea urchin sperm flagella. *J. Cell Biol.* **81**, 355–360.
PORTER, M. E., SCHOLEY, J. M., STEMPLE, D. L., VIGERS, G. P. A., VALE, R. D., SHEETZ, M. P. AND MCINTOSH, J. R. (1986). Characterization of the microtubule movement produced by sea urchin egg kinesin. *J. biol. Chem.* **262**, 2764–2802.
SCHIFF, P. B., FANT, J. AND HORWITZ, S. B. (1979). Promotion of microtubule assembly *in vitro* by taxol. *Nature (Lond.)* **277**, 665–667.
SUMMERS, K. E. AND GIBBONS, I. R. (1971). ATP-induced sliding of tubules in trypsin-treated flagella of sea urchin sperm. *Proc. natn. Acad. Sci. U.S.A.* **68**, 3092–3096.
VALE, R. D., REESE, T. S. AND SHEETZ, M. P. (1985). Identification of a novel force-generating protein, kinesin, involved in microtubule-based motility. *Cell* **42**, 39–50.
VALE, R. D. AND TOYOSHIMA, Y. Y. (1988). Rotation and translocation of microtubules *in vitro* induced by dynein from *Tetrahymena* cilia. *Cell* **52**, 459–469.
VALLEE, R. B. (1982). A taxol-dependent procedure for the isolation of microtubules and microtubule-associated proteins (MAPs). *J. Cell Biol.* **92**, 435–442.
VENTILLA, M., CANTOR, C. R. AND SHELANSKI, M. (1972). A circular dichroism study of microtubule protein. *Biochem.* **11**, 1554–1561.
WADE, R. H., CHRETIEN, D. AND JOB, D. (1990). Characterization of microtubule protofilament numbers: how does the surface lattice accommodate? *J. molec. Biol.* **212**, 775–786.
WHITE, J. G., AMOS, W. B. AND FORDHAM, M. (1987). An evaluation of confocal *versus* conventional imaging of biological structures by fluorescence light microscopy, *J. Cell Biol.* **105**, 41–48.

Kinesin interactions with membrane bounded organelles *in vivo* and *in vitro*

SCOTT T. BRADY*

Department of Cell Biology and Neuroscience, University of Texas Southwestern Medical Center, Dallas TX 75235, USA

and K. KEVIN PFISTER

Department of Anatomy and Cell Biology, University of Virginia School of Medicine, Charlottesville, VA 22908, USA

*To whom correspondence should be addressed

Summary

The ability of kinesin to interact with microtubules in a nucleotide-dependent manner and mediate microtubule-based motility has received the greatest amount of attention to date. Several lines of experimentation are now beginning to examine the interaction with membrane-bounded organelles. Immunochemical, biochemical and morphological approaches have shown that kinesin is associated with some, but not all, classes of membrane-bounded organelles found in cells. Similarly, evidence suggests that the distal portion of the rod and the tail portions of the kinesin heavy chain as well as the kinesin light chains may be important for the interaction with membrane surfaces. As a substantial amount of information about the molecular structure and biochemistry of kinesin has become available, the functional implications of interactions with membrane structures *in vivo* are being addressed.

Key words: kinesin, membrane-bounded organelles, immunocytochemistry, synaptic vesicles, mitochondria, microtubules, light chains, isoforms.

Introduction

Perfusion of a nonhydrolyzable analogue of ATP, adenylyl imidodiphosphate [AMP-PNP], immobilizes all of the membrane bounded organelles in isolated cytoplasm from the squid giant axon, by inducing an apparent rigor complex of organelle, motor and microtubule, even in the presence of significant amounts of ATP (Lasek and Brady, 1985). The discovery of this effect led directly to the conclusion that a new class of microtubule-based mechanochemical ATPases existed in eukaryotic cells (Lasek and Brady, 1984, 1985). Subsequent work resulted in the identification of a third class of motor proteins, the kinesins (Brady, 1985; Vale *et al.* 1985). Since that time, most investigators have focused on determining the polypeptide composition and structure, enzymatic properties and sequence of kinesins from a variety of species and tissues. The initial cloning of a *Drosophila* kinesin heavy chain (Yang *et al.* 1988, 1989) defined several important features of the kinesin ATPase. Using this information, studies in molecular genetics led to the identification of several additional members of the kinesin family in yeast (Meluh and Rose, 1990), *Aspergillis* (Enos and Morris, 1990) and *Drosophila* (Endow *et al.* 1990; MacDonald and Goldstein, 1990; Zhang *et al.* 1990). Mutations in these kinesin-like polypeptides lead to defects in nuclear migration and cell division. However, despite identification of additional kinesin-like polypeptides, which may play a variety of roles in cells, the most widely accepted function for the kinesin family of motor proteins in translocation of membrane bounded organelles along cytoplasmic microtubules.

The fact that kinesin was initially identified and purified by its nucleotide-sensitive binding to microtubules (Brady, 1985; Scholey *et al.* 1985; Vale *et al.* 1985), and the early determination that kinesin was an ATPase (Brady, 1985; Kutznetsov and Gelfand, 1986; Cohn *et al.* 1987), focused attention on the interaction between kinesin and microtubules. By contrast, relatively little information has been available about the interaction with membrane bounded organelles. Thus, while considerable progress has been made in defining the properties of the microtubule-activated ATPase of kinesins, the interaction of the kinesins with membrane bounded organelles and the intracellular distribution of kinesins has been largely a matter of speculation. Many questions remain about the physiological functions and subcellular distribution of kinesin. Work in a number of laboratories has begun to provide insights into these interactions both *in vivo* and *in vitro*. Our own laboratories have utilized video, confocal and electron microscopy, immunochemistry, molecular genetics and biochemistry in order to address questions about kinesin–membrane interactions.

Microscopy and immunocytochemistry

A library of monoclonal antibodies suitable for immunocytochemical studies were generated against the heavy and light chains of native kinesin (Pfister *et al.* 1989). Each antibody was shown to recognize a distinct epitope on kinesin. Two antibodies recognized the full range of isoforms resolved in one and two dimensional electrophoresis for the heavy chain (H2) and light chain (L1), while the remaining antibodies recognized a subset of these isoforms. Immunocytochemical studies at the light level using this library of anti-kinesin antibodies established that both light and heavy chains are most heavily concentrated on membrane bounded organelles in a variety of cell types (Fig. 1). As seen in Figs 1 and 2, the

Journal of Cell Science, Supplement 14, 103–108 (1991)

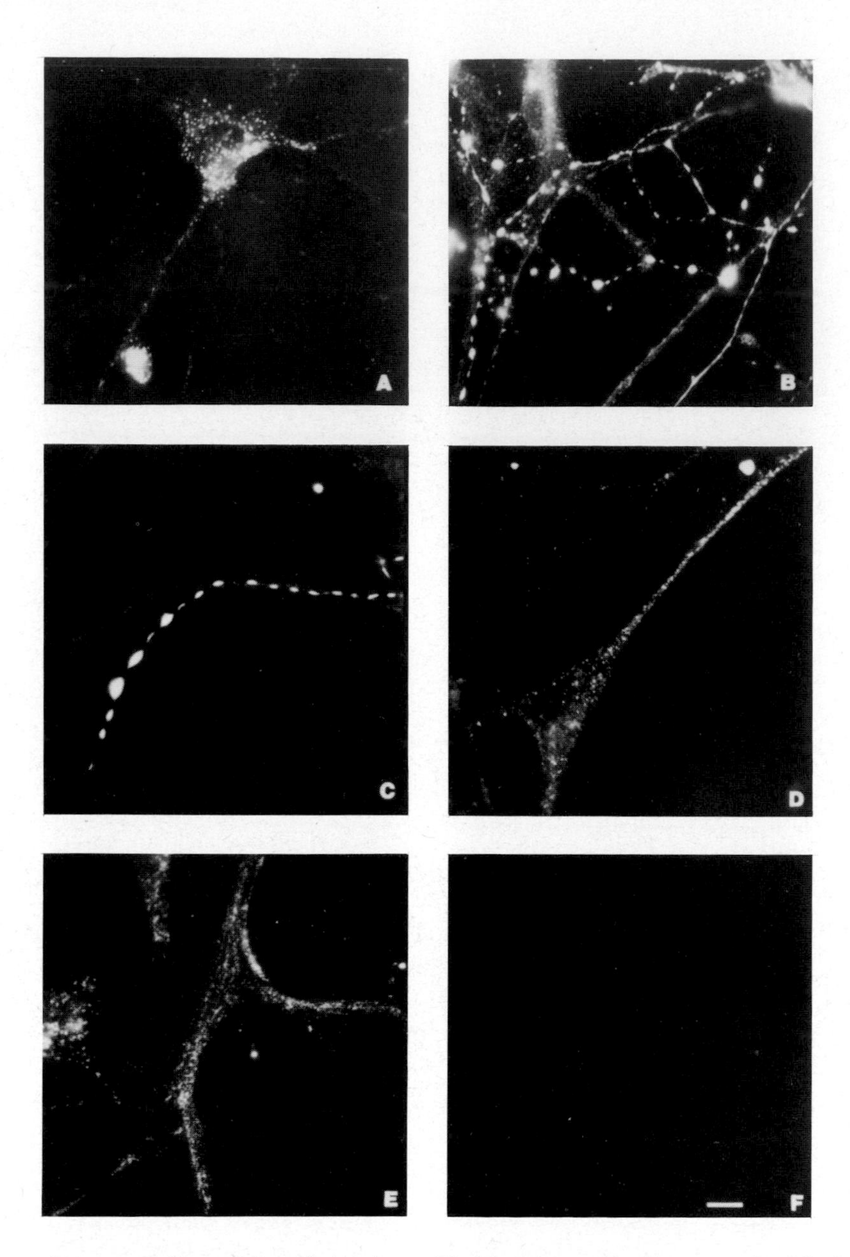

Fig. 1. Immunofluorescence patterns in neonatal rat brain cultures with antibodies to kinesin heavy and light chains. Cells were cultured and stained as described in Pfister *et al.* (1989) for immunofluorescence microscopy. Cells were stained with antibodies specific for kinesin heavy chains H1 (A), H2 (B), H3 (C,D), antibodies specific for kinesin light chain L2 (E) and normal mouse IgG antibodies (F). In all cell types, the characteristic staining was punctate, consistent with the staining of membrane-bounded organelles in the cytoplasm. In the axons of cultured neurons, varicosities known to contain large numbers of membrane-bounded organelles were very brightly stained by H2 and H3 (B,C). Little or no staining was detectable with control antibodies (F). Punctate staining with kinesin antibodies was abolished by extraction of the cells with Triton X-100 prior to fixation. The figure is reproduced from Pfister *et al.* (1989) by permission of the Journal of Cell Biology and the Rockefeller University Press. Bar, 10 μm.

characteristic pattern of staining with kinesin antibodies was a punctate pattern enriched in microtubule-containing domains of cells, but not corresponding to microtubules. Detergent extraction of cells under conditions that stabilized cytoplasmic microtubules eliminated the punctate pattern without significantly altering the appearance of the microtubular cytoskeleton in differential interference contrast microscopy (Pfister *et al.* 1989). Using several different approaches, it is becoming possible to identify the specific classes of organelles that have kinesin association.

Laser scanning confocal microscopy has a number of advantages over conventional immunofluorescence microscopy, particularly for double label studies. While the resolution remains limited by the physics of light and technical considerations for the instruments, judicious use of this method can provide substantially more detail about the precise locations of specific epitopes within a cell. When compared with conventional epi-illumination fluorescence light microscopy, the signal to noise ratio is substantially improved and effective resolution is increased (down to 400–500 nm), and this is particularly valuable for adding the information along the z-axis for moderately thick specimens. When combined with suitable analytical paradigms, significantly more information about associations may be derived from confocal images than from conventional immunofluorescence images.

Choice of a suitable antigen for colocalization studies is critical for effective analyses. Ideally, antigens meeting two criteria are needed. Firstly, for the broadest comparisons, an integral membrane protein associated with all membranes is needed; secondly, protein antigens specific for different classes of membrane bounded organelles are needed to permit positive identification of the organelle with which the kinesin is associated. An integral membrane protein, influenza virus hemagglutinin (HA), exists that can meet both of these criteria. A genetically engineered form of HA has been generated (Lazarovitis and Roth, 1988) that, when transfected into tissue culture cells, can be found in Golgi complex, endoplasmic reticulum, plasmalemma and endosomes. With proper experimental manipulations, HA can also be restricted to specific classes of membrane structures. Preliminary confocal microscopic studies of cells transfected with HA,

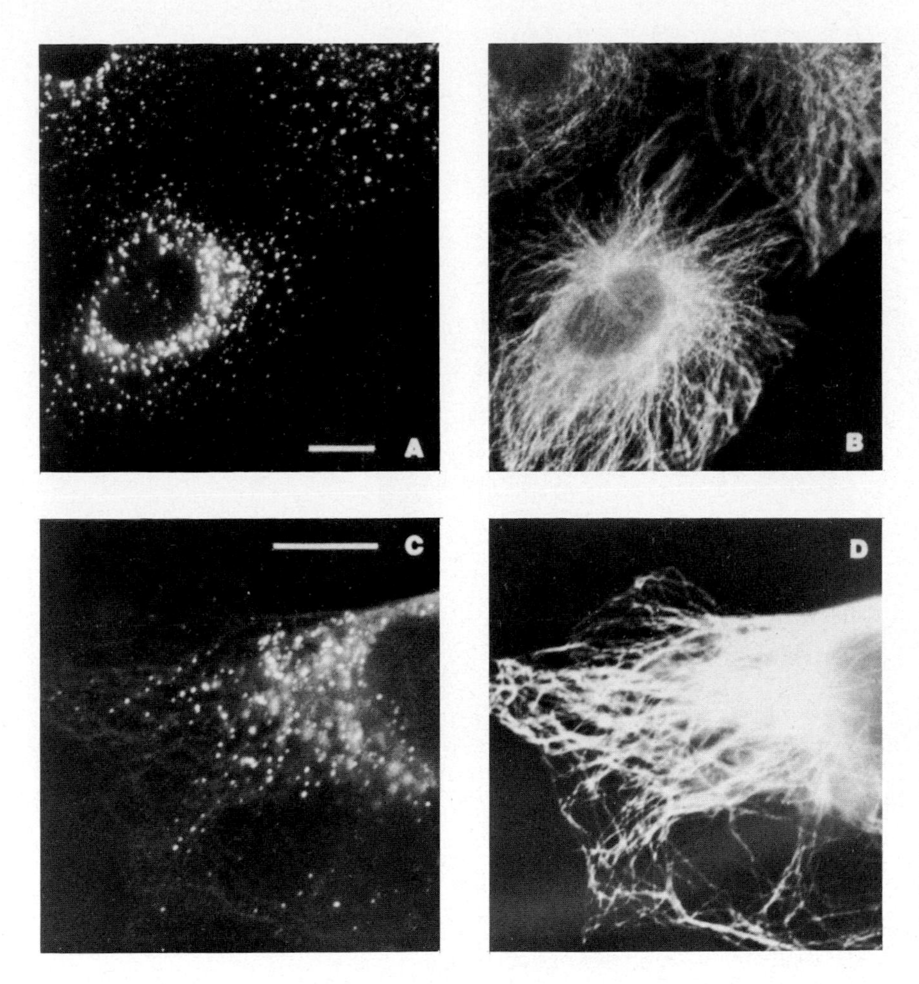

Fig. 2. Double immunofluorescence patterns in cultured cells using antibodies to kinesin and tubulin. PtK_1 cells were cultured and prepared from immunofluorescence microscopy as described in Pfister *et al.* (1989). Antibodies used were the kinesin H1 murine monoclonal antibody and a rabbit polyclonal antibody to tubulin. TRITC goat anti-mouse and FITC goat anti-rabbit were employed as secondary antibodies for visualization. In (A) and (B), very low dilutions of the anti-tubulin antibody were used to minimize crossover between rhodamine and fluorescein channels. In (C) and (D), higher concentrations of the tubulin antibody that resulted in some crossover were used to facilitate comparisons of the distributions for the two antigens. The kinesin reactive structures appear to align along tracks formed by the faintly visible microtubules. The figure is reproduced from Pfister *et al.* (1989) by permission of the Journal of Cell Biology and the Rockefeller University Press. Bar, 10 μm.

using antibodies to kinesin and HA, confirm that kinesin is associated with a variety of different classes of membrane bounded organelles (K. K. Pfister, C. Brewer, M. Roth, G. S. Bloom and S. T. Brady, unpublished observations). However, not all membranes have kinesin associated with them. For example, both the Golgi complex and nuclear membranes appear to have little or no kinesin immunoreactivity.

A parallel study using an antigen to a mitochondrial antigen indicates that kinesin is also associated with mitochondria. Interestingly, initial images suggest that the distribution of kinesin is restricted to discrete patches and is not uniformly distributed on the surface of the mitochondria. Although the immunofluorescence data is still preliminary, such a patchy distribution is consistent with observations of mitochondrial movement in video microscopy (Martz *et al.* 1984) and with immunogold electron microscopic localization of kinesin on isolated mitochondria (Leopold *et al.* 1990).

Confocal microscopy has also provided additional information about the relationship of kinesin with microtubules. The distribution of kinesin with tubulin in double label immunofluorescence experiments in both interphase and mitotic cells has been examined. In Fig. 3, the pattern in interphase cells for both kinesin (A) and tubulin (B) was similar to that obtained with conventional immunofluorescence (see Fig. 2 and Pfister *et al.* 1989). The punctate kinesin structures are located in the vicinity of microtubules, but do not co-localize with the microtubules. In rounded, mitotic cells, little kinesin can be detected in optical sections and no kinesin appears to be associated with the mitotic spindle itself (Fig. 3).

The physiological properties of the organelle can also be utilized as a means of identifying specific organelles. Immunocytochemistry of ligated nerves with conventional and confocal microscopy indicates that kinesin is preferentially associated with anterograde moving organelles in the axon, although some kinesin is also present on retrograde moving organelles (Dahlstrom *et al.* 1991; Hirokawa *et al.* unpublished). Immunoelectron microscopy of similar nerves also indicates that kinesin is primarily on anterograde moving organelles, while an antibody to cytoplasmic dynein is present at similar levels for both anterograde and retrograde moving organelles (Hirokawa *et al.* 1990; unpublished). Light and electron microscopic immunocytochemistry thus suggest that the level of kinesin in a freely diffusible form is low in axons and probably in other cell types as well.

Biochemical and immunochemical evidence

If, as indicated by the immunocytochemistry, kinesin is primarily associated with membrane bounded organelles in the cell, then purified organelles should contain significant amounts of kinesin. Using quantitative immunoblots and immunogold electron microscopy of purified organelle fractions, a variety of organelle classes have been purified with kinesin bound to the membrane surface (Leopold *et al.* 1989). Kinesin-containing classes of membrane organelles include synaptic vesicles, a more

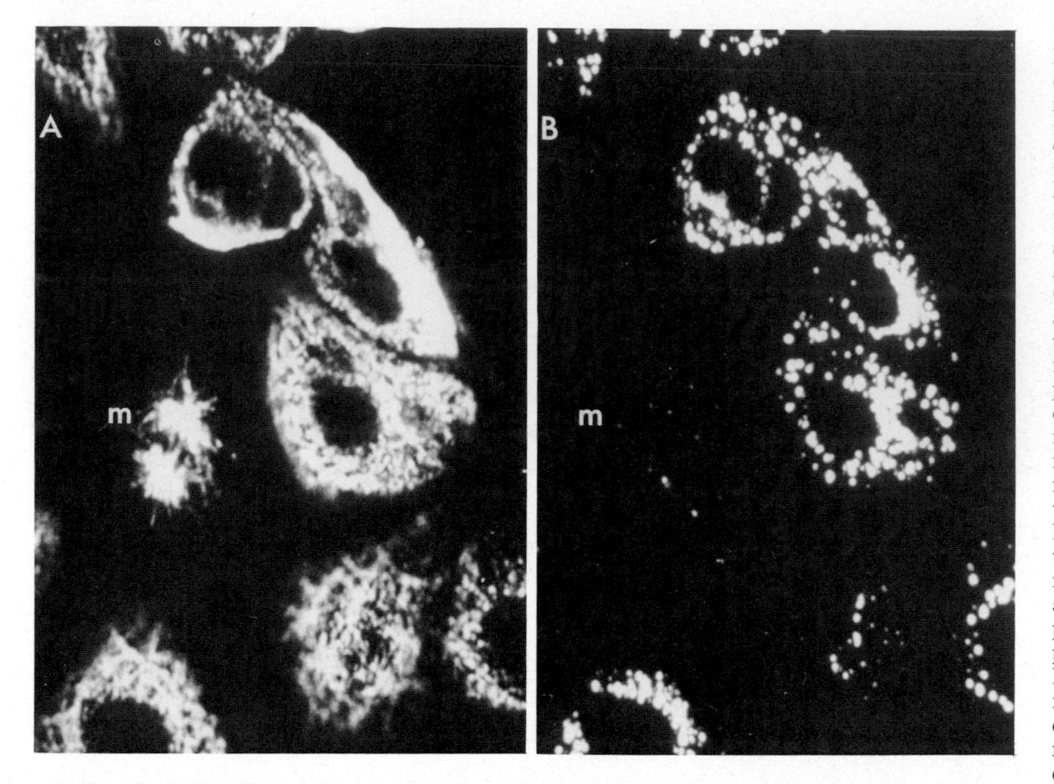

Fig. 3. Confocal microscopy of double immunofluorescence for kinesin and tubulin in interphase and mitotic cells. The cells were prepared as in Fig. 2, but examined using the Biorad MRC-600 scanning laser confocal microscope system attached to a Zeiss Axioskop microscope. (A) shows the tubulin pattern and (B) the same field with the kinesin pattern. The interphase cells show patterns in both A and B comparable to what was seen in conventional immunofluorescence microscopy. A mitotic cell (m) with spindle is visible on the left side of A and illustrates the characteristic array of microtubules on both sides of a presumptive metaphase plate. In the corresponding position of the field in B, little kinesin immunoreactivity can be detected in this plane of focus. Little or no kinesin is detectable in the mitotic spindles of mitotic cells and the overall amount of kinesin immunoreactivity appears reduced in mitotic cells. Confocal microscopy has the potential to provide information about the identity of kinesin reactive structures in the cell using double label immunofluorescence.

heterogeneous microsomal fraction, coated vesicles and mitochondria. In contrast, certain other classes of membrane bounded organelles such as the Golgi apparatus and nuclei do not contain detectable levels of kinesin, consistent with the results of immunofluorescence studies. Interaction with cellular membranes therefore appears to be selective to specific organelle surfaces. Moreover, the organization of kinesin on the surface of membrane bounded organelles in immunoelectron microscopic studies suggests that kinesin may form supramolecular complexes on the surface of isolated organelles, possibly mediated by specific receptors or by cooperative interactions between neighboring kinesins.

Studies on the axonal transport of kinesin itself have been conducted in the rat optic nerve, following metabolic labelling of retinal ganglion cell proteins with ^{35}S-methionine. Using our antibodies to kinesin, both heavy and light chains of kinesin were labelled in immunoprecipitates from nerves containing labelled fast axonal transport proteins. A peak of radiolabelled kinesin moves along the nerve, along with the peak of fast axonal transport. Thus, a substantial fraction of axonal kinesin is moving with membrane bounded organelles in fast axonal transport (Elluru *et al.* 1990). By contrast, very little labeled kinesin was detectable in nerves at times consistent with labeling of cytoplasmic proteins of the axon. The small amount of labelled kinesin detectable in the nerve at these times was uniformly distributed along the nerve and did not move down the nerve with slow axonal transport. This fraction of kinesin may be moving back to the cell body in retrograde axonal transport. These observations reinforce the suggestion that the bulk of kinesin in the axon is present on membrane bounded organelles and little is found in a cytoplasmic form.

Kinesin structure and interactions with membrane surfaces

The molecular architecture of kinesin has been explored using biophysical, immunochemical, electron microscopic and molecular genetic approaches. Kinesin holoenzyme is a tetramer containing two heavy chains and two light chains (Bloom *et al.* 1988; Kutznetsov *et al.* 1988). The protein is highly elongated, approximately 80 nm in length, with a globular head domain, a rod-like shaft and a fan-shaped tail region (Hirokawa *et al.* 1989) (Fig. 4A). The ATP binding site(s) and the microtubule binding sites are associated with the globular portion of the heavy chain, as shown by biochemical (Bloom *et al.* 1988; Penningroth *et al.* 1987; Kutznetsov *et al.* 1989; Scholey *et al.* 1989), electron microscopic (Hirokawa *et al.* 1989; Scholey *et al.* 1989) and molecular genetic (Yang *et al.* 1989; 1990) approaches.

Surprisingly, the location of the kinesin light chains is not on the globular head, as in the case of myosin. Immunolocalization of kinesin light chains in rotary shadowed electron microscopic studies indicate that they are found in the region of the fan-shaped tail (Hirokawa *et al.* 1989) (Fig. 4A). The function of the light chains are still open to speculation, although the assumption is that they serve some regulatory function. Kinesin heavy chains or kinesin heavy chain fragments retain the ability to bind to microtubules, hydrolyze ATP and induce microtubule gliding on coverslips in the absence of light chains (Kutznetsov *et al.* 1989; Scholey *et al.* 1989; Yang *et al.* 1990). The position of light chains on the tail do not exclude the possibility that they may act to regulate ATPase activity. A precedent for such 'long-range actions' can be found in the modulation of myosin ATPase by

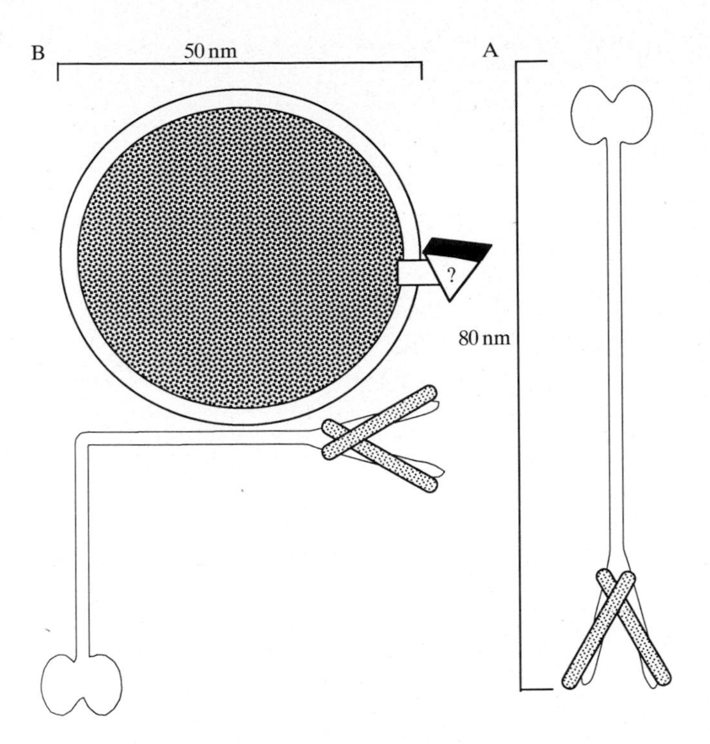

Fig. 4. Schematic diagram illustrating the approximate dimensions of kinesin and synaptic vesicle-sized membrane-bounded organelles. The length of the kinesin molecule (80 nm) (A) and the size of the crossbridges between membrane-bounded organelles and microtubules (20–30 nm) *in vivo* help establish limits for the interaction of kinesins with membrane-bounded organelles. A substantial length of the kinesin shaft and the tail domain with associated light chains (stippled rods) of kinesin must be interacting with the membrane surface (B). The existence of a receptor (labeled ?) for the kinesin on membrane-bounded organelles has been hypothesized, but has not yet been demonstrated. The functional implications of this substantial interaction are currently under investigation.

phosphorylation or proteolysis of the myosin heavy chain tail. However, no direct evidence for this mechanism is yet available for the kinesin ATPase, although Hackney has suggested this possibility from kinetic studies (1988 and elsewhere in this volume).

Several features of kinesin structure may be relevant for the interaction of kinesin with membranes (Fig. 4B). The localization of light chains on the fan-shaped tail, far removed from the microtubule and ATP binding regions in the head, is consistent with such a role for kinesin light chains. A hinged region is frequently noted on the shaft of rotary shadowed preparations of kinesin, located approximately halfway between the head and tail domains (Hirokawa *et al.* 1989). The length of crossbridges between vesicles and microtubules *in vivo* is of the order of 15–30 nm (Miller and Lasek, 1985; Hirokawa *et al.* 1989), which is less than half of the 80 nm length of kinesin. One possibility is that the hinge may be related to the interaction of kinesin with membrane surfaces (Hirokawa *et al.* 1989). This implies that a substantial stretch of the kinesin ATPase must be associated with the surface of the membrane (see Fig. 4B). The mechanisms by which this process may be mediated or regulated are currently unknown.

The existence of isoforms for both heavy and light chains of kinesin (Wagner *et al.* 1989; Pfister *et al.* 1989) raise the possibility that these may be either cell type-specific or organelle-specific forms of kinesin. These isoforms may be due either to post-translational modifications or to differences in primary sequence. Recent studies on the primary sequences of kinesin light chains suggest that some of the variability in light chains from mammalian brain can be related to primary sequence differences (Cyr *et al.* 1990; J. Cyr and S. T. Brady, unpublished data). Further investigation should allow us to distinguish between these two possibilities.

Functional implications of kinesin–membrane associations

Kinesin is readily isolated from a soluble fraction of brain or other tissue (see for example, Wagner *et al.* 1989; Kutznetsov and Gelfand, 1986; Cohn *et al.* 1987; Porter *et al.* 1987). Following differential extraction studies on chick fibroblasts, Hollenbeck (1989) reported that nearly 70 % of the kinesin was extracted into the soluble fraction by standard cell fractionation methods. However, the properties of kinesin are not consistent with a substantial, soluble pool of kinesin *in vivo*. The activity of soluble kinesin as a microtubule-activated ATPase should lead to a substantial amount of ATP hydrolysis in a cell with a

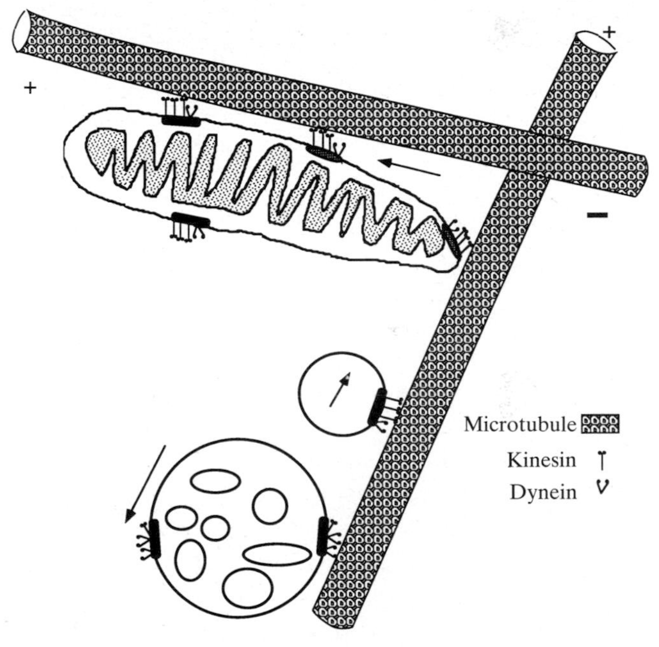

Fig. 5. Hypothetical organization of kinesin on the surface of various classes of membrane-bounded organelles *in vivo* and *in vitro*. Synthesis of results from immunofluorescence localization at the light microscopic level and immunogold electron microscopy suggests that the kinesin is not distributed uniformly on the surface of the organelle. Instead, the kinesin appears organized into 'patches' that may contain several kinesins and possibly cytoplasmic dynein motors as well. Receptors for the kinesin could be responsible for the clustering of kinesins and possibly of the dyneins as well. Alternatively, the substantial lengths of kinesin tail and shaft on the surface of the membrane could be involved in organizing the kinesins in a cooperative fashion. In either case, this organization may have a substantial impact on the functional properties of molecular motors *in vivo*. Although this scheme is speculative, the features illustrated are consistent with demonstrated aspects of kinesin distribution in cells and properties of organelle transport in the axon.

cytoplasmic microtubule cytoskeleton. Such ATP hydrolysis would not be coupled to translocation of cellular structures and would presumably constitute a drain on cellular energy resources. While putative regulatory elements could prevent such uncoupled ATP hydrolysis by soluble kinesin, no evidence is currently available to suggest that such a regulatory element exists. Alternatively, cellular kinesin could be associated with membrane bounded structures and the soluble pool may represent an artifact of homogenization.

Immunocytochemical (Pfister *et al.* 1989) and axonal transport (Elluru *et al.* 1990; Dahlstrom *et al.* 1991; Hirokawa *et al.* unpublished) studies do indicate that the bulk of cellular kinesin is associated with membrane bounded organelles. A fraction of cellular kinesin also cofractionates with purified membrane organelles from brain, such as synaptic vesicles and mitochondria (Leopold *et al.* 1989; 1990). The reason that most kinesin is recovered from the soluble fraction is uncertain at present, but may be related to the effects of standard biochemical buffer solutions on kinesin structure.

Conclusions

Studies on the cell biology of kinesins in the nervous system complement our continuing investigation of the biochemistry and molecular architecture of these motor molecules. The existence of multiple isoforms of kinesin in nervous tissue has been demonstrated by biochemical and molecular genetic methods. The importance of these variations in kinesin structure for enzymatic activity and motility are being evaluated and the functional significance of these differences are currently under investigation (see Fig. 5). Information about the interaction of kinesin with membrane surfaces is now also becoming available. By combining various approaches, we hope to obtain a more complete understanding of the cellular and molecular dynamics of the kinesins.

References

Bloom, G. S., Wagner, M. C., Pfister, K. K. and Brady, S. T. (1988). Native structure and physical properties of bovine brain kinesin and identification of the ATP-binding subunit polypeptide. *Biochemistry* **27**, 3409–3416.

Brady, S. T. (1985). A novel brain ATPase with properties expected for the fast axonal transport motor. *Nature* **317**, 73–75.

Cohn, S. A., Ingold, A. L. and Scholey, J. M. (1987). Correlation between the ATPase and microtubule translocating activities of sea urchin egg kinesin. *Nature* **328**, 160–163.

Cyr, J., Bloom, G. S. and Brady, S. T. (1990). Kinesin light chains: the generation of protein isoforms by multiple mRNA species. *J. Cell Biol.* **111**, 416a.

Dahlstrom, A., Pfister, K. K. and Brady, S. T. (1991). The axonal transport motor kinesin is bound to anterogradely transported organelles: quantitative studies of fast anterograde and retrograde axonal transport in the rat. *Acta Physiol. Skand.* (Manuscript in press).

Elluru, R. G., Bloom, G. S. and Brady, S. T. (1990). Axonal transport of kinesin in the rat optic nerve/tract. *J. Cell Biol.* **111**, 417a.

Endow, S. A., Henikoff, S. and Soler-Niedziela, L. (1990). A kinesin-related protein mediates meiotic and early mitotic chromosome segregation in *Drosophila*. *Nature* **345**, 81–83.

Enos, A. P. and Morris, N. R. (1990). Mutation of a gene that encodes a kinesin-like protein blocks nuclear division in *A. nidulans*. *Cell* **60**, 1019–1027.

Hackney, D. D. (1988). Kinesin ATPase: rate-limiting ADP release. *Proc. natn. Acad. Sci. U.S.A.* **85**, 6314–6318.

Hirokawa, N., Pfister, K. K., Yorifuji, H., Wagner, M. C., Brady, S. T. and Bloom, G. S. (1989). Submolecular domains of bovine brain kinesin identified electron microscopy and monoclonal antibody decoration. *Cell* **56**, 867–878.

Hirokawa, N., Yoshida, Y., Sato-Yoshitake, R. and Kawashima, T. (1990). Brain dynein localizes on both anterogradely and retrogradely transported membranous organelles. *J. Cell Biol.* **111**, 1027–1037.

Hollenbeck, P. J. (1989). The distribution, abundance, and subcellular localization of kinesin. *J. Cell Biol.* **108**, 2335–2342.

Kutznetsov, S. A. and Gelfand, V. I. (1986). Bovine brain kinesin is a microtubule-activated ATPase. *Proc. natn. Acad. Sci. U.S.A.* **83**, 8530–8534.

Kutznetsov, S. A., Vaisberg, E. A., Shanina, N. A., Magretova, N. A., Chernyak, N. M. and Gelfand, V. I. (1988). The quartenary structure of bovine brain kinesin. *EMBO J.* **7**, 353–356.

Kutznetsov, S. A., Vaisberg, E. A., Rothwell, S. W., Murphy, D. B. and Gelfand, V. I. (1989). Isolation of a 45 kda fragment from kinesin heavy chain with enhanced ATPase and microtubule-binding activities. *J. biol. Chem.* **264**, 589–595.

Lasek, R. J. and Brady, S. T. (1984). Adenylyl imidodiphosphate (AMPPNP), a nonhydrolyzable analogue of ATP, produces a stable intermediate in the motility cycle of fast axonal transport. *Biol. Bull.* **167**, 503.

Lasek, R. J. and Brady, S. T. (1985). AMP-PNP facilitates attachment of transported vesicles to microtubules in axoplasm. *Nature* **316**, 645–647.

Lazarovitis, J. and Roth, M. (1988). A single amino acid change in the cytoplasmic domain allows the influenza virus hemagglutinin to endocytose through coated pits. *Cell* **53**, 743–752.

Leopold, P. L., Pfister, K. K., Bloom, G. S. and Brady, S. T. (1989). Association of kinesin heavy and light chains with purified membrane bounded organelles from mammalian cells. *J. Cell Biol.* **109**, 81a.

Leopold, P. L., McDowall, A. W., Pfister, K. K., Bloom, G. S. and Brady, S. T. (1990). Immunogold localization of kinesin on purified synaptic vesicles and mitochondria. *J. Cell Biol.* **111**, 417a.

MacDonald, H. and Goldstein, L. S. B. (1990). Identification and characterization of a gene encoding a kinesin-like gene in *Drosophila*. *Cell* **61**, 991–1000.

Martz, D., Lasek, R. J., Brady, S. T. and Allen, R. D. (1983). Mitochondrial motility in axons: membranous organelles may interact with the force generating system through multiple surface binding sites. *Cell Motility* **4**, 89–102.

Meluh, P. B. and Rose, M. D. (1990). KAR3, a kinesin-related gene required for yeast nuclear fusion. *Cell* **60**, 1029–1041.

Miller, R. H. and Lasek, R. J. (1985). Crossbridges mediate anterograde and retrograde vesicle transport along microtubules in squid axoplasm. *J. Cell Biol.* **101**, 2181–2193.

Penningroth, S. M., Rose, P. M. and Peterson, D. D. (1987). Evidence that the 116 kda component of kinesin binds and hydrolyzes ATP. *FEBS Lett.* **222**, 204–210.

Pfister, K. K., Wagner, M. C., Stenoien, D. S., Brady, S. T. and Bloom, G. S. (1989). Monoclonal antibodies to kinesin heavy and light chains stain vesicle-like structures, but not microtubules, in cultured cells. *J. Cell Biol.* **108**, 1453–1463.

Porter, M. E., Scholey, J. M., Stemple, D. L., Vigers, G. P. A., Vale, R. D., Sheetz, M. P. and McIntosh, J. R. (1987). Characterization of microtubule movement produced by sea urchin egg kinesin. *J. biol. Chem.* **262**, 2794–2802.

Scholey, J. M., Porter, M. E., Grissom, P. M. and McIntosh, J. R. (1985). Identification of kinesin in sea urchin eggs, and evidence for its localization in the mitotic spindle. *Nature* **318**, 483–486.

Scholey, J. M., Heuser, J., Yang, J. T. and Goldstein, L. S. B. (1989). Identification of globular mechanochemical heads of kinesin. *Nature* **338**, 355–357.

Vale, R. D., Reese, T. S. and Sheetz, M. S. (1985). Identification of a novel force-generating protein, kinesin, involved in microtubule-based motility. *Cell* **42**, 39–50.

Wagner, M. C., Pfister, K. K., Bloom, G. S. and Brady, S. T. (1989). Copurification of kinesin polypeptides with microtubule-stimulated MgATPase and kinetic analysis of enzymatic properties. *Cell Motil. Cytoskel.* **12**, 195–215.

Yang, J. T., Saxton, W. M. and Goldstein, L. S. B. (1988). Isolation and characterization of the gene encoding the heavy chain of *Drosophila* kinesin. *Proc. natn. Acad. Sci. U.S.A.* **85**, 1864–1868.

Yang, J. T., Laymon, R. A. and Goldstein, L. S. B. (1989). A three domain structure of kinesin heavy chain revealed by DNA sequence and microtubule binding analyses. *Cell* **56**, 879–889.

Yang, J. T., Saxton, W. M., Stewart, R. J., Raff, E. C. and Goldstein, L. S. B. (1990). Evidence that the head of kinesin is sufficient for force generation and motility *in vitro*. *Science* **249**, 42–47.

Zhang, P., Knowles, B. A., Goldstein, L. S. B. and Hawley, R. S. (1990). A kinesin-like protein required for distributive chromosome segregation in *Drosophila*. *Cell* **62**, 1053–1062.

Purification and assay of kinesin from sea urchin eggs and early embryos

DAN BUSTER and JONATHAN M. SCHOLEY

University of California at Davis, Department of Zoology and Graduate Groups in Cell and Developmental Biology, and Biochemistry, Davis, CA 95616, USA

Summary

This paper describes the procedures used to purify the microtubule motor, kinesin, from mitotic cells, namely sea urchin eggs and cleavage stage embryos, and describes methods for assaying its motor activity.

Key words: kinesin purification, motility and ATPase, sea urchins, microtubule-based transport.

Introduction

Intracellular transport systems that move and position cytoplasmic particles such as organelles, chromosomes and perhaps macromolecules, play essential roles in the functioning of eukaryotic cells. Many particles are thought to be transported along microtubule 'tracks' by enzymes called 'motors'. One such motor, kinesin, is a microtubule-activated ATPase that transports particles towards the plus ends of microtubules (Vale *et al.* 1985*b*; Porter *et al.* 1987; Cohn *et al.* 1989; Howard *et al.* 1989). Kinesin has been isolated from a variety of eukaryotic organisms (reviewed by McIntosh and Porter, 1989) and several studies suggest that kinesin may perform a number of critical functions within the cell (Vale *et al.* 1986). Possible physiological functions include: anterograde transport of vesicles in neurons (Vale *et al.* 1985*a*; Schroer *et al.* 1988; Brady *et al.* 1990) and other cell types (Dabora and Sheetz, 1988*a*; Pfister *et al.* 1989), spreading of tubular membrane structures of lysosomes and the endoplasmic reticulum along microtubules (Dabora and Sheetz, 1988*b*; Vale and Hotani, 1988; Hollenbeck, 1989; Hollenbeck and Swanson, 1990), mitosis (Scholey *et al.* 1985; Leslie *et al.* 1987; Endow *et al.* 1990; Enos and Morris, 1990; Meluh and Rose, 1990) and using microtubule tracks to move and organize intracellular membranes associated with interphase and mitotic asters of early sea urchin embryos (Wright *et al.* 1991).

Sea urchin eggs and early embryos are well suited for the study of kinesin-driven motile events *in vitro* and *in vivo*. Sufficiently large quantities of eggs and synchronized dividing embryos can easily be obtained for biochemical and cytological studies of early embryogenic events such as chromosome movement, directed transport of vesicles and organelles and pronuclear migration (see Bloom and Vallee, 1990 for a review). In addition, microtubules and motors can be isolated from eggs and embryos and used to reconstitute microtubule-based motility (e.g. Porter *et al.* 1987; Cohn *et al.* 1987, 1989), and antibody microinjections into cells of the early embryo can be used to learn the function of motor proteins *in vivo* (Mabuchi and Okuno, 1977; Kiehart *et al.* 1982).

We have isolated and characterized kinesin from sea urchin eggs and early embryos (see Table 1 for a summary of the structural and functional properties of sea urchin kinesin). The motor binds microtubules in a nucleotide-sensitive fashion, and this property has been exploited in the first two purification procedures described below, which make use of the fact that kinesin binds to MTs (see Appendix for abbreviations) in the absence of ATP and ADP, in the presence of the nucleotide analogue, AMP-PNP, or in the presence of the divalent cation chelators, EDTA and PPP_i; subsequently, MT-bound kinesin can be released from MTs by adding MgATP. An alternative purification scheme makes use of anti-kinesin monoclonal antibodies that have been raised against sea urchin kinesin (Ingold *et al.* 1988); the latter were used in the last purification scheme described below which is a rapid immunoaffinity procedure using one of our anti-kinesin monoclonal antibodies covalently coupled to a Sepharose matrix. To assay kinesin mechanochemical activity, our lab routinely measures two properties of kinesin function: the MT-activated hydrolysis of ATP and the translocation of MTs over a surface coated with immobilized kinesin. The procedures used by our lab to purify and assay sea urchin kinesin will be described in detail below. Additional information can be obtained from articles by Scholey *et al.* (1984, 1985, 1989); Porter *et al.* (1987); Cohn *et al.* (1987, 1989); Leslie *et al.* (1987); Ingold *et al.* (1988); Johnson *et al.* (1990).

Collection of sea urchin eggs and early embryos, and preparation of cytosolic extracts

Most of our studies are performed with sea urchins, *Strongylocentrotus purpuratus*, obtained from the University of California Bodega Bay Marine Biology Laboratory or from Marinus, Inc. (Long Beach, CA). We sometimes also use *S. franciscanus*, *S. droebrachiensis*, *Lytechinus pictus*, and *L. variegatus*. *S. purpuratus* are transported as rapidly as possible in styrofoam boxes, packed in layers with paper well-soaked with sea water and cooled with ice packs. The urchins can be processed for gamete production immediately upon their arrival in the lab or they may be placed in well-aerated tanks and processed at leisure. Alternatively, the gametes of sea urchins (*S. franciscanus* or *S. purpuratus*) are collected at Bodega Bay and transported in jars suspended in large containers of 10–15 °C sea water.

1. Sea urchins are induced to shed their gametes by injection of approx. 2 ml of 0.56 M KCl solution into the coelomic cavity. The yellow eggs are collected by inverting the shedding females onto beakers filled with sea water for 1–2 h. The white sperm are collected by inverting males

Journal of Cell Science, Supplement 14, 109–115 (1991)

onto dry petri dishes. For production of HSS, perform this and subsequent steps at 0–4 °C; for developmental studies, keep the eggs at their physiological temperature of approx. 15 °C.

Note: Contamination of eggs with even traces of sperm must be rigorously avoided to prevent unexpected fertilization of the eggs. Thorough rinsing of hands, containers, and tools with tap water will kill unwanted sperm.

2. Decant excess sea water from the egg-containing breakers, pool the eggs, and pass the eggs 10–15 times through sea water-moistened Nitex screen (150 μm mesh; Small Parts, Inc.) to remove debris and the eggs' jelly coats.

Production of cytosolic extracts (HSS) (steps 3–5)

3. To produce HSS, transfer the eggs to 50 ml plastic tubes and gently pellet using a low speed, short duration (1 min) spin in a clinical centrifuge. Aspirate off the supernatant and gently resuspend the egg pellets in approx. 7 vols '19:1 buffer' (see Appendix for buffer recipes). Repeat the centrifugation and resuspension steps twice more to remove completely the eggs' jelly coats and to minimize the Ca^{2+} concentration (many sea urchin proteases are Ca^{2+}-activated and MT polymerization is Ca^{2+} sensitive). Then resuspend the eggs in sufficient PMEG buffer to reduce the NaCl concentration to less than 0.1 M and gently pellet again.

4. Resuspend the eggs in 2 vols PMEG buffer and homogenize on ice using a chilled Dounce homogenizer until intact eggs are no longer visible under a stereomicroscope (about 10–20 strokes).

5. Centrifuge the homogenate at 85 000 ***g*** for 30 min at 4°C. Discard the large pellet and the orange, top lipid layer; save and centrifuge the clear, intermediate supernatant at 175 000 ***g*** for 1 h at 4°C. Once again, save the clear, intermediate, high speed supernatant (HSS). This material is a cytosolic extract and can be stored for extended periods if immediately frozen in liquid nitrogen and then transferred to −80 °C.

Embryo production (steps 3a–6a)

3a. For developmental studies, resuspend the dejellied eggs in approx. 5 vols sea water (pH 8.1, 15 °C) containing 10 mM *p*-aminobenzoic acid (PABA) to soften fertilization membranes. Aspirate off the supernatant when the eggs have settled. Repeat this step 2–3 times.

4a. Resuspend the eggs in at least 10 vol of previously well-aerated sea water (containing 10 mM PABA, pH 8.1, 15 °C) and *very gently* stir with a paddle driven by an overhead electrical motor. The metabolic rate of eggs greatly increases upon fertilization. Adequate dilution of the eggs into well-aerated sea water is necessary to prevent asphyxiation.

5a. Activate sperm by diluting 2 drops of sperm into 50 ml sea water (use within 15 min). To fertilize eggs, add approx. 0.0025 vol. of diluted sperm to the diluted eggs. Caution: different sea urchin species are differentially prone to polyspermy and the quantity of sperm used must be varied accordingly (see Leslie and Wilson, 1989).

6a. The extent of successful fertilization can be assessed within a few minutes by observing an aliquot of eggs under a stereomicroscope: fertilized eggs will be ringed by a clearly visible fertilization membrane. Egg development is improved if the dilute egg suspension is occasionally *gently* aerated. Well-fertilized and aerated eggs remain largely synchronized in their mitotic cycles through two or three divisions.

7a. Aliquots of developing embryos can be taken at selected points in the mitotic cycle for HSS production or histological studies. For HSS production, the embryos must be passed through a sea water-moistened Nitex screen (~10 times) to remove the fertilization membranes. The embryos can then be processed into HSS as described above (steps 3–5).

Purification procedures

Kinesin purification by AMP-PNP-induced MT affinity binding (Scholey *et al.* 1985; Cohn *et al.* 1987; Ingold *et al.* 1988)

Depletion of actomyosin and denatured protein. 1. Thaw frozen HSS (stored at −80 °C in 50 ml plastic tubes) in a beaker of tap-fed cool water. Move the thawed HSS to ice immediately.

2. Incubate HSS with hexokinase (10 units ml^{-1}) and 50 mM glucose (to convert ATP to ADP) at room temperature for 30 min. Then centrifuge at 100 000 ***g*** (30 min, 4°C) to pellet actomyosin and denatured protein, leaving kinesin and unpolymerized MT protein in the high speed supernatant.

Induced MT binding by kinesin. 3. Kinesin can be removed from the supernatant if kinesin is induced to bind to MT polymers, and then the kinesin–MT complex is pelleted. First, add 0.5–1 mM GTP and 20 μM taxol and incubate at room temperature (RT) for 15 min to promote MT assembly in the supernatant. Supplement the solution with ⩾1 mM AMP-PNP to enhance MT binding by kinesin and incubate at RT for 20 min. Sometimes we have used 5 units ml^{-1} apyrase in place of AMP-PNP to induce MT-binding by kinesin.

4. Centrifuge the solution over a 15 % sucrose cushion (containing 1 mM GTP, 2.5 μM taxol may be added to stabilize MTs) in a swinging bucket rotor at 23 000 ***g***, 60 min, 10°C (or 57 000 ***g***, 20 min, 10°C, in fixed angle rotor) to pellet the kinesin–MT complexes.

5. Wash the pellet by homogenizing in 0.1 mM GTP in PMEG buffer (or in 20 mM EDTA, 2 mM ATP, 0.1 mM GTP, 10 μM taxol in PMEG buffer with magnesium sulfate omitted) and let stand 10 min at 4°C. Centrifuge in a swinging bucket rotor at 45 000 ***g***, 20 min, 4°C (or at 57 000 ***g***, 20 min, 4°C, in a fixed angle rotor).

6. Release kinesin from MTs in the pellet by homogenizing the pellet in 100 mM KCl, 10 mM MgATP (0.1 mM GTP, 5 μM taxol may be included) and holding overnight at 0–4°C. Add a further 5 mM MgATP before centrifuging at 45 000 ***g***, 20 min, 4°C to pellet MTs and leave MAPs in supernatant.

Biogel A5M chromatography. 7. Fractionate the extracted MAPs supernatant by gel filtration chromatography (Biogel A5M (Bio-Rad), 1.8 cm×12 cm) in PMEG buffer containing 1 mM ATP (ATP successfully omitted in some experiments).

8. Fractions are tested for MT-translocating activity by video microscopy (see below), and analysed by SDS–PAGE and Western blotting (using anti-kinesin monoclonal antibodies). The peak of kinesin activity constitutes those fractions that will support MT movement at ⩾0.4 μm s^{-1}. These fractions are pooled for further analysis.

MT binding and release. 9. The pooled kinesin fractions are further purified by addition of 1.5 mg ml^{-1} P11 phosphocellulose-purified bovine brain MTs (polymerized by incubation with >10 μM taxol and 0.5 mM GTP for 15 min at RT; tubulin purified according to the method of

Table 1. *Biochemical properties of sea urchin kinesin*

1. Polypeptide subunits:	►2 mol 130×10^3 M_r heavy chain ►2 mol $78/84\times10^3$ M_r light chain
2. Size/morphology:	►Sedimentation coefficient≃9.6 S ►2 heads, stalk and tail in EM ►Total length≃75 nm ►Stalk diameter≃3–5 nm ►Globular head diameter≃9–10 nm ►Tails≃heterogeneous
3. Globular heads:	►Released by proteolysis as 45×10^3 M_r fragments ►Formed from N-terminal portion of heavy chain ►Contain ATP and MT binding sites ►Motor domains
4. MT-activated MgATPase: (nmol min^{-1} mg^{-1})	►Intact kinesin typically≃50 ►Kinesin proteolysed to 45×10^3 M_r≃520–615
5. Motility:	►Plus-end directed ►Blocked by some monoclonal antibodies that bind heads ►*Substrates*; broad specificity for Mg nucleotides (ATP>GTP>TTP≅UTP>CTP>ITP) MgATP; K_m≃60 μM, V_{max}≃0.6 μm s^{-1} MgGTP; K_m≃2 mM, V_{max}≃0.4 μm s^{-1} ►*Inhibitors*; Mg-free ATP and Mg^{2+} chelators (EDTA, PPPi) MgATPγS; competitive inhibitor, K_i≃15 μM MgADP; competitive inhibitor, K_i≃150 μM AMP-PNP, vanadate; potent/complex inhibitors AMP-PCP, NEM; weak/complex inhibitors

Williams and Lee, 1982), 20 mM EDTA, and 2 mM ATP, and incubation for 20 min at RT. Pellet the resulting kinesin–MT complex by centrifugation (45 000 ***g***, 20 min, 20°C).

10. Resuspend the kinesin–MT pellet in PMEG buffer containing 10 mM ATP, 12.5 mM magnesium sulfate, 30 μM taxol, and 0.1 mM GTP for 30 min, RT, to release kinesin from MTs. Centrifuge the solution again at 45 000 ***g***, 20 min, 20°C to pellet MTs and leave purified kinesin in the supernatant.

P11 phosphocellulose chromatography. 11. Tubulin and ATP can be removed from the kinesin-containing supernatant by P11 phosphocellulose chromatography; kinesin (but not tubulin or ATP) binds to P11 phosphocellulose resin (Whatman). Activate P11 phosphocellulose resin according to manufacturer's instructions and equilibrate in PMEG buffer. Mix 1 vol P11 to 10 vols of supernatant from step 10 (or from step 8 if required) for 30 min, 4°C.

12. Wash the resin batch-wise (by centrifuging in a clinical centrifuge) 2 times with 5 vols PMEG buffer, then 2 times with 5 vols PMEG containing 0.1 M KCl to remove non-binding material. Pack the resin into a column and then elute kinesin using PMEG buffer containing ⩾0.5 M NaCl or KCl. Collect 25 drop fractions and use SDS–PAGE to locate kinesin-containing fractions. Dialyze the pooled fractions against PMEG buffer (0–4°C) to remove salt.

The purified kinesin has both ATPase and MT translocating activities and can be stored for extended periods in liquid nitrogen. Using this method, we obtain 50–500 μg kinesin from 100 ml HSS (see Fig. 1A).

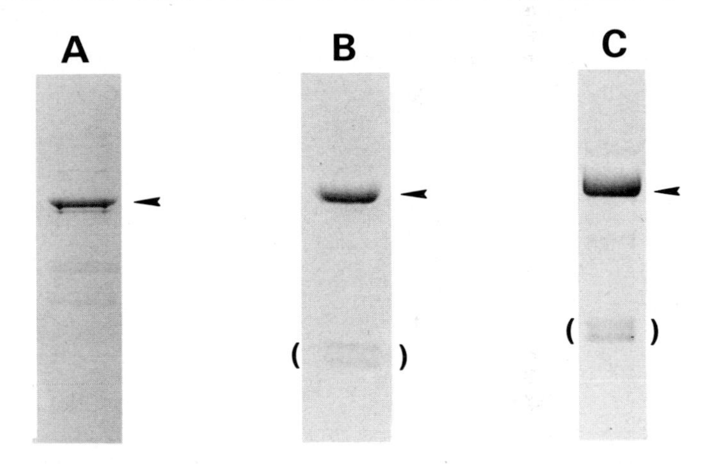

Fig. 1. SDS polyacrylamide gel electrophoresis of purified sea urchin kinesin. The three lanes are of kinesin samples analyzed on different, Coomassie-stained gels. Kinesin was purified by (A) AMP-PNP-induced MT binding (procedure I), (B) PPP_i-induced MT binding (procedure II), and (C) immunoaffinity chromatography (procedure III). Arrows indicate the 130×10^3 M_r kinesin heavy chain polypeptides. Brackets mark the 84 and 78×10^3 M_r kinesin light chain polypeptides. Procedure I yields active kinesin with variable light chain content, whereas procedure II yields kinesin with similar MT-translocating properties and a consistent light chain content (1 mole heavy chain: 1 mole light chain doublet). Procedure III yields kinesin that is usually inactive (because of contaminating SUK4) but has consistent light chain content (1 mole heavy chain: 1 mole light chain doublet). Contaminating polypeptides can be recognised because they also cosediment with control IgG resins.

Kinesin purification by PPP_i-induced MT affinity binding (Johnson *et al.* 1990)

P11 phosphocellulose chromatography of HSS. 1. Thaw 150 ml of HSS (stored at −80°C) and supplement with fresh protease inhibitors (see Appendix). Add HSS to approx. 20 ml of activated P11 phosphocellulose resin and gently mix the slurry for 1 h (all steps at 0–4°C). Remove non-binding proteins by washing the P11 resin 3 times with about 4 vols of PMEG buffer, then once with 4 vols of PMEG buffer containing 0.1 M KCl. A clinical centrifuge is used to pellet the resin after each wash.

2. Bound proteins (including kinesin) are eluted with 2 vols of PMEG buffer containing 0.75 M KCl. The eluate is passed through glass wool to remove any residual P11 resin – the presence of P11 resin in subsequent steps

greatly decreases kinesin yield. Dilute eluate with 1 vol of PMEG buffer.

Ammonium sulfate precipitation. 3. Precipitate kinesin from the eluate by adding 0.39 g ammonium sulfate per ml eluate (60 % final concentration) and stirring overnight at 4 °C.

4. Precipitated material is pelleted by centrifugation at 17 250 **g** for 10 min, 4 °C. Resuspend the pellet in 3 ml PMEG buffer and clarify by centrifugation at 17 250 **g**, 10 min, 4 °C.

Biogel A1.5M chromatography. 5. The clarified supernatant is fractionated and desalted by gel filtration chromatography on a Biogel A1.5M (Bio-Rad) column (2.5 cm × 15 cm) equilibrated with PMEG buffer containing 2 mM ATP and 100 μM (instead of 2.5 mM) magnesium sulfate. Kinesin-containing fractions are identified by SDS–PAGE and immunoblotting, pooled, and then supplemented with 20 μM (final concentration) taxol.

PPP_i-induced MT binding: 6. Polymerize MTs by mixing a previously prepared, concentrated solution of P11 phosphocellulose-purified bovine brain tubulin (sufficient tubulin is added to make the tubulin final concentration 1–2 mg ml^{-1} after the pooled Biogel fractions are added at step 7), with roughly equimolar taxol (10 μM per 1 mg ml^{-1} tubulin), and 0.5 mM GTP. Incubate at 37 °C for 30 min.

7. Chill and then mix the MT solution with the pooled, taxol- and kinesin-containing Biogel fractions and incubate for 10 min, 4 °C. Add 20–30 mM (final concentration) PPP_i (or EDTA) to promote kinesin binding to MTs and incubate for a further 10 min, 4 °C. The kinesin–MT complexes are pelleted by centrifugation at 85 000 **g**, 20 min, 4 °C.

8. Kinesin is released from the MTs by resuspending the pellet in 2 ml of cold resuspension buffer using a chilled Dounce homogenizer and incubating for 1 h at 4 °C. Pellet MTs by centrifugation at 85 000 **g**, 20 min, 4 °C.

P11 phosphocellulose chromatography. 9. Mix the kinesin-containing supernatant with 1–2 ml of activated P11 phosphocellulose resin and gently rock the slurry at 4 °C for 1 h. Wash the P11 resin 2 times with about 8 vols of PMEG buffer, and wash twice more with PMEG buffer containing 0.1 M KCl to remove non- and weakly-binding proteins. After each wash, gently pellet the resin in a clinical centrifuge before resuspending in the next wash buffer.

10. Pack the washed P11 resin into a disposable 11 ml polystyrene column (Bio-Rad), and elute kinesin by developing the column with PMEG buffer containing 0.6 M KCl and collecting 25 drop fractions.

Kinesin-containing fractions can be identified by SDS–PAGE and dialyzed against PMEG buffer to remove KCl. This procedure yields ~0.1 mg active kinesin from 100 ml HSS (see Fig. 1B).

Kinesin purification by monoclonal antibody immunoaffinity chromatography (Johnson *et al.* 1990)

For immunoaffinity chromatography, we routinely use the monoclonal anti-kinesin, SUK4, prepared from mouse ascitic fluid using Protein A Affigel and the MAPs II kit (Bio-Rad) as described previously (Ingold *et al.* 1988; Johnson *et al.* 1990). Nonspecific mouse IgG and rat IgG purchased from Sigma are routinely used as controls for non-specifically sedimenting polypeptides. We usually immobilize the SUK4 to cyanogen bromide-activated Sepharose 4B (Pharmacia) to produce an effective immunoadsorbent. We have used SUK4 Sepharose immunoadsorbents on fresh and frozen HSS from eggs and embryos, and we have observed no significant differences in the kinesin preparations obtained. Our current immunoaffinity protocol is described below.

Preparation of HSS. 1. Thaw frozen HSS and supplement with fresh protease inhibitors (see Appendix). Clarify by centrifugation at 150 000 **g**, 1 h, 4 °C, to remove any denatured protein.

Elimination of non-specific IgG-binding proteins. 2. Gently mix the HSS with 1/20 vol rat IgG-Sepharose for 1–2 h to remove nonspecific IgG-binding proteins (all steps at 4 °C). The resin is pelleted by brief centrifugation in a clinical centrifuge and the HSS saved. The resin is regenerated with successive washes of 1 M NaCl in 10 mM phosphate buffer, pH 8.5, then 1 M NaCl in 0.1 M diethylamine buffer pH 11.5, and then PMEG buffer.

Immunoprecipitation. 3. The preadsorbed HSS is divided and mixed with either control (non-specific mouse IgG) or SUK4 resin that has been incubated with an equal volume of 10 mg ml^{-1} SBTI for 1–2 h at 4 °C to saturate nonspecific protein binding sites. Approximately 1 vol of resin is incubated with 50 vols HSS for 1–2 h (4 °C) with gentle mixing.

4. Wash the resins 3 times with 30 vol PMEG buffer. Then wash with 30 vol of 1 M NaCl, 10 mM ATP (pH 7) to remove kinesin-associated proteins (see Schroer *et al.* 1988), and then with 30 vol 1 M NaCl in 10 mM phosphate buffer pH 8.5 (to get the resin into a higher pH solution in preparation for elution).

5. A 50 μl sample of a resin may be boiled in SDS sample buffer (Laemmli, 1970) for PAGE analysis. The remainder of the resin (approx. 400–450 μl) is treated with at least 2 vols of 1 M NaCl, 0.1 M diethylamine pH 11.5 to elute the bound kinesin. The eluate can be neutralized by immediate addition of 0.1 M sodium acetate (pH 4) or 1 M dibasic phosphate. Alternatively, kinesin can be precipitated from the eluate using 20 % trichloroacetic acid and washed with cold acetone.

Kinesin purified by this procedure lacks MT translocating activity, possibly as a result of a small amount of the function-blocking monoclonal, SUK4, leaching from the column into the eluate or because the elution conditions denature the kinesin. In some experiments using the same elution conditions, we were able to recover active kinesin using immunoadsorption with SUK2-Sepharose, although the efficiency of adsorption was lower than with SUK4-Sepharose. Approximately 0.5–1 mg kinesin can be immunoaffinity purified from 100 ml HSS (Fig. 1C). The resins can be re-used if they are immediately neutralized with PMEG buffer and regenerated with the same washing protocol as is used with the rat IgG-Sepharose (above). We have found that some kinesin (presumably denatured) becomes irreversibly bound to re-used SUK4-Sepharose resin. This residual kinesin does not elute from the resin and, therefore, presumably does not contaminate subsequent immunoaffinity purification experiments.

Assay procedures

Video microscopic assay of sea urchin kinesin-driven microtubule motility

We have adapted the original MT gliding assay described by Vale *et al.* (1985*a*) to allow us to measure conveniently the speed of kinesin-driven MT movement over glass surfaces in real time. This is a highly reproducible, quantitative assay that allows us to characterize motor enzymology and screen for probes that inhibit kinesin-driven transport using very small quantities of the motor (on the order of 1 μg per assay).

The motility assay (Porter *et al.* 1987; Cohn *et al.* 1987, 1989)

1. Spread a solution of purified kinesin (15 μl) onto a No. 1 coverslip and allow it to adsorb for 20 min in a humidified chamber.
2. After adsorption, 2 μl of approx. 0.2 mg ml^{-1} MTs (bovine brain tubulin purified by P11 phosphocellulose chromatography and polymerized in 1 mM GTP and 20 μM taxol) are added, along with 0.5–2.0 μl each of nucleotide and/or inhibitor solution (made up as stock solutions of 1–100 mM) to obtain a final volume of approx. 20 μl.
3. Invert the coverslip and seal the edges to a microscope slide with molten VALAP (vaseline/lanolin/paraffin, 1:1:1 (w/w/w)).
4. The sealed slide is mounted on the microscope, taking care to avoid bubbles or dust in the immersion oil. Focus onto the coverslip surface looking through the ocular, adjust field iris and condenser with lamp set for critical illumination and differential interference contrast (DIC) optics set at extinction. Redirect image to monitor, then turn slider away from extinction until MT-covered coverslip surface is visible. 'Fine tune' contrast on monitor using optics and camera control box by trial and error. Use the computer mouse to track and measure velocities of 10–30 MTs, and finally command the computer to print out the velocity histogram, with the mean MT velocity and the standard deviation.

In our lab, kinesin-driven MT translocation is visualized by use of a video-enhanced microscope system consisting of a Zeiss standard research microscope with a ×100 (NA 1.25) objective, ×2 optavar, a wide-band-pass green interference filter (wavelength 546 nm), and standard DIC optics. The microscope is equipped with a 100 W mercury arc lamp set at a critical illumination and is connected to a MTI Series 68 Newvicon video camera (equipped with remote gain and black offset level) whose output is directed to a RCA monochrome monitor. Using this set up, we find that no computer background subtraction is needed to visualize the MTs (Porter *et al.* 1987; Cohn *et al.* 1987, 1989). The velocities of MTs 'gliding' across the coverslip are measured in real time using computer-assisted analysis (Cohn *et al.* 1987, 1989). The video output of an Amiga 1000 computer is mixed with the output of the MTI video camera using a video mixer installed by DAGE Inc. – allowing the image of the computer's mouse-controlled cursor (an arrow) to be superimposed on the image of the MTs. Alternatively, a Genlock-type video mixer can be connected to the Amiga computer directly, for better resolution and greater versatility. A computer program (Appendix) monitors the x and y coordinates of the cursor, as well as the time measured by an internal computer clock. In practice, the cursor is moved to the site of a MT end and the mouse input button triggered to mark its position and the time of the observation. After an arbitrary period (usually 10–30 s), the cursor is moved to the new site of the same MT end and again the mouse button triggered. The program calculates the distance between the two x,y coordinates (using scalar factors between x,y coordinates and actual distances measured previously with a stage micrometer) and divides this value by the time difference to obtain the velocity of the translocating MT. After the measurement of an arbitrary number of MT velocities (usually 10–30), the program will calculate the mean velocity and standard deviation and also print out a histogram of MT velocities.

We have also used the assay to monitor the velocity of MT-gliding induced by 21S dynein prepared by sucrose density gradient centrifugation from sea urchin sperm flagellar axonemes. We find, however, that the dynein often moves MTs too fast for us to accurately find MT ends, causing greater errors in the measured velocity. Therefore, we would recommend the restricted use of this assay for accurately measuring velocities ≤2 μm s^{-1}. Other advantages and limitations of this assay are discussed by Cohn *et al.* (1989).

Use of MT motility assays for analysis of steady state kinetic parameters of kinesin (Cohn *et al.* 1989)

Using the procedure described above, MT motility assays and their analyses can be performed to determine the kinetic parameters of kinesin-driven MT translocation.

1. Adsorb 14–15 μl kinesin to a glass coverslip for 20 min, then add the desired concentrations of substrates, inhibitors and phosphocellulose-purified bovine brain MTs (to 20 μg ml^{-1}) to a final volume of 20 μl. Nucleotide stock solutions are made up as 100 mM solutions of 0.1 M Tris, pH 7. Inhibitor solutions are made up as 100 mM stocks in either 0.1 M Tris pH 7 or PMEG buffer. Mg-nucleotides (or analogs) are prepared by adding 100 mM $MgSO_4$ to the stock solutions. Dilutions of working solutions are made in PMEG buffer.
2. Measure the velocities of at least 10 MTs for each data point and each experiment should be performed on at least two different preparations of kinesin. For determinations of K_m and K_i, kinesin is used at a concentration of 250–300 μg ml^{-1}, which is at least 4–5 times the critical concentration necessary for MT motility (Cohn *et al.* 1987, 1989).

   Note: In order to eliminate any decrease in ATP concentrations during the motility assays due to hydrolysis, perform all experiments using ATP at concentrations ≤100 μM (except those involving measurements of ADP) in the presence of an ATP regenerating system of 10 units ml^{-1} creatine-phosphokinase and 50 mM phosphocreatine. Samples containing creatine-phosphokinase and phosphocreatine retain MT motility for several hours, even when using ATP concentrations as low as 10–25 μM.
3. The V_{max} and K_m of substrates that support kinesin-driven MT-motion can be obtained from Lineweaver-Burk plots (Fig. 2). To study compounds interfering with MT translocation, the type of inhibition is determined by plotting the reciprocal of MT velocity *versus* the reciprocal of substrate concentration at a constant inhibitor concentration (Lineweaver and Burk, 1934) or the reciprocal of MT velocity *versus* inhibitor concentration at constant substrate concentration (Dixon, 1953).

Use of motility assays for identifying function-blocking kinesin antibodies (Ingold *et al.* 1988)

The ability of antibodies to inhibit the MT-translocating activity of kinesin can be assayed using video-enhanced microscopy and computer-assisted velocity analysis as described above.

1. Adsorb 14 μl of 100–300 μg ml^{-1} kinesin solutions onto coverslips for 10–20 min at RT in a humidified chamber before addition of the test antibodies. Then add antibody solutions to the coverslips for an additional 10–20 min. Routinely we use solutions of monoclonal antibodies purified from ascitic fluid. These can be

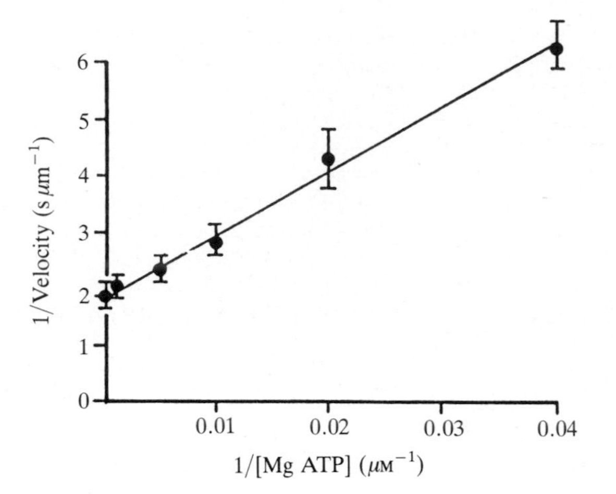

Fig. 2. A Lineweaver-Burk plot of kinesin-driven microtubule gliding rate as a function of MgATP concentration. Values for the maximal velocity (V_{max}) and the Michaelis constant (K_m) can be obtained from this plot (Table 1; see Cohn *et al.* 1989 for details).

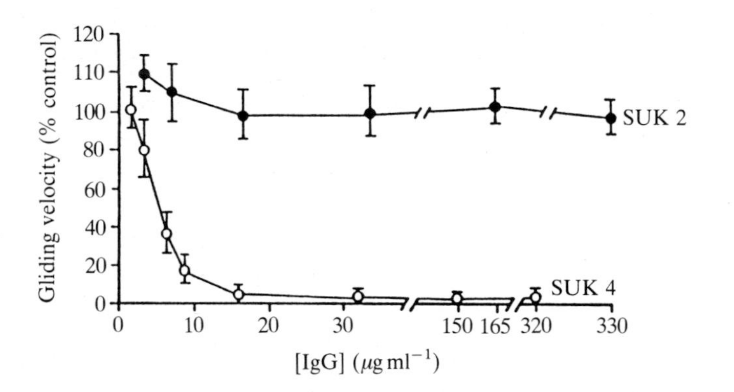

Fig. 3. Effects of two monoclonal antibodies on kinesin-driven microtubule gliding. SUK2 is an IgG that binds the stalk of kinesin and does not significantly inhibit microtubule movement, whereas SUK4 is an IgG that binds to the kinesin head and blocks kinesin-driven motility. (See Ingold *et al.* 1988 for details).

obtained at sufficiently high concentrations that only small (1–2 μl) volumes are needed.

2. Add polymerized, P11 phosphocellulose-purified, bovine brain MTs (20 μg ml^{-1}) and MgATP (10 mM) to the coverslip, and seal the coverslip to a slide using VALAP.

3. Determine the velocity of a gliding MT over an interval of at least 10 s and randomly select 10–20 MTs from several microscopic fields to be used for each data point. Each set of data points and standard deviations in Fig. 3 is from single typical experiments that have been reproduced many times.

The ATPase assay (Cohn *et al.* 1987)

ATPase assays are performed at RT by modification of the method of Seals *et al.* (1978). The assays themselves are quantitative and reproducible (even though we have observed variability in the specific ATPase activity of our kinesin preparations; see Cohn *et al.* 1987), but require larger quantities (~100 μg per assay) of kinesin. The specific ATPase activities of our tubulo–kinesin preparations (<100 nmol min^{-1} mg^{-1} at 25°C) are usually much lower than those found by other workers, but the activity can be stimulated markedly by addition of function blocking antibodies (Ingold *et al.* 1988) or proteolysis (Table 1).

1. Kinesin (prepared by purification procedure 1 or 2, above) in 200–300 μl PMEG buffer is mixed with 40 μl phosphocellulose-purified bovine brain MTs (10 mg ml^{-1}) or with 40 μl buffer plus 1 μl of 10 mM taxol (as a control).

2. Add sufficient PMEG buffer to each assay tube to give a final volume of 380 μl, and incubate the mixture for approx. 10 min at RT.

3. To start the reaction, add 20 μl of 50 mM [γ-^{32}P]MgATP (specific activity 10 000–40 000 cpm nmol^{-1} ATP). Remove aliquots (100 μl) at a minimum of three time points (0, 5, 10, 15, 20, 30 or 60 min), and quench the reaction in the sample by addition of 30 μl 10 % SDS.

4. Add 75 μl phosphate reagent (4 vols 5 N sulfuric acid/5 % ammonium molybdate: 1 vol 0.1 M silicotungstic acid) to react with liberated phosphate; extract the resulting ^{32}P-labelled phosphomolybdate complex with 1 ml of a 65 %:35 % (v/v) xylene:isobutanol mixture.

5. After centrifugation in a microfuge to separate organic and aqueous phases, analyze a 0.7 ml aliquot of the organic phase (containing the ^{32}P-phosphomolybdate) of each sample by scintillation counting.

6. To calculate ATPase specific activity, aliquots of the [γ-^{32}P]MgATP stock solution are scintillation counted to determine cpm nmol^{-1} ATP for the stock solution. The concentration of liberated phosphate for each sample from step 5 can be calculated by dividing the sample's cpm by the value for cpm nmol^{-1} ATP. The ATPase specific activity (nmol phosphate min^{-1} mg^{-1}) is calculated by dividing the slope of plots of free phosphate concentration as a function of time by the protein concentration of the kinesin sample.

Conclusion

It is clear that kinesin performs a number of vital functions within the cell, all of which are a consequence of kinesin's ability to translocate cellular structures along MT tracks. Sea urchin eggs and early embryos have proved to be a rich source of kinesin and also serve as amenable experimental subjects for cytological investigations of kinesin function in fixed cells and *in vivo*. Our lab has concentrated on isolating and characterizing sea urchin kinesin, and has developed the purification and assay procedures described here. This work has provided us with basic information on the structure and *in vitro* functions of the sea urchin kinesin motor molecule; Table 1 summarizes our current understanding of these properties of sea urchin kinesin. The isolation and assay protocols described in this paper may assist other workers who wish to study kinesin-like proteins isolated directly from their natural host cells, or expressed *in vitro* or in transformed cells from cloned and manipulated genes.

Appendix

Abbreviations

AMP-PNP: 5′-adenylyl imidodiphosphate.
HSS: high speed supernatant of cytosolic extract.
MAP: microtubule associated protein.
MT: microtubule.
PAGE: polyacrylamide gel electrophoresis.
PPP_i: tripolyphosphate.

SBTI: soybean trypsin inhibitor.

SUK: sea urchin kinesin (notation used for monoclonal antibodies).

Buffer solutions (all concentrations are final)

1. Sea water: 760 g Instant Ocean sea salts brought to 20 liters with tap water (check specific activity= 1.025–1.026, 4°C).

2. 19:1 buffer: 530 mM NaCl, 28 mM KCl, 1 mM EDTA, 5 mM Tris HCl in distilled water, pH 7.

3. PMEG buffer: 0.1 M potassium Pipes, 2.5 mM magnesium sulfate (promotes MT assembly), 0.1 mM EDTA (chelates divalent metal ions which catalyze sulfhydryl oxidation), 5 mM EGTA (chelates Ca^{2+} which inhibits MT assembly and activates some proteases), 0.9 M glycerol (establishes correct osmolarity for eggs) in distilled water, pH 6.9. Stored at 4°C.

Protease inhibitors: 1 μg ml^{-1} pepstatin and leupeptin, 2 μg ml^{-1} aprotinin, 100 μg ml^{-1} SBTI, 0.1 mM phenyl methyl sulfonyl fluoride, 1 mg ml^{-1} tosyl arginine methyl ester, 1 mM dithiothreitol, 20 μg ml^{-1} benzamidine, 1 mM sodium azide. Protease inhibitors are added to PMEG buffer immediately before use.

4. Resuspension buffer: 0.5 mM GTP, 30 μM taxol, 10 mM ATP, 15 mM magnesium sulfate, 0.33 mg ml^{-1} bovine serum albumin, 0.1 M KCl in PMEG buffer. Made immediately before use.

The computer programs used to analyze MT translocation were developed by Dr S. A. Cohn while working in the lab of J.M.S. Current updates of the programs can be obtained from SAC (current address: De Paul Univ., Dept of Biological Sciences, 1036 W. Belden Ave, Chicago, IL 60614, USA). SAC was recipient of an American Cancer Society Postdoctoral Fellowship no. PF-2925.

This work was supported by American Cancer Society Grant no. BE-46C and March of Dimes Birth Defects Foundation Grant no. 1-1188 to J.M.S. Thanks to Dr S. A. Cohn for reviewing an early version of this paper.

References

Bloom, G. S. and Vallee, R. B. (1989). Microtubule-associated proteins in the sea-urchin egg mitotic spindle. In *Mitosis, Molecules and Mechanisms*. (Ed. J. S. Hyams and B. R. Brinkley), pp. 183–201. Academic Press: New York and London.

Brady, S. T., Pfister, K. K. and Bloom, G. S. (1990). A monoclonal antibody against kinesin inhibits both anterograde and retrograde fast axonal transport in squid axoplasm. *Proc. natn. Acad. Sci. U.S.A.* **87**, 1061–1065.

Cohn, S. A., Ingold, A. L. and Scholey, J. M. (1987). Correlation between the ATPase and microtubule translocating activities of sea urchin egg kinesin. *Nature* **328**, 160–163.

Cohn, S. A., Ingold, A. L. and Scholey, J. M. (1989). Quantitative analysis of sea urchin egg kinesin-driven microtubule motility. *J. biol. Chem.* **264**, 4290–4297.

Dabora, S. L. and Sheetz, M. P. (1988*a*). Cultured cell extracts support organelle movement on microtubules *in vitro*. *Cell Motil. Cytoskel.* **10**, 482–495.

Dabora, S. L. and Sheetz, M. P. (1988*b*). The microtubule-dependent formation of a tubulovesicular network with characteristics of the ER from cultured cell extracts. *Cell* **54**, 27–35.

Dixon, M. (1953). The determination of enzyme inhibitor constants. *Biochem. J.* **55**, 170–171.

Endow, S. A., Henikoff, S. and Soler-Niedziela, L. (1990). Mediation of meiotic and early mitotic chromosome segregation in *Drosophila* by a protein related to kinesin. *Nature* **345**, 81–83.

Enos, A. P. and Morris, N. R. (1990). Mutation of a gene that encodes a kinesin-like protein blocks nuclear division in *A. nidulans*. *Cell* **60**, 1019–1027.

Hollenbeck, P. J. (1989). The distribution, abundance and subcellular localization of kinesin. *J. Cell Biol.* **108**, 2335–2342.

Hollenbeck, P. J. and Swanson, J. A. (1990). Kinesin supports the radial extension of macrophage tubular lysosomes. *Nature* **346**, 864–866.

Howard, J., Hudspeth, A. J. and Vale, R. D. (1989). Movement of microtubules by single kinesin molecules. *Nature* **342**, 154–158.

Ingold, A. L., Cohn, S. A. and Scholey, J. M. (1988). Inhibition of kinesin-driven microtubule motility by monoclonal antibodies to kinesin heavy chains. *J. Cell Biol.* **107**, 2657–2667.

Johnson, C. S., Buster, D. and Scholey, J. M. (1990). Light chains of sea urchin kinesin identified by immunoadsorption. *Cell Motil. Cytoskel.* **16**, 204–213.

Kiehart, D. P., Mabuchi, I. and Inoue, S. (1982). Evidence that myosin does not contribute to force production in chromosome movement. *J. Cell Biol.* **94**, 165–178.

Laemmli, U. K. (1970). Cleavage of structural proteins during the assembly of the head of bacteriophage T4. *Nature* **227**, 680–685.

Leslie, R. J., Hird, R. B., Wilson, L., McIntosh, J. R. and Scholey, J. M. (1987). Kinesin is associated with a nonmicrotubule component of sea urchin mitotic spindles. *Proc. natn. Acad. Sci. U.S.A.* **84**, 2771–2775.

Leslie, R. J. and Wilson, L. (1989). Preparation and characterization of mitotic cytoskeletons from embryos of the sea urchin *Strongylocentrotus franciscanus*. *Analyt. Biochem.* **181**, 51–58.

Lineweaver, H. and Burk, D. (1934). The determination of enzyme dissociation constants. *J. Am. chem. Soc.* **56**, 658–666.

Mabuchi, I. and Okuno, M. (1977). The effect of myosin antibody on the division of starfish blastomeres. *J. Cell Biol.* **74**, 251–263.

McIntosh, J. R. and Porter, M. E. (1989). Enzymes for microtubule-dependent motility. *J. biol. Chem.* **264**, 6001–6004.

Meluh, P. B. and Rose, M. D. (1990). KAR3, a kinesin-related gene required for yeast nuclear fusion. *Cell* **60**, 1029–1041.

Pfister, K. K., Wagner, M. C., Stenoien, D. L., Brady, S. T. and Bloom, G. S. (1989). Monoclonal antibodies to kinesin heavy and light chains stain vesicle-like structures, but not microtubules, in cultured cells. *J. Cell Biol.* **108**, 1453–1463.

Porter, M. E., Scholey, J. M., Stemple, D. L., Vigers, G. P. A., Vale, R. D., Sheetz, M. P. and McIntosh, J. R. (1987). Characterization of the microtubule movement produced by sea urchin egg kinesin. *J. biol. Chem.* **262**, 2794–2802.

Scholey, J. M., Neighbors, B., McIntosh, J. R. and Salmon, E. D. (1984). Isolation of microtubules and a dynein-like MgATPase from unfertilized sea urchin eggs. *J. biol. Chem.* **259**, 6516–6525.

Scholey, J. M., Heuser, J., Yang, J. T. and Goldstein, L. S. B. (1989). Identification of globular mechanochemical heads of kinesin. *Nature* **338**, 355–357.

Scholey, J. M., Porter, M. E., Grissom, P. M. and McIntosh, J. R. (1985). Identification of kinesin in sea urchin eggs, and evidence for its localization in the mitotic spindle. *Nature* **318**, 483–486.

Schroer, T. A., Schnapp, B. J., Reese, T. S. and Sheetz, M. P. (1988). The role of kinesin and other soluble factors in organelle movement along microtubules. *J. Cell Biol.* **107**, 1785–1792.

Seals, J. R., McDonald, J. M., Bruns, D. and Jarett, L. (1978). A sensitive and precise isotopic assay of ATPase activity. *Analyt. Biochem.* **90**, 785–795.

Vale, R. D. and Hotani, H. (1988). Formation of membrane networks *in vitro* by kinesin-driven microtubule movement. *J. Cell Biol.* **107**, 2233–2241.

Vale, R. D., Reese, T. S. and Sheetz, M. P. (1985*a*). Identification of a novel force-generating protein, kinesin, involved in microtubule-based motility. *Cell* **42**, 39–50.

Vale, R. D., Schnapp, B. J., Mitchison, T., Steuer, E., Reese, T. S. and Sheetz, M. P. (1985*b*). Different axoplasmic proteins generate movement in opposite directions along microtubules *in vitro*. *Cell* **43**, 623–632.

Vale, R. D., Scholey, J. M. and Sheetz, M. P. (1986). Kinesin: possible biological roles for a new microtubule motor. *Trends Biol. Sci.* **11**, 464–468.

Williams, R. C. Jr. and Lee, J. C. (1982). Preparation of tubulin from brain. *Meth. Enzym.* **85**, 376–385.

Wright, B., Henson, J. H., Wedaman, K. P., Willy, P. J., Morand, J. N. and Scholey, J. M. (1991). Subcellular localization and sequence of sea urchin kinesin heavy chain: evidence for its association with membranes in the mitotic apparatus and interphase cytoplasm. *J. Cell Biol.* (in press).

Axonemal dynein from *Tetrahymena*

E. M. CROSSLEY[1], S. C. HYMAN[2] and C. WELLS[1,*]

[1]*Department of Biochemistry, University of Leicester, Leicester LE1 7RH, UK*
[2]*Electron Microscopy Laboratory, School of Biological Sciences, University of Leicester, Leicester LE1 7RH, UK*

**Author for correspondence*

Summary

Axonemal dynein from *Tetrahymena* cilia can be separated on a sucrose gradient into two fractions, at least one of which appears to be polymorphic. We have been using immuno-electron microscopy in order to try and locate the different types of dynein molecules within the axonemal structure.

Key words: dynein, *Tetrahymena*, immuno-electron microscopy.

Introduction

The motility of eukaryotic cilia and flagella is driven by the microtubule–dynein system. The axonemal structure, which is responsible for this movement, typically comprises the characteristic 9+2 arrangement of microtubules, where 9 outer doublets surround 2 central single microtubules (Fig. 1). Inner and outer dynein arms emanate from the A-tubule of each doublet and interact, in an ATP-sensitive manner, with the B-tubule of the adjacent pair. This causes the relative sliding of the microtubule doublets, which forms the basis of axonemal motility. How the localized sliding between microtubule doublets is translated into the propagated bending waves (which often include 3-dimensional, helical and reversal movements) is largely unknown. Undoubtedly, other axonemal structures will be involved, such as the radial spokes and the central pair, the rotation of which has been demonstrated during cilia beating (Omoto and Kung, 1979). It is interesting to speculate, however, on the direct influence of dynein on the wide repertoire of movements of eukaryotic cilia and flagella, as this protein displays a large degree of polymorphism. At least two different dynein molecules exist within a given system and in *Chlamydomonas* axonemes, six different ATPase-containing dynein subunits have been isolated (Piperno, 1988). It is also not inconceivable that dynein may be located in a region of the axoneme other than the outer and inner arms and thus affect motility by an alternative mechanism to that traditionally envisaged.

In recent years, a much broader role for dynein in a wide range of microtubule-based motility systems has been implicated (Gibbons, 1987; Shpetner *et al.* 1988). Nevertheless, many questions concerning the function of dynein in axonemal motility remain unanswered. This article primarily addresses the question of the location of dynein within *Tetrahymena* cilia. In order to understand the molecular role of dynein in this system, it is necessary to relate the structural and functional properties of the isolated proteins *in vitro* to the situation that exists within the cilia.

Dynein from *Tetrahymena* cilia

Dynein was originally isolated from the ciliated protozoan, *Tetrahymena*, by Gibbons and Rowe (1965) in the form of a 30 S and a 14 S particle. It was assumed then that the latter species was probably a breakdown product, but subsequent work using SDS gel analysis indicated that they were distinct proteins (Porter and Johnson, 1983). More recent estimations of the sedimentation coefficient of the 30 S species gave a value of 21 S (Mitchell and Warner, 1981; Clutter *et al.* 1983). Subsequent papers referred to this fast sedimenting species as 22 S dynein, with reference to the aforementioned papers and Clutter and coworkers' unpublished results. We have recently shown that the sedimentation coefficient obtained for this species is dependent on ionic strength (Wells *et al.* 1990). This parameter increases with decreasing ionic strength and the diffusion coefficient decreases under the same conditions. By combining these data and also using sedimentation equilibrium analysis, we conclude that 22 S dynein self-associates forming dimers under low salt conditions. There may be a relationship between this self-association and the reduced ATPase activity observed under these conditions. As the s-value varies with ionic strength, it seems inappropriate to use this parameter to name this protein and we propose (Wells *et al.* 1990) that the larger dynein species be referred to as simply, dynein, as recently adopted by Shimizu *et al.* (1989). This species has been shown by STEM analysis to be a three-headed molecule, with flexible strands connecting the heads to a common basal region (Johnson and Wall, 1983).

Dynein can be isolated from *Tetrahymena* cilia essentially as described by Porter and Johnson (1983). Following extraction of the dynein (with 0.6 M NaCl) and purification on a DEAE-Sephacel column, dynein and 14 S dynein can be separated on a sucrose gradient. The advantages of studying dynein from *Tetrahymena* include the ability to produce sufficiently large quantities of protein for biochemical studies and the relative stability of the protein. This system, however, does not possess some useful characteristics which have been used in other systems. The ability to form mutants in *Chlamydomonas* lacking either outer or inner arms has been used to locate the different dynein species and to provide insight into their relative functions (Huang *et al.* 1979). In sea urchin sperm, the outer arms can be selectively extracted, thus enabling the dynein species in this location to be identified (Tang *et al.* 1982). In *Tetrahymena*, however, selective extraction does not occur. The larger dynein species is assumed to be present in the outer arms, by analogy with

Journal of Cell Science, Supplement 14, 117–120 (1991)

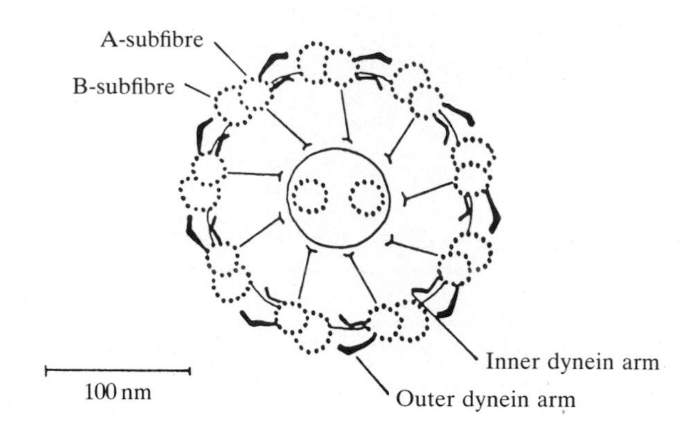

Fig. 1. A diagrammatic representation of a transverse section through a demembranated axoneme.

the sea urchin sperm system, but this has not been rigorously shown.

14 S dynein from Tetrahymena *cilia*

The location and nature of the 14 S dynein fraction is much more uncertain. Marchese-Ragona *et al.* (1988) have shown by electron microscopy that 14 S dynein displays heterogeneity, comprising two types of molecules. SDS gel electrophoresis shows that this fraction contains two immunologically distinct heavy chains. The location of 14 S dynein within the axoneme has not been established. ATPase activity has been associated with a microtubule–membrane link (Dentler *et al.* 1981). Rotation of the central pair has also been reported (Omoto and Kung, 1980) and with the demonstration, by Vale and Toyoshima (1988) that 14 S dynein can cause rotation of microtubules, this may be a possible site for 14 S components. Also, by analogy with other systems, 14 S dynein may be located at the inner arms. In view of the apparent heterogeneity of the 14 S fraction, it may contain protein components located in different positions within the axoneme.

By Fast Protein Liquid Chromatography, using a mono-Q column and a 200–400 mM NaCl gradient, we can separate 14 S dynein into three fractions (Fig. 2). All three fractions have ATPase activity, the specific activity of fraction 2 being greater than the other two fractions (0.6 μmol min^{-1} mg^{-1} as compared to 0.2–0.3 μmol min^{-1} mg^{-1}).

Immuno-electron microscopy studies

In order to try to answer fundamental questions concerning the localization of dyneins within the axoneme, we have raised polyclonal antibodies to dynein (formally 22 S dynein), the whole 14 S dynein and the three different 14 S fractions isolated by FPLC and used them for immuno-electron microscopy studies.

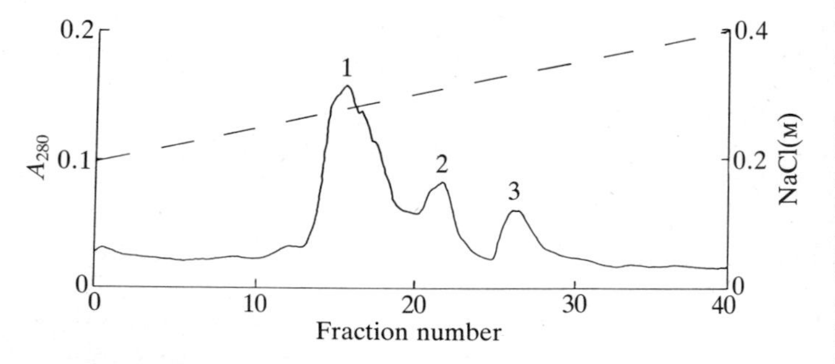

Fig. 2. Fast Protein Liquid Chromatography profile, on a Mono-Q column, of 14 S dynein fractionated on a 0.2–0.4 M NaCl gradient in 20 mM Hepes, 4 mM $MgCl_2$, 0.1 mM EGTA, pH 7.4. Three fractions are separated, as indicated.

Western blots using these antisera confirmed that there was no cross-reactivity between dynein and 14 S dynein. Furthermore, there appears to be no cross-reactivity between the three 14 S fractions, even though the SDS gel patterns suggest that fractions 2 and 3 could be breakdown products of fraction 1.

The original strategy we used to carry out immuno-electron microscopy was to react demembranated cilia with antisera, followed by reaction with a secondary antibody conjugated to gold particles (usually IgG–gold). This preparation was then fixed and embedded for sectioning. Although we obtained labelling by this method, we were uncertain as to whether the antibodies were freely diffusable into the axonemal structure. Furthermore, this procedure was expensive on time and materials.

Therefore we adopted the procedure whereby cilia were first lightly fixed (2 % glutaraldehyde for 30 min), embedded (in 3 % agar, followed by slow dehydration through an ethanol series into Lowicryl HM23 resin) and sectioned (80–100 nm thickness). The sections were then placed on grids and reacted with antibodies using the following procedure.

The sections were blocked in 0.6 % bovine serum albumin, 0.2 % Tween 20 in phosphate-buffered saline, pH 7.2 for 30 min at room temperature. (The technique used for this and all subsequent procedures was to submerge the grid in 30 μl droplets of the appropriate solution). The grid was then placed in primary antibody-containing solution, either whole antisera, or purified antibody to a specific band eluted from a Western blot. Incubation with antibody was overnight, usually at room temperature, in the presence of 5 mM sodium azide. Gold grids were used, as copper grids were found to react with the antisera. The grids were then washed 4 times with blocking buffer and then reacted with IgG–gold (usually a 1:50 dilution, Biocell) for 4 h at room temperature. The grids were then washed 3 times with block buffer and 3 times with phosphate-buffered saline. This was followed by fixing in 1 % glutaraldehyde and then a distilled water wash to remove all traces of phosphate before staining with saturated aqueous uranyl acetate and Reynold's lead citrate. The grids were carbon-coated for stabilization.

Controls were carried out which included reacting the sections on grids with anti-tubulin as a primary antibody (which gave extensive labelling of all the microtubules with gold particles) and omitting primary antibody altogether (which gave essentially no labelling at all).

Fig. 3 shows the distribution of 10 nm gold particles on electron micrographs of membranated (A,B) and demembranated (C–F) cilia following incubation with purified anti-dynein (specific for the heavy chain components of the large dynein species). The vast majority of the gold particles localize in the region of the outer arms. Some of the labelling seen would be consistent with the position of a microtubule–membrane link and likewise, other gold particles coincide with the inner arm position. Consideration must be given, however, to the degree of flexibility introduced by this labelling method. The length of the dynein molecule has been calculated as 35 nm (Johnson and Wall, 1983) and this is attached by two antibody molecules (9–10 nm each) to a 10 nm gold particle.

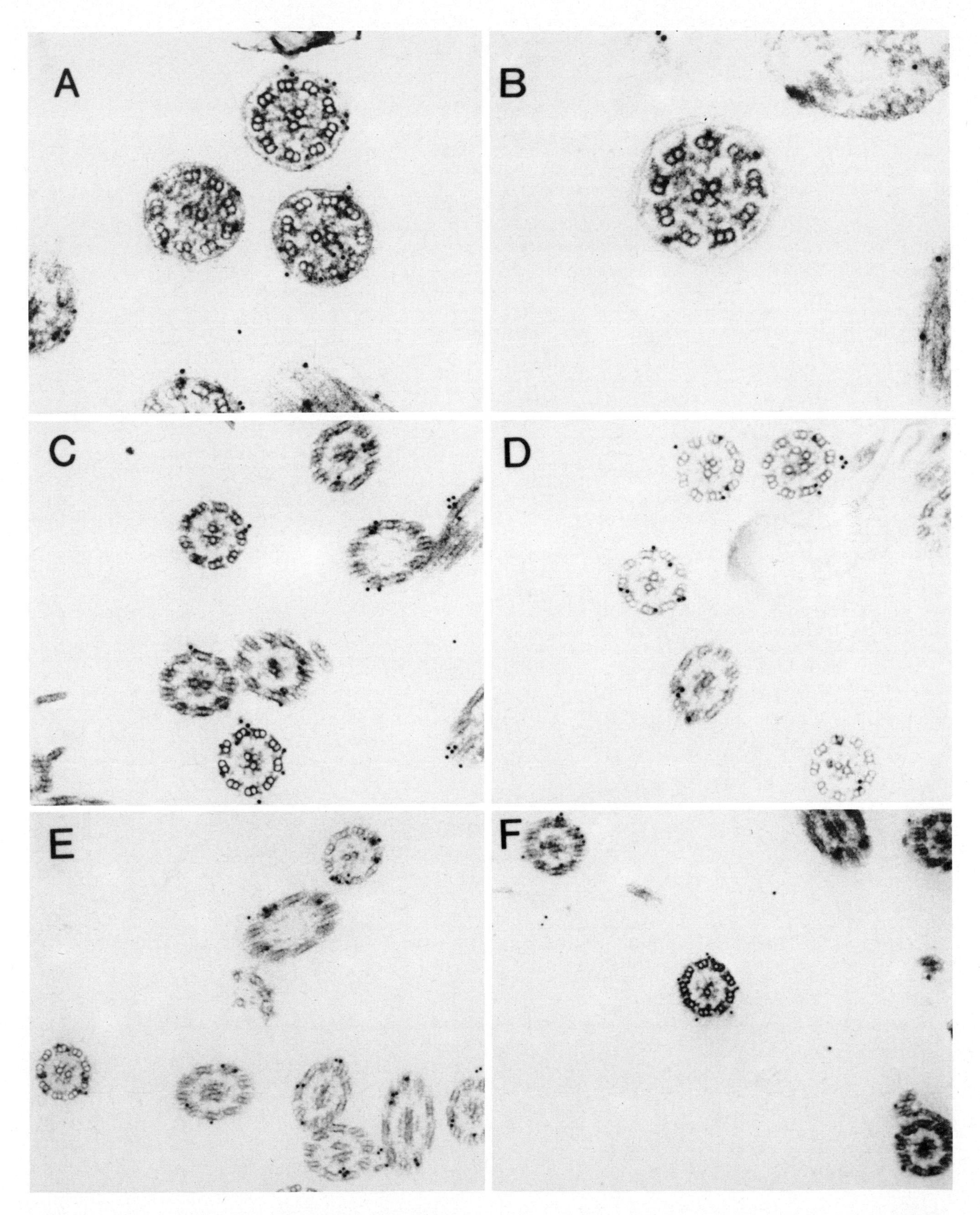

Fig. 3. Electron micrographs of (A,B) membranated and (C–F) demembranated *Tetrahymena* axonemes which have been reacted with purified antibody to the heavy chains of dynein (formally 22 S dynein) and then reacted with IgG–gold. The gold particles are 10 nm in diameter for scale.

Therefore, a gold particle could potentially be located more than 60 nm from the attachment site of the labelled dynein molecule. If this fact is taken into account, then virtually all the labelling observed in Fig. 3 can be associated with an outer arm location.

Immuno-electron microscopy using antibodies raised against 14 S and fractions thereof, however, has produced a less clear picture. The data so far indicates that the 14 S dynein is probably located at the inner arm. Essentially the same labelling pattern was observed using antibodies raised against whole 14 S dynein (which shows no cross-reactivity with the larger dynein species) and purified antibodies raised against the heavy chains of the separate fractions (which show very little cross-reactivity with the other fractions).

Conclusions

Axonemal dynein was first discovered 25 years ago and yet some fundamental questions still remain unanswered concerning its structure and localization within the axoneme. The difficulty experienced in obtaining this information is likely, in part, to be due to the polymorphic nature of dynein within a given system.

The situation is further complicated by the fact that dyneins are very susceptible to endogenous protease digestion during purification from axonemal sources, which could be contributing to the polymorphic profile observed. The work reported here, on the localization of *Tetrahymena* dyneins, is consistent with dynein (formally 22 S dynein) being located at the outer arm and 14 S dynein probably being located at the inner arm. As previously alluded to, eukaryotic cilia and flagella exhibit a broad range of motile properties. Whether there is any direct relationship between this diversity and the heterogeneous nature of the dynein population has yet to be determined.

We acknowledge the SERC for financial support.

References

CLUTTER, D., STIMPSON, D., BLOOMFIELD, V. AND JOHNSON, K. A. (1983). The structure of *Tetrahymena* dynein in solution. *J. Cell Biol.* **97**, 197a.

DENTLER, W. L., PRATT, M. M. AND STEPHENS, R. E. (1981). Microtubule–membrane interactions in cilia. *J. Cell Biol.* **84**, 381–403.

GIBBONS, I. R. (1987). New jobs for dynein ATPases. *Nature* **330**, 600.

GIBBONS, I. R. AND ROWE, A. J. (1965). Dynein: a protein with adenosine triphosphatase activity from cilia. *Science* **149**, 424–425.

HUANG, B., PIPERNO, G. AND LUCK, D. J. L. (1979). Paralyzed flagella mutants of *Chlamydomonas reinhardtii* defective for axonemal doublet microtubule arms. *J. biol. Chem.* **254**, 3091–3099.

JOHNSON, K. A. AND WALL, J. S. (1983). Structure and molecular weight of the dynein ATPase. *J. Cell Biol.* **96**, 669–678.

MARCHESE-RAGONA, S. P., WALL, J. S. AND JOHNSON, K. A. (1988). Structure and mass analysis of 14 S dynein obtained from *Tetrahymena* cilia. *J. Cell Biol.* **106**, 127–132.

MITCHELL, D. R. AND WARNER, F. D. (1981). Binding of 21 S dynein ATPase to microtubules: effects of ionic conditions and substrate analogs. *J. biol. Chem.* **256**, 12 535–12 544.

OMOTO, C. K. AND KUNG, C. (1980). Rotation and twist of the central-pair microtubules in the cilia of *Paramecium*. *J. Cell Biol.* **87**, 33–46.

PIPERNO, G. (1988). Isolation of a sixth dynein subunit adenosine triphosphatase of *Chlamydomonas* axonemes. *J. Cell Biol.* **106**, 133–140.

PORTER, M. E. AND JOHNSON, K. A. (1983). Characterization of the ATP-sensitive binding of *Tetrahymena* 30 S dynein to bovine brain microtubules. *J. biol. Chem.* **258**, 6575–6581.

SHIMIZU, T., MARCHESE-RAGONA, S. P. AND JOHNSON, K. A. (1989). Activation of dynein adenosine triphosphatase by cross-linking to microtubules. *Biochemistry* **28**, 7016–7021.

SHPETNER, H. S., PASCHAL, B. M. AND VALLEE, R. B. (1988). Characterization of the microtubule-activated ATPase of brain cytoplasmic dynein (MAP-1C). *J. Cell Biol.* **107**, 1001–1009.

TANG, W. Y., BELL, C. W., SALE, W. S. AND GIBBONS, I. R. (1982). Structure of the dynein-1 outer arm in sea urchin sperm flagella. *J. biol. Chem.* **257**, 508–515.

VALE, R. D. AND TOYOSHIMA, Y. Y. (1988). Rotation and translocation of microtubules *in vitro* induced by dyneins from *Tetrahymena* cilia. *Cell* **52**, 459–469.

WELLS, C., MOLINA-GARCIA, A., HARDING, S. E. AND ROWE, A. J. (1990). Self-interaction of dynein from *Tetrahymena* cilia. *J. Muscle Res. and Cell Motil.* **11**, 344–350.

Effect of MAP2, MAP2c, and tau on kinesin-dependent microtubule motility

SUSANNE HEINS, YOUNG-HWA SONG, HOLGER WILLE, ECKHARD MANDELKOW
and EVA-MARIA MANDELKOW*

Max-Planck-Unit for Structural Molecular Biology, c/o DESY, Notkestrasse 85, D-2000 Hamburg 52, Germany

* Corresponding author

Summary

By making use of DIC video microscopy to monitor microtubule motility we have studied the effect of several MAPs (MAP2, MAP2c, tau) on microtubule–kinesin interactions and microtubule gliding. Of the three MAPs tested, MAP2 interferes most strongly with kinesin-dependent microtubule motility.

Key words: kinesin, microtubules, motility, MAPs.

Introduction

The transport of vesicles and organelles in nerve cells depends on microtubules and their motor proteins (Allen *et al.* 1985; Brady *et al.* 1985; Vale *et al.* 1985). These cells also contain a variety of microtubule-associated proteins (MAPs) which tightly bind to and stabilize the microtubules. Some of the MAPs (e.g. MAP2) are fairly large, have elongated shapes and can protrude for several tens of nanometres from the microtubule surface. The crowding of proteins in the axoplasm poses some problems; for example, how can a motor protein move along a microtubule in spite of the MAPs present? What is the influence of MAPs on axonal transport?

The analogue of anterograde axonal transport can be studied *in vitro* by observing microtubules gliding over a surface covered with the motor protein kinesin, using video-enhanced DIC microscopy (Allen *et al.* 1985; Vale *et al.* 1985). The visualization of the microtubules depends not only on the optical parameters and image enhancement, but also, because of the small depth of field, on the confinement of the microtubules to a region just above the surface. If the microtubules detach from the surface and diffuse into the solution they become essentially invisible. Thus the image contains information both on microtubule attachment and on motility. It can therefore be used to study factors that affect these two parameters.

We have recently reported on the interference of MAP2 with microtubule attachment and kinesin-dependent motility (von Massow *et al.* 1989). The experiments (see diagrams in Fig. 1) showed that (1) pure microtubules attach to the surface and are therefore visible, (2) a surface covered with MAP2 prevents the attachment of microtubules which are therefore invisible and (3) kinesin attached to the surface binds and moves microtubules. However (4) when the surface is covered with both kinesin and MAP2, the effect of MAP2 dominates, i.e. microtubules neither attach nor move, and thus are not visible; (5) when MAP2 is present both on the surface and on microtubules, microtubules attach to the surface again, in contrast to (2), as if the MAP2 molecules coming from opposite directions interlocked with one another; in this case kinesin cannot induce movement either.

These effects depend on certain concentration ratios between kinesin, MAPs, and microtubules, which we have now studied in more detail. In addition we were interested to determine if other MAPs had effects similar to MAP2. We have now tested two proteins which are homologous with MAP2 (1828 residues) in the C-terminal region, which contains the internal repeats responsible for microtubule binding (Lewis *et al.* 1988). They are tau (around 400 residues, depending on isoform; Lee *et al.* 1988; Goedert *et al.* 1988), which has a much shorter N-terminal domain than MAP2 and shows no homology to it, and MAP2c (467 residues), a juvenile form of MAP2, arising from alternative splicing, which is homologous to MAP2 but lacks most of the N-terminal region (Papandrikopoulou *et al.* 1989). The results reported below show that their effects are much less pronounced than those of MAP2 in terms of the above assays, suggesting that only the full length MAP2 contains special regions, presumably in the N-terminal domain, that are responsible for the repulsion of microtubules from the surface or the MAP-2–MAP2 interlocking.

Materials and methods

Preparation of microtubules

PC–tubulin was prepared as previously described (Mandelkow *et al.* 1985). MAP-free microtubules were polymerized in assembly buffer (0.1 M PIPES, pH 6.9, 1 mM each of $MgSO_4$, EGTA, DTT, and GTP) and stabilized by 10 μM taxol.

Preparation of kinesin

Kinesin was prepared after Kuznetsov and Gelfand (1986) or Vale *et al.* (1985), with modifications (von Massow *et al.* 1989; Song, unpublished).

Preparation of MAPs

MAP2 and tau were prepared from porcine brain as described (Hagestedt *et al.* 1989), involving a boiling step, separation on Mono S HPLC (Pharmacia), and their differential solubility in perchloric acid. MAP2c was prepared similarly and separated from MAP2 by gel filtration (Superose 12 column, Pharmacia).

Motility assay and sample preparation

Microtubules and their movement by kinesin were observed by DIC video microscopy following the method of Allen *et al.* (1985). The effects of MAPs on microtubule attachment and gliding were performed as previously described (von Massow *et al.* 1989). Briefly, the samples are prepared in two steps: (1) coating of the glass surface by pre-incubation with kinesin, MAPs, or both, (2) addition of microtubules, made either from PC–tubulin and stabilized by taxol, or with re-attached purified MAPs. The pre-

Journal of Cell Science, Supplement 14, 121–124 (1991)

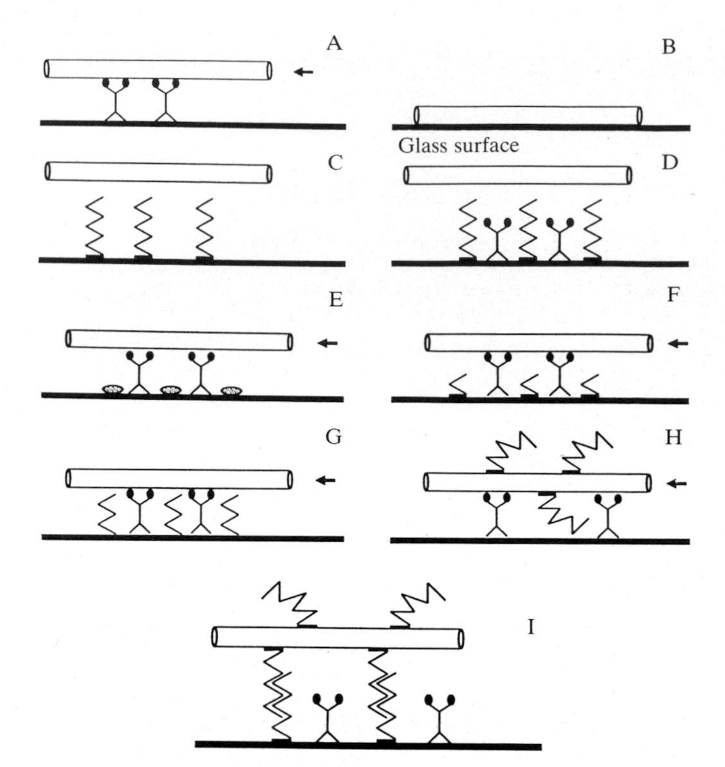

Fig. 1. Diagrams illustrating possible mechanisms of interaction between kinesin, MAP2, and microtubules. (A) Kinesin adsorbed to the glass surface is capable of moving microtubules (arrow), irrespective of whether the added microtubules have attached MAP2 or not (see H, below). (B) Microtubules added without prior adsorption of kinesin or MAP2 attach to the glass and can thus be seen easily by focussing the microscope to the level of the glass surface. (C) MAP2 adsorbed to the glass surface repels microtubules so that they tend to remain in the bulk solution and are difficult to visualize. (D) MAP2 plus kinesin adsorbed to the glass does not allow attachment or motility of MAP-free microtubules. (E) As a control, kinesin adsorbed to the glass together with BSA shows uninhibited motility, i.e. BSA does not interfere with the interaction between microtubules and kinesin. (F) The same result as with BSA is obtained when kinesin and the assembly fragment of MAP2 are adsorbed to the glass, showing that this fragment does not prevent the kinesin-based motility. (G) Similar to (F), but using the projection fragment of MAP2 which leaves the kinesin-induced motility largely unaffected. (H) Kinesin adsorbed to the glass by itself supports motility of MAP2-containing microtubules, provided that no free MAP2 is present on the glass surface; i.e. microtubule-bound MAPs do not interfere with the microtubule–kinesin interaction. This is indicated by the altered configuration of MAP2. (I) Kinesin and MAP2 pre-adsorbed to the glass plus microtubules with attached MAP2. The MAP2 molecules projecting from the glass and from the microtubule become entangled and/or crosslinked and keep the microtubule near the surface, but prevent their gliding. From von Massow *et al.* 1989.

incubation was typically done by applying a volume of 8 μl of kinesin/MAP solution to the coverslip, spread over an area of about 0.4 cm^2, and letting the proteins adsorb for about 15 min. This area was marked with a felt pen for later identification of the boundary. After that, 2 μl of microtubules and ATP were added, and the sample was covered with a $1.5\times1.5\,cm^2$ coverslip. The 10 μl solution was distributed over the whole area of 2.25 cm^2 (thickness of layer about 40 μm), but the pre-adsorbed proteins diffused only very slowly outside the initial area so that the boundary could be identified by the microtubule attachment and motility criteria described below. All protein concentrations refer to the final 10 μl volume.

The upper limit of the surface density of pre-adsorbed proteins (number of particles nm^{-2}) can be estimated by assuming that all particles attach to the surface area so that:

$$\delta = (V/A)(C_w/M) \times 6 \times 10^6 ,$$

and the mean separation between particles is d (nm)$=1/\sqrt{\delta}$. C_w is the weight concentration in mg ml^{-1}, M is the molecular weight, V is the applied volume in ml, and A the area in cm^2. For example, for a typical solution of kinesin (M_r=360 000) with V=0.01 ml, A=0.4 cm^2, C_w=0.05 mg ml^{-1} we get $\delta\approx0.02\,nm^{-2}$ and d≈7 nm.

Results

(1) *Effect of MAP2/kinesin ratios on microtubule gliding*

Coating with kinesin alone leads to visible and motile microtubules, irrespective of whether the microtubules are covered with MAPs or not, as previously described (von Massow *et al.* 1989; see Fig. 1A,H). Pre-adsorption of the surface with MAP2 leads to the repulsion of microtubules which are therefore not visible (Fig. 2; Fig. 1C). The same effect occurs in the presence of kinesin, even at very low molecular ratios of MAP2:kinesin of 0.02:1 (Fig. 1D). Thus the repulsion of microtubules by MAP2 dominates over the attraction by kinesin. This can be nicely visualized by pre-incubating two adjacent areas in different ways, one with kinesin only, the other with kinesin and MAPs. Microtubules glide over the surface containing only kinesin; when they cross the boundary into the MAP2-containing region, they lift off and disappear out of focus (Fig. 3).

The repulsion of microtubules by surface-bound MAP2 is routinely observed for MAP2-free microtubules; when the added microtubules are also saturated with MAP2 the behavior is more complex. Repulsion still dominates when the MAP2:kinesin ratio is above 0.15:1. However, between ratios of 0.15:1 and 0.02:1 the microtubules are attached to the surface but do not glide. This can be interpreted as an 'interlocking' of MAP2 molecules protruding from the glass and microtubule surfaces (Fig. 1I). At even lower concentrations of MAP2 on the surface, gliding takes over again, as in Fig. 1H. In the intermediate range where interlocking takes place the mean distance between MAP2 molecules is of the order of 25 nm or more, compared with only 7 nm for kinesin. In contrast to intact MAP2, neither the assembly nor the projection fragment of MAP2 (obtained by limited chymotryptic cleavage, as described by Vallee, 1980) has the repulsive or interlocking effect when used at comparable stoichiometric ratios (Fig. 1F and G).

(2) *Influence of MAP2c or tau on kinesin–microtubule interactions*

Coating of the surface with kinesin plus MAP2c or tau does not show the same strong repulsive effect as with MAP2. In the case of MAP2c, the MAP2c:kinesin ratio has to be raised to 0.9:1 in order to observe repulsion; this is six times higher than the corresponding concentration of MAP2. For tau the minimum stoichiometry for repulsion is even higher, namely 8:1. In these conditions the calculated mean distances between MAP molecules would be about 7 and 2.5 nm for MAP2c and tau, respectively.

In contrast to MAP2, neither MAP2c nor tau produces an 'interlocking' effect, i.e. when the glass and microtubule surfaces are both coated with the MAPs one

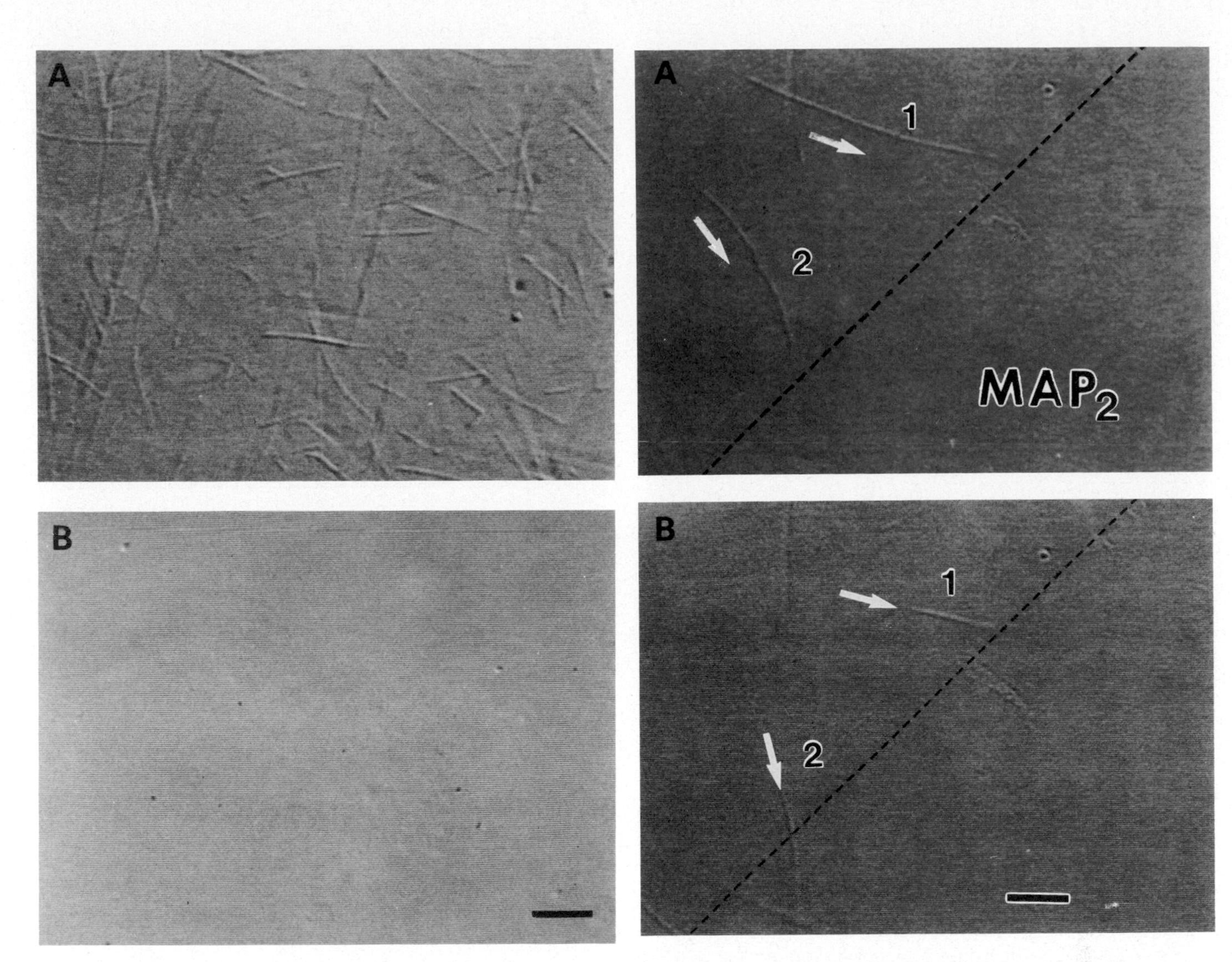

Fig. 2. Effect of MAP2 on the attachment of microtubules to the glass surface in the absence of kinesin. (A) Without preincubation with MAP2, microtubules (prepared from PC–tubulin and containing no MAPs) attach to the glass surface and can therefore be visualized by video microscopy. (B) When the glass surface is first covered with MAP2 and then microtubules are added, the microtubules are repelled from the surface and remain invisible in solution. For concentration ranges see text. Bar, 5 μm.

Fig. 3. Effect of MAP2 on kinesin-induced microtubule motility across a boundary. The coverslip was first coated with kinesin (50 $\mu g\,ml^{-1}$), in addition a part of it was coated with MAP2 (50 $\mu g\,ml^{-1}$). The approximate position of the boundary is indicated by the dashed line (upper left, only kinesin; lower right, kinesin plus MAP2). After adding microtubules and ATP they are visible and moving across the MAP2-free area (two of them are labeled 1 and 2; direction of movement indicated by white arrows). When they cross into the MAP2-containing area their leading part lifts off the surface and disappears from the image while the trailing part is still seen moving (compare (A) and (B), separated by a 22 s interval). The converse case can be observed as well (not shown). Microtubules that are initially invisible become attached in the MAP2-free area with their leading part; as they move further into it they gradually become visible along their whole length. Superficially they appear to be growing while moving, up to the point where they are fully visible. This effect is typical of MAP2 but not of MAP2c or tau. Bar, 5 μm.

observes either repulsion (at high MAP concentrations, analogous to Fig. 1D) or microtubule gliding (at low concentrations, analogous to Fig. 1G). In other words, whether or not a microtubule is coated with these MAPs has no obvious effect on the microtubule behavior. This is true even in hybrid situations, e.g. when the surface contains MAP2 and kinesin and the microtubules are coated with MAP2c or tau. In this case the strong repulsion of microtubules by MAP2 again dominates, and one observes no 'interlocking', i.e. stationary microtubules attached to the surface.

Discussion

These studies were originally prompted by the observation that crude kinesin preparations did not support microtubule gliding. A search for inhibitory factors led to the discovery of the effects of MAP2, which was present in these preparations and could be removed by gel filtration. One conclusion was that microtubule-bound MAP2 does not interfere with microtubule gliding while MAP2 preadsorbed to the glass does. Our working model (von Massow *et al.* 1989) is that MAP2 adsorbed to the glass surface forms a lawn which either repels microtubules from approaching the surface and thus prevents the interaction with kinesin, or, at lower concentrations, interacts somehow with MAP2 bound to microtubules, thereby keeping them close to the surface yet immobile. The effects occur at surprisingly low concentrations of glass-bound MAP2 (at least six times less than that of kinesin), where the mean separation of MAP2 molecules is expected to be of the order of 20 nm or more. In contrast, MAP2 bound to microtubules appears to have no effect on their motility *per se*, in spite of their high density on a microtubule. Saturated microtubules contain about 20% MAP2 which translates into roughly one MAP2 molecule every 6 nm along the length. Thus far we have found only one condition where microtubule-bound MAPs affect gliding, that is, when the MAP2–MAP2 interlocking takes place (at low MAP2 densities on the surface).

We have now studied two other MAPs which are homologous to MAP2 in their C-terminal (microtubule-binding) region. Neither of them shows the pronounced repelling effect of MAP2, that is, in order to observe repulsion one requires six to fifty times higher molar concentrations than with MAP2 (for MAP2c and tau, respectively). Nevertheless, these effects appear to be specific, since controls with BSA show no effect at even much higher concentrations (100-fold and more). Moreover, the interlocking effect is not observed with either MAP2c or tau.

One interpretation of these observations is suggested by the size of MAP2; it is four to five times larger than MAP2c or tau which makes the stronger repulsion of microtubules understandable. This view is probably oversimplified since the projection fragment of MAP2 lacks the repelling or interlocking capacity even though it is only 10–20% smaller than intact MAP2, Moreover, arguments based on size alone fail to explain the MAP2–MAP2 interlocking. It seems therefore likely that MAP2 contains specific sequences responsible for its effect. These sequences are unlikely to be in the C-terminal microtubule-binding domain where the three MAPs are largely homologous. However, the N-terminal projection domain of MAP2 contains about 1400 unique amino acid residues not found in MAP2c or tau which are therefore candidates for MAP2-specific functions. These considerations are derived from the published cDNA sequences of MAP2, MAP2c, and tau (see Lewis *et al.* 1988; Papandrikopoulou *et al.* 1989; Lee *et al.* 1988; Goedert *et al.* 1988), but they can only be approximate since we used proteins purified from brain tissue.

Finally, we note that one common feature of all MAPs tested is their capacity to repel microtubules and to overcome the attractive force of kinesin. There has been an ongoing debate as to whether MAPs act as crosslinkers or as spacers of microtubules. Recent reports have emphasized the crosslinking function, since cells transfected with MAP2 or tau developed microtubule bundles (Kanai *et al.* 1989; Lewis *et al.* 1989). In contrast, biophysical experiments such as the ones discussed here tend to support the view of MAPs as spacers, and X-ray experiments (to be described elsewhere) point in the same direction. This would mean that one function of MAPs (apart from microtubule stabilization) is to keep the space around microtubules free from structures (microtubules or other cytoplasmic components) that might interfere with kinesin-based transport processes.

We thank Dr M. Suffness (US National Cancer Institute, Bethesda, MD) for providing taxol. This project was supported by the Bundesministerium für Forschung und Technologie.

References

Allen, R., Weiss, D., Hayden, J., Brown, D., Fujiwake, H. and Simpson, M. (1985). Gliding movement of and bidirectional transport along single native microtubules from squid axoplasm: evidence for an active role of microtubules in cytoplasmic transport. *J. Cell Biol.* **100**, 1736–1752.

Brady, S. T., Lasek, R. J. and Allen, R. D. (1985). Video microscopy of fast axonal transport is extruded axoplasm: a new model for study of molecular mechanisms. *Cell Motility* **5**, 81–101.

Goedert, M., Wischik, C., Crowther, R., Walker, J. and Klug, A. (1988). Cloning and sequencing of the cDNA encoding a core protein of the paired helical filament of Alzheimer disease: identification as the microtubule-associated protein tau. *Proc. natn. Acad. Sci. U.S.A.* **85**, 4051–4055.

Hagestedt, T., Lichtenberg, B., Wille, H., Mandelkow, E.-M. and Mandelkow, E. (1989). Tau protein becomes long and stiff upon phosphorylation: correlation between paracrystalline structure and degree of phosphorylation. *J. Cell Biol.* **109**, 1643–1651.

Kanai, Y., Takemura, R., Oshima, T., Mori, H., Ihara, Y., Yanagisawa, M., Masaki, T. and Hirokawa, N. (1989). Expression of multiple tau isoforms and microtubule bundle formation in fibroblasts transfected with a single tau cDNA. *J. Cell Biol.* **109**, 1173–1184.

Kuznetsov, S. A. and Gelfand, V. I. (1986). Bovine brain kinesin is a microtubule-activated ATPase. *Proc. natn. Acad. Sci. U.S.A.* **83**, 8530–8534.

Lee, G., Cowan, N. and Kirschner, M. (1988). The primary structure and heterogeneity of tau protein from mouse brain. *Science* **239**, 285–288.

Lewis, S. A., Wang, D. and Cowan, N. J. (1988). Microtubule-associated protein MAP2 shares a microtubule binding motif with tau protein. *Science* **242**, 936–939.

Lewis, S. A., Ivanov, I. E., Lee, G. H. and Cowan, N. J. (1989). Organization of microtubules in dendrites and axons is determined by a short hydrophobic zipper in microtubule-associated proteins MAP2 and tau. *Nature* **342**, 498–505.

Mandelkow, E.-M., Herrmann, M. and Rühl, U. (1985). Tubulin domains probed by subunit-specific antibodies and limited proteolysis. *J. molec. Biol.* **185**, 311–327.

Papandrikopoulou, A., Doll, T., Tucker, R., Garner, C. and Matus, A. (1989). Embryonic MAP2 lacks the cross-linking sidearm sequences and dendritic targeting signal of adult MAP2. *Nature* **340**, 650–652.

Vale, R. D., Reese, T. S. and Sheetz, M. P. (1985). Identification of a novel force-generating protein, Kinesin, involved in microtubule-based motility. *Cell* **42**, 39–50.

Vallee, R. B. (1980). Structure and phosphorylation of microtubule-associated protein 2 (MAP2). *Proc. natn. Acad. Sci. U.S.A.* **77**, 3206–3210.

Von Massow, A., Mandelkow, E.-M. and Mandelkow, E. (1989). Interaction between kinesin, microtubules, and microtubule-associated protein 2. *Cell Mot. and Cytoskel.* **14**, 562–571.

Preparation of marked microtubules for the assay of the polarity of microtubule-based motors by fluorescence

A. A. HYMAN

Department of Pharmacology, University of California San Francisco, San Francisco, California 94143-0450, USA

Summary

Short microtubule seeds are constructed using heavily rhodamine-labeled tubulin. Polymerisation off the ends of these seeds is initiated using a mixture of 1:10 labeled and unlabeled tubulin, so that the new polymerisation is only dimly labeled. This is done in the presence of NEM tubulin, which inhibits growth from the microtubule minus ends. The polarity-marked microtubules are fixed at a desired length by adding taxol.

Key words: microtubule polarity, microtubule polymerisation, dynein, kinesin, fluorescent tubulin.

Introduction

Since the original discovery of kinesin as a plus-end directed microtubule-based motor, there has been an explosion in the discovery of other microtubule-based motors. Microtubules are polar structures, with 'plus' and 'minus' ends, and different motors can be either plus-end directed, like kinesin (Vale *et al.* 1985), or minus-end directed, like dynein (Paschal *et al.* 1987). Prospective motors may be identified by amino acid sequence similarity (Endow *et al.* 1990; Enos and Morris, 1990; McDonald and Goldstein, 1990; Meluh and Rose, 1990), so it is essential to have a fast and reliable assay for the polarity of newly discovered microtubule-based motors, in order to understand their physiological function. Previously a number of methods have been used to determine polarity. Centrosomes, which nucleate microtubules with a defined polarity, were used to demonstrate the polarity of kinesin-based movement in a bead assay (Vale *et al.* 1985). One can also visualize the gliding of microtubules on a motor-bound substrate, using transmitted light techniques such as dark-field or differential interference contrast microscopy, to follow the movement of microtubules which have been nucleated by structures such as isolated axoneme fragments or sperm heads with attached axonemes (Yang *et al.* 1988). However, because of the possibility of contamination by axonemal microtubule motors in these assays, it would be useful to have a simple assay in which the polarity of movement could be followed using single microtubules polymerized from pure tubulin. In this paper, I describe methods of preparing polarity-marked microtubules which can be seen by fluorescence. Fluorescence has the advantage that (1) it is easy to extend existing fluorescent-labeling techniques (Hyman *et al.* 1990) to prepare marked microtubules, and (2) fluorescence can be used in situations where cytoplasmic components might obscure resolution in transmitted light microscopy (Belmont *et al.* 1990).

Preparation of labeled microtubules

Briefly, the strategy for making polarity-marked microtubules is shown in Fig. 1 and is as follows: short microtubule seeds are constructed by polymerizing heavily rhodamine-labeled tubulin (Hyman *et al.* 1990), and these brightly labeled microtubules are then diluted into a mixture of rhodamine and unlabeled tubulin so that new polymerization from seeds has a much dimmer fluorescence. These microtubules are then stabilized with taxol. Plus and minus ends are identified as described below.

Short, brightly labeled microtubules can be prepared in two ways. (1) Stable seeds can be made by polymerizing tubulin with a non-hydrolyzable analogue of GTP, GMPCPP (Hyman *et al.* 1990). These have the advantage of rapid nucleation, using up all tubulin monomer and providing short microtubules of a fairly uniform length. Unfortunately, currently there is no available commercial source of GMPCPP. As an alternative, seeds polymerized with GTP can be stabilized by chemical cross-linking (Koshland *et al.* 1988). (2) Alternatively, one can use standard short microtubules as seeds. These are prepared by polymerizing 8 mg ml^{-1} rhodamine-labeled tubulin in BRB80 (80 mM K-Pipes, 1 mM GTP, 1 mM $MgCl_2$, pH 6.8)+ 4 mM $MgCl_2$+40 % glycerol, for 5 min.

The highly labeled seeds are then diluted into a mixture of unlabeled and labeled tubulin monomer. This mixture is prepared at a 10:1 molar ratio of unlabeled to labeled tubulin. Polarity is identified by including NEM-treated tubulin in the assembly mixture, which prevents assembly from microtubule minus ends (Hyman *et al.* 1990). NEM-tubulin is included in the monomer mixture at a molar ratio of 0.7:1.0 (NEM-tubulin:untreated tubulin). The mixture is then spun at 30 *psi* in the airfuge at 4 °C for 5 min and frozen in aliquots in liquid nitrogen. Just prior to use, the aliquot is thawed and kept at 4 °C. Care should be taken to ensure that the tubulin does not go above 4 °C at this stage, or more of the tubulin will be inactivated. This becomes a problem because the unpolymerized, dead tubulin increases background fluorescence in the assay. The tubulin aliquot is diluted into BRB80+1 mM GTP to a final concentration of 15 μM, and transfered to 37 °C for 1 min. After 1 min, seeds are diluted into the tubulin aliquot at a 1:5 ratio. Under these conditions, microtubules will now polymerize only from the ends of pre-existing microtubules and not spontaneously (Mitchison and Kirschner, 1984; Mitchison, 1984). The microtubules

Journal of Cell Science, Supplement 14, 125–127 (1991)

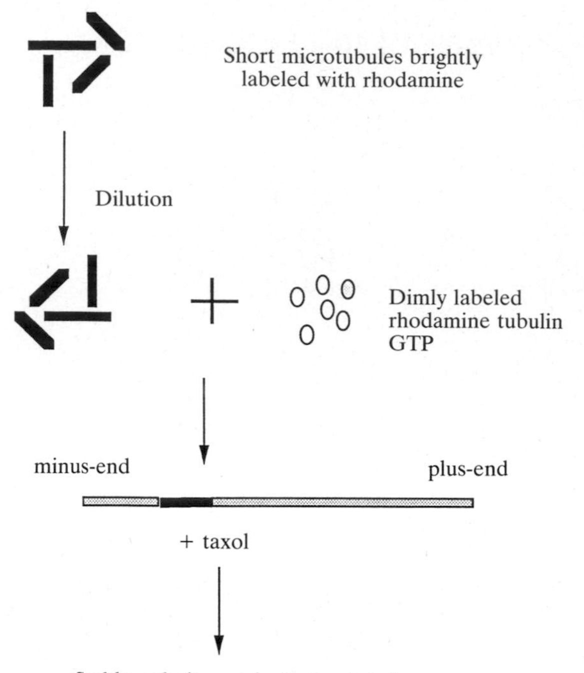

Fig. 1. Schematic diagram illustrating the design of polarity marked microtubules.

are allowed to polymerize until they reach a desired length, generally about 15 min. The longer they are polymerized, the less free rhodamine–tubulin remains. The microtubules are then stabilized by adding three volumes of BRB80+10 μM taxol and pre-warmed to 37 °C, taking care to avoid shear. The microtubules can now be used in the assay. To remove any remaining free rhodamine–tubulin monomer, microtubules can be spun and washed in the airfuge, if desired.

Visualization of marked microtubules in motor polarity assays

The major disadvantage of fluorescence is that it bleaches rapidly, generally within 10 s. In addition, microtubules will break upon observation with sufficiently high illumination (Vigers *et al.* 1988). To prevent photobleaching we use the oxygen scavenging system developed for the observation of single actin filaments (Kishino and Yanigida, 1988). In the motility buffer to be used, are added, 0.1 mg ml^{-1} catalase, 0.03 mg ml^{-1} glucose oxidase, 10 mM glucose, and 0.3 % 2-mercaptoethanol. Glucose oxidase contains a chromophore and will therefore tend to quench the fluorescence at high concentrations. With this system, microtubules have been successfully recorded for 10 min under full mercury illumination, and even when they fade, no case of a microtubule break has ever been observed. Nevertheless, it is recommended to reduce the light dose by shuttering if extended recordings of a single microtubule are planned. A SIT camera is used to record the images.

Rhodamine microtubules can be used to follow the movement of kinesin and dynein

Kinesin or dynein were perfused into a chamber constructed with two strips of double-sided tape, with a 22 mm coverslip on top. The chamber was then washed once with 10 % BSA in BRB80 and left for 3 min. Labeled microtubules were diluted into the anti-fade buffer in BRB80. These were then washed out leaving those bound to the motor proteins by a rigor state. 1 mM ATP was washed in and the movement of microtubules followed. The results with kinesin are shown in Fig. 2. No difference was seen in

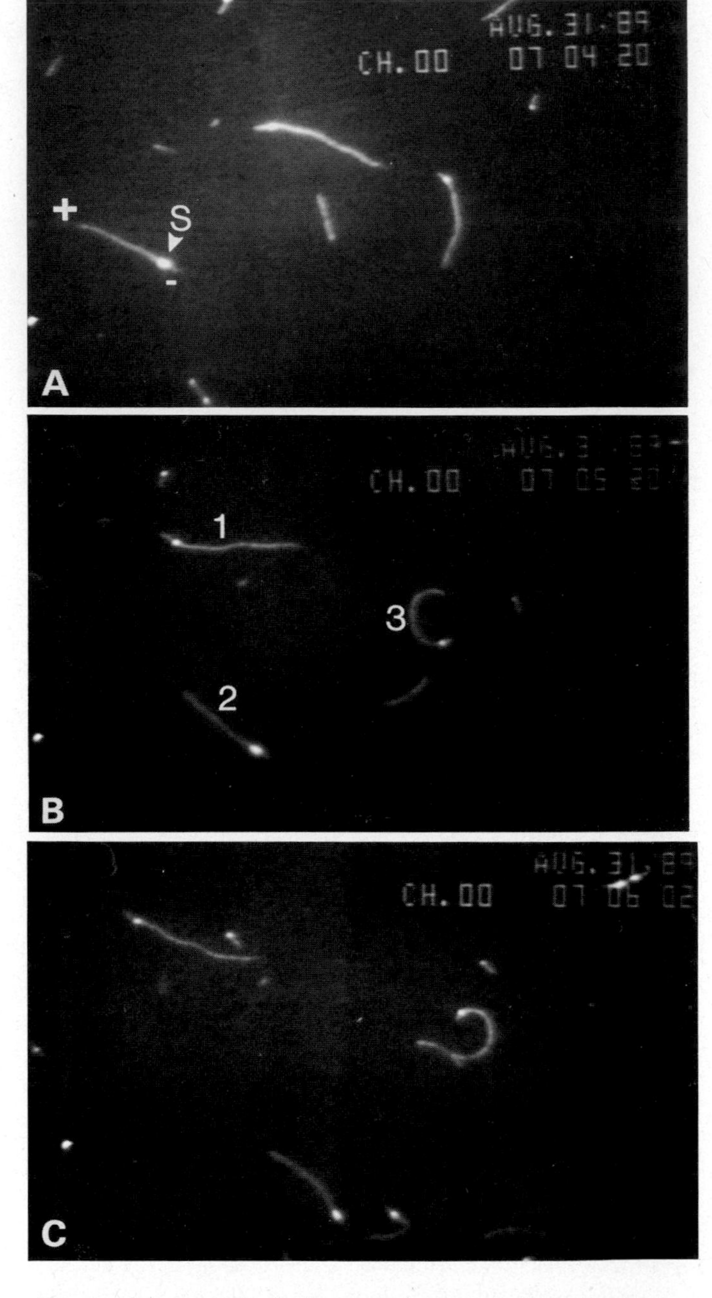

Fig. 2. Movement of polarity marked microtubules on bovine brain kinesin. (A) The plus-end is marked +, the minus end − and the seed S. There is a short minus-end because a 0.5:1 molar ratio of NEM treated:untreated tubulin was included in the tubulin polymerization (Hyman *et al.* 1990). Kinesin is plus-end directed, therefore the seed is leading as the microtubule translocates across the coverslip. (B) Shows three microtubules that can be followed moving. (C) Shows the three microtubules moving plus-end directed with the bright seed leading.

the velocity of microtubule movement when compared with transmitted light techniques, and the observation had no effect on the activity of the motor proteins, since the same area could be followed for at least 20 min. This demonstrates that the polarity-marked rhodamine microtubules provide a simple and fast way to assay the polarity of new proteins.

I thank Fady Malik for providing the kinesin and dynein, Tim Mitchison for extensive help and support, Paul Tvedten for building my room, and Doug McVay for invaluable engineering assistance and Ken Sawin for comments on the manuscript. Tony Hyman is a Lucille P. Markey Visiting Fellow.

References

Belmont, L. D., Hyman, A. A., Sawin, K. E. and Mitchison, T. J. (1990). Real-time visualization of cell cycle dependent changes in microtubule dynamics in cytoplasmic extracts. *Cell* (in press).

Endow, S. A., Henikoff, S. and Niedziela, L. S. (1990). Mediation of meiotic and early mitotic chromosome segregation in *Drosophila* by a protein related to kinesin. *Nature* **345**, 81–83.

Enos, A. P. and Morris, N. R. (1990). Mutation of a gene that encodes a kinesin-like protein blocks nuclear division in *A. nidulans*. *Cell* **60**, 1019–1027.

Hyman, A. A., Drechsel, D., Kellog, D., Salser, S., Sawin, K., Steffen, P., Wordeman, L. and Mitchison, T. J. (1990). Preparation of modified tubulins. *Meth. Enzym.* (in press).

Kishino, A. and Yanigida, T. (1988). Force measurements by manipulation of a single actin filament. *Nature* **334**, 74–76.

Koshland, D., Mitchison, T. J. and Kirschner, M. W. (1988). Chromosome movement driven by microtubule depolymerization *in vitro*. *Nature* **311**, 499–504.

McDonald, H. B. and Goldstein, L. S. B. (1990). Identification and characterization of a gene encoding a kinesin-like protein in *Drosophila*. *Cell* **61**, 991–1000.

Meluh, P. B. and Rose, M. D. (1990). KAR3, a kinesin-related gene required for yeast nuclear fusion. *Cell* **60**, 1029–1041.

Mitchison, T. J. (1984). Microtubule assembly nucleated by isolated centrosomes. *Nature* **312**, 232–236.

Mitchison, T. J. and Kirschner, M. W. (1984). Dynamic Instability of microtubule growth. *Nature* **312**, 237–242.

Paschal, B. M., Shpetner, H. S. and Vallee, R. B. (1987). MAP 1C is a microtubule-activated ATPase which translocates microtubules *in vitro* and has dynein-like properties. *J. Cell Biol.* **105**, 1273–82.

Vale, R. D., Reese, T. S. and Sheetz, M. P. (1985). Identification of a novel force-generating protein, kinesin, involved in microtubule-based motility. *Cell* **42**, 39–50.

Vale, R. D., Schnapp, B. J., Mitchison, T., Steuer, E., Reese, T. S. and Sheetz, M. P. (1985). Different axoplasmic proteins generate movement in opposite directions along microtubules *in vitro*. *Cell* **43**, 623–632.

Vigers, G. P. A., Coue, M. and McIntosh, J. R. (1988). Fluorescent microtubules break up under illumination. *J. Cell Biol.* **107**, 1011–1024.

Yang, J. T., Saxon, W. M. and Goldstein, L. S. B. (1988). Isolation and characterization of the gene encoding the heavy chain of *Drosophila* kinesin. *Proc. natn. Acad. Sci. U.S.A.* **85**, 1846–1886.

An approach to reconstituting motility of single myosin molecules

STEPHEN J. KRON*, TARO Q. P. UYEDA, HANS M. WARRICK and JAMES A. SPUDICH

Departments of Cell Biology and Developmental Biology, Stanford University School of Medicine, Stanford, CA 94305, USA

* Present address: Whitehead Institute for Biomedical Research, Cambridge, MA 02142, USA

Summary

Over the last five years, the value of *in vitro* motility assays as probes of the mechanical properties of the actin–myosin interaction has been amply demonstrated. Motility assays in which single fluorescent actin filaments are observed moving over surfaces coated with myosin or its soluble fragments are now used in many laboratories. They have been applied to a wide range of problems including the study of structure–function relationships in the myosin molecule and measurement of fundamental properties of the myosin head. However, one limitation of these assays has been uncertainty over the number of myosin heads interacting with each sliding filament, that frustrates attempts to determine properties of individual heads. In order to address this limitation, we have modified the conditions of the actin sliding filament assay to reduce the number of heads interacting with each filament. Our goal is to establish an assay in which the motor function of a single myosin head can be characterized from the movement of a single actin filament.

Key words: motility assay, actin filament, myosin head, sliding velocity, methylcellulose.

Introduction

The transformation of ATPγP bond energy to mechanical work by muscle proteins has been studied by a wide range of chemical and physical techniques. It has long been recognized that functional reconstitution from purified components is a key experimental approach. However, for forty years after Szent Gyorgyi and colleagues (1942) first studied the contraction of threads of actomyosin, progress in reconstitution of motility trailed advances in the solution biochemistry of molecular motors. This situation changed with the development of the *Nitella*-based assay for myosin movement (Sheetz and Spudich, 1983). In this simple quantitative assay, myosin-coated particles move along parallel tracks of actin bundles. Though a powerful tool, this assay was found to be unsuitable for reconstitution from purified components as it relied upon a biochemically uncharacterized actin filament substrate.

Attempts to develop a purified actin substratum met with mixed success (Spudich *et al.* 1985), but a successful alternative approach was adopted from the microtubule motility field. Following the observation of sliding movement of microtubules in axoplasm deposited on glass surfaces (Allen *et al.* 1985), Vale and colleagues (1985) used microtubule sliding as a functional assay in their purification of soluble factors mediating fast axonal transport. Similarly, visualization of single, fluorescently labelled actin filaments (Yanagida *et al.* 1985) allowed ATP-dependent sliding movement to be observed over surfaces coated with myosin filaments (Kron and Spudich, 1986), and proteolytic fragments of myosin (Toyoshima *et al.* 1987). These assays have been applied to studies of properties of myosin energy transduction. Two definitive results from such experiments were experimental demonstrations of movement activity (Toyoshima *et al.* 1987) and force-production (Kishino and Yanagida, 1988) by the isolated myosin head, subfragment 1.

The sliding filament assay has been used to measure the step size or power stroke of the myosin molecule, the displacement of an actin filament associated with a single ATP hydrolysis. Toyoshima *et al.* (1990) measured the rates of sliding movement and ATP hydrolysis in the myosin-coated surface assay. By normalizing the ratio of filament velocity and ATP hydrolysis to the number of myosin heads interacting with an actin filament, a step size of less than 20 nm ATP^{-1} was estimated. Using a similar approach, Harada *et al.* (1990) estimated a step size of more than 100 nm ATP^{-1}. This difference may be explained by considerations of measurement of ATPase or other parameters. Nonetheless, one prominent difficulty with the myosin-coated surface assay which affects the step size calculation of these studies is uncertainty in the number of myosin heads interacting per unit length of sliding actin filament.

We (Uyeda *et al.* 1990) have taken a second approach to measuring the step size, where we have modified the sliding assay to examine the relationship between sliding velocity and filament length at limiting densities of myosin heads. Methylcellulose (MC) was found to inhibit diffusion of actin filaments away from surfaces coated even very sparsely with myosin. MC is a high molecular weight, inert polymer used in industry for its activity as a viscosity enhancer in dilute aqueous solutions. When the motility assay was performed in the presence of MC, slow movement over sparsely coated surfaces could be observed. This was interpreted to be movement of actin filaments not in continuous contact with active myosin heads. The proportional decrease in sliding speed with decrease in the probability of interacting with an active myosin head was used to measure a power stroke. While, as in our previous study, the conclusions of Uyeda *et al.* (1990) are weakened by uncertainty of numbers of interacting heads, this work suggests a way of eliminating the statistical problems posed by multiple interacting heads. Our current approach is to investigate whether actin filament movement can be observed in the presence of MC when driven by single myosin heads. Here, we discuss progress toward observing the movement activity of single myosin heads and a proposed experiment to measure the force of a single myosin head.

Materials and methods

Instrumentation

The major consideration for non-destructive observation of single tetramethylrhodamine–phalloidin labelled (Rh–Ph labelled) actin filaments in real time at high magnification is to optimize fluorescence imaging. We have used an upright epifluorescence microscope, stripped of all unnecessary optical components, which may absorb or scatter light (Fig. 1). The standard epi-illumination light source is the 546 nm emission line of a 100 W mercury high pressure arc lamp, attenuated with neutral density filters or an iris diaphragm to control photodamage. Rh–Ph labelled actin has maximal fluorescence emission at 575 nm, requiring relatively close excitation and emission bands in the filter set which must also display high transmittance in the pass band and excellent rejection of the excitation light by the emission filter.

In general, magnification and image brightness are at cross purposes. Video recording of sliding movement at the magnification used for single myosin head movement (screen diameter 25 μm) demands careful selection of the optical components and camera system. Because the brightness of fluorescence in an epi-illumination microscope depends on the objective N.A. to the fourth power, the most important optical component is the objective, which must be an immersion lens with N.A.$>$1.2. Even with a bright image of labelled actin filaments, a microchannel-plate intensified Newvicon or CCD camera (e.g. GENIISYS, Dage-MTI, KS1381, Videoscope or C2400-09, Hamamatsu), rather than a SIT camera, may be required.

Good quality images have plenty of contrast, making real-time digital image enhancement unnecessary. Noise reduction by frame averaging should be avoided because of inevitable distortion of motion. Currently, the best alternative for data storage is video recording on optical disc.

Buffers

MSB: 25 mM imidazole–HCl pH 7.4, 0.6 M KCl, 1 mM ethylene diamine tetraacetic acid (EDTA), 1 mM dithiothreitol (DTT).

BED: 0.1 mM $NaHCO_3$, 0.1 mM ethyleneglycol-*bis*-(2-amino-ethylether) tetracetic acid (EGTA), 1 mM DTT.

2×CHB: 20 nM imidazole-HCl pH 7.0, 1 M KCl, 4 mM $MgCl_2$, 10 mM DTT.

Phenylmethyl sulfonyl fluoride (PMSF) stock: 0.2 M PMSF in ethanol.

PMB: 25 mM imidazole–HCl pH 7.4, 100 mM NaCl, 5 mM $MgCl_2$, 1 mM DTT.

E64 stock: 1 mg ml^{-1} E64 in dimethylsulfoxide.

SB: 10 mM imidazole–HCl pH 7.4, 0.5 M KCl, 1 mM DTT.

AB: 25 mM imidazole–HCl pH 7.4, 25 mM KCl, 4 mM $MgCl_2$, 1 mM EGTA, 1 mM DTT.

AB/BSA: AB, 0.5 mg ml^{-1} bovine serum albumin (BSA).

AB/BSA/GOC: AB/BSA, 0.018 mg ml^{-1} catalase, 0.1 mg ml^{-1} glucose oxidase, 3 mg ml^{-1} glucose.

AB/BSA/GOC/ATP: AB/BSA/GOC, 2 mM Na_2-ATP.

Protein preparation

For these experiments, we have used both chymotryptic heavy meromyosin (HMM) and papain–Mg^{2+} subfragment 1 (S1) made from proteolytic digestion of skeletal muscle myosin (Hynes *et al.* 1987), stored at −20°C in MSB and 50% glycerol at a final concentration greater than 10 mg ml^{-1}.

To make HMM, we use a modification of the method of Okamoto and Sekine (1985). To about 20 mg of stock myosin in the centrifuge tube, add 9 volumes cold BED and mix gently and completely. After at least 10 min incubation on ice, centrifuge to sediment the myosin filaments. Dissolve the pellet in 2×CHB and BED as needed to achieve a final concentration of 15 mg ml^{-1} myosin in CHB. Incubate the myosin solution for 10 min at 25°C. Add TLCK-treated (alpha)-chymotrypsin to 12.5 μg ml^{-1}, gently mix, and incubate for 7.5 to 10 min at 25°C. To stop the reaction, add 9 volumes cold BED with 3 mM $MgCl_2$ and 0.1 mM PMSF to the reaction. After more than 1 h on ice, clarify the suspension at high speed. The supernatant is typically 0.7 mg ml^{-1} HMM.

To make S1, prepare a pellet of myosin filaments as above. Dissolve the pellet in an equal volume (about 0.25 ml) of 1.2 M KCl and 20 mM DTT in BED, and incubate for 10 min at 25°. Dilute with 19 volumes ice cold BED (to about 10 ml final volume), incubate on ice for at least 10 min, and again sediment the filaments. Resuspend the pellet in PMB to a final concentration of 10–12 mg ml^{-1} and incubate for 10 min at 25°. Add papain to 12.5 μg ml^{-1}, mix gently and incubate for 7.5–10 min at 25°. Stop

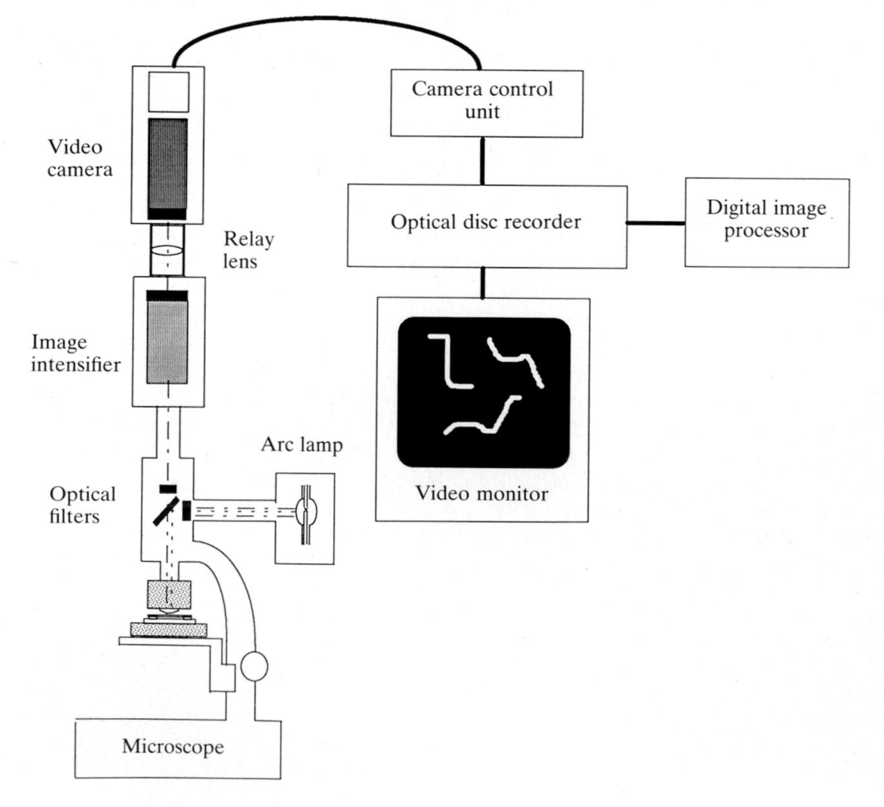

Fig. 1. Diagrammatic representation of the experimental apparatus. The myosin-coated surface assay is performed on a glass coverslip mounted on a microscope slide. Fluorescently labelled actin filaments sliding over the myosin-coated surface are visualized by intensified-video fluorescence microscopy using an upright epifluorescence microscope coupled to a microchannel-plate intensified video camera. The video signal is recorded on an optical memory disc recorder to facilitate high-resolution motion analysis. The analog video data are later digitized and analyzed to determine the position of each filament in each video frame during a movement record.

the reaction with an equal volume of ice-cold BED with 5 mM $MgCl_2$ and 5 $\mu g\,ml^{-1}$ E64. Incubate for at least 1 h on ice before high speed centrifugation. Recover the supernatant, typically 1.1 $mg\,ml^{-1}$ S1, and store on ice.

Actin affinity purification is performed to remove irreversible rigor heads from either preparation. Add filamentous actin to 0.15 $mg\,ml^{-1}$ and ATP to 1 mM to an aliquot of the stock myosin fragment. After a short incubation on ice, centrifuge the mixture at high speed to sediment the actin. Actin preparations can have an associated proteolytic activity. Thus, use the supernatant within 2 to 3 h.

Gel filtration (e.g. on a Superose 12 HR FPLC column (Pharmacia LKB) for S1 or Superose 6 HR for HMM, flow rate 0.25 $ml\,min^{-1}$ SB) yields further purification. Concentrate the actin-purified supernatant in a Centricon 30 (Amicon) ultrafiltration device to less than 0.5 ml. Clarify the retentate by high speed centrifugation and load the supernatant onto the column. Pool the major protein peak by A_{280} to yield the final myosin preparation, which is stable on ice for several days.

Muscle actin (Pardee and Spudich, 1982) is stored on ice as filaments at about 4 $mg\,ml^{-1}$ in AB with 0.2 mM ATP. To label actin fluorescently, first dry 94 μl of a 3.3 μM solution of tetramethylrhodamine-phalloidin (Rh–Ph) in methanol (Molecular Probes) to a residue, in a Speed Vac concentrator (Savant) or by a N_2 stream. Redissolve the phalloidin in 2 μg ethanol. Add 290 μl AB and vortex for approximately 30 s. Add 10 μl of 1 $mg\,ml^{-1}$ actin (freshly diluted in AB), mix well and incubate overnight on ice in the dark. Test the extent of labelling by fluorescence microscopy. Rh–Ph labelled actin is stable on ice in the dark for several weeks.

Motility assay

Construction of the experimental flow cell requires slides, coverslips (typically No. 1, 18 mm square), a diamond scribe and a glass syringe fitted with a blunted 21 French needle and filled with Apiezon M grease. To coat the coverslips with nitrocellulose (NC, 1% in amyl acetate, Fullam), fix several on a slide with small drops of water. Apply a drop of NC diluted to 0.1% in amyl acetate to each of the coverslips and spread the film using a side of a pasteur pipette tip. Touch the edge of the slide to filter paper to absorb the excess NC. Allow the film to air dry. The coverslips can be detached from the slide and used for up to two days.

Using the syringe, place two parallel beads of grease several cm in length about 10 mm apart onto a slide. Using a slide as a guide, cut several 2 mm wide strips of untreated coverslip with the scribe. Position two strips outside the pair of grease lines. Place an NC-coated coverslip, film side down, onto the grease and press down with the forceps until it rests on the strips. The resulting flow cell has an internal volume of about 50 μl. Place a bead of grease across the coverslip close to the inlet of the flow channel, forming a dam to prevent flow of methylcellulose solutions over the coverslip rather than through the channel.

Prepare a 2% (w/v) methylcellulose (Sigma M-0512) stock solution and dialyze it overnight against AB with 0.05% NaN_3. All solutions are freshly diluted from stocks and kept on ice. To inhibit photobleaching, dissolved oxygen is removed from AB and AB/BSA by degassing under vacuum.

Prop the dry flow cell up at an angle of approximately 30°, with its 'inlet' down, and fill the channel with a dilution of stock myosin fragments in AB/BSA. Immediately flip the flow cell upright and introduce a second aliquot of the diluted myosin into the flow channel and incubate for 60 s. (Initially, start with about 50 $\mu g\,ml^{-1}$ S1 in AB/BSA in order to oberve smooth sliding movement. Then, titrate the concentration of S1 by dilution in AB/BSA to achieve optimal motility. Stock S1 may need to be diluted 500 fold to observe single-head motility.)

Beginning with 100 μl AB/BSA, infuse solutions in approximately 50 μl aliquots (one flow cell volume) forming a small pool at the inlet to the flow channel. Infuse 100 μl of a 1 to 70 dilution of stock Rh–Ph actin in AB/BSA. After 60 s, infuse 100 μl AB/BSA, followed by 100 μl of 0.7–1.0% MC in AB/BSA/GOC, which provides both the viscous medium for low myosin density movement (Uyeda *et al.* 1990) and an enzymatic scavenger of dissolved oxygen (Kishino and Yanagida, 1988). Immediately follow with 50 μl of AB/BSA/GOC/ATP with 0.7–1.0% methylcellulose.

Once the ATP solution has been infused, dry the exposed parts of the slide, oil the coverslip and place the flow cell onto the microscope stage. With the actin filaments bound to the underside of the coverslip in focus, sliding movement should be immediately visible.

Results

We use methylcellulose (MC) as a high molecular weight, inert solute to increase the 'macro-viscosity' of the assay solution (Uyeda *et al.* 1990). Use of MC restricts actin filaments to a characteristic form of brownian motion, reptation. Actin filaments suspended in 0.5–2.0% MC do not undergo rapid lateral displacements like those of filaments suspended in assay buffer. However, the axial component of brownian motion is largely preserved (Fig. 2). When the motility assay is performed in the presence of MC, concentrations of myosin heads which otherwise could not hold filaments to the surface can effect net actin sliding movement. Yet the sliding speed of actin filaments on surfaces sparsely coated with myosin are unchanged in the presence of 0.5%–2.0% MC, suggesting little if any viscous loading of myosin heads by the methylcellulose.

When high concentrations of S1 are applied to the surface, smooth sliding movement occurs at 1–2 $\mu m\,s^{-1}$ in the presence or absence of MC. However, when surfaces are only sparsely coated with myosin heads, sliding motility can only be observed in the presence of MC. As the concentration of S1 applied is decreased, the sliding movement slows and becomes intermittent, consisting of smooth sliding interrupted by short reptating movements. At the lowest applied concentrations, the movement is no

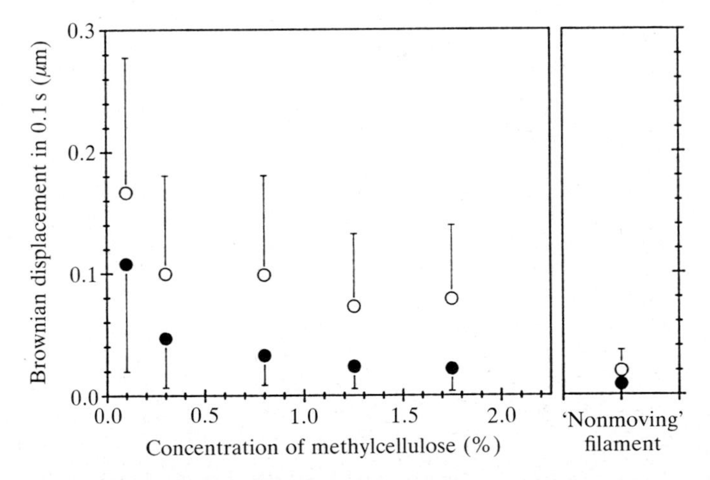

Fig. 2. Plot of axial *versus* lateral components of brownian motion of actin filaments in methylcellulose solutions. Video records of the motions of single fluorescent actin filaments of length 4.5 to 6.0 μm in the presence of 0.1 to 1.75% methylcellulose in assay buffer were analyzed to determine displacement of the centroid of the filament at intervals of 3 frames (0.1 s). The displacements are shown along the filament axis (open circle, mean±S.D.) and perpendicular to the filament axis (closed circle, mean±S.D.). Characteristic reptating movements of actin filaments in methylcellulose solutions reflect the signficant difference between the degree of inhibition of lateral motions and axial motions. Apparent motion of the 'nonmoving' filament, an actin filament fixed to a myosin-coated surface by rigor bonds, reflects in part system noise as well as any flexibility in the molecular components.

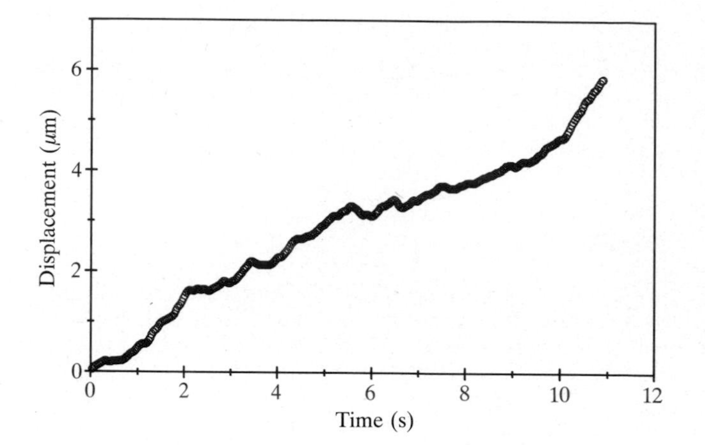

Fig. 3. Record of displacement of a single actin filament sliding over a surface sparsely coated with myosin subfragment 1. The position of the centroid of a single actin filament (length approx. 2 μm) was determined in each video frame (30 frames s^{-1}) for 11 s. Here, a 3 frame rolling average was used to smooth rapid transients. Displacements along the filament path were determined for each frame and plotted *versus* elapsed time. Note that the record consists of runs of slow forward sliding interrupted by rapid random displacements. At least two distinct speeds of sliding are demonstrated in this record.

longer smooth, even for short intervals. In this regime of limiting myosin head density, individual filaments undergo limited brownian movement at all times, yet make slow but significant progress in a single direction along their axis. Such filaments move in runs and pauses, periods of slow sliding which may move a filament the distance of its length or greater, punctuated by periods of free reptation (Fig. 3). When no myosin is applied to the surface, actin filaments are found near the surface only rarely. These filaments appear to undergo free reptation.

Conclusions

Kinetic evidence suggests that the tightly bound states of myosin may represent only a fraction of the kinetic cycle. Thus, an actin filament undergoing sliding movement in the presence of ATP must interact simultaneously with many myosin heads in order to remain continuously attached to a myosin-coated surface. The model-dependent analysis necessary to extract the properties of the individual myosin head from sliding movement due to large ensembles of myosin heads is subject to criticism. The stochastic nature of the actin–myosin interaction is itself confounding to such studies. However, a new class of approaches can now be considered, that involve mechanical measurements of the function of single myosin heads.

Prior to this work with myosin, sliding movement of microtubules over single, two-headed, kinesin molecules was reported (Howard *et al.* 1989). Comparison of the activites of these two motors in single molecule movement assays may be illuminating. Considering the kinetic properties of muscle myosin, a motility assay for single myosin head movement may provide a particularly useful tool for the analysis of molecular motor function.

The system for assaying the movement described is an extension of our recent studies of sliding movement over surfaces sparsely coated with heavy meromyosin (Uyeda *et al.* 1990). We found that the limit on sliding movement at low densities of myosin could be overcome by including in the assay buffer a high molecular weight polymer, methylcellulose (MC), to increase the macroviscosity of the solution. Using methylcellulose and surfaces coated very sparsely with myosin, we have now observed filaments undergoing a characteristic form of sliding movement. We have interpreted this to result from intermittent impulses from a small number of single myosin heads superimposed on the brownian motion of the filaments.

During intervals when a single head interacts with an actin filament, its sliding movement in MC solutions is readily predicted. According to the simplest form of the swinging crossbridge model (Huxley, 1969), the average speed should be determined by the product of the step size and the turnover rate. In order to extract a value for the step size, we wish to measure the average sliding speed of an actin filament in contact with a single myosin head. Toward this end, we have begun to analyze video records of filament movement, using digital image processing to track the centroids of individual filament images from frame to frame. Preliminary analysis suggests that the sliding of filaments over surfaces coated very sparsely with S1 or HMM indeed consists of intermittent runs, with sliding speeds of integral multiples of a minimum velocity.

This new system constitutes one of the desired geometries for analysis of the function of individual motors – a single myosin head driving a single actin filament. Based upon current models of chemo-mechanical transduction, high temporal and spatial resolution microscopy of such a system would be expected to detect binding of the head to the filament, the subsequent step and then the release. Several approaches to 'super-resolution' in the light microscope have been reported (as reviewed by Inoue, 1989). Using quantitative imaging and numerical analysis, the position of the centroid of an image can be detected to nm precision (Gelles *et al.* 1988). A similar analysis could be used with appropriately tagged actin filaments (e.g. see Harris and Weeds, 1984).

Force measurements from single myosin molecules may

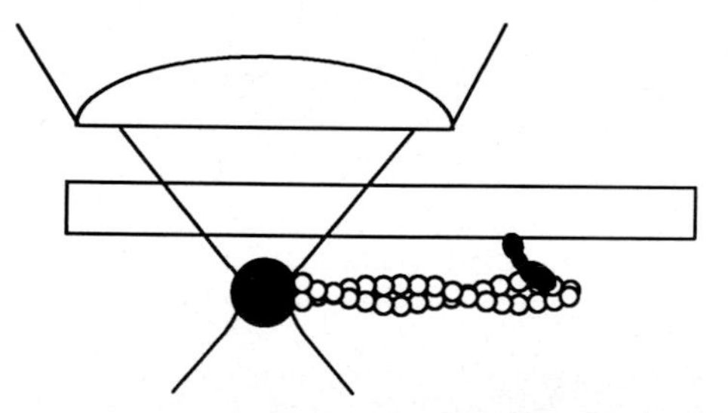

Fig. 4. Diagrammatic representation of a proposed experiment to use a single-beam, gradient–force, optical trap to perform mechanical experiments with single myosin heads interacting with single actin filaments. A polystyrene bead (<1 μm diameter) is attached to the barbed end of a single fluorescent actin filament. The actin filament is bound to a single myosin head on the coverslip surface. A laser beam is focussed by a microscope objective, forming a light trap in the focal plane. Within the optical trap, the polystyrene bead is held in a roughly spherical, potential well with diameter near the optical wavelength. Flash photolysis of caged ATP will result in repetitive cycling of the myosin head and movement of the polystyrene bead to the edge of the trap. The position of the bead is monitored continuously, with nm precision. Modulation of the force of the optical trap is used to observe the effect of increasing tension in the actin filament on the motion of the polystyrene bead, yielding a force–velocity curve for a single myosin head.

also be possible. Flexible glass needles (Kamimura and Takahashi, 1981), as used by Kishino and Yanagida (1988), probably lack the necessary sensitivity. A novel approach for force measurement in the microscope has recently been applied to studies of microtubule–kinesin interactions (Block *et al.* 1990; Kuo and Sheetz, 1990). The single-beam, gradient force, optical trap (Ashkin *et al.* 1986; Ashkin and Dziedzic, 1987) provides a micromanipulator capable of exerting large forces on dielectric materials, such as glass or plastic beads. A possible experiment with an optical trap is to hold in place, *via* an attached polystyrene bead, a single actin filament during its interaction with a single myosin head (Fig. 4). The function of the trap is two-fold in this experiment. It not only provides a way to impose a variable load on the myosin head, but also damps brownian motion of the actin filament, facilitating high precision imaging of position during cycling of the myosin head. Thus, the force–velocity relation, the step size and the mechanical duty cycle should each be able to be determined in this geometry.

We thank Dr Steven Chu for helpful discussions. S.J.K. was a trainee of the Medical Scientist Training Program at Stanford University. T.Q.P.U. was supported by a fellowship from the Japanese Ministry of Health, Technology and Culture. This work was supported by a grant from the NIH to J.A.S. (GM33289).

References

ALLEN, R. D., WEISS, D. G., HAYDEN, J. H., BROWN, D. T., FUJIWAKE, H. AND SIMPSON, M. (1985). Gliding movement of and bidirectional transport along native microtubules from squid axoplasm: evidence for an active role of microtubules in cytoplasmic transport. *J. Cell Biol.* **100**, 1736–1752.

ASHKIN, A. AND DZIEDZIC, J. M. (1987). Optical trapping and manipulation of viruses and bacteria. *Science* **235**, 1517–1520.

ASHKIN, A., DZIEDZIC, J. M., BJORKHOLM, J. E. AND CHU, S. (1986). Observation of a single-beam gradient force optical trap for dielectric particles. *Optics Lett.* **11**, 288–290.

BLOCK, S. M., GOLDSTEIN, L. S. B. AND SCHNAPP, B. J. (1990). Bead movement by single kinesin molecules studied with optical tweezers. *Nature* **348**, 348–352.

GELLES, J., SCHNAPP, B. J. AND SHEETZ, M. P. (1988). Tracking kinesin-driven movements with nanometre-scale precision. *Nature* **331**, 450–453.

HARADA, Y., SAKURADA, K., AOKI, T., THOMAS, D. D. AND YANAGIDA, T. (1990). Mechanochemical coupling in actomyosin energy transduction studied by *in vitro* movement assay. *J. molec. Biol.* **216**, 49–68.

HARRIS, H. E. AND WEEDS, A. G. (1984). Plasma gelsolin caps and severs actin filaments. *FEBS Lett.* **177**, 184–188.

HOWARD, J., HUDSPETH, A. J. AND VALE, R. D. (1989). Movement of microtubules by single kinesin molecules. *Nature* **342**, 154–158.

HUXLEY, H. F. (1969). The mechanism of muscular contraction. *Science* **164**, 1356–1356.

HYNES, T. R., BLOCK, S. M., WHITE, B. T. AND SPUDICH, J. A. (1987). Movement of myosin fragments *in vitro*: domains involved in force production. *Cell* **48**, 953–963.

INOUE, S. (1989). Imaging of unresolved objects, superresolution, and precision of distance measurement with video microscopy. *Meth. Cell Biol.* **30**, 85–112.

KAMIMURA, S. AND TAKAHASHI, K. (1981). Direct measurement of the force of microtubule sliding in flagella. *Nature* **293**, 566–568.

KISHINO, A. AND YANAGIDA, T. (1988). Force measurements by micromanipulation of a single actin filament by glass needles. *Nature* **334**, 74–76.

KRON, S. J. AND SPUDICH, J. A. (1986). Fluorescent actin filaments move on myosin fixed to a glass surface. *Proc. natn. Acad. Sci. U.S.A.* **83**, 6272–6276.

KUO, S. C. AND SHEETZ, M. P. (1990). Force of kinesin-dependent microtubule translocation measured by optical trapping. *Biophys. J.* **57**, 399a.

OKAMOTO, Y. AND SEKINE, T. (1985). A streamlined method of subfragment one preparation from myosin. *J. Biochem. (Tokyo)* **98**, 1143–1145.

PARDEE, J. D. AND SPUDICH, J. A. (1982). Purification of muscle actin. *Meth. Cell Biol.* **24**, 271–289.

SHEETZ, M. P. AND SPUDICH, J. A. (1983). Movement of myosin-coated fluorescent beads on actin cables *in vitro*. *Nature* **303**, 31–35.

SPUDICH, J. A., KRON, S. J. AND SHEETZ, M. P. (1985). Movement of myosin-coated beads on oriented filaments reconstituted from purified actin. *Nature* **315**, 584–586.

SZENT-GYORGYI, A. (1942). *Studies Inst. Med. Chem. U. Szeged* **1**, 17–26.

TOYOSHIMA, Y. Y., KRON, S. J., MCNALLY, E. M., NIEBLING, K. R., TOYOSHIMA, C. AND SPUDICH, J. A. (1987). Myosin subfragment-1 is sufficient to move actin filaments *in vitro*. *Nature* **328**, 536–539.

TOYOSHIMA, Y. Y., KRON, S. J. AND SPUDICH, J. A. (1990). The myosin step size: Measurement of the unit displacement per ATP hydrolyzed in an *in vitro* assay. *Proc. natn. Acad. Sci. U.S.A.* **87**, 7130–7134.

UYEDA, T. Q. P., KRON, S. J. AND SPUDICH, J. A. (1990). Myosin step size: Estimation from slow sliding movement of actin over low densities of heavy meromyosin. *J. molec. Biol.* **214**, 699–710.

VALE, R. D., SCHNAPP, B. J., REESE, T. S. AND SHEETZ, M. P. (1985). Organelle, bead, and microtubule translocations promoted by soluble factors from the squid giant axon. *Cell* **40**, 559–569.

YANAGIDA, T., ARATA, T. AND OOSAWA, F. (1985). Sliding distance of actin filament induced by a myosin crossbridge during one ATP hydrolysis cycle. *Nature* **316**, 366–369.

A model for kinesin movement from nanometer-level movements of kinesin and cytoplasmic dynein and force measurements

SCOT C. KUO, JEFF GELLES*, ERIC STEUER, and MICHAEL P. SHEETZ

Department of Cell Biology, Box 3709, Duke University Medical Center, Durham, NC 27710, USA

* Present address: Department of Biochemistry, Brandeis University, Waltham MA 02254, USA

Summary

Our detailed measurements of the movements of kinesin- and dynein-coated latex beads have revealed several important features of the motors which underlie basic mechanical aspects of the mechanisms of motor movements. Kinesin-coated beads will move along the paths of individual microtubule protofilaments with high fidelity and will pause at 4 nm intervals along the microtubule axis under low ATP conditions. In contrast, cytoplasmic dynein-coated beads move laterally across many protofilaments as they travel along the microtubule, without any regular pauses, suggesting that the movements of kinesin-coated beads are not an artefact of the method. These kinesin bead movements suggest a model for kinesin movement in which the two heads walk along an individual protofilament in a hand-over-hand fashion. A free head would only be able to bind to the next forward tubulin subunit on the protofilament and its binding would pull off the trailing head to start the cycle again. This model is consistent with the observed cooperativity between the heads and with the movement by single dimeric molecules. Several testable predictions of the model are that kinesin should be able to bind to both alpha and beta tubulin and that the length of the neck region of the molecule should control the off-axis motility. In this article, we describe the technology for measuring nanometer-level movements and the force generated by the kinesin molecule.

Key words: kinesin, microtubule motility, optical trapping, particle tracking, video-enhanced microscopy.

Introduction

Recently, a number of proteins have been purified which are microtubule-activated ATPases and sufficient for *in vitro* motility of microtubules. Two of these proteins, kinesin and cytoplasmic dynein, are believed to be responsible for fast axonal transport of organelles (Vale *et al.* 1985; Scholey *et al.* 1985; Lye *et al.* 1987; Paschal *et al.* 1987), with kinesin moving towards the plus-end of microtubules (anterograde) and cytoplasmic dynein towards the minus-end (retrograde). Both proteins have two globular heads, containing both microtubule-binding and ATPase activities, attached to a rod-like structure (Amos, 1987; Hirokawa *et al.* 1989; Hisanaga *et al.* 1989; Scholey *et al.* 1989; Vallee *et al.* 1988), overall reminiscent of myosin structure and flagellar dynein. When using the cloned *Drosophila melanogaster* kinesin gene (Yang *et al.* 1989) and genes for various rod-like proteins, protein chimeras demonstrate that the kinesin globular heads are sufficient for heterologous rod-like structures, presumably acting as rigid spacer arms, to move microtubules (Yang *et al.* 1990).

Video-enhanced differential interference contrast microscopy (VDICM) has been instrumental in identifying microtubule-dependent motors. VDICM can visualize unstained microtubules (Allen *et al.* 1985), and allows the initial purification of kinesin by monitoring microtubule gliding on kinesin-coated coverslips (Vale *et al.* 1985). A modification of the gliding assay has shown that microtubules can be translocated by a single molecule of kinesin (Howard *et al.* 1989). Furthermore, the position of motor-coated latex beads can be measured with nanometer precision (Gelles *et al.* 1988) from VDICM images. At limiting ATP concentrations, kinesin bead tracks follow single protofilaments of the microtubule and pause at 4 nm intervals, corresponding to the tubulin monomer spacing in the microtubule polymer. Comparable experiments with cytoplasmic dynein beads do not exhibit regular pauses and bead paths move across all accessible protofilaments on the surface of the microtubule. The contrast between kinesin and cytoplasmic dynein bead motions indicates that the kinesin results are not artifacts of the methodology.

Since kinesin motors travel along single protofilaments with great fidelity and single kinesin molecules can translocate microtubules, kinesin offers technical advantages over actomyosin in biophysical characterization of single force-generating proteins. However, examining the force of a single or small number of motor molecules requires novel technology. Kishino and Yanagida (1988) and Chaen *et al.* (1989) used the deflection of a thin glass needle to measure the force of 30–50 (or more) myosin heads on actin filaments, with either actin filaments or myosin filaments, respectively, attached to the glass needle. Rather than a glass needle, we have used an optical trap ('laser tweezers') to manipulate latex beads attached to microtubules. Optical traps, a recently developed technology, use the intensity gradient of laser light formed from a high numerical aperture lens to capture microscopic particles (Ashkin *et al.* 1987*a*). If near-infrared laser light (1064 nm from Nd:YAG laser) is used, bacteria and yeast can be trapped and can reproduce for generations in the optical trap (Ashkin *et al.* 1987*b*).

Materials and methods

Video microscope and optical trap

Our video microscope and optical trap is based on the Zeiss IM-35

Journal of Cell Science, Supplement 14, 135–138 (1991)

inverted microscope. We use the 1.4 NA condenser and the 63×/1.4 Plan-Apo objective from Zeiss for all our video microscopy; the amount of extinction is controlled with a quarter-wave plate and a rotating analyzing polarizer, as described by Allen *et al.* (1981) and images are projected with a 6.3×lens onto a Newvicon camera with manual gain and black-level controls (Dage MTI model 70).

Two major modifications have been made to improve the optical qualities of this microscope. First, all film polarizers (polarizer and analyzer) were replaced with birefringent crystal polarizers, which transmit much more light and greater polarization purity. To mount the analyzer crystal and quarter-wave plate, a special holder was constructed to fit into the slot which usually holds the fluorescence dichroic filters. Second, the Zeiss illuminator and condenser tube was replaced with a custom illuminator that uses a fiber-optic scrambler to flatten out the inhomogeneities of the mercury arc lamp, as described by Ellis (1985). All components, including the microscope, are clamped or bolted to 8-inch thick honeycomb optical table on pneumatic isolators (Newport Corp.).

The schematic diagram for the optical trap is shown in Fig. 1. The polarized output of a 1W Nd:YAG laser (CVI Corp) operating in TEM_{00} mode is directed through an 80 mm focal length achromat doublet lens into the epifluorescence port of the IM-35. The epifluorescence insert was modified, removing the field stop and heat-blocking filter. A half-wave plate mounted in a motorized rotator and a beam-splitting crystal polarizer allows continuous attenuation of the laser illumination during experiments. A beam splitter on the attenuated beam illuminates a power meter to monitor the instantaneous variations in beam power. Custom video overlay electronics allows recording of the power meter readings onto the video image. All transmittive optics of the laser trap external to the microscope have been dielectric-coated to minimize laser light reflection.

Calibration of optical trap

The maximum retention force of the laser optical trap was calibrated by holding latex spheres stationary against the viscous flow of aqueous solution through a miniature flow cell. The flow cell (Berg and Block, 1984) exhibits laminar flow in the area of observation, and the force of viscous drag is simply Stokes law for a sphere: $\boldsymbol{F}=6\pi\eta rv$, where r is the radius of the sphere, v the velocity of fluid flow, and η the viscosity of the medium. The fluid viscosity was determined using an Ostwald viscometer, and the fluid velocity was determined using video microscopy to track the latex spheres which are unaffected by the laser trap and are moving with the fluid flow. The limiting velocity at which beads are retained in the optical trap indicates the maximum force of the laser trap. Both water and dilute Ficoll 400 solutions (a Newtonian fluid, Berg and Turner, 1979) have been used. The trapping force is linear with the amount of laser illumination.

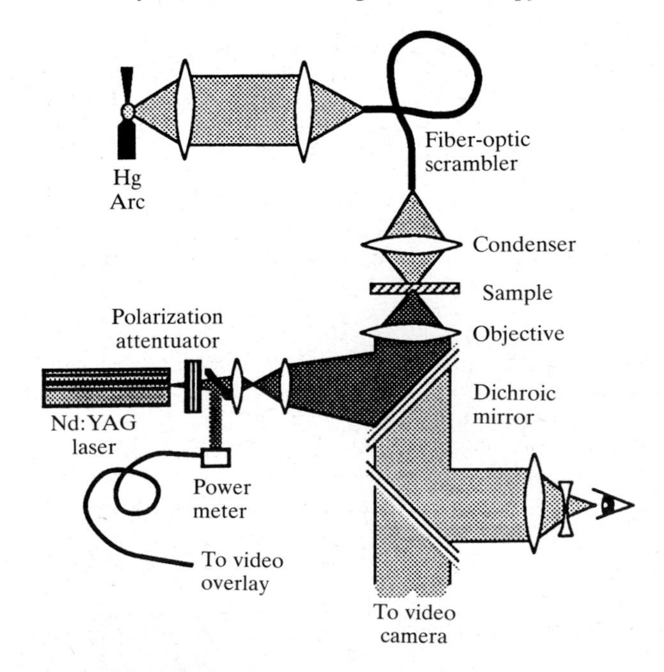

Fig. 1. Schematic diagram of optical trap on the video microscope.

Computational equipment

Although we have developed the software to perform positional measurements on a number of computers and image processors, our current system is the most affordable and has the widest applicability for other researchers. This configuration uses a Zenith Z386/25 computer and the Series 151 Image Processor from Imaging Technology, Inc. The minimum necessary configuration of the Series 151 Image Processor includes the ADI, FB, ALU, and BTM boards in the enclosure box. Video images are stored in the Panasonic Optical Memory Disk Recorder (OMDR, model TQ20285F) which is controlled by the computer *via* an RS-232 serial line. For applications that do not require full nanometer precision, adequate images can be stored on a super-VHS videocassette recorder (Panasonic Model AG7300), but a customized controller must be built to allow computer control of the VCR for frame-by-frame analysis. With either recording medium, a time-base corrector, such as the Fortel model DHP-525S, cleans up video image timing signals prior to digitization by the image processor.

Cross-correlation analysis of particle position

Motion of an object in the microscope field can be reconstructed from sequential positional measurements of the diffraction image of the object. We use the cross-correlation algorithm to try to maximize the amount of positional information extracted from the bead image. Intuitively, the cross-correlation algorithm can be considered a process of finding the best match of a template image, called the kernel, in the image area. As the kernel is slid across the image area, the cross-correlation calculation yields a measure of the degree of similarity of the kernel and the portion of the image that it overlaps. Since the image and the kernel are digital, the point of best alignment is extrapolated by weighted averaging (centroid) of the cross-correlation values around the cross-correlation peak. The complete algorithm and equations have been described elsewhere (Gelles *et al.* 1988).

Factors affecting measurements

Optical factors

The general optical limitations in microscopy have been covered in other texts (e.g. text by Inoué, 1986). Practical aspects of microscopy can clearly affect the precision and accuracy of positional calculations.

Optical surfaces. The microscope should be adjusted for the best images possible. All optical surfaces should be clean; we use many passages with lint-free cotton swabs made from long-fiber surgical cotton dipped into 70:20:10 ether:acetone:methanol to clean lenses. The camera is most sensitive to smudges on the projection lens.

Condenser adjustment. Though often neglected, re-adjusting the condenser for every sample is *extremely* important for best microscope performance. The condenser should be adjusted for Köehler illumination, and the image of the field diaphragm moved or adjusted in size to achieve the most homogeneous illumination across the video camera face plate.

Camera alignment. Even with the best objectives, there will be 'pincushion' or 'barrel distortion' effects. These distortions are least in the center of the field of view, so the camera should be centered.

Mechanical factors

Since our tracking algorithm relies on fiducial markers to

provide an internal frame of reference, it is relatively insensitive to low frequency, small amplitude vibrations. As discussed later, optical and electronic considerations have greater effects on the precision of positional measurements. However, the position of the laser trap would vary with respect to the sample under mechanical stress, so we try to minimize mechanical flexibility of the apparatus. We use a stiff optical table (see above) to mount the microscope, laser, laser optics and camera. The microscope is clamped to the table, and the camera is mounted on a small table supported by four stainless steel 6×0.56 inch rods, arranged to minimize the compliance of the small camera table. Similar considerations were used to mount the laser and laser optics. Finally, the whole apparatus is used at thermal equilibrium to minimize variations of focus and field of view during the recording session.

Electronic factors

Camera. Since we operate our video-enhanced DIC microscope optimized for best contrast of microtubules, there is a large amount of background light. Therefore a video camera with a good dynamic range, specifically good linearity of response even with large amounts of light, is needed. Appropriate cameras and factors in choosing cameras have been discussed elsewhere (e.g. Inoué, 1986).

Timing and digitization. The biggest problems we encountered with recorded images (both OMDR and sVHS) are pixel jitter and poor synchronization pulses in the video signal. The time-base corrector restores clean synchronization pulses, but the specification for standard RS-170 video allows a maximum 190 ns variation in H-sync pulse timing, which translates to almost two full pixels in the x direction, or 90 nm, typically. Commercial video equipment is rarely this bad and repeatability is more important, but cascading such equipment could accumulate noticeable timing variations. With timing variations on the order of half a pixel, we try to minimize its effects by using the video digitizer to average 256 video frames of the same frame from the OMDR prior to positional calculations.

Characteristics of sample and particles

Both the centroid and cross-correlation methods assume that the shape of the particle image remains constant throughout the tracking period. The image shape may change if the particle is asymmetric and rotates, or if the particle goes out of focus. To reduce changes in focus, we use a microscope with stable focus and samples at thermal equilibrium with the microscope. For *in vitro* work, we try to choose cells that are flat, extended, and well adhered to the substratum.

A final factor that affects the tracking algorithms is the image quality of the background. Homogeneous illumination helps (e.g. fiber-optic illuminator), but more limiting are the optical qualities of the sample. A very particulate or textured background makes it very difficult to distinguish particles from the background. Also the density of particles should be low enough that particles are well separated, allowing us to measure the movement of a single isolated object with high precision.

Although the laser trap can manipulate various cellular structures, including whole cells, organelles, chromosomes and even patches of plasma membranes, the force of the trap on these structures is difficult to calibrate because of their irregular and deformable shapes. We prefer to use artificial particles, such as latex spheres, for force

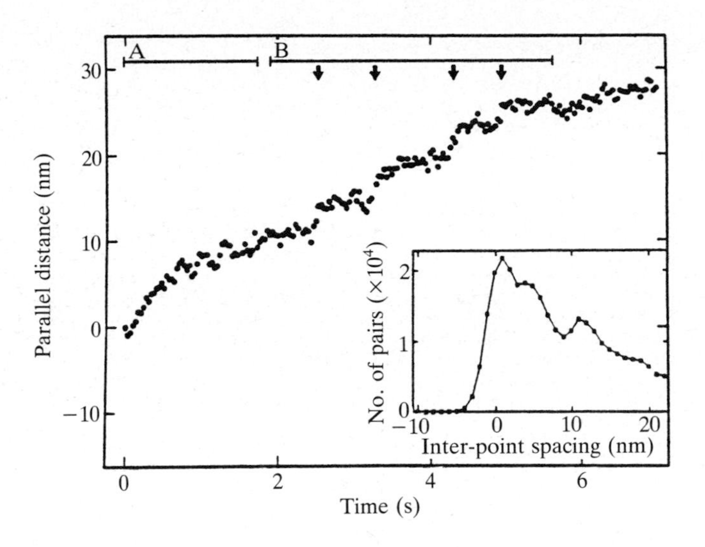

Fig. 2. Motion of kinesin-coated beads in 2.5 μM ATP (from Gelles *et al.* 1988).

measurements. Even with these particles, the optical quality of the sample can interfere; in aqueous buffer alone, there is a maximum working depth to which the laser trap will hold particles. Although we are unsure of the mechanism for this limitation, all our calibrations are performed close to the coverslip surface.

Applications

Nanometer-precision analysis of kinesin movements

Kinesin-coated latex beads will move unidirectionally along microtubules; in limiting concentrations of ATP (2.5 μM), periods of both smooth and discontinuous movements can be seen. As shown in Fig. 2, region A shows smooth translocation with a mean velocity of 5.6 nm s^{-1}. Movement in region B occurs in four discrete jumps (arrows) of length 3.7±1.7 nm. The inset of Fig. 2 shows the inter-point spacing of the bead position, enhancing the 4 nm spacing of the pauses in translocation.

Conclusions and future directions

The confluence of video microscopy and optical trapping promises to be very fruitful. Our image processing algorithms can follow movements in the DIC video microscope with almost nanometer precision; the laser trap provides an extremely fine, and calibratable, micromanipulator. We have done our initial force and tracking measurements on kinesin, but applications to cytoplasmic dynein are obvious. Other force generating systems, such as actomyosin or nucleotide polymerization reactions, are being considered, but more development is needed to improve *in vitro* motility. Compared to mechanical micromanipulators, the optical trap offers a distinct advantage that when the particle is released, it is free to diffuse unhindered and undamaged. This makes the technology particularly attractive for studies of the dynamic conversion between active transport and diffusion of individual membrane proteins tagged with submicron particles. If precise, quantitative measurements are not required, a whole host of applications could make use of these technologies. We suspect that optical traps will become quite commonplace in the near future.

References

ALLEN, R. D., ALLEN, N. S. AND TRAVIS, J. L. (1981). Video-enhanced contrast, differential interference contrast (AVEC-DIC) microscopy: A new method capable of analyzing microtubule-related motility in the reticulopodial network of *Allogromia laticollaris*. *Cell Motil. Cytoskel.* **1**, 291.

ALLEN, R. D., WEISS, D. G., HAYDEN, J. H., BROWN, D. T., FUJIWAKE, H. AND SIMPSON, M. (1985). Gliding movement of and bidirectional organelle transport along single native microtubules from squid axoplasm. Evidence for an active role of microtubules in cytoplasmic transport. *J. Cell Biol* **100**, 1736–1752.

AMOS, L. A. (1987). Kinesin from pig brain studied by electron microscopy. *J. Cell Sci.* **87**, 105–111.

ASHKIN, A. AND DZIEDZIC, J. M. (1987*a*). Optical trapping and manipulation of viruses and bacteria. *Science* **235**, 1517–1520.

ASHKIN, A., DZIEDZIC, J. M. AND YAMANE, T. (1987*b*). Optical trapping and manipulation of single cells using infrared laser beams. *Nature* **330**, 769–771.

BERG, H. C. AND BLOCK, S. M. (1984). A miniature flow cell designed for rapid exchange of media under high-power microscope objectives. *J. Gen. Microbiol.* **130**, 2915–2920.

BERG, H. C. AND TURNER, L. (1979). Movement of microorganisms in viscous environments. *Nature* **278**, 349–351.

CHAEN, S., OIWA, K., SHIMMEN, T., IWAMOTO, H. AND SUGI, H. (1989). Simultaneous recordings of force and sliding movement between a myosin-coated glass microneedle and actin cables *in vitro*. *Proc. natn. Acad. Sci. U.S.A.* **86**, 1510–1514.

ELLIS, G. W. (1985). Microscope illuminator with fiber optic source integrator. *J. Cell Biol.* **101**, 83a.

GELLES, J., SCHNAPP, B. J. AND SHEETZ, M. P. (1988). Tracking kinesin-driven movements with nanometre-scale precision. *Nature* **331**, 450–453.

HIROKAWA, N., PFISTER, K. K., YORIFUJI, H., WAGNER, M. C., BRADY, S. T. AND BLOOM, G. S. (1989). Submolecular domains of bovine brain kinesin identified by electron microscopy and monoclonal antibody decoration. *Cell* **56**, 867–878.

HISANAGA, S., MUROFUSHI, H., OKUHARA, K., SATO, R., MASUDA, Y., SAKAI, H. AND HIROKAWA, H. (1989). The molecular structure of adrenal medulla kinesin. *Cell Motil. Cytoskel.* **12**, 264–272.

HOWARD, J., VALE, R. D. AND HUDSPETH, A. J. (1989). Movement of microtubules by single kinesin molecules. *Nature* **342**, 154–158.

INOUÉ, S. (1986). *Video Microscopy*. Plenum Press, New York.

KISHINO, A. AND YANAGIDA, T. (1988). Force measurements by micromanipulation of a single actin filament by glass needles. *Nature* **334**, 74–76.

LYE, R. J., PORTER, M. E., SCHOLEY, J. M. AND MCINTOSH, J. R. (1987). Identification of a microtubule-based cytoplasmic motor in the nematode *C. elegans*. *Cell* **51**, 309–318.

PASCHAL, B. M., SHPETNER, H. S. AND VALLEE, R. B. (1987). MAP 1C is a microtubule-activated ATPase which translocates microtubules *in vitro* and has dynein-like properties. *J. Cell Biol.* **105**, 1273–1282.

SCHOLEY, J. M., HEUSER, J., YANG, J. T. AND GOLDSTEIN, L. S. B. (1989). Identification of globular mechanochemical heads of kinesin. *Nature* **338**, 355–357.

SCHOLEY, J. M., PORTER, M. E., GRISSOM, P. M. AND MCINTOSH, J. R. (1985). Identification of kinesin in sea urchin eggs, and evidence for its localization in the mitotic spindle. *Nature* **318**, 483–486.

VALE, R. D., REESE, T. S. AND SHEETZ, M. P. (1985). Identification of a novel force-generating protein, kinesin involved in microtubule-based motility. *Cell* **42**, 39–50.

VALLEE, R. B., WALL, J. S., PASCHAL, B. M. AND SHPETNER, H. S. (1988). Microtubule-associated protein 1C from brain is a two-headed cytosolic dynein. *Nature* **332**, 561–563.

YANG, J. T. AND GOLDSTEIN, L. S. B. (1990). Evidence that the head of kinesin is sufficient for force generation and motility *in vitro*. *Science* **249**, 42–47.

YANG, J. T., LAYMON, R. A. AND GOLDSTEIN, L. S. B. (1989). A three-domain structure of kinesin heavy chain revealed by DNA sequence and microtubule binding analyses. *Cell* **56**, 879–889.

Studies using a fluorescent analogue of kinesin

P. K. MARYA, P. E. FRAYLICH, C. R. FLOOD, R. RAO and P. A. M. EAGLES

Department of Biophysics, King's College, London, UK

Summary

The microtubule motor protein kinesin has been conjugated with 5-iodoacetamido fluorescein (5-IAF). The analogue, AF–kinesin, supports organelle motility and the movement of microtubules.

Key words: kinesin, fluorescein, microtubules, photodamage.

Introduction

Kinesin is a microtubule motor (Vale *et al.* 1985*a*,*b*,*c*; Brady, 1985) though its precise role in moving organelles inside cells is still unclear. As a direct approach in the study of the function and cellular localization of kinesin we have prepared a fluorescent analogue of the molecule. Here we describe a procedure for its preparation using kinesin isolated from bovine intra-dural nerve roots, and report on its characterization.

Methods

Kinesin was isolated from ox spinal nerve roots and purified by methods similar to those already described (Amos, 1987; Bloom *et al.* 1988; Kuznetsov and Gelfand, 1986). The initial nerve root homogenate was spun and the supernatant taken to polymerize the tubulin present. Kinesin was attached to the microtubules by lowering the endogenous ATP concentration with hexokinase and glucose, and by adding AMP-PNP to 50 μM. The microtubules were then spun out of solution, washed and respun through a sucrose pad. To the final pellet 10 mM ATP was added for releasing kinesin.

The kinesin-containing supernatant was dialysed against 25 mM Pipes, 5 mM $MgCl_2$, 0.5 mM EDTA, 0.5 mM EGTA, pH 7.0 (buffer A) at 4°C. 5-IAF (20 mM 5-IAF dissolved in dimethylformamide) was then added to 30 μM in order to label the reactive sulphydryl groups on the proteins and the reaction mixture was incubated for 2 h in the dark at 25°C. Dithiothreitol (70× molar excess) was used to inactivate unreacted 5-IAF. Sodium chloride was added to 50 mM and the mixture was applied to a small DEAE-cellulose 52 column (1 cm×0.5 cm) pre-equilibrated in buffer A supplemented with dithiothreitol (2 mM) to separate labelled kinesin, which was collected in the flow-through, from free 5-IAF, which bound to the resin. Purification of labelled kinesin from other proteins was by HPLC. After purification, AF–kinesin was stored in 50 % glycerol at −20°C.

Results

Analysis by SDS–PAGE of samples containing purified AF–kinesin (Fig. 1) revealed the presence of only one major fluorescent polypeptide, which had a size around 125×10^3 M_r, equivalent to the kinesin heavy chain. No labelling of the light chains was seen. In gels where the marker dye had not migrated off the end, no free 5-IAF could be detected at the dye front; the fluorescence intensity in this region was the same as in the empty lanes (Fig. 1, lanes d and e).

The time course for the reaction of 5-IAF with kinesin was determined. The reaction nears completion after about 30 min and no further labelling of kinesin is seen if samples are left for 24 h. The fluorescein-to-protein ratio of AF–kinesin was estimated to be 1.05±0.16 (S.D.) assuming a molecular mass for kinesin of 380×10^3 M_r.

AF–kinesin was active in causing organelle movement and microtubule motility. Gliding of microtubules was on linear paths across the cover slip surface. Motility, dependent on MgGTP, was seen with AF–kinesin, and the velocities were about 30 % less when compared with those obtained by adding MgATP at the same concentration. Motility was also observed in the presence of anti-photobleaching reagents *p*-phenylenediamine (PPD, 5 mg ml^{-1}) and ascorbic acid (5 mg ml^{-1}). We tested the ability of AF–kinesin to translocate organelles. The crude kinesin fraction corresponding to lane (a) of Fig. 1 was used. In addition to kinesin, this fraction contains abundant small membranous organelles, which can be identified by electron microscopy. Addition of taxol-stabilized microtubules and ATP to the 5-IAF labelled fraction caused the organelles to bind to microtubules and move along them.

We thought it possible that the labelling of kinesin with 5-IAF could result in an inactive molecule, and that the motility which we were observing was in fact due to unlabelled kinesin in the sample. We investigated ways of testing for this. Firstly, we divided a crude preparation of kinesin into two, then labelled half of it with 5-IAF, and afterwards purified both halves identically. Assaying the amount of active kinesin, by means of limiting dilution, indicated that the same amount of active kinesin was present in each of the purified preparations.

Secondly, we used antibodies to inhibit selectively the activity of AF–kinesin. A rabbit antifluorescein serum was employed which reacted with AF–kinesin after immunoblotting, but not with unlabelled kinesin. At a serum dilution that had no effect on microtubule gliding due to unlabelled kinesin, the serum completely abolished the gliding activity caused by AF–kinesin and abolished motility of organelles in crude preparations. These experiments support the idea that fluorescently labelled kinesin molecules are responsible for microtubule movement.

We tested the effect of illumination conditions on the ability of purified AF–kinesin to translocate microtubules. Illumination of AF–kinesin using the 100 W Hg arc and 546 nm filter for DIC microscopy did not perturb its activity in this assay, though the addition of light that excited the fluorophore caused microtubule motility to

Journal of Cell Science, Supplement 14, 139–142 (1991)

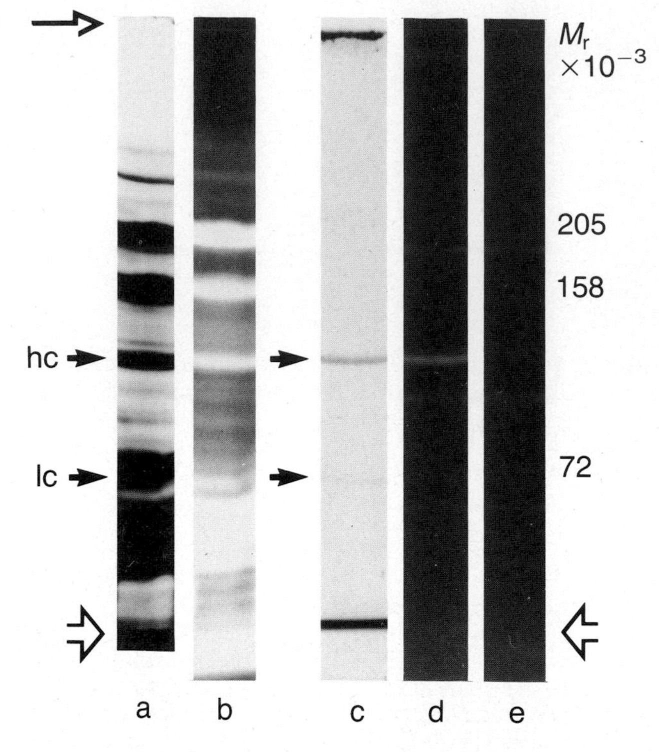

Fig. 1. Photographs, obtained after SDS–PAGE, of samples containing AF–kinesin, when observed under ultra violet light and before staining with Coomassie Blue (b,d,e) or under natural light and after staining (a,c). (a) and (b) show the crude kinesin fraction following labelling with 5-IAF. The heavy chains (hc) and light chains (lc) of kinesin are marked. (c) and (d) show the sample of kinesin after purification by HPLC. The heavy chain only is conjugated with 5-IAF. (e) shows an empty lane without sample. An open arrow marks the tops of each gel. The dye front is marked with open arrow heads and indicates that free 5-IAF is not detectable in the preparation of purified AF–kinesin, compare (d) with (e). Molecular masses ($\times 10^{-3}$) are marked at the side and they correspond to the neurofilament polypeptides from intradural nerve roots (Carden and Eagles, 1983).

cease within about 30 s in the area being irradiated. Excitation was produced by a 100 W halogen lamp and 450–490 nm filter which gave 0.35 μW cm^{-2} to the specimen. The effect was highly localized and not transmitted to neighbouring regions where motility was unaffected. Microtubules stayed attached to the coverslip after being stopped. When unlabelled kinesin was substituted for AF–kinesin and irradiated in a like manner with 450–490 nm radiation, no change in gliding velocities was noticed over 30 min. The addition to AF–kinesin of either PPD or ascorbate mitigated the damaging effect of radiation at 450–490 nm and in their presence normal motility continued with undiminished velocities for at least 10 min. At a lower power setting (0.02 μW cm^{-2}) motility was observed for at least 15 min in the absence of any free radical scavenging compounds (Marya *et al.* 1990).

We also carried out experiments to determine whether we could detect a relationship between the power level of the epifluorescence illumination that is needed to cause cessation of microtubule gliding and the level required for measurable photobleaching to occur to the AF–kinesin molecule when it was bound to microtubules with AMP-PNP. Continuous illumination at 0.35 μW cm^{-2} caused a rapid loss in intensity, being complete in about 80 s, whereas at a setting of 0.02 μW cm^{-2} the decrease over about 5 min was reduced to less than 50 %. These results are consistent with the idea that photodamage to AF–kinesin by 450–490 nm radiation not only destroys its activity in motility assays but also causes a related photobleaching effect on the attached fluorescein molecule.

Kinesin forms a high affinity complex with microtubules in the presence not only of AMP-PNP but also with ADP (Wagner *et al.* 1989; Cohn *et al.* 1989). We investigated the ability of our labelled kinesin to form such a complex. AF–kinesin was pre-incubated with microtubules in the presence of 2.5 mM ADP, and the free nucleotide was washed out. AF–kinesin bound to microtubules under these conditions (Fig. 2). Control samples, where 5 mM ATP was present instead, showed only a diffuse background fluorescence with no filament-like structures visible. Binding of AF–kinesin to microtubules by 2.5 mM AMP-PNP resulted in the microtubules appearing much brighter. Binding could also be detected sometimes, in the absence of nucleotides, with very weak intensity.

AF–kinesin was released from microtubules, to which it had been bound with ADP, when ATP was added to the sample and allowed to diffuse into the area under observation (Fig. 2). Addition of buffer alone caused no change in the pattern. The decrease in fluorescent signal could not be attributed to loss of microtubules – there being no change in the microtubule distribution at the end of the experiment (Fig. 2B) compared with the starting situation (Fig. 2A); nor could photobleaching account for the decrease – the conditions used result in minimal photobleaching. Thus the loss in signal is a consequence of the release of kinesin from the microtubule. As well as ATP, GTP could be used to remove AF–kinesin from microtubules, and at 10 mM was as effective as a similar concentration of ATP. When kinesin was attached by AMP-PNP instead of by ADP, the time for complete loss of signal on addition of ATP was much longer, generally in the order of 3–5 min as opposed to a few seconds.

Fig. 2 also illustrates another important point. When ATP is used to remove AF–kinesin from microtubules, to which it has been attached either with ADP or AMP-PNP, there is a total loss of signal. From this experiment it is

Table 1. *The binding of AF–kinesin to latex beads and glass particles*

	Appearance under DIC with kinesin	Appearance under DIC with AF–kinesin	Assessment of binding AF–kinesin
Latex beads	mainly aggregated	mainly aggregated	+++
Carboxylated beads	non-aggregated	non-aggregated	+
Amino-beads	mainly aggregated	mainly aggregated	+++
Glass particles	non-aggregated	non-aggregated	+

All beads were between 0.1 μm and 0.15 μm diameter. Glass particles were ground down to this size. The assessment of AF–kinesin binding to the fractions was made visually under fluorescence, keeping conditions as reproducible as possible from one sample to the next.

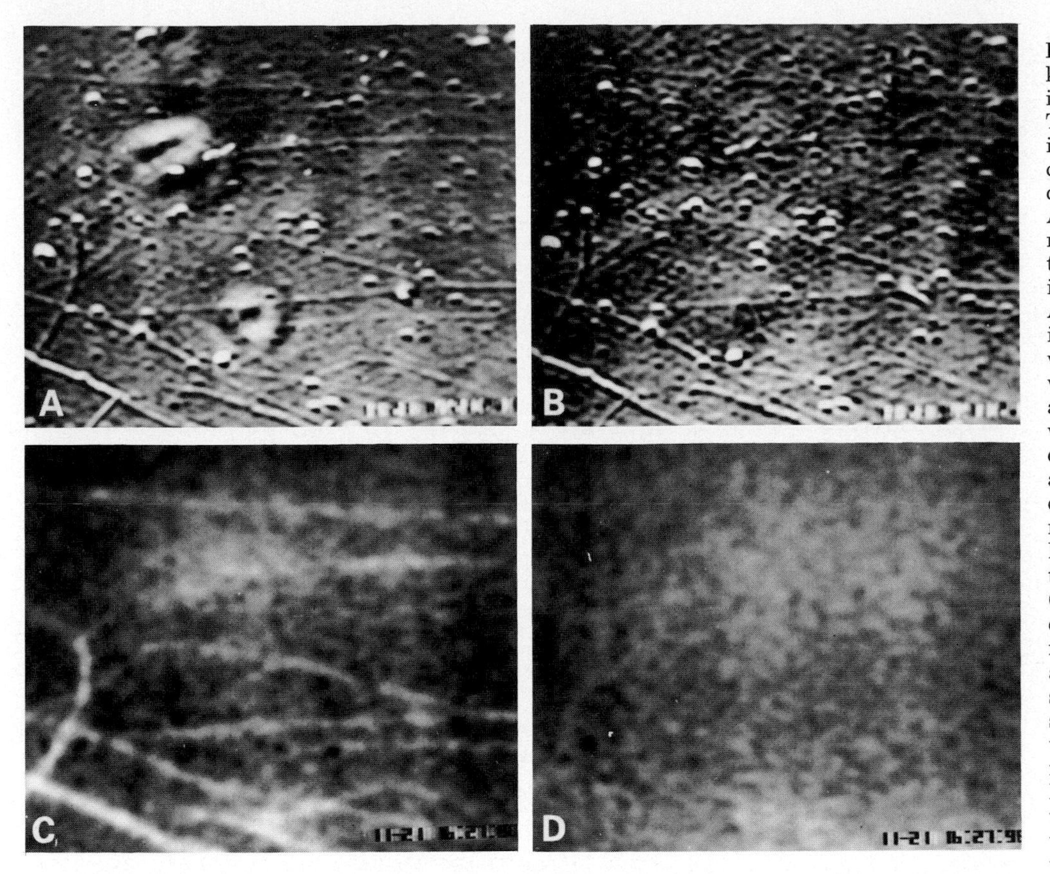

Fig. 2. The binding of AF–kinesin to microtubules and its ATP-dependent removal. The figure shows a display of images taken during the course of an experiment designed to release bound AF–kinesin from microtubules by adding ATP to the coverslip and allowing it to diffuse into the system. AF–kinesin was bound initially with 2.5 mM ADP, which was subsequently washed out before ATP was added. Ascorbate (5 mg ml^{-1}) was included in the sample during the experiment. (A) and (B) show pictures obtained under DIC microscopy of microtubules labelled with AF–kinesin at the start (A), and at the end (B) of the experiment. They demonstrate that the field of microtubules has not changed as a result of adding ATP. C shows the same area at the start of the experiment but viewed under fluorescence microscopy (0.12 $\mu W\,cm^{-2}$). Most of the microtubules in the field have been labelled with AF–kinesin. (D) shows the fluorescent image 20 s after the addition to the system of 10 μl 10 mM MgATP containing ascorbate. In the course of the experiment there is a complete loss of fluorescent signal which cannot be accounted for by photobleaching effects or by microtubule removal. The sample received the equivalent of about 30 s of continuous illumination.

clear that those protein molecules that have been labelled with fluorescein also bind to microtubules in an ATP-dependent fashion and rules out the possibility of a significant population of labelled molecules that bind non-specifically to microtubules in our preparation of AF–kinesin.

As far as we could judge, the fluorescent images mirrored the microtubule profiles as seen by DIC microscopy. An example of this is shown in Fig. 3. The fluorescent images of microtubules labelled with AF–kinesin were, however, quite variable and in many cases a continuous distribution of even intensity was not found; this was true for both microtubules labelled in the presence of ADP and those labelled in the presence of AMP-PNP. Detailed analysis of the fluorescence intensity distribution confirms this (Fig. 3).

One remarkable property of kinesin is its ability to bind to latex particles, and ferry them along microtubules (Vale *et al.* 1985*b*). A requirement for this motility is that the beads possess a carboxylated surface. We tested the ability of AF–kinesin to bind to underivatized latex beads, latex beads carrying a carboxyl or amino group, and glass particles. They all bound the analogue (Table 1).

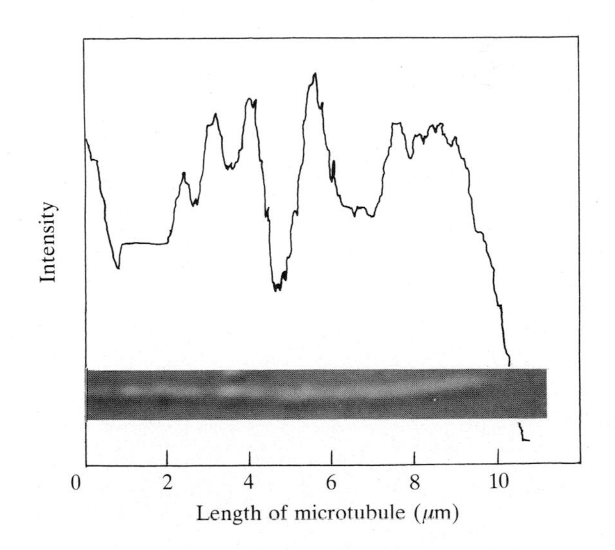

Fig. 3. The fluorescence intensity profile measured along a microtubule that has been labelled with AF–kinesin. The intensity units are arbitrary and have been normalised to remove background variations. The corresponding DIC picture of the microtubule, magnified to conform with the X-axis is shown in the inset. The intensity distribution mirrors quite well the changing appearance of the microtubule along its length, which indicates that some regions of the tubule bind less kinesin than others.

Discussion

We have shown in this paper that a fluorescent derivative of kinesin can be produced by conjugating ox nerve root kinesin with 5-IAF. The analogue, AF–kinesin, is indistinguishable from unlabelled kinesin in many functional aspects: it is active in moving microtubules and in moving organelles; motile rates are comparable to those obtained with the unconjugated protein; AF–kinesin utilises MgGTP as well as MgATP for moving microtubules and the ratio of velocities with the two substrates is similar to that found for unlabelled kinesin. AF–kinesin also binds to microtubules in the presence of AMP-PNP and ADP, providing further evidence for its similarity with the unlabelled molecule. We conclude from these observations that AF–kinesin differs little from unconjugated kinesin in many aspects relating to its ability to move microtubules and organelles.

An important issue is where the fluorophore is binding on the kinesin molecule. It is presumably interacting near the ATP-binding head region, because photobleaching causes modification of the microtubule/AF–kinesin association (Marya *et al.* 1990). It seems unlikely that the fluorophore is binding to the very reactive, NEM-sensitive sulphydryl groups (Pfister *et al.* 1989) because labelling with NEM results in inactivation of the kinesin.

Future work is aimed at characterizing the location of the fluorophore on the kinesin molecule and in using AF–kinesin to study its distribution on microtubules and organelles in cultured cells.

We thank the MRC, the Wellcome Trust, and Research into Ageing for support throughout this work.

References

AMOS, L. A. (1987). Kinesin from pig brain studied by electron-microscopy. *J. Cell Sci.* **87**, 105–111.

BLOOM, G. S., WAGNER, M. C., PFISTER, K. K. AND BRADY, S. T. (1988). Native structure and physical properties of bovine brain kinesin and identification of the ATP-binding subunit polypeptide. *Biochemistry* **27**, 3409–3416.

BRADY, S. T. (1985). A novel brain ATPase with properties expected for the fast axonal transport motor. *Nature* **317**, 73–75.

CARDEN, M. J. AND EAGLES, P. A. M. (1983). Neurofilaments from ox spinal nerves. *Biochem. J.* **215**, 227–237.

COHN, S. A., INGOLD, A. L. AND SCHOLEY, J. M. (1989). Quantitative analysis of sea urchin egg kinesin-driven microtubule motility. *J. biol. Chem.* **264**, 4290–4297.

KUZNETSOV, S. A. AND GELFAND, V. I. (1986). Bovine brain kinesin is a microtubule-activated ATPase. *Proc. natn. Acad. Sci. U.S.A.* **83**, 8530–8534.

MARYA, P. K., FRAYLICH, P. E. AND EAGLES, P. A. M. (1990). Characterization of an active, fluorescein-labelled kinesin. *Eur. J. Biochem.* **193**, 39–45.

PFISTER, K. K., WAGNER, M. C., BLOOM, G. S. AND BRADY, S. T. (1989). Modification of the microtubule-binding and ATPase activities of kinesin by N-ethylmaleimide (NEM) suggests a role for sulfhydryls in fast axonal transport. *Biochemistry* **28**, 9006–9012.

VALE, R. D., REESE, T. S. AND SHEETZ, M. P. (1985*a*). Identification of a novel force generating protein, kinesin, involved in microtubule based motility. *Cell* **42**, 39–50.

VALE, R. D., SCHNAPP, B. J., MITCHISON, T., STEUER, E., REESE, T. S. AND SHEETZ, M. P. (1985*b*). Different axoplasmic proteins generate movement in opposite directions along microtubules *in vitro*. *Cell* **43**, 623–632.

VALE, R. D., SCHNAPP, B. J., REESE, T. S. AND SHEETZ, M. P. (1985*c*). Organelle, bead and microtubule translocations promoted by soluble factors from the giant squid axon. *Cell* **40**, 559–569.

WAGNER, M. C., PFISTER, K. K., BLOOM, G. S. AND BRADY, S. T. (1989). Copurification of kinesin polypeptides with microtubule-stimulated Mg-ATPase activity and kinetic analysis of enzymatic properties. *Cell Motil. Cytoskel.* **12**, 195–215.

Dynamin: A microtubule-associated GTP-binding protein

ROBERT A. OBAR, HOWARD S. SHPETNER and RICHARD B. VALLEE

Cell Biology Group, The Worcester Foundation for Experimental Biology, Shrewsbury, MA 01545, USA

Summary

We recently identified dynamin as a third nucleotide-sensitive microtubule-associated protein in brain tissue, in addition to kinesin and cytoplasmic dynein. Molecular cloning analysis has revealed that dynamin contains the three consensus elements characteristic of GTP-binding proteins, and biochemical results support a role for GTP in dynamin function. Dynamin is also homologous to the Mx proteins, involved in interferon-induced viral resistance, and the product of the yeast VPS1 gene, involved in vacuolar protein sorting. These results identify a novel class of GTP-utilizing proteins, with apparently diverse functions.

Key words: microtubule, GTPase, interferon, Mx, protein sorting.

Introduction

In recent years two distinct molecular motors have been identified which are capable of producing force in opposite directions along microtubules – kinesin and cytoplasmic dynein (reviewed by Vallee and Shpetner, 1990; Vallee and Bloom, 1991). In the course of our initial characterization of cytoplasmic dynein (Paschal *et al.* 1987; Paschal and Vallee, 1987), a third nucleotide-sensitive microtubule binding protein was identified in brain tissue, which was termed dynamin (Shpetner and Vallee, 1989). On the basis of cDNA cloning analysis, we have now obtained the predicted primary sequence of the principal polypeptide component of dynamin (Obar *et al.* 1990). In this chapter we briefly review the biochemical properties of dynamin, and discuss the implications of the primary structure data.

Biochemical properties of dynamin

Dynamin was first seen as a $100\times10^3 M_r$ polypeptide in brain microtubules prepared in the absence of added nucleotide (Paschal *et al.* 1987). Like kinesin and cytoplasmic dynein, it could be extracted from the microtubules using ATP, and, like kinesin, it could also be extracted using GTP (op. cit). A combination of GTP and AMP-PNP extracted only dynamin, providing a useful first step in its purification from microtubules (Shpetner and Vallee, 1989). The dynamin extracts contained an ATPase activity which could be activated 6–7 fold by microtubules. Upon rebinding to microtubules for further purification, the ATPase activity separated into two components, the $100\times10^3 M_r$ polypeptide and an additional factor which remained soluble.

Both the GTP/AMP-PNP extract and the purified $100\times10^3 M_r$ dynamin polypeptide caused microtubules to bundle. Addition of ATP induced fragmentation of the bundles, some of which showed evidence of elongation as well. This behavior, like the ATPase activity, required both the $100\times10^3 M_r$ dynamin polypeptide and the soluble activating fraction. These observations combined with the biochemical data led to the proposal that dynamin was a third microtubule-related mechanochemical enzyme, with an apparent role in inter-microtubule sliding.

Primary structure of dynamin

To obtain further insight into the mechanism of action of dynamin and its relationship to other proteins, a rat brain cDNA library was screened with anti-dynamin antibodies. A 3.2 kbp cDNA clone was obtained which appeared to encode the entire $100\times10^3 M_r$ polypeptide (Obar *et al.* 1990). Located near the N terminus (Figs 1, 2) was the nucleotide binding consensus element $GX_4GKS(T)$ (Walker *et al.* 1982; Fry *et al.* 1986). This element is common to both ATP and GTP binding proteins. Surprisingly, two additional elements common only to GTP-binding proteins were found downstream (Figs 1, 2). The third element is considered to be of particular importance in specifying guanine nucleotide binding (Dever *et al.* 1987). Curiously, kinesin heavy chain was found to contain the second GTP-binding consensus element (Obar *et al.* 1990; Fig. 2), but lacked the third element. Kinesin can hydrolyze GTP and use it for force production, though based on the relative K_m values for ATP and GTP (Cohn *et al.* 1989; reviewed by Vallee and Shpetner, 1990) it is thought to use ATP as its physiological substrate. Because some additional sequence similarity was observed between kinesin heavy chain and dynamin surrounding the second GTP-binding consensus element, it is possible that this region has been conserved for other functional purposes such as microtubule binding. Hence, conservation of the second GTP-binding consensus element in kinesin could represent a secondary evolutionary effect.

Examination of several protein and nucleic acid sequence databases revealed extensive sequence homology between dynamin and a class of interferon-inducible proteins termed Mx. Expression of Mx in transfected cells (Staeheli *et al.* 1986) or transgenic mice (Arnheiter *et al.* 1990) has been found to be sufficient to induce resistance to viral infection, though the detailed mechanism of this process is not understood.

Recent analysis of yeast protein sorting mutants has revealed another dynamin homologue, VPS1p (Rothman *et al.* 1990). Mutations in the *VPS1* gene interfere with the normal transport of proteins from the Golgi apparatus to the vacuole, a large, degradative organelle. What may be the same or a very closely related gene, *SPO15*, has also been identified as the locus of a sporulation mutation (Yeh *et al.* unpublished data). This mutation causes arrest in meiosis, and mutant cells have been found to be defective in meiotic spindle pole body separation. It is not yet known whether the protein sorting and meiotic phenotypes result

Journal of Cell Science, Supplement 14, 143–145 (1991)

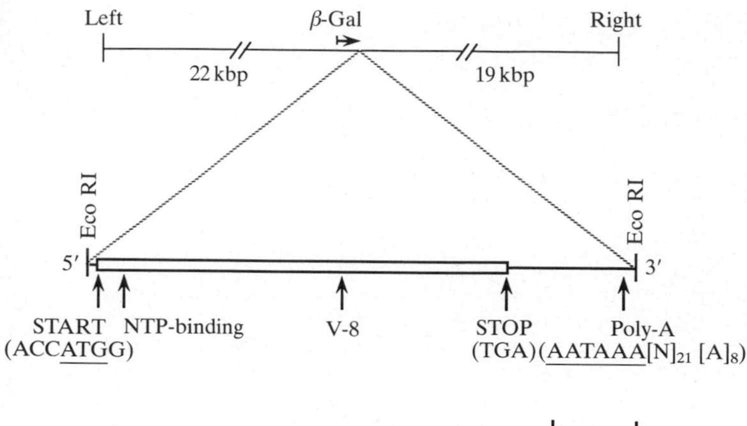

Fig. 1. Diagram of dynamin clone. The 3.2 kbp cDNA is shown inserted within the *lacZ′* gene of lambda-ZAP, which encodes 38 amino acids of the alpha-peptide of beta-galactosidase upstream of the fusion site. Arrows indicate the positions of translational initiation and termination sites, a consensus polyadenylation signal (poly(A)), the proposed nucleotide triphosphate- (NTP-) binding site, and *Staphylococcus aureus* V8 protease cleavage site (V-8). Amino acid microsequence data from a calf brain V8 fragment matched the amino acid sequence predicted from the rat brain cDNA at 32 of 33 positions.

from distinct mutations in the same gene, or whether they are different manifestations of the same mutation.

Homology between dynamin, VPS1p, and the Mx proteins is most striking within a 288 amino acid N-terminal domain, which contains the three GTP-binding consensus elements. This is in contrast to the more limited sequence similarity between dynamin, VPS1p, Mx, and other classes of GTP-binding protein such as ras and the G proteins, which is restricted to the three short GTP-binding consensus elements. Within the 288 amino acid N-terminal domain, Dynamin and VPS1p show 66 % amino acid identity. Sequence conservation between dynamin and Mx is lower within this domain (43 % amino acid identity), as is conservation between VPS1p and Mx (44 % amino acid identity). This is despite the fact that the available sequence data for dynamin and the Mx proteins are derived from vertebrate species (rat in the case of dynamin *vs.* mouse, human and fish in the case of Mx) in contrast to VPS1p, which is known so far known only in yeast. Thus, dynamin is more likely to be functionally related to VPS1p than to Mx.

Nucleotide specificity

The identification of GTP-binding consensus elements in dynamin has prompted further analysis of its substrate specificity. Previous evidence indicated that GTP, at least at high concentrations, was as effective as ATP in extracting dynamin from microtubules (Shpetner and Vallee, 1989). Analysis of the substrate preference of dynamin has now revealed that the purified 100×10^3 M_r polypeptide is a potent GTPase (Shpetner and Vallee, 1990). While both ATPase and GTPase activities were low in the absence of microtubules (less than 10 nmol min^{-1} mg^{-1}), hydrolysis of GTP, but not ATP, was dramatically stimulated by microtubules. Microtubule-stimulated GTPase activity was not inhibited by ATP, a further indication that GTP is likely to be the substrate for dynamin in the cell.

Why, then, is ATPase activity detected in dynamin preparations? One possible scenario involves nucleoside diphosphokinase (NDPK) activity, which has been reported to co-purify with brain microtubules (Penningroth and Kirschner, 1977; Burns and Islam, 1981). NDPK in the soluble 'activating' fraction, which is separated from dynamin during purification (Shpetner and Vallee, 1989), could use ATP to phosphorylate residual GDP, which would, in turn, be hydrolysed by dynamin. The GDP might well be present as a remnant of the high levels of GTP added during microtubule extraction, or it could be released from tubulin or even dynamin itself. In this scenario, the activator could be a combination of NDPK and GDP, though other regulatory factors may also be present in this fraction.

Functional implications

The pattern of homology between dynamin, Mx, and

A

	Element I		Element II		Element III	
Dynamin-1	I A V V G G Q S A G K S S V L E	(34-49)	L T L V D L P G M T K V	(132-143)	T K L D	(205-208)
ras	L V V V G A G G V G K S A L T I	(6-21)	L D I L D L A G Q E E Y	(53-64)	N K C D	(116-119)
Gs alpha	L L L L G A G E S G K S T I V K	(43-58)	F H M F D V G G Q R D E	(219-230)	N K K D	(292-295)
SRP54	I M F V G L Q G S G K T T T C S	(104-119)	I I I V D T S G R H K Q	(186-197)	T K L D	(248-251)
EF-Tu	V G T I G H V D H G K T T L T A	(14-29)	A H V D D C P G H A D Y	(76-87)	N K C D	(135-138

B

	Element I		Element II	
Dynamin-1	I A V V G G Q S A G K S S V L E	(34-49)	L T L V D L P G M T K V	(132-143)
Kinesin HC	I F A Y G Q T S S G K T H T M E	(88-103)	L Y L V D L A G S E K V	(234-245)
ncd	I F A Y G Q T G S G K T Y T M G	(430-445)	I N L V D L A G S E S P	(576-587)
KAR3	I F A Y G Q T G S G K T F T M L	(470-485)	L N L V D L A G S E R I	(622-633)
BimC	I F A Y G Q T G T G K T Y T M S	(163-178)	L N L V D L A G S E N I	(318-329)

Fig. 2. (A) Alignment of the three consensus elements of the nucleotide-binding motif of D100 with those from GTP-binding proteins representative of several families. Dynamin-1, rat brain dynamin derived from the ZAP dynamin-1 cDNA clone; *ras*, human H-ras p21; Gs alpha, the alpha subunit of the bovine brain Gs complex; SRP54, the M_r 54 000 subunit of the endoplasmic reticulum signal recognition complex; EF-Tu, *E. coli* elongation factor Tu. (B) Alignment of Elements I and II from the D100 sequence with those of the kinesin heavy chain (HC) and three kinesin-heavy chain-like proteins identified in *Drosophila* (ncd), yeast (KAR3), and *Aspergillus* (BimC). (From Obar *et al.* 1990.)

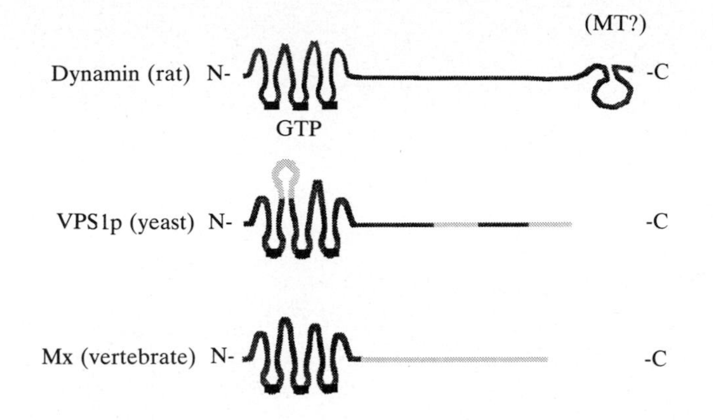

Fig. 3. Hypothetical structural relationship between members of the dynamin gene family. Conserved domains are heavily shaded with GTP-binding consensus elements as filled rectangles. Less well-conserved domains are more lightly shaded. The conserved amino-terminal domains are assumed to be globular as in more well-characterized GTP-binding proteins, such as Gs alpha. The remainder of the molecules are arbitrarily drawn. The carboxy-terminal $10\times10^3 M_r$ of the dynamin polypeptide is proline rich (32%) and very basic (pI=12.5). Such a region would be expected to interact with microtubules, though this remains to be tested experimentally.

VPS1p is strikingly reminiscent of that among the members of the myosin and kinesin families (Kiehart, 1990; Vale and Goldstein, 1990). It is tempting to speculate that the 288 amino acid conserved N-terminal domain represents a force-producing 'head' as in the case of the myosins and kinesins. This region contains the apparent site for nucleotide binding, as in the other protein classes. It remains to be seen whether it also interacts with microtubules.

This is an important issue to resolve, as the mechanism of action of the three proteins is incompletely understood. Dynamin has been most extensively analysed biochemically, and its properties indicate a role in microtubule mechanochemistry. However, its specific cellular role remains to be determined.

It is not yet certain whether the product of the VPS1 gene interacts with microtubules. There is no direct evidence for an involvement of microtubules in vacuolar protein sorting. However, the vacuole has been reported to vesiculate under microtubule-disrupting conditions (Guthrie and Wickner, 1988). This suggests at minimum a role for microtubules in maintaining the structural integrity of the vacuole. The defect in meiotic spindle pole separation seen in the SPO15 mutant is reminiscent of the mitotic phenotype seen for the BimC mutant in *Aspergillus* (Enos and Morris, 1990). An attractive hypothesis is that both cases represent a failure in the mechanochemical machinery associated with the spindle microtubules.

An interaction of the Mx proteins with microtubules is more difficult to imagine. While some are cytoplasmic, others are found in the nucleus (Meier *et al.* 1988; Staeheli and Haller, 1985). Expression of the mouse Mx1 protein was found to block the infective cycle of influenza virus at a step between uncoating and translation, though the precise mechanism of inhibition remains unknown. It is not obvious how these observations can be reconciled with a role for microtubules. Perhaps some members of the Mx family interact with other cellular components, such as the nuclear matrix. Alternatively, the observed sequence conservation between Mx, dynamin, and VPS1p could signify a common feature of their mechanism of action that has yet to be identified (Fig. 3). Finally, it is possible that while the three classes of protein evolved from a common ancestor, their functions have diverged. This could well be the case for the Mx proteins *versus* the other family members. However, it seems less likely for dynamin *versus* VPS1p considering their remarkable sequence conservation. One goal of further work on these proteins will be to identify common features in their mechanism of action.

References

Arnheiter, H., Skuntz, S., Noteborn, M., Chang, S. and Meier, E. (1990). Transgenic mice with intracellular immunity to influenze virus. *Cell* **62**, 51–61.

Burns, R. G. and Islam, K. (1981). Nucleoside diphosphate kinase associates with rings but not with assembled microtubules. *Eur. J. Biochem.* **117**, 515–519.

Cohn, S. A., Ingold, A. I. and Scholey, J. M. (1989). Quantitative analysis of sea urchin egg kinesin-driven microtubule motility. *J. biol. Chem.* **264**, 4290–4297.

Dever, T. E., Glynias, M. J. and Merrick, W. C. (1987). GTP-binding domain: three consensus sequence elements with distinct spacing. *Proc. natn. Acad. Sci. U.S.A.* **84**, 1814–1818.

Enos, A. P. and Morris, N. R. (1990). Mutation of a gene that encodes a kinesin-like protein blocks nuclear division in *A. nidulans*. *Cell* **60**, 1019–1027.

Fry, D. C., Kuby, S. A. and Mildvan, A. S. (1986). ATP-binding site of adenylate kinase: mechanistic implications of its homology with *ras*-encoded p21, F1-ATPase, and other nucleotide-binding proteins. *Proc. natn. Acad. Sci. U.S.A.* **83**, 907–911.

Guthrie, B. A. and Wickner, W. (1988). Yeast vacuoles fragment when microtubules are disrupted. *J. Cell Biol.* **107**, 115–120.

Kiehart, D. (1990). Molecular genetic dissection of myosin heavy chain function. *Cell* **60**, 347–350.

Meier, E., Fah, J., Grob, M. S., End, R., Staeheli, P. and Haller, O. (1988). A family of interferon-induced Mx-related mRNAs encodes cytoplasmic and nuclear proteins in rat cells. *J. Virol.* **62**, 2386–2393.

Obar, R., Collins, C. A., Hammarback, J. A., Shpetner, H. S. and Vallee, R. B. (1990). Molecular cloning of the microtubule-associated mechanochemical enzyme dynamin reveals homology with a new family of GTP-binding proteins. *Nature* **347**, 256–261.

Paschal, B. M., Shpetner, H. S. and Vallee, R. B. (1987). MAP 1C is a microtubule-activated ATPase which translocates microtubules *in vitro* and has dynein-like properties. *J. Cell Biol.* **105**, 1273–1282.

Paschal, B. M. and Vallee, R. B. (1987). Retrograde transport by the microtubule-associated protein MAP 1C. *Nature* **330**, 181–183.

Penningroth, S. M. and Kirschner, M. W. (1977). Nucleotide binding and phosphorylation in microtubule binding *in vitro*. *J. molec. Biol.* **115**, 643–673.

Rothman, J. H., Raymond, C. K., Gilbert, T., O'Hara, P. J. and Stevens, T. H. (1990). A putative GTP binding protein homologous to interferon-inducible Mx proteins performs an essential function in yeast protein sorting. *Cell* **61**, 1063–1074.

Shpetner, H. S. and Vallee, R. B. (1989). Identification of dynamin, a novel mechanochemical enzyme that mediates interactions between microtubules. *Cell* **59**, 421–432.

Shpetner, H. S. and Vallee, R. B. (1990). GTP utilization by the microtubule-activated nucleotidase of dynamin. *J. Cell Biol.* **111**, 290a.

Staeheli, P. and Haller, O. (1985). Interferon-induced human protein with homology to protein Mx of influenza virus-resistant mice. *Molec. cell. Biol.* **5**, 2150–2153.

Staeheli, P., Haller, O., Boll, W., Lindenmann, J. and Weissmann, C. (1986). Mx protein: constitutive expression in 3T3 cells transformed with cloned Mx cDNA confers selective resistance to influenza virus. *Cell* **44**, 147–158.

Vale, R. D. and Goldstein, L. S. B. (1990). One motor, many tails: an expanding repertoire of force-generating enzymes. *Cell* **60**, 883–885.

Vallee, R. B. and Bloom, G. S. (1991). Mechanisms of fast and slow axonal transport. *A. Rev. Neurosci.* **14**, 59–92.

Vallee, R. B. and Shpetner, H. S. (1990). Motor proteins of cytoplasmic microtubules. *A. Rev. Biochem.* **59**, 909–932.

Walker, J. E., Saraste, M., Runswick, M. J. and Gay, N. J. (1982). Distantly related sequences in the alpha- and beta-subunits of ATP-synthase, myosin, kinases and other ATP-requiring enzymes and a common nucleotide binding fold. *EMBO J.* **1**, 945–951.

Aluminum fluoride, microtubule stability, and kinesin rigor

YOUNG-HWA SONG, SUSANNE HEINS, ECKHARD MANDELKOW and EVA-MARIA MANDELKOW

Max-Planck-Unit for Structural Molecular Biology, c/o DESY, Notkestrasse 85, D-2000 Hamburg 52, FRG

Summary

Aluminum fluoride may be used both to stabilize microtubules and to induce strong binding of kinesin, thus circumventing the need for taxol and AMP-PNP in kinesin preparations.

Key words: kinesin, microtubules, aluminum fluoride.

Introduction

The preparation of kinesin is usually based on a step which involves the tight binding of kinesin to microtubules. This requires that microtubules be stabilized, which is usually achieved by taxol. The tight binding of kinesin to these microtubules takes place in the presence of the ATP analogue AMP-PNP, but not with ATP (Vale *et al.* 1985; Lasek and Brady, 1985). Both taxol and AMP-PNP are rather expensive, thus limiting the scale of the preparation.

In a previous report we noted that the requirements for AMP-PNP and taxol can both be reduced by adding fairly high amounts (100 mM) of NaF to the preparation. This leads to both the stabilization of microtubules and to the strong binding of kinesin to microtubules (von Massow *et al.* 1989).

We have now investigated this effect in more detail and report that it is actually due to aluminum fluoride, possibly in the form of the AlF_4^- complex. Microtubules are known to be rather labile, but in the presence of aluminum fluoride and ATP they are stable for days. Likewise, kinesin binds to microtubules in the presence of aluminum fluoride and ATP without requiring AMP-PNP. Thus, kinesin forms a rigor complex with microtubules. The effects take place at low concentrations of aluminum fluoride, in the mM range. Both are compatible with the assumption that aluminum fluoride stabilizes tubulin and kinesin in a tight binding state by binding to the position of the gamma phosphate of GTP or ATP, respectively. The effect can be exploited to prepare kinesin without taxol and AMP-PNP.

Materials and methods

Preparation of microtubules

PC–tubulin was prepared as previously described (Mandelkow *et al.* 1985). Microtubules were polymerized from PC–tubulin in assembly buffer (0.1 M PIPES pH 6.9, 1 mM each of $MgSO_4$, EGTA, DTT and GTP) with varying concentrations of $AlCl_3$ and NaF which were added either during assembly or afterwards. Alternatively microtubules were polymerized in the presence of 10 μM taxol.

Preparation of kinesin

Kinesin was prepared either following the method of Kuznetsov and Gelfand (1986) involving 50 mM imidazole buffer and PPP_i, or that of Vale *et al.* (1985) involving 100 mM PIPES buffer and AMP-PNP, with modifications as described by von Massow *et al.* (1989) and below.

Motility assay

Microtubules and their movement by kinesin were observed by DIC video microscopy using a Zeiss IM35 microscope, a Hamamatsu Newvicon TV camera, and Leutron Vision image processor, follwing the procedures of Allen *et al.* (1981).

Results and discussion

Stability of microtubules in the presence of aluminum fluoride

One prerequisite for the binding of kinesin to microtubules is that the microtubules are stable. The reason is basically that kinesin is affinity purified, by using microtubules as a matrix which must not dissolve during the kinesin release step, so that after pelleting the supernatant is enriched in kinesin (Vale *et al.* 1985; Kuznetsov and Gelfand, 1986). Stabilization is traditionally achieved by taxol whose supply is limited.

We have tried to circumvent the use of taxol by stabilizing microtubules with aluminum fluoride. To optimize conditions we polymerized microtubules from PC–tubulin at 10 mg ml^{-1} and varied the concentrations of NaF and $AlCl_3$ as well as their ratio; MgGTP was kept at 1 mM. The stability assay consisted of observing the microtubules by video microscopy immediately after assembly and at later time points, up to three days. Since the microscopy was at room temperature, microtubules tended to disassemble and typically disappeared within 10–15 min. NaF alone at higher concentrations (up to 100 mM) had a moderate stabilizing effect; the same is true for $AlCl_3$ (up to 100 mM). However, neither of these conditions stabilized microtubules for three days, which we chose as a reference period since taxol-treated microtubules are stable for this length of time or more (Fig. 1A).

By contrast, combinations of NaF and $AlCl_3$ had a pronounced stabilizing effect, even at low concentrations. We found the optimal conditions to be 4 mM NaF and 1 mM $AlCl_3$. In this case the stability was indistinguishable from that induced by taxol, and even after three days numerous microtubules were visible (Fig. 1B). This was true even when the microtubule solution was cooled to 4°C. This means that, with regard to stability, taxol can be replaced by aluminum fluoride. The effect is reversible; when aluminum fluoride is removed, the microtubules become labile again.

The optimal ratio of $AlCl_3$:NaF was about 1:4, suggesting that the stabilizing activity arises from AlF_4, but this interpretation should be regarded with caution. Other

Journal of Cell Science, Supplement 14, 147–150 (1991)

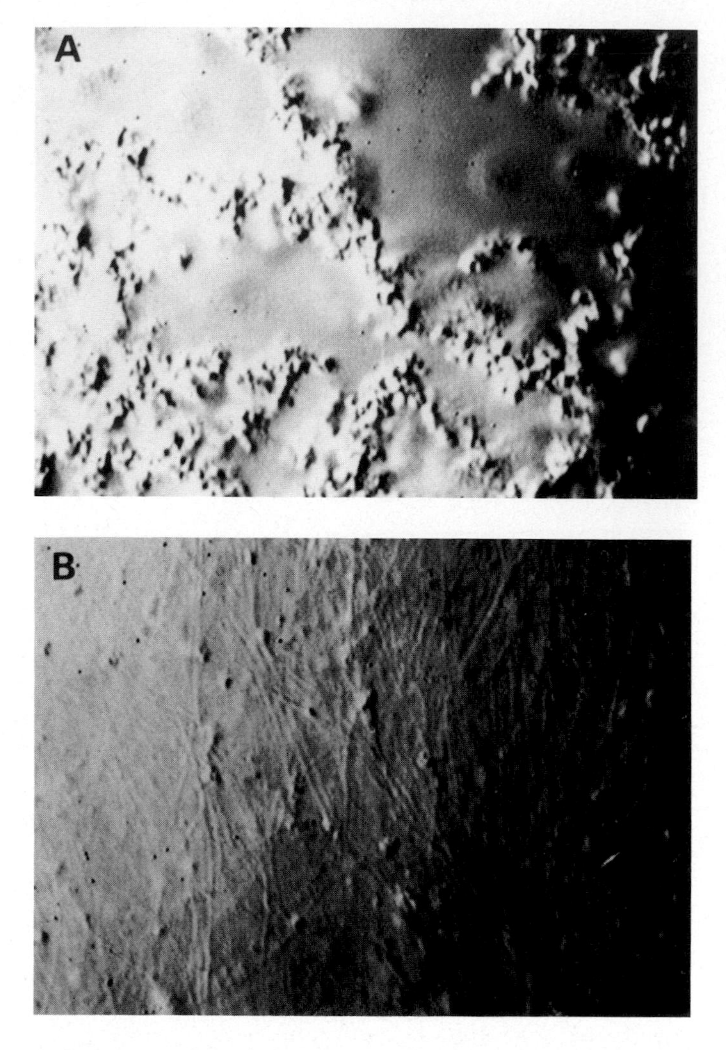

Fig. 1. Stability of microtubules (assembled from PC–tubulin with 1 mM GTP) in the absence (A) and presence (B) of 1 mM $AlCl_3$ and 4 mM NaF. Observations were made by DIC microscopy. The protein was polymerized at 37 °C and then kept in solution for three days at room temperature. Initially both samples contained roughly equal amounts of microtubules. Without aluminum fluoride the microtubules disappear within 10–20 min, leaving only amorphous precipitate (A); with aluminum fluoride they keep for several days (B). The stabilization is optimal when the ratio of $AlCl_3$ to NaF is about 1:4. The aluminum fluoride-stabilized microtubules are stable even at 4 °C, which is comparable to the effect of taxol. The stabilizing effect occurs regardless of whether $AlCl_3$ and NaF are added during the assembly or afterwards, and it is reversible.

possibilities have to be considered as well. For example, elevated Al^{3+} can substitute for Mg^{2+} in the metal–GTP complex, thereby reducing the GTPase and enhancing microtubule stability (Macdonald *et al.* 1987). Secondly, aluminum can form several complexes with fluoride (AlF_x), and it is at present not clear which of these is the active one (Humphreys and Macdonald, 1988). Thirdly, elevated fluoride also stabilizes microtubules. This could be due to a general chaotropic effect influencing the water structure and thus microtubule assembly (Humphreys and Macdonald, 1988); in addition it could be due to a reaction of the fluoride with trace amounts of aluminum. Finally, it has been proposed that AlF_4 acts as an analogue of P_i, substituting for the gamma-phosphate of GTP and thereby stabilizing microtubules (Carlier *et al.* 1988), similar to the mechanism proposed for G-proteins (Bigay *et al.* 1987). The hypothesis is attractive but has been a matter of debate (Caplow *et al.* 1989).

These options should be kept in mind when considering the basis of kinesin rigor to be discussed below. From a practical point of view, the important fact is that mM concentrations of $AlCl_3$ and NaF in combination will circumvent the need for taxol. We will refer to this mixture as aluminum fluoride without specifying a molecular interpretation.

Effect of aluminum and fluoride on kinesin-induced microtubule motility

In the next set of experiments we tested whether mM concentrations of NaF and of $AlCl_3$ had an influence on kinesin-dependent microtubule gliding. NaF by itself has no noticeable effect on motility unless one raises the concentration, say to 100 mM. $AlCl_3$ at mM concentrations has no influence either. However, mixtures of the two were strongly inhibitory. As with microtubule stability, the effect was most pronounced at concentrations of about 1 mM $AlCl_3$ and 4 mM NaF, where the microtubules stuck to the glass without moving (not shown).

Kinesin rigor induced by AMP-PNP, PPP_i, or aluminum fluoride

From the above experiments our working hypothesis is that kinesin binds to microtubules in a state analogous to the rigor state of actomyosin. However, other interpretations are possible, for example, kinesin might not be bound to microtubules at all and would thus not generate motility. To distinguish between these possibilities we performed binding studies between taxol-stabilized microtubules and kinesin in different conditions. In particular we compared the effect of aluminum fluoride with two other compounds known to induce strong binding of kinesin to microtubules, namely AMP-PNP (Vale *et al.* 1985) and PPP_i (Kuznetsov and Gelfand, 1986).

Microtubules (from tubulin at 1 mg ml^{-1}, stabilized with 20 μM taxol) and kinesin (150 $\mu g\,ml^{-1}$) were mixed in different buffer conditions. Microtubules were pelleted, the pellet was washed once, the kinesin was released by adding 10 mM MgATP, and the quantity of kinesin in the pellet and supernatant was estimated by silver stained SDS–PAGE gels (Fig. 2). The left four lanes show the effect of 2.5 mM PPP_i. During the first binding step, essentially all the kinesin binds to the microtubules and can be pelleted with them so that no kinesin remains in the supernatant (S_1). Washing the pellet with buffer does not release the kinesin into the supernatant (S_2). When kinesin is released with ATP most of it appears in the supernatant (S_3), although a minor amount still remains bound to the microtubules (pellet P). The next four lanes show the same type of experiment with 1 mM AMP-PNP (Fig. 2, middle four lanes). As before, kinesin is initially fully bound to the microtubules (S_1). ATP releases a fraction of kinesin (S_3), but another substantial fraction remains bound to the microtubules (P), showing that in this series of experiments AMP-PNP is less efficient than PPP_i with regard to kinesin recovery. Finally, we supplemented the buffer with 1 mM $AlCl_3$ and 4 mM NaF (Fig. 2, right four lanes). Again the S_1 supernatant contains no kinesin, the S_3 supernatant after release with ATP contains the major fraction, and very little remains bound to microtubules (P).

These experiments show that aluminum fluoride indeed

causes kinesin to bind to microtubules, and the efficiency of kinesin recovery is comparable to the one achieved with PPP_i and AMP-PNP.

Next we repeated these experiments, stabilizing the microtubules not with taxol but with aluminum fluoride (Fig. 3). The results were essentially the same, with some differences in the efficiency of kinesin recovery. This means that the effect of aluminum fluoride on kinesin binding and on microtubule stability are largely independent of one another.

Conclusions

The analysis of the effect of aluminum fluoride on

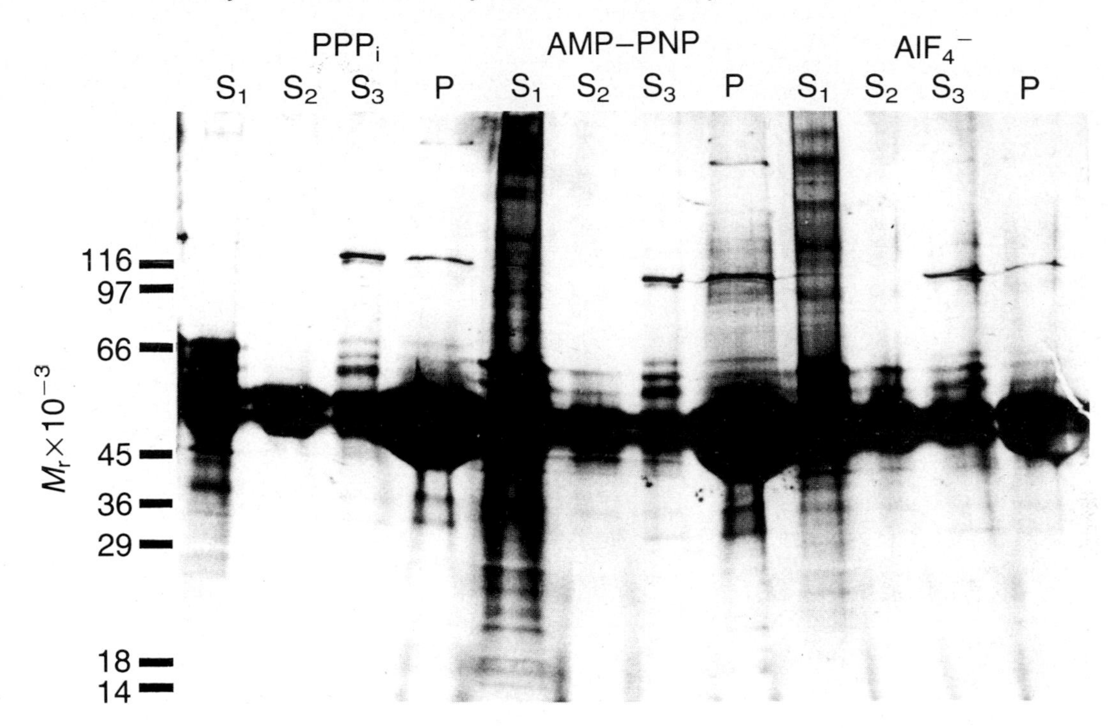

Fig. 2. Binding of kinesin to taxol-stabilized microtubules in the presence of different agents: 2.5 mM PPP_i (left), 1 mM AMP-PNP (middle), and 1 mM $AlCl_3$, 4 mM NaF and 1 mM ATP (right). The buffer (H) contained 50 mM imidazol pH 6.7, 0.4 mM $MgCl_2$, 0.1 mM EDTA, 1 mM EGTA, 1 mM mercaptoethanol, 1 M glycerol, 5 mM $MgSO_4$, 50 mM KCl and 0.1 mM PMSF. In each case four lanes are shown in the silver-stained SDS–PAGE gel. S_1, supernatant after the first microtubule–kinesin binding step and pelleting. No kinesin remains in the supernatant. S_2, supernatant after washing the pellet with buffer H; kinesin remains bound to microtubules and is not released into the supernatant. S_3, supernatant after releasing the kinesin with buffer containing additional 10 mM MgATP. P, pellet after the releasing step, showing that some kinesin is still bound to the microtubules.

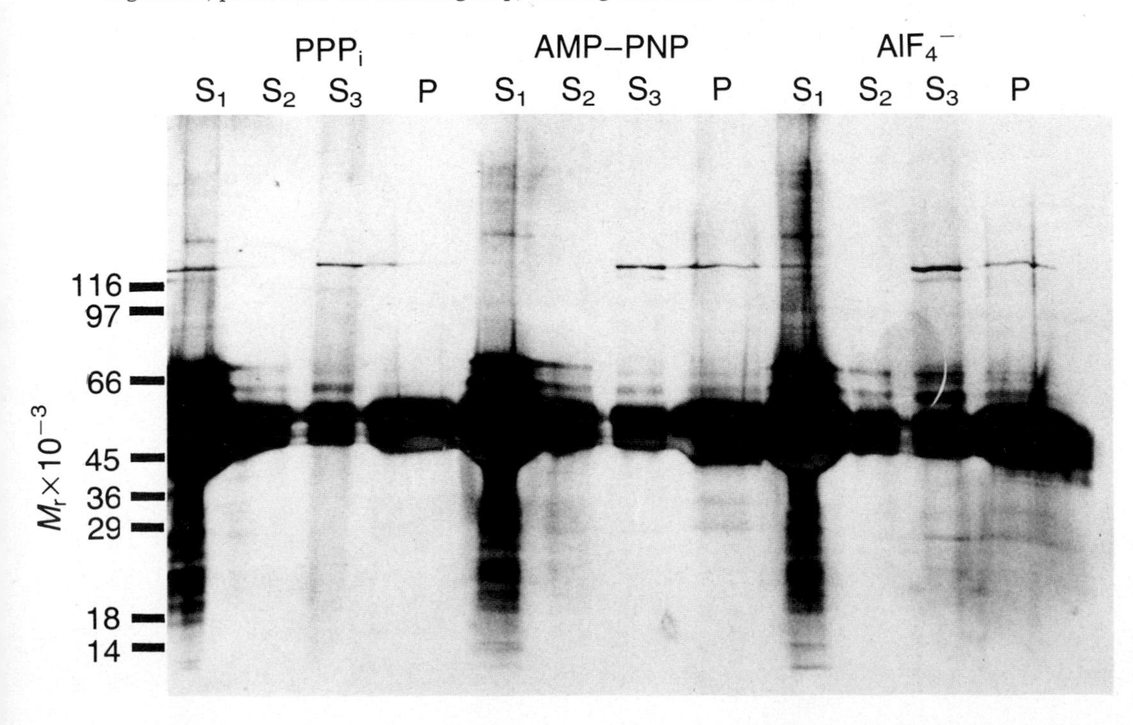

Fig. 3. Same experiment as in Fig. 2, except that microtubules were stabilized with 1 mM $AlCl_3$ and 4 mM NaF. The results are essentially the same, showing that the stabilization of microtubules by AlF and the rigor complex of kinesin occur independently of one another.

microtubules and kinesin is interesting for two reasons. The first is a practical one; aluminum fluoride helps to circumvent the need for taxol and AMP-PNP, reagents that are on short supply or expensive. This is important for scaling up the preparation when one wants to produce sufficient amounts of kinesin for structural studies, e.g. crystallization. The second reason is a theoretical one related to the reaction mechanism of kinesin. The present data opens the possibility that aluminum fluoride induces kinesin rigor *via* the AlF_4^- complex which could act as an analogue of a transition state NDP–P_i complex, substituting for the terminal phosphate of GTP (for tubulin) or ATP (for kinesin). Such a mechanism seems to hold for G-proteins (Bigay *et al.* 1987) and may apply to microtubules as well (Carlier *et al.* 1988), although here the case is less clear-cut (see Hymphreys and Macdonald, 1988; Caplow *et al.* 1989). At any rate, analogs that arrest the reaction cycle of kinesin are valuable tools for studying its reaction mechanism. This has been demonstrated in the case of AMP-PNP, and aluminum fluoride may turn out to be similarly useful in the future.

We thank Dr M. Suffness (US National Cancer Institute, Bethesda, MD) for providing taxol. The project was supported by a grant from the Bundesministerium für Forschung und Technologie.

References

ALLEN, R. D., ALLEN, N. S. AND TRAVIS, J. L. (1981). Video-enhanced contrast, differential interference contrast (AVEC-DIC) microscopy: a new method capable of analysing microtubule-related motility in the reticulopodial network of *Allogromia laticollaris*. *Cell Motility* **1**, 291–302.

BIGAY, J., DETERRE, P., PFISTER, C. AND CHABRE, M. (1987). Fluoride complexes of aluminium or beryllium act on G-proteins as reversibly bound analogues of the gamma phosphate of GTP. *EMBO J.* **6**, 2907–2913.

CAPLOW, M., RUHLEN, R., SHANKS, J., WALKER, R. A. AND SALMON, E. D. (1989). Stabilization of microtubules by tubulin-GDP-P_i subunits. *Biochemistry* **28**, 8136–8141.

CARLIER, M. F., DIDRY, D., MELKI, R., CHABRE, M. AND PANTALONI, D. (1988). Stabilization of microtubules by inorganic phosphate and its structural analogues, the fluoride complexes of aluminum and beryllium. *Biochemistry* **27**, 3555–3559.

HUMPHREYS, W. G. AND MACDONALD, T. L. (1988). The effects on tubulin polymerization and associated guanosine triphosphate hydrolysis of aluminum ion, fluoride and fluoroaluminate species. *Biochem. biophys. Res. Comm.* **151**, 1025–1032.

KUZNETSOV, S. A. AND GELFAND, V. I. (1986). Bovine brain kinesin is a microtubule-activated ATPase. *Proc. natn. Acad. Sci. U.S.A.* **83**, 8530–8534.

LASEK, R. J. AND BRADY, S. T. (1985). Attachment of transported vesicles to microtubules in axoplasm is facilitated by AMP-PNP. *Nature* **316**, 645–647.

MACDONALD, T. L., HUMPHREYS, W. G. AND MARTIN, R. B. (1987). Promotion of tubulin assembly by aluminum ion *in vitro*. *Science* **236**, 183–186.

MANDELKOW, E.-M., HERRMANN, M. AND RÜHL, U. (1985). Tubulin domains probed by subunit-specific antibodies and limited proteolysis. *J. molec. Biol.* **185**, 311–327.

VALE, R. D., REESE, T. S. AND SHEETZ, M. P. (1985). Identification of a novel force-generating protein, Kinesin, involved in microtubule-based motility. *Cell* **42**, 39–50.

VON MASSOW, A., MANDELKOW, E.-M. AND MANDELKOW, E. (1989). Interaction between kinesin, microtubules, and microtubule-associated protein 2. *Cell Mot. and Cytoskel.* **14**, 562–571.

MAPs and motors in insect ovaries

HOWARD STEBBINGS, CHERRYL HUNT and ANGELA ANASTASI

Department of Biological Sciences, Washington Singer Laboratories, University of Exeter, Exeter, UK

Summary

MAPs and microtubule motor proteins from the massive microtubule translocation complexes within the ovaries of hemipteran insects have been identified and characterized. Both classes of proteins have been compared with those of other systems, and the function of both in the insect ovaries is speculated upon.

Key words: microtubules, MAPs, hemipteran ovaries, intracellular translocation, motors.

The insect ovarian microtubule system

Since microtubules were first identified they have been equated with assymetrical cell shape and intracellular translocation – and the ovaries of one order of insects, the hemipterans (bugs), exhibit both properties in an exaggerated form.

Hemipterans are unlike other insects in having telotrophic meroistic ovaries. The term meroistic refers to the fact that the ovaries possess nutritive cells which supply the developing oocytes, and telotrophic to the nutritive cells being situated at the anterior end of the organ (see Bonhag, 1958). The geometry of the system is such that the nutritive cells pass materials to the oocytes by way of connections known as nutritive tubes – cell extensions (Valdimarsson and Heubner, 1989) which may be some 20–30 μm in diameter and many millimetres in length. Each nutritive tube is packed with a parallel arrangement of microtubules which typically number tens of thousands.

The microtubules have a common polarity, with their plus or fast-growing ends towards the anterior, nutritive cell end of the nutritive tubes (Stebbings and Hunt, 1983); their spacings within the nutritive tubes vary between different hemipteran species, possibly being linked to differences in the size of components translocated in different species (Hyams and Stebbings, 1977). As well as continually increasing in length throughout oogenesis, the nutritive tubes also increase in diameter, so that new microtubules must be added continuously to the bundle (Hyams and Stebbings, 1979). This process has recently been clarified by immunocytochemical studies which show different distributions of α-tubulin isotypes in a single developing nutritive tube (Harrison *et al.* manuscript in preparation). Eventually, when an oocyte reaches the vitellogenic stage of oogenesis, the nutritive tube supplying it becomes redundant, and the typical microtubule spacing gives way to a bundling of the microtubules (Hyams and Stebbings, 1979) followed by their depolymerization.

Throughout their functional existence, nutritive tubes act as conduits for the passage of nutritive cell components to the oocytes. These consist of RNA and cytoplasm and, in some cases, organelles as large as mitochondria – the importation of the latter apparently being typical of species where the period of oogenesis is short, *i.e.* a few weeks, and not occurring where oogenesis lasts many months. Translocation of components along the microtubule-packed nutritive tubes has been shown to take place at different rates even within a single tube (see Stebbings, 1986), with slow rates of movement being monitored by autoradiography (Macgregor and Stebbings, 1970; Mays, 1972) and faster rates by video-enhanced contrast microscopy (Dittmann *et al.* 1987). Moreover unidirectional translocation of mitochondria along nutritive tube microtubules has been reactivated *in vitro* (Stebbings and Hunt, 1987), showing it to be an active process.

The insect ovarian nutritive tube provides an excellent model for the study of microtubules and microtubule-associated components generally. It complements studies which have been carried out mainly with mammalian nerve, and has a number of advantages over the latter. Microtubules are the sole cytoskeletal elements involved, whereas nerve axons are complicated by neurofilaments. Many more microtubules are found in nutritive tubes than nerve axons. Translocation along nutritive tubes is unidirectional, as opposed to bidirectional movement in nerves. Most importantly though, nutritive tubes can be microdissected from ovaries intact (Fig. 1) for biochemical, structural and motility studies. This has allowed the investigation of (microtubule-associated proteins) MAPs and motors in the insect system and provided a model for the study of microtubule–MAP/ motor interactions generally.

Insect ovarian microtubule motors

Using techniques that involve the addition of AMP-PNP and the depletion of ATP, together with purified MAP-free taxol-stabilized pig brain microtubules (methods which have been developed by others for the isolation of microtubule motors), we have identified motors in hemipteran ovary homogenates (Anastasi *et al.* 1990). Under such conditions a high molecular weight protein, similar in size to mammalian brain MAP1, bound specifically to the microtubules, as did a $116\times10^3\,M_r$ polypeptide (Fig. 2). Both were extracted from the microtubules with MgATP. The material released was sedimented on a sucrose density gradient (Fig. 3). The high molecular weight protein sedimented at 20S, co-electrophoresed with MAP1C and, like dynein heavy chains, showed susceptibility to cleavage at a single site when irradiated with u.v. light in the presence of vanadate and ATP, yielding two polypeptides of molecular weights $190\times10^3\,M_r$ and $230\times10^3\,M_r$. The $116\times10^3\,M_r$ species sedimented at

Journal of Cell Science, Supplement 14, 151–155 (1991)

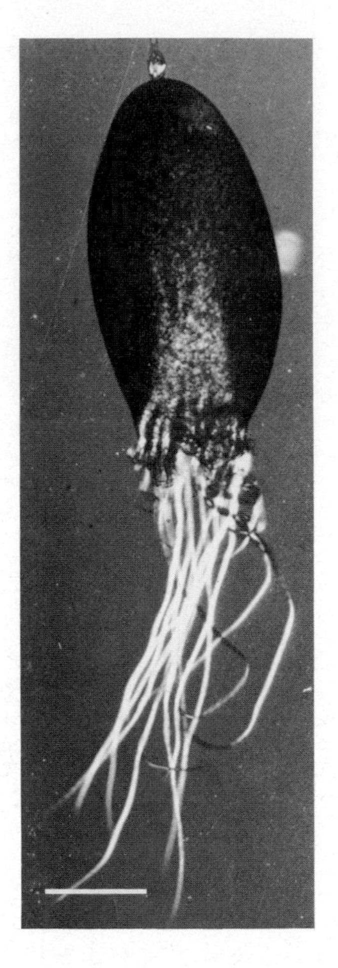

Fig. 1. Insect ovariole viewed in polarized light and microdissected to free the birefringent nutritive tubes. Bar, 0.25 mm.

around 9S and showed immunoreactivity with anti-chicken brain kinesin.

The hemipteran ovarian material released from pig brain microtubules with MgATP was tested to see whether it produced movement of latex beads on microtubules. Bundles of parallel microtubules of known polarity from isolated, detergent and salt-extracted nutritive tubes provided a novel and highly efficient motility substrate (Fig. 4).

Polybead carboxylated microspheres coated in the ATP extract were applied to the substrate in the presence of 10 mM ATP, whereupon many such beads attached to the microtubule bundles (Fig. 5) and about half of these appeared to be moving along the bundle at any one time. Bidirectional bead movements were observed, but movements in an anterograde direction predominated. Purification of the proteins in the ATP extract showed that kinesin was responsible for bead movement in the anterograde direction, but the retrograde movement was not attributable to any particular fraction. The nature of the movements in the two directions promoted by the ATP extract were quite different in terms of velocity and duration, and were comparable to the movements generated by kinesin and dynein from a range of systems (see Hollenbeck, 1988; Vallee *et al.* 1989).

Intriguingly kinesin appears not to be present in nutritive tubes, while a dynein-like protein is seen in these translocation channels – a finding which correlates with the retrograde translocation which occurs along their length *in situ* throughout oogenesis, and suggests that dynein may be responsible for the translocation *in vivo*.

$M_r \times 10^{-3}$ 205 116 97 66 45 29 A B C

Fig. 2. Binding of polypeptides from insect ovary homogenates to microtubules in the absence of ATP, and their subsequent release from microtubules by MgATP. (A) Clarified ovary homogenate to which hexokinase/glucose had been added. (B) Polypeptides that bind to taxol-stabilized, purified brain microtubules in the absence of ATP and the presence of AMP–PNP. (C) Polypeptides released by 10 mM MgATP from microtubule pellets as prepared in B. Samples were run on a 5–10 % gradient gel.

Insect ovarian MAPs

Whereas the motor proteins in insect ovaries appear to be the same as those in mammalian and other systems, this is not true of the MAPs. When microtubules are isolated from ovaries of hemipterans by cycles of assembly and disassembly, as well as containing tubulins they are seen to comprise a spectrum of MAPs (Fig. 6), and a similar collection of MAPs is extracted from taxol-assembled microtubules with high salt. The predominant MAPs are of high molecular weight (greater than $200 \times 10^3 M_r$) but these differ in molecular weight from mammalian brain MAPs and immunological studies show no cross-reactivity with them. Insect ovarian high molecular weight MAPs do however have comparable properties to brain MAPs and have been shown to promote microtubule assembly, to be heat stable (Anastasi *et al.* manuscript in preparation) as is brain MAP2 (Vallee, 1985), and to show a high degree of turnover (Stebbings *et al.* 1985) as do mammalian brain MAPs (Okabe and Hirokawa, 1987; Nixon *et al.* 1990).

Unusually the same high molecular weight MAPs in insects show nucleotide-sensitive binding to microtubules (Fig. 7) (Anastasi *et al.* manuscript in preparation). Such MAPs appear to bind to microtubules in the presence of low levels of nucleotide (e.g. 2 mM GTP), but are removed progressively to about 50 % with increasing concentrations of both GTP and ATP (to 10 mM). To this extent

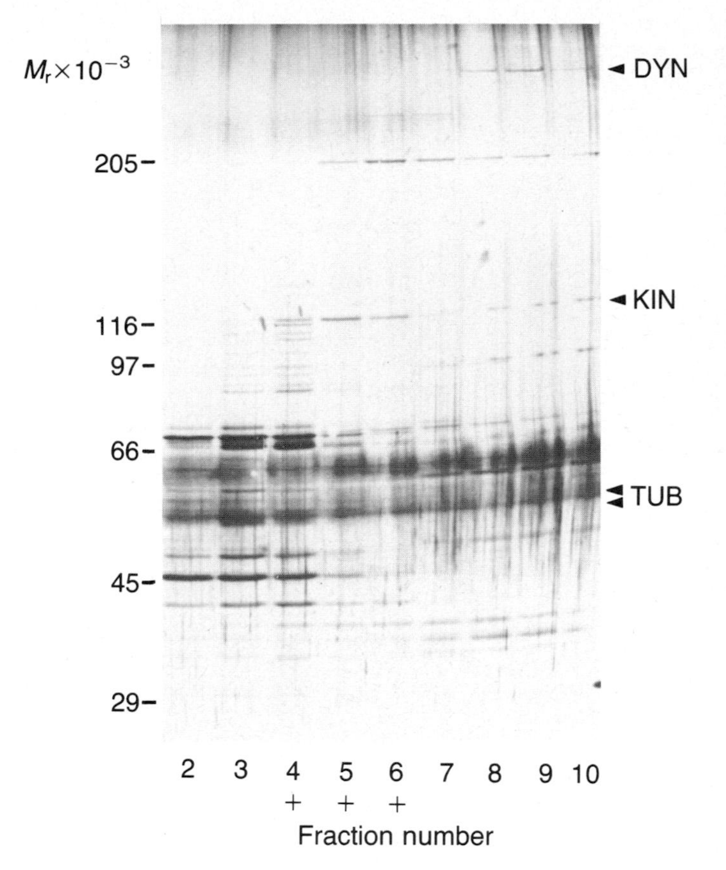

Fig. 3. Fractions from a 5 % to 30 % sucrose density gradient loaded with an ATP-released supernatant (as in Fig. 2C) and centrifuged, were analysed by gel electrophoresis and silver stained. Fractions are numbered from the top to the bottom of the gradient. (Kinesin is separated on the gradient from dynein and tubulin. A doublet at approx. $64-66\times10^3$ M_r co-sedimented with the kinesin). Sedimentation standards thyroglobulin, catalase and bovine albumin, sedimented at fractions 9, 6 and 3, respectively. Motile activity in a fraction is indicated by +. DYN, dynein; KIN, kinesin; TUB, tubulin.

the high molecular weight MAPs resemble the previously discussed cellular motor proteins; but no ATPase activity, indicative of a motor, was found, and neither do they generate motility. Such properties are not unique to the insect ovarian high molecular weight MAPs, however, as MAPs with comparable properties have also been identified recently in HeLa cells (Rickard and Kreis, 1990).

Fascinatingly, the high molecular weight MAPs of insect ovaries show species specificity, having slightly different molecular weights (Fig. 8) in all of the species studied so far. Also, antibodies raised against a high molecular weight MAP of one species do not cross-react with MAPs of other species. Furthermore, apart from being species-specific the high molecular weight MAPs from insect ovaries also show some tissue specificity, and certainly do not occur in the neurones of the same species.

Such MAP specificities are of considerable interest as they may point to the MAP function. First, the fact that the high molecular weight MAPs occur in one translocation system (the ovarian nutritive tubes) but not in others (nerve axons) suggests that they are not involved in the translocation process generally. Secondly, their occurrence in the ovarian nutritive tubes (though only in microtubules) and not the nerve axons (microtubules and neurofilaments) means that they appear not to be involved in linking microtubules and neurofilaments, as has been proposed for other MAPs. Rather their existence in purely microtubule systems would suggest a microtubule-based function.

In this regard, the variability of the high molecular weight MAPs in the ovaries of different species is particularly telling. There are two well-known and very variable factors in the microtubule systems of ovarian nutritive tubes of different hemipterans. Firstly, differences exist in the stability of the microtubules in different species, which is perhaps not surprising since insects inhabit a wide range of environments, some species experiencing temperatures of below 0°C and others equatorial conditions. Secondly, differences have been found in the inter-microtubule spacings in the nutritive tubes of different species, such that a particular spacing is typical of a particular species and is also related to the components travelling along and between the microtubules comprising the nutritive tubes of that species. Certainly MAPs in a wide range of systems have been shown to confer stability on microtubules, and there is also previous data which suggests that MAPs may determine microtubule arrangements. However, direct evidence for insect ovarian high molecular weight MAPs being involved in either function remains to be obtained.

The significance of the nucleotide-sensitive binding of the insect ovarian high molecular weight MAPs to microtubules also requires to be resolved. There have been many demonstrations of microtubules associating *via*

Fig. 4. Detergent-extracted nutritive tube viewed using dark-field optics. Bar, 20 μm.

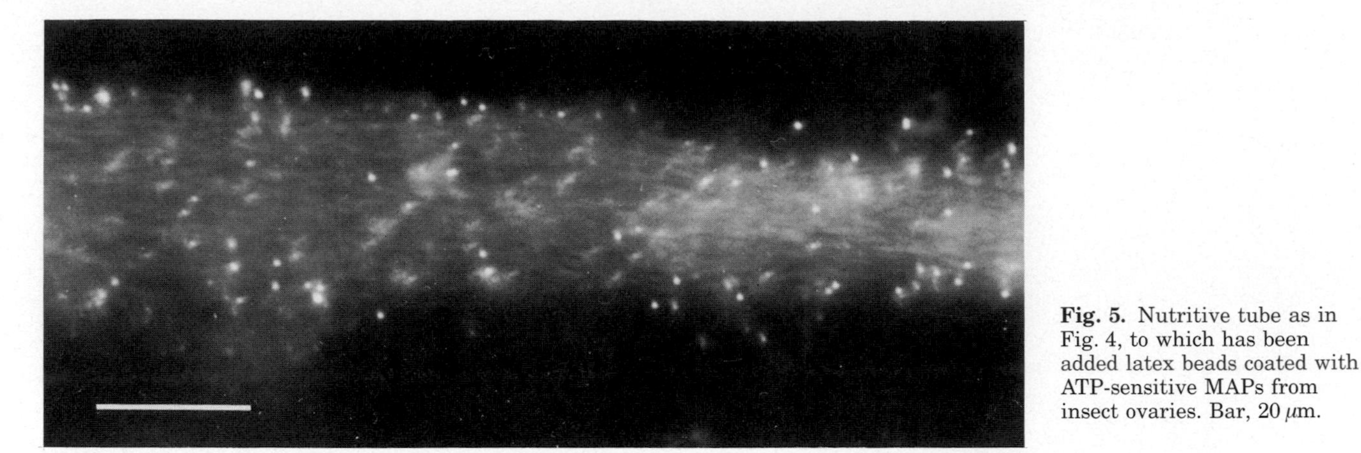

Fig. 5. Nutritive tube as in Fig. 4, to which has been added latex beads coated with ATP-sensitive MAPs from insect ovaries. Bar, 20 μm.

various polypeptides with membranous organelles, cytoskeletal filaments, and probably more relevant to this discussion, with each other. In most cases ATP has been shown to bring about their dissociation, and a view is emerging that such relatively stable associations might interpose periods of motor-driven interaction (Linden *et al.* 1989; Mithieux and Rousset, 1989). Whether insect ovarian MAPs do serve to form associations between microtubules, during the establishment of the initial microtubule pattern and the continuous addition of

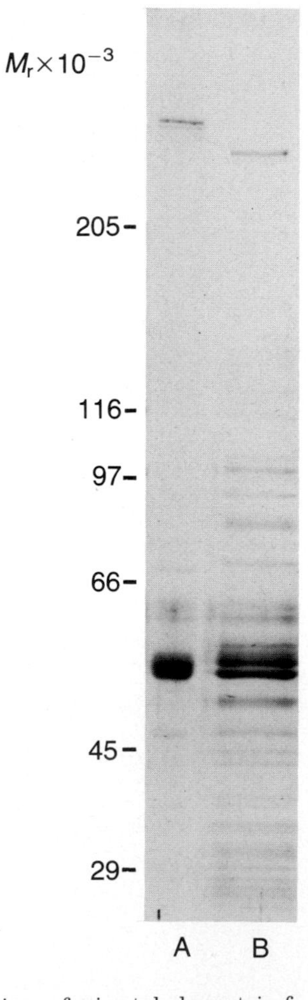

Fig. 6. Comparison of microtubule protein from (A) pig brain and (B) insect ovary, by SDS–PAGE.

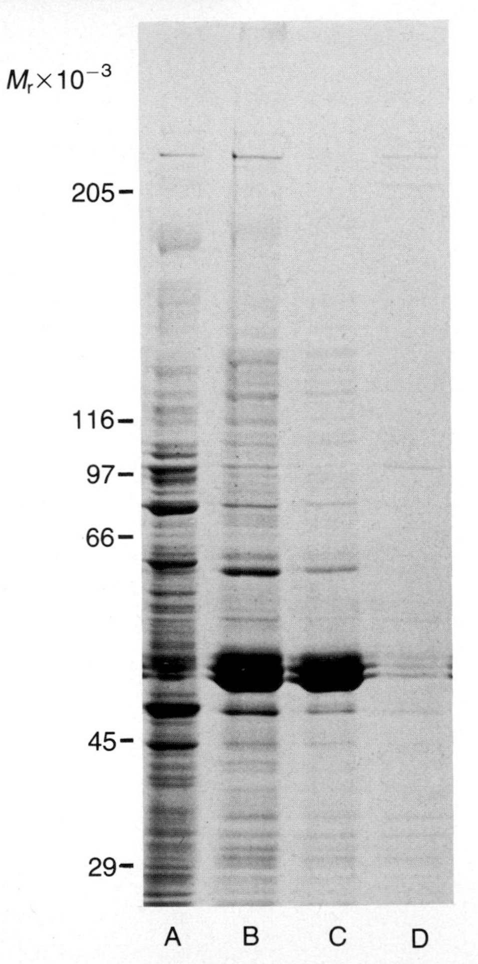

Fig. 7. Gel to show nucleotide sensitivity of some insect ovary MAPs. (A) Ovary homogenate; (B), microtubule protein polymerised from (A) in the presence of 20 μM taxol; (C), microtubule protein polymerised from (A) in the presence of 20 μM taxol, 5 mM GTP; D, polypeptides extracted from (B) by 10 mM ATP.

similarly-spaced microtubules to such aggregates, is not known. What is clear is that their removal results in a breakdown of their characteristic spacing.

The suggestion that certain MAPs may act similarly to axonemal dynein, which binds cytoplasmic microtubules *in vitro* by a structural site at one end and more weakly to

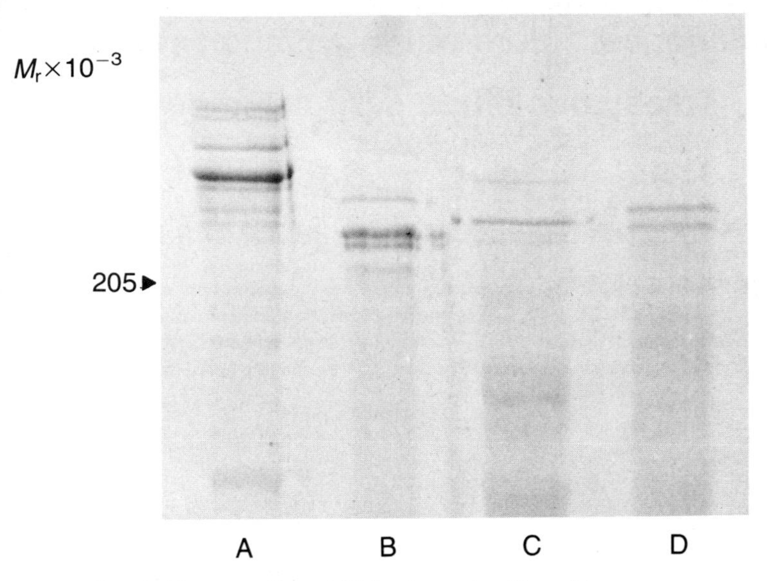

Fig. 8. Upper portion of SDS-polyacrylamide gel to compare high molecular weight MAPs of taxol-polymerised microtubules from (A) pig brain, (B) *Notonecta* ovary, (C) *Oncopeltus* ovary and (D) *Corixa* ovary.

an ATP-sensitive site at the other (Porter and Johnson, 1983; Haimo and Fenton, 1984) to result in cross-bridging, is not a new one (Suprenant and Dentler, 1982; Hollenbeck and Chapman, 1986). Similar behaviour by the high molecular weight MAPs in insect ovaries could explain their possession of characteristics of both microtubule-associated *and* microtubule-binding proteins.

This work has been supported by grants from the SERC and the Wellcome Trust to H.S.

References

Anastasi, A., Hunt, C. and Stebbings, H. (1990). Isolation of microtubule motors from an insect ovarian system: characterization using a novel motility substratum. *J. Cell Sci.* **96**, 63–69.

Bonhag, P. F. (1958). Ovarian structure and vitellogenesis in insects. *A. Rev. Entomol.* **3**, 137–160.

Dittmann, F., Weiss, D. G. and Münz, A. (1987). Movement of mitochondria in the ovarian trophic cord of *Dysdercus intermedius* (Heteroptera) resembles nerve axonal transport. *Roux's Arch. Devl Biol.* **196**, 407–413.

Haimo, L. T. and Fenton, R. D. (1984). Microtubule crossbridging by *Chlamydomonas* dynein. *Cell Motil.* **4**, 371–385.

Hollenbeck, P. J. (1988). Kinesin: its properties and possible functions. *Protoplasma* **145**, 145–152.

Hollenbeck, P. J. and Chapman, K. (1986). A novel microtubule-associated protein from mammalian nerve shows ATP-sensitive binding to microtubules. *J. Cell Biol.* **103**, 1539–1545.

Hyams, J. S. and Stebbings, H. (1977). The distribution and function of microtubules in nutritive tubes. *Tissue Cell* **9**, 537–545.

Hyams, J. S. and Stebbings, H. (1979). The formation and breakdown of nutritive tubes – massive microtubular organelles associated with cytoplasmic transport. *J. Ultrastruct. Res.* **68**, 46–57.

Linden, M., Nelson, B. D. and Leterrier, J. F. (1989). The specific binding of the microtubule-associated protein 2 (MAP2) to the outer membrane of rat brain mitochondria. *Biochem. J.* **261**, 167–173.

Macgregor, H. C. and Stebbings, H. (1970). A massive system of microtubules associated with cytoplasmic movement in telotrophic ovarioles. *J. Cell Sci.* **6**, 431–449.

Mays, U. (1972). Stofftransport in ovar von *Pyrrhocoris apterus* L. *Z. Zellforsch. mikrosk-Anat.* **123**, 395–410.

Mithieux, G. and Rousset, B. (1989). Identification of a lysosome membrane protein which could mediate ATP-dependent stable association of lysosomes to microtubules. *J. biol. Chem.* **264**, 4664–4668.

Nixon, R. A., Fischer, I. and Lewis, S. E. (1990). Synthesis, axonal transport, and turnover of the high molecular weight microtubule-associated protein MAP 1A in mouse retinal ganglion cells: tubulin and MAP 1A display distinct transport kinetics. *J. Cell Biol.* **110**, 437–448.

Okabe, S. and Hirokawa, N. (1989). Rapid turnover of microtubule-associated MAP2 in the axon revealed by microinjection of biotinylated MAP2 into cultured neurons. *Proc. natn Acad. Sci. U.S.A.* **86**, 4127–4131.

Porter, M. E. and Johnson, K. A. (1983). Characterization of the ATP-sensitive binding of *Tetrahymena* 30S dynein to bovine brain microtubule. *J. biol. Chem.* **258**, 6575–6581.

Rickard, J. E. and Kreis, T. E. (1990). Identification of a novel nucleotide-sensitive microtubule-binding protein in HeLa cells. *J. Cell Biol.* **110**, 1623–1633.

Stebbing, H. (1986). Cytoplasmic transport and microtubules in teletrophic ovarioles of hemipteran insects. *Int. Rev. Cytol.* **101**, 101–123.

Stebbings, H. and Hunt, C. (1983). Microtubule polarity in the nutritive tubes of insect ovarioles. *Cell Tiss. Res.* **233**, 133–141.

Stebbings, H. and Hunt, C. (1987). The translocation of mitochondria along insect ovarian microtubules from isolated nutritive tubes: a simple reactivated model. *J. Cell Sci.* **88**, 641–648.

Stebbings, H., Sharma, K. and Hunt, C. (1985). Protein turnover in the cytoplasmic transport system within an insect ovary – a clue to the mechanism of microtubule-associated transport. *FEBS Lett.* **193**, 22–26.

Suprenant, K. A. and Dentler, W. L. (1982). Association between endocrine pancreatic secretory granules and *in vitro*-assembled microtubules is dependent upon microtubule-associated proteins. *J. Cell Biol.* **93**, 164–174.

Valdimarsson, G. and Huebner, E. (1989). The development of microtubular arrays in the germ tissue of an insect teletrophic ovary. *Tissue Cell* **21**, 123–138.

Vallee, R. B. (1985). On the use of heat stability as a criterion for the identification of microtubule associated proteins (MAPs). *Biochem. biophys. Res. Commun.* **133**, 128–133.

Vallee, R. B., Shpetner, H. S. and Paschal, B. M. (1989). The role of dynein in retrograde axonal transport. *Trends Neurosci.* **12**, 66–70.

ported organelles, however, such a correlation was absent (Seitz-Tutter, 1990; D. Seitz-Tutter, D. G. Weiss and G. M. Langford, unpublished observations).

The pharmacological properties of the gliding behaviour of nMTs (Seitz-Tutter, 1990; D. Seitz-Tutter, D. G. Weiss and G. M. Langford, unpublished observations) show the following similarities with MT-motility in reconstituted systems containing purified kinesin (Cohn *et al.* 1987; Scholey *et al.* 1985; Vale *et al.* 1985*a*,*c*): gliding of nMTs is inhibited by apyrase ($0.5\,mg\,ml^{-1}$), AMP-PNP and PPP_i

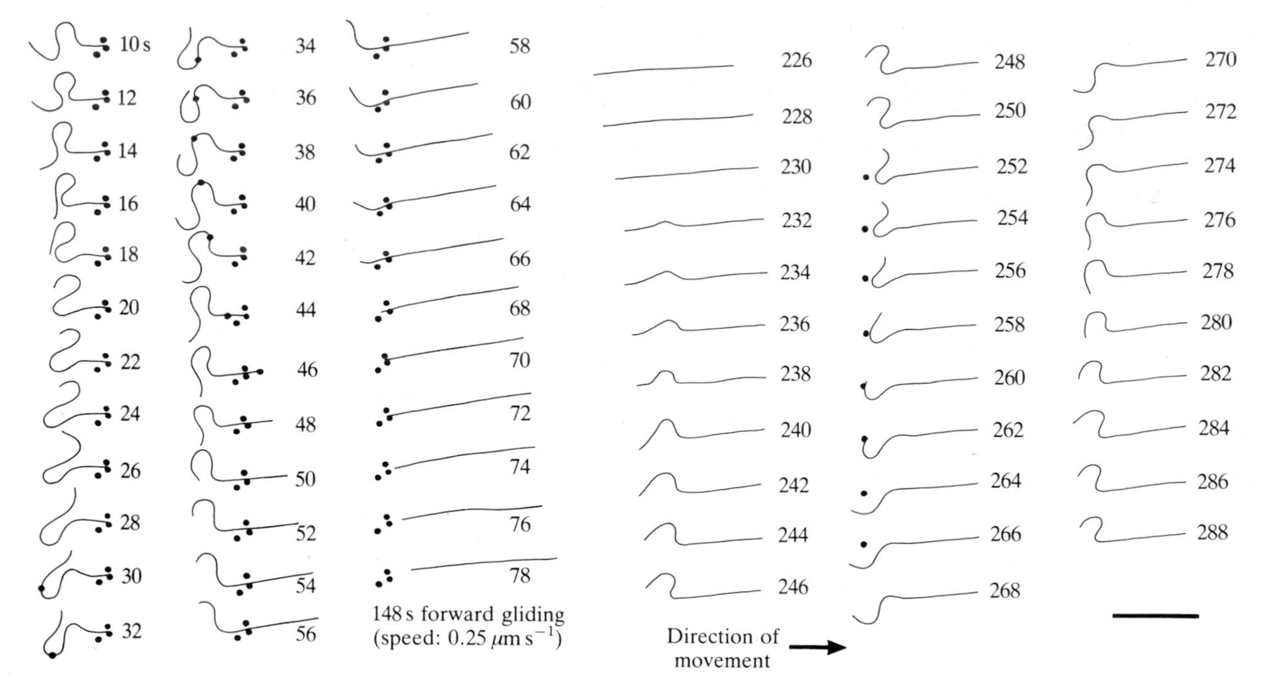

Fig. 1. Gliding and fishtailing behaviour of nMT. For the first 34 s the straight gliding movement of this nMT is hindered by an invisible obstacle. The force-generating enzymes cause the fishtailing motion until the MT becomes free from the obstacle (44 s). The MT straightens and glides between second 78 and 226 toward the right (speed $0.25\,\mu m\,s^{-1}$) until the frontal end becomes again attached by an obstacle or a 'sticky' site on the surface and the tail end displays the fishtailing behaviour again. The particle which is transported along the microtubule between time points 30 and 44 does not interfere with the fishtailing motion, but probably causes the microtubule to resume gliding by releasing the front end from its attachment. The numbers indicate the elapsed time in seconds. Bar $5\,\mu m$.

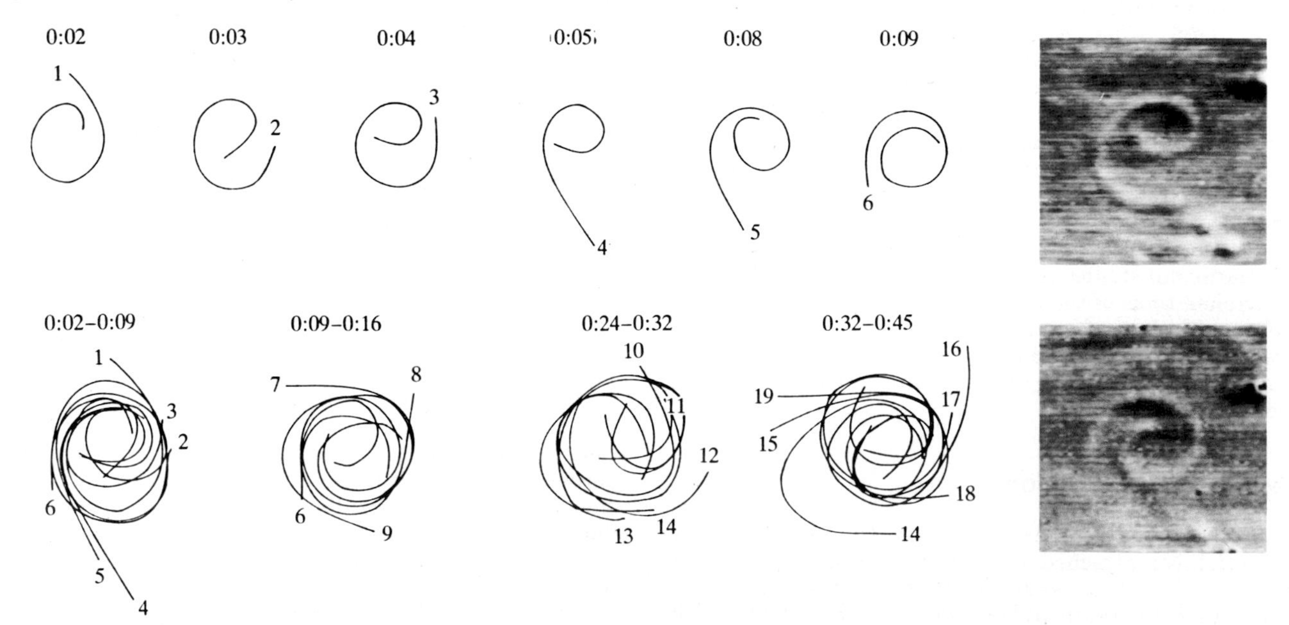

Fig. 2. Spiraling motion of nMT. A sequence of shapes assumed by a microtubule which is bent into a planar spiral and rotates. The length of this microtubule is $11\,\mu m$. The consecutive shapes are drawn at 1 s intervals and were superimposed in the lower panels to show that the microtubule is not moving on a defined track on the glass surface (the small numbers indicate the rear, pushing end). Instead, it appears as if the microtubule can move equally well over the entire area. The observation time in the top row is 8 s corresponding to about 3/4 of a full turn (time is given in min:s). Two representative videomicrographs of this MT are included.

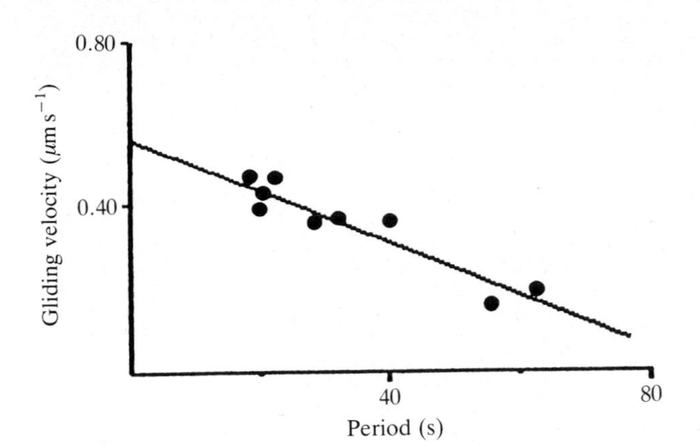

Fig. 3. Correlation between beating frequency and gliding velocity. The period of the fishtailing motion is independent of the length of the motile part of the MT but highly correlated with the mean gliding velocities of other MTs in the same preparations (r=0.94, P=0.01, Pearson-Bravais test). This result supports the notion that gliding, fishtailing, and rotating movements are all driven by the same mechanism. The point of intersection of the regression line with the ordinate (0.58 μm s^{-1}) matches the maximum gliding velocity of MTs observed in the extruded axoplasm preparations (mean: 0.45±0.04 μm s^{-1}, max.: 0.65 μm s^{-1}).

when in excess of ATP. But, in contrast to the movement of reassembled MTs induced by purified kinesin, the gliding of nMTs was partially inhibited by 1 mM *N*-ethylmaleimide (NEM) and completely blocked by 2 mM (Table 1). Furthermore, the straight gliding of nMTs was significantly inhibited by relatively low concentrations of vanadate (20 μM) that specifically inhibit dynein-like, but not kinesin-like, ATPases. Gliding was blocked completely in the presence of 50 μM vanadate.

These findings are in contrast to reports on kinesin-induced gliding in *in vitro* reconstituted systems, where inhibition occurs usually only at 50–400 μM vanadate (for references see Table 1). In the presence of 50 μM vanadate, retrograde organelle transport along nMTs was already completely blocked, whereas anterograde organelle transport continued almost undisturbed for about 30 min before a decline in velocity was detected. After replacement of the vanadate-containing buffer by vanadate-free medium only, anterograde organelle transport was reactivated. Gliding of nMTs is therefore sensitive to even lower concentrations of vanadate than retrograde transport (Table 1).

The sensitivity to various inhibitors of the ATPase activity and of the microtubule translocation is well known, for both purified kinesin and cytoplasmic dynein. In such preparations kinesin was found to be less sensitive than dynein for NEM inhibition (2–5 mM *versus*

Table 1. *Properties of motors and MT motility* in vivo *and* in vitro

	Purified enzyme	Native MTs	Intact axons
Kinesin			
Gliding (retrograde)			
Vanadate inhibition	50–100 μM[b,d,e]	20 μM[a]	Gliding absent, MTs anchored
NEM inhibition	3–5 mM[b], >5 mM[d]	1–2 mM[a]	Gliding absent, MTs anchored
Velocity	0.4–0.6 μm s^{-1}[d]	0.45 μm s^{-1}[a,i]	Gliding absent, MTs anchored
Organelle movement (anterograde)			
Vanadate inhibition	No published report	100 μM[a]	100 μM[f,g], 55 μM[n]
NEM inhibition	No published report	1 mM[a,c]	2 mM[f], 1 mM[a]
Velocity	No published report	1–3 μm s^{-1}[a,i]	1–3 μm s^{-1}[a]
ATPase activity			
Vanadate inhibition	>10 μM[m], 50 μM[h], 420 μM[e],	n.a.	n.a.
NEM inhibition	2–5 mM[c,d,h,l,m]	n.a.	n.a.
Cytoplasmic dynein			
Gliding (anterograde)			
Vanadate inhibition	10 μM[p]	Not observed	Gliding absent, MTs anchored
NEM inhibition	0.1 mM[p]	Not observed	Gliding absent, MTs anchored
Velocity	1.25 μm s^{-1}[q]	Not observed	Gliding absent, MTs anchored
Organelle movement (retrograde)			
Vanadate inhibition	No published report	50 μM[a]	3 μM[n]
NEM inhibition	No published report	1 mM[a,c]	1 mM[n]
Velocity	0.87 μm s^{-1}[r]	0.2–1 μm s^{-1}[a,i]	0.2–1 μm s^{-1}[a]
ATPase activity			
Vanadate inhibition	5–10 μM[o]	n.a.	n.a.
NEM inhibition	1 mM[o]	n.a.	n.a.

[a] Seitz-Tutter, 1990.
[b] Porter *et al.* 1987.
[c] Pfister *et al.* 1989.
[d] Vale *et al.* 1985c.
[e] Cohn *et al.* 1987.
[f] Lasek and Brady, 1985.
[g] Forman *et al.* 1983.
[h] Wagner *et al.* 1989.
[i] Allen *et al.* 1985.
[k] Pryer *et al.* 1986.
[l] Penningroth *et al.* 1987.
[m] Kuznetsov and Gelfand, 1986.
[n] Smith, 1988.
[o] Shpetner *et al.* 1988.
[p] Paschal and Vallee, 1987.
[q] Paschal *et al.* 1987.
[r] Gilbert and Sloboda, 1989.
n.a., not analysed; NEM, *N*-ethylmaleimide.

0.1–1 mM, respectively) and for vanadate inhibition (20–400 μM *versus* 5–10 μM, respectively) (Table 1). In intact axons the same difference was observed for the effect of vanadate on organelle transport in the two directions (retrograde transport is inhibited by 3 μM vanadate, anterograde transport by 50–100 μM), but it was found that NEM inhibited both directions similarly (1–2 mM) (Table 1). This situation is very similar to our findings on the pharmacology of organelle transport along nMTs (Table 1). The major surprise is, however, the finding that gliding of nMTs is even more sensitive than retrograde organelle movement.

Taken together, gliding of nMTs has mainly kinesin-like properties: (1) the direction is the same as that of the motor for anterograde vesicle transport, (2) the velocity of anterograde transport and nMT gliding correlate closely (although the absolute values are different) (Allen *et al.* 1985). Not fully compatible with the hypothesis that kinesin is the motor for nMT motility and yet to be resolved are the following findings: (1) the velocity of purified kinesin is much slower than anterograde vesicle transport and nMT gliding, (2) the sensitivity to NEM and vanadate of nMT gliding is higher than that of purified kinesin and resembles to some extent that of purified dynein or retrograde organelle transport, (3) Leopold *et al.* (1990) reported that the nucleotide specificity of anterograde organelle motility in the native system differs considerably from the properties of purified kinesin.

However, the majority of the findings support the notion that the motility of nMTs is driven by kinesin, the motor of anterograde transport. Although retrograde motors were certainly present, since we observed minus end-directed organelle transport in all our preparations, there was no indication of plus end-directed MT gliding caused by cytoplasmic dynein. A similar situation is known from *in vitro* experiments in which both motors were simultaneously present (Lye *et al.* 1989).

The discrepancies in the pharmacological properties of nMT gliding and anterograde organelle transport may indicate different states of the motor. In the case of gliding the kinesin molecules are attached to the glass surface perhaps in an unspecific way and may even be partially denatured. In the case of vesicle transport the conformational state of kinesin may be well defined and determined by its interaction with specific receptors or perhaps with dynein residing nearby at the organelle membrane (compare Brady *et al.* 1990; Sheetz *et al.* this volume).

The study of nMTs can therefore be used to bridge the gap between the relatively different results obtained with intact axons and with purified motors. Moreover, it appears that knowledge of the detailed properties of nMT motility is necessary, in addition to the study of purified motors, in order to understand fully the molecular mechanism of MT-dependent organelle transport.

The support from NSF Grant BNS-9004526, NATO Research Grant CRG(30)0874/86 (GML), and DFG grant We 790/12 (DGW) is gratefully acknowledged. We thank W. Maile for his help in providing Figs 1 and 2.

References

ALLEN, R. D. AND WEISS, D. G. (1985). An experimental analysis of the mechanisms of fast axonal transport in the squid giant axon. In *Cell Motility: Mechanism and Regulation* (ed. H. Ishikawa, S. Hatano and H. Sato), vol. 10, pp. 327–333. Yamada Conference, Sept. 1984, Tokyo: University of Tokyo Press.

ALLEN, R. D., WEISS, D. G., HAYDEN, J. H., BROWN, D. T., FUJIWAKE, H. AND SIMPSON, M. (1985). Gliding movement of and bidirectional organelle transport along single native microtubules from squid axoplasm: evidence for an active role of microtubules in cytoplasmic transport. *J. Cell Biol.* **100**, 1736–1752.

BRADY, S. T., PFISTER, K. K. AND BLOOM, G. S. (1990). A monoclonal antibody against kinesin inhibits both anterograde and retrograde fast axonal transport in squid axoplasm. *Proc. natn. Acad. Sci. U.S.A.* **87**, 1061–1065.

BROKAW, C. J. (1986). Computer simulation of bend propagation by axoplasmic microtubules. *Cell Motil. Cytoskel.* **6**, 347–353.

COHN, S. A., INGOLD, A. L. AND SCHOLEY, J. M. (1987). Correlation between the ATPase and microtubule translocating activities of sea urchin kinesin. *Nature* **328**, 1160–1163.

FORMAN, D. S., BROWN, K. J. AND LIVENGOOD, D. R. (1983). Fast axonal transport in permeabilized lobster giant axon is inhibited by vanadate. *J. Neurosci.* **3**, 1279–1288.

GILBERT, S. P. AND SLOBODA, R. D. (1989). A squid dynein isoform promotes axoplasmic vesicle translocation. *J. Cell Biol.* **109**, 2379–2394.

KUZNETSOV, S. A. AND GELFAND, V. I. (1986). Bovine brain kinesin is a microtubule-activated ATPase. *Proc. natn. Acad. Sci. U.S.A.* **83**, 8530–8534.

LASEK, R. J. AND BRADY, S. T. (1985). Attachment of transported vesicles to microtubules in axoplasm is facilitated by AMP–PNP. *Nature* **316**, 645–647.

LEOPOLD, P. L., SNYDER, R., BLOOM, G. S. AND BRADY, S. T. (1990). Nucleotide specificity for the bidirectional transport of membrane-bounded organelles in isolated axoplasm. *Cell Motil. Cytoskel.* **15**, 210–219.

LYE, R. J., PFARR, C. M. AND PORTER, M. E. (1989). Cytoplasmic dynein and microtubule translocators. In *Cell Movement, Volume 2: Kinesin, Dynein, and Microtubule Dynamics* (ed. F. Warner and R. McIntosh), pp. 141–201. A. R. Liss, New York.

PASCHAL, B. M., SHPETNER, H. S. AND VALLEE, R. B. (1987). MAP 1C is a microtubule-activated ATPase which translocates microtubules *in vitro* and has dynein-like properties. *J. Cell Biol.* **105**, 1273–1283.

PASCHAL, B. M. AND VALLEE, R. B. (1987). Retrograde transport by the microtubule-associated protein MAP 1C. *Nature* **330**, 181–183.

PENNINGROTH, S. M., ROSE, P. M. AND PETERSON, D. D. (1987). Evidence that the 116 kDa component of kinesin binds and hydrolyzes ATP. *FEBS Lett.* **222**, 204–210.

PFISTER, K. K., WAGNER, M. C., BLOOM, G. S. AND BRADY, S. T. (1989). Modification of the microtubule-binding and ATPase activities of kinesin by *N*-Ethylmaleimide (NEM) suggests a role for sulfhydryls in fast axonal transport. *Biochemistry* **28**, 9006–9012.

PORTER, M. E., SCHOLEY, J. M., STEMPLE, D. L., VIGERS, G. P. A., VALE, R. D., SHEETZ, M. P. AND MCINTOSH, J. R. (1987). Characterization of the microtubule movement produced by sea urchin egg kinesin. *J. biol. Chem.* **262**, 2794–2802.

PRYER, N. K., WADSWORTH, P. AND SALMON, E. D. (1986). Polarized microtubule gliding and particle saltation produced by soluble factors from sea urchin eggs and embryos. *Cell Motil. Cytoskel.* **6**, 537–548.

SCHOLEY, J. M., PORTER, M. E., GRISSOM, P. M. AND MCINTOSH, J. R. (1985). Identification of kinesin in sea urchin eggs and evidence for its localization in the mitotic spindle. *Nature* **318**, 483–486.

SEITZ-TUTTER, D. (1990). Organellentransport und Eigenbewegung der Mikrotubuli im Axoplasma des Tintenfisch-Riesenaxons. Ph. D. Thesis, University of Munich.

SEITZ-TUTTER, D., LANGFORD, G. M. AND WEISS, D. G. (1988). Dynamic instability of native microtubules from squid axon is rare and independent of gliding and vesicle transport. *Expl Cell Res.* **178**, 504–511.

SHPETNER, H. S., PASCHAL, B. M. AND VALLEE, R. B. (1988). Characterization of the microtubule-activated ATPase of brain cytoplasmic dynein (MAP1C). *J. Cell Biol.* **107**, 1001–1009.

SMITH, R. S. (1988). Studies on the mechanism of the reversal of rapid organelle transport in myelinated axons of *Xenopus laevis*. *Cell Motil. Cytoskel.* **10**, 296–308.

VALE, R. D., REESE, T. S. AND SHEETZ, M. P. (1985*c*). Identification of a novel, force-generating protein, kinesin, involved in microtubule-based motility. *Cell* **42**, 39–50.

VALE, R. D., SCHNAPP, B. J., MITCHISON, T., STEUER, E., REESE, T. S. AND SHEETZ, M. P. (1985*b*). Different axoplasmic proteins generate movement in opposite directions along microtubules *in vitro*. *Cell* **43**, 623–632.

VALE, R. D., SCHNAPP, B. J., REESE, T. S. AND SHEETZ, M. P. (1985*a*). Organelle, bead and microtubule translocations promoted by soluble factors from the squid giant axon. *Cell* **40**, 559–569.

WAGNER, M. C., PFISTER, K. K., BLOOM, G. S. AND BRADY, S. T. (1989). Copurification of kinesin polypeptides with microtubule-stimulated

Mg-ATPase activity and kinetic analysis of enzymatic properties. *Cell Motil. Cytoskel.* **12**, 195–215.

Weiss, D. G., Langford, G. M., Seitz-Tutter, D. and Keller, F. (1988). Dynamic instability and motile events of native microtubules from squid axoplasm. *Cell Motil. Cytoskel.* **10**, 285–296.

Weiss, D. G., Maile, W. and Wick, R. A. (1989). Video microscopy. In *Light Microscopy in Biology. A Practical Approach* (ed. A. J. Lacey), pp. 221–278. IRL Press, Oxford.

Weiss, D. G., Meyer, M. A. and Langford, G. M. (1990). Studying axoplasmic transport using video microscopy and the squid giant axon. In *Squid as Experimental Animals* (ed. D. L. Gilbert, W. J. Adelman and J. Arnold), pp. 303–321. Plenum Press, New York.

Index

EMBO WORKSHOP ON MOTOR PROTEINS

Group photograph by Stuart Ingham

Robinson College, Cambridge, England   8th–13th September 1990

The Company of Biologists Limited is a non-profit-making organization whose directors are active professional biologists. The Company, which was founded in 1925, is the owner and publisher of this and *The Journal of Experimental Biology* and *Development* (formerly *Journal of Embryology and Experimental Morphology*).

Journal of Cell Science is devoted to the study of cell organization. Papers will be published dealing with the structure and function of plant and animal cells and their extracellular products, and with such topics as cell growth and division, cell movements and interactions, and cell genetics. Accounts of advances in the relevant techniques will also be published. Contributions concerned with morphogenesis at the cellular and sub-cellular level will be acceptable, as will studies of micro-organisms and viruses, in so far as they are relevant to an understanding of cell organization. Theoretical articles and occasional review articles will be published.

Subscriptions

Journal of Cell Science will be published 14 times in 1991 in the form of 3 volumes, each of 4 parts, and 2 Supplements. The subscription price of volumes 98, 99, 100 plus Supplements 14 and 15 is £500 (USA and Canada, US $870; Japan, £550) post free. Supplements may be purchased individually – prices on application to the Portland Press Ltd. Orders for 1991 may be sent to any bookseller or subscription agent, or to Portland Press Ltd, PO Box 32, Commerce Way, Colchester CO2 8HP, UK. Copies of the journal for subscribers in the USA and Canada are sent by air to New Jersey for delivery with the minimum delay.

Back numbers of the *Journal of Cell Science* may be ordered through Portland Press Ltd. This journal is the successor to the *Quarterly Journal of Microscopical Science*, back numbers of which are obtainable from Messrs William Dawson & Sons, Cannon House, Park Farm Road, Folkestone, Kent CT19 5EE, UK.

Copyright and reproduction

1. Authors may make copies of their own papers in this journal without seeking permission from The Company of Biologists Limited, *provided that such copies are for free distribution only: they must not be sold.*

2. Authors may re-use their own illustrations in other publications appearing under their own name, without seeking permission.

3. Specific permission will *not* be required for photocopying copyright material in the following circumstances.

(a) For private study, provided the copying is done by the person requiring its use, or by an employee of the institution to which he/she belongs, without charge beyond the actual cost of copying.

(b) For the production of multiple copies of such material, to be used for bona fide educational purposes, provided this is done by a member of the staff of the university, school or other comparable institution, for distribution without profit to student members of that institution, provided the copy is made from the original journal.

4. For all other matters relating to the reproduction of copyright material written application must be made to:

Dr R. J. Skaer,
Company Secretary,
The Company of Biologists Limited,
Department of Zoology,
Downing Street,
Cambridge CB2 3EJ,
UK.

Journal of Cell Science Supplements

No. 1	**Higher Order Structure in the Nucleus** Edited by P. R. Cook and R. A. Laskey ISBN: 0 9508709 4 3 234 pp. Proceedings of 1st BSCB–Company of Biologists (COB) Symposium	£12.00	U.S.$23.00	1984
No. 2	**The Cell Surface in Plant Growth and Development** Edited by K. Roberts, A. W. B. Johnston, C. W. Lloyd, P. Shaw and H. W. Woolhouse ISBN: 0 9508709 7 8 350 pp. The 6th John Innes Symposium	£15.00	U.S.$30.00	1985
No. 3	**Growth Factors: Structure and Function** Edited by C. R. Hopkins and R. C. Hughes ISBN: 0 9508709 9 4 242 pp. BSCB–COB Symposium	£15.00	U.S.$30.00	1985
No. 4	**Prospects in Cell Biology** Edited by A. V. Grimstone, Henry Harris and R. T. Johnson ISBN: 0 948601 01 9 458 pp. An essay volume to mark the journal's 20th anniversary	**SOLD OUT** ~~£15.00~~	~~U.S.$30.00~~	1986
No. 5	**The Cytoskeleton: Cell Function and Organization** Edited by C. W. Lloyd, J. S. Hyams and R. M. Warn ISBN: 0 948601 04 3 360 pp. BSCB–COB Symposium	**SOLD OUT** ~~£15.00~~	~~U.S.$30.00~~	1986
No. 6	**The Molecular Biology of DNA Repair** Edited by A. R. S. Collins, R. T. Johnson and J. M. Boyle ISBN: 0 948601 06 X 353 pp.	£40.00	U.S.$70.00	1987
No. 7	**Virus Replication and Genome Interactions** Edited by J. W. Davies *et al.* ISBN: 0 948601 10 8 350 pp. The 7th John Innes Symposium	£40.00	U.S.$70.00	1987
No. 8	**Cell Behaviour: Shape, Adhesion and Motility** Edited by J. Heaysman, A. Middleton and F. Watts ISBN: 0 948601 12 4 449 pp. BSCB–COB Symposium	£35.00	U.S.$60.00	1987
No. 9	**Macrophage Plasma Membrane Receptors: Structure and Function** Edited by S. Gordon ISBN: 0 948601 13 2 200 pp.	£29.00	U.S.$50.00	1988
No. 10	**Stem Cells** Edited by Brian I. Lord and T. Michael Dexter ISBN: 0 948601 16 7 280 pp.	£35.00	U.S.$65.00	1988
No. 11	**Protein Targeting** Edited by K. F. Chater, N. J. Brewin, R. Casey, K. Roberts, T. M. A. Wilson and R. B. Flavell ISBN: 0 948601 21 3 270 pp. The 8th John Innes Symposium	£35.00	U.S.$65.00	1989
No. 12	**The Cell Cycle** Edited by Robert Brooks, Peter Fantes, Tim Hunt and Denys Wheatley ISBN: 0 948601 23 X 300 pp. BSCB–Journal of Cell Science Symposium	£40.00	U.S.$60.00	1989
No. 13	**Growth Factors in Cell and Developmental Biology** Edited by M. D. Waterfield ISBN: 0 948601 27 2 210 pp. BSCB–Journal of Cell Science Symposium	£30.00	U.S.$55.00	1990

This series of supplementary casebound volumes deals with topics of outstanding interest to cell and molecular biologists

These are provided free to subscribers to *Journal of Cell Science*. They may be purchased separately from:

Portland Press Ltd, PO Box 32, Commerce Way, Colchester CO2 8HP, UK

Do It Yourself Life Plan Astrology

How Planetary Cycles Affect Your Whole Life

LYN BIRKBECK

ELEMENT

Shaftesbury, Dorset • Boston, Massachusetts • Melbourne, Victoria

First published in the UK in 2000 by
Element Books Limited
Shaftesbury, Dorset SP7 8BP

Published in the USA in 2000 by
Element Books, Inc.
160 North Washington Street
Boston, MA 02114

Published in Australia in 2000 by
Element Books and distributed
by Penguin Australia Limited
487 Maroondah Highway, Ringwood,
Victoria 3134

Front cover illustration © Stephen Sturgess
Cover design by Slatter-Anderson
Designed and Typeset by PPG Design & Print Limited
Printed and bound in Great Britain by Butler & Tanner, Frome

British Library Cataloguing in Publication
data available

Library of Congress Cataloging in Publication
data available

ISBN 1 86204 733 2

CONTENTS

To the Awakening of the Cosmic Nature of Humanity

'True astrology is not so much in aid of predicting events as understanding what those events mean when they happen or happened, or preparing yourself for certain possibilities'

from this book, page 94

HOW THIS BOOK WORKS

DO IT YOURSELF LIFE PLAN ASTROLOGY is multi-layered and its various parts are inter-connecting – just like life itself. Because of this you may wonder where to start and which way to go next – unless you are one of those people who always read a book from cover to cover. You are perfectly welcome to use the book any way you like, starting at the back if you wish, dipping in here and there, or using 'What *Do it Yourself Life Plan Astrology* Can Show You' on page 3. However, if you want a 'guided tour' there are three suggested 'Reading Routes' given on page 2.

Whatever way you use this book, make sure that sooner rather than later you read the general descriptions of all the Planetary influences that introduce each of the relevant sections. Not familiarizing yourself with these descriptions can be somewhat like trying to do a jigsaw without seeing the picture on the box!

Every so often throughout this book you'll see a symbol which is

'A Memo from Mercury'.

This points out useful and important information that enables you to get the best out of Do it Yourself Life Plan Astrology, *Mercury being the Messenger of the Planetary Gods.*

Most importantly, it helps to appreciate that this book has been used as a vehicle to convey an awareness that life, and your life in particular, is a 'cosmically natural' event that has an inherent path and pattern like the life cycle of anything – be that a flower, a frog or a human being. Because of this, by 'dipping in' or looking at your past and future, or looking at any time for anyone, you will see more and more of the bigger picture, thereby clarifying where you are now.

Choices of 'Reading Route'

First, begin by reading 'How This Book Works' on page 1, the Introduction on page 5 and 'Your Birth, the Start of all Planetary Cycles' on page 9, and then choose and use one of the following routes.

Route 'A'

1 Go to 'The Planetary Cycles of Your Life' beginning on page 21. Read these, Planet by Planet, including the Introduction for each. Look for any Planetary influences you were, are or will be under at any age of your life that interests you. Remember that Jupiter and the Moon are always active, while Saturn, Uranus, Neptune and Pluto are only active at certain ages depending on your month and year of birth. These are given in the tables beginning on page 289.

2 Go to 'Your Personal Astro-Life Plan' on page 243 to plot the key events of the whole of your life so far into the 'schedule' of all the Planetary influences occurring throughout that time. You'll find that this exercise really puts you 'on the map'.

3 In whatever order you like, read the remaining chapters 'Managing Your Planetary Cycles and Your Life' (page 257) and 'The Zodiacal Life-Stream' (page 283). Take in 'The Age Index' starting on page 235 as an extra, if you wish.

Route 'B'

1 Go to 'The Age Index' on page 235 to find out when all the various Planetary Cycles were active for you for any particular period, be it a year or two, or the whole of your life. You can then read up what they mean, one by one, by going to the page indicated for that influence. Alternatively you can enter them into 'Your Personal Astro-Life Plan' chart on page 243 and then read them all up together. It is recommended that you read the Introduction to the Planet whose influence you are concerned about, as this gives you a fuller and more accurate idea of what that influence means in your life.

2 Go to 'Your Personal Astro-Life Plan' on page 243 to plot the key events of the whole of your life so far into the 'schedule' of all the Planetary influences (which you may have entered previously in Step 1) occurring throughout that time. You'll find that this exercise really puts you 'on the map'.

3 In whatever order you like, read the remaining chapters 'Managing Your Planetary Cycles and Your Life' (page 257) and 'The Zodiacal Life-Stream' (page 283).

Route 'C'

Go to 'What *Do it Yourself Life Plan Astrology* Can Show You' on the next page, and take it from there.

What *Do it Yourself Life Plan Astrology* Can Show You

INTRODUCTION

WE ARE ALL CYCLICAL BEINGS. OUR LIVES are not only governed by diurnal (daily), biorhythmic, menstrual and seasonal cycles, but also by planetary ones – as this book will show you. When we are born, the Sun, Moon and Planets are positioned in the Signs of the Zodiac – a depiction of which is called your Birth Chart. The Planets (from hereon in, we will refer to the Sun, Moon and planets as simply the Planets, with a capital 'P') will then move on through the Signs, creating their respective cycles, and at certain times in your life make certain significant connections with those Planets in your Birth Chart (*see* figure 1), until they eventually return to the position they were in at birth.

Figure 1

Planet B (in the sky) at a given time in your life influencing Planet A (in your birth chart).

There are two basic types of Planetary Cycle that influence us as we proceed along our life path.

Universal Planetary Cycles

These are the subject of this book and are concerned with the significant periods within any one Planet's Cycle that occur at more or less the same ages for everyone. Put another way, a particular Planet in the sky is periodically seen to interact with the same Planet in your Birth Chart. (Figure 1 illustrates this with the Planet Saturn arriving at the half-way point in its cycle, in Opposition to where it was placed at birth). The most well known of the Universal Planetary Cycles is that of the Sun, the cycle we call a year. Its most significant personal point for nearly everybody is their birthday, when the Sun returns to where it was in the Zodiac when you were born. We attach particular significance to certain of those birthdays – for example, the age of 13 when we enter our teens, the legal ages for being able to vote, smoke, drink, drive or have sex, the 'big five-0' and retirement.

But the Universal Cycles which we look at in this book are those which reveal your life to be a story with a plot, chapters, highs and lows, stops and goes – and most importantly, meaning. Astrologically, these Planetary Cycles are those of Jupiter, Saturn, Uranus, Neptune, Pluto and what is called the Progressed Moon. The 84-year cycle of the Planet which governs astrology itself, Uranus, is the symbolic astrological life-span. The other Planets – Mercury, Venus and Mars – also have their cycles, but they are too short to be of practical use here. The Sun, as mentioned above, also has too short a cycle for the context of this book, but its significance in terms of Sun Signs is known to most people in the world. However, starting on page 283 I give a new perspective on the Sun as a cycle in terms of what I call the Zodiacal Life-Stream, and this shows you how your particular Sun Sign is part of a meaningful cycle.

Individual Planetary Cycles

These are to do with the influence that a Planet in the sky at a given time has upon any Planet in your Birth Chart *other* than the Planet that is doing the influencing. For example, at the time of writing, the Planet Pluto is travelling through the early part of Sagittarius. So this means that anybody with their Sun, Moon or any of their Planets placed in that part of Sagittarius will be under the powerful influence of Pluto. If you are not familiar with your Birth Chart and the Sign positions of your Moon and Planets, then I refer you to my books *Do it Yourself Astrology* and *Do it Yourself Relationship Astrology* (both Element Books) or any other astrology book that goes beyond mere Sun Signs.

By their very nature, Individual Planetary Cycles have to be worked out for each individual. At the back of the book there is a list of addresses to which you can write if you wish to discover your Individual Planetary Cycles and what they mean.

Your Template of Fate

Universal Planetary Cycles could be likened to your basic template of fate, a bit like a 'scheduled flight', which every human being is subject to. An anagram of the word Planets is a fascinating testament to this – SET PLAN. On the other hand, Individual Planetary Cycles can be viewed as the unique, personal excursions that you make within that basic 'flight plan'. Neither kind of cycle on its own can give you the full story of your life all of the time – you would need both for that. But the Universal Cycles have the advantage of being readily accessible to everyone, and this book exclusively gives them for the whole of your life. Furthermore, I have worded the descriptions of all these cycles so that you may use their influence to shed light upon what you are experiencing as the result of any Individual Planetary Cycles that might be happening at the same time, whether you are aware of them or not.

Transits and Progressions

Transits and Progressions are the astrological terms used to describe the ongoing influences of both Universal and Individual Planetary Cycles that occur from the beginning of your life until the end of it. Put simply, Transits refer to the actual Planets in the sky at any given moment and where they are in relationship to the Planets in your Birth Chart (which are called Natal Planets). Progressions, on the other hand, are worked out in a symbolic way and, like Transits, their significance lies in where they are at any time in relationship to your Birth Chart. All of the Universal Planetary Cycles used in this book are Transits except for the Moon, which is Progressed and already worked out for you.

Fate and Free Will

Possibly the most common question or criticism levelled at astrology is one to do with the perceived implication that our fates are predetermined. The above section 'Your Template of Fate' could be interpreted as stating this, but it is not. Ongoing Planetary influences can be likened to the life cycle of any living thing. A frog, for example, has a definite life cycle which goes from spawn to tadpole to frog, and within its life as a frog there are also certain other stages, mating being the most important of them.

Yet not all frogs have exactly the same lives – even though they are not deemed as having free will. Human beings, on the other hand, do have free will, which makes our life cycles a lot more complex. We still have those basic biological stages, but we are free to choose what to do with them and during them. This is somewhat like the weather's influence over us. It might be pouring with rain outside, but that does not mean that everyone will stay in. A person who had just had their hair done probably wouldn't go outside, but then a fisherman probably would. Furthermore, we humans, unlike animals, can wear protective clothing. Yet another way of looking at this can be demonstrated with a little 'astro-parable'. A man is told by an astrologer that if he goes outside today the sky will fall in on him. The man is so alarmed by this that he stays in bed – only to have the ceiling fall in on him.

The human picture gets even more complex when you consider how much our past experiences affect the way we think, feel and live in the present. In other words, our past experiences tend to constitute our fate – especially if you believe in reincarnation. And so we are very much creatures of both fate and free will, like a little boat on the sea. The little boat is your personality which is free to go where it chooses, but it is subject to the wind, current, tide and weather. You are also influenced by your predispositions and predilections, which have been predetermined by your voyage so far, and the intentions you set out with. Both these sets of influences are further influenced by whether or not you have a chart or a weather forecast to guide you. Astrology is the subject which is the study of all these things: the wind, current, tide and weather (your Transits and Progressions), your predispositions and predilections (your Natal Planets or personality), your voyage so far, and the intentions you set out with (your Transits and Progressions, *and* Your Natal Planets). The Transits and Progressions that are called Universal Planetary Cycles are the subject of this book. Natal Planets are the subject of my books *Do it Yourself Astrology* and *Do it Yourself Relationship Astrology* or any other astrology book concerned with your Birth Chart itself, as distinct from those influences that come to bear upon your Birth Chart (Transits and Progressions). In the context of the above nautical metaphor then, astrology is both your chart and your weather forecast – and thereby an indicator of both your fate *and* your free will. Consequently, it can help you travel more smoothly and securely, as well as further, and in a more informed way.

Chapter One

YOUR BIRTH, THE START OF ALL PLANETARY CYCLES

In the second section of this book, 'The Planetary Cycles of Your Life', beginning on page 21, I have set out the Universal Cycles, Planet by Planet, from the Cycle of Jupiter to that of Pluto, in chronological order. First of all, though, we look at that moment when all cycles begin – your birth, and the significant manner in which all the Planetary Cycles unfold from that point in time.

As I described in the Introduction, at the moment of your birth the Sun, Moon and all the Planets are positioned and seen to be in various parts of the sky relative to the Earth and your birth place. We also saw that the diagrammatic image of this is called your Birth Chart, and that the Sun, Moon and Planets will all be placed in Zodiacal Signs. Nearly everyone knows their Sun Sign, the Sign their Sun is placed in at birth, your Birth Sign, but your Moon and all your Planets also have a 'Birth Sign'.

Starting from their positions in your Birth Chart then, the Planets proceed along their courses as seen from the Earth, our point of view. Each Planet has a journey or cycle of its own, from the year that the Sun takes to the epic 248-year voyage of Pluto.

The Divine Playwright

'All the world's a stage' as Shakespeare so aptly put it, and we can liken the Planets in your Birth Chart to the players that perform that production called *Your Life*. The author of this play that is your life has some idea or vision of what the play is going to be about: its plot, what point it's going to make, the effect it will have upon the audience, etc. And as every playwright knows, the storyline is ultimately created by the characters in that play. Or as astrologers simply say, 'Character *is* destiny'.

Our birth is but a sleep and a forgetting:
The soul that rises with us, our life's star,
Hath had elsewhere its setting,
And cometh from afar

from 'Intimations of Immortality' from *Recollections of Early Childhood* by William Wordsworth

The author, who is whatever you regard as having created you in the first place, gives all these characters, the Planets, a role each to play. The astrological image of each individual role unfolding is symbolized by each Planet's own particular nature and Cycle.

However, there is a subtle distinction between the Player and the Planet. At birth they are, or appear to be, one and the same – much as the cast of a play states that so-and-so will be playing the part of such-and-such – and ideally a really good actor becomes the very part he or she is playing. The actor will even 'flesh out' the part and develop it even more than the author had in mind. In other words, the individual lives up to their potential.

On the other hand, the actor might not live up to the part they're playing. So at certain times the writer/director has to step forward and say, in so many words, 'Hey, that's not the part you are supposed to be playing' – telling the actor that they are not doing credit to the part, or conversely, that they are over-acting or ad-libbing too much. In astrological terms, this is precisely when a particular Planet arrives at a significant point in its Cycle, and asserts the need for a 'Course Correction'. Then again, there will also be times when the production is going well – enjoying a 'good run', as they say. Underlying all of this, remember, is the notion that every character has an energy and life of its own as well, thereby continually influencing, like the Planet, the overall plot – of your life, that is.

The Nature of Planetary Cycles

So what are the 'significant points' in a Planet's Cycle when a 'Course Correction' occurs or when there is a 'good run'?

The Course Corrections

The four *most* significant points of any Planet's cycle are when it reaches the quarter-way point (called the **Waxing Square**, which is at right angles to where it started), the half-way point (called the **Opposition**, directly opposite where it began), the three-quarter way point (the **Waning Square**, again at right angles to where it started), and finally, and most significant of all, when it arrives back to the position in the Zodiac it was at the time of birth (called the **Conjunction** or **Return**). A simple and very visible example of this four-part cycle is the phases of the Moon: New = (Conjunction to the Sun); First Quarter or Half Moon (right angles to the Sun) = Waxing Square; Full Moon = Opposition (to the Sun); Last Quarter or Half Moon (right angles to the Sun) = Waning Square; then Return to New and another Cycle.

Not all Planets will complete their Cycle in a lifetime, and Neptune and Pluto cannot ever do so because they take 165 years and 248 years respectively. One of the most important Planetary Cycles of all is that of Saturn, which is 29½ years long. This means that during its first orbit or cycle its Waxing Square occurs at around seven years of age, its Opposition at 14 to 15 years, its Waning Square at around 21 to 22, and its Conjunction or Return at 29½ years – at which time it then begins its second cycle. You can see from Saturn's Cycle where the idea of the 'seven-year itch' came from, and also the well-known significance of the times of the Opposition (onset of adolescence) and the Waning Square (traditionally, coming of age). The Saturn Return itself is not so well-known but it ought to be! More will be described later about all these Saturnian influences, but for the time being ask yourself (if you're old enough) what happened, ended or began, on or around your 30th year. Also, keep this age in mind whenever you read or hear the news, and you will notice that this age (29 to 30 – and the Second Saturn Return at 58 to 59) will crop up time and again regarding people going through some critical phase, either in a positive or negative sense.

It is also important to realize that Course Corrections are not always 'bad' in the sense that there is something dire happening. If you have been 'playing your part well' and living up to your potential, then you can find yourself being rewarded, going up in the world, learning something of value, or anything else which has a positive quality about it. But failing this, Course Corrections will hold you up or even regress you. It's all remarkably similar to the game of 'Snakes and Ladders', except that with the aid of astrology and this book you will be prepared and informed to deal with Course Corrections.

Throughout this book, to help you see how powerful any Planetary influence is, I have given them one of three ratings, denoted by various shadings:

Course Corrections are always of a Major or Medium influence.

With regard to these Course Corrections, it is important to emphasize, in this scientifically-led age of ours, that these four points are quite 'real' and not just astrological notions. It is connected with electro-magnetic and resonant fields, and if you wish to know more about this I refer you to Dr Percy Seymour's book *Astrology: The Evidence of Science*.

The Good Runs

There are two other significant points in a Planet's Cycle, but they are nowhere near as dynamic or crucial as the above Course Corrections, which is why their influence is classified as Mild. They are more flowing and harmonious in quality, and open the way but neither push you down it – nor block it. The first one is when a Planet reaches a third of the way through its Cycle (called the **Waxing Trine**) and the second when it gets two-thirds of the way through its Cycle (called the **Waning Trine**). It could also be said that the Return is a Good Run, as well as a Course Correction, but as you will see later, the nature of being human usually means that we go off-course more often than not! Taking the first Cycle of Saturn again as our example (0–29½ years), it would reach its first Waxing Trine at around 10 years, and arrive at its first Waning Trine at around 19 years. By and large, you can get the best out of Trines when you use them consciously, which is why when you look back at times of Trines they often do not seem to have been especially 'good'. Another reason why they may not have felt particularly positive is because some more difficult part of the Universal or Individual Cycle of another Planet could also have been prevailing, which would have been even harder without that Trine occurring simultaneously. In any event, I will be showing you how to get the most out of Trines, turn them into Medium or even Major influences, and into the advantages they potentially are.

Conflicting Influences

Sometimes you will find you are experiencing the influence of one Planet that contradicts the influence of another. When two influences are simply 'one helpful' and 'one challenging' then it is quite evident how one serves the other. However, sometimes you might have one influence saying 'do this' and another saying 'do the other'. I have tried to cater for this as much as possible, but in some cases it is part and parcel of going through a difficult patch. Conflicting influences are more likely between the influences of Jupiter and the Moon because they are continuous. For example, Jupiter might be saying 'let it be' while the Moon could be saying 'take decisive action'. So a possible response to this would be to take that decisive action and then allow matters to develop from there, without any further action until it was deemed necessary.

Timing your Cycles

Although Universal Planetary Cycles are regular and their significant points mostly occur at the same ages for everyone, there are some variations owing to what is called Retrograde Motion. Retrograde Motion is when a Planet *appears* to travel backwards for a while. This illusion is created by our living on a planet that is also moving in its own cycle – somewhat similar to two trains leaving a station at different speeds, so that to a passenger in the faster one the slower one seems to be going backwards.

With the Cycles of Uranus, Neptune and Pluto, these variations can be considerable, especially with Pluto as it has a very irregular orbit anyway. So there are some simple Tables (page 289), which I will refer you to every so often, that tell you at what time in your life you will personally experience the influence of any one of these three Outer Planets', depending on your month and year of birth. During your Saturn Cycle you may find that some Transits or influences occur some months, even a year, before or after the age given – but not both. With the Progressed Moon and Jupiter we do not have this problem because the Moon does not have Retrograde Motion as it goes straight around the Earth, whereas with Jupiter we are using the Jupiter 'Years', the meaning of which is explained on page 23. However, the Moon does have fluctuations in its cycle, so the influences of her 'Phases' can begin some weeks before or after the age given.

Your Birth is a Microcosm of Your Life

From the above, it could be understood that it is the more difficult and pressuresome Planetary influences that are most noticeable, and that easy passages are only there if you are aware of them. Most of us would agree, if we are being absolutely honest, that life is difficult much of the time – and the natural world certainly reflects this fight for survival. But it also has to be said that it is through attuning oneself to ongoing Planetary Cycles and influences that one is able to 'feel the force' more and more and get in flow with the greater current of one's own life and of life as a whole.

It is as if we come to this world to prove or manifest something, and, at birth, the first thing we prove is our right to exist, and the first thing we manifest is a separate physical body. Birth is a painful and anxious process, yet it can of course also be joyful and exhilarating. This is why a Birth Chart is the most significant entity in astrology. At that moment is registered the joy and the pain of the life to be, and in effect every day is a 'birth' as we yet again emerge into the world from out of the 'sleep and forgetting' of our night's rest. Insomnia, by the way, is often caused by that anxiety about being 'born'.

Furthermore, the actual nature of our birth itself can be seen to encapsulate the nature of our life to come. An extreme example of this was of a client of mine who was always very afraid of not surviving, and had lived for 66 years 'in the shadow of death'. On studying her Birth Chart, I saw that doom-laden Pluto had been cast in the role of making the first appearance upon the stage of her life. Upon this (and what it might mean) being pointed out, she at last 'remembered' that she had been born almost dead with the umbilical cord wrapped three times around her neck! Interestingly, at the time of her consulting me, Pluto itself was Trining that Pluto (*see* page 231) in her Birth Chart; Trines often iron out or clarify problems associated with the Planet in question.

So if you investigate and look into the details of your own birth, you will see how much it has affected your life as a whole. And as the Planetary Cycles unfold, you will see how, like a bud blossoms into a flower, you can bloom into what you potentially are, stage by stage. Furthermore, your life in your mother's womb – that is, what was happening to you as an embryo or in your parents' lives at that time – is also full of clues about the life to come. And again, the widespread complaint of chronic insomnia can be indicative of anxieties felt by your parents during gestation, or possibly because of some deeper karmic reason. It is as if the embryo picks this up and interprets it as life ahead being fraught with difficulties. Another example of womb-life that comes to mind is a client who was beset with inexplicable feelings of guilt. Upon inquiring into what was happening while in the womb, he discovered that he was one of twins, his younger brother being born dead. His parents were told by the doctor who delivered him that it was as if he had 'taken all the sustenance and left none for his brother'. (*See* also 'First Emerging Moon Phase' on page 49 for more about your birth's effect on your life to come.)

No One Player Makes a Whole Play; No One Planet Makes a Whole Life

Last but by no means least, it is important to see that any one Planetary Cycle is not happening on its own, but in concert with all the others, be it concurrently or consecutively. This is particularly true with regard to what Moon 'Phase' you are going through, and also with regard to the combination of Jupiter's and Saturn's Cycles. These two form a pair in that Jupiter is Growth and Saturn is Status – the one developing and the other maintaining – a bit like director and producer. So whatever Saturn, or for that matter any other Planet, is doing, look to the year in your Jupiter Cycle that you are currently in, for this will show you what positive and expansive principle of Growth and Understanding, what 'Growth

Mode', is occurring at that time (*see* page 25). There is also further information on what I call the Jupiter–Saturn balance of power on page 265.

All of this has given you some idea of how the 'cast of characters', the Planets, play and interplay on the stage of your life, but we now need to examine the Play itself.

The 'Story' of Your Life

After some time studying and using astrology, it becomes evident that this wonderful tool for looking at life is asking you a singularly important question: what is the actual nature of this thing called 'life' that you are looking at from an astrological viewpoint? More particularly it asks: what is the basic nature and purpose of this creature, the human being, who is living this 'life', where did it come from, and where is it going?

In order to answer these questions in a way that is generally appreciable and useable, and not merely some scientific theory or abstract philosophy, I have found that it is best put in the form of a metaphor for life.

In the last chapter we saw how we can view human life as in Jaques' metaphor in *As You Like It* by Shakespeare – that 'All the world's a stage' – and apply it to the meaning of the Planets, firstly as they symbolize the various parts of your personality, and subsequently as the various ongoing influences throughout your life. However, this metaphor only takes us so far, merely describing the journey in terms of the cast of characters and the fortunes of a theatrical production, rather than in terms of the 'Play' or 'Story' itself. So it is here we introduce the second metaphor, that also serves as a title for this Play.

A Voyage Home Via The Unknown

We can now view the Planets in terms of the main features of the story that is Life. At the start of each of the chapters on the individual Planets in the second part of this book, entitled 'The Planetary Cycles of Your Life', these features will be explored more in the context of you and your own individual life. But as you will see, the individual's life and Life itself are ultimately one and the same.

Jupiter

This is the Joy of the Journey itself. Jupiter is the meaning and experience of travelling, and as such it says 'Life is what you make it!'. Jupiter tells you how to enjoy life, how to grow and profit from it. It is also the Way or Path that you are travelling along, which suggests that there is some Philosophy, Belief or Law to go by. By the same token, because the 'stone' you are standing on is ultimately connected to all the 'stones' ahead and behind (like the Yellow Brick Road), this way contains a Prophecy of what is to come. Jupiter *is* the Prophet. And here's a little Word Magic – 'journey' is an anagram of 'joy' and 'rune', a rune being one of a number of stones or bones used to shed light on what is happening and where events are leading.

Jupiter's ongoing influence, as described on page 25, therefore tells you how to maintain a feeling that you are going somewhere, that there is a meaning to your life, that each stage of the journey is rewarding and encouraging according to the nature of that stage.

The Moon

This is the Traveller on his or her way Home, with all the feelings and memories associated with where you have come from, and the needs, fears and longings relating to where you are bound – which in fact have more to do with your past. The Moon therefore suggests that you left Home once upon a time, and are either still outward bound, or are on what is called your 'Return Path'. All this begs the question of what the word 'Home' means to you. It could be your family home, your place of birth, whatever you regard as a place of safety, where you go to after being out or at work, or the place and people you care about and care about you. Then again, it may be something more abstract or imaginary, like somewhere you long to be and which is definitely not where you are right now, or conversely, simply feeling at home with yourself. Viewed more spiritually, it could be the place from whence your Soul originally came. Your Soul on its Journey then got lost, went astray, until at last it had a strong feeling – possibly through meeting someone you felt you already knew – that there *is* a Way Home, and that you have to find it. Perhaps this is another meaning of the expression 'Home is where your heart is'.

Whatever the case, the ongoing influence of the Moon carries for you the feelings of being 'near or far from home' at certain points along your way. With her 'Phases', like waves, bearing you back and forth with their ebb and flow, such emotions can therefore mean anything from feeling you are genuinely where you belong, to feeling you are having an identity crisis; from feeling safe and secure with whatever or whoever is family to you, to feeling that those same places and people do not really know or have much to do with the real you, and are holding you back.

At some point on your Lunar Voyage you may realize that Home for you is something other than what you had been led to believe, that Home is a feeling inside you which has its origins in the deepest and darkest recesses of your Soul. It is at this point that you would begin to take that Return Path, that path which would have you realize that everyone and everything in your life was entirely a reflection of your own story so far. So, at such a point, you have progressively to take responsibility for your Soul's state, which includes making any necessary 'repairs' to it and taking stock of how others have been affected by it, and are a reflection of it, for good or ill. The Moon governs both memory and mirrors, and the Return Path is ultimately where astrology will lead you. It is the travelling of the Return Path that blesses your Soul, simply because you have remembered, or are in the process of remembering, who you truly are. Your Lunar Cycle serves to remind you of where you emotionally stand through putting you in touch with where you came from, familiarizing you with what 'Phase' you are going through, and beckoning you ever on towards your Home.

Saturn

This is Ground Control. Saturn can mean the 'real world' in the sense of seeing life as being only what it physically appears to be. As such, we are controlled by the conditions of the status quo, and limited by what we see ourselves to be in the physical, financial or political circumstances in which we find ourselves. Alternatively – or eventually – it can come to mean a place where you have established yourself firmly enough in order to use it as a 'base camp' from which to explore beyond the 'known', 'conventional' or 'normal' world and discover a life and reality that is exciting, new and different. However, Saturn still has the 'control' because it is that 'consensus reality' which virtually everyone agrees on as being that 'real world'. This means to say that whatever new ideas or inventions you might get or be given, they have to conform to what is currently regarded as practical and acceptable. For example, no-one would have wanted a can opener before cans were invented, would they? The same rule would apply to your emotional condition. Although the 'real world' allows you a certain amount of privacy to be what you want to be, there can come a point where one has to measure up and grow up. Saturn is the adult world in that childish emotions are not tolerated for long unless you actually are a child – and maybe not even then, it would seem. Saturn is the 'hard, cold, cruel world' into which our little Lunar selves are thrown. Indeed, the restrictive and 'only what you can touch and see is real' side of Saturnian reality is the *only* reality for some, or arguably most, people.

So, Saturn is Ground Control, either in the sense of being the material conditions and conventions that limit and control our lives, but to a certain extent maintain our stability, or in the sense of having one's feet planted firmly enough on the ground to be able to take a step beyond, into the Unknown, that which will one day be the 'real world'. In both cases, however, it is a place of learning. Also, there are many shades between these two extremes, one being the person who is firmly established in the 'real world' but is sick or bored with it, yet still finds it hard to get out of it because it is the 'devil they know' (Saturn = Satan).

Uranus

This is the Rocket Ship that launches and carries you into the Unknown, beyond the emotional constraints of the Moon and the physical limitations of Saturn. Uranus first 'ignites' itself in your life whenever you initially feel that there is 'something else' beyond what the Lunar/Saturnian world is presenting you with. Usually, such a feeling is characterized by a sense of restlessness, alienation and/or rebelliousness – all of which are symptomatic of one thing: the desire to liberate yourself from whatever appears to be holding you down or back. Restlessness is basically a feeling that there is something you ought to know or be doing and feeling – the cosmic itch. Feeling alienated is more likely to be a negative state because by its very nature it precludes support for one's unusual ideas or feelings from others (like friends or teachers). As such, your Uranian 'thrust' can become inverted – that is, neurotic or even present itself as some complaint like stuttering, hyperactivity, epilepsy or autism. Hopefully, such an 'abort mission' will get a second chance later on in life. When rebelliousness characterizes your sense of Uranian thrust, you probably have the inherent will, intelligence and sense of your own uniqueness that will find a 'launch window'. Then again it could launch you 'outside' rather than into the beyond – meaning you become an outsider and possibly an outlaw, rather than actually being free to be yourself and free from what is truly imprisoning you. Although there exist true liberators who fight for the rights of anything from humans to trees, from children to animals, all too often rebels are more Lunar than Uranian in that they are hurt innocents rather than behaving as genuine socio-political activists. In a large number of cases, Uranian thrust only shows itself momentarily in the growing being, for they swiftly revert to being one of the safe and conventional mass – the silent majority. Yet again, later in life – under the influence of Uranus itself – they might again try to relaunch themselves, to rediscover their specialness and freedom. When all is said and done, Uranus makes you *aware* by taking you *unawares*, for shock or surprise are the heralds of change. And change is, after all, what Uranus is all about. However, you will still need to be in touch with Ground Control, that practical sense needed for the 'real world'.

Whenever you do consciously lift off from your launch-pad of the Known, Uranus takes you up to a viewpoint from which you can see life on Earth from a different perspective. It does this through such means as technology, psychology, astrology, mythology, or any 'ology' which streamlines, elevates or revolutionizes your way of seeing and being.

Sooner or later everyone has to encounter the Unknown, whether they do so deliberately or accidentally. You then find out that the Unknown is something which is simply not yet known, the future in the making, the invention waiting to be invented, the 'you' waiting to be awakened, the truth waiting to be found, the realm waiting to be explored, the emotions needing to be liberated, a right waiting to be asserted, or anything that sets you or others free.

Neptune

The Sea of Mystery. If we view the story so far in a global context, with a little imagination we can begin to appreciate that between the various expressions and interpretations of Jupiter, Moon, Saturn and Uranus an awful lot of problems and differences have arisen. For example, the Way of Jupiter could have got set down by one group of people as a system of Laws and Beliefs that intruded upon the Lunar security and customs of another group. With such different Saturnian realities and conventions, they then build a wall between each other. Through time, this pattern then gets repeated by other groups everywhere until we have a lot of groups who don't see eye to eye, giving rise to war and strife, not to mention a 'have and have not' situation which ultimately threatens the stability of the whole planet. To add to the instability, there would then be Uranian uprisings in certain groups in support of certain 'visionaries' or other groups as a result of them feeling trapped by their own respective Saturnian systems.

The story so far is that of Humanity, and as an individual you cannot easily exclude yourself from it and all the misinformation and disinformation that has been generated by it. This is the negative side of the Neptunian Sea of Mystery (or Sea Mist, more like) – which has been called the Grand Illusion. It is called this because we are all infected, to one degree or another, with someone else's similarly confused version of what is going on. This is mostly a case of the blind leading the blind.

When you are still dwelling exclusively in the realms of Lunar racial and familial security, or of Saturnian conventionality and governmental system, then the Grand Illusion is not that clearly perceived, simply because you're in it! When, however, you have lifted off in your Uranian Rocket Ship, you see the whole scene with an increasingly clearer eye. Although Uranian blows for freedom usually start out as something entirely selfish or unconscious, in the end such a desire for truth and liberation is in aid of piercing the veil of that Grand Illusion, or dispelling that Sea Mist, so that the One People that we are can again be seen for the reality it originally was.

As you seek and sail upon the Neptunian Sea of Mystery, you will be pulled this way and that by the swell of political and religious propaganda on the one side, and incoming tide of genuinely spiritual, healing and creative inspiration on the other. After many lifetimes or 'days at sea' we will all ultimately 'cross over', depart from the Grand Illusion, and return to the 'Heaven which is our Home' (Wordsworth). In addition to all of this, there is another dimension to your Journey that cannot be avoided, for it is the call of ...

Pluto

Your Destiny. Someone once said that when you discover where you are really supposed to be, you find that you have been there all along. This is the link between the Moon and Pluto, between your Soul and your Destiny. What calls from your personal unconscious as a subjective longing for getting back Home is really a call to discover the ultimate reason for your being in existence at all. And remember, the ultimate reason for your being is to get back Home again.

The important point is that in taking this circular journey we become a better and better expression of Life itself, a Voice in the Wilderness, a Light in the Darkness, a Contrail of Significance. And when we have completed one round, we do it again, but at a deeper level, taking us one layer nearer towards the nucleus of our being. Pursuing your Destiny is like burrowing into your own centre, wherein dwells the most condensed and convinced you. But this journey through Pluto's Underworld can be gloomy, doomy and scary, as it purges you of superficial or lesser thoughts and feelings. To become more concentrated, you have to concentrate upon what are your deepest feelings and convictions, via your passions and obsessions.

So Pluto is Destiny in the sense of a compulsion to unite, or reunite, with whomever or whatever is genuinely one's own. Ultimately, after many, many rounds, we will all realize and feel we belong to the One Core, which means One Heart, in the same way that the Earth upon which we live has only one core, one heart.

Here's a Synopsis of this Planetary Plot, this Story of Your Life

Planet	Metaphor	Meaning
JUPITER	*The Journey*	The outward- and homeward-bound Voyage. A Prophecy of the way ahead. Fellow travellers. The Rules of the Road. The Vision that beckons.
MOON	*The Traveller Home*	The Soul, the Hero/Heroine, who sets out from Home, seeking Experience. One day pulled back to from whence he or she came, which means being torn away from what he or she has grown accustomed to. What amounts to a two-way pull Home.
SATURN	*Ground Control*	The 'Real World' where the drama apparently takes place. Can become a prison where everyone has forgotten who they truly are or has given up hope of finding out (the Rat Race or System), or the test-bed and launch-pad for the Rocket Ship.
URANUS	*The Rocket Ship*	The Process of Individuation, through which the Soul at first dimly intuits its true identity. It then at last breaks free from the Rat Race, and strikes out on the Journey Home – the Return Path.
NEPTUNE	*The Sea of Mystery*	Encountering the Illusions which got the Soul lost in the first place, and that it can still get you lost in. The Sublime Release that calls you forever on. The 'Paradise Lost' that you want to regain.
PLUTO	*Destiny*	The hunger of the Soul that tugs and drags it deeper and deeper into an intimacy with whatever and whoever connects it with its most genuine feelings, its only homing instinct. The 'Dark Night of Soul' that purges the Soul of everything that it cannot take with it on its Journey Home. This genuine inner truth is your true personal Power, jealously guarded by your own Fear of Death, which is the Great Unknown.

The Wave, the Wave the Dolphin rides –
Therein lies your destiny.
Nearer than the Moon, yet
As far as you can see –
Therein lies your Destiny.

Chapter Two

THE PLANETARY CYCLES OF

YOUR JUPITER CYCLE

Growth and Understanding

Traveller's Joy! Traveller's Joy!
I see your Star above me
Shine on down and love me!
Traveller's Joy! Traveller's Joy!

Jupiter can be regarded as the Director of your Life's play. As such, it knows your 'story' inside-out, what it means and where it's going. This is why Jupiter is the key to understanding what your life, and life in general, are about. It will also expand upon the original story, that is, your basic potential and fate. In fact, its prime aim is to make you 'more than you are or were' – this is why Jupiter is the Planet of Growth. Jupiter also teaches you how to be philosophical about life (whereas Saturn makes it a necessity – as you will see later!).

Although, appropriately enough, it is the largest of all the Planets, Jupiter's cycle is the shortest of all the Planetary Cycles discussed in this book. It takes approximately 12 years to go around the Sun – or around *us* from the Earth's point of view. Because of the relative shortness of its cycle, dividing it into those points (Waxing Square, etc) we use with the other Planets is not so appropriate, simply because such influences are usually too fleeting to be noticed. The 'Director' would leave such minor details to its underlings!

Instead, we look at Jupiter's Cycle in a way that is far more suitable to its role as Director, the one with a good grasp of the story being told – your Life. This is neatly done by taking each year of its 12-year cycle and making it into a 'chapter' ascribed to one of each of the 12 Signs of the Zodiac. And so each year of your life can be understood in terms of developing according to the qualities of a particular Sign, through challenge, opportunity or sheer good fortune. I call these Sign qualities of your Jupiterian process of Growth and Understanding 'Growth Modes'. But do not confuse them with Sun Signs or Star Signs (or with the Sign which Jupiter is actually placed in at birth or transitting at any given time) – which is why I place each Sign name in inverted commas.

However, you may well find that you come into your own during Jupiter years that are the same Sign as your Sun, Moon or Rising Sign (*see* my books *Do it Yourself Astrology* and *Do it Yourself Relationship Astrology*, to discover what these are and mean). Also, by the same token, you will see that the meanings given in the Zodiacal Life-Stream on page 283 are applicable to your Jupiterian years, and you will be able to see each of those years in the context of the whole story or cycle. Incidentally, there is an astronomical link between the Sun and Jupiter, for it is the only Planet that generates more energy than it receives from the Sun itself. So in a way it is like a second Sun.

So, whatever Planetary influences are occurring at any particular age, look to your Jupiterian Year or 'Sign' because its Growth Mode will give you positive ways of looking at and developing through that time. The qualities of Jupiter's 'Sign' will also contribute to the various scenes and characters through which that year will be acted out. For example, you may well find that the Sign qualities (Sun, Moon or Rising Sign) of significant others correspond to the Jupiter Year that you became involved with them or parted with them.

Here follow all the Jupiter Years, headed by the actual ages they occur on the left, and the 'Sign' name of those years on the right. Don't expect each year to begin and end exactly on your birthday – give a week or so 'cusp' either way. With each Jupiter Year interpretation, I give examples of what each individual age of that year could mean, but please note that these are not the only possible manifestations for those ages. Finally, each year description ends with a 'Go-Zone' guide to fruitful areas of appreciation and understanding. And Jupiter being Jupiter – that is, wisdom – Go-Zone information for any Year is useful at any time. You can either flip through the pages of this section and look for the age in which you're interested, or use the following table as an index.

Incidentally, for the meaning of the glyph of Jupiter (♃), *see*, 'The Jupiter–Saturn Balance of Power' on page 265.

Jupiter Years

Age	'Sign' Year	Page	Age	'Sign' Year	Page	Age	'Sign' Year	Page	Age	'Sign' Year	Page
0	'Aries'	27	**22**	'Aquarius'	37	**44**	'Sagittarius'	35	**66**	'Libra'	33
1	'Taurus'	28	**23**	'Pisces'	38	**45**	'Capricorn'	36	**67**	'Scorpio'	34
2	'Gemini'	29	**24**	'Aries'	27	**46**	'Aquarius'	37	**68**	'Sagittarius'	35
3	'Cancer'	30	**25**	'Taurus'	28	**47**	'Pisces'	38	**69**	'Capricorn'	36
4	'Leo'	31	**26**	'Gemini'	29	**48**	'Aries'	27	**70**	'Aquarius'	37
5	'Virgo'	32	**27**	'Cancer'	30	**49**	'Taurus'	28	**71**	'Pisces'	38
6	'Libra'	33	**28**	'Leo'	31	**50**	'Gemini'	29	**72**	'Aries'	27
7	'Scorpio'	34	**29**	'Virgo'	32	**51**	'Cancer'	30	**73**	'Taurus'	28
8	'Sagittarius'	35	**30**	'Libra'	33	**52**	'Leo'	31	**74**	'Gemini'	29
9	'Capricorn'	36	**31**	'Scorpio'	34	**53**	'Virgo'	32	**75**	'Cancer'	30
10	'Aquarius'	37	**32**	'Sagittarius'	35	**54**	'Libra'	33	**76**	'Leo'	31
11	'Pisces'	38	**33**	'Capricorn'	36	**55**	'Scorpio'	34	**77**	'Virgo'	32
12	'Aries'	27	**34**	'Aquarius'	37	**56**	'Sagittarius'	35	**78**	'Libra'	33
13	'Taurus'	28	**35**	'Pisces'	38	**57**	'Capricorn'	36	**79**	'Scorpio'	34
14	'Gemini'	29	**36**	'Aries'	27	**58**	'Aquarius'	37	**80**	'Sagittarius'	35
15	'Cancer'	30	**37**	'Taurus'	28	**59**	'Pisces'	38	**81**	'Capricorn'	36
16	'Leo'	31	**38**	'Gemini'	29	**60**	'Aries'	27	**82**	'Aquarius'	37
17	'Virgo'	32	**39**	'Cancer'	30	**61**	'Taurus'	28	**83**	'Pisces'	38
18	'Libra'	33	**40**	'Leo'	31	**62**	'Gemini'	29	**84**	'Aries'	27
19	'Scorpio'	34	**41**	'Virgo'	32	**63**	'Cancer'	30			
20	'Sagittarius'	35	**42**	'Libra'	33	**64**	'Leo'	31			
21	'Capricorn'	36	**43**	'Scorpio'	34	**65**	'Virgo'	32			

'ARIES' YEARS — AGES: 0 12 24 36 48 60 and 72

'GROWTH IN BUD' – These are the years when each 12-year Jupiter Cycle of Growth begins, and this is highly significant. The reason why the beginning of this, or any Planetary Cycle, is so significant is because it means that the nature and extent of the objective you have, and effort you put into setting out to achieve it, totally determine how successful you are going to be in attaining that objective. Such years are similar to Spring when the sap is rising and everything is bursting forth for a new cycle of growth. Like any good farmer or gardener, you plant your seeds now in order to maximize growth through utilizing this important time. Included in all of this is what you have left over from the previous Jupiter Cycle (and particularly the last 'Pisces' Year), because 'good seed' (ideas, experience and minimal claims upon you from the past) will obviously promise better 'fruit' during this cycle, whereas 'bad seed' (debts, preoccupations, doubt and little or no experience) are likely to not produce much, slow you down, and complicate matters.

For understanding the significance of year zero or birth, I refer you back to the last chapter 'Birth, the Start of All Planetary Cycles' and to the next chapter, 'Your Moon Cycle'. At the age of 12, some new departure is very likely (at school) which sets the scene for your (educational) progress for the remainder of this cycle, and possibly others too. The age of 24, along with all the other 'Aries' Years to come, offers you the chance to 'start again' or 'muster' yourself with respect to whatever you feel never got properly off the ground hitherto, or simply to launch something brand new.

Looking back at the 'Aries' Years of your life so far, you are bound to find there the 'Spring' of something. You might find that you started something beforehand at half-cock, only to do it properly during an 'Aries' Year. I was forced to decide to become a professional astrologer during an 'Aquarius' Year and 'Deciding Moon Phase', only to find that I hadn't really got my act together until two years later. I was not aware of Jupiter Years at that time, but when you are, it makes a great deal of difference to the energy and focus you have available for the launch of a new project.

GO-ZONES for your 'Aries' Jupiter Years

Areas of encouragement, growth and wisdom – and how to make the most of them.

GO-ZONE	APPRECIATE AND UNDERSTAND:
Action	That the time has come to walk your talk, rather than just keep things at the ideas or intention stage. That action really does speak louder than words.
Initiative	How, as described above, these Years are times to launch any project, which would also include any private or secret undertakings.
Leadership	That if you wish to step out in front and take a leading position in the field of your choice, you will do so more effectively now than most other times.
Presentation	That you now have a natural sense of your own personal presence, at least relatively speaking, and so are more likely to make a good impression. This is 'selling yourself' time! Overdoing it could have the reverse effect though.
Self-assertion	That this is a great time to improve your powers of assertion, either in the sense of making an important statement to a significant other in your life, or in the form of making an affirmation to yourself regarding some facet of your personality that you want to be more confident about.
Championing	If there is any cause or person that you want to uphold or win over then your are in the right time for it, but make sure you are in the right place too – that is, it is for the right reasons and you are not just being gung-ho.

'TAURUS' YEARS — AGES: 1 13 25 37 49 61 and 73

'GROWTH IN STABILITY' – Following 'Aries' Years as they do, these 'Taurus' Years see to it that whatever you then set in motion is now stable, in the sense of being physically productive, profitable or realizable. They basically make it clear what you have going for you – and depending on what that is, you are liable to attract something or someone of value to you. They can also be times when you become more than usually aware of your body's state and needs with regard to sustenance and/or sensuality. 'Taurus' Years are ideally when you are able to stand, four-square, on the earth – stand your ground, in fact. Making an effort in this respect is positive even if you do not manage to hold your position, or if you overdo it and find yourself stuck. Money is rather more essential to continued growth than at other times, and is either forthcoming or a lack of funds might force you to reconsider or cut back. The Jupiterian point here is that you can improve your material position one way or the other, either through attracting funds/making a profit or learning how to adapt your 'product' to suit the 'market' (*see* Self-Worth below).

At one year old, you are discovering your body in a quite literal sense (walking, taste, etc); furthermore, this age used to be critical in terms of physical survival before modern healthcare. The age of 13 coincides biologically with the state of puberty, sexual interest and the ability to reproduce – and astrologically with the 'Emotional Discovery' of your 'First Realizing Moon Phase' (*see* page 75). So what you have going for you comes into sharp focus at this time. The age of 25 is also a time when some kind of sexual realization can occur – be it having a baby or enjoying hitherto unknown pleasures of the flesh. In any event, in true 'Taurus' Year fashion, you discover things by either having or not having. The other ages do not appear to have anything intrinsically significant about them – probably because money and physical pleasure are perennial issues for us all. It is just that they will find focus and a chance for advancement, or indulgence, during these 'Taurus' Years.

GO-ZONES for your 'Taurus' Jupiter Years

Areas of encouragement, growth and wisdom – and how to make the most of them.

GO-ZONE	APPRECIATE AND UNDERSTAND:
Sensuality	That being in touch with your five senses can make you glad to be alive. Involvement with your own or another's body, or Nature herself, is 'go' now.
Steadiness	That now you are 'going steady' or, at least, there is a potential for steadiness with whatever or whomever you're involved. This is a time when it can be useful to maintain a 'steady as she goes' attitude, and not allow yourself to be fazed by the up/down or on/off nature of life. With Jupiter there is always the possibility of excess which, in this case, would amount to being 'too steady', that is, stuck in the mud of some safe but boring job or relationship or belief. *See* Self-worth below, for such is the issue behind your being in a situation that is doing you no credit.
Self-worth	That, like it or not, the most essential issue for anyone in today's world is that of what they are worth in terms of physical attractiveness, talent and earning power. Now is the time and opportunity to increase, enjoy or realize what you have going for you in any or all of these respects.
Productivity	Having something to show for yourself is what makes you feel that you are worth something inside. Productivity can mean the obvious in terms of some saleable item you've made or helped to make, or a baby or piece of artwork, or in the sense of a fruitful relationship or activity.

'GEMINI' YEARS — AGES: 2 14 26 38 50 62 and 74

'GROWTH IN COMMUNICATION' – These Jupiter Years are highly significant in that they are times when we reach out to our environment out of curiosity, eagerness for contact and response, or to simply feel the contrast between 'me' and 'not me'. This is the famous Geminian duality that is part of everyone's world, not just Geminis. This all begins at the age of two. In fact this time is known as the 'terrible twos' because this is when a child is mobile and into everything, and verbal and asking everything. The child then splits its reality further still by classifying reality into two parts: people and things who interact with it, on the one hand, and those that do not, on the other. This Geminian splitting occurs again at a more sophisticated level at the age of 14, which coincides with the First Saturn Opposition, 'Me *v* the Rest' (*see* page 111). We can see how adolescent awkwardness is characterized by the inclination to divide the world into 'us' and 'them' in order to create a sense of identity in the face of a million things that appear to oppose it. It can be seen that at both these stages of the Jupiterian Growth Process, the individual can take a step toward integration through interaction, or alienation through a lack of it.

Subsequent ages are also marked by the inclination to feel divided within oneself, in the sense of being in two minds, or confronted equally by two external opposing forces that represent that split within one's own psyche. Apart from being caught in a 'valley of decision', there can also be the need to split off from whatever or whoever one is involved with. These 'schizophrenic' years do not seem very Jupiterian or advantageous. The reason for this is that Jupiter, the Planetary energy that likes to organize things and people into wholes, is not very at home in a Sign that wants to divide them into two (this is because Gemini is opposite to Sagittarius, the Sign Jupiter rules). The growth that takes place is, as the heading above states, one of Communication, because it is only through getting to know and intelligently interacting with one's opposite number that progress can be made. So whether it takes place between yourself and another, or between one part of yourself and another, or between your mind and some other source of intelligence, all such communication goes towards the growth of one's contacts and information banks.

GO-ZONES for your 'Gemini' Jupiter Years

Areas of encouragement, growth and wisdom – and how to make the most of them.

GO-ZONE	APPRECIATE AND UNDERSTAND:
Levity	How a light touch works wonders now – while not taking things *too* lightly! That being too heavy or earnest definitely alienates and misses the point.
Humour	That a sense of humour is a sense of proportion; being able to see the funny side of things is actually a brand of wisdom for it diminishes self-importance, thereby allowing you and others to see things in a clearer and healthier light. 'Laugh and world laughs with you, weep and you weep alone' – though beware excessive irreverence and sarcasm that is mocking born of cynicism.
Eloquence	How the ability to make yourself clearly understood now attracts good fortune both socially and professionally, as distinct from gossiping and chattering, which if not avoided will find you *being* avoided.
Knowledge	That indeed, 'knowledge *is* power' – especially in the sense of knowing that there are two sides to everything and everyone. Being able now to see life 'stereoscopically' is very balancing and promotes healthy relationships. That an informed mind is better able to find work and a place in the world.

'CANCER' YEARS AGES: 3 15 27 39 51 63 and 75

'GROWTH IN NURTURING' – These are years when you need to feather your nest and find a place and situation that is secure enough to create a root from which to grow. Sometimes, however, such times can seem anything but secure while they are happening, for 'emotional development' could also describe these Jupiter Years. How you emerge from the 'terrible twos' *see* ('Gemini' Year) and what happens at three years of age (also the 'First Striving Lunar Phase', *see* page 58) can set an 'emotional security tempo' for years to come. Likewise, the emotional volatility of 15 is notorious, and not surprising considering the Saturn and possibly Neptune influences also prevailing then (*see* Age Index on page 235). The 'Cancer' Year that is probably the most significant is 27 when you also reach your First Lunar Return – so this is a very Lunar time (Cancer being the Sign ruled by the Moon). Issues concerning the past, home, family, mother, motherhood and emotional ties in general are uppermost in your life at this time. These issues can also come to the fore at 39 or 51, but, at these times, very possibly with regard to feeling trapped or compromised by them. Security and family concerns can take on a particularly poignant hue at 63 and 75, or indeed at 27, 39 or 51.

Generally, these are years when you create a foundation upon which to build whatever you have in mind, or whatever is going to unfold throughout the remainder of the Jupiter Cycle (another eight years). So make such a foundation a priority and do not get too distracted or dazzled by more ambitious or glamorous prospects or projects. As you do so, determine what elements in your life are more emotional habit or blackmail than security or nurture. Jupiter's expansive and future-oriented energy can appear incongruous in safety- and past-oriented Cancer. This means to say that the 'emotional development' mentioned above can force you into emotionally upsetting situations in order that you grow emotionally, possibly pulling you away from past attachments, but with some difficulty. There is an image here of a plant being transplanted (therefore uprooted) into a soil that is more conducive to its future flowering and fruiting.

GO-ZONES for your 'Cancer' Jupiter Years

Areas of encouragement, growth and wisdom – and how to make the most of them.

GO-ZONE	APPRECIATE AND UNDERSTAND:
Care	That any field of endeavour that is concerned with the care of yourself, others or the natural environment is well-starred now.
Home	That domestic improvements want to happen, which can mean anything from making a positive move (ultimately), to expanding/redecorating your living space, to possibly even living in a mobile home. Whatever the case, your home-life will figure strongly at this time.
Family	That one or a number of aspects relating to your family – blood relatives or otherwise – are in a process of development and/or increased awareness. This means that the 'life' of your family/mother is creating changes and realizations in you.
Dreams	How, by recording and researching your dreams during this time, there will be a vast improvement in your emotional wellbeing and understanding.
The Past	How by looking into the past you can now discover many things that will aid you in the present, and eventually propel you into a more secure future. Going down 'memory lane' is a strong likelihood during these years, and this can give you a stronger sense of your life's continuum, that is, a sense that you came from somewhere and are therefore going somewhere.

'LEO' YEARS — AGES: 4 16 28 40 52 64 and 76

'GROWTH IN CREATIVE SELF-EXPRESSION' – After hopefully securing yourself in the previous 'Cancer' Years, these are times when you grow in terms of making more of yourself. This can mean many things, from discovering that you have a personality all of your own at the age of four, to discovering some unique talent at 16 that will serve you in your adult career. At 28 such self-expression could peak as creating children, art or music, while at 52 seeing your creative efforts more in spiritual terms such as what you are giving to society. The age of 64, being possibly a time of, or prior to, retirement could mean finding some creative pastime to give meaning to a life after full-time work, and 76 would hopefully be the glow of a life lived to the full. Whatever the age, a 'Leo' Jupiter Year is all about being celebrating the 'I' in you, the centre that creates its own reality, that is sovereign in its own realm.

Because Leo and Jupiter are both expansive by nature, such years as these can be a field day for simply enjoying life and creating or attracting a better lifestyle and class of people. However, the bias towards excess which is the nature of Jupiter can also make you overstep the mark and become undone by the very thing being used to further yourself, be it childish outburst, market speculation, snobbish, overblown ego-expression or passionate sexual involvement. Avoiding the pride, vanity or arrogance that comes before a fall is an elementary guideline here.

Having said all of this, it is important not to check too much the natural flowering of personality that occurs during these years, for it is through such 'shows of self' that a feel for your life-force and creativity is discovered. Leo is the sign of gambling, and it is always a case of 'win some, lose some' when it comes to thrusting yourself into the field of play in a bid to succeed.

GO-ZONES for your 'Leo' Jupiter Years

Areas of encouragement, growth and wisdom – and how to make the most of them.

GO-ZONE	APPRECIATE AND UNDERSTAND:
Play	That becoming as a little child is now the way to breathe freshness, spontaneity and vitality into your life. Playing is the art of living. The actor in you.
Romance	That 'love is in the air' in the romantic sense of there being opportunities to make yourself feel special because you find/have someone that thinks you *are* special, and you think they are too. All the world loves a lover.
Creativity	It is the natural inclination of a human being to create. We have been created and therefore we must create – that is why we are called 'creatures'. The pursuit or promotion of any skill, especially an artistic one, is well-starred during such years. That (your) children are also (your) creations.
Enterprise	That you were born to make more of yourself, and that the key to this is having a sense that there is a spark within you that has its own significance, and that you owe it to yourself to kindle that spark into a glow, that you then fan into a flame. This is a time to shine and show who's boss.
Generosity	Giving of yourself – physically, emotionally, mentally or spiritually – is the surest way to feel worthwhile in yourself. That virtue is its own reward.
Recognition	The need to have recognized whatever worth you possess is a vital human requirement, without which self-worth problems will arise. You may first have to recognize this fact, then discover what it actually is in you that you want recognized. Whatever the case, make hay – for the Sun is now shining.

'VIRGO' YEARS — AGES: 5 17 29 41 53 65 and 77

'GROWTH IN EFFICIENCY' – These are years when progress is difficult and painstaking. The attention that you have to give to myriad details and complications might give a feeling that you are getting nowhere fast. However, you should not let this get to you – try to see everything you are doing in the spirit of preparation for or service to something higher and better. 'Work' is the keynote now. A 'Virgo' year can be especially disheartening when compared to the high hopes and extravagant displays of the preceding 'Leo' year. But really Virgo follows Leo specifically in order to trim sails and cut the suit to fit the cloth – whether it's having to knuckle down to primary school at 5 years, exams or revision at 17, your Saturn Return at 29 (*see* page 115), mid-life crisis at 41, health issues at 53, retirement at 65, or the limitations imposed by old age at 77. And creative visions (Leo) are always followed by having to get down to the practical details (Virgo).

Generally, an issue that can arise now is one of health – probably from working too hard or of letting the slog get to you. Also a change of diet or some other health regimen may be imposed or called for, as well as finding a more efficient way of living that cuts down wear and tear. Try to get emotional issues into perspective so that they do not overwhelm you on the one hand or that you lose sight of them on the other. Much of this perspective will depend upon and be implemented by devising or discovering 'techniques for living'. Such techniques are methods one can use to deal with people and things in a way that is clever and effective. For example, a difficult person could be best handled by pre-empting whatever it is that makes them difficult. So if they are indecisive, giving them too much choice would be asking for trouble. (Also *see* 'Protection along the Way' on page 271.)

All such efforts will bear fruit come your following 'Libra' year, but a 'Virgo' year just spent worrying and carping will breed very little. Now is the time for working and studying hard in order to improve yourself and the life you lead.

GO-ZONES for your 'Virgo' Jupiter Years

Areas of encouragement, growth and wisdom – and how to make the most of them.

GO-ZONE	APPRECIATE AND UNDERSTAND:
Work	That work is the key to success. That 'the devil makes work for idle hands' in that being busy at anything is preferable to feeling slothful, useless, vacuous or pointless – or simply neurotic about what you think you 'should' be doing. Just get on with it and find out as you go.
Health	How to listen to, respect and care for your body and mind. How your psychological and physical well-being are absolutely dependent upon one another. How health is a priority and recipe for success in all areas of life.
Service	That being of help to others has a deeply satisfying effect upon your state of being, and also stops you from fretting too much over personal issues.
Analysis	How you and others tick as individuals. How to be clear in yourself by simplifying your life and prioritizing your involvements, and vice versa.
Method	How much a job done in the right way is very satisfying, and that the right way must consist of a plan, preparation, order and the right tools and technique, or you'll just botch things and become frustrated.
Study/Retraining	How it will improve your mind and/or career prospects. How it will resolve financial problems, increase self-esteem.

'LIBRA' YEARS — AGES: 6 18 30 42 54 66 and 78

'GROWTH IN RELATIONSHIP' – These are years when you grow towards becoming more of a 'member of society', depending on your age and social inclinations. At the youngest end, at six years of age one begins to become aware that there is something called 'Other'. Other is everything that you are not, be it an individual or a group, human or otherwise. At 18 you are probably in the throes of taking relationships more seriously; at 30, during or just after your Saturn Return (*see* page 115), you may well have made some lasting commitment to a relationship, such as marriage itself, or be at some kind of turning-point with regard to a relationship. At 42 your circumstances and experiences, being in the midst of certain mid-life crisis factors (marked by specific Planetary influences – *see* Age Index on page 235), challenge you to relate in a far more sophisticated manner than ever before. At 54, being the time of your Second Lunar Return (*see* page 54), apart from a new emotional focus coming into view, domestic/family relationships could well be a major issue. The last two ages for 'Libra' Years possibly involve the ending of some longstanding relationships, and of generally relating to others in the light of a wealth of experience.

All in all, 'Libra' Jupiter Years offer you the opportunity to be more socially aware and involved. How you go to meet such an offer is entirely up to you and your current relationship status, social life, and ability to relate. But it must be borne in mind that a 'Libra' Year is a highly important one in the Jupiter Cycle because it marks the point of culmination in your endeavours to establish yourself in the world of Other – which includes business and market relationships, and the public at large. In your adult years then, you stand to see the world as being far more available to you as it offers up its resources, agencies and avenues of expansion.

On a more psychological level, a 'Libra' Year, being that of Other, means that you can learn a great deal about why certain people are in your life, through understanding that they are a projection of your own Self. In other words, through realising that everyone in your life is there for a good reason, and what that reason is, you then appreciate just what you are and have as an individual.

GO-ZONES for your 'Libra' Jupiter Years

Areas of encouragement, growth and wisdom – and how to make the most of them.

GO-ZONE	APPRECIATE AND UNDERSTAND:
Relating	That whatever your relationship situation is, you stand to improve your skill in relating and thereby have a more rewarding love and social life.
Harmony	That the role of 'peacemaker' may well fall to you in that you are the one who brings accord and understanding to any situation that requires it. Any innate or learned skills regarding fields such as counselling, diplomacy or public relations now come to the fore and enable you to make something of them. 'Libran' indecisiveness can and must be overcome now.
Attractiveness	How through being more socially and/or fashionably aware, in terms of skills and grooming, you become more in demand, personally and professionally.
Art and Aesthetics	That you are more than usually aware of what pleases your own values and senses, and those of others generally, and can therefore become more successful materially and/or in terms of acclaim and self-esteem.
Justice and Balance	That you could be called to sue for justice in some respect. This will mean weighing the odds and coming to a just decision, sticking to your principles.

'SCORPIO' YEARS AGES: 7 19 31 43 55 67 and 79

'GROWTH THROUGH INTIMACY' – This means that during these years, intimate or sexual relationships, be they your own or those of someone close to you, play an important part in your growth process. It can also mean that any relationship of any weight or significance reaches critical mass and either breaks apart or breaks through to a new level. Alternatively, the unsatisfied need for intimacy can also reach critical mass and demand some sort of desperate release, which may not properly come until the following 'Sagittarius' Year. In any event, 'Scorpio' Jupiter Years can be quite cathartic as they force a piece of your soul to the surface – which is what intimacy is ultimately all about.

Because of the encountering of certain taboos, Scorpionic features such as obsession, manipulation, secretiveness, wrangling and intrigue can arise, giving grist for the mill of your emotional development. The biggest taboo of all, death, may also play a part in taking your life to a new level of involvement, but it is not something you should morbidly await. At any rate, if a death does occur during a 'Scorpio' Year, you can be sure that it marks the end of an important chapter in your life, and the birth of an equally important new one, and again, act as a fillip to intimacy.

The individual 'Scorpio' Jupiter Years do not seem to coincide with conventional ages of any significance, other than perhaps seven years which could be regarded as a time when sexual interest first makes itself felt. Perhaps this is because such Scorpionic extremes as sex and death can rear their powerful heads at any time, with dynamically transformative effect. But if they do so during a 'Scorpio' Jupiter Year you can be sure that such events are major turning points in your life's plot.

Another way of looking at a 'Scorpio' Year is seeing it as the intensification necessary to resolve any stalemate or deadlock created during the over-diplomatic or indecisive 'Libra' Year that preceded it. Shared resources or legacies could also be an area of good fortune or development.

GO-ZONES for your 'Scorpio' Jupiter Years

Areas of encouragement, growth and wisdom – and how to make the most of them.

GO-ZONE	APPRECIATE AND UNDERSTAND:
Sex	What sex actually is and thereby take it to new levels of pleasure, health and general effectiveness in the world. That being or getting (more) sexually involved now is a strong possibility, and that this can either take you like a wave (mostly in the younger 'Scorpio' Jupiter Years) or be something you consciously or ritually embark upon, like the practice of Tantra or Taoism.
Power	What power actually is in your life, and life generally, will be taken to a new level. It is through concentrating on what is most genuine within you or about you, and not flinching from any challenge, that you will most empower you. That if you see power as something that is outside or over you, then that is where it will remain, and possibly become an issue.
Dealing	That investing your time and energy in business and generally making a financial killing could be your Growth Mode now. Taxation can be an issue.
Delving	That you will be drawn (even more) into such fields as psychology and the occult as you seek to discover the hidden or root causes beneath outer phenomena. Growth in such inner knowledge is deeply influential.
Inevitability	That as one door closes, another one opens. An awareness such as this could now well be your most productive, informative or consoling truth.

'SAGITTARIUS' YEARS — AGES: 8 20 32 44 56 68 and 80

'GROWTH IN ITSELF' – Because Sagittarius is the Sign that Jupiter 'rules' – meaning that they are both the same sort of energy – Jupiterian Growth functions particularly well in the Sign of the Archer-Centaur. Consequently, 'Sagittarius' Jupiter Years are usually upbeat, expansive, advantageous or positive in some way. At the very least, they will be times when it is a case of it being 'an ill-wind that blows nobody any good'. In fact, the whole principle or philosophy of growth is something worth focusing upon at these times.

As the Archer, Sagittarius is Growth in the sense of having a target to aim for, or simply letting loose an intention and following it, encountering whatever adventures you meet along the way as part and parcel of that growth process. So, during these years, such enthusiasm for some goal or path is par for the course, along with the positive thinking that is integral to it.

As the Centaur, Sagittarian growth is greatly assisted by having the 'horse power' and 'horse sense' to get where you want to get to. So if you are going to make a lift-off, now is the time. The ground that you cover can be literal in the sense of travelling, cerebral in the sense of education towards some qualification that equips you and others even better for the journey ahead, or seeking in the sense of looking for a meaning to life, such as a philosophy or a set of beliefs or laws, or simply what you are after. The significant point here is that you have reached a significant point. Consequently, you are confident and eager about making a move and going for it.

So, furtherance is the byword now – into whatever field your aspirations or your loins lead you. The power of optimism and enthusiasm is what counts now, whether it is your entertaining dreams at eight years old of what you are going to be when you grow up, university life at age 20, a new path of learning at 32 or 44, or more religious or spiritual paths at 56, 68 or 80. Indeed, at any age, but especially the later ones, being able to see the bigger picture is the important and encouraging thing.

GO-ZONES for your 'Sagittarius' Jupiter Years

Areas of encouragement, growth and wisdom – and how to make the most of them.

GO-ZONE	APPRECIATE AND UNDERSTAND:
Sport	How now your physical body is able to prove itself, whatever your 'Sagittarian' Jupiter Year, notwithstanding your age and condition. Some form of physical culture is now in the offing or is a path to further success.
Travel	That travel and foreign issues not only broaden the mind but can be a lot of fun too, and they give you a sense of the adventure that life essentially is.
Friction	How, through engaging with whatever it is that opposes, threatens or frightens you, you will make your greatest advances. Anything or anyone that now confronts you is probably a great teacher for you.
Higher Mind	How a sense and awareness that there is some higher reason for life will now transport you to a new level of social and/or professional involvement. Such could entail your pursuit of any subject that is never-ending in its content, such as law, philosophy, religion and astrology.
Higher Education	That at any one of these stages of your life you stand to advance through improving your mind or (working towards) winning educational honours, be it in a formal or informal way.

'CAPRICORN' YEARS AGES: 9 21 33 45 57 69 and 81

'GROWTH IN AUTHORITY' – After the expansiveness and opportunity of a 'Sagittarius' Year, the Growth Principle of Jupiter is limited or conditioned in the following 'Capricorn' Year. This is because anything that goes through Capricorn has to pass its tests of practicality and authority. So at nine years of age there is the possibility of hitting some sort of crisis with respect to those who have or had authority over you, like teachers and parents. In other words, this is a critical time because it determines how you're measuring up in the eyes of those authority figures and the systems that they set up and control. Consequently, if you pass their tests you go up a step, but if you fail you go back one, or 'miss a turn'. Similar circumstances force themselves upon you at 21, when you 'come of age', particularly considering your Saturn Waxing Square occurs then too. (Saturn has a similar influence to Capricorn because it rules it, that is the Planet and the Sign have the same energies.) Thirty-three can be critical because you may go up the career ladder then, or, conversely, you may have to knuckle down to the real world (again), or at worst find yourself out of favour with those in authority. Interestingly, from a Christian viewpoint, Jesus was crucified at this age, and one could say that at that time he ran foul of the existing authorities, but the authority he gained in being nailed to the cross (a Saturnian symbol, by the way) was inestimable. The age of 45 occurs during your second Saturn Opposition, so this can be very telling with respect to status and authority, whereas 57 should be an age when you really do know where you stand. With regard to 69 and 81, it should be borne in mind that Capricorn and Saturn both rule old age, and so the pluses and minuses that go with this are strongly in evidence at these times.

GO-ZONES for your 'Capricorn' Jupiter Years

Areas of encouragement, growth and wisdom – and how to make the most of them.

GO-ZONE	APPRECIATE AND UNDERSTAND:
Status	That your professional and/or social position will now be increased in proportion to the efforts you have made previously.
Discipline	That through adhering to some course or regimen you will build for your future and create an inner sense of stability as you do so – this would have to include making certain sacrifices, hopefully in the sure awareness of necessity.
Authority	What you think authority is for you. Is it something which is created by the status quo and the powers that be, or something more of your own making or on a higher spiritual level? Such years are times when you can, or rather must, find this out. You will have to do whatever you have to in order to measure up to what that authority is demanding of you. On a more personal level, you can now build a sense of authority that dispels an unnecessary need for approval from certain figures in your emotional life.
Objectivity	That you now have the ability or opportunity to see things as they really are. From looking hard at what is going on in your life as dispassionately and fairly as possible, you then know what is, and what is not, required of you. Knowing where you stand is the big advantage now.
Organization	These are times to create structure, functionality and order in your life, both business-wise and personally. This is when you create systems or are beaten by them. If you are stuck in a system, then now you should organize yourself either out of it as a 'system' in your own right (ready for the 'Aquarian' Year that follows), or into a better position within it.

'AQUARIUS' YEARS AGES: 10 22 34 46 58 70 and 82

'GROWTH IN PRINCIPLES' – If the 'Capricorn' Year was about status and ambitions as the means and object of growth, then the 'Aquarius' Year sees ideals and aspirations as what need to be expanded and expounded. All this means that some kind of reform, rebellion, splintering or breaking away is what has to happen in order to make way for new growth. As ever with Aquarius, there is the danger of 'throwing the baby out with the bath-water' in the sense that the structures and associates you have previously acquired become ditched or affronted as you make your Aquarian bid for freedom. Ideally, you should aim for a happy medium, but not sacrifice your principles in the process.

At the age of 10, the Aquarian drive not to be 'one of the many' is largely unconscious, and so any signs of rebelliousness then should be interpreted as the child connecting with some deeply instilled programme to be true to themselves rather than to the system in which they find themselves. The age of 22 can be a time of groups and friends splintering as you determine what your true course is. For many people, 34 and 46 can be times of being 'settled' and as such the Aquarian Principle of Growth may just manifest as boredom with existing involvements. For the more individualistic types though, new horizons beckon or are sought. The age of 58 is a contradictory one because it occurs on or near the Second Saturn Return, an influence which inclines to conforming in some way. So, 'Be what you are, my friend!' is the definite convention for this, or any Jupiter year. Being the allotted three score years and ten, 70 must have some unique significance for the individual human being. At 82, you would be sliding into your Uranus Return, a highly Aquarian 'moment of truth' as Uranus governs (is of the same energy as) that Sign.

GO-ZONES for your 'Aquarius' Jupiter Years

Areas of encouragement, growth and wisdom – and how to make the most of them.

GO-ZONE	APPRECIATE AND UNDERSTAND:
Friends	Friends are an integral part of your growth process now. New friends will be particularly instrumental in affecting your life course and development as an individual. Old friends and associates may need to be given a shot in the arm – or be left behind as being no longer suitable to your evolution.
Idealism	That your growth and furtherance as a unique individual has everything to do with being true to your ideals or vision, which necessitates knowing what they are. The clearer your sense of *dharma* or individual purpose, then the clearer will be your path ahead. What is unusual about you.
Humanity	An increase in your awareness of what actually being human is, will now greatly encourage you and simplify your life too. That you are human but may never have really been shown how to be so, and that through being more or merely human you discover who you are as a one-off that has never occurred before or will ever exist again. All this could lead you to help others to discover their humanness, and uphold their right to be so.
Revolution	The Chinese symbol for Revolution is an old skin being shed, which means that these 'Aquarius' Years are times of 'sloughing' for you. You could find yourself out of sorts at these times because you could feel you are neither one thing or the other – the old you hasn't gone and the new you has not yet grown to replace it. But that is what sloughing is, a new skin (life or persona) growing beneath and displacing the old one.

'PISCES' YEARS — AGES: 11 23 35 47 59 71 and 83

'GROWTH THROUGH ACCEPTANCE' – Through simply allowing things to happen during these times, you find that, eventually, you move forward and on. And, considering that this is the last stage of a Jupiter Cycle, things 'happening' can mean that something ceases to happen, often in the sense of 'fading away'. A 'Pisces' Jupiter Year could therefore be regarded as a time of dissolution or disillusionment, or both, depending upon your emotional state or philosophy of life.

By the same token, such a year can be best used as an opportunity to relinquish anything that is proving too difficult, damaging or draining to keep up. Such a thing could be a relationship, a style of living, a job, an appearance, or a habit. Then again, you may not have any choice in the matter as whatever it is just seems to slip from your grasp. But it is at just such a time that this relinquishing must take place, for otherwise you could end up like someone who is about to embark on a new journey (in the following 'Aries' Year), but is still carrying the baggage from the last one. This would be particularly true of emotional baggage in the form of unresolved relationship issues.

At the age of 11 you have to leave behind your childhood as you go to 'big' school with its more serious curriculum; there may also be other moves away in that period. At 23 an adolescent hangover or some feckless lifestyle may have to be shaken off in order to move forward, or then again you could find yourself in some backwater as you attempt to retreat from life's harsher realities. The ages of 35 or 47 could also be times when the pressures of life tempt you to find a way out, say into alcohol or some other distraction, or to find a transcendental path in some spiritual discipline. The age of 59 coincides with your Second Saturn Return (*see* page 126), making this time a major watershed in your life, with the necessary bill of acceptance of the way things are. The ages of 71 and/or 83 should hopefully see 'Growth through Acceptance' in the light of some kind of submission to one's mortality and any thoughts on an afterlife, as well as a realization that compassion is the ultimate response to the human condition. In typical Piscean fashion, none of the above suggestions should be strictly ascribed to each age, for such endings and submissions as a passport to the next round are common to them all. 'Pisces' Years are essentially times of coming to an understanding of what (your) karma is, and endeavouring to clear it.

GO-ZONES for your 'Pisces' Jupiter Years

Areas of encouragement, growth and wisdom – and how to make the most of them.

GO-ZONE	APPRECIATE AND UNDERSTAND:
Mystery	This means that basically we know nothing in the face of the mystery that is life, and that the best and only way forward is to succumb to one's fate and the beautiful and sublime feelings that characterize this mystery. Mysticism and mythology could also show you the way through and ahead now. The power of sacrifice.
Devotion	That some task or ministry to which you can give yourself body and soul would be a wonderfully appropriate way of showing your acceptance that there is some greater cause to serve. Devotion may also be to some commitment that is to your own salvation. Making amends is advantageous.
Sensitivity	Any higher or psychic sense or ability that you have, and attune it to relieving or enlightening others. The healing arts now call.
Imagination	That art and music can be the channels for the expression of the visions and finer feelings that you currently experience, taking you and/or others to new heights, in preparation for the next stage of growth.

YOUR MOON CYCLE

Your Emotional Life

I Must Go
Down by the River
Down to the Sea
See if the Fishes
Agree with Me
I'm Born of Water
Born of Water
I Must Flow

In the end there is no better single word to sum up the complex significance of the Moon than Soul. Your Soul is the feeling deep within you, a longing for home and security, to be where you belong – and the instinct for what is safe and familiar, or dangerous and alien. The Soul is also the emotions that bubble up from the well spring of your being when enraptured by love or art; the emotions that dwell still, sometimes stagnant and fearful, within you: that course consistently through your life and being, or at times rage like a torrent. You can see from this kind of imagery that the Soul and the Moon are decidedly watery – and indeed the great majority of our physical make-up is water (around 70 per cent). The tides of the waters that comprise most of the surface of our home planet Earth are governed by the Moon. So, your body, and the thoughts and feelings that it experiences, are also subject to 'tides', the nature of which are what your Moon Cycle describes. One way or another, such thoughts and feelings can then seek to be given form or take place as events.

The Moon is our own satellite, and as such it is our 'governor' (that which regulates flow) in that it gives us that monthly cycle and rhythm that is short enough to be familiar, and long enough to digest or regulate the experience of something, and, in the case of a woman, to produce a new egg and shed it if not fertilized, along with the old lining of her womb. This last vital point tells us that this 'governor' is really the Mother, doing her rounds, looking after her children, or creating or making ready for them. So, Mother is arguably a word as valid for summing up the Moon as Soul – be it your biological mother, a mother-like figure or institution, a Moon Goddess, or anything or anyone that protects or nurtures. The connection between Mother and Soul is as close as the words 'Nature and Nurture', or 'Chicken and Egg'.

By the same token, you can never have a Mother without there being a Child, a third contender for the Moon Keyword title! And no matter how old you are, there is always that Child within you, with its need for security, its longing for a 'welcome in the hillsides', and its yearning to express and give form to its feelings and memories. Most importantly, that child – which we could call your Receptivity – is in need of protection, otherwise it can drown in a sea of emotional input, or conversely, can dry up through denying or being denied that emotionality. (To help you with the protection and control of your 'satellite dish of receptivity' I refer you to the chapter 'Protection Along the Way' on page 271.) In this light, we can regard the passage of the Moon through the skies around our planetary home as being like a shepherdess to her flock, gathering and guiding us towards that 'home', be it far or near. She is the measure of our emotional development and experience. As such, her Cycle is more of a continuous flow or succession of waves than a number of significant points like the Cycles of Saturn, Uranus, Neptune and Pluto.

Consequently, we look at the Cycle of the Moon in terms of a number of 'Phases' that flow in and out of one another, although there still are 'significant points' at the times when any 'Phase' actually begins. And I put the word 'Phase' in inverted commas because they are not the actual phases of the Moon which we see in the sky through the month; rather symbolize them. This is because the Moon that we use to see the Cycles of our Emotional Life is the *Progressed* Moon. The Progressed Moon takes the lunar month of an average of approximately 27⅓ days, and projects it onto a wider screen of approximately 27⅓ years – that is, 27 years and four months. This period is then divided up into eight 'Phases' of our emotional life and development, as depicted in the illustration on the following page. These 'Phases' use the same principle of waxing and waning, flowing and ebbing, as the actual Lunar Phases we see, but they are not exactly the same because the actual Lunar Phases are produced by the

Moon's angle of relationship to the Sun. Such Phases *are* used in astrology as something called the 'progressed soli-lunar cycle', but this is far too complex and technical to include here.

The Cycle of Moon 'Phases'

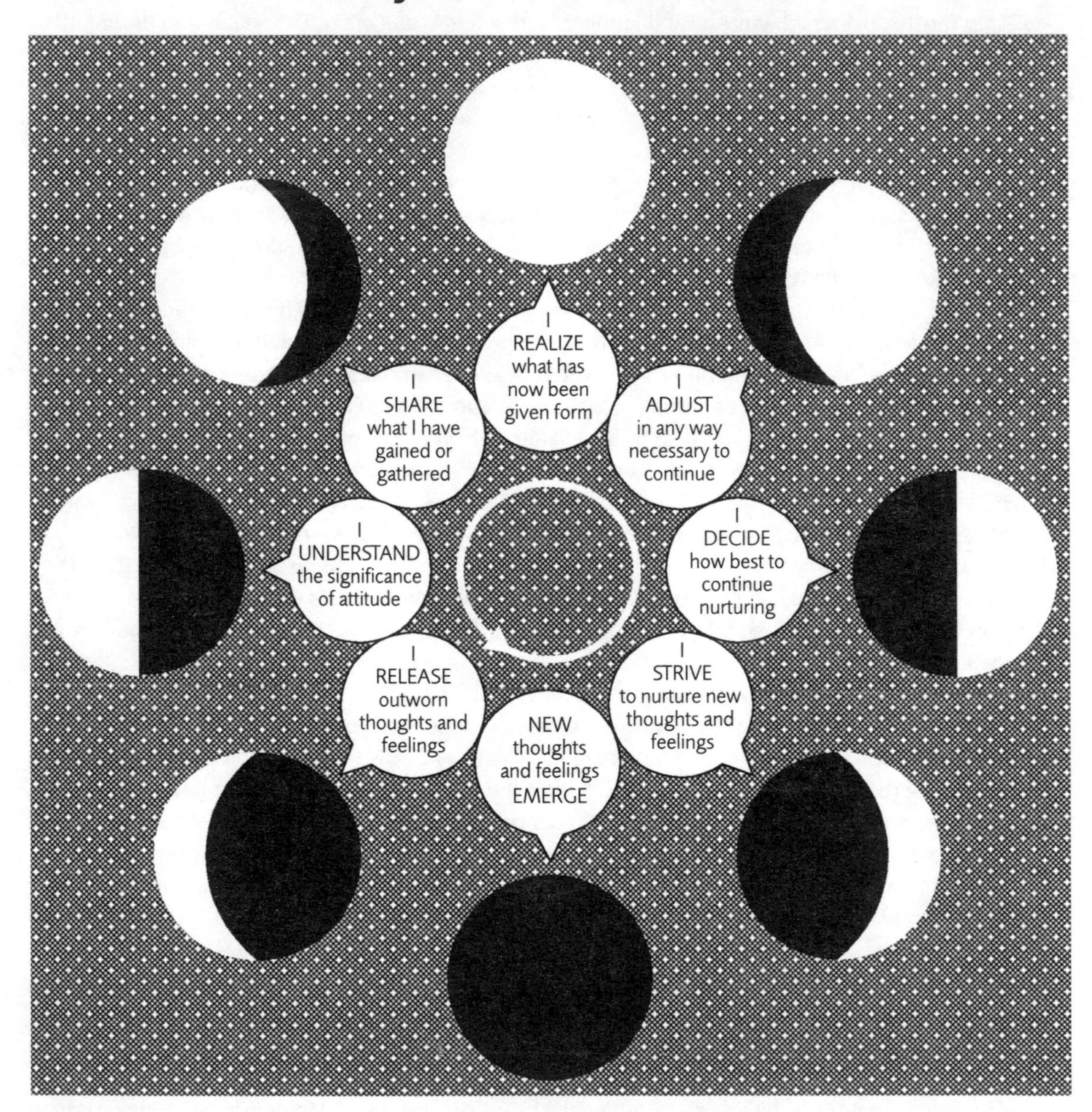

Each Cycle, which lasts 27 years and four months, begins with the 'Emerging Phase', at the bottom of the picture, with the 'Phases' unfolding in an anti-clockwise direction. The Name and a Key phrase or Affirmation for each 'Phase', which lasts three years and five months, are given in the bubbles, the diameter of which also denotes the actual beginning and end of that 'Phase'.
Note that 'thoughts and feelings' can take form as products or events. Also observe how the right Waxing half of the picture/cycle mirrors the left 'Waning' half. (*See* page 46 for your Moon 'Phase'.)

Beginning on page 48, I give the eight Lunar 'Phases' that symbolize and measure the natural flow of your emotional life and evolution over each period of 27⅓ years of your life. Each 'Phase' therefore lasts three years and five months, and preceding the general description of what they each mean individually is shown:

- an image of the Lunar Phase of which it is symbolic
- its actual name
- the periods in your life when it occurs.

Following this general description of each 'Phase', I then give individual descriptions of the meanings of them as they occur at the different times in your life, both in the context of where you are according to your age and of other Universal Planetary Cycles that are possibly or definitely active at the same time. Generally speaking, the influence of each 'Phase' is at its *strongest* at the time it begins, and this lasts for about nine months. Its influence then fades until it gives way to the onset of the next 'Phase'. This 'hit and fade' process is graphically depicted in the graduated shading of each 'Phase' as shown. Do not, however, take the this fading too literally, as in some cases the initial strength of a 'Phase' can persist. Your Moon Cycle is hereby seen as a succession of eight waves, four 'flowing in' or 'waxing' for the first half of the cycle, and four 'ebbing out' or 'waning' for the second. The Waxing half of the cycle is the 'sowing' time when we bring something to Realization, whereas the Waning half is when we 'reap' or harvest what we have sown. Also, at the heading of key individual 'Phases' I give a 'title' for it (which is also featured in 'The Stages of Your Life' on page 242).

All of this serves to show how our life unfolds, like a written tale that we can also edit as it does so. And the Progressed Moon's Cycle is marvellously revealing in this respect. However, you may sometimes see that major changes or moves (often with regard to home and emotional relationships) happen up to a year before or after they are 'supposed' to at the time a particular 'Phase' begins. Apart from there being another Universal or Individual Planetary Cycle prevailing at that time which would account for it, there seems to be either an inclination to pre-empt a change by jumping before it jumps on you, or delays due to the time it takes for it to sink in that the change is unavoidable, or to the time it takes, say, for a house sale to go through. Also, the notion springs to mind of Moon-ruled gestation, in that an event occurs as being conceived nine months *before* it's due, or nine months *after* it was conceived at the given time. Even so, be careful not to force an event to 'fit' the astrological influence. Concentrate upon what really happened or is happening at the time and you will see what it really meant and is telling you – this would include your Birth and the period prior to it (*see* the 'First Emerging Phase' on page 49 and the chapter entitled 'Your Birth, the Start of All Planetary Cycles' on page 9). For example, an elderly client of mine made a major move a year after the start of her 'Second Emerging Phase', but where she stayed for a short time whilst it was actually occurring is where she regularly visited all her life from then on, and where eventually she returned to live and has remained for the last 25 years.

Possibly more than any other astrological influence, the Moon can show you where you are coming from and where you are going to – and more importantly, give you a definite sense that your life has emotional meaning, direction and purpose. So it is worth taking the time and making the effort to remember and ponder the events and feelings that comprise the course and the flow of the River of your Life. It can also help to see each Cycle of eight 'Phases' as a kind of 'gestation process' through which you are formed as you were in your mother's womb, to be born again at each 'Emerging Phase'.

A 'Filter for Experience'

As the Moon is what gives form to things rather than what actually creates them (which is the role of the Sun) the 'Phases' are not actually making events or experiences happen but informing you of the way in which you should naturally emotionally respond to those events and experiences. So for instance, during an 'Emerging Phase' you are emotionally in tune with that time when you are living it in the spirit of something Emerging or New in your life. The Moon being the Realm of the Unconscious and a case of 'chicken and egg' means that something may indeed Emerge before you even think about it, but being consciously attuned to whatever 'Phase' it might be ensures that you are feeling as positive as possible about whatever is happening.

Moon Cycle Index

You can just flip through the pages of this section and look for the ages in which you're interested, but here is an Index if you wish to use it. Simply find the age you're looking for, and the 'Phase' that is occurring then is shown, along with the page number on which it is described. But first of all read the general description of that 'Phase', given next to the actual picture of it a few pages beforehand.

The Mansion of the Moon

Your Moon Cycle can be likened to a building of three storeys – or stories, for that matter. You are conceived and gestate at 'basement' level from which you are born or Emerge into the outer physical realm where you feel your way through the 'rooms' – that is, the 'Phases' of your Lunar Cycle – of the first storey or 'Ground Floor'. This is the first orbit of the Moon, lasting 27⅓ years. You then go up to the next storey or 'First Floor', then through that and up on to the last storey or 'Second Floor'. The 'Roof' is symbolic of departing the physical realm, and of course, it could be 'accessed' at any time during the Cycle, at any storey or room when life ends, as your Soul chooses.

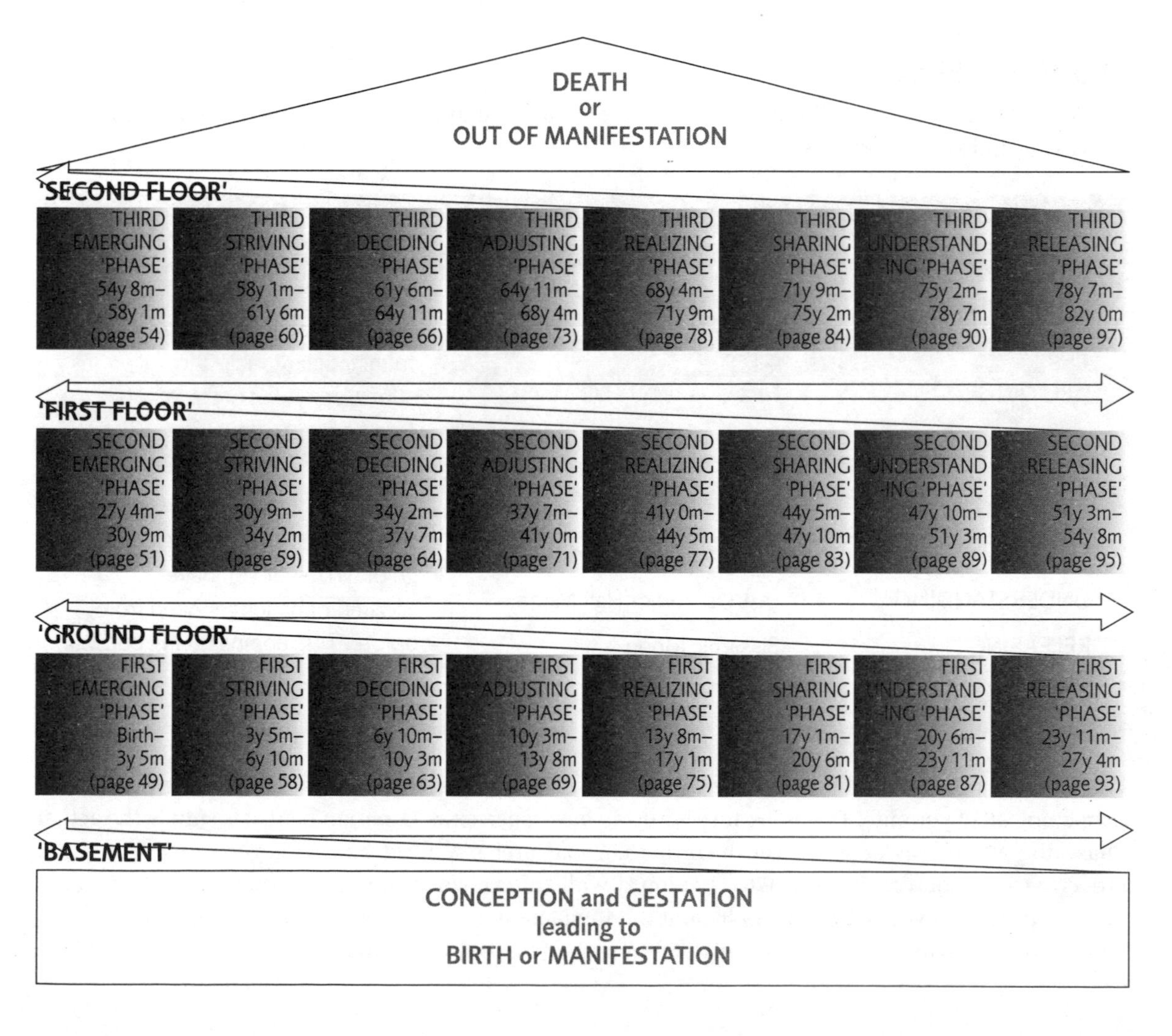

General Guidelines for Reading your Moon Cycle

Key Affected Areas

Your Moon Cycle will affect and be concerned with any number of the following areas of life:

Family
Food and catering
Home and security
Babies and children
Mother and motherhood
Anyone or anything female
Stomach, breasts and womb

Fluctuations
Fluids and flow
The Unconscious
Water in all its forms
Giving anything form
Feelings and emotions
Nature and naturalness

Habits
Receptivity
Private matters
Health (all forms)
The Sign of Cancer
Background and past
Dreams and memories

The Moon 'Phases'

The Moon 'Phases' are given here with their names as applied to actual Moon Phases, and the meaning of Waxing and Waning Phases:

Symbolic 'Phase'	Actual Phase	Waxing and Waning
EMERGING STRIVING DECIDING ADJUSTING	New Moon Crescent Moon First Quarter or Half Moon Gibbous Moon	**Waxing** = Sowing: Building from birth or inception to some culmination or realization of your efforts or natural unfoldment. Tide flowing/coming in.
REALIZING SHARING UNDERSTANDING RELEASING	Full Moon Disseminating Moon Last Quarter or Half Moon Balsamic Moon	**Waning** = Reaping: Giving out, shedding, distributing or letting go of whatever you have accrued or no longer need from before. Tide ebbing/going out.

When reading the general descriptions of the Moon 'Phases' you may feel that one of them has been happening all of your life. This is because we do all have a personal 'Lunation' which is simply the actual Phase the Moon was at when you were born. You'd probably need an astrologer to work this out properly (*see* Resources on page 376). Equipped with this knowledge, you will then find that you are more in tune with yourself emotionally and the fundamental orientation of your personality when the 'Phase' you are living through at any given time is the same as your 'Lunation'.

When reading through the Moon Cycles, it is worth bearing in mind the following points.

- **Pregnant?** If you are pregnant you will find that the 'Phase' (or 'Phases') you are going through at the time characterizes your term quite accurately.
- **Examples** that I cite are often personal or drawn from my clients because the Moon, by its very nature, engenders private or subjective experiences.
- **The key** to understanding the Moon's influences is focusing upon whatever is of emotional significance in your life at any given time.
- **Jupiter Years** or Growth Modes occurring at the same time as any 'Phase' are indicated at the end of the description of that 'Phase'. The ones written in capitals are the more significant ones.
- **Other Universal Planetary Cycles active** at the same time as any 'Phase' are indicated after the Jupiter Years, with some rare exceptions regarding pre-1930 births.
- **Lunar Links** Where you see ◐ at the end of each 'Phase' description, you will be given the page number for the next 'Phase' to occur in your life. If you haven't already, be sure to read the general introduction to that 'Phase' a few pages beforehand, beneath the illustration of the 'Phase' itself.

'Emerging Phases'

BIRTH – 3y 5m
27y 4m – 30y 9m
54y 8m – 58y 1m
82y 0m – 85y 5m

THIS 'PHASE' WE COULD SEE AS THE SYMBOLIC equivalent to the New Moon, the start of a new emotional Cycle of $27\frac{1}{3}$ years, and also the end of the previous one. In fact, it is the previous Cycle, and particularly the last 'Releasing Phase', that bear greatly upon what you make of the birth of this new Cycle, and the Cycle as a whole.

If the previous Cycle ended with certain emotional or family issues still unresolved, then the new Cycle could be hampered or confused by this. Then again, the whole nature of the new Cycle, and beyond even, could have the express purpose of sorting out that 'unfinished business' – whatever it may be. Furthermore, that previous Cycle has probably provided you with the experience and insight to forge ahead with the new one. But whether or not you are 'clear' of the undertow of previous encumbrances or actually well-equipped because of them, the beginnings of any new Cycle are always somewhat indistinct – just like the image of the New Moon itself (*see* above).

But the main point is that it is the start of a new Cycle of emotional experience and development, a bit like getting seven new letters in a game of Scrabble with which to a make a brand new word, except that one usually has some odd letters left over from the previous go! The clues as to what this new Cycle is going to be about are essentially written into the 'set and setting' of the time during which it begins. This means that the emotional and mental attitude that you have now, along with the physical and relationship circumstances in which you find yourself, have everything to do with what that new Cycle is going to be concerned – and again with what could be influencing it from the previous Cycle and 'Phase', for good or ill.

So whether it is a past or present situation that you are investigating here, look closely at the circumstances of that time (and what immediately preceded it) and you will find some wonderful clues as to where the River of your Life is wanting to go. Because of the nature of the Moon, these instances can be very personal, so it is hard to find examples of this among public figures; for this reason I will give my own experience of the Second Emerging Phase which begins at around $27\frac{1}{3}$ years of age.

About four months earlier, the woman with whom I had a long-standing relationship that had been winding down over the previous few years (very much in accord with the 'Releasing Phase' that was occurring then) was reading my 'horoscope' from a magazine. At the time I had no interest in astrology or knowledge of it other than my Sun Sign, and so as the forecast for the coming year said that I was going to leave my current job, find a new home, and, unless a miracle happened, come to the end of a relationship, I could be forgiven for having raised an eyebrow or two! But it did in fact all happen – even though 'horoscope' columns cannot take into account anyone's Lunar Return as the 'Emerging Phase' is also called. This so impressed me that I began my study of astrology.

Apart from this being the rebirth of my professional and personal life, and the main theme of the whole 27⅓ year cycle to come, the point is that the reality of this was not that evident to me as it was happening – I was still busy trying to become a successful songwriter. I did not actually decide to take up astrology as a profession until the 'Deciding Phase' almost seven years later, encouraged, believe it or not, by my then meeting the author of that 'horoscope', the late Patric Walker!

This personal anecdote indicates how at 'Emerging Phases' we have this strong sense of something 'sprouting' or 'germinating'. One eminent astrologer coined the term 'pregnant with futurity' which aptly describes the feeling or state of this 'Phase'. As such, at this time you have to be quite self-possessed, and not consider too much the opinions or even feelings of others. This does not mean to say that you should be insensitive, but that the little sprouting seed of your Emerging Self should not be held back too much because of what someone else says or does that disagrees with it. Another metaphor that springs to mind here is that of a flower in bud; it is totally concerned with its own unfolding and reaching its destiny as a full bloom. It cannot afford to be otherwise.

During any 'Emerging Phase' you can manage, learn and benefit from it by allowing and nurturing whatever is wanting to Emerge.

Here are each of the individual 'Emerging Phases' that can occur during your lifetime.

1st Emerging Phase Birth–3y 5m

'YOUR BIRTHING' – It cannot be stressed enough as to how important and significant this 'Phase' is. I have already focused upon this under 'Your Birth is a Microcosm of Your Life' on page 14 where I showed how much the nature of your birth and your time in your mother's womb affects your personality and life to come. Events and experiences during that time were 'in at the basement', not just the 'ground floor' (*see* 'The House of Saturn' on page 108). This means that such influences got to you before you had any means of vetting them whatsoever, and that they remain entirely unconscious until you are somehow made aware of them or take steps to familiarize yourself with them. And even when you have found out or have been told what your birth was like, it often does not sink in as to what it actually felt like or indicated. The truth is that whatever it did feel like has coloured and determined your most basic emotional responses and patterns of behaviour. So ask your mother, father or anyone around during your time in the womb and at birth, what was going on. Then allow yourself to remember what this felt like. As you do so, you will then see how much such experiences have, and probably still are, affecting you. There are an infinite number of possible experiences, but a common one is that the unborn child is so supersensitive to its mother's state of being that it will pick up on her hopes or fears for her child to be. After all, her state is described as 'expecting'. Often I have had clients who have found out that their mother was expecting or hoping for a child of the opposite sex to what they turned out to be. This sows a seed of confusion, rejection or disappointment from the start – what I call 'bombing at birth'. Another common one is the mother's fear of giving birth, possibly owing to previous difficulties or to a miscarriage. A common result here is that the child is born anxious, with a sense of dread. Yet another is a premature birth with the baby not being expected to survive. This confers upon the infant a feeling that their existence makes no difference, that whatever they do is going to be ineffectual – all giving rise to a rather victimlike personality. However, when the 'Second or Third Emerging Phase' comes round (or preferably before) they should tell themselves that they *are* still alive and that they *have* made a difference.

On the positive side, good influences at this time are probably taken for granted – which is possibly because that was how Mother Nature originally intended pregnancy and birth to be. Even so, if you find positive experiences around that time for you, then it would be wise to draw upon them to offset your current cares, or to encourage and care for others with the feelings those positive experiences have bestowed upon you.

We cannot really consider this subject of pre-birth experience without taking reincarnation into consideration. Whether you believe in it or not, the fact is that more people in the world believe in it than not. Suffice to say, that purely from a cause-and-effect standpoint, whatever has happened before birth has an enormous bearing upon the birth itself and upon the one being born. The whole concept of reincarnation can get extremely complex; there are many complete books written on the subject, so I won't expand upon it here. However, there is one particular astrological idea concerning reincarnation which has always appealed to me, perhaps because it is so simple: whatever Moon Sign you were born with is what your Sun Sign was in your last life. So if, like me, you were born with your Moon in Aries then you were an Arian in your previous incarnation. This means that your immediate memory (Moon) is Arian in nature, in that your last life would have been concerned with issues of war, soldiering, fighting for a cause, self-assertion, pioneering, etc – what I call 'battle karma'. And so your main predisposition (Moon) in this life would centre around these things, either literally or metaphorically. Furthermore, your 'Emerging Phase' would tend to be coloured by whatever your Moon Sign is. (To determine your Moon Sign, if you do not already know it, see my books *Do it Yourself Astrology* and *Do it Yourself Relationship Astrology*.

Whatever your birth and pre-birth scenario, the remainder of the 'First Emerging Phase' is also going to strongly affect your behaviour and sense of what life is like because you learn and experience some very elementary functions during this time, such as walking, talking, potty-training and breast-feeding. How well we do with these things, and any problems we encounter, form our basic attitude and patterns of behaviour. Walking initiates our sense of how to get on in life; did you streak ahead or have a bad fall? How easily did you learn to talk, and was anyone listening and responding? Potty-training symbolizes our ability to process our experiences and not hold on to them and poison our system – how did that go? Breast-feeding reflects our sense of supply in terms of quality and quantity – from having little or no breast-feeding to being fed that way until around or beyond the end of this 'Phase'. Society's current preoccupation with breast size probably has something to do with all this in that the Child in many men needs 'feeding' and the Mother in many women wants to oblige. Sex, in the usual sense of the word, has little real bearing on it, but merely distracts the attention away from the embarrassing truth of the matter.

Many of us can hardly remember this 'Phase' of our lives at all, particularly the birth and pre-birth parts, and wonder why we are sometimes so much at the mercy of uncontrollable feelings. This 'Phase' really does set the scene – so looking into it, remembering it, and reliving it (possibly through some therapy such as hypnosis or rebirthing), is rather like at last seeing the picture on the box of the jigsaw puzzle of your life. It may not be the picture you expected or wanted, and what you'd pieced together so far may resemble it somewhat or just be a work of fantasy, but at last you'll know and feel what you truly set out as. The fact is, we *do* start out as we mean to carry on – but that does not mean responding forever to natal rejection with feelings of guilt or apology, or to natal disappointment with self-defeating behaviour, or, that by way of compensation to pretend that everything is all right until we find that we are getting nowhere with a relationship or our career because we are not being true to ourselves. The idea is to regard the experiences and impressions of the 'First Emerging Phase' and immediately before as a challenge to overcome the negative feelings that they engendered. If you start out feeling rejected then you may have to keep 're-rooting' yourself (changing homes and relationships) until you find somewhere you feel you truly belong and leave behind the 'no room at the inn' feeling you encountered at birth and/or before. Additionally, you would have to superimpose a positive belief about yourself (for instance, through affirmations and refusing to accept bad treatment from others).

This description of this 'Emerging Phase' would not be complete without stressing how important it is for parents to be aware of what is actually going on in their child's life, or children's lives, at this time. Through being keenly, and empathetically, aware of your infant's state with respect to all the above, you will give them a start in life that is as powerfully positive as it is subtle. The phrase 'the soul of sensitivity' suggests itself here in a slightly different light. The 'soul' is the child is 'sensitivity' is what is being conveyed here. A baby who knows that it is known at this deep level will eventually better know itself as youth and adult, and therefore be more confident, happy and successful. As one child I know said of a sleeping baby, 'Look, it's dreaming of its past lives'.

Jupiter Growth Modes – 'ARIES – TAURUS – GEMINI – CANCER' (*see* pages 27–30)

There are no other Universal Planetary Cycles active at this time

Your First Moon Cycle continues with your 'First Striving Phase' on page 58. Be sure to read the general description of this 'Phase' on page 57, if you have not done so already.

2nd Emerging Phase, 1st Lunar Return 27y 4m–30y 9m

'YOUR FIRST EMOTIONAL ROUND-UP' – The first nine months to a year of this 'Phase' is one of the most significant points in anyone's life, and it is highly unlikely for there not to be some important change or milestone marking it. The reason for this is that the 'Second Emerging Phase' is like a form of rebirth. This is akin to the transformation from caterpillar to butterfly in that your 'caterpillar stage', the first Lunar Cycle of 27⅓ years, is when you undergo a range of emotional experiences that serves to make you aware of your life's potential and possibilities. And so during this 'Phase' you are moved towards whatever events, places and relationships – along with the opportunities and experiences they engender – to develop whatever you have discovered as most valid about your personality so far. This discovery may or may not be conscious. Remember that the Moon is symbolic of the unconscious, so it usually moves you for reasons you do not at first recognize, that is unless you are in particular touch with your emotions. Then again, you may think you know where you are going but discover later that your unconscious mind had something else in store for you. Or you might only get it part right. But you'll only know for sure in retrospect, because it could take several further 'Phases' of your Second Lunar Cycle to notice what is now trying to take shape in your life, or *as* your life. Having said this, with the equally important First Saturn Return at around 29 years of age (*see* page 115), there could be a more definite sign of what that shape is supposed to be – or *not* supposed to be. Added to all this, at 28 you enter your Second of the Three Ages (*see* page 238), where you move from being more or less entirely, and necessarily selfishly, concerned with the development of your 'basic self' to becoming a 'social being'. The 'basic self' is what we develop in order to function in the world in a fundamental way – rather like taking a foundation course before deciding exactly what 'subjects' you wish to specialize in. Some people are well on their way career-wise at this time, but the Moon has more to do with how you feel on the inside than what you are doing on the outside.

However – and it's a big however – unlike the natural world where creatures grow and develop according to a ruthlessly efficient and unbending schedule, we as human beings do not. And so at this time our level of security in the world becomes very evident – personal and family life being the most likely focus of concern. Any children born at this time usually have some significance over and above what is usual. In fact, issues around birth and babies are common now, but they are emotionally unique for each mother or father. However, just as in the natural world, it would seem that many people have children at this time when they are only just 'grown' themselves, so it is not surprising that children born during this time in a parent's life turn out to be a greater than usual responsibility, because they are somehow an extension of what still needs nurturing in that parent. It is as if we produce in a child that which is the child in ourself. This is not unnatural or neurotic, but quite the reverse. Nature uses subtle means to achieve Her ends – any half-decent natural history TV programme will show you that.

Returning to the 'caterpillar to butterfly' metaphor, there is also a 'pupation' period when we may not be sure what we are. This period would occur during the preceding 'First Releasing Phase' (*see* page 93) when we have to succumb to the close of the First Cycle. The 'chrysalis' that we live in can be quite comforting or isolating, until we emerge from it – or until someone emerges from us – during this 'Second Emerging Phase'. Certain people, for some reason close to their own Soul, choose not to Emerge from that 'chrysalis' at all. Rather than shuffle it off, they shuffle off the mortal coil altogether. Famous examples of this are rock icons Brian Jones, Jimi Hendrix, Jim Morrison and Janis Joplin – all dying within six months of the start of their 'Second Emerging Phase' (Brian was dead on time, so to speak, at $27\frac{1}{3}$ years). I call these 'once around the bay' lifetimes, for it seems that this single emotional circuit was sufficient for them to make their point, either to the world or themselves. And yet it would appear from contemporary reports that Brian, Jimi, Jim and Janis were rather lost and in-between prior to their deaths (like motherless children), and to a lesser extent such an emotional state can characterize this time for any of us. Someone else who made a monumental mark at this point in his life, but then kind of shrivelled up and died not long after, was Russian cosmonaut Yuri Gagarin. He didn't so much go 'once around the bay' at $27\frac{1}{2}$, but was the first man to go once around the Earth – just like the Moon! He literally couldn't follow that, being killed in an air crash seven years later. I also give an uncelebrated example for this kind of lifetime as part of the description of the 'First Releasing Phase' on page 93. This concerned one of my clients who informed me that he had 'got what he had come for' and then left.

Here are a four more celebrated examples of the 'Second Emerging Phase' – they all just happen to be male and definite fighters for certain causes. South African President, Nelson Mandela, was appointed as National Secretary of the African National Congress, which led through the events at and between his 'Second and Third Realizing Phases' (*see* pages 77 and 78) up to becoming his country's first black president at the start of his 'Third Understanding Phase' (*see* page 90). Palestinian leader, Yasser Arafat, co-founded the Al Fatah movement, which was his first step towards becoming Chairman of the Palestine Liberation Organization during his following 'Realizing Phase' (*see* page 78). *See* also page 126 concerning his renunciation of terrorism and recognition of Israel. Civil rights leader, Martin Luther King, in his first high profile act concerning civil rights, led a 381-day boycott of Montgomery city buses at the very start of his 'Second Emerging Phase'. It was also during this Phase that he founded a national vehicle for civil rights reform, the SCLC. *See* also page 171 regarding the assassination of Martin. John Lennon married Yoko Ono (very much the mother figure to him) and then the Beatles split up – all at the beginning of his 'Second Emerging Phase'. *See* also page 359 to be directed to other astro-events concerning John's life and death.

Now here are three women whose lives took a significant turn at the very beginning of their 'Second Emerging Phase'. In fact, for the first two, the respective events took place a few months before the

27 years and four months mark, which indicates that they were a little 'premature', something that happens quite often with 'Emerging Phases' because they are somewhat like a birth.

Queen Elizabeth II was crowned on 2 June 1953. As the Moon governs womanhood, Elizabeth becoming the country's 'leading lady' and Head of the Church of England at this time was very much 'on time'. *See* page 359 to be directed to other astro-events regarding this sovereign. Another, this time theatrical, leading lady, Elizabeth Taylor, converted to Judaism at the start of this 'Phase'. This little-known fact demonstrates how private and personal the Moon's influence can be, even for someone so much in the public eye. Having lost her third husband Mike Todd in a plane crash the previous year, she felt in dire need of some inner support, and this was her predilection. *See* page 359 to be directed to other astro-events regarding this star. Marilyn Monroe married Joe DiMaggio in January 1954. This is sadly significant because, in a way, she had 'come home' in marrying him because he was reputed to be the one male in her life who truly cared about her. He could have given her the Lunar protection that she so sorely needed – but glamour and a confused sense of independence tempted her away, and she divorced and left him after being, as it happened, nine months in the womb of his care. *See* page 359 to be directed to other astro-events regarding Marilyn's life.

So the 'Second Emerging Phase' is very much a 'birth and death' kind of time – but obviously the event of a literal death is quite rare. All the same, that caterpillar, that old skin or pupal stage, is what has to die, for hanging on to it would be suffocating (which is what Marilyn, ironically, was feeling in her marriage to Joe). It is all part of the process of 'putting away childish things' – and significantly, the word 'pupa' means 'doll'. It is also an emotional landmark, with home- and family-building, not to mention marriage, being high on the list of life priorities – but as such it could eventually affect your outer or work life as well. Moreover, it can be helpful to refer back to your emotional orientation at this time to get a bearing on your life at a later date. This highly important 'Phase' could also be dubbed 'Bringing It All Back Home'.

Jupiter Growth Modes – 'CANCER (very appropriately!) – Leo – Virgo – Libra' (*see* pages 30–33)

Other Universal Planetary Cycles possibly active at this time

Check the tables on page 289 to see if these Planetary Cycles are active for you at this time. Remember that any Trine influences may have to be actualized to make them that effective.

URANUS WAXING TRINE (*see* page 166) Should it be also happening now, this useful and refreshing influence will guide and oil the wheels for you. New ideas and intuitive reckoning resonate with new horizons beckoning. Yuri Gagarin had this going for him at this time – and Uranus governs any kind of flight, as well as technology, pioneering and record-setting.

PLUTO CONJOINING NEPTUNE (*see* page 221) Something quite special is scheduled to emerge at this time. The 'death of one thing/birth of another' is very much on the cards for you, so be as attuned as you can to this process and let the 'tail' of one thing disappear while you deliver or give birth to what you see appearing as the 'head' of something else. John Lennon was experiencing this during his Second Emerging Phase.

Your Second Moon Cycle continues with your 'Second Striving Phase' on page 59. Be sure to read the general description of this 'Phase' on page 57, if you have not done so already.

3rd Emerging Phase, 2nd Lunar Return 54y 8m–58y 1m

'YOUR SECOND EMOTIONAL ROUND-UP' – This can be quite a extraordinary time for you, but a great deal depends upon where your preceding 'Second Releasing Phase' has left you. If you fought against the tide that was trying to get you to let go, which was the meaning of the previous 'Phase', and hung on to certain attitudes and feelings because they were the devil you know, then the 'extraordinary' is not going to enter your life. An 'Emerging Phase' is quite simply the tide turning, coming in again. But, if you are caught on the reef of some old attachment then it will be to no avail, and you could eventually drown in whatever comprises that attachment. Hopefully, though, you have shaken off the cobwebs of your last Moon Cycle and are primed to set sail for new horizons. If my voluminous client files are anything to go by, new departures are more the norm than getting stuck on that reef.

True to the Moon, home moves are a common expression of breathing some fresh air into your emotional lungs. And such home moves are not, or should not be, merely a change of address. Most moves at this time involve a shift in career emphasis or lifestyle too. One possibility is moving into a larger house which can be used as a guest house or for functions of some kind. Conversely, sizing down to make life simpler or less expensive, and leaving more time for one's inner life could be the desirable route. Apart from these contingencies, a home move could be forced by some occurrence such as the death of a parent or loved one, illness or divorce/separation. Then again, changes in the domestic set-up could be on the cards, as distinct from actually moving.

The important thing now is to be alive to the fact that Nature herself does not regard human beings as over the hill just because they have reached their mid to late fifties. On the contrary, most who have had the responsibility of parenthood, or caring for elderly parents, are free from it now. If they are not, then their 'Second Emerging Phase' is presenting them with a new beginning with respect to such responsibilities. A son or daughter who still needs parental support, or conversely, the parent who needs the support of one or more of their children, are also possibilities now. It all depends on what the Moon deems fit for you with regard to caring or being cared for.

If the previous 'Second Releasing Phase' found or left you somewhat confused with regard to what you are doing with your life, then this 'Phase' could present you with the breakthrough you need. However, this is dependent upon not being attached to some habitual attitude and the situation that goes with it. A classic example of this is the person who for some time has had a partner outside of a stale marriage but for some reason just cannot seem to make the break. Come this 'Phase', they can emerge from out of that marriage, like a stranded boat being floated free by the incoming tide. However, if they still stay in that old marriage, then their extra-marital partner should be careful of not being left high and dry. But the reverse could be the case, with this 'Phase' making one the person being cuckolded.

With the Moon being the Moon, another strong possibility now is that of going, or being taken, to some new level of caring. This could entail embarking upon some course of healing – for oneself or others, or both. An illness at this time can shift one's priorities or emotional orientation quite dramatically, causing one to review one's old ways and attempt to be more open and tolerant with both oneself and others. In fact, any such illness or complaint at this time could well be ascribed to either a lack of self-care or the insufficient care of others. Any long-standing imbalances in one's system (caused by lifestyle or habits) really have to be addressed now because the body or mind can no longer take such strain. Returning to our nautical analogy, this is a bit like a vessel that lists heavily to one side being capsized by a bigger wave than usual and/or because its structure is no longer quite as sound or elastic. Celebrated examples of such capsizing are Peter Sellers who died of a heart attack brought on by drug abuse at 54 years 10 months; Humphrey Bogart who died of cancer at 57 after years of heavy drinking and smoking; Adolf Hitler who killed himself at 56 after it finally struck him that he really had got things wrong; and Andy Warhol who also died at 56 following a gall-bladder operation after what could be described as a rather unnatural and impure mode of living (the gall-bladder deals with purifying the body).

Unlike Peter, Bogey, Adolf and Andy, here are two examples of characters who believed they were doing rather well at this time of their lives, but in fact such dynamic events actually set them up for their future demise. Benito Mussolini led Italy into war at 56 with great hopes of becoming a world-class leader, only to be shot dead by partisans five years later, four days before his above-mentioned partner-in-crime Hitler committed suicide (and note how they both died at 56, the end of their Second Age – *see* page 238). Richard Nixon became the 37th US president at 55 but resigned in disgrace six years later. It is interesting to note that both 'Il Duce' and 'Tricky Dickie' met their downfall during the first few months of their 'Third Deciding Phase' when indeed their fate was Decided (*see* page 66).

At the very close of this 'Phase' you experience an equally important Planetary influence in the form of your Second Saturn Return. This will probably affect the transition from this 'Phase' to the next, your 'Third Striving Phase'. This implies that whatever was changed or produced during your 'Third Emerging Phase' will then be tested for its 'roadworthiness', or rather whether or not it holds water. According to what the result is, you will have to strive to make things work, or you will be happily thriving while striving. (*See* page 126 for a more thorough description of that highly significant Second Saturn Return.) One famous person who is a good example of the connection between these two important influences is Boris Yeltsin. At 54, at the very start of his 'Third Emerging Phase', he was appointed to the Politburo of the then USSR by Mikhail Gorbachev. Come his Second Saturn Return, however, and following the intervening demise of the Soviet Union, he resigned from the Communist Party at 59, and was elected President of the Russian Republic the following year (*see* page 359 for more on Boris Yeltsin).

Still on the subject of politics, Margaret Thatcher became Prime Minister of Great Britain just prior to her 'Second Emerging Phase' (she couldn't wait!), and came out fighting during it by declaring war on Argentina, following its invasion of the Falkland Islands. She did this with the ferocity and zeal of a lioness protecting one of her cubs – not surprising considering she was born with her Moon in Leo! Be directed to more about Margaret Thatcher on page 359.

Finally, with regard to four of the above example personalities, note how their Lunar experiences had to do with their Homeland or Motherland, which are ruled by the Moon.

Jupiter Growth Modes – 'LIBRA – SCORPIO – Sagittarius – Capricorn – Aquarius' (*see* pages 33–37)

Other Universal Planetary Cycles possibly active at this time

Check the tables on page 289 to see if these Planetary Cycles are active for you at this time. Remember that any Trine influences may have to be actualized to make them that effective.

URANUS WANING TRINE (*see* page 175) If you are into making a new start out of your 'Third Emerging Phase' then this will give you the ideas, intuition and willingness to experiment that will help it along nicely. The unique in you gets a big chance to live for itself, or even make a long-awaited mark.

NEPTUNE WAXING TRINE (*see* page 204) This particularly favours any new beginning if it entails a healing or spiritual path. The inspiration and sensitivity necessary for such a course will be at hand. Apart from this, you are able to sense more easily the way your life wants to go, and to accept whatever that might mean.

NEPTUNE OPPOSING PLUTO (*see* page 207) As this influence tends to give you the feeling that you are not going to live forever, the opportunity of a new beginning that your 'Third Emerging Phase' offers should be grabbed. The combination of a sense of what really matters and of changing into a higher gear should prove to promote you into a very interesting period of your life. Conversely, feeling maudlin or morbid would be the kiss of death itself, with heavy and meaningless obligations being too much with you.

PLUTO WAXING TRINE (*see* page 231) There is a power available to you here in the form of money, influential people or simply your depth of experience that allows you to embark on some new venture or way of life. The chances of finding a higher and better level of existence is strong now.

Your Third Moon Cycle continues with your 'Third Striving Phase' on page 60. Read the general description of this 'Phase' on page 57, if you have not done so already.

4th Emerging Phase, 3rd Lunar Return 82y 0m–85y 5m

'YOUR LAST EMOTIONAL ROUND-UP' – Apart from being seen as the emergence from life altogether – either literally, symbolically or psychologically, this 'Phase' is almost entirely bound up with the significance of your Uranus Return, which symbolically marks the end of an astrological human lifetime. So I refer you to page 179 where this momentous, but at present seldom reached, time of life is described and interpreted.

Jupiter Growth Modes – 'AQUARIUS – Pisces – Aries' (*see* pages 37, 38 and 27)

There are no other Universal Planetary Cycles active at this time apart from the Uranus Return.

'Striving Phases'

3y 5m–6y 10m
30y 9m–34y 2m
58y 1m–61y 6m

THIS 'PHASE' WE COULD SEE AS THE SYMBOLIC equivalent to the Crescent Moon, which literally means 'Growing One'. Whatever has been set in motion at or since the start of the cycle at the 'Emerging Phase' you can now see signs of what is to come. The trouble is, so can everyone else, and this means that whatever the world thinks, or whatever you feel the world thinks, of what you are developing at this stage is taken on board very personally by you. And again, as with all other 'Phases', the sense of what has gone before is also highly influential but particularly the preceding 'Emerging Phase'. In effect, you are now having to make an effort to establish a trend that can be maintained for some years to come, in the face of the competition, conditions and opinions that you feel are being put forward by the outside world on the one hand, and any inner doubts or need to withdraw or tread water on the other. It is during this 'Phase' that something can occur that challenges or confuses our progress. This could be an external event or an internal feeling, or a combination of the two. If the 'Emerging Phase' was akin to being a 'seed', the 'Striving Phase' is like a 'seedling' in that it suggests what is trying to grow or take shape. And, just because of this 'tenuous visibility', we can be easily put off by such outer or inner influences. However, all being well, one can usually feel more confident, or is made to be so, around half-way through this 'Phase'.

But when all is said and done, it is called the 'Striving Phase' with good reason. You have to strive against any possible discouragement – active or passive – from that outer or inner world, strive to discern what really is trying to grow, and strive to nurture it in whatever way you can. By way of example, let me continue with the events and experiences in my own life that began with the onset of my 'Second Emerging Phase' (*see* page 51) when I moved from city to country and began my study of astrology (but with the initial intention of becoming a professional songwriter.) With the 'Striving Phase' I moved back into the city, which meant initially having to leave behind the peace and quiet amidst which I had been able to write songs. Effectively, my striving was firstly to leave behind that life and my ambition to be successful in the music field. My striving was also greatly involved with having to make my way in the city again, which included coping with the difficulties and distractions that had originally caused me to leave. But all the time I was striving to learn more and more about astrology, and now, because there was a larger and more varied social input, I was able to be far more practical in terms of helping others through astrology, as well as getting a lot more feedback. And all this was having to be done when I had time free from my day job and normal social engagements. Also worth noting is that about half-way through this 'Phase' I received a great deal of encouragement from my girlfriend of the time.

During a 'Striving Phase' then, the vital issue is to detect what is trying to take form, and to nourish it in whatever way you can. At the same time, be wary of inclinations to revert to the 'devil you know' in the form of old habits and haunts, attitudes and activities. Whatever it is that is growing, it will then assuredly grow even more.

During any 'Striving Phase' you can Manage, Learn and Benefit from it by Striving with and through whatever is happening.

Here are each of the individual 'Striving Phases' that can occur during your lifetime.

1st Striving Phase 3y 5m–6y 10m

During this 'Striving Phase' the most strikingly significant event that occurs for everyone in the civilized world is that of going to school for the first time. The 'seed' that was originally planted and germinated in Mother Earth before and during the 'First Emerging Phase' is now that 'seedling self' which has to contend with the elements that lie beyond hearth and home. Events may occur now that destabilize it in terms of changes to the security system that the infant has just come to accept as permanent. For instance, apart from the well-known rigours of early schooldays, home moves at this time can be a common cause of later feelings of being unsettled. There can be other events, such as younger siblings being born or a much older brother or sister going away to college. Such events are not bound to happen, by any means, and those that do are significant because of one's subjective experience of them – subjectivity ever being the case with the Moon.

The point to grasp is that if there are changes that destabilize, they indicate that the child's fate has somehow called for it. The need to strive has been implanted early on. This can go one of two basic ways: it can make the child grow up as, or into, a real tryer, or else the child feels like a leaf in the wind, believing that there is not much point in trying because something else is bound to come along and render such effort useless. Psychologically this has to do with the first type having what is called an 'internal locus of control' and the second type an 'external locus of control'. It doesn't take too much imagination to see what a difference falling into one category or the other makes to a person's life as a whole. Tracking back to what happened for you, and how it affected you, can be enlightening and, ultimately, healing in this respect. This applies mainly to the second type, because such can hamper one's progress through life. However, people with a strong internal locus of control can suffer the more subtle complaint of not being able to understand how others tick, with subsequent relationship problems. As ever, that Lunar riddle of 'Nature or Nurture' arises here. The answer is both, because people are born with either an internal or external locus of control, and it 'officially registers' itself during this 'Phase'. But, if as a parent you are aware of it at the time, then you can take steps to make sure that your child is not unduly disturbed by any unavoidable changes between the ages of three years, five months, and six years, ten months.

If you are viewing all this in retrospect, then endeavour to 'recast the past' by putting yourself back in that subjective state and viewing it with the objectivity of an adult, seeing what alterations and rationalizations you could have made in order to see how it was 'good', or at least 'right', for you. You can also try to rectify it now.

Jupiter Growth Modes – 'CANCER – LEO – Virgo – Libra' (*see* pages 30–33)

There are no other Universal Planetary Cycles active at this time .

Your First Moon Cycle continues with your 'First Deciding Phase', *see* page 63. Before or after, be sure to read the general description of this 'Phase' on page 62, if you have not done so already.

2nd Striving Phase 30y 9m–34y 2m

Considering that this 'Phase' follows upon the most powerful and important points of two Universal Planetary Cycles, your First Lunar Return (or 'Second Emerging Phase') and First Saturn Return, then it is not surprising that the 'Second Striving Phase' lives up to its middle name. Whatever changes, pressures or events were experienced between 27 and 30 years of age, you are now more than likely dealing with the aftermath, backlash or consequences of them. Remembering that a 'Striving Phase' is all about nurturing the growth of something that has just begun to take shape, your particular effort could relate to any number of things. However, whatever it might be is very likely traceable or connected to whatever happened back then, particularly when you were 27.

Looking through my client files, two things I note is that both relationships and the body can start showing evidence of fragility now. As a result of this, affairs and illnesses are in evidence. If yours happens to be the first case, I suggest that an affair is possibly a symptom of trying to find something you believe you have lost. The fact is that you have probably lost track of whatever was trying to spring forth while you were in your 'Second Emerging Phase', and either you gave up on that or did not even notice what it was. If this raises a frown, then go back and read about it on page 51 and the general description of 'Striving' on page 57. Anyway, whatever it is that you have not discovered in yourself is not going to be found in someone else – at least, not directly speaking. By this I mean that your cage might need rattling to bring home the fact that you've placed yourself in one, that you settled for a seemingly secure life with someone only to find that it cost you the freedom to pursue your journey of self-discovery. So until you retrace and resurrect that 'seedling' of future growth, you are in danger of chasing your own tail, or of jumping from the frying-pan into the fire. Apart from all this, bear in mind that striving to make your relationship(s) work now is very much on the cards – so do not let a relationship 'go to seed' for lack of attention.

On the other hand, you may well have discovered, clearly or vaguely, what you want to make something of in your life —and this 'Striving Phase' demands that you put your money where your mouth is. A couple of famous examples of this are Muhammed Ali and Fidel Castro. Ali strove to regain his World Heavyweight boxing title at this time, ultimately doing so with a knock-out punch, while Castro defeated the CIA-backed invasion of Cuba – *see* page 359 to be directed to more concerning Ali's and Castro's astrological life. Going back to my files, another thing I notice are clients who are Striving to build some kind of professional practice at this time – often having to prove their independence at the same time. But, as ever, the secret is not to lose sight of that 'seedling' which is trying to grow into a healthy plant. As any gardener knows, seedlings have to survive spells of bad weather, being trodden on, or even trying to grow too quickly. They also have to be fed and watered regularly. This means that any infant project will be challenged by market forces, encounter attacks from others, or take on too much too soon. And you have to have encouragement, information and experience. Most of all, keep in touch with whatever spurred you to start that project in the first place – for that will be the seed itself.

Talking of infants and growth, children born or raised at this time in a parent's life can prove more demanding than usual. Again, as I have pointed out, this would have everything to do with something that needed nurturing or healing in yourself re-presenting itself to you in the form of your 'offspring' – something that has sprung from you. Everything is natural and meant to be as far as the Moon is concerned; but so often we have been driven off-course by a society divorced from Nature. Remember that a child is saying 'I am more yours than anything apart from your own body, mind and soul'. A child is also a symbol and a messenger, telling you something about you as parent and parents, as a couple, as a human being. And like the literal meaning of the word 'crescent', the shape of this 'Phase', a child is also literally a 'growing one'.

As regards health, this is not necessarily a time when you are likely to get sick more than usual, but striving can be a strain on the system. A good way to cut down wear and tear is to draw up a list of essentials and non-essentials, on either side of a sheet of paper. For instance, an 'essential' could be to make sure that you reach a goal you have set yourself, whereas a 'non-essential' would be to entertain doubts of ever getting there. Then again, another 'essential' could be to entertain enough doubts to make sure you organize yourself well and do not drop any stitches. You will find that by drawing up such a list your 'natural intelligence' is able to access these essentials and non-essentials quite easily. This is because within you is a wealth of answers – you just have to ask the right questions. Having sorted this out, stresses and strains are marvellously diminished. However, you may still have to strive to root out the cause of any serious complaint, should there be one. In fact, if there is one, it is there expressly to get you to track down the root cause of the problem. One paradox here is the possibility of the complaint being caused by striving too much or in the wrong way or direction. In fact, the correct kind of striving is the 'essential essential'.

Jupiter Growth Modes – 'LIBRA – SCORPIO – Sagittarius – Capricorn – Aquarius' (*see* pages 33–37)

Other Universal Planetary Cycles possibly active at this time

Check the tables on page 289 to see if this Planetary Cycle is active for you at this time. Remember that any Trine influences may have to be actualized to make them that effective.

URANUS WAXING TRINE (*see* page 166) Having this occur during your 'Second Striving Phase' is fortunate because this influence can bring you all manner of unusual, alternative or technological assistance just when you need it. So read up on this, and use the flowchart on page 167.

Your Second Moon Cycle continues with your 'Second Deciding Phase', *see* page 64. Be sure to read the general description of this 'Phase' on page 62, if you have not done so already.

3rd Striving Phase 58y 1m–61y 6m

This 'Phase' takes on great significance because it coincides with one of the most important of all Planetary influences, your Second Saturn Return (*see* page 126). Basically, Saturn's effect at this time emphasizes the difference between, on the one hand, a life that has reached a state of maturity bordering on the wise, and on the other, a lifestyle that has grown so rigid and resistant to any change, either externally or internally, that circumstances and prospects appear bleak and lonely. Its effects can also range between these two extremes.

The striving at this time is one of making the effort to see life from a more spiritual standpoint and to avoid going to seed by being weighed down by seeing life from a solely material viewpoint. The words 'wise' and 'spiritual' are linked here because they both pose a response to life and others that is strongly coloured by the sense that everything and everyone are here for a purpose – and a higher purpose, at that. As a consequence of this, one's striving is in aid of somehow teaching others from one's experience rather than boring them with it; or simply carrying on learning rather than believing that you can't teach an old dog new tricks. In this respect it can be helpful to see life from the perspective of your being a spiritual entity having material experiences, rather than that of being a material entity having occasional spiritual experiences.

For many, this is a time of possible retirement, but in astrological symbolism such a thing does not really exist, since the Planetary Cycles persist from birth and before, up to and beyond one's physical death. So-called retirement at this time is really just emphasizing the transition from one style of life and outlook to another as a result of leaving behind the material world of career and day-to-day labour.

Alternatively, this could be a time when your striving could be academic in the sense that it is the striving you have exercised down through the years that is now at last appreciated, rather than just come to an end with 'retirement'. Or this could be a time when what has been so long in growing through your efforts now needs that last push to take it that vital stage further, perhaps to that final vital stage. It is also quite possible that your creative efforts of the past are appreciated by those born a generation before you. Whatever the case, this all stresses the significance of striving to make some kind of positive mark in life – or as the Vikings saw it, not to leave this world without being well remembered for your time here. The emphasis is upon what you have created, or are creating in life, rather than how long it took or takes to do so.

Jupiter Growth Modes – 'AQUARIUS – Pisces – Aries – Taurus' (*see* pages 37, 38, 27, 28)

Other Universal Planetary Cycles possibly active at this time

Check the tables on page 289 to see if these Planetary Cycles are active for you at this time. Remember that any Trine influences may have to be actualized to make them that effective.

URANUS WANING SQUARE (*see* page 177) This Planetary influence forces some kind of change, and avoiding this could cause chaos. Striving not to resist change but rather to see the necessity for it is the overall theme, even if it is just accepting, or better still, adopting some positive youthful viewpoint.

NEPTUNE OPPOSING PLUTO (*see* page 207) This Planetary influence is called 'Intimations of Mortality' and so if it occurs during your 'Third Striving Phase' it can mean anything from trying to making sense of life in the light of the prospect of some afterlife, or giving up on there being any such thing. In fact, this influence sharply polarizes the negative and positive possibilities outlined above for this 'Phase'. If one wants to avoid feeling depressed and pointless, it should go without saying that it is preferable striving to see beyond the material and obvious towards a more profound and eternal concept of existence.

PLUTO WAXING TRINE (*see* page 231) All being well, you should be on a bit of a roll now, so striving is merely about staying on course rather than pushing uphill. If, however, there are difficulties and resistance, this influence should help you strive to get to the fundamental reason for such trouble – 'If thou wouldst only delve' as Roman Emperor Marcus Aurelius said.

Your Third Moon Cycle continues with your 'Third Deciding Phase', *see* page 66. Read the general description of this 'Phase' on page 62, if you have not done so already.

'Deciding Phases'

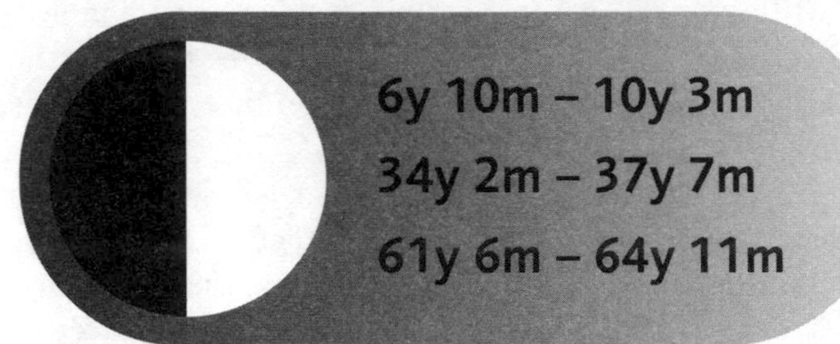

THIS 'PHASE' WE COULD SEE AS THE symbolic equivalent to the Waxing Half Moon, which can be viewed as a critical point in your emotional life in that whatever you are involved in is now demanding some form of decision. Alternatively, it could be a case of a decision being made for you.

In the first case, making a crucial decision is one of the hardest things to do if you are having to deal with an emotional issue where the rights and wrongs are not so clearly defined. Yet the 'tide' is coming in fast now and you cannot afford to dither or procrastinate – the time for action has come. Otherwise you could be become overwhelmed by whatever the existing situation is. So this is a case of 'publish and be damned'. The important point is to set something definite in motion, one way or the other. You can 'fake it till you make it' and make any necessary alterations come the ensuing 'Adjusting Phase'. But if you sit on the fence you could get torn in two or feel compromised for a long time to come. Everything depends ultimately upon you having the courage of your convictions – and having convictions in the first place. What is really helpful is gaining a sense of where your fate, where that River of your Life, actually wants you to go at this time. To do this you will probably have to distinguish between essentials and non-essentials, and accept the course that your life wishes to take.

In the second case, your decision may be made for you. To continue with my own case, during my 'Second Deciding Phase' I was made redundant from the job I had with a publisher. It was decided for me that I was not supposed to be going where that job and everything it involved was taking me. I then had to decide what to do as this way was closed to me – and I decided to go professional as an astrologer. For many people, there are decisions like this that are made by someone or something in their lives who is more 'powerful' than them. Although this is probably par for the course at the 'First Deciding Phase' when you are a child, later on such situations may point to the necessity of deciding to be your own power in your life. It is then that it can be discovered why people *are* in power – they're good at making decisions! They are also prepared to take the flak if what they decided was wrong – or they ought to be.

It has been said that it is not making decisions that is so difficult – it is keeping to them. This implies that we can think some course of action is a good idea until it gets hard to maintain, and then we look for reasons to get out of it – like dieting, for instance. In order to keep to a decision we need to have a very firm sense that it is the way we must go, can go, and want to go. If you don't get an honest 'yes' to all three – I must, I can and I want to – then the chances are that your decision will come to nothing.

During any 'Deciding Phase' you can manage, learn and benefit from being decisive about whatever is happening.

Here are each of the individual 'Deciding Phases' that can occur during your lifetime.

1st Deciding Phase 6y 10m–10y 3m

'THE DIE IS CAST' – This really is a 'Decider' because any decisive event occurring now, strongly affects the course your life takes for a large part of it – if not all of it. This is emphasized by the concurrence of your First Waxing Square of Saturn between six and seven (*see* page 110), the time when this 'Phase' begins. All this echoes the Jesuit saying 'Give me a child for the first seven years and you may do what you like with him afterwards'. Essentially though, the crucial decision or decisions being made now are more or less bound to be ones made by someone other than yourself. And so this implies that it is purely your fate in action, not some conscious choice on your part. It could be argued that a ten-year-old can make decisions, but it is debatable whether the ten-year-old can be held totally responsible for them. The difficult case of young offenders, especially of serious crimes like murder, raises this issue. Did they make utterly their own choice, or were they some unconscious expression of the questionable moral standards of the society in which they grew up, and in which they are still growing up? And then there is the further decision by the judge who sentences them as to what their fate should be.

But apart from such dire cases, your life is bound to take a decisive turn now. Because of this, it pays to focus upon this time in your life, or that of a child with whom you are involved as a parent, relative, carer, etc. Look very closely at whatever crops up or arises, or did so, during this period, for it sets a trend that triggers certain personality traits, and therefore life patterns, that have hitherto been dormant. Such events may be quite internal in the sense that they are on an emotional level and are not that evident to others. A client of mine's mother began an affair with another man at this time. This eroded the family structure generally, and relegated the father to a kind of loser status. But the effect upon my client was that she then began to harbour feelings of deep rejection and insecurity, for her mother had made a priority of her time spent with her lover, and left her youngest daughter to suffer in silence. It was only during her 'Second Emerging Phase' at 27 years of age that she had a chance to begin to leave behind that pattern of rejection that had been decided at that earlier time and get out of a relationship where she was tied to someone who had kept her where he wanted her (that is, like her father, not rejecting *him*) through her own fear of being rejected.

From the above it should be apparent just how much events at this time affect us, and how it is not that easy to spot which those influential events were. My client had always glamorized her mother in an attempt to deny the unacceptable fact that her mother's sexual desires were more important than she was. It could also be said that her mother probably felt rejected by her husband and looked for love and acceptance in her lover. But this just further emphasizes the vital importance of identifying such events at this time in one's own life and in the lives of others, coming to terms with them, and either healing whatever damage they inflicted or actually pre-empting any damage occurring at all. In this way the rot, the so-called 'sins of our fathers' (and mothers) is decidedly stopped.

In endeavouring to discover how one may have been affected at this time, especially at around six to seven years old, it does help to bear in mind that the Moon is mainly to do with 'Mother'. My own mother saw her mother for the last time at this age, giving rise to emotional complexities and denials that boggle the mind – hers in particular, and mine by referral.

That other Lunar issue, 'Home', can also be a main player with regard to how the emotional course of one's life's can be set during your 'First Deciding Phase'. For example, a move at this time to a better house and neighbourhood would give a general sense of uplift and progression to a child which would stay with him or her through life, perhaps in the feeling that change is generally positive. On the other hand, a downward move of this kind could infect the child with a sense of change being a negative trend altogether.

Fortunately, when the First Waxing Trine of Saturn (*see* page 111) occurs between nine and ten years of age, there can be an upswing in our fate. This can either ameliorate any misfortune that happened earlier, or lead to a far more stable life trend. Making an important friend or having a good teacher at school could be what happens now.

Once again, if it is your own child you are looking at concerning this time in their life, I must stress how sensitive they are to what is happening to them and around them, and the importance of not merely interpreting things from their outward behaviour, but to check how they are feeling inside. Then again, just looking at what is going on in your child's life now could be the simple but decisive action that forestalls any possibility of an emotionally afflicted life.

Jupiter Growth Modes – 'LIBRA – SCORPIO – Sagittarius – Capricorn – Aquarius'(*see* pages 33–37)

There are no other Universal Planetary Cycles active at this time

Your First Moon Cycle continues with your 'First Adjusting Phase', *see* page 69. Before or after, be sure to read the general description of this 'Phase' on page 68, if you have not done so already.

2nd Deciding Phase 34y 2m–37y 7m

After the striving of the last 'Phase', whatever happens during this period has to have a decisive effect on the direction your life now takes. You may regard events now as not that significant or just plain hard, but the truth of the matter is that something is trying to push your life into a new chapter that had its beginnings during your 'Second Emerging Phase' – even though you may not have recognized it. So if you are not aware of these significant events and how they are connected, then take time out to become so. Stresses and strains in your life are simply 'growing pains', not just bad luck or occurring merely to depress you. The crisis now can be small or great, but whatever the case, it is intensifying feelings to the point of forcing you to take radical steps to pull away from negative or fearful feelings about where you are, what you are doing, and who you are with – or *not*, with respect to any or all of these three cases.

Anything negative or unpleasant in your life at this time is rearing its head because it is telling you that you must act in order to resolve the difficulty you perceive yourself to be experiencing. Occurrences such as physical complaints, new involvements or children are messages from your unconscious which are saying 'What does this mean, because it represents a call to decisive action', something which may simply mean saying your piece. A famous example of this was Fidel Castro who, having striven to defeat the invasion at the Bay of Pigs (*see* page 59), decided to mobilize his Cuban forces in the Soviet Missile

Crisis, which apart from strengthening his own power base effectively led to the Soviet Union deciding to make him a hero of that nation.

A crisis of decision could also take the form of the loss of someone or something in your life. During this 'Deciding Phase' such a loss is making a decision for you, and forcing you to be more decisive in yourself. The heavier the crisis or loss, the greater the need must be for you to take control of your life. The chances are that if you have been stronger on the inner or emotional side of life, then the decision is probably one that has to take place on a more practical and material level. Conversely, if you have been more capable in the outer world, then it means you must now decide to be more emotionally aware, committed and involved.

Between 36 and 37 years of age, your Second Saturn Waxing Square kicks in, and this will add further weight to the above described imperatives. If all this sounds rather grim, it has to be so because this time could be described as the run-up to your 'mid-life crisis'. In astrological terms, your mid-life crisis is mainly your Uranus Opposition (*see* page 168), but also probably your Neptune Waxing Square (*see* page 198) and Pluto Waxing Square (*see* page 226 and below). The Universal Planetary Cycles appear to be quite concerted and adamant about your late thirties to early forties being a major shift in your life's direction and/or meaning. And this, your 'Second Deciding Phase' is setting you up for it – or rather, it is forcing you to prepare yourself for it. Such preparation is essentially about your life becoming more consciously constructive. This means that it is no longer appropriate to adhere to any conditioning you have which promotes the notion that life is merely about toeing the line, responding to immediate needs, conforming to what you think everyone else does, or just getting from A to B unscathed. 'Deciding Phases' – and particularly this one – are expressly about getting you to appreciate that life is a challenge, and that it has a unique and definite purpose. It may have to jolt you out of any complacency – or just living automatically and going along with crowd – into an awareness of this. To put all this another way, the more of a leaf in the wind you are, the harder will be the decision forced upon you. On the other hand, the more self-actualizing you are, then the more you will view all this as healthy grist to the mill, and eventually achieve something.

The classic and tragic example of how badly a leaf in the wind can fare during this 'Phase' is the death of Marilyn Monroe. Or it could be said, according to some theories, that her fate was decided for her. Whatever the case, Monroe was no more. (Be directed to more about Marilyn on page 359).

Notwithstanding this 'leaf in the wind' syndrome, during this 'Phase' you stand to become a stronger, more effective person – someone who is in the driving seat of their own vehicle.

Jupiter Growth Modes – 'AQUARIUS – Pisces – Aries – Taurus' (*see* pages 37, 38, 27, 28)

Other Universal Planetary Cycles possibly active at this time

Check the tables on page 289 to see if this Planetary Cycle is active for you at this time.

PLUTO WAXING SQUARE (*see* page 226) If this is occurring during your 'Second Deciding Phase' you can be sure that something has got to go in order for something new to take shape in your life. Pluto is very uncompromising. Hanging on to what is outworn – be it an external or internal matter, or both – would be rather like not putting out your rubbish; something starts to stink and get in the way. So decide what this is, and ruthlessly cut it out. Fate will then, and only then, be on your side.

Your Second Moon Cycle continues with your 'Second Adjusting Phase', *see* page 71. Be sure to read the general description of this 'Phase' on page 68, if you have not done so already.

3rd Deciding Phase 61y 6m–64y 11m

Like the 'Third Striving Phase' (*see* page 60) that came before it, this 'Phase' may precede or coincide with retirement from regular employment or for your profession. One could say simply that this was the decision made for you or by you at this time, but the real decision is what to do now that there is not that daily routine to give form and meaning to your life. As this is an utterly personal issue which I have not yet experienced myself, I cannot do anything other than give a general astrological point of view or guidance for this point in human life.

In addition to what I said regarding retirement under your 'Third Striving Phase', it can be helpful to look at what was going on or taking your interest during your 'Third Emerging Phase' between 54 years and 58 years of age (*see* page 54) or even your 'Second Emerging Phase' in your late twenties (*see* page 51). This is because you may now find that you want to use your time and experience to resurrect or expand upon what was happening at either of these times in your life. Conventional opinion is rather linear in that it fails to recognize that if something does not develop or take off at a certain time, then it never will. Astrological thought, on the other hand, appreciates that, as in the natural world, seeds can lie dormant until they sprout many years later.

Apart from events such as retirement, this 'Phase' can also present you with situations or changes that call for some form of decision. Certain endings in the lives of close friends or relatives may create a shift or alteration in your life too. Whatever the case, observing the 'I must, I can, I want to' rule of all 'Deciding Phases' will be useful (*see* the general description beginning on page 62).

Finally, by way of negative example of how a 'Third Emerging Phase' can be misused with dire consequences in the 'Third Deciding Phase', I cite Benito Mussolini and Richard Nixon. During their 'Third Emerging Phase' Mussolini led his country into war and Nixon became the 37th US president. Come their 'Third Deciding Phase' though, 'Il Duce' was shot dead by his own countrymen, and 'Tricky Dickie' resigned following the Watergate scandal. (*See* also 'Third Emerging Phase' on page 54, if you haven't already.)

Jupiter Growth Modes – 'TAURUS – Gemini – Cancer – Leo' (*see* pages 28–31)

Other Universal Planetary Cycles possibly active at this time

Check the tables on page 289 to see if these Planetary Cycles are active for you at this time. Remember that any Trine influences may have to be actualized to make them that effective.

URANUS WANING SQUARE (see page 177) If this Planetary influence is happening now, then some change is probably occuring that is forcing you to decide on some level, about some issue. The influence of young people, or even unruly ones, could be a decisive factor too.

NEPTUNE OPPOSING PLUTO (see page 207) This Planetary influence is called 'Intimations Of Mortality' and if it is happening during your 'Third Deciding Phase' it can mean once and for all arriving at some decisive belief with regard to what happens after death, and what your life has been all about from a spiritual standpoint. The latter decision should involve the resolve to leave this world having accomplished at least one worthwhile task – such as learning truly to love one person or to rid oneself of a bad habit or attitude.

PLUTO WAXING TRINE (*see* page 231) Wealth of experience is what should serve you well now in whatever decision you have to make. Having a good sense of what does and does not matter, and rejecting trivial or superficial considerations, is of the essence now.

◐ Your Third Moon Cycle continues with your 'Third Adjusting Phase', *see* page 73. Read the general description of this 'Phase' on page 68, if you have not done so already.

'Adjusting Phases'

10y 3m – 13y 8m
37y 7m – 41y 0m
64y 11m – 68y 4m

THIS 'PHASE' WE COULD SEE AS THE symbolic equivalent to the strangely named Gibbous Moon which occurs just before the Full Moon. So during this 'Phase' we are between the decision that was made in the previous 'Deciding Phase' and the form of what was set in motion reaching fruition at the 'Realizing Phase'. The word Gibbous literally means 'hump-backed', so this 'Phase' is like a hump-backed bridge you have to cross in order to attain that peak of Realization.

This indicates that we have to adjust our speed and trajectory to negotiate this bridge from the one 'Phase' to the other. It is human nature to feel that once you are past a crucial stage, such as the previous 'Deciding Phase', you can relax and freewheel a bit. But the fact is that this adjustment has to be made carefully, otherwise, if you are unable or unwilling to do so, you could well 'get the hump' – and as any driver knows, freewheeling gives you the least control over your vehicle, especially when control is just what is required.

In terms of the events that can characterize the 'Adjusting Phase', we are mainly confronted with a change or changes that demand some kind of acclimatization. Such a change can be quite subtle, and not necessarily externally obvious. Remember that the Moon symbolizes our subjective experience of things; what can come and go without hardly being noticed by one person can have a major emotional effect upon another person. Again, as with the 'Deciding Phase', an adjustment may be made for you, and have long-lasting repercussions. Personally, during my 'First Adjusting Phase', I was made a prefect at school, only to be demoted a few months later for 'behaviour unbecoming someone in authority'. For many years after this, in order to fit in or make any impression on the world around me, I was forever having to adjust my radical 'outsider' stance to the demands of the status quo and the authorities that control it – in fact I still have to!

So what equips you for handling this 'Phase' successfully is firstly, as ever, an awareness and acceptance of there being this necessity to make adjustments. Following upon this, flexibility and a sense of accommodation are the vital qualities. Conversely, obstinacy or a know-it-all attitude can be fatal – or at least, have you finding that your principles are a lonely place to be. During these 'Phases', one can expect to experience any contrast between what we essentially are and the nature of the environment in which we find ourselves. Such a contrast can be interpreted as a useful sense of getting the measure of our own character, or as an indictment against it. As a rule, this is not a time to dig your heels in, but paradoxically it is a time to persevere. The difference between the two being that the former can be arrogant rigidity asking for a rocky ride, while the latter is sure enough of its long-term goals to bend enough in order to negotiate that hump.

During any 'Adjusting Phase' you can manage, learn and benefit from it by Adjusting to whatever is happening.

Here are each of the individual 'Adjusting Phases' that can occur during your lifetime.

1st Adjusting Phase 10y 3m–13y 8m

Apart from having to adjust to any decisions made for you or by you during the previous 'Phase', there are two obvious adjustments that have to be made during this 'Phase'. The first one is biological/hormonal – puberty. Many books have been written about this subject so I won't expand upon it much here. Suffice to say that we go from being children to being capable of producing them ourselves during this period, and that discovering sexual urges, body hair and emissions, not to mention such visible signs as menstruation, breasts and spots and puppy fat, can be exciting to some and intimidating to others. At one extreme, if you feel vibrant and attractive at this time then adjusting to things will be relatively easy, and will probably persist as an underlying sexual self-assurance throughout life. At the other extreme, feeling awkward, suppressed or unattractive now can cast one big shadow over your sexual development and experiences for years to come. So check out your experience of puberty and discover that what occurred then was the prototype of the kind of sexual being you are now. If you are bringing up or relating to kids of this age, then again refer to your own experience if you wish to understand them better – that is unless you were one of the chosen few that encountered no problems at all! A common problem can be that as you are attempting to adjust to increased hormonal activity and all that it poses, one or both parents are busy trying to keep you where you are, thereby making relating a lot harder now and in the future.

The other obvious adjustment that has to be made, especially at the significant beginning of this 'Phase', at least in the civilized world, is the result of changing from junior to senior school, from an educational establishment that treats you as a child to one that regards you as an adult in the making. In some educational systems, owing to place or time, there may not be an actual change of school at this time (as in my case, where I went to English preparatory school at age nine – *see* general description above). However, whatever the case, some kind of change or shift of emphasis at this time creates some need to adjust your 'natural' inclinations to the set-up in which you find yourself. As ever with the Moon, quite what is meant by 'natural' inclinations is open to question, because whatever they are, you are now faced with a 'when in Rome do as the Romans do' type of situation. On the face of it, because you also have your First Saturn Waxing Trine happening at this time which helps you fit in with the status quo (*see* page 111), this should not present too much of a problem. However, the strength of your individuality is going to be one of two major contributing factors now – something which is further emphasized by having an individuality-provoking 'Aquarian' Jupiter Year (*see* page 37) happening at the start of this 'Phase'. How the other individuals surrounding you respond to your individualism is going to be the other major contributing factor.

Looking through my client files, I am made to recall an Aquarian female whose experience at this time was to be bullied in class by some of the other pupils (because she looked unusual), to be ridiculed by a school set-up frighteningly different from her previous one, *and* to be subjected to an experimental educational technique which was eventually scrapped because it did not work.

Basically then, she could not adjust to all this and simply found herself wanting, installing a long-lasting inferiority complex, not to mention resentment. One is inclined to think that it is the system that should have adjusted to the individual here! But the Moon shows us the Soul's course, the River of your Life, and as such, the reality it encounters at every step or 'Phase' is there to form its course – just as a river is contained by its banks. At this stage the 'river' is very young, a mere trickle, and so the reality of the 'earth' it encounters dictates where it has to go. The trickle has to adjust to the lie of the land. Years later, during her 'Second Emerging Phase' and the start of a whole new Lunar Cycle, my client began to see the significance of these experiences, and to turn the resentment into an understandable sense of injustice, which in turn fuelled a pledge to help others similarly afflicted. Incidentally, she made her greatest friend and 'protector' at school at this time, thanks to that 'Aquarian' Jupiter Year (matching her Sun Sign) and the stabilizing assistance of her First Saturn Waxing Trine.

Quite often the need for adjusting can be 'serial' in its effect. Another client, this time elderly, was first fostered out to another family at ten years, then they had to move to a 'not so good' neighbourhood owing to the father of the house being sent to jail. Then during the third year of this 'Phase', the father returned and made good, moving them upwards again. However, he started drinking and my client had to keep a very low profile – something against her Leonine nature. Not surprisingly, this woman has been 'adjusting' all her life because these experiences triggered an innate inclination of hers to fit in with the circumstances, yet all the while her suppressed need to be her own person kept bubbling under and awkwardly asserting itself, as she forever felt put upon.

Yet another case, which is rather relevant to our current obsession with physical looks and to the sexual aspect of the 'Phase', was a male who at this time was teased for being plump at a new school. Needless to say, he could hardly adjust himself by suddenly not being overweight, and grew up feeling very self-conscious, especially with the opposite sex. He is now of 'normal' weight and size, but the ongoing feeling of not fitting in persists, even though (or possibly because) he is an extremely nice guy.

Finally, a famous example of how critical this 'Adjusting Phase' can be is Elizabeth Taylor. As a child star of 11 years of age she not only had to adjust to fame (like many other child stars), but she also injured her back filming *National Velvet* which created a lifelong need to adjust herself to the ongoing pain of this injury. This example also shows how Moon 'Phases' can show how we deal with things emotionally in accordance with whatever that 'Phase' is. In turn, our manner of dealing with them at that time can persist for years, as in her case, for her continual ill-health is still a call for help that is not really being heard. In typical Piscean fashion she is too good at adjusting as opposed to contacting and expressing her pain more directly (for more on Elizabeth Taylor *see* page 359).

It can be seen that the 'First Adjusting Phase', as with all the 'Phases' of the First Lunar Cycle, is subtle and profound in its effect upon the growing child, and therefore upon the adult. If only parents or carers were more aware of the potential trials and perils of adjustment being imposed at this time, perhaps they'd take time out to help their children adjust with the benefit of their adult experience and objectivity. For example, in the case of the Aquarian female, if her parents had enquired into what was going on perhaps some allowance and provision could have been made, and/or her own self-image reinforced.

Jupiter Growth Modes – 'AQUARIUS – Pisces – Aries – Taurus' (*see* pages 37, 38, 27, 28)

Other Universal Planetary Cycles possibly active at this time

Check the tables on page 289 to see if this Planetary Cycle was active for you at this time.

NEPTUNE SQUARING PLUTO (*see* page 193) Having this 'Adolescence Strikes' influence happening now could be regarded as a bit steep! It is a bit like learning to walk on a moving pavement that is itself in need of some maintenance. But you cannot argue with Planetary Cycles when they deem it necessary for a couple of generations to be plunged into the deep end by combining the above described needs for adjustment with the decidedly fruity ingredients of compulsion and imagination, not to mention sex and drugs and rock 'n' roll, supplied by Neptune interacting with Pluto. From an evolutionary standpoint these two generations (many but not all of those born between the close of World War Two and around 1990) could be said to be undergoing not only a crash course in sexual and psychological development, but, by dint of being forced to express themselves quite strongly, they are forcing others to become more aware of what psychic and sexual energies can get up to – for good or ill. As these two generations mature, they progressively shed more and more light upon these essential powers that govern our existence. This would have started out with Beat Generation of the late 1950s, through the taboo-breaking Sixties, the AIDS years, the increase in numbers of mothers under 14 years of age and teenage suicides, and beyond. (Note that these generations were also products of those of you who experienced Neptune Squaring Pluto during your 'First Realizing Phase' – *see* page 75.)

◐ Your First Moon Cycle continues with your 'First Realizing Phase' on page 75. Before or after, be sure to read the general description of this 'Phase' on page 74, if you have not done so already.

2nd Adjusting Phase 37y 7m–41y 0m

Whether your Second Moon Cycle has so far made you a lot more mature, confident or experienced, or only a little, this 'Phase' still demands that you bend with the wind, whether you're in pursuit of something or someone, or even if you're not doing much in particular. Although this 'Phase' starts off with a change-resistant 'Taurus' Jupiter Year (which is just as well, for you won't bend *too* readily), when you reach your 'Gemini' Years at 38, being in two minds about something will make being firm and unequivocal decidedly elusive. What is now in need of adjusting is whatever it is in your current situation that is getting in the way, whereas more technically speaking it will be something you begun at your 'Second Emerging Phase' (*see* page 51). Or the adjusting may simply take the form of putting things on hold or carrying on in the face of circumstances that are far from ideal. Fortunately, between 38 and 39, your Second Saturn Waxing Trine arrives to assist you (*see* page 121), providing you with a steadying influence or opportunity.

Having said all of this though, the very nature of adjusting at this time of your life implies several things. Firstly, by now you probably have a number of things going on in your life – like family, a busy work and/or social schedule, ageing parents, etc – so the chances are that you have not got that much room to manoeuvre. Secondly, there is also the possibility that you are beginning to think you can settle into a comfortable rhythm of life, that you have and know your priorities, that you have a degree of control or authority in your life – so why should you have to adjust? The short answer is that this is the way Nature intended it – that human beings should not get too set in their ways just because they've reached the halfway point. Our culture's idea of life's progression can very often be at odds with the greater scheme of things.

This 'scheduled' need to adjust can be prompted by more or less anything, whether self-induced or apparently beyond one's control. Having an affair can be one method of stirring up your life, to shatter any complacency or emotional inertia that may have crept in. Because this need to adjust is on the agenda, be very careful in how you assess your feelings. Maybe the real adjustment was needed in your marriage or ongoing relationship, and an affair is just a red herring or is teaching you a lesson. You find you have to adjust, whatever the case – so allow yourself to go through the process of doing so, and learn what you need to know as you go, without making any hard and fast decisions until you reach or are some way into your 'Second Realizing Phase' (*see* page 77).

Then again, you may be having to adjust to a divorce or the demise of a relationship, or to your partner having an affair. But the same 'Lunar logic' would apply – keep adjusting, bending with the wind of change, and thus finding out that you are far more flexible than you thought, or paradoxically, that your adjustment is one of *not* being so flexible. The 'Second Adjusting Phase' can be very demanding, but it will only seem impossible if you will not adjust to the basic fact that you are *able* to adjust! Thinking that your life should be free of the need to make adjustments is what would be asking for the most trouble. 'Bend or break' is the call of the day. Trust your unconscious mind, for it is throwing things at you which will force you into becoming a stronger, better and more capable and aware person. Impress yourself with that famous human aptitude for adapting to circumstances.

Some other classic events at this age that are Nature's way of getting you to adjust are having children; reaching a crisis of confidence because your beliefs are not suited to the reality you live in; losing your job and having to retrain, rethink or go freelance; having to live on less money; experiencing changes in your domestic set-up; health issues; or whatever fate chooses to throw at you. They are all in aid of getting you to adjust, and it won't last forever – unless you *refuse* to adjust.

Jupiter Growth Modes – 'TAURUS – GEMINI – Cancer – Leo' (*see* pages 28–31)

Other Universal Planetary Cycles active at this time

Check the tables on page 289 to see if this Planetary Cycle was active for you at this time.

SECOND SATURN WAXING SQUARE (*see* page 119) This influence emphasizes the need for maturity described above. What needs adjusting should be plain to see – if not that easy.

URANUS OPPOSITION (*see* page 168) This is the classic astrological indicator of your 'Mid-Life Crisis' and having it occur during this 'Phase' could be regarded as very appropriate because both influences are demanding that you change and adjust. Events or involvements that call for adjustment are liable to be more dramatic and obvious – as too is the inclination to overreact. So adjust your speed and intensity of reaction as well.

NEPTUNE WAXING SQUARE (*see* page 198) As this influence has a lot to do with not knowing which way to go until you realize how lost you are, having this 'Adjusting Phase' at the same time could find you prevaricating and dissembling to an almost certifiable extent. Or such adjusting could be more positively expressed as a going with the flow, adopting a more spiritual lifestyle, and accepting the way of the world so that you go around obstacles rather than waste time and energy taking issue with them.

PLUTO WAXING SQUARE (*see* page 226) The heavier, more inevitable, things in life are what you are having to adjust to. Therefore it is made even more imperative that you *do* adjust. You may well find that you are dealing with a very stubborn, entrenched part of your personality – or someone else who reflects that part of you. Recognizing this will be half the battle won.

Your Second Moon Cycle continues with your 'Second Realizing Phase', *see* page 77. Be sure to read the general description of this 'Phase' on page 74, if you have not done so already.

3rd Adjusting Phase 64y 11m–68y 4m

This 'Phase' coincides with the conventional time of retirement from routine work, as do the 'Third Striving Phase' and 'Third Deciding Phase', which, if you haven't already, I recommend you read with regard to this issue. This time could also be viewed as the onset of old age, although in some cases this would not be seen that way. In any case though, there is definitely some adjusting to be done at this time.

The feeling or prospect of becoming elderly – or the denial of it – is further emphasized by there being two Saturn influences occurring during the 'Phase'. The first, your Third Saturn Waxing Square, affects you between 65 and 66 years of age, and like many Saturn Transits, it can make you feel your age, either physically, psychologically or both. The physical aspect is often all too obvious, with stiff joints, aches and pains, and less elasticity being well-known afflictions, even for people considerably younger that this. A balanced diet and exercise, now if not sooner, are called for. Additionally, it is one's psychological condition which greatly dictates the physical one. A rigid attitude creates or worsens a rigid body, and an inactive life on a mental and/or emotional level can cause the body to be inactive too. And if one is closed to new ideas and circumstances then loneliness and uselessness can tip one down the slippery slope.

As ever, a lot depends upon the previous 'Phase', which in this case was that of deciding to do something worthwhile, new or renewed with one's life. A positive decision to make something out of life with an 'it's never too late' philosophy will have pre-empted any feeling that there is no place in the world for you. Apart from being able independently to make a new life for yourself, there are always things like clubs and charity shops to keep one from apathy or relying solely upon family for company and meaning. Such practical and useful activities become readily available come the Third Saturn Waxing Trine around 68 (continuing into 69 and the next 'Phase').

What you are mainly adjusting to at this time is those conventional ideas of getting older, either in the sense of accepting that you have to withdraw and be less in the land of the living, or in feeling old and infirm when you have to do such things as other retired people do. Of course, if you do not retire but carry on with some creative avenue of expression, then none of this will apply, except perhaps for having to adjust what you are creating or doing so that it is ultimately successful – and to any negative attitudes you might encounter from others simply because they cannot positively adjust themselves.

Two famous examples of the 'Third Adjusting Phase' are Queen Elizabeth II and Margaret Thatcher – both 'leading ladies' of Great Britain. During this 'Phase' the Queen finally had to adjust to public opinion and agree both to taxation of her personal income and to opening parts of Buckingham Palace to the public. (*See* page 359 to be directed to more about Elizabeth.) Four years earlier in November 1990, Thatcher had to make probably her biggest adjustment so far when she had to resign as Prime Minister after failing to win her party's leadership contest (*see* page 359).

Jupiter growth modes – 'LEO – VIRGO – Libra – Scorpio – Sagittarius' (*see* pages 31–35)

There are no other Universal Planetary Cycles active at this time

Your Third Moon Cycle continues with your 'Third Realizing Phase', *see* page 78. Read the general description of this 'Phase' on page 74, if you have not done so already.

'Realizing Phases'

13y 8m – 17y 1m
41y 0m – 44y 5m
68y 4m – 71y 9m

THIS 'PHASE' WE COULD SEE AS THE symbolic equivalent to the Full Moon, which in itself is symbolic of the maximum illumination of the emotional state, and a revelation of what the foregoing period has amounted to.

So it is the 'Realizing Phase' both in the sense of becoming aware of something, and of something being made real or coming to a point of fruition. This 'Phase' is highly significant because, *if we look*, we can now get a far clearer idea of what is going on and how we're doing. Often we are reluctant to look because we fear that what we see might tell us something we did not want to know. But to this I can only say that astrology is not a subject for ostriches!

Furthermore, if you do see something at the time of one of these 'Phases' that disturbs you, then hold your gaze a little longer and you will then have revealed to you something very valuable that will better equip you for the future. Admittedly, this is a bit like going for a medical scan – but what you don't know not only can, but will, hurt you. If there is anything festering in your unconscious, now is your chance to spot and eradicate it.

Positively speaking, it now becomes plain if you are doing well with whatever your endeavour might be, giving rise to confidence and further success. In any event though, your 'Realizing Phase' should not be regarded as a judgement any more than the Sun itself could be seen as judging any darkness it is dispelling. Apart from this, it is through owning your darker side that you make your brightness shine. Or as the poet Danté said, 'Take away my Demons and you take away my Angels'.

Remember that the Moon is essentially about your emotional and private life, so do not expect such realizations always to be material ones, although they can coincide with such events brought about by other Planetary influences. Anyway, it is through keeping track of and in touch with our feelings that outer success is made more possible. It is our emotions that motivate us, not ambitions as is commonly believed, for ambitions are the externalization of our emotional needs. My own 'Second Realizing Phase' saw a time when it was made evident to me that the confident image I presented to the world gave the lie to the insecure emotional being behind that image. Most of this came about with the help of a male friend, through a new sexual involvement, and as the result of a belated and vain attempt at bonding with my mother. None of this was easy but it was a vital realization for me in terms of making me infinitely more aware of my inner state, and consequently far more strongly motivated from within regarding my career and social relationships. Interestingly, that 'confident image' of mine is down to having Leo rising, and the male friend, the lover, and my mother, are all Leos!

So, although a 'Realizing Phase' can be rather exposing – a kind of emotional 'outing' – it does have the advantage of finding out what's in your 'basement'. In there, maybe all you can see at first is the rubbish, but clear some of it away and you can then see and lay claim to something valuable that you forgot existed. It also means that you no longer have to hide behind some mask that is possibly suffocating you. In fact, the 'Realizing Phase' is so useful and clarifying that personally I sometimes wish I had it on tap!

During any 'Realizing Phase' you can manage, learn and benefit from it by Realizing what is or should be happening.

Here are each of the individual 'Realizing Phases' that can occur during your lifetime.

1st Realizing Phase 13y 8m–17y 1m

'EMOTIONAL DISCOVERY' – Now we are right in the thick of adolescence, so realizing here can be somewhat of a rude awakening. The essential realization here is that life is now polarized into Self and Other, you and somebody or something else, what you feel yourself to be against what others want you to be – peers versus parents, for instance. Perhaps in order to simplify matters at this time, we are inclined to see things in black and white, along with the brittle confidence that is common to adolescents. So it is not that your environment is necessarily divided – it is just that you insist on seeing it that way in order to establish a sense of your own identity. If there is a 'them', there has to be a 'me'. This 'splitting off' is particularly noticeable at the age of 14 as it is a 'Gemini' Jupiter Year (*see* page 29).

Because hormones are still raging around your system, your sexual state is especially aware of polarities – male/female being the main one. However, the input received from each of your parents also gets processed now. This means that if they don't get on that well, then unconsciously you will pick partners with whom you cannot have a successful relationship, or you avoid getting involved at all, thus setting a future pattern. Conversely, a stable parental relationship would mean that you enter a similarly 'normal' relationship, or feel secure enough to take your time, and be choosy. Any latent homosexuality, which may have been triggered by a dominant mother or father, now comes to the fore – either confidently or as a confusing contrast between what you are supposed to feel and what you actually feel. However, through being aware of these processes you can diminish, if not wholly overcome, their effects.

Another 'sexual scenario' that can arise now is that of one parent or the other seeing you either as sexual competition or temptation, or as being sexually significant in some other way. For example, a mother can focus upon her son as being a man now, and as more malleable than her husband who she has given up trying to 'reform'. And so a boy can get sexually 'manipulated' into the image of what his mother thinks her ideal male to be – which is probably in sharp distinction to what he is actually like. So what he is realizing here is the contrast between the two versions of himself. But the astrologically significant fact here is that such a 'split' or identity confusion would have been latent at birth – it is just that the 'Realizing Phase' now makes it real. A female client of mine, whose parents had previously split up, had her father making a love object out of her, but never touching her. This realized her innate desire to be looked at and admired but not actually touched or got close to.

It should be pointed out that such 'heavy' realizations as described in the previous paragraph usually only take place during the First Saturn Opposition between 14 and 15 years of age. But whatever the case, we can view our experiences at this time as creating a 'role model' for the way we relate to others, especially sexually. Anything that is said or done, or not said or not done, to us at this time has a dynamic and lasting effect upon our ability to relate and the type of relationships we attract – or do not attract. So if you are at this stage, be aware of what you are choosing to be in relation to others – for it'll carry through into the future. Are you reacting or interacting? Are you developing your relating skills with a person or with a computer? Now is the time to express yourself openly and receive back a sharp impression of how that other experiences you. You are interacting now in order to find something out, not to prove anything other than that you are a person in your own right. Interaction leads to a stable personality and relationship, whereas reaction leads to alienation.

In retrospect, whether your life was full and active at this time, or unavoidably awkward or barren – the realization was that whatever was going on then was a 'read-out' of who or what you started out as. If your 'seed' is that of a 'daisy' it shows itself as such during this 'Realizing Phase', not as a 'rose' or whatever. What is more, there is now an 'ego flash' which creates a model that persists until you are able to refashion it somehow. One client's father said to her at this time, 'I don't like clever girls'. She was and still is most definitely clever – and such has been both the bane and the boon of her life for she still attracts men who like but do not like her mental superiority. And talking of mental ability, this 'Phase' is also a time when we realize what we are made of at school during and leading up to examination time. Generally speaking, at this time we get some kind of feedback as to what kind of person we are in the making – be it good, bad or indifferent.

Not surprisingly, during this 'Phase' differences and conflicts can flare up. These can literally show up like skin disease. Acne, that notorious affliction of adolescence, can be interpreted as the eruption of differences or poisonous feelings between, say, father and son.

The classic 'splitting' at this time is literally a case of leaving home or running away from school. This could be interpreted as a straightforward way of discovering who you are by putting yourself outside of the set of people that think they know you, and putting yourself in the midst of one that definitely does not know you. Then again, it might set a trend for running away from difficult or confining relationships. Whatever the case, you can be sure that this 'Realizing Phase' may be viewed as preview of how you relate and see relationships.

Jupiter Growth Modes – 'TAURUS – GEMINI – Cancer – Leo – Virgo' (*see* pages 28–32)

Other Universal Planetary Cycles active at this time

Check the tables on page 289 to see if this Planetary Cycle is active for you at this time.

NEPTUNE SQUARING PLUTO (*see* page 193) Having this 'Adolescence Strikes' influence occurring now could be said to add spice to an already spicy dish. However, most of you born between roughly the mid 1930s and mid/late 1960s , and some after the mid/late 1970s would experience this double whammy. One could say that this astro-combination punch is the flavour of adolescence for the greater part of the 20th century – along with the truly outlandish and taboo-breaking statements and styles that characterize it, not to mention the 'youth culture'. Any youthful social statements of the pre-mid 30s generation were generally safer, or more confined to the privileged classes. This though is the Beat and Rock 'n' Roll Generation, or the generation that is growing up in their aftermath. (Note that these generations were also the products of those of you who experienced Neptune Squaring Pluto during your First Adjusting Phase – *see* page 69.)

Your First Moon Cycle continues with your 'First Sharing Phase', *see* page 81. Before or after, be sure to read the general description of this 'Phase' on page 80, if you have not done so already.

2nd Realizing Phase 41y 0m–44y 5m

'EMOTIONAL AWARENESS' – At this time, certain truths related to your overall state of emotional awareness are shown to you, or at least are begun to be. Such truths can bear upon your life history as a whole, or stem particularly from whatever was happening or whatever you set out to achieve during your 'Second Emerging Phase' between 27 and 30 (*see* page 51).

How this 'awareness' comes upon you is usually the result of one or usually more emotional confrontations – or even showdowns. As ever, the scenarios that can bring these about are as varied as experience itself, but emotional relationships are the prime ones. It is as if the significant other in your life now shows themselves to you 'full frontal', as it were. Consequently, the mirror to your own emotional state is now directly opposite you, not at some more oblique angle.

In the purest psychological terms, such 'confrontations and reflections' are an excellent opportunity to gain a far clearer idea of where you are coming from emotionally, to see your patterns of behaviour and how they either serve you or dominate you. However, in most people's realities there is quite a degree of need and desire involved, so seeing the emotional score for what it actually is can be distorted by those ever-present fears and wishes. So, although this is a 'Realizing Phase', the realizing is in strict proportion to how much you are able or willing to do so. Because of this, the content of such realization is probably going to unfold over several, even many, years to come. And this will happen whether or not the relationship still exists. In fact, in retrospect this time can be viewed as a kind of emotional landmark, a registering of one's most basic and essential emotional make-up. At its most extreme, it is rather like some deep and complex subject which is a lifetime's work and study. In any event, it is probably a good idea to view it this way, for thinking that whatever goes on now is of passing significance could be classed as flippancy at best, or grave emotional denial at worst.

Other areas of focus for such emotional revelation, apart from one-to-one relationships, can be those involving one's parents – especially your mother. Becoming considerably more aware of her influence upon you is what can take place now. Again, what you can draw, learn and benefit from this is entirely up to you. Although this is by no means a stock expression of this 'Phase', losing a parent (or anyone close) can obviously act as a powerful emotional awakening.

A more straightforward, and therefore advisable, way of experiencing your 'Second Realizing Phase' is through some form of therapy, counselling or group work. If you are intentionally looking for increased emotional and self-awareness, then now is the time to get it.

Although all Moon 'Phases' are most powerful during their first year, this one can end as it begun because your Second Saturn Opposition occurs at that time (between 43 and 44). This period also makes it clear that there is ultimately no side-stepping or rationalizing away what is emotionally on the cards. If there is any music to face, then this is the time you are forced to do so. This may entail the 'real world' of external authority or material conditions, or as some health condition.

This, possibly one of the most difficult 'Phases', is basically about 'contrast' – between what you want and what you have, between what you feel and what you think, between reality and the dream, between yourself and someone or something else. Although he never reached this age, the poet John Keats lived his life according to the principle of contrast because he found that in such a way he became as acutely aware of existence as it is possible to be. This gave rise to some of the most moving and profound poetry ever written, and reading it is recommended to anyone suffering the extremes of emotional life, at this or any other time.

Finally, a famous example of someone who experienced an ugly but highly significant contrast and realization at this time in their life was Nelson Mandela, who was arrested and imprisoned at the very end his Second Realizing Phase' and spent a few months over an entire Moon Cycle of 27⅓ years in jail, being released at the very end of his 'Third Realizing Phase'. It was as if he needed that whole cycle to go from one realization to another, and come to a complete realization. When he became South Africa's first black president in May 1994, his government immediately established the Truth and Reconciliation Commission to investigate and come to terms with the country's violent and apartheid past, such being a perfect and positive expression of his then current 'Third Understanding Phase' (*see* page 90).

Jupiter Growth Modes – 'VIRGO – LIBRA – SCORPIO – SAGITTARIUS' (*see* pages 32–35)

Other Universal Planetary Cycles active at this time

Check the tables on page 289 to see if this Planetary Cycle was active for you at this time.

URANUS OPPOSITION (*see* page 168) This is the classic astrological indicator of your 'Mid-life Crisis' and having it occur during this 'Phase' is doubly exposing and revealing. The contrast between what you think life to be and what it actually is can be particularly acute. The more you have a fondness for the unadulterated truth (or can cultivate such), the more likely you are to progress as a result of this time of truth – for that is what it is. Nelson Mandela, mentioned above, was experiencing this when he was arrested in 1962 for sabotage and terrorism, which are ruled by freedom- and rights-seeking Uranus.

NEPTUNE WAXING SQUARE (*see* page 198) As this influence makes your illusions go critical, having this 'Realizing Phase' happening at the same time should prove, shall we say, 'interesting'. So much depends upon whether you regard your illusions as precious and in need of preserving, or as something to identify and be rid of, and feel lighter and clearer for doing so. This is not a time for half-measures or being on the run from your own reality. It is best to see things through to some kind of conclusion, trusting that the pain along the way is simply the price you pay for the realization. Failing this, a game of cat-and-mouse could be your lot.

PLUTO WAXING SQUARE (*see* page 226) The heavier, more inevitable things in life are what you are having to realize and become more aware of. An ostrich and his feathers are soon parted – meaning that if you put your head in the sand you could get a kick in the rear! The truth never hurts anyone for very long except those who persist in hiding from it.

Your Second Moon Cycle continues with your 'Second Sharing Phase', *see* page 83. Be sure to read the general description of this 'Phase' on page 80, if you have not done so already.

3rd Realizing Phase 68y 4m–71y 9m

'EMOTIONAL WISDOM' – Here we get to the legendary 'three score and ten' mark which is traditionally regarded as the allotted human life-span, although not as the astrologically regarded one, which is 84 years of age. Furthermore, this time is seen astrologically as a period in your life when some great realization can be made, brought about by a wealth of experience and an acceptance of life's great mystery as being ultimately positive. If, however, cynicism or narrow-mindedness have been allowed to take hold, then such a realization is not going to take place. Instead, such realization would be seen as 'through a glass darkly', that life was some bad and pointless joke, with only regret and resentment for constant company. A third alternative could be the realization that life was *not* (or just could not be accepted as) such a negative affair – giving rise to the first type of realization, that life is something great and that you are now in a better position than most to appreciate this.

Two perfect examples of how the realization of this time of life can be a crowning glory is seen in the examples of Nelson Mandela and Emmeline Pankhurst. After having been imprisoned for the rebellious expression of his beliefs and principles for just over the average Lunar Cycle of 27 years and four months, Mandela returned to public life, and was recognized by the world as a human rights hero. (*See also* page 77 for the 'Second Realizing Phase' when he was incarcerated.) Emmeline Pankhurst, that well-known champion of Female Rights, saw the realization of all her dreams and labours when the full and equal vote for women was at last introduced in Great Britain on the day she died in June 1928 at 69 years of age.

In a strictly relative fashion, you too can rise to the realization of being someone unique and useful in this world. Whether this is in a small or large way, the Third Saturn Waxing Trine between 68 and 69 years of age should help you to make this realization through utilizing your age and experience, the opportunities and agencies that society offers, or a combination of both.

Jupiter Growth Modes – 'SAGITTARIUS – CAPRICORN – Aquarius – Pisces' (*see* pages 35–38)

There are no other Universal Planetary Cycles active at this time

Your Third Moon Cycle continues with your 'Third Sharing Phase', *see* page 84. Read the general description of this 'Phase' on page 80, if you have not done so already.

'Sharing Phases'

17y 1m – 20y 6m
44y 5m – 47y 10m
71y 9m – 75y 2m

THIS 'PHASE' WE COULD SEE AS THE symbolic equivalent to the Disseminating Moon, which is all about letting others know and experience what you have accomplished, learnt, and become interested in so far – but especially what came into flower or focus during the preceding 'Realizing Phase'. So this is called the 'Sharing Phase' because it is about imparting to others something which is felt to be worth having or knowing, or of giving away that which is deemed as not being exclusively your property, and the circulation of which is regarded as being mutually beneficial.

Furthermore, it is through sharing in this way that we become more sure of whatever it is that we are feeling, demonstrating or communicating – all of which contributes to a better 'understanding', the 'Phase' following this one. It is as if we are 'road-testing' what we have emotionally and mentally assimilated previously, seeing if it needs tweaking this way or that in order to make it more effective and cohesive. It can be seen from the picture of this 'Phase' that it is the mirror image of the 'Adjusting Phase'. This time, however, you are adjusting what has already taken shape, rather than adjusting what is on the way to realization. This is rather like the actor who 'wears in' or edits his part of the script so as to appeal more to his audience, as distinct from the playwright who edits and fine-tunes his script as a whole into the finished article before it is actually performed. So the 'Sharing Phase' poses the necessity of shaping your personality and its expressions so that they appeal to others rather than just one's own pet ideas. It is quite possible that you do not feel able to share yourself with others at this time, or that what's inside of you should 'cook' for a while longer. But this is not a good idea because it is through this 'wearing in' or 'weathering' that you substantiate your personality and become clearer about its nature. Letting things fester inside can give rise to emotional and mental problems later on. 'So let it out and let it in.'

By way of example, I will share my own experiences of this 'Phase'. My 'First Sharing Phase' was marked by making the harsh transition from an English public school to working in a job on a factory floor where I had desperately to 'edit my script' to fit in with people who had prejudices about people of my social and educational background – even though the very reason I was working there was because I did not have the social or financial privileges of that background. But in retrospect I can see that this was my initial basic training in becoming able to reach the 'man in the street' through being merely one human being amongst others, free of any socio-cultural stereotyping. Come my 'Second Sharing Phase' 27 years later, I had my first book published, sharing what I had learnt from astrology and life.

Like all the other 'Phases', the meaning of this one can be expressed passively rather than actively – people sharing their thoughts and feelings with you as well as, or instead of, you doing so. This may take the form of some significant emotional exchange, being a sympathetic listener, or simply being taught something academically. If you are of a passive disposition, this is far more likely to be the case, yet at the same time you could also interpret sharing as a physical or bodily act. After all, the way we sensually or sexually express ourselves is as much a sharing of ourselves as demonstrating what we think and feel in other ways. In fact, it is probably doing so at a far more fundamental level.

Then again, the sharing could be more in the usual sense of sharing what is yours with someone else – whether willingly and gladly, or not. Consciously and genuinely sharing in this way can be profound or spiritual in the sense of being a quintessentially human 'What's mine is yours' kind of thing. Inasmuch as there is no sense of 'property', then there is no sense of loss either.

During any 'Sharing Phase' you can manage, learn and benefit from it by Sharing what is happening or has happened.

Here are each of the individual 'Sharing Phases' that can occur during your lifetime.

1st Sharing Phase 17y 1m–20y 6m

Now at last we reach this period which could be regarded as late adolescence, with the accent upon showing others how 'in the know' you are with regard to whatever it is your peer group regards as worth knowing. If you're maybe a bit of a wallflower and usually find yourself on the receiving end of this sharing, then there has to be something that has happened (or not happened) previously that has made you shy of sharing yourself. Because of this, you feel that there is something about you that you'd rather not share. This is particularly likely at 17 when going through a self-critical 'Virgo' Jupiter Year. But this is rather like what I call the 'Venous Blood Syndrome' which is analogous to the process whereby deoxygenated blood in your body, a dirty purple in colour, is returned to the lungs for re-oxygenation. People are often surprised to discover this, saying that they have never bled any purple blood – and certainly not the dirty sort. This is because if you cut a vein, the blood immediately hits the air and is re-oxygenated, turning bright red. The metaphor here is that what we regard as negative in ourselves (the deoxygenated blood in need of regeneration), we are inclined to keep to ourselves until we are attacked, wounded or get angry ('see red'). If we consciously and deliberately let it, the 'blood', out, it changes into something (more) positive because we have 'aired' it. At a 'Sharing Phase', especially this one, do not wait to be wounded; let it out anyway and you will be amazed at the transformation. If anyone is shocked or repelled to see what you're made of, then that is their problem – you're simply reminding them of the 'dirty purple blood' they themselves are holding inside.

Between 19 and 20 years of age, your First Saturn Waning Trine occurs (*see* page 113), which should somehow give you a greater sense of being in tune with the world around you, or that someone in authority approves of you – or that the time has come when you cannot repress yourself any longer. Either way, this is a good time to 'let it bleed'. The feeling now of having found someone you can connect with can be so strong that you should be careful not to overreact – a tendency if you are of an insecure disposition and think this could be your last chance. Remember to get the measure of any significant other's sharing of themselves. If it seems more like a self-advertisement, or a lack of any self-disclosure, rather than an honest emotional expression or exchange, then be cautious, and do not share too much of yourself with them. There are far too many cases on my books of clients marrying in haste at this time and repenting at leisure later. With the Saturn Waning Trine particularly, you have time on your side, so don't be pushed.

Sometimes, the sharing can be of the more obvious kind – like sharing resources and living space, intentionally or otherwise. Several clients of mine have had late siblings born at this time, or had a parent remarrying, meaning that they feel they now have to share their mother's and/or father's attention and resources. This can even give rise to a reluctance to share emotionally and mentally as an adult, with negative consequences. Then again, such experiences can teach you as a youth just how to share, with positive consequences. As ever, neither the Moon nor any other Planet creates a situation, but simply activates what is already there within the individual. Sharing of this kind can also be strongly imposed if you should become a parent now. The sharing of living space is the most common and elementary experience of sharing that most people have upon moving away from the family home. Or possibly the sharing of one's family home becomes an issue. But sharing is still the keyword.

Another event that can now be peculiarly influential is the death or serious illness of someone close. If you are not able or allowed to share your feelings and grief with someone, it can inhibit a sharing of your deeper self later, especially on a sexual level. Having said this, sexual sharing to excess during this 'Phase' can desensitize you later. Note that such issues as sex and death are more likely to arise during your 'Scorpio' Jupiter Year at 19. One client of mine tragically lost her mother at this time, but the only way she could share the intensity of her emotions was through sex, which led to further complications that she still could not share.

So if you have not got the message by now, the priority during this 'Phase' is to find someone with whom you can be and share yourself. If you cannot find a person, then keep a journal where you freely allow your thoughts and feelings on to paper – or talk to a pet or a tree. But share you must!

Jupiter Growth Modes – 'VIRGO – LIBRA – Scorpio – Sagittarius' (*see* pages 32–35)

Other Universal Planetary Cycles active at this time

Check the tables on page 289 to see if these Planetary Cycles are active for you at this time.

URANUS WAXING SQUARE (*see* page 162) This is 'Breaking Away and Breaking Through' and so lends an element of surprise and experimentation to your process of sharing. Occasionally you might bite off more than you can chew, but then that will be something to share as well! As this influence so often forces leaving the family home or being involved in something new like a first job or university, or having no job and little money, sharing takes on a very special meaning.

NEPTUNE SQUARING PLUTO (*see* page 193) Having this 'Adolescence Strikes' influence occurring now makes it even more imperative that you share what you are feeling and thinking. The power in this influence may even inspire you to say or do something very creative. In any event, the point is that there should be plenty *to* share because this Transit can be quite dramatic in its effect and the events it brings. Be particularly careful of hasty emotional commitments now, because although you might start out with a lot to share on a physical level (sex), you may later find that there is little else. A mutual sharing of thoughts and feelings should effectively filter out such delusions during this heady, even dizzy, time.

Your First Moon Cycle continues with your 'First Understanding Phase', *see* page 87. Be sure to read the general description of this 'Phase' on page 86, if you have not done so already.

2nd Sharing Phase 44y 5m–47y 10m

As this 'Phase' begins with your Second Saturn Opposition, it is quite likely that there is a reluctance to share or exchange your thoughts and feelings with someone. This is because Saturn's influence can emphasize our inhibitions, and/or find us in the midst of what we regard as an emotionally unsympathetic environment. Yet looked at another way, Saturn could be regarded as forcing us to look at our weaknesses by sharing them with someone else. In so doing, you would loosen up blocked feelings and generally clarify your emotional situation. Furthermore, such sharing would lead to the understanding necessary for the next 'Phase'. As ever, in order to share just when it is most difficult to do so would require courage – and as someone once said, 'Courage is not merely one of the virtues, but every virtue at its testing point'.

Looked at in yet another way, as Saturn governs the objective or professional world, a sharing of yourself with someone on that level – like a counsellor, therapist or impartial friend – is advisable. Whatever happens, though, you are going to be finding yourself sharing in some way or other, consciously or unconsciously, willingly or unwillingly. As our lives at this time are usually quite full or busy with the accumulated commitments that come with middle age, sharing can take on a number of meanings, depending, as all Planetary influences do, on the context of experience in which they take place.

If in the midst of a separation or divorce at this time, the issue of sharing can be one of having no-one to share with any more, or having (had) to share your partner with someone else. In either case the importance of sharing is driven home. This poses further issues, or rather questions you might need to ask yourself. Have I been receptive, open or honest enough for Other to share with me? Have I been a good or bad example of sharing? Have I been too possessive – or not possessive enough? Have I shared myself too much in that I leave no mystery? Answering any of these questions may entail sharing what you think are the answers with someone else. This is because sharing ultimately involves being very honest emotionally, something which few people are to that great an extent, and so a 'mirror' in the form of someone else is necessary. I am made mindful of how powerfully influential self-honesty, and the lack of it, is, by a client of mine who had kept his feelings close to his chest all his life. Following the death of his father during the preceding 'Realization Phase', during the beginning of this 'Sharing Phase' he got so much 'off his chest' that he ceased to suffer from the asthma which he had had all his life.

This brings us to another important experience that would bear upon the importance of sharing – grief. Apart from the more obvious, but no less important, aspect of grief which is that of simply missing someone very deeply, there is also the case where resentments and other negative feelings towards the deceased can give rise to very complicated and confusing states of being. Without sharing these convoluted feelings of guilt and grief, all manner of quite dangerous emotional and ultimately physical complaints can ensue. Remember that after the Full Moon of the preceding 'Realizing Phase', we are supposed to 'empty' ourselves more and more through the Waning half of the Moon's Cycle. Without doing so, such thoughts and feelings can begin to fester. So now is the time to *begin* to unload negative feelings, eventually leading to 'making your peace'. (*See* 'The Venous Blood Syndrome' described under the 'First Sharing Phase' on page 81.)

New relationships that begin during this time should, apart from other things, be viewed as being opportunities to share like you have never shared before. If it doesn't work out, the chances are that it was down to not seeing the writing on the wall, or that it was simply a part of your sharing process. Within existing relationships, difficulties in sharing can take the form of going from one form of sharing to another, as in the case of physical sex being no longer (so) available or possible, meaning that a new level of sharing must be attained that either replaces your sex life or revitalizes it.

Other likely scenarios of sharing may include having children at this relatively late stage, being a member of a less boundary-conscious form of culture, making breakthroughs in work or creative life by sharing what you might have previously kept to yourself, or getting blocked or stuck because of a reluctance to share your ideas.

Jupiter Growth Modes – 'SAGITTARIUS – Capricorn – Aquarius – Pisces' (*see* pages 35–38)

Other Universal Planetary Cycles possibly active at this time

Check the tables on page 289 to see if this Planetary Cycle was active for you at this time.

URANUS OPPOSITION (*see* page 168) This is the classic astrological indicator of your 'Mid-life Crisis', and having it occur during this 'Phase' is a strong indication of the need and opportunity to share, with one or many, what you have learned, felt and experienced – and what is going on inside you right now. Whatever is or has been bottled up can explode if not shared.

PLUTO WAXING SQUARE (*see* page 226) The heavier, more taboo-ridden aspects of life and yourself are what you are having to share and thereby become more aware of. Carrying around dark secrets or heavy and negative feelings has now reached saturation point, so get unloading! 'They that drink (or hang on to) the old wine, have no room for the new'.

Your Second Moon Cycle continues with your 'Second Understanding Phase', *see* page 89. Be sure to read the general description of this 'Phase' on page 86, if you have not done so already.

3rd Sharing Phase 71y 9m–75y 2m

As the essence of 'Sharing Phases' is that of imparting to others what you have learned and experienced, this one has the advantage of a lifetime of gathering such things. But as with all the 'Phases' that occur in later life, in what can be retirement years, there can be a disconnection from the world at large for one reason or another. Such isolation, which is more likely to become an issue during your Third Saturn Opposition between 73 and 74 years of age, could have a legion of causes – from the loss of a lifelong companion to an existing habit of withdrawing yourself, or some other inclination born of resentment, bitterness or simply subscribing to the conventional belief that being elderly is supposed to be a second-class state.

Whatever the case, sharing is the key to making the most of this time in your life, and of resolving any issues or problems. I recommend that you read the general description on page 86, if you have not already, in order to appreciate the full significance of sharing. One might even say that sharing is the opposite of being on your own, lacking anyone to share with. By this it is implied that *not* sharing can be the very thing that creates that lack. So, if you are lonely, put aside any reservations and muster the courage to share whatever you can, be it emotionally, mentally or physically. You will then find how Nature, abhorring a vacuum as it does, will bring you what you need. But you have to open the door in order to let in Nature's gifts.

All being well, this can a wonderful time of helping, teaching and entertaining others with the ripeness and wisdom of your personality and its experiences.

Jupiter Growth Modes – 'PISCES – ARIES – Taurus – Gemini – Cancer' (*see* pages 38, 27–30)

There are no other Universal Planetary Cycles active at this time

Your Third Moon Cycle continues with your 'Third Understanding Phase', *see* page 90. Read the general description of this 'Phase' on page 86, if you have not done so already.

'Understanding Phases'

20y 6m – 23y 11m
47y 10m – 51y 3m
75y 2m – 78y 7m

THIS 'PHASE' IS THE SYMBOLIC EQUIVALENT TO the Waning Half Moon, which we could see as representing a period requiring re-orientation as involvements and certain attitudes of mind that are no longer necessary to our life must fall away, like the husk of a fruit. Seeing that the Waning part of the Moon Cycle is like a harvest-time, at this, the critical point in it, we have to separate the wheat from the chaff – that is, retain the 'good seed' and let go of what is useless. But the main issue here is that we understand why this letting go must happen, otherwise we are inclined to hang on to attitudes which we do not need. What is basically demanded from us now is an understanding of the simple truth that cycles are all about renewal, and that we only hang on to the unnecessary out of the mistaken sense that life only deals you one hand. It does not. The player remains what he or she essentially is, but new scripts and parts keep coming round – be they good, bad or mediocre.

This 'Phase' is a critical one because although at the time there may be some great or small crisis occurring, we are not easily inclined to understand what it means. All we can focus upon is what or who has departed from our life, or is threatening to do so. Again this harks back to that ignorance of the fact that our lives are cycles within cycles. Nature has Her purpose. Like the Lord, She giveth and taketh away – and giveth again, and so on and so on. But what is *really* critical now is the *attitude* that we have to whatever is happening now, and to life in general, because 'as you think, so shall it be'. To put it another way, whatever difficulties you are experiencing emotionally now, they are there because of an attitude or philosophy that is no longer appropriate. It is as if your attitude is like the combination that unlocks the door to the next chapter of your life – and it is different to the one you have been using. If you still hang on to some role or outlook that you have grown too attached to, then your mind is closed to whatever life is trying to present you with, and whatever that is simply becomes unavailable to you – possibly you'd never know what it was.

If you look at the above picture of the 'Understanding Phase', you will notice that it is the mirror image of the 'Deciding Phase' on page 62. This too is a time of decision, but now it is a decision about how to view what is happening, not necessarily a decision about what to do or how to act. You have to know what you are looking at before you can do anything effective to, with, or about it. When you do know, you then have the new combination, and the way ahead becomes clear.

In terms of what kind of events can characterize this 'Phase' and its need to reform your attitude, they are bound to have one thing in common – the old departing to make way for the new. This could be a mother giving birth to her first child at the start of World War Two (a personal revolution and a global one), the end of a relationship that frees you up for a fresher and better one, the death of a parent forcing you to become someone in your own right or reappraise your feelings, the end of a personal era of life that ultimately ushers in a whole new chapter, a new relationship or form of relationship, or the loss of a job that forces you to look again at what you are doing with your head and hands. Whatever it might be, the point is that you must understand it in this light of the old dying and the new being made room for, conceived, and ultimately born. The timing of this can be quite straightforward, or decidedly subtle, or a combination of both. Moreover, it may not be fully appreciated or understood until viewed in retrospect from the next cycle – and as ever, its meaning could be extremely personal to you. In my own case, during my 'First Understanding Phase' I let go of a whole group of people who were previously my whole life; retreated even more from them during the following 'Releasing Phase', and was positively reborn when I left for a new life somewhere else come the beginning of the next Moon Cycle. But to emphasize the point once more, at the time I was not aware of quite what I was doing or where I was going – I was only aware of what could not or should not be in my life, not what was going to take their place. In the light of this awareness, my attitude to what was happening, or rather ceasing to happen, was what changed. A new attitude to life not only paves the way for the future, but actually heralds it in. What that is could arrive at any time if you've adopted the right attitude, but in terms of this, your Moon Cycle, it is 'scheduled' to come at or during your next 'Emerging Phase'.

During any 'Understanding Phase', you can manage, learn and benefit from it through endeavouring to Understand what is happening and thereby adopting the right attitude towards it.

Here are each of the individual 'Understanding Phases' that can occur during your lifetime.

1st Understanding Phase 20y 6m–23y 11m

After the roller-coaster of the three 'Phases' of your adolescence, upon entering your twenties, you begin to formulate your own code for living – that 'new combination' mentioned in the general introduction to this 'Phase'. But it would seem pretty unavoidable that confusion will arise as your old attitude to life conflicts with the new one that is trying to take shape. Whatever the content of your confusion though, it will have a key bearing upon the direction your life wants to go, emotionally and attitudinally speaking. For example, during this time one client of mine was obsessed with the fear of catching AIDS (with no good reason other than a brief infidelity with someone of virtually non-existent risk) and could only find one person to confide in. To cut a long story short, the fear of AIDS passed but then she felt hostage to this friend's continuing confidentiality. This feeling of having the threat of exposure hanging over her was an emotional blackmail scenario that had been haunting her since she was very young. The reason was that from the age of seven she had to keep her *mother's* extramarital affair secret from her father. In her seven-year-old head she believed that if she did not keep the secret, her mother would run off with her lover. Now this pattern was trying to come to an end, but true to the nature of this 'Phase', some form of crisis had to arise to force the issue. The Moon, that is her unconscious, was saying to her 'this paranoid belief that you have a guilty secret is driving you into a very tight corner'.

Now my client did not understand what was actually happening to her at the time, and her only change in attitude was to tread even more carefully and consequently intensify her paranoia. But one of the beauties of this Moon Cycle 'Phase' is that through back-tracking in this way, we can then 'recast the past', which is a bit like editing the videotape of your life. My client managed to do this during the following 'Releasing Phase' when she released all this material to me. Then with her 'Second Emerging

Phase' when she was able to put these shadows of the past behind her as she moved away from the area and changed to a partner with whom she could be open.

All of this hopefully conveys both how critical and subtle the 'Understanding Phase' can be, especially during your early twenties when you are having to devise your own moral standards. Such pressures to alter your attitude are most likely to occur during your First Waning Square of Saturn (*see* page 114) between 21 and 22 years of age. This was when my poor paranoid client divulged her dark secret to her new-found friend (note that 22 is an 'Aquarius' Jupiter Year, the Sign which governs friendships.)

Such inner conflicts created by this 'Understanding Phase' are common to people in their early twenties. Often all the individual can feel is depression because they cannot begin actually to look at what comprises the conflict. This is probably because, by its very nature, it involves a lack of self-awareness. During such a difficult time, possibly the key lies in that 'Aquarius' Jupiter Year. To unburden yourself to a friend, or a professional therapist, is to lend some vital objectivity to what can be a horribly subjective state. Incidentally, my client's friend (an Aquarian) never did break her confidence – but that's another story.

Jupiter Growth Modes – 'SAGITTARIUS – CAPRICORN – Aquarius – Pisces' (*see* pages 35–38)

Other Universal Planetary Cycles possibly active at this time

Check the tables on page 289 to see if these Planetary Cycles are active for you at this time. Remember that any Trine influences may have to be actualized to make them that effective.

FIRST SATURN WANING TRINE (*see* page 113) This influence should help you to come to whatever understanding is needed in a relatively controlled and stable fashion. Such help is quite likely to come from someone older or more experienced, whom you respect or like.

URANUS WAXING SQUARE (*see* page 162) If you are experiencing this during your 'First Understanding Phase', then more likely than not you will encounter some event that demands understanding in some way or other. The experience of my fearful client described above is a good example of this for Uranus is always out to expose us (to the truth). So you might say that Truth and Understanding go together very well – something which you could find well worth appreciating during this combination of influences, and in the light of my client's salutary experience. In other words, if you wish to find or give understanding now, let the Truth be your friend and guide – particularly through the Uranus-ruled 'Aquarian' Jupiter Year or the candid 'Sagittarian' Year.

NEPTUNE SQUARING PLUTO (*see* page 193) With this cycle active at the same time as your 'First Understanding Phase', you are more or less bound to have some sort of crisis – be it of conscience, consciousness or both. Essentially, this is all saying that you must re-orientate yourself drastically. Being wilful and sticking to some teenage myth or rebellious role could become quite destructive – either now or later. This is a time to grow up fast.

PLUTO CONJOINING NEPTUNE (*see* page 221) Should this extraordinary influence, called the 'Illusion Buster' and 'Inspiration Booster', happen to occur so early as during your 'First Understanding Phase', then you will be experiencing and hopefully learning rather more than is usual for this time of an individual's life. Either understanding is essential now, or else whatever is happening will lay down the foundations for it later.

◐ Your First Moon Cycle continues with your 'First Releasing Phase', *see* page 93. Before or after, be sure to read the general description of this 'Phase' on page 92, if you have not done so already.

2nd Understanding Phase 47y 10m–51y 3m

This 'Phase' is all about putting together an attitude of mind that allows you to manage and relate to the world in a way that could be regarded as 'mature' and in keeping with what should be expected of someone of your age and experience. One reality could be that you experience situations that demand this, and how well you fare has everything to do with the current state of your individual personality. The chances are that the part of your life and/or character that finds itself in the frame will be what is in need of being brought up to par. Then again, you may have arrived at a point where your emotional understanding has developed to a degree where some new level of living and relating comes about in the form of a new or renewed relationship, job or lifestyle. Seeing that this 'Phase' gets underway at 48 during an 'Aries' Jupiter year and a Saturn Waning Trine, the odds are that some definite forward push is there to get you moving, or at least, give you some sense of being in control of your life and its direction. Taken together, these two possibilities pose circumstances that are both encouraging and demanding. This could be a career move that is upward but more stressful, or a love interest that both delights and confuses you, complements and compromises you. But it is still all down to attitude, that is, the way you look at it and consequently, the way you present yourself to it.

If you look at whatever is demanding your attention in a fashion that is inappropriate, then it will demand even more of your attention. For example, I had a client who was newly involved with someone and she was wanting more from him than he felt he could afford. But this attitude was the very thing at fault, for in putting his usual limit upon emotional commitment, he made his new partner put even more pressure upon him, eventually compromising his professional position. He failed to understand that in giving a little more quality time and attention to her than was his custom, he would satisfy her needs and diminish the pressure. Instead, by sticking to his old but unintentional attitude of 'treat 'em mean and keep 'em keen' he was making the situation into the very one he feared – of feeling emotionally put upon. At a deeper level, he had to understand why he felt that he could not give enough emotionally, and why he felt so easily crowded. Giving her a loving look rather than a dismissive one took exactly the same amount of time, but his faulty attitude had him believe that a loving look would incur some kind of unwelcome complication or obligation, while a dismissive one would dispel it. Not surprisingly, such an attitude got formed earlier on in his love-life at the time of his 'First Understanding Phase' when he was rejected by a girlfriend after having shown his feelings.

All this very much bears witness to the esoteric dictum that 'Energy follows thought', that how you think determines what will happen to you. After all, everything in the world began with a feeling or idea. This is particularly applicable to cases where you could be giving yourself a hard time over some issue, particularly a relationship. Either with some concentrated introspection, or preferably with the help of a good friend, counsellor or therapist, you will find that what you thought was a problem was not a problem – but just the way you were looking at it. This is analogous to getting headaches when all that is required is a new pair of spectacles.

Apart from your attitude affecting the circumstances, most circumstances occurring during this 'Phase' are there to change your attitude. Resisting this simple formula can make a hard time harder – and last longer. This is especially the case if an inappropriate attitude that the difficult situation is reflecting has been around for a long time. Simply facing the truth is often the first step to a solution, whereas arguing the toss is asking for the 'lawsuit' to persist, perhaps for years.

Considering that this is a critical 'Phase' and that it occurs at a time in life when one is 'supposed to know better', it is not too surprising that it has a sting in its tail. This is the Second Saturn Waning Square at the very end of it. This is saying, 'If you haven't got the message – take that!' In other words, the pressure increases as reality impinges upon that unsuitable attitude, compounded by the inclination to cling on created by the concurrent 'Cancer' Jupiter Year. However, that same pressure can lift instantaneously when you finally look at the situation in the right way. It may help if you cast your mind back to your 'Second Realizing Phase' (*see* page 77) and determine what you learnt or what became clear then. Whatever it was – and it should have been something – it contains the key to your correct understanding now.

Jupiter Growth Modes – 'PISCES – ARIES – Taurus – Gemini – Cancer' (*see* pages 38, 27–30)

Other Universal Planetary Cycles possibly active at this time

Check the tables on page 289 to see if this Planetary Cycle was active for you at this time. Remember that any Trine influences may have to be actualized to make them that effective.

PLUTO WAXING SQUARE (*see* page 226) This 'eliminatory' influence really does stress the need to let go of an old and outworn attitude that will otherwise cause you all manner of difficulty. Reading the general description that introduces the Pluto section on page 213, especially the part headed 'Obsessions or Convictions', will prove very helpful.

PLUTO WAXING TRINE (*see* page 231) Having this occur now will alleviate or preclude any pressures caused by your viewing difficulties in the wrong way. This will be because your understanding of what does and does not matter has developed enough for you to drop whatever needs dropping as or before it becomes an issue.

◐ Your Second Moon Cycle continues with your 'Second Releasing Phase', *see* page 95. Be sure to read the general description of this 'Phase' on page 92, if you have not done so already.

3rd Understanding Phase 75y 2m–78y 7m

This is when ripeness and maturity should turn to wisdom, if they have not done so already. Nelson Mandela, whom I used as a positive example of the Second and Third 'Realizing Phases' (*see* pages 77 and 78), also came up trumps with his 'Third Understanding Phase'. At the very beginning of this 'Phase' he was inaugurated as the first black president of South Africa. Most significantly for this 'Phase', though, was the immediate implementation of his government's Truth and Reconciliation Commission. At the time of writing, this is still seeking to heal the country's past of apartheid and all that it entailed through inviting perpetrators of crimes to come forward and expose themselves to their victims or the friends or relatives of those victims. Rather than create more violence and animosity through revenge and punishment, this wise and understanding method can only create more Understanding. Astrologically, it is also worth noting that this first year was a 'Cancer' Jupiter Year, the Sign of emotional understanding and also Nelson's Sun Sign.

This example is a beautiful testament to how harsh experience, when transmuted with and into understanding, makes the true human being, which is surely the prime purpose of a human life.

Jupiter growth modes – 'CANCER – Leo – Virgo – Libra' (*see* pages 30–33)

There are no other Universal Planetary Cycles active at this time apart from your Third Saturn Waning Trine (*see* page 131).

Your Third Moon Cycle continues with your 'Third Releasing Phase', *see* page 97. Read the general description of this 'Phase' on page 92, if you have not done so already.

'Releasing Phases'

23y 11m – 27y 4m
51y 3m – 54y 8m
78y 7m – 82y 0m

THIS 'PHASE' IS THE FINAL PHASE OF the 27⅓ year cycle. Being the symbolic equivalent to the Waning Crescent or Balsamic Moon, it is a time of healing in the sense of letting go as much as possible of the very last of anything that could impede or compromise you during the next cycle, especially at its beginning. As well as having to do with healing, Balsam is also an oily substance that can smooth the way if properly applied.

Ideally then, this is a time when we can empty ourselves of feelings, memories, attachments and habits that are inappropriate to our current circumstances, and in so doing smooth the way for the future. But in reality, we can still feel that if we give something up, even that which we find uncomfortable, there won't be anything to replace it, or that something untoward will happen. In Western culture at least, we are conditioned to avoid feeling empty. This is because we are achievement-orientated, which is more to do with the Waxing Crescent or 'Striving Phase'. If you look at the picture of this 'Phase' above, you will see that it is the mirror opposite of the 'Striving Phase' on page 57. And the opposite of 'Striving' is 'Releasing'. We can also release ourselves from pressures and conflicts with certain others by not reacting to them, but simply 'letting it go'. So a 'Releasing Phase' could be regarded as making room for what is arriving with the 'Emerging Phase' and new Moon Cycle that follow it. The less you let go of stale or redundant thoughts, feelings and reactions – or any involvement that goes with them – then the less room there will be for the new. Hanging on to the outworn in this way could amount to a self-fulfilling prophecy because you might lose someone or something through doing so, and mistakenly think that you should have hung on more rather than let go.

So consciously or unconsciously, deliberately or involuntarily, at this time we release from ourselves anything that is 'imprisoning' us. Remember we are talking about internal things, like thoughts, feelings, attitudes and habits – not necessarily external things like relationships, jobs and domestic set-ups. Nonetheless, releasing such internal things can, and does, prepare the way for a release from any external ones that have passed their 'sell-by date'. All this can mean a surrendering to whatever forces are prevailing – and, to be more precise, we also encounter the need for making some kind of sacrifice for the good of the whole. Indeed, the 'good of the whole' is in many ways the essence of the 'Releasing Phase'. This is because the 'emptying' that takes place during this 'Phase' is rather like a river emptying into the sea. This means that individual feelings are having to be sacrificed for the good of the feelings of the group – be it a large group or just a couple – or for the good of the whole of the rest of your life. You may well experience a feeling of being lost or 'at sea' during one or more of these 'Phases', and this would be a sign that you are trying to hang on to your sense of separateness and self-containment, rather than surrendering to the 'outgoing tide'. Any such egotism, for that is what it is, could be regarded as the enemy now – although the exception to this could be the 'First Releasing Phase'.

The kinds of events that one could experience during this time are ones where it can appear that you are losing out, or that life is passing you by. To complete the example of my own experiences through the Moon Cycle, during my 'First Releasing Phase', after having accepted quite a few endings in the previous 'Understanding Phase', I then seemed to lose everything that I had had going for me formerly in terms of profession and social life. From being a record producer in London during the Swinging Sixties I went to vegetating in a boring office job, and living a very uneventful private life. Amazingly, this period of surrendering to my rather unprepossessing fate lasted three years and five months, the exact length of this 'Phase'. Then, having emptied myself completely, the new Cycle began with absolutely everything changing.

From reading the above, you may have felt that this 'Phase' is quite similar to the foregoing 'Understanding Phase' in that they are both to do with letting go of what it is in your life and personality that is no longer appropriate. But while the 'Understanding Phase' is akin to the stage where a vessel goes from being half empty to quarter empty, the 'Releasing Phase' is when it goes from there to being totally empty. Apart from anything else, this means that whatever outdated thoughts, feelings and attitudes that were not revised or eliminated during the previous 'Phase' are rotten and must be rooted out during this one. In fact, considering that 'balsamic' viscous quality, this 'Phase' is perfectly analogous to an 'oil change'. You cannot fill the tank with fresh oil until it has been emptied, and when it is being drained you cannot go anywhere. But once it has been filled again you are refreshed and can go on your way, travelling smoothly again, ready for the next complete Moon Cycle. None of this means to say that during the whole of this 'Phase' you are going to be 'off the road'. This is because, firstly, it would be labouring the metaphor (we still carry on living), and secondly, that notorious thing, the human ego, can ignore all these psychological traffic-lights and needs for emotional servicing, and steam on regardless – only to grind to a halt somewhat further down the road.

During any 'Releasing Phase', you can manage, learn and benefit from it by Releasing yourself and others from whatever is happening or has happened.

Here are each of the individual 'Releasing Phases' that can occur during your lifetime.

1st Releasing Phase 23y 11m–27y 4m

The surrendering or sacrificing quality of this 'Phase' is not so likely to take place voluntarily or consciously. This is because at this age you are still in the process of building and developing a life and career, and such passive and selfless responses to whatever is happening to you or around you could be regarded as inappropriate and ineffectual. But this is precisely what makes the 'First Releasing Phase' a potentially difficult one. Having said that though, statistically there is a great likelihood of having one or more children at this time, and this is probably the most sacrifice-demanding thing that a human being can experience. I dare say that if one looked into the figures, one would find that this is precisely the event that characterizes this 'Phase' for most people. It could also be said that for a pregnant woman, the experience of at last giving birth is a classic form of Release!

Even if you don't experience early parenthood, it is at this time that you really have to 'put away childish things'. Such 'things' can be girlish or laddish behaviour, or any pursuit that was really just an extension of adolescent fun and experimentation. This does not mean to say that having a good time and trying new things should come to an end – far from it, for such letting off steam is a releasing in itself – but it does mean that there now has to be some consideration of how you fit into the greater scheme of things. Although the Moon does not have anything directly to do with socialization, the 'Releasing Phase' does have this quality of making you feel one among the many. But, as I say, during your twenties you are quite likely to be still looking out for number one. Yet it all depends upon what this entails. One client of mine was at this time working hard to make it as a professional healer, but the fact of the matter was that she did not have enough sense or experience of people, or of being one of the people, to get very far, despite gaining the academic qualifications.

Another client came to see me during this 'Phase', bearing all the trademarks of a sad social misfit. His releasing took the form of unloading all the peculiarities of his personality and life experience so far. This was very intensive, involving his strange thoughts, feelings and activities that no-one on Earth knew about, let alone his alcoholic parents with whom he still lived. After numerous appointments spreading over three years he came to see me for the last time. He said that he did not need to come any more because he at last felt that one person now knew him. He went down my drive looking almost ethereal. He died in his sleep a few weeks later, at the very end of his First Moon Cycle. (*See* page 51 for more of what I call 'once around the bay' lifetimes.)

Now don't go thinking that you too are likely to come to a premature end at the close of this 'Phase' and Cycle. True astrology is not so much in aid of predicting events as understanding what those events mean, when they happen or happened, or preparing yourself for certain possibilities. This is particularly the case with these Moon 'Phases'.

Then there was the young mother who during this 'Phase' did not feel 'at sea' – until her husband went away *to* sea! She also had Pluto Conjoining her Neptune, however, which would intensify such a state (*see* below). I also have a few clients who actually lost or aborted children during this 'Phase', giving rise to intense feelings, not just of loss, but also of being lost, and that the release was unnatural or premature. Also, being newly married during this period and then feeling totally cut off can be a problem, possibly because you do not feel ready yet to make that kind of sacrifice of your individuality – but for the time being that is what your unconscious is saying you must try to do.

A final and famous example of this 'Phase' – and how during it one can appear to be losing out when in truth it is just that there is a bigger issue at stake – is Muhammed Ali. At this time he had his World Heavyweight boxing title stripped from him as a result of being convicted for refusing induction into the US Army (because of his own religious convictions). It was not until his Saturn Return (*see* page 115) and Pluto Conjoining Neptune (*see* page 221) that this conviction was reversed, and he could eventually go on to regain his title. (*See* also page 227 concerning Ali's retirement and illness.)

All in all, you can see that the experiencing of the 'Releasing Phase' of your mid to late twenties is an apparently contradictory set-up. But in Nature's terms, which is the Moon's domain, it probably all makes sense – it is just that we have departed from Nature to a significant degree.

Jupiter Growth Modes – 'PISCES – ARIES – Taurus – Gemini – Cancer' (*see* pages 38, 27–30)

Other Universal Planetary Cycles possibly active at this time

Check the tables on page 289 to see if these Planetary Cycles are active for you at this time. Remember that any Trine influences may have to be actualized to make them that effective.

URANUS WAXING TRINE (*see* page 166) As this can assist your intuition, you are able to plot your way through the possible subtleties and confusions of this 'Phase'. Muhammed Ali, mentioned above, was quite clear about the course he was taking through the reefs and disappointments that he encountered then.

■ **PLUTO CONJOINING NEPTUNE** (*see* page 221) This extraordinary influence is called the 'Illusion Buster' and 'Inspiration Booster' and its occurrence now strongly intensifies the 'Releasing Phase' sense of being caught up in an ebb tide where no matter what you do something seems to be governing your fate. This can be a thrilling, uplifting and a kind of 'white-water' experience if you go with flow, or very disorientating and upsetting if you are trying to control things by swimming against the tide and the natural order of things. Apart from a thousand other fascinating things, my own simultaneous experience of both these influences found me cut off by the actual tide until it started to ebb! Here, Pluto Conjoining Neptune was felt quite simply as the power (Pluto) of the sea (Neptune), or the inevitability (Pluto) of Fate (Neptune).

◐ Your First Moon Cycle, having now come to a close, continues with the first 'Phase' of your Second Cycle, your 'Second Emerging Phase' or 'First Lunar Return' *see* page 51. Before or after, be sure to read the general description of this 'Phase' on page 48, if you have not done so already.

2nd Releasing Phase 51y 3m–54y 8m

This time in your life can either be experienced as the beginning of the (bitter) end, or as a surrender to a life lived in a more spiritually or psychologically aware fashion. In reality, it could well oscillate between these two extremes. Everything depends upon whether you experience fate in general, and your own fate in particular, as something you feel offended by or trapped in, or as something that you should trust as knowing best. On the face of it, it would appear that the latter choice of surrendering to your fate (and possibly another's) is far more desirable. The trouble is that the ego always likes to think that its lifestyle should live up to its romantic, glamorous, self-indulgent, comfortable or vain idea of itself. Consequently, seeing that your situation during this 'Phase' has the in-built necessity of making some kind of sacrifice, your lifestyle is going to appear wanting from the ego's point of view. Such an 'inadequate lifestyle' may include a restricted or even non-existent social or love life, a materially or emotionally unrewarding job, or any situation that seems to be out of your control or beyond your influence. In 1998, the headlines revealed the marital crisis of Rod Stewart, just as he turned 54. It is evident that a deal of this kind of releasing needs to be done for his marriage to Rachel – who is going through her Saturn Return – to survive, or more likely, for Rod to progress on his own.

Returning to our River of your Life metaphor, this stretch is the estuary where the river inexorably empties into the sea – especially as the tide is now ebbing (this is the last 'Phase' of the Waning or second half of the Moon cycle, remember). This symbolizes that time and place where your individual path has to succumb to universal, collective or karmic forces – fate in other words. And unlike your 'First Releasing Phase', which was a more unconscious mingling with the masses during your mid-twenties, this one means having to live for others. This engenders a most practical definition of that much misused word 'spiritual': meeting a collective need. This can be anything from looking after an ailing parent to helping someone close to you and getting nothing (or worse) in return; from swallowing your pride and doing what has to be done, to surrendering what seems most precious to you personally for the good of the whole. Unless you are consciously committed to and involved with some form of healing process or spiritual discipline (the best expression of this 'Phase'), the trouble is that the ego regards all these courses as a surrender of control or what it wants, and therefore possibly inviting chaos (the word 'estuary' means *commotion*, by the way). But the truth of the matter is that if the ego tries to get its way then there *will* be commotion – not least of all inside of you. This could be physical (health problems), emotional (relationship problems) or mental (work problems). And the reason for any commotion would be that you were trying to control the situation in the face of uncontrollable forces.

This is why this 'Phase' is termed 'Releasing', for you need to release physical, emotional and mental blockages – or rather what is creating them. These can be very old issues. I had a client who was experiencing all manner of skin and hair problems as something was trying to break out and release itself through the only channels being allowed it. Astrologically, I was able to locate a point in her early life that contained a trauma about which she had told nobody. It was a great relief to her there and then to release this matter at last, but I emphasized that it would be highly compressed after more than 40 years, and encouraged her to release it more and more, whenever, however and to whomever she could. Whether or nor my client continued to empty her old hurt, I do not yet know – but truth will out as surely as the river finds the sea. If one accepts and lives through the 'commotion' of the struggle between one's fate and one's ego, peace and resolution are released into your life.

In my own case, I had a 'message' from my motor car, that wonderful symbol of one's own physical-emotional-mental vehicle (the personality). For the first time ever I ran out of petrol – on my way to seeing my doctor. This was symbolic of an unconscious need to empty myself of whatever I regarded as essential to my ego's idea of progress, something which was interfering with my physical health. I also had to surrender to the need to be helped out by someone or something else. However, I did not get this message at the time – my radiator had to blow up a month later and empty out all the water and coolant!

This 'Releasing Phase' could be described as one of 'enforced selflessness' as you are obliged to live up to something higher or more inclusive than an egocentric sense of life. A symptom of this is that of feeling more sensitive than usual. This is owing to sensing the whole more keenly, with a view to attuning yourself to your role in it. Such increased sensitivity can express itself as irritability and feeling 'drowned' by others or circumstances. As well as letting go of your ego barriers, in the way that is being suggested here, it is also a good idea to protect yourself (*see* page 271 – Protection Along the Way) because there is no point in being incapacitated. One exercise given that can be particularly useful with regard to allowing inside what's on the outside is what I call 'Listening to the World'. This is a simple meditation in which you sit somewhere where you are not going to be disturbed (preferably outside) and listen in an intent but relaxed way to whatever sounds are going on around you. Do not try to analyse or identify the noises, just aurally acknowledge them. After a short while you will hear sounds that you were not even aware of, until eventually you have a sense of 'listening to the world'. This is quite integrating, and automatically instils a trust in the great 'out there'.

Seeing that this 'Phase' begins during your Second Saturn Waxing Square (*see* page 119), you can be forgiven for initially wrestling with and balking at whatever assails you from without. But this should be regarded as just a sensible stance to adopt as you assess the situation, with a view to seeing what or who it is in your life that you are possibly going to have to meet more than half way. This 'Phase ' is a time to go with the flow – not for maintaining a stiff upper lip or being rigidly non-compliant, for this will sink you.

Jupiter Growth Modes – 'CANCER – Leo – Virgo – Libra' (*see* pages 30–33)

Other Universal Planetary Cycles possibly active at this time

Check the tables on page 289 to see if these Planetary Cycles are active for you at this time. Remember that any Trine influences may have to be actualized to make them that effective.

URANUS WAXING TRINE (*see* page 166) This offers you the chance and awareness to find an unusual path through the 'estuary' of this 'Phase'. You are more likely to see and appreciate the wisdom of taking the spiritual course rather than struggling with the ego's chosen path.

NEPTUNE WAXING TRINE (*see* page 204) To have this occurring along with this 'Phase' implies a certain 'Planetary poetry' because the one serves the other. The Neptune Waxing Trine offers the ultimate opening of the way of the spirit, while the 'Second Releasing Phase' is what demands that you best follow that way.

PLUTO WAXING SQUARE (*see* page 226) This strongly emphasizes the necessity of surrendering to fate and giving up any fixed or egotistical ideas that are presently being denied. Failure to succumb in this way could lead to far more serious problems.

PLUTO WAXING TRINE (*see* page 231) Under the influence of this, your ego is less likely to insist on having its own way, simply because age and experience tell you that there are more important things than that. You could even be aware that surrendering to the power of fate and ministering to the needs of others are the most empowering things of all.

◐ And now with the end of this 'Phase', your Second Moon Cycle has come to a close. You now move on to your Third Moon Cycle, commencing with your 'Third Emerging Phase' or 'Second Lunar Return' – *see* page 54. Be sure to read the general description of this 'Phase' on page 48, if you have not done so already.

3rd Releasing Phase 78y 7m–82y 0m

Without being too dramatic about it, and in the awareness that I personally am nowhere near this age, the experiencing of your 'Third Releasing Phase' has about it the ring of surrender and redemption in the face of what could be near the end of life. Whatever the case, it is most definitely a time to make your peace with everyone through releasing anything that you are holding on to which could be regarded as weighing you down. Thus emptied and enlightened, you can proceed toward whatever is in store for you with a transparency that is the passport for crossing the Great Divide.

Jupiter Growth Modes – 'LIBRA – SCORPIO – Sagittarius – Capricorn' (*see* pages 33–36)

There are no other Universal Planetary Cycles active at this time apart from your Third Saturn Waning Trine and Waning Square (*see* page 131)

Your Third Moon Cycle concludes with your 'Fourth Emerging Phase' and 'Third Lunar Return' – *see* page 56. Read the general description of this 'Phase' on page 48, if you have not done so already.

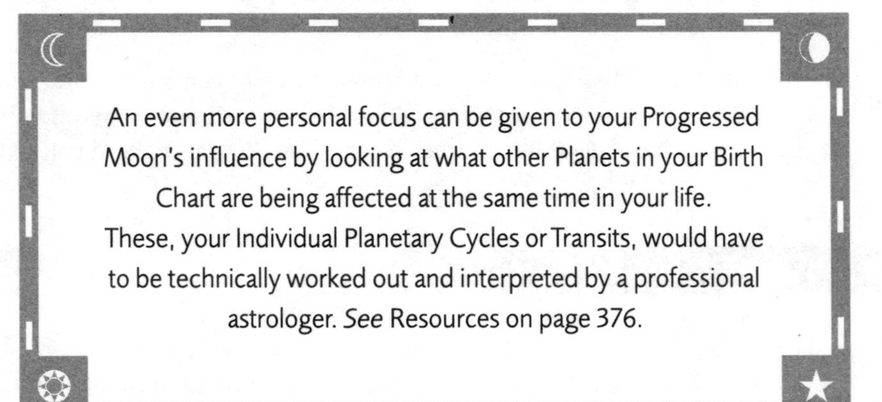

An even more personal focus can be given to your Progressed Moon's influence by looking at what other Planets in your Birth Chart are being affected at the same time in your life. These, your Individual Planetary Cycles or Transits, would have to be technically worked out and interpreted by a professional astrologer. *See* Resources on page 376.

YOUR SATURN CYCLE

Status and Responsibility

'Know this, in all the ages past,
I am the one who set your task;
Since first you came unto this Earth
As spirits young to prove your worth'

Key Influences

+	-
Consolidating	Pressurizing
Teaching	Punishing
Proving	Testing
Cautioning	Thwarting

As far as Saturn is concerned, there is most definitely a specific point and purpose to your being here, alive on Planet Earth. Possibly, we are most hard-put by Saturn's influence when we are failing to grasp that very point – that there is one. It is as if many of us are born thinking it's going to be easier than it actually is. Consequently, upon the shock of finding that there is someone or something making it difficult for us, we overreact and commence looking for some way of denying or getting around whatever Saturn appears to be plaguing us with – with the consequence that it then piles on even more pressure. Then again, there are some people who naturally take to the fact that discipline and hard work are central to existence. However, they too may be missing a point.

This is because there are *two* points inherent in Saturn's dictum, and I describe them below. Interestingly, this dual nature is reflected in the two-faced god, Janus, from whom the name of the month January is derived, the month which is mainly that of the Sign Capricorn, which Saturn rules. Janus is the god of Doorways and Thresholds and, as you will see, is very relevant to Saturn. Yet another mythological connection between the two is that when Jupiter banished Saturn from the sky, Janus welcomed him in, for Saturn must have resonated with him. The two faces of Saturn allow him to look outwards across the Threshold, the First Point(of view) and inwards across the Threshold, the Second Point (of view): the Threshold simply being the boundary between Inner Reality and Outer Reality. In human physiological terms, this would be your skin. Incidentally, Saturn is also called the Lord of Boundaries and the Dweller on the Threshold.

The Two Points of Saturn

Saturn's First Point is the point and purpose that the world at large, the status quo, would have us accept. This is 'Ground Control' (*see* 'The "Story" of Your Life' on page 15) in the sense of being the conventions and controls of the society in which one lives – the 'real world'. These mainly take the form of any external authorities and structures, such as the government, teachers, officials, elders, traditions, ancestors, how one is supposed to behave, and, initially at least, one's parents (or whoever was responsible for you as a child). All these things – rightly or wrongly, well or badly – provide us with a foundation upon which to build our lives and achieve the purpose that this status quo says is valid. Caution and conservatism are fundamental to the nature of the First Point because tradition and material stability depend upon such things. The First Point is also that of the concrete, physical world by which we are, by and large, bound and limited – the world of form. This Point sensibly states that one has to fit into society and be materially stable, to have a 'job definition' and status in the eyes of that status quo.

Saturn's Second Point has more to do with what you are learning in terms of your more fundamental and internal reason for being here, the overseer of which is your conscience. This is Ground Control in that there are certain forces and circumstances which keep us to that path of learning, doing what our conscience deems to be 'right', whether we like it or not. This learning entails certain lessons about life itself in the sense that Planet Earth is like one enormous school. The key lesson here is that of learning to take full responsibility for what one says, does and even thinks. In turn, this engenders the idea that we have a grade, a curriculum, lessons and examinations, a timetable, and the teachers themselves.

But then the First Point also engenders these educational factors, but on a different level, involving different criteria. Then again, an important aspect of Saturn is that there is a measure of the First Point within the Second Point, for you cannot very easily start to improve yourself on an inner level if you have nothing to depend upon in the outer world. Conversely, someone who is successful in the outer world has, sooner or later, to come to terms with their inner life. Furthermore, one is probably pursuing more than just one Point at a time. Possibly some people are exclusively involved with the First Point of material status and concrete reality, but there is usually someone they depend upon who is more concerned with the Second Point. And there are relatively few who subscribe exclusively to the Second Point – like mystics and non-conformists who deny the material world – but even they have to eat and have somewhere to sleep. The vast majority of us, however, simultaneously subscribe to both the First Point, the conventional values of making something of yourself in the material world, and the Second Point, the values which state that true success has something to do with personal integrity and spiritual discipline. However, it has to be said that in today's world, the First Point holds sway. The Monica Lewinsky Affair is a testament to the USA's majority allegiance to the First Point. Many would agree that the First Point is dangerously over-stressed in our world today, and that the personal integrity of the Second Point is in need of a resurgence. This is why the Clinton crisis created such a furore, division and confrontation. In February 1999, after having been impeached, he was cleared, but it is hard to say how it will turn out – especially in the long run. From both of Saturn's Points of view, there is more in the balance than just Clinton's reputation and position.

Your Grade

This is the calibre or 'form' you have at any given time. From the Second Point of view, Saturn is the 'Lord of Karma' keeping the score of how well or badly you are doing with regard to your integrity and spiritual standing. Being the Lord of Karma implies that Saturn has something to do with reincarnation – and it does. It dictates the circumstances and limitations we find ourselves in until we have learnt whatever we are supposed to be learning. For instance, if one is learning about the proper value of money, then one would either have a difficult time getting hold of the stuff, or one could have too much of it and have to learn the value of things that money cannot buy – like love or health. In this case, your Saturn Cycle will test or reward you in this respect; when you have learned something about the true value of money, for example, then you would probably not care about it so much, and one way or the other it would cease to be a problem. If you had failed to learn that lesson, money would continue to be a problem. 'What good is it for a man to gain the world yet forfeit his soul?' (Mark 8, 36).

On the other hand, those who subscribe more heavily to Saturn's First Point often prefer to disregard or even ridicule such things as reincarnation and karma. Synchronistically, on the very day after writing this, England's football team manager, Glenn Hoddle, spoke out about his belief in reincarnation, saying (in private) that disabled people were paying for sins committed in previous lifetimes. The next day he was forced to resign amidst howls of derision. (*See* also pages 169–170 to read about Glenn's gaffe.) This is because the First Point is heavily invested in keeping things 'real' in so far as it is deemed that the only things that exist are those which can be proved to exist, using the very rules for existence laid down by those who preside over the First Point rules – governments and the scientific establishment. And so Saturn's influence in your life would be on that level. In other words, you are tested and rewarded in the context of what are conventionally regarded as being socially and materially desirable. Using the same example, one would learn how to get, earn and manage money – or not, as the case may be. And from Saturn's First Point of view, making money *is* making the grade.

The Curriculum

This, the course which Saturn is running, and upon which you enrolled at birth (or before, if you subscribe to the Second Point), basically has two Levels which correspond to the two Points. A simple way of looking at it is to say that one starts off with the First Point at Level One and eventually graduates to the Second Point and goes on to Level Two. But really it is more complex, because one Point is inextricably bound up with the other. For example, some people might regard themselves as spiritual seekers, or just drop-outs, belonging to the Second Point, when really they are just woolly-headed dreamers and escapists. You could say they have jumped the queue and enrolled themselves on the Second Level of the course without having gone through the 'foundation course' of Level One. It can take quite a lot of time, pain and confusion before such individuals realize that they have to put first things first and learn how to manage the material world. And yet it is ironic that the governments of First World countries, the ones who to a great degree run the First Level of the course, subsidize many people who are floundering in this way. But then, as I have said, everyone is dependent upon the First Point to a certain extent. Even genuine Second Point types who have integrated their personalities enough to be virtually self-governing depend somewhat upon the State or upon the First Point.

However, those who have genuinely embarked upon Level Two of the course have qualified by proving themselves able to function in the material world, even if only just. They then set about learning how the world really works. This part of the curriculum involves the study and practice of right thinking and right relating, and Saturn's ongoing influence is there to guide and cajole them towards this end. Going back to our Ground Control metaphor, First Point people at Level One are like the ground crew which creates and manages, or in most cases simply works for, what is happening at ground level – the so-called 'real world'. Second Point people at Level Two are training to be air crew who eventually qualify to take off in that Rocket Ship which is symbolic of Uranus (*see* "Your Uranus Cycle", page 149). But, as previously pointed out, even they are still dependent upon the ground crew and practical reality. The 'star' who has risen to fame and fortune still has to pay their taxes and maintain their health. It is only when they are in 'free flight' – utterly self-governing – that they are truly liberated; and like all astronauts and flyers, they are relatively few in number. All the same, consciously or unconsciously, these are the heights to which everyone sooner or late aspires, whether it is to be free of the rat race by achieving material success or by rising above it. Yet again, though, there are those who sneak a ride in that Rocket Ship only to come crashing down soon after. This is like someone winning the lottery and having it ruin their life. Still, we often learn valuable lessons that serve us later on, but more significantly, it is these excursions that can eventually (and sometimes haphazardly) upgrade us from Level One to Level Two. But this, as I say, is more to do with Uranus, which we come to in the next section.

The Lessons and the Examinations

These are the actual influences that occur every so many years during your Saturn Cycle, to which I have briefly referred already, and which the last part of this section, the Timetable, is about. In other words, these are the various ways and areas in which Saturn can manifest in your life.

Material Stability – At Level One, money and possessions are seen as the index of how well you are doing in the world, and as a rule, having enough of them is supposed to make one happier. This is one of Saturn's favourite zones of influence because your material or financial state is not something you can pretend about – at least, not for very long. Hard cash is hard cash, and bricks and mortar are bricks and mortar. It also reflects your earning power which in turn reflects what you are worth – at least, in the eyes of the world of Level One. Another very important aspect of material stability here is structure. For something to stand up and continue to stand up, it must have a sound structure. Structural strength is in the foundations and framework, which are laid and made before everything else, and are not as a rule visible from the outside.

At Level Two, material stability is nothing more than a means to an end. If you wish to live a 'good' life through establishing that right thinking and right relating, then in the process of learning to do so you will automatically attract a reasonable material situation. The Devil (Saturn) may look after his own, but so too does God. They say one cannot serve two masters – but it is probably more a case of learning both lessons that are being taught here. As one old Saturnian esotericist put it when asked by a student if he should do more on the spiritual side or the material side of life: 'Do more on both!' Structure at Level Two *is* the right thinking and right relating that ensures that ideas and relationships stand up and continue to stand up. The greatest strength of all is usually hidden, and is essential to Level Two.

Time and Timing – At Level One, time is an artificial thing in that it has little or no connection to natural and cosmic cycles. For instance, the numbers 24 and 60 have no correlation to anything. They are simply figures used to mark the passing of time in order to make sure that things in the material, manmade world, happen *on* time. This artificial time is possibly the most insidious controlling factor over people because nearly all of us all keep to it without really noticing it – other than saying there is too much or not enough of it. Be that as it may, time keeping and punctuality are essential to order at Level One.

At Level Two, time is the essence of life itself. Astrology itself is a true measure of time because it observes how True Time moves and influences us in unconscious or superconscious ways. This book is expressly about True Time. Timing, as understanding and using the process of time, is absolutely essential with regard to achieving at either this Level or at Level One. Listening to and obeying the physical body's cycles is also essential to health. Synchronicity is an important and liberating aspect of True Time, and this is examined in the Uranus section. The Sacred Calendar of the Mayans of Southern Mexico is also an accurate expression of the movement and measure of True Time – a time which is radial as opposed to linear.

At both Level One and Level Two, the maxims 'Everything happens in the fullness of time', 'Time alone heals' and 'Time alone will tell' are all perfectly true – great comfort and guidance can be drawn from them; 'all' one needs is faith and patience.

Discipline and Control – At Level One, discipline and control of people through rules, regulations, and their enforcers are unfortunately essential because some people cannot control or discipline themselves. However, because the ones who have the control often cannot control themselves either, it has all got rather *out* of control, with the controllers abusing their power over the controlled – a sorry state of affairs (*see* also Authority below). Another negative aspect of control is rigidity, which eventually causes things to seize up, including parts of the physical body.

At Level Two, we definitely learn self-control and self-discipline. Not only are we learning these, but what really controls us in the form of guilt and fear (*see* below) and how we seek to control others for the same reasons. Rigidity as a negative form of discipline is recognized to be a great enemy to well-being on this Level.

♄ **The Way of the World** – At Level One this *is* reality. The Level One world is one where you have to watch your back, not be too trusting, read the fine print, be hard-nosed and objective, not take things too personally, and know that there is no such thing as a free lunch. Life is seen as a jungle where the fittest survives, becomes the strongest, gets the lion's share, and where there is little room for sentiment or sympathy. There is no 'divine justice' in the sense of there being any payback for something you have done wrong but got away with. Pragmatism holds sway over idealism, and Saturn's influence here will make sure that this order prevails.

At Level Two the way of the world is still, to quite a degree, as Level One maintains it, except for the crucial difference that you know you do not get away with anything. Thinking ill of someone will eventually have the effect of making you think ill of yourself, or having to depend upon that person for something so that you are forced to be nice to them – or some such other form of poetic justice. The strong shall protect the weak, and give them what they really need in the knowledge that the security of the one is ultimately only as good as the security of all.

♄ **Purpose** – At Level One your purpose is to achieve something in the world, whether it be as a parent of fine children or as a successful professional or businessperson. Whatever earns one stàtus is seen as the purpose of life. Saturn confronts you with this necessity, and consolidates or promotes when you are getting it right. It liquidates or demotes when you are not.

At Level Two your purpose has to do with your ultimate reason for being here, with accomplishing your 'mission'. This may not be recognized or praised by the Level One world at large, just so long as that purpose is achieved. Sometimes, by the very nature of the mission, it will have to be recognized and approved of by the world at large. In the first case, such a mission could be, say, to come more from your heart and less from your head; or in the second case, to teach people how to do this. With both these examples, Saturn will let you know how you are doing; things will feel better if you/they are learning to come from the heart, or feel worse if you/they are not.

♄ **Relationships** – At Level One the conventional approach rules the manner in which relationships are created and managed: the man does this and the woman does that. Or you get married and that should make it all right. But of late, possibly more than any other Saturnian lesson or examination, relationships have the toughest time just keeping to Level One. The reasons for this are complex and explored elsewhere in this book, and also in my book *Do it Yourself Relationship Astrology*. Here it suffices to say that too much conformity, staying too long at Level One, gave rise to millions of people going to two World Wars because they were told to do so, and consequently changed the face of society. As a result of this, to a large degree neither men nor women are quite clear as to what their roles in relationship, or in the world in general, are any more.

At Level Two one is learning to reintroduce some Level One standards of clearly defined roles and rights, the big difference being that such roles and rights are consciously created by the couple and not by the State and Church who have been seen to let them down. Level Two first has to involve the fervour and confusion of having hardly any boundaries of behaviour at all, most manifest in the 1960s, giving rise to increased separation and divorce. This allows us to make our own mistakes and then rewrite the rule book out of our own experience.

♄ **Restriction and Delay** – At both Level One and Level Two, Saturn states that if you are not ready for something, then you will be kept back until you are ready – possibly for your own good. That 'something' will be one or more of the other lessons or examinations mentioned here. Esoterically, Saturn is called the Ring-Pass-Not or the Dweller on the Threshold, which simply states that 'Until you pass my test you cannot go any further'. Inhibitions would also come under this heading, but consciously dealing with them would be more the subject matter of Level Two. At Level One, inhibitions are hardly questioned because they get too easily mixed up with being something or being somewhere that cannot or should not be changed – like a lot of things at Level One.

♄ **Authority** – At Level One, authority is simply one's boss, the government, officialdom, parents, elders, teachers, or whatever or whoever has more power than you do. If you transgress the rules laid down, clearly or otherwise, then you are punished in some way. If you keep to these rules then you will be furthered, looked after, or at least left alone.

At Level Two, although Level One authority can still hold sway, the real authority is something like God or one's conscience. Such a thing, which we can call the Higher Power (or one's link with it) guides us in a way that is considerably more subtle and far-seeing than Level One authority. The more attuned you are to such a Higher Power, the freer you become from conventional authority. In legend, Robin Hood is symbolic of this internal authority for he sustains the inner world of hopes and dreams, which was all the serfs could cling to in the face of the evil barons who abused the authority they had over them in the outer world or Level One. This is an important reality of Saturn, for it is a bridge, possibly *the* bridge, to the freedom of Uranus.

♄ **Physical Condition** – This is to do with the working order of the world of things, which includes the human body and its health. At Level One, 'things' are seen as objects that need to be serviced and maintained. This is perfectly all right as far as inanimate or man-made objects are concerned for they are the very stuff of Level One and the First Point – that is, they *are* mechanical. But things that are not mechanical, or entirely mechanical, are something that Level One reality refuses to see as anything other than *purely* mechanical – be it the human body or the Universe itself. And so Level One reality is filled with sick bodies and a strange animal that could be called 'a limited version of the greater reality'. Or, as Arthur Koestler described the scientist, 'Someone who looks at infinity through a keyhole.' At Level One a person with, say, a bad back is asked how they 'did it' and they are then treated with some corrective behaviour or posture (which can be quite effective), or failing that, bits of the back are surgically fused or replaced with synthetic mechanical parts as if the body was a car put in for repair. Another common complaint, anxiety, is seen here as simply not measuring up to, or fitting into, the System – which is another name for Level One or the First Point. Suppressant drugs are then administered for this kind of complaint in order to keep one 'in place'. Psychologist Carl Jung described the neurotic as 'someone who is non-cognisant of society'. Level One sees the physical condition as being a result of something that physically happened to it, and that it should be dealt with in the same way.

At Level Two the way things work takes into account dimensions that cannot be measured or defined in the usual Level One way. Neither can they (yet) be completely understood in Level Two reality, but it is recognized that there is something subtle happening that is *not* necessarily known or mechanical. So, in this case, the origin of the bad back is first ascertained, in the sense of the sufferer not feeling enough emotional or material 'support' in their life, or of some earlier emotional trauma causing them to hold themselves in an unhealthy fashion, or putting themselves in a position where at Level One something physical *does* happen to them. Anxiety, in Level Two terms, is seen to be caused by not believing enough in that Higher Power referred to under 'Authority'. This means that if one chooses to rely entirely upon the external and tangible powers of Level One, then the Higher Power is bound by Cosmic Law to leave you to the mercy *of* those powers. Level Two sees the physical condition as being a symptom or expression of one's inner being, which is where its true cause lies and where it must be healed.

♄ **Adversity** – At Level One this is regarded as either the very reason for striving to make sure that one is stronger and richer than most of the others, or it is endured and cursed as 'bad luck' and one is branded as a failure. This necessitates a life of hard work and discipline until one makes it, or one of drudgery, depression and dependency that ensures that one never does make it. To push this along, Saturn will heap extra responsibilities on one's shoulders – usually of a work or material nature – but ultimately one is forced up to Level Two.

At Level Two, adversity is seen as a direct result of your karma – that something you did or didn't do in the past is demanding that you make the effort to determine what it is and redeem yourself. The only way you can 'fail' here is through refusing to take this on, because, by definition, no one else is calling the shots but you. In the process, your character is strengthened and inner discipline is acquired. At this Level, in order to force this issue, Saturn will increase your load of responsibilities, but they will be more emotional or psychological in nature.

♄ **Guilt and Fear** – At Level One, as with many things, such states are regarded as something you just put up with, are controlled by, and do not question too closely. The Level One system is heavily invested in keeping people in these states because they effectively keep you under control *at* Level One. The continued criminalization of drug abuse is a sinister symptom of this 'policy', and as short-sighted and damaging as Prohibition was in the United States in that it gave rise to the organized crime that bedevils us to this day. The rules of Level One say that such-and-such should make you feel guilty, and if you do not feel guilty you will be cast out into an area where there are no longer the support systems of Level One – being cast out is one of the very things you are made to fear. The point here is that such guilt and fear keep one ignorant of the existence of Level Two.

At Level Two you learn that guilt and fear are emotions which mask what you are really feeling. This entails possibly the hardest lessons of Level Two because you have to wean yourself off an addiction to blaming someone or something else for the way you feel and are. This is when you learn to take full responsibility for your own thoughts, feelings, words and actions through facing that Dweller on the Threshold – the accumulation of all your illusions and misdeeds. Incidentally, you may also be blaming yourself in a self-incapacitating fashion which is again masking that need for being responsible for yourself. All of this is what is called, in esoteric terms, Initiation.

♄ **Ageing** – This is the hourglass that Saturn holds, along with his scythe with which he reaps his grim harvest. No matter who you are, the Grim Reaper will one day come for you. At Level One, ageing and death are feared above all else because little is seen beyond one's physical existence or physical condition. The search for eternal youth, living by appearance only, cryogenics and forestalling the wisdom of old age (the very thing that lies beyond Level One) are all symptoms of this.

At Level Two, ageing is regarded as a noble process whereby you become an elder who knows the true way of the world – both at Level One and Level Two. The body is seen as a temporary vehicle that should be taken care of but not worshipped, and which you cannot take with you.

♄ **Saturn Types** – These are people who are Capricornians, or who have a Capricornian/Tenth House emphasis or strongly placed Saturn in their Birth Charts. In ordinary terms, such types (at both Level One and Level Two) would be orderly, responsible, conservative, industrious and achieving on the positive side, or rigid, controlling, cold, workaholic and depressive on the negative side. Their professions at Level One could include such things as dentist, architect, builder, politician, businessperson, official, teacher or any position that is either in charge or subservient. At Level Two, professions would include anything that wielded authority *on* Level Two. As such they would be the 'agency' or figure that brought Saturn into your life through possessing the qualities associated with this Planet and the Sign that it rules. (*See* 'Planetary Effects, Figures and Responses' on page 259).

♄ – For the meaning of the actual glyph of Saturn, *see* 'The Jupiter–Saturn Balance of Power' on page 265.

The Timetable

BELOW IS THE ACTUAL SCHEDULE THAT LAYS DOWN when Saturn's lessons and examinations occur in your life and when possible honours are bestowed. As you read these the active points in your Saturn Cycle, consider how they are affecting you at Level One or Level Two, that is, on an external level and/or an internal one.

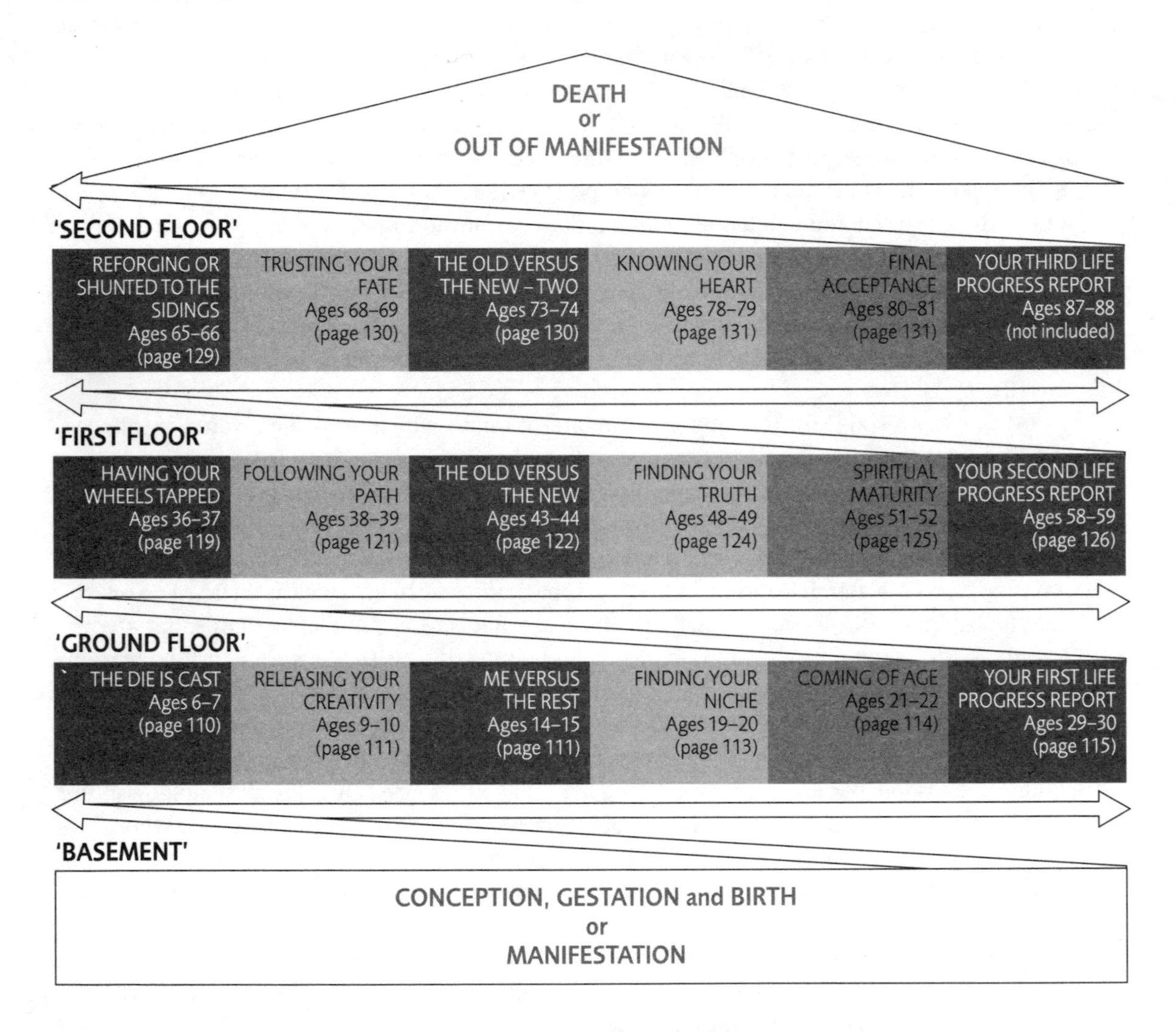

THE HOUSE OF SATURN

Like your Moon Cycle, your Saturn Cycle can be likened to a building of three storeys or 'stories'. You are conceived and gestate at 'basement' level (Moon) from which you are born into the outer physical realm where you work your way through the 'rooms' (or significant periods of your Saturn Cycle) of the 'Ground Floor', which is the first orbit of Saturn, lasting 29½ years. You then go up to the next storey or 'First Floor', then through that and up on to the 'Second Floor'. Note how each 'room' relates to the one above and/or below it, and that each floor can be split between Level One and Level Two. The 'Roof' is symbolic of departing the physical realm (Neptune), and of course, it could be 'accessed' at any time during the cycle, at any storey or room when life ends, as Saturn sees fit.

Here follow the significant periods and meanings of your Saturn Cycle

★ Saturn's influence will occur during the period of your life given at the heading of each influence. Its unmistakeable effects, as described in this section, will be felt or experienced by you for as little as a month or two, or for up to a year off and on. Very occasionally, it can sometimes be felt for up to a year before or after (but not both) the age given. An astrologer could easily determine for you exactly when your Saturn Cycle is active, but for the scope of this book this would involve tables far too long and complex. However, as I say, you will know and feel when it is active.

★ After you have used the timetable, you can then go on to meet 'The Teachers' on page 133. This is rather like having a personal tuition where you gain a more detailed idea of how Saturn affects you during any particular influence described in the timetable. The Teachers can also help you get a more accurate fix on the time any influence occurs.

★ Please remember that there is also the Individual Cycle of Saturn to be considered (that is, Saturn's Transits to other Planets in your Birth Chart that occur throughout your life), but these would have to be calculated by an astrologer or with an astrological computer program. If some of what has been said in the foregoing pages has struck any chords, then this is possibly because such Saturn Transits are what you are or were experiencing.

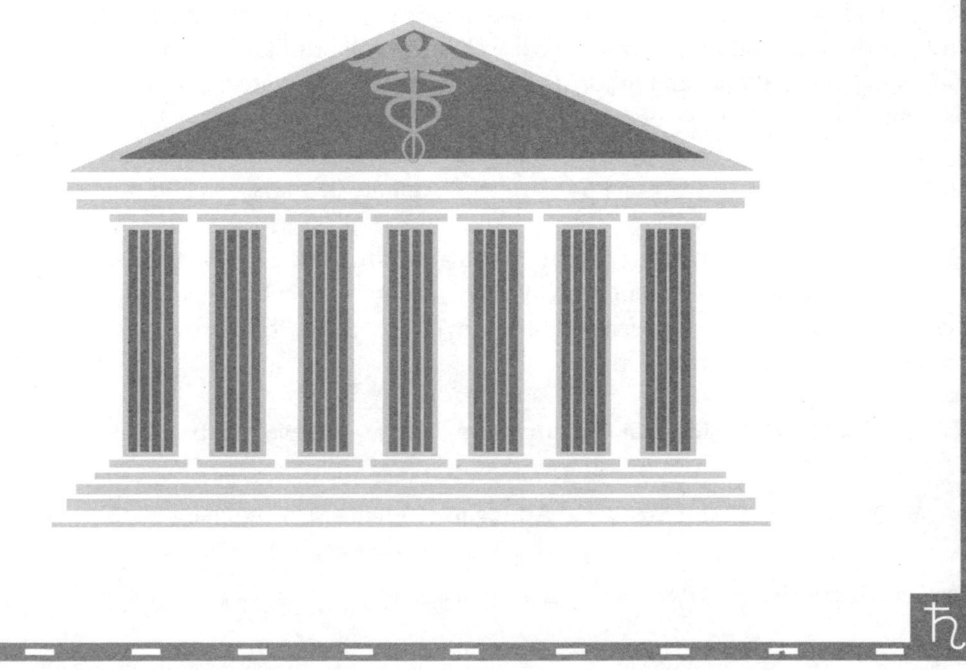

First Waxing Square | Between 6 and 7 years of age

'THE DIE IS CAST' – The soft and pliable being that you were as a new-born baby was moulded and formed into a product in the image of the expectations of your parents, family, first schooling, and, to a lesser or greater degree, of society as it impinged upon the protection of those guardians. Countering or enhancing these expectations and influences that sought to form you would have been your own innate personality, your soul. And now that being must find its adult teeth and begin to contend with the world as a more independent and accountable ego, all with the assistance or encumbrance that its guardians and teachers have laid upon it.

Until this age we often only refer to ourselves by our actual name or by a pet-name or baby-name. But with the onset of your First Saturn Waxing Square, something urges you more and more to say 'I'. In this way you begin that long road of establishing yourself as a separate entity with its very own set of thoughts and feelings. Consequently, at this age you begin to show the calibre and nature of your will. The reaction that you get to this from parents, siblings, teachers and peers causes your will to adjust itself or to assert itself even more, depending upon its innate qualities.

Looking back at this time can be very useful and enlightening with respect to noting the experiences and feelings that happened then. Your importance and potency, whether it shows or not, are the most pressing and sensitive issues at this age. At six years old, a client of mine was not told that a house move was about to happen, and one day she came home from school to nothing and nobody. It was only a matter of a few minutes before a neighbour sorted her out but in that time her sense of personal significance was severely damaged. So parents take heed – of this and of all the astrology of growing up! At and up to this time, the 'die is cast'.

More than anything else, the encouraging of a child to be creative and playful is what really counts now. During these one or two years it forms its more conscious idea of the nature of the world it lives in – especially in terms of the nature of authority. If it finds that authority – in the form or parents, teachers and all adults – is fair-minded, with a balanced sense of the need for work and play, of consideration for itself and others, and stresses the importance of being good at expressing itself as distinct from merely getting it right, then that 'die' will serve it well into a colourful future. If it is made to see authority as rigid, overly conventional and hypocritically 'right/wrong' orientated, then it can be doomed to a black and white (or grey) future – or else it is caused to rebel. If it experiences very little sense of authority at all, then such an absence of structure will bedevil it, first with first a false sense of freedom and then a very real sense of insecurity and no sense of discipline. Of course, a child might experience and be a product of a blend of these.

Possibly more than anything else, at this time you are made aware of your sense of choice: to choose to look at something in a certain way, to contribute or not to what is going on around you, to choose what kind of world you have been born into – in other words, what you as a unique individual make of it and what you think about is crucial.

JUPITER 'YEAR' for your Growth Mode – 'Libra – Scorpio' (*see* pages 33 and 34)

MOON 'PHASE' for Emotional Attunement – 'First Striving Phase' to 'First Deciding Phase' (*see* pages 58–63)

There are no other Universal Planetary Cycles active at this time

First Waxing Trine Between 9 and 10 years of age

'RELEASING YOUR CREATIVITY' – The nature of the creative impulse that is the birthright of all human beings now has a chance to burst forth. A great deal, though, depends upon the degree to which it has been nurtured previously. If as a child you were at first dammed up by the ignorance and narrow expectations of others, then this release will be to little avail. But where originality of mind and personality has been encouraged, the child positively flourishes, tumbling over adversity with enthusiasm. In either case, however, there is still the possibility of some helpful and sympathetic teacher or friend appearing on the scene. The very best should be made of any such occurrence, because the inhibited child can be given a chance to loosen up its repressed creativity, whereas the uninhibited one can be shown how to be more creative still – and even generous with its gifts.

This can be a time when the authority of parents, teachers, or adults generally, can be a positive influence just when it is needed – for instance during a change of school. There is no guarantee here, as the Trine influence needs to be used consciously – something which the child itself is obviously not that able to do.

Parents should take serious note of the above, not least of all because at this time the growing child, rather that just accepting or reacting to authority as before, is actually on the look-out for a form of authority that it can respect and appreciate, but will reject one that it cannot. This is further emphasized by the ninth year being a 'Capricorn' Jupiter Year.

JUPITER 'YEAR' for your Growth Mode – 'Capricorn – Aquarius' (*see* pages 36–37)

MOON 'PHASE' for Emotional Attunement – 'First Deciding Phase' to 'First Adjusting Phase' (*see* pages 63–69)

There are no other Universal Planetary Cycles active at this time apart from Neptune Squaring Pluto (see page 193) for those born in the 1960s.

First Opposition Between 14 and 15 years of age

'ME *VERSUS* THE REST' – During this time, the impressions you get of how life works for you can be very far-reaching. On a purely natural level, puberty is forcing you to erupt from the matrix of parents and family, whether the matrix is like a warm, secure nest or a cold and sticky mud-pit. At the same time, you are also probably having your first 'serious' interaction with others on a sexual or emotional level. On top of all this, or more likely underlying it, there can occur some kind of 'karmic confrontation' now. Thrilling and energetic as this time can be for many, courtesy mainly of sex hormones, its impact is often experienced as anything but a groovy teenage picnic. If it *was* for you, then count yourself very lucky or karmically clean, meaning that your soul's course was still aligned to its natural one.

However, astrologically speaking, and if my friends, acquaintances and clients are anything to go by, the First Saturn Opposition is when you are basically made aware of what you are in for – particularly on a social and emotional level. It is rather as if one embarks on the roller-coaster of life and freewheels through the first downhill half with the sheer impetus of being born. Come this, the halfway point of the first round, kinetic energy ebbs and one is made aware of an equal and opposite force coming in the other direction. The world and life comes at you like an uphill gradient.

How you relate and receive this has everything to do with the type of person you essentially are. For me, it was a time when I was just beginning to gain confidence with the opposite sex, in quite a pleasant neighbourhood. Then all of a sudden, for financial reasons, my father had to sell up and move away to a cheaper area 60 miles away. This place was a social wasteland, and my social and sexual development was put on ice, with repercussions that I shan't go into here. Also at that time, my mother and father were both having affairs, and had little time for poor little desolate me. But suffice to say, these years were 'formative' – a good old Saturnian adjective that describes for many the effect of this time only too well.

Unless you happened to have, or had, a 'Happy Days' type of First Saturn Opposition, you will find that a lot can be gleaned from your experiences now in terms of that grade I referred to in the introduction to this section. This is that 'karmic confrontation' in that you are made aware of what you have to deal with in this lifetime. For me, I was driven into my interior which ultimately bred in me an intense awareness of how myself and others tick – but I paid the price of such introversion by being very hard to reach emotionally for a large part of my life. At that time I also learnt to play the guitar, rather than with girls, and that too formed an important part of my curriculum vitae.

This time can, not surprisingly, be rather alienating – hence the 'Me *Versus* the Rest' tag. The basic thrust of this time, as I pointed out, is to pull away from parental influence and become a healthy adolescent. Unfortunately, owing to the alienating and unnatural quality of modern society, more often than not such a natural, hormonal process backfires on itself. Teenage gang warfare is an example of this alienation looking for something it can identify with. A group other than family that also embodies this alienation and one's emotional denial, *and* has a similar group (the opposing gang) to project one's hate and frustration upon, fits the bill nicely.

Every picture tells its own story, and I'm sure each teenage gangster has his or her own reason for feeling split within themselves in this way. A client of mine, for example, got to kiss with the 'catch' of her group, but the only thing she caught from him was glandular fever which did wonders for her sexual confidence. Thereupon she entered into a relationship with someone who abused her, and she put up with it because it seemed to be all she deserved in terms of love and attention. Breaking out of this negative sexual self-image and an obsession with health was the main part of her 'form' which she had to contend with in life. Another client of mine went to work for someone who found her very attractive and so she unwittingly antagonized his wife. This is a pattern of behaviour she continues to this day, some 60 years later – that is, she always relates to men in a way which offends the women they are with, evidencing a karma of unsure sexuality.

Whatever is a focal issue at this time is going to be a focal issue in your life as a whole. It does not necessarily have to be sexual in nature, but it usually is. For instance, it can also concern issues of belief, school, family life, a younger sibling being born, shyness or self-consciousness, or any event, person, change or thing that makes you aware of how different you feel from those around you. For this reason, running away from home is a popular expression of this time, as is anything that gives you an intense impression of yourself, whether you want it, like it, or not.

Your Saturn Opposition is so typical of Saturn itself because it makes you starkly aware of the difficulties and challenges of life at a time when you are supposed to be having fun, according to popular belief. Then again, 'having fun' can be the antidote or counterbalance to Saturn's heaviness, and, as such, is a positive and natural response that you would be wise to adopt if you are experiencing it now. If you are viewing this time in retrospect, then understanding the hard bits and what they posed, and not forgetting the good bits, would be your antidote against it still dogging you to this day. Also, on a more esoteric note, this age can be a time when you come very close to finding your mission in life, hopefully grasping it fully later on. This is reflected in the myth of Parsifal who on his quest for the Holy

Grail came upon the Grail Castle in his youth. The trouble was he did not know what to say or do when he got there and had to wander off to gain the experience he needed before he eventually came back properly equipped years later. So, if you are having your First Saturn Opposition now or are looking back on it, try to focus upon any near misses, for they could hold a valuable piece of information.

Finally, if you are a parent reading this with regard to your teenage offspring – then read it again!

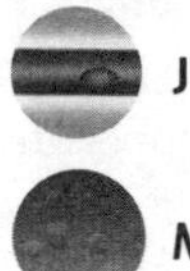

JUPITER 'YEAR' for your Growth Mode – 'Gemini – Cancer' (*see* pages 29 and 30)

MOON 'PHASE' for Emotional Attunement – 'First Realizing Phase' (*see* page 75)

Other Universal Planetary Cycles possibly active at this time.

Check the tables on page 289 to see if this Planetary Cycle is active for you at the same time as your First Saturn Opposition.

NEPTUNE SQUARING PLUTO (see page 193) The temptations and hairy, but typical, teenage experiences that this influence can bring, magnify the effects of your Saturn Opposition quite dramatically. I suggest you read the description of it (if you haven't already, or again if you have) with a serious eye for its implications on the one hand, and with some relish for tasty experiences on the other. If you are reading it in retrospect, then seriously relish the consolation, or rather reward, of understanding.

First Waning Trine **Between 19 and 20 years of age**

'FINDING YOUR NICHE' – This is a time when you are inclined to settle into some form of situation or relationship. This is not to say that it is ideal or that it will last forever, but that it is more like a friendly port where you can feel relatively 'at home'. Statistically, this is a popular time for getting married, probably because this Saturn Trine gives an underlying feeling that emotional stability is at hand. Similarly, the kind of work you are doing now reflects a feeling of material stability – or it simply feels like the right road to be on. Personally, at this time I left my steady job to go on the road with my band; this was not particularly secure, but it felt right. I also started my first proper relationship which lasted seven years. Others, probably most others, will settle for job security now, especially if they are tying the knot as well.

Another common feature of this period is that you can attract an older person or older people who support or guide you in some way. This may be a trainer or teacher, someone who houses you, or simply a workmate. They may also be someone in authority who gives you that vital step up.

JUPITER 'YEAR' for your Growth Mode – 'Scorpio – Sagittarius' (*see* pages 34 and 35)

MOON 'PHASE' for Emotional Attunement – 'First Sharing Phase' to 'First Understanding Phase' (*see* pages 81 and 87)

Other Universal Planetary Cycles possibly active at this time

Check the tables on page 289 to see if any of these Planetary Cycles are active for you at the same time as your First Saturn Waning Trine.

URANUS WAXING SQUARE (*see* page 162) The overall necessity of 'Breaking Away and Breaking Through' is made considerably easier by Saturn's supportive influence – it may even nullify it altogether. What changes you make are more likely to be sensible and well-organized. Looked at the other way around, the buffeting you get now could force you to make good use of the positive opportunities that Saturn may bring.

NEPTUNE SQUARING PLUTO (*see* page 193) Saturn's influence can be totally over-ridden by the powerful urges and temptations provided by this influence. Then again, helpful advice and support could come at a time when you really need it. It would pay not to be too wilful – and you know it.

First Waning Square Between 21 and 22 years of age

'COMING OF AGE' – Although traditionally this is when the world confers 'adulthood' upon us, it is really more a case of the world challenging us to cut the mustard, to shape up in the face of its tests and opportunities. This can take many forms and be the consequence of any number of situations. Leaving or going to university or college, the ending of something, leaving home, changed circumstances – these are some of the things that can manifest Saturn's call to 'get it together'. But, whatever the case may be, you are probably going to have to make more of an effort, endure harder circumstances, do without in some way, acclimatize yourself to an alien environment, or anything that tests your mettle as a young adult setting out on the hard road of life. The types you encounter may well be unobliging or unsympathetic. Beware of Scrooges, too.

However, despite the fact that this is a Square influence, it is also a time of significant honours and events. Celebrated examples of this abound, but here are a few. Grace Kelly, Greta Garbo and Brigitte Bardot all made their successful film debuts at this time in their lives, while Howard Hughes first began to invest in films then. Another kind of debut was that of Queen Elizabeth II's maiden speech to the nation on television. There are also significant graduations: Mahatma Gandhi qualified as a barrister then, Margaret Thatcher and Jackie Onassis graduated from college, and Yuri Gagarin got his pilot's licence. Mikhail Gorbachev joined the Communist Party. An important point here is that all these honours or events are underwritten by effort, commitment or tradition. Saturn never rewards when these elements are absent.

JUPITER 'YEAR' for your Growth Mode – 'Capricorn – Aquarius' (*see* pages 36 and 37)

MOON 'PHASE' for Emotional Attunement – 'First Understanding Phase' (*see* page 87)

Other Universal Planetary Cycles possibly active at this time

Check the tables on page 289 to see if any of these Planetary Cycles are active for you at the same time as your First Saturn Waning Square.

URANUS WAXING SQUARE (*see* page 162) The overall necessity of 'Breaking Away and Breaking Through' is particularly pressing at certain critical times during this part of your Saturn Cycle. Beware of fears and conditions that can beset you and stop you from doing your own thing or from taking advantage of opportunities to spread your wings. Taking a calculated risk is the call of the day, as opposed to doing something too impulsive or not doing anything because you are being too cautious – and regretting it for a long time into the future.

NEPTUNE SQUARING PLUTO (*see* page 193) The temptations of this influence can cause you to do something foolish which you later regret, so the reserve and caution of Saturn can be a useful brake. However, be careful that you do not trade fruitful experience for mere safety. Caution means checking out your options, not rejecting them out of hand because they seem odd or weird. Listening to your own sense of sobriety is important now, but then so is listening to your urge for experiencing the unknown.

PLUTO CONJOINING NEPTUNE (*see* page 221) This extraordinary influence is called the 'Illusion Buster' and 'Inspiration Booster' and when it occurs alongside your First Saturn Waning Square it means that you are having to get into a higher gear emotionally. This entails having to move swiftly with the times, while at the same time exercising enough caution so as not to be fooled or seduced.

First Saturn Return | Between 29 and 30 years of age

'YOUR FIRST LIFE PROGRESS REPORT' – Saturn is the shape or mould that your life is supposed to take. Its first impression is made at birth, and when it has completed an orbit around the Sun it comes back, impresses itself upon you again, and checks out whether or not you are realizing your potential and living up to your responsibilities. Depending on how you are doing, it will impose tests and/or promote or consolidate your position in life. Over the previous few years you may well have been experiencing endings and beginnings as your fate set you up for this time of reckoning – especially as decreed by your 'Second Emerging Moon Phase' or First Lunar Return (*see* page 51).

Using our 'all the world's a stage' metaphor, Saturn is seen as the Producer – that is, the one who makes sure that the material wherewithal is in hand and that practical considerations and demands are met and maintained. Saturn is all about the shape and form we give to life and to ourselves, not just about notions and ideas. So, if you have some dream or ambition, then you must really make the effort to act upon it now and take the time to see if it works. If it does not work, then you will have to rethink and try again – and again, if necessary. At the same time, it is important to stress that the fantasies, ideals and 'crazy ambitions' of your youth and twenties are also necessary in order to withstand the pressures and doubts that the world at large presents you with during this time. Without that dream, Saturn can make things crushingly dull and meaningless. Just think of a well-financed and well-produced play that has a poor script and second-rate acting! This is a case of Level One pressures being withstood, balanced out, and eventually influenced by Level Two energies (*see* the introduction to this Saturn section concerning Level One and Level Two).

If after several good attempts to realize an ambition (possibly over a period of years) there is, under Saturn's influence, still no success, then you can be sure that your dream is only that, or that the efforts you have made will come in useful at some later date. But if no effort is or has been made to fulfil some ambition or to live up to some inner belief or standard, then you will obviously gain neither achievement nor experience, and possibly become unstuck.

On the other hand, if you have been making the effort *and* are made of the right stuff, then Saturn will see to it that you are crowned with the laurels of success. An example of striving and succeeding came at the Saturn Return of Jana Novotna (aged 29) who, after years of competing, won the Wimbledon 1998 Women's Tennis Championship, and, also an example of striving and failing at the Saturn Return of her opponent, Nathalie Tauziat (30). Saturn puts us to the test, and rewards or otherwise.

But your Saturn Return is not just about material success or failure on a professional level. Much as the Producer might deem it necessary to inject the play with a new character or element, at this time life may well present you with an event which imposes greater responsibility (= ability to respond), forcing you to measure up or develop in some way. The classic event for this is having a baby. Other such events could be moving into or buying a new home, getting married, starting a demanding job, or some other responsibility that affects you for years to come. It could also mean losing or having your job threatened in some way. Again, rather than just experiencing Level One events such as these, you would also probably be given extra responsibilities on Level Two – that is, responsibilities involving moral decisions and emotional judgements.

In any event, Saturn forces you to establish your correct position, be it professionally, socially or emotionally, often by having to make your boundaries a lot more definite. This means determining a balance between what you do and do not owe others, and vice versa. More particularly, it means having to know your own ground (a Level Two situation) as distinct from the ways, views and territories of others (possibly Level One situations). Not knowing your own space – that is, where responsibilities and expectations should begin and end – could bring you quite low and make you feel very insecure. So, not least of all, relationships can at this time go through the mill in some way, meaning that stronger commitments have to be made – either to stay in a relationship or get out of one once and for all. If you attract a new relationship now, you can be sure that it will test you emotionally and materially, giving you experiences that will serve and/or affect you for the rest of your life.

Speaking of 'boundaries' points us to another Saturnian issue: structure. During this period of your life it is vitally important how you organize your activities, and that you formulate a plan or task. A lack of structure becomes very evident now, because the pressures of life can implode upon you if you have little or none. In turn, this all leads to another Saturnian factor, that of time. How you use or misuse your time is another critical issue. Time is *the* instrument of Saturn, for it is through time well spent that something is achieved, and through it being wasted that nothing is accomplished. You may well feel that there is very little time to do what you have to do – but this would really be that heavy Saturnian sense of importance and urgency making itself felt. So read the writing on the wall. Also, bear in mind that not respecting or wasting others' time could attract coldness and a lack of response.

At a more karmic or psychological level, the Saturn Return can bring something or someone into your life that challenges you to confront or overcome some major issue that does not involve just you and those close to you. Again you may fail or succeed – or, what may have appeared a failure could at a later date prove essential to success. Examples of each of these are Lt. Paula Coughlin who single-handedly took on the US Navy in a sexual harassment case and won, and Karen Silkwood who blew the whistle on a nuclear scandal only to be mysteriously killed in a car 'accident' on the exact day of her Saturn Return. Both these cases could be described as Level Two values clashing with Level One values, as individuals standing testament to superior moral standards in the face of the System which is inclined to be guided by purely financial considerations and the lowest common denominator of morality.

Another crisis that Saturn can bring, as the 'jelly mould' of your fate hits the 'jelly' of your life so far, is one of health. But this would usually be because you had evaded or not consciously taken on some challenge, only for it to re-present itself on a cellular level. It could also be the result of some event from the distant past, that still had not been addressed, surfacing as some complaint. Accidents and other physical traumas would also fall into this category. (*See* 'Physical Condition' in the introduction to this section.)

Naturally, your Saturn Return is not too likely to be as dramatic or drastic as the examples given above, but, relatively speaking, it will feel very important and of weighty significance to you. As such, it will be taken seriously – or it had better be – for in many respects your Saturn Return is the beginning of the rest of your life. At this time you are more yourself than at any other time because you are feeling the difference or similarity between what you are meant to be and what you have so far turned out to be. Taking the time and exercising the discipline to accomplish something is highly important. But possibly more important is that you determine quite what 'success' really means to you personally (Level Two), rather than in the eyes of the world (Level One) or according to some naïve fantasy (Level Two 'posing'). At my own Saturn Return I was striving to become a singer/songwriter, having been in the music business nearly all my working life. But it was at this time that I took up astrology – which is now my profession, while music is now a pastime. But I was not really aware of this ending and beginning until some years later when I became a fully-fledged professional astrologer. So have your ear to the ground because during your First Saturn Return you should be picking up the essential plot of your life. And, like me, you can in retrospect see how you did this.

People have often asked me, after their Saturn Return has technically ended, things like, 'Why do I still feel Saturn's effect?' or 'Have I handled it okay?'. The answer here is that any Planet's effect, but especially Saturn's, is a bit like punching a pillow. Take your fist away and the dent in the pillow is there for some time, or until it is plumped up. In other words, you will feel Saturn's pressure as long as you need to, until you have achieved, learnt or straightened out something. In many ways, your First Saturn Return is when you truly become, or have to become, an adult and 'put away childish things'. You have gone through the 'grounding' of the 'Ground Floor' of the House of Saturn (*see* page 108), and now you have to recognize the existence of Level Two and make the effort to aspire to its values. A positive statement that I heard from someone soon after the end of their First Saturn Return was 'I am now building myself up to make a better person'.

To help you manage and make the most of your First Saturn Return, overleaf you will see a flowchart of 'Your First Life Progress Report', which gives you a graphic idea of what Saturn is bringing into your life, and what the various ways of responding to it can bring you. Just follow the instructions in the top panel in order to steer your way onward and upward, rather than standing still or going down.

The following is a 'Saturn Return Gallery' of celebrated examples of this super-significant Planetary event.

People who gained honours at this time include Marlon Brando winning his first Oscar for *On the Waterfront* and Elizabeth Taylor winning hers for *Butterfield Eight* a month after almost dying from pneumonia (*see* page 359 to be directed to more about Brando and Taylor). Howard Hughes set a new air speed record of 351mph (*see also* page 166). Marie Curie discovered a new element, Polonium. Muhammed Ali regained his honours as World Heavyweight Champion when the US Supreme Court reversed his conviction for draft-dodging (*see also* page 359). Sir Edmund Hillary literally peaked by becoming the first person to climb Mount Everest, and went on to climb it for a second time at his Second Saturn Return (*see* page 126); interestingly, Hillary is a Saturn-ruled Capricorn, the Sign of the Mountain Goat.

Another individual who made a highly significant career move at both his First and Second Saturn Returns was Boris Yeltsin; he joined the Soviet Communist Party at the first one and resigned from it at his second. Other singularly significant professional landmarks were Che Guevara when he led his group against the troops of General Batista in Cuba (*see* also page 171), and a big one for John Lennon – the Beatles broke up (*see* page 359 for more concerning John).

A MAJOR INFLUENCE occurs from the age of 29 to 30

YOUR FIRST SATURN RETURN

Your First Life Progress Report

The flowchart below shows how the Positive and Negative Saturn Traits *(left)*, that you may or may not have, encounter Saturn's influence *(arrow on right)*, which in turn consolidates the Positive Traits into a Positive Outcome *(top)* and/or relegates the Negative Traits to a Negative Outcome *(bottom)*. Negative Outcomes can be made Positive through developing the Positive Traits.

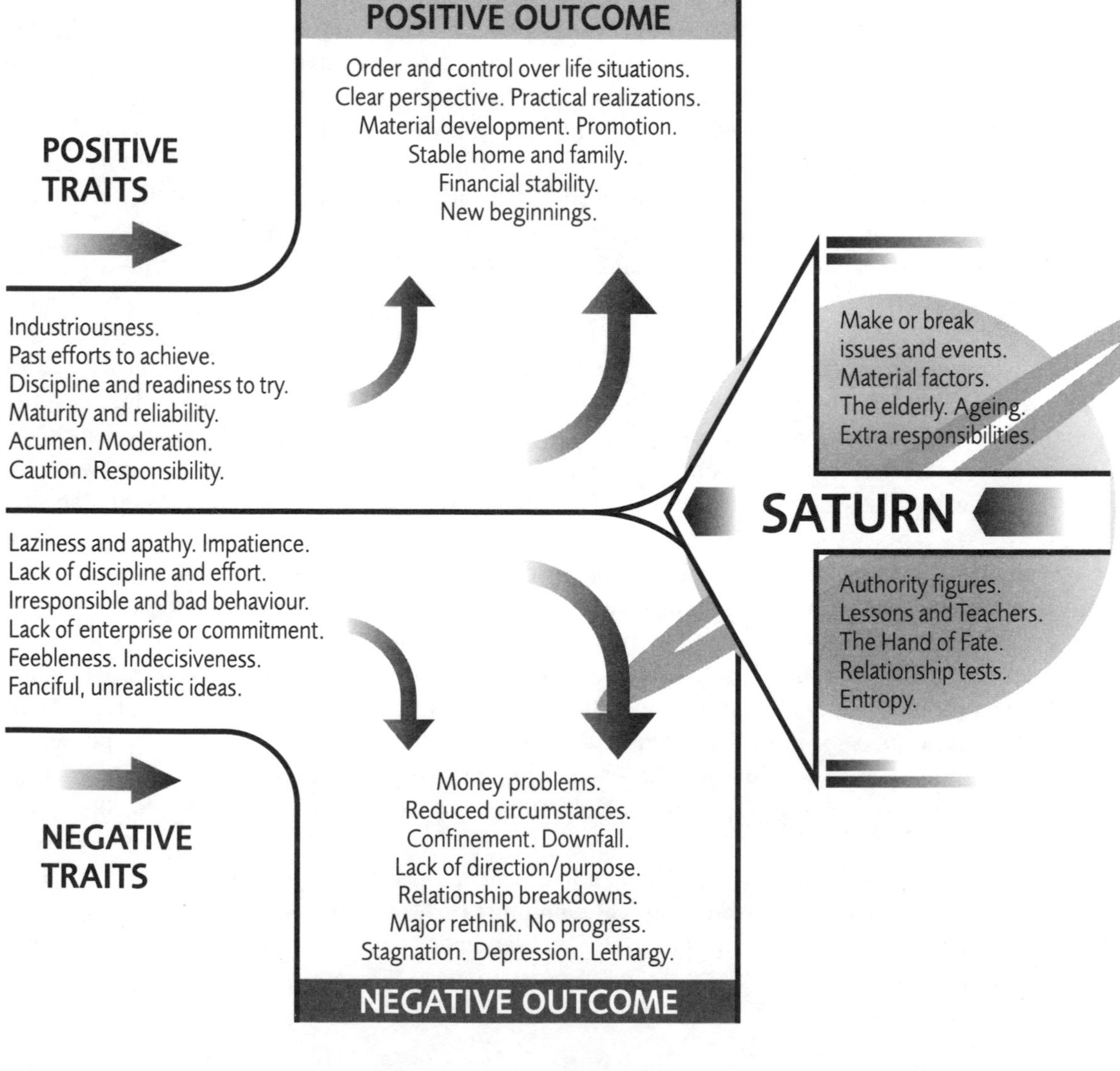

Marilyn Monroe married the much older Arthur Miller, probably in the hope of gaining the father figure she never had (Saturn rules father figures) but was divorced five years later – *see* page 359 for more about Marilyn.

A notorious example of the luck of the Devil (Satan = Saturn) looking after his own, was Al Capone who peaked at 30 million dollars from his various illegal activities that year. This peak of his evil was also marked by his ordering the infamous St Valentine's Day Massacre. But as is the nature of peaks, it was downhill all the way after this for Capone.

Finally, a significant but tragic event concerning ordinary people; in December 1998 six friends were involved in a climbing accident. The 29- and 30-year-olds were killed along with two 28-year-olds (*see* pages 51 and 238), whereas the ones that survived were 'unassailed' 24- and 25-year-olds.

JUPITER 'YEAR' for your Growth Mode – 'VIRGO – Libra' (*see* pages 32 and 33)

MOON 'PHASE' for Emotional Attunement – 'Second Emerging Phase' to 'Second Striving Phase' (*see* pages 51 and 59)

Other Universal Planetary Cycles possibly active at this time.

Check the tables on page 289 to see if any of these Planetary Cycles are active for you at the same time as your First Saturn Return. Remember that any Trine influences may have to be actualized to make them that effective.

URANUS WAXING TRINE (*see* page 166) This useful and refreshing influence happens for a year or two between the ages of 24 and 31. If it coincides with your First Saturn Return it can be distinctly helpful during this critical time in your life. If a peak is achieved during your Saturn Return then Uranus can give it that extra sparkle, as with Howard Hughes and his air speed record described above (note that Uranus rules all forms of flight). Al Capone also had the concurrence of both these transits, but this could be ascribed to the *luck* of the Devil!

PLUTO CONJOINING NEPTUNE (*see* page 221) This extraordinary influence is called the 'Illusion Buster' and 'Inspiration Booster', and when it occurs alongside your Saturn Return then it makes it even more important that you manage it properly. Saturn always lends the weight of reality to whatever is going on in your life, and as Pluto Conjoining Neptune can intensify one's illusions, Saturn will make sure that you spot them. This means that if you are labouring under some false impression or aspiring towards some impossible goal, it is all too likely that satisfaction will elude you and frustration dog you. On the other hand, ideas and feelings of very real value can also enter your life, and as such stay with you for years to come as a true guide and companion. Muhammed Ali was under the influence of both these influences, proving himself to be the giant that actually *was* the 'Illusion Buster' and 'Inspiration Booster' for blacks, draft-dodgers, and many more besides – a credit to Level Two, in fact. And despite his present condition, he still *is*.

Second Waxing Square | Between 36 and 37 years of age

'HAVING YOUR WHEELS TAPPED' – As this marks the end of the first stage of your Second Saturn Cycle after that all-important First Saturn Return, we could see this as a kind of minor test of how well you handled the major test. So this 'seven year itch' of Saturn's Cycle can be particularly itchy. Basically, this test is there to see what and how well you learnt your lessons from seven years back, and whether you are still on the right road, like the railway engineer who taps the wheels of trains to see if they are sound. Did you manage to stand on your own two feet then – or begin to – and how are you doing now? Or are you again unconsciously looking for someone or something to lean on? Are you still looking for a scapegoat whom to blame any difficulties?

This influence could be regarded as a time when one really should go from Level One to Level Two, if you haven't already (*see* Saturn introduction). This means that the time has come to begin taking full responsibility for your thoughts, feelings and actions. And so Saturn, in order to make this point, begins to emphasize certain issues in your life that reflect this and challenge you. Personally, I suffered a major collapse in the form of a relationship ending and losing tenure on a house, both of which I had felt very secure in. Saturn was saying to me, 'You are getting too comfortable with all this and you're not really getting your act together.' And so they had to go. In retrospect I can see Saturn's wisdom in this – but at the time I was bereft, and had to limp along for a year or two until my Uranus Opposition forced a more radical and conscious move (*see* page 168).

A client of mine thought she had 'arrived' after having moved into a far better house and neighbourhood than she'd ever had previously, only to get involved in an extra-marital affair with a younger and insecure man. This was her own unlooked-at insecurities being visited upon her via another person to whom she became attracted. This was Saturn's way of 'tapping her wheels', showing that her issue was not just one of material stability (Level One), but also emotional stability (Level Two).

Two famous examples of people who failed to make this transition from Level One to Level Two were Marilyn Monroe and Princess Diana. Marilyn's dependency upon barbiturates and unreliable men was amounting to a quite spectacular flunking of Saturn's test, and she knew it – or rather felt it. She could not understand it or deal with it, and withdrew from it in the most complete way that is possible. Suicide is a very 'Level One' phenomenon, because one fails to see beyond the physical realm and the material world. So there seems to be only one way out. And if, as conspiracy theorists have it, she was murdered, then it amounts to the same thing for such assassins could be regarded as Level One enforcers.

The same could be said, more or less, for that other female icon of the 20th century, Princess Diana, who also died at 36 years of age. Diana also had a well-known security problem, but despite appearances of becoming stronger and more self-directing shortly before her death, she was from an astrological viewpoint simply falling into another security trap. This is borne out by the fact that is was the very shaky security systems around her at the time that were responsible for her death: not wearing a seat-belt, a drunk chauffeur unfamiliar with the car, the car hire firm not ensuring that there would be a suitable driver, and other factors we will probably never be aware of. And again, like Marilyn, if there was some sort of sinister conspiracy going on, it amounts to the same thing: Level One enforcers removing what they see as a loose cannon.

But none of this is intended to imply that you can only expect negative events, and certainly not of the fatal variety, but in most cases Saturn is putting one to a test which one is not likely to pass with flying colours. Two notable exceptions are Marie Curie when she won the Nobel Prize for Physics, and Albert Einstein when he published his Theory of Relativity. Needless to say these two individuals were exceptional in themselves, very disciplined and actually seeing scientifically beyond Level One to Level Two – although of course that is not how they expressed this.

Tests along the way for two other notables were Martin Luther King organizing difficult and dangerous demonstrations for the right of Blacks to vote in Alabama, and Che Guevara protesting against his erstwhile companion-in-arms Fidel Castro, *and* being under attack from Cuban exiles, while speaking at the UN headquarters.

This influence also introduces the beginning of a new Jupiter Cycle, indicating that your next cycle of growth has to be rooted in far firmer ground than ever before, so that there is no going back, and that you might know how well you are running along the track of your life.

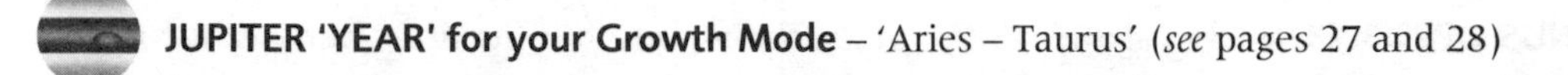

JUPITER 'YEAR' for your Growth Mode – 'Aries – Taurus' (*see* pages 27 and 28)

MOON 'PHASE' for Emotional Attunement – 'Second Deciding Phase' to 'Second Adjusting Phase' (*see* pages 64 and 71)

Other Universal Planetary Cycles possibly active at this time.

Check the tables on page 289 to see if any of these Planetary Cycles are active for you at the same time as your Second Saturn Waxing Square.

URANUS OPPOSITION (*see* page 168) Very few people have this influence early enough (at 37) for it to coincide with their Second Saturn Square, but if you do, then the balancing of the radical and conservative sides of life and how you personally live it, is the main issue.

PLUTO WAXING SQUARE (*see* page 226) If this influence is also happening now, then you are being given notice that the above-described imperatives are even more critical. Saturn and Pluto are the most ruthless of all Planetary influences for they have to be very thorough, otherwise the power structure that is the physical universe would collapse – after all, they symbolize fate itself. Nothing and no-one is let off this hook. Princess Diana was in sight of both these influences when she died, implying that her death was fated for reasons that go beyond the subject matter of this book. Suffice to say that she did not die in vain, but it naturally would not have been her conscious choice to die at all at that stage in her life. But as Einstein himself is supposed to have said, 'God does not play dice with the Cosmos'; there is a deep design in all things that goes beyond the awareness of our little selves.

Second Waxing Trine — Between 38 and 39 years of age

'FOLLOWING YOUR PATH' – This Transit is really a continuation of the last one, your Second Waxing Square, in that it could give you a much needed respite to make any necessary adjustments, get your bearings, and select the best path ahead. But really it would be more accurate to say that this Trine works overtime to contend with your Mid-life Crisis or the years leading up to it, for the chances are that you will be simultaneously experiencing at least one of the other influences given below under 'Other Universal Cycles Possibly Active At This Time'. If you are not, then your Second Waxing Trine of Saturn should make anything run more smoothly. In either case, what achieves this is either your living skills born of experience or somebody else either in authority or in a position to assist you.

In any event, this is a good time to plan or make a move, possibly because a move is being made for you, or is unavoidable. Saturn is saying that you're a big girl or boy now, and you should be able to manage anything life throws at you – to bend with the wind, to follow the path as the road unwinds.

JUPITER 'YEAR' for your Growth Mode – 'Gemini – Cancer' (*see* pages 29 and 30)

MOON 'PHASE' for Emotional Attunement – 'Second Adjusting Phase' (*see* page 71)

Other Universal Planetary Cycles possibly active at this time.

Check the tables on page 289 to see if any of these Planetary Cycles are active for you at the same time as your Second Saturn Waxing Trine.

URANUS OPPOSITION (*see* page 168) This singularly important influence, your 'Midlife Crisis', can rock your boat considerably, so having this Saturn Trine occurring at the same time comes as a useful and welcome stabilizing factor. You may even be able to elect for a safe and predictable course, but it is up to you whether such a thing makes your life worth living.

NEPTUNE WAXING SQUARE (*see* page 198) As this is called 'Fascinations and Illusions', having Saturn around to help you tell the true from the false is more than a bit useful. No matter what sea of confusion Neptune may sink you in, you'll still have dry land in sight. On the other hand, though, do not use it to avoid adventure altogether; that would be like putting to sea and staying in your cabin all the time.

PLUTO WAXING SQUARE (*see* page 226) With this influence occurring as well, the changes are definitely being rung, and you would be a fool to hang on to anything or anyone that is remotely past their sell-by date. Mind you, if the issue is one of determining quite what or who is or is not past their sell-by date, then Saturn will grant you the experience and/or assistance to establish just this. But establish it and act on it you must.

Second Opposition | Between 43 and 44 years of age

'THE OLD *VERSUS* THE NEW' – During or following upon the super-significant Uranus Opposition (*see* page 168), Saturn steps in to see whether or not you have revolutionized yourself and your life. It is saying that some kind of realization should be taking (or has taken) place that effectively ushers in the new while discarding the old. One could even call this juncture in life the 'Last Chance Saloon', it being probably the last real opportunity you get to turn your life around. Or rather, your Uranus Opposition is or was that; Saturn is effectively rounding up any stragglers. But then again, it's never too late if your heart's in the right place and you're willing to learn.

However, this Transit could also have been called 'Level One *Versus* Level Two' (*see* Saturn introduction) because from an astrological or spiritual point of view, the time has come to recognize the reality and significance of your inner rules, strengths and values (the New) and leave behind any unnecessary dependence upon the rules and values of the System, Level One or any other external factor (the Old).

Because of this important aspect of the Second Saturn Opposition, it is difficult to find examples of this 'Level One to Level Two' version of 'The Old *Versus* The New'. A shift of emphasis from outer to inner values is not, by its very nature, that noticeable from the outside – at least, not while that shift is going on, for the rewards (or desserts) tend to come *after* this Transit is over. Instead I can only find glaring instances of this Opposition in figures who were ushering in the new and doing away with the old in a political, world-stage manner, rather than attending to their inner, Level Two, affairs. First we have Adolf Hitler who became German Chancellor at 43 in 1933, putting him in a position to bring in his 'new order' of the Third Reich, and literally do away with the old, thereby displaying an absolute lack of Level Two, inner integrity. As things turned out, the new order was a new Europe (and World) following the disaster and holocaust of World War Two, and it was the Third Reich that went the way of the dinosaurs. And then there was John F. Kennedy, the face and voice of a new, young and discrimination-free United States, who was elected President in 1960 at the age of 43. Unfortunately, his Level Two condition of moral rectitude was greatly wanting, and it is open to discussion as to how much that contributed to his assassination three years later (*see* page 359 for more about Kennedy).

A third example of this kind of Second Saturn Opposition, but with an important difference, is Nelson Mandela. In 1962, at 43 years of age, he was arrested for allegedly conspiring to overthrow the government of South Africa – the 'old'. However, he then had over 27 years in gaol to ponder and look within and discover the 'new' in himself – Level Two. As a consequence of this, the outer world manifested support and opportunity and ultimately he did usher in the new South Africa as its first black president (*see* page 359 for more about Mandela).

All the above examples serve to show the vital importance of understanding what 'The Old *Versus* The New' really means. In more everyday terms, on the outside events can occur that make the 'old' apparent, such as life seeming stale and meaningless, or being tired and lonely, or being stuck in old habits. Work and relationships very much reflect and play out these dilemmas. Your body too can start to show signs of wear and tear now, and exercise and diet could well be important issues. Things seem to want to die, either literally or figuratively – emphasized by the 'Scorpio' Jupiter Year at 43. One asks, 'What have I got going for me?' – possibly in desperation. But at the 'Last Chance Saloon' it is definitely a case of 'they that drink of the old wine have no place for the new'. What you really have to identify is the way in which you evaluate your life and experience. Is it cynical, escapist and resistant to inner, and subsequently outer, change? And is this characterized by watching too much TV, not meeting anyone new or different, overworking to avoid emotional situations, or drinking and smoking too much? If you are a man, are you becoming boring? if you are a woman, are you becoming neurotic about your looks? Such things or something like them are warning signs from Saturn. You have become a casualty of Level One living, of evaluating and therefore living your life by standards set for you by something external to you, and that is not really interested in how you feel – just so long as you 'fit in'. You are a creative being that can discover its own values and live by them – just so long as they are rooted in something good, true and beautiful.

For myself, at this time I became involved in a relationship that was spelling out the death of a certain way of relating and emotionally responding. I did not know this at first, of course, because I was still on the same old tack of seeking sexual highs and not being aware of how I really felt. It took an actual physical disaster and being unceremoniously rejected to get me painfully to see the light. My father also died at this time. A manifestation of my seeing the light was getting my first book published, and then meeting someone who really 'knew' me. The 'Last Chance Saloon' is quite a dive, but you could say that it gives value for money if you get the message that you are creating your own reality on Level Two – what's 'new' – and that it is psychological suicide to be dictated too much by those 'old' external standards of Level One – what's done and gone.

JUPITER 'YEAR' for your Growth Mode – 'Scorpio – Sagittarius (*see* pages 34 and 35)

MOON 'PHASE' for Emotional Attunement – 'Second Realizing Phase' to 'Second Sharing Phase' (*see* pages 77 and 83)

Other Universal Planetary Cycles possibly active at this time.

Check the tables on page 289 to see if any of these Planetary Cycles are active for you at the same time as your Second Saturn Opposition.

URANUS OPPOSITION (*see* page 168) This singularly important influence, your 'Mid-life Crisis', is all too suitable an accompaniment to your Second Saturn Opposition because both of them are about a plateau, a realization and a revolution – or they'd better be. Mandela and Kennedy, in two different ways, bear witness to this astro-fact.

NEPTUNE WAXING SQUARE (*see* page 198) As this is called 'Fascinations and Illusions', there is a distinct danger of missing the point of your Second Saturn Opposition, as described above. But on the other hand, simply because it is a delusory influence, it can make it all too obvious to you how ghastly hanging on to your illusions can be. Neptune, after all, is a Planet closely connected with the inner world of Level Two.

PLUTO WAXING SQUARE (*see* page 226) 'Having to Let Go' is what this influence has been dubbed, so in the light of what your Second Saturn Opposition means, nothing more need be said. I had both these happening at 43 to 44, and the rod was not spared!

Second Waning Trine **Between 48 and 49 years of age**

'FINDING YOUR TRUTH' – These are years when you can expect some kind of 'profit' from any efforts you made during the preceding years that included your Second Saturn Opposition. Failing any attempt at all to have revitalized or developed your inner being, this Transit is not going to confer much upon you other than a sense of ease materially that is spoiled by frustration and boredom on an emotional or psychological level.

But assuming that you have turned inward and begun to recognize that the world really is what you make it by the way you're looking at it, and thereby how you respond to it, then this influence can give you a sense of your age, experience and maturity that amounts to a *genuine* confidence and an incipient wisdom. And the more you recognize and feed this truth, the more you find that it is *your* truth. You then slowly wake up to how powerful is that inner truth which is your inner being. All being well – which you also realize is something that is entirely up to you – your life becomes more and more a quest for increased wisdom, something which is simply a knowledge of how life and human nature actually interact. As a result of this you become more and more the true teacher and/or benefactor of others. The 'Aries' Jupiter Year that occurs at 48 also marks a new beginning, but how new it is depends upon how much you previously put an end to any stale or rigid ideas and ways.

JUPITER 'YEAR' for your Growth Mode – 'Aries – Taurus' (*see* pages 27 and 28)

MOON 'PHASE' for Emotional Attunement – 'Second Understanding Phase' (*see* page 89)

*There are no other Universal Planetary Cycles active at this time apart from the Pluto Waxing Trine (*see *page 231) for some born between 1954 and 1972.*

Second Waning Square | Between 51 and 52 years of age

'SPIRITUAL MATURITY' – The Saturnian imperative to consider seriously that you have to adopt a new outlook on life in order eventually to grow old gracefully, that begun with the Second Saturn Opposition at 43 to 44 years of age, now has further weight lent to it. You should continue to identify and discard any old and entrenched attitudes in order to reach a wisdom that befits your age. The one old idea that you *should* cling to is that which says that if human beings were not supposed to grow wiser as they got older, then in terms of our species' survival, what would be the point of having old human beings?

Yet the irony here is that at this age it is very likely that you have to bear with ageing parents who are well-entrenched in old and inappropriate patterns of behaviour, and are far from wise. It could also be younger people that you have to bear with, possibly your own children. It would seem that at this time you are challenged to contend with all manner of annoying or demanding circumstances that test your own centredness and self-control. But failing to bear with others and maintain your own integrity is likely at this time, simply because this is the very way in which you are being tested. This means that the test is also about being able to bear with yourself.

Losing your rag and feeling guilty are signs that you need to reconcile yourself with those around you, especially the ones who are close to you. To have a parent die when you are on bad terms can leave a stain that is hard to remove and pollutes your life to come. If you have long since removed yourself from such family members or friends, then Saturn could bring a heavy feeling of emptiness or hollowness down upon you. Whatever the case here, that imperative still holds sway: to come to an understanding of how you may have inadvertently cultivated a negative or faulty attitude, and to find a way of releasing any bitterness, anger, resentment or regret. Pleading that you are too old to change now simply does not wash, because the negative feelings are merely shuffled to the bottom of the deck, to be dealt later when you are even less well-equipped.

A positive starting point is to discover for yourself a nobler way of living and a more wonderful reason for being, and try at all costs to live up to them. On an esoteric note, the star-cluster called the Pleiades takes 52 years to return to the same spot in the sky. According to that enigmatic and highly advanced civilization, the Mayans, it is the Pleiades that are the source of love and wisdom in our galaxy, the Milky Way. And so this, your Pleiadian Return, is a time when you can tune into this source, and elevate your manner of living and loving. In one way or another, the challenge to do so will certainly be there – in the form of someone or something demanding selfless care and devotion as the best means to manage your life.

JUPITER 'YEAR' for your Growth Mode – 'Cancer – Leo' (*see* pages 30 and 31)

MOON 'PHASE' for Emotional Attunement – 'Second Understanding Phase' to 'Second Releasing Phase' (*see* pages 89 and 95)

Other Universal Planetary Cycles possibly active at this time

Check the tables on page 289 to see if any of these Planetary Cycles are active for you at the same time as your Second Saturn Waning Square. Remember that any Trine influences may have to be actualized to make them that effective.

URANUS WANING TRINE (*see* page 175) If this Transit occurs during your Second Saturn Waning Square then it would be extremely appropriate because it gives you the opportunity to renew and refresh whatever needs it.

NEPTUNE WAXING TRINE (*see* page 204) This Transit is very spiritual in that it allows you to accept things more easily and go with the flow, thereby countering any tendencies to let stubbornness or contentiousness get the better of you.

PLUTO WAXING TRINE (*see* page 231) This regenerative influence is just what is needed during the Saturn Transit, for it enables you to see what matters and what doesn't matter. As such, any petty values or opinions you have can be let go of, for such would inhibit the continued acquisition of the spiritual maturity that is the main objective of this time.

Second Saturn Return | Between 58 and 59 years of age

'YOUR SECOND LIFE PROGRESS REPORT' – This is a time when you may quite rightly ask yourself what your life has amounted to. A good clue or starting point to answering this question is to cast your mind back to your First Saturn Return at 29 to 30 years of age (*see* page 115). This is because whatever happened or began then was the seed of where you have got to now. In strictly Saturnian terms, this is the time when you should be able to see what has developed with regard to what you started then. There is no rule of thumb here, because whatever has happened over those intervening years should have an entirely personal message for you.

Two celebrated examples that were cited for the First Saturn Return were Sir Edmund Hillary who was the first person to climb Mount Everest at that time, and who climbed it for the second time at his Second; and Boris Yeltsin who joined the Russian Communist Party at that time and then resigned from it at the Second. One can only guess at what these corresponding events meant to these two men. I should imagine that Sir Edmund felt not only a great sense of achievement the second time, but somehow also felt a register of all his experiences since the first time, and not least of all, the difference in his bodily state and the equipment available. Yeltsin perhaps felt a sense of disillusionment mingled with a sense of having taken something as far as it could go. But I am sure there were far more intricate feelings involved for both of them.

Another interesting case was Indira Gandhi who, during her First Saturn Return at 29, enjoyed her father becoming the first Prime Minister of India, while she acted as his political hostess. Come her Second Saturn Return, though, she was defeated in a general election after having been Prime Minister herself for 11 years, during which she was convicted for corrupt electoral practices. She was re-elected three years later only to be assassinated four years after that. Here Saturn's judgement seems quite straightforward in one respect, but quite complex in another.

Yasser Arafat had his First Saturn Return shortly after co-founding the Al Fatah terrorist movement. Come his Second Saturn Return he formally renounced terrorism and recognized the state of Israel.

And a poignant mix of experiences befell Mikhail Gorbachev during his Second Saturn Return when at 59 he won the Nobel peace prize and then resigned as leader of the Soviet Union four days after its abolition.

The above are what could be called monumental occurrences for 'Second Life Progress Reports', but for so-called ordinary people, events at this time are in their own way just as significant. Seeing what your life has added up to is, as I have said, a key feature of this time. From a philosophical point of view – which is really the essential point of view to adopt now – appraising one's life and personality in terms of having stood for something worthwhile is what is recommended. Failure should possibly be seen as having kow-towed too much to the values and figures of the System or Level One (*see* introduction to the Saturn section), whereas success may best be framed in terms of having been true to yourself and your own individual sense of choice (Level Two), and of having managed to remain relatively free from crystallized attitudes born of convention. Not least of all, having been of help, support and guidance to others should figure highest of all in one's self-assessment.

For many, retirement, or the prospect of it, can be a major issue now. Whether such is seen as a new beginning or a meaningless ending has everything to do with the above-described criteria. Being one's own person would grace you with the former, but having sold one's soul to the System for the sake of a limited form of security would be only too likely to inherit the latter.

Overleaf you will see a flowchart of your Second Saturn Return, 'Your Second Life Progress Report', which gives you a graphic idea of what Saturn is bringing into your life, and what the various ways of responding to it can bring you.

JUPITER 'YEAR' for your Growth Mode – 'Aquarius – Pisces' (*see* pages 37 and 38)

MOON 'PHASE' for Emotional Attunement – 'Third Emerging Phase' to 'Third Striving Phase' (*see* pages 54 and 60)

Other Universal Planetary Cycles possibly active at this time.

Check the tables on page 289 to see if any of these Planetary Cycles are active for you at the same time as your Second Saturn Return. Remember that any Trine influences may have to be actualised to make them that effective.

URANUS WANING TRINE (*see* page 175) With this Transit occurring during your Second Saturn Return, you have the opportunity to liberate yourself and keep abreast of the times – and make use of modern methods and values to propel you further down your life's road. If you are 'young at heart', Uranus will reward you for it.

URANUS WANING SQUARE (*see* page 177) This Transit could in some way challenge you to break away from old attachments and attitudes. Whether such challenges are seen to come in the form of what you regard as an effrontery to your entrenched lifestyle and values, or in the form of opportunities to reform and rejuvenate you, is entirely up to you.

NEPTUNE WAXING TRINE (*see* page 204) With this Transit occurring during your Second Saturn Return you are, at least, more able to accept any difficulties. If you are working on your creative or spiritual development, then this will carry you forward smoothly and firmly.

NEPTUNE OPPOSING PLUTO (*see* page 207) This Transit has been dubbed 'Intimations of Mortality' and so the sense of being in or on the verge of the Autumn of your years is going to be to the fore. Quite how you view this – as a time of mellowness and wisdom born of experience, or as one of loss and regret – is wholly dependent upon your philosophy of life (or the lack of one).

PLUTO WAXING TRINE (*see* page 231) This is a regenerative influence and therefore 'just what the doctor ordered'. But then again, if you have refused to take the 'medicine' of letting go of negative and fixed ideas, do not expect any miracle cures.

A MAJOR INFLUENCE occurs from the age of 58 to 59

YOUR SECOND SATURN RETURN

Your Second Life Progress Report

The flowchart below shows how the Positive and Negative Saturn Traits *(left)*, that you may or may not have, encounter Saturn's influence *(arrow on right)*, which in turn consolidates the Positive Traits into a Positive Outcome *(top)* and/or relegates the Negative Traits to a Negative Outcome *(bottom)*. Negative Outcomes can be made Positive through developing the Positive Traits.

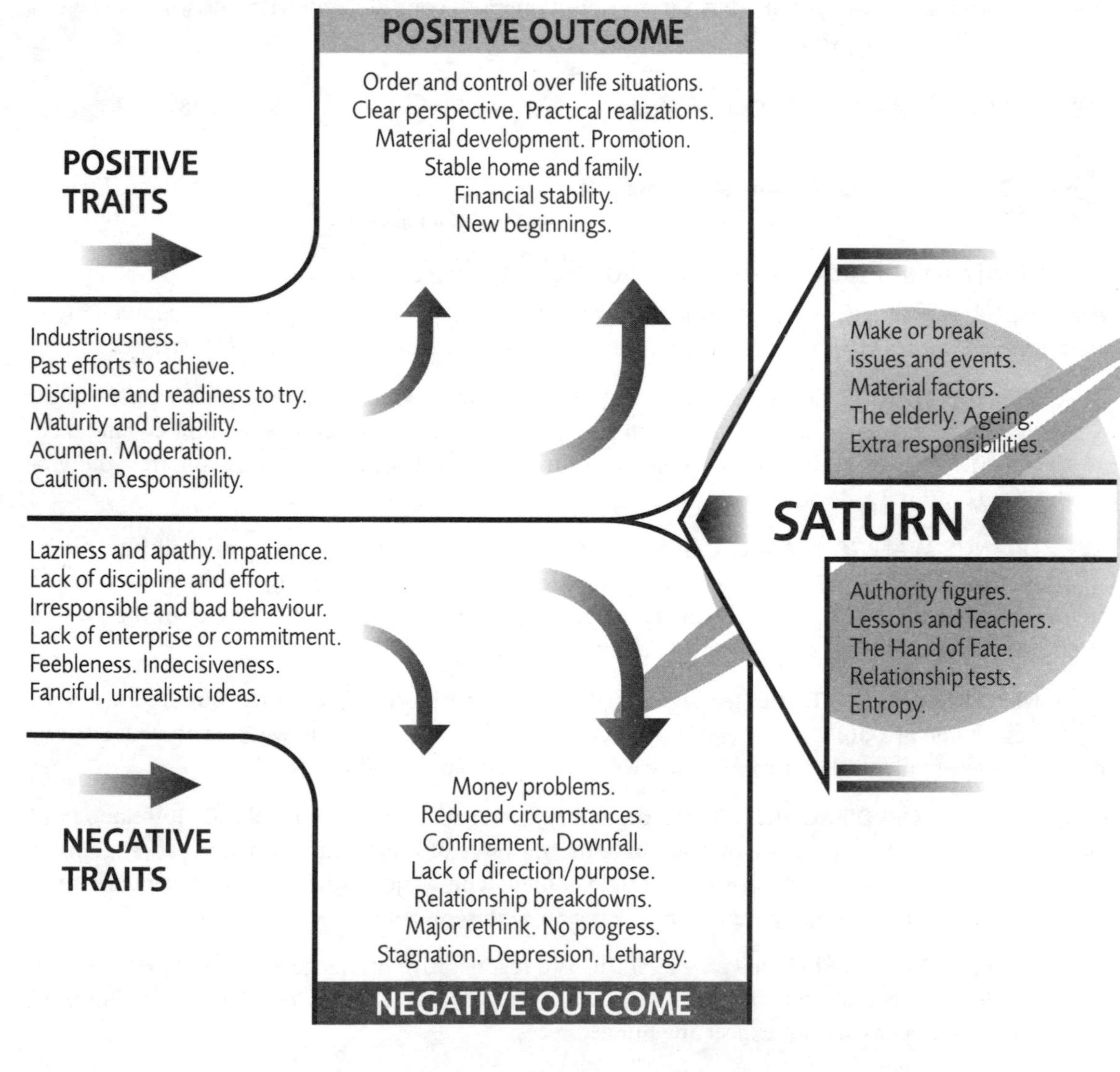

Proclamation of Limitation

Saturn being the Planet that defines all limitations, the remaining Transits of Saturn have mostly limited and theoretical descriptions, simply because I personally have no experience of these ages, and know only a limited number of friends, relatives, clients or celebrities that have. Useful information can still be gained from the respective Jupiter 'Years' and Moon 'Phases'. Also note that as Saturn has to do with the ageing process, health from hereon can become an issue, especially during the Squares and Oppositions.

Third Waxing Square | Between 65 and 66 years of age

'RE-FORGING OR SHUNTED TO THE SIDINGS' – Other than those who are self-employed or have some ongoing work commitment, by now or at this time, most people have retired from regular work – or at least, their partner has. In any event, in terms of what this influence poses, it is time to take stock of what you do with your time and whether or not you can continue to perform efficiently or rewardingly enough. One public example of this was Margaret Thatcher, the 'Iron Lady' who seemed pretty immovable, but who was beaten in the leadership contest for her political party at this time.

There could arise some challenge to take stock of yourself on a more personal level. A dramatic example of this was Marlon Brando, whose son Christian was convicted for manslaughter of his sister Cheyenne's boyfriend, leading to her suicide soon after.

These examples are not meant to encourage you to expect disaster, but serve to show how a person's 'record' can be visited upon them. Whatever may happen on these fronts, your choice is that of welcoming any change or arrival of something or someone different breathing new life into your veins, or of giving in to a quiet but possibly meaningless existence.

JUPITER 'YEAR' for your Growth Mode – 'Virgo – Libra' (*see* pages 32 and 33)

MOON 'PHASE' for Emotional Attunement – 'Third Adjusting Phase' (*see* page 73)

Other Universal Planetary Cycles possibly active at this time.

Check the tables on page 289 to see if this Planetary Cycle is active for you at the same time as your Third Saturn Waxing Square.

URANUS WANING SQUARE (*see* page 177) This Transit could in some way challenge you to break away from old attachments and attitudes. Whether such challenges are seen to come in the form of what you regard as an effrontery to your entrenched lifestyle and values, or in the form of opportunities to reform and rejuvenate you, is entirely up to you.

Third Waxing Trine Between 68 and 69 years of age

'TRUSTING YOUR FATE' – Although this is a Trine and therefore an easy Saturn period, because it is nudging the proverbial 'three-score years and ten' that a human being has been traditionally allotted, it can give rise to anxiety and doubt. Of course, this is not necessarily so, but of *this* you *can* be sure: it is your fate to die – but the question of what follows that is the tricky one. If you truly believe that there is no hereafter, that when you die you are no more and that's it, then any anxiety would obviously be totally unfounded – some regrets maybe, but no anxiety. If there is doubt, however (or doubt posing as anxiety), this implies that you are no longer so sure of there being nothing when you die – which leads us on to looking at what 'hereafter' means. To approach it scientifically, according to Newton's Second Law of Physics, the Law of the Conservation of Energy, energy is never born or dies – it only changes the form it is takes. In Quantum Theory, everything is energy, including you and your body. When you die, the energy that was you is going to become something, or part of something, or someone else. But this still leaves us with the prospect of the Great Unknown – which is the real cause for anxiety.

As far as Saturn is concerned, a 'score' has been kept, and when you die there is a reckoning – just as there has been in life, but this time that score will be determined by Level Two and not just Level One criteria (*see* Saturn introduction). And so I venture to suggest that any anxiety is similar to what most people feel who have sat an examination and are waiting for their results. But the highly significant point is that you are not dead yet – you are still alive in the examination room! And so your Third and very probably final Saturn Waxing Trine offers you the opportunity to still 'put in a good paper' or at least admit to making a bit of a mess of it and promise yourself you'll do better next time, for this makes all the difference to the form your energy will take on after you die.

JUPITER 'YEAR' for your Growth Mode – 'Sagittarius – Capricorn' (*see* pages 35 and 36)

MOON 'PHASE' for Emotional Attunement – 'Third Adjusting Phase' to 'Third Realizing Phase' (*see* pages 73 and 78)

There are no other Universal Planetary Cycles active at this time

Third Opposition Between 73 and 74 years of age

'THE OLD *VERSUS* THE NEW – *TWO*' – One can have relationship difficulties at any age, and this is one of them. This time, however, they could quite likely be because you have lost a partner, or that you have no-one to relate to and be with, or that there are people around you that you would rather not have to relate to. In any event, Saturn is confronting you with your social and emotional state. If all is well, in that you are still in a positive relationship, then this Opposition will confer a feeling of satisfaction upon you, possibly enabling you to help those less fortunate or wise than you, be they younger, older or the same age. If, though, you feel emotionally or socially ill at ease, then either you must search inside yourself for the reason, and reconcile yourself to it, or you can carry on bemoaning your fate, probably worsening what relationships you do have in the process.

Relationship and relating are the stuff of life. Having the curiosity and humility to interact equally with others, whoever they are, will keep you interested in life, and, more's the point, keep life interested in you.

JUPITER 'YEAR' for your Growth Mode – 'Taurus – Gemini' (*see* pages 28 and 29)

MOON 'PHASE' for Emotional Attunement – 'Third Sharing Phase' (*see* page 84)

There are no other Universal Planetary Cycles active at this time

Third Waning Trine Between 78 and 79 years of age

'KNOWING YOUR HEART' – The greatest positive power a human being can possess is the power to help other living creatures, be this one or many, in a small or big way. In the introduction to this Saturn section, I put forward the idea that life was like a big school and periodically we have to sit examinations. Under your Third Waxing Square at 65 to 66 and your Third Opposition at 73 to 74 (*see* pages 129 and 130), we saw how one is confronted with a sense of how well one has or has not done in these examinations, and that as long as one was alive there was always a chance to 'improve one's marks', so to speak. This, the time of your Third and probably last Waxing Trine, is one of those chances. And the chance comes in the form of someone or something you can help, possibly because they remind you of yourself at some time in your past when you could have done with some help. Some people meet such an opportunity merely with scorn and bitterness, simply because it does remind them of how no-one helped them when they needed it. This attitude, needless to say, does not earn any marks – quite the contrary in fact. On the other hand, recognizing life's poetry in motion, and responding to it with a generous heart that expects no reward other than the pleasure of giving, means you can rest assured.

JUPITER 'YEAR' for your Growth Mode – 'Libra – Scorpio' (*see* pages 33 and 34)

MOON 'PHASE' for Emotional Attunement – 'Third Understanding Phase' to 'Third Releasing Phase' (*see* pages 90 and 97)

There are no other Universal Planetary Cycles active at this time

Third Waning Square Between 80 and 81 years of age

'FINAL ACCEPTANCE' – All of what has gone before has been done, and nothing can alter that. However, the way you now think and feel about it can. And how you feel and think about it should be focused solely upon an acceptance of it all – that it was all for the best. Reading through the various influences of the whole of your Saturn Cycle, but especially the more recent ones, and accepting what is written there, may help you in doing this. Doing the same with your Moon Cycle is also a good idea, for this will help you recall and clear away unwanted memories, and polish up the good ones.

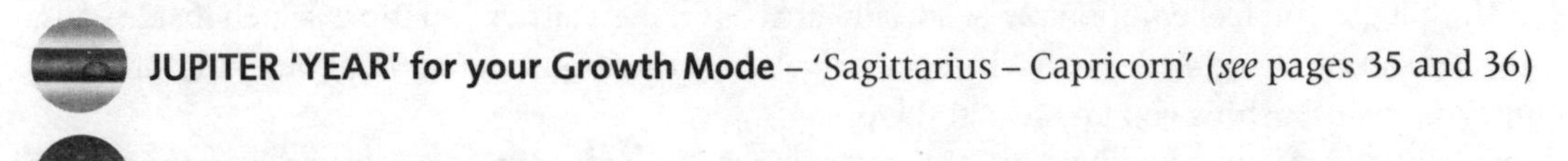

JUPITER 'YEAR' for your Growth Mode – 'Sagittarius – Capricorn' (*see* pages 35 and 36)

MOON 'PHASE' for Emotional Attunement – 'Third Releasing Phase' (*see* page 97)

There are no other Universal Planetary Cycles active at this time

The Teachers

GETTING TO KNOW THE TEACHERS THEMSELVES IS not absolutely essential to understanding and making use of your Saturn Cycle, but with a little more concentration (that's Saturn for you!) far more detailed information will be available to you.

These teachers are the people, things and situations which, intentionally or otherwise, give us our lessons, set our examinations, and then confer honours upon us – or not, as the case may be. It is the actual period in time that you are living through which determines what kind of teacher will be in the 'classroom' of your life. Just like the Sun, Saturn is also passing through a certain Zodiacal Sign at any given time, and the way it teaches its lessons or lays down its laws will bear the same qualities as the Sign it is passing through – or transitting, to use the correct astrological term. It is as if the teacher is of a different character, with differing teaching methods and priorities, depending which one is 'on duty' – just like an actual teacher.

So if you are experiencing an active part of your Saturn Cycle in one of the ways described above in the 'timetable', then further insights can be gained concerning that time through knowing the Sign of the teacher on duty – the Sign that Saturn is passing through at that time. And even if you are not in an active part of your Saturn Cycle, that teacher will still be on duty, and its 'character' or Sign qualities will be prevailing generally. This can be better understood with an example. At the time of writing this, Saturn is transitting or travelling through the Sign of Aries. Now I am not personally experiencing an active part of my Universal Saturn Cycle at present, but many people are. Furthermore, people who are born with the Sun (or Moon, Ascendant or a number of Planets) placed in Aries will be feeling Saturn's influence. And so the Arian influence in the world (that is Arians or people or things with Aries emphasized in their Birth Charts) will be particularly Saturnian – or to put it the other way around, Saturn's influence will be particularly Arian. (If you can, check when Saturn was/is passing through any Sign that is occupied by the Sun, Moon or a number of Planets in your Birth Chart, and you will appreciate that at such times you feel or felt Saturn's influence as described here.)

In order to know what the Sign of the teacher on duty is, ie what Sign Saturn is passing through at any given time, simply consult 'The Teachers' Duty Roster' on the following page. Each entry shows the year, month and day when Saturn left one Sign and entered another. As with our example above, you can see that the date of writing this, 1 February 1999, fell between 25 October 1998 at 18.55 when Saturn entered the Sign of Aries, and 1 March 1999 at 1.38 when Saturn left Aries and entered Taurus. So Saturn right then was in Aries, and the teacher is Arian by nature. Note that all times are given as 24-hour clock time. And if the time you are concerned with is near the time Saturn changes Signs, then make sure you are using Greenwich Mean Time, which is what the times in this Table are given in. It is not as if you'll suddenly feel Saturn changing Signs – although you might.

Having determined from the Roster the Sign that Saturn is transitting, turn to 'The Twelve Teachers – Saturn's Transits through the Zodiac' on page 135, where are described the meanings of Saturn travelling through each of the Signs – that is, the kind of teacher who is teaching the lessons and setting the examinations at that time.

The Teachers' Duty Roster

year	month	day	time	sign	year	month	day	time	sign	year	month	day	time	sign
1900	JAN	01	00.00	SAG	1935	FEB	14	14.24	PIS	1975	SEP	17	04.44	LEO
1900	JAN	21	08.21	CAP	1937	APR	25	06.45	ARI	1976	JAN	14	13.30	CAN
1900	JUL	18	17.43	SAG	1937	OCT	18	03.53	PIS	1976	JUN	05	05.28	LEO
1900	OCT	17	05.20	CAP	1938	JAN	14	10.22	ARI	1977	NOV	17	02.28	VIR
1903	JAN	19	22.01	AQU	1939	JUL	06	15.39	TAU	1978	JAN	05	00.58	LEO
1905	APR	13	08.23	PIS	1939	SEP	22	05.11	ARI	1978	JUL	26	11.46	VIR
1905	AUG	17	00.22	AQU	1940	MAR	20	09.35	TAU	1980	SEP	21	10.37	LIB
1906	JAN	08	12.38	PIS	1942	MAY	08	19.32	GEM	1982	NOV	29	10.13	SCO
1908	MAR	19	14.10	ARI	1944	JUN	20	07.55	CAN	1983	MAY	06	19.24	LIB
1910	MAY	17	07.13	TAU	1946	AUG	02	14.31	LEO	1983	AUG	24	11.43	SCO
1910	DEC	14	23.02	ARI	1948	SEP	19	04.23	VIR	1985	NOV	17	02.23	SAG
1911	JAN	20	09.02	TAU	1949	APR	03	03.44	LEO	1988	FEB	13	23.37	CAP
1912	JUL	07	06.23	GEM	1949	MAY	29	12.49	VIR	1988	JUN	10	05.11	SAG
1912	NOV	30	18.26	TAU	1950	NOV	20	15.45	LIB	1988	NOV	12	09.32	CAP
1913	MAR	26	13.18	GEM	1951	MAR	07	12.13	VIR	1991	FEB	06	18.40	AQU
1914	AUG	24	17.38	CAN	1951	AUG	13	16.36	LIB	1993	MAY	21	04.44	PIS
1914	DEC	07	06.34	GEM	1953	OCT	22	15.26	SCO	1993	JUN	30	08.13	AQU
1915	MAY	11	21.36	CAN	1956	JAN	12	18.44	SAG	1994	JAN	28	23.29	PIS
1916	OCT	17	15.26	LEO	1956	MAY	14	03.40	SCO	1996	APR	07	08.35	ARI
1916	DEC	07	19.03	CAN	1956	OCT	10	15.24	SAG	1998	JUN	09	06.24	TAU
1917	JUN	24	13.38	LEO	1959	JAN	05	13.24	CAP	1998	OCT	25	18.55	ARI
1919	AUG	12	13.43	VIR	1962	JAN	03	19.15	AQU	1999	MAR	01	01.38	TAU
1921	OCT	07	17.35	LIB	1964	MAR	24	04.17	PIS	2000	AUG	10	02.36	GEM
1923	DEC	20	04.09	SCO	1964	SEP	16	21.04	AQU	2000	OCT	16	00.55	TAU
1924	APR	06	08.33	LIB	1964	DEC	16	05.39	PIS	2001	APR	20	22.06	GEM
1924	SEP	13	22.19	SCO	1967	MAR	03	21.13	ARI	2003	JUN	04	01.33	CAN
1926	DEC	02	22.49	SAG	1969	APR	29	22.35	TAU	2005	JUL	16	12.36	LEO
1929	MAR	15	13.59	CAP	1971	JUN	18	16.22	GEM	2007	SEP	02	13.53	VIR
1929	MAY	05	04.28	SAG	1972	JAN	10	03.28	TAU	2009	OCT	29	17.13	LIB
1929	NOV	30	04.03	CAP	1972	FEB	21	14.33	GEM	2010	APR	07	18.45	VIR
1932	FEB	24	02.22	AQU	1973	AUG	01	22.40	CAN	2010	JUL	21	15.18	LIB
1932	AUG	13	11.12	CAP	1974	JAN	07	20.19	GEM	2012	OCT	05	20.39	SCO
1932	NOV	20	02.25	AQU	1974	APR	18	22.22	CAN					

Sign Abbreviations

Aries ....ARI	Leo ....LEO	Sagittarius ....SAG
Taurus ....TAU	Virgo.................VIR	Capricorn ....CAP
Gemini ....GEM	Libra ....LIB	Aquarius...........AQU
Cancer.............CAN	Scorpio ....SCO	Pisces ....PIS

The Twelve Teachers according to Saturn's Transits through the Signs of the Zodiac

How to Understand your Teachers

1 Each of your teachers is named according to the Sign that Saturn is travelling through, that is, 'When Saturn is travelling through Aries', 'When Saturn is travelling through Taurus', etc.

2 In the table below, the words Waxing Square, Waxing Trine, etc, refer to the various influences or Transits of your Saturn Cycle, the time and meaning of which are given in the timetable on page 108. The word 'Natal' refers to the Sign position of Saturn in your Birth Chart. For example, during my First Waning Square at 22 years of age during mid-to-late 1967, Saturn was travelling through **Aries**, making a **Waning Square** to my **Natal** Saturn in Cancer. At that time I was actually making an immense effort to get myself started (Aries) as a professional record producer while at the same time having to live out of a suitcase at a friend's apartment (Cancer).

REALITY CHECK: Except in the case of your Saturn Return, always check that transitting Saturn is in the right Sign to be making the influence in question by using the table below. Usually it will be, but first determine the Sign of your Natal Saturn by looking in the 'Teachers' Duty Roster' for what Sign it was in for your date of birth. Continuing with my example, for my birth date in September 1944, I see that my Natal Saturn, was in Cancer. Now I actually started my production job in January 1967, but if you look at that time in the 'Teachers' Duty Roster' you will see that Saturn was in Pisces then, and so for a Waning Square to be happening, according to the table below, I would have to have my Natal Saturn in Gemini, which I do not. So I now know that Waning Square must have begun a bit later, after Saturn entered Aries, which was 3 March according to the Roster. Doing this little exercise also helps you to get a more accurate fix upon when you are or were under any particular influence of Saturn.

If your SATURN Sign at birth is ... ↓	... then SATURN would have to be transitting the Sign given below under the type of influence you are experiencing.					
	WAXING SQUARE	WAXING TRINE	OPPOSITION	WANING TRINE	WANING SQUARE	RETURN
ARIES	Cancer	Leo	Libra	Sagittarius	Capricorn	Aries
TAURUS	Leo	Virgo	Scorpio	Capricorn	Aquarius	Taurus
GEMINI	Virgo	Libra	Sagittarius	Aquarius	Pisces	Gemini
CANCER	Libra	Scorpio	Capricorn	Pisces	Aries	Cancer
LEO	Scorpio	Sagittarius	Aquarius	Aries	Taurus	Leo
VIRGO	Sagittarius	Capricorn	Pisces	Taurus	Gemini	Virgo
LIBRA	Capricorn	Aquarius	Aries	Gemini	Cancer	Libra
SCORPIO	Aquarius	Pisces	Taurus	Cancer	Leo	Scorpio
SAGITTARIUS	Pisces	Aries	Gemini	Leo	Virgo	Sagittarius
CAPRICORN	Aries	Taurus	Cancer	Virgo	Libra	Capricorn
AQUARIUS	Taurus	Gemini	Leo	Libra	Scorpio	Aquarius
PISCES	Gemini	Cancer	Virgo	Scorpio	Libra	Pisces

3 'UNDER SATURN' means being under test, giving rise to promotion or demotion, to advancement or adversity, to review and possible reform, or being at a standstill until such has happened.

Here come the Teachers . . .

♄

SATURN travelling through ARIES

This teacher stresses the necessity of independence, leadership, getting something started, push, persistence and patience.

So when during your Saturn Cycle you are experiencing a:

WAXING SQUARE – this would mean that your Natal Saturn is in Capricorn, and so it tests your practical ability to deal with career and external authorities and pressures whilst also having to deal with your more individual pursuits and concerns, and not to let the one jeopardize or impinge on the other. 'If you can't beat 'em, join 'em' could become a key issue – but having the courage not to be anyone's puppet would be too.

WAXING TRINE – this would mean that your Natal Saturn is in Sagittarius, and so your faith, enterprise and/or adventurousness gain support, reward and/or guidance from some forceful person or leading light, or from your own sense of initiative and powers of self-assertion.

OPPOSITION – this would mean that your Natal Saturn is in Libra, so it makes it apparent to you that a balance must be struck between your social and relationship needs and duties on the one hand, and your more individual pursuits and concerns on the other. You have to maintain these two areas equally, and appreciate how one reflects the other.

WANING TRINE – this would mean that your Natal Saturn is in Leo, and so your style and creativity gain support, reward and/or guidance from some forceful person or leading light, or from your own sense of initiative and powers of self-assertion.

WANING SQUARE – this would mean that your Natal Saturn is in Cancer, and so it tests your practical ability to deal with your private, family and domestic affairs whilst also having to attend to your more individual pursuits and concerns, and not to let the one jeopardize or impinge on the other. Learning to adopt or exercising the right demeanour for each situation is a key issue. If you can successfully make it the same for both, then all the better.

RETURN – this would mean that your Natal Saturn is also in Aries, and so it stresses that the necessity of independence, leadership, getting something started, push, persistence and patience is a critical issue for you – on both an external, material and professional level (Level One), and an internal, emotional and private level (Level Two).

The following people and matters are 'under Saturn' in the world at large, and could be what bring Saturn's influence to you:

Leaders, the military, sport and sportspeople, violence, forceful or pushy people, Arian types.

♄ ♄

SATURN travelling through TAURUS

This teacher stresses the necessity of self-worth, material wherewithal, and making only necessary changes while conserving the natural and the good.

So when during your Saturn Cycle you are experiencing a:

WAXING SQUARE – this would mean that your Natal Saturn is in Aquarius, and so it tests your practical ability to express your ideals and originality whilst also having to attend to your material and physical needs and concerns, and not to let the one jeopardize or impinge on the other. Bearing in mind that today's radicals are tomorrow's conservatives could be very useful.

WAXING TRINE – this would mean that your Natal Saturn is in Capricorn, and so your discipline, hard work and worldliness gain support, reward and/or guidance from some financial quarter or person of substance, or from your own talents, practical abilities and sense of economy.

OPPOSITION – this would mean that your Natal Saturn is in Scorpio, so it makes it apparent to you that a balance must be struck between your emotional desires and dependency upon the resources of others on the one hand, and your own physical attractiveness and earning power on the other. You become aware of how much what you want is dependent upon what you have, and that material factors are as important as emotional ones.

WANING TRINE – this would mean that your Natal Saturn is in Virgo, and so your attention to detail and good services rendered gain support, reward and/or guidance from some financial quarter or person of substance, or from your own talents, practical abilities and sense of economy.

WANING SQUARE – this would mean that your Natal Saturn is in Leo, and so it tests your practical ability to deal with your romantic, creative and/or speculative interests whilst also having to attend to your material and physical needs and concerns – and not to let the one jeopardize or impinge on the other. Controlling your appetites and knowing the true worth of things are key issues.

RETURN – this would mean that your Natal Saturn is also in Taurus, and so it stresses that the necessity of self-worth, material wherewithal, and making only necessary changes while conserving the natural and the good, is a critical issue for you – on both an external, material and professional level (Level One), and an internal, emotional and private level (Level Two).

The following people and matters are 'under Saturn' in the world at large, and could be what bring Saturn's influence to you:

Money, banks and bankers, the market, conservatism, agriculture/farming, production, produce and producers, Taurean types.

SATURN travelling through GEMINI

This teacher stresses the necessity of good communication, having the facts and figures, and of not being cynical as a result of being too logical.

So when during your Saturn Cycle you are experiencing a:

WAXING SQUARE – this would mean that your Natal Saturn is in Pisces, and so it tests your practical ability to express your sensitivity, imagination and spiritual interests whilst also having to attend to more everyday affairs and rational concerns, and not to let the one jeopardize or impinge on the other. Controlling any inclination to escape or dissemble is a key issue.

WAXING TRINE – this would mean that your Natal Saturn is in Aquarius, and so your ideals and originality gain support, reward and/or guidance from the media or your own or someone else's good contacts and communication skills.

OPPOSITION – this would mean that your Natal Saturn is in Sagittarius, so it makes it apparent to you that there is a difference between the letter of the law and the spirit of it, between ideals and everyday affairs – and that you must create a balance between the two. You become aware of how much what you believe is dependent upon your ability to express that belief in everyday terms.

WANING TRINE – this would mean that your Natal Saturn is in Libra, and so your sense of harmony and diplomacy gains support, reward and/or guidance from the media or your own or someone else's good contacts and communication skills.

WANING SQUARE – this would mean that your Natal Saturn is in Virgo, and so it tests your practical ability to work efficiently and be of service whilst also taking time-out for more light-hearted pursuits and amusements, and not to let the one jeopardize or impinge on the other. Balancing out earnestness with humour could be a key issue.

RETURN – this would mean that your Natal Saturn is also in Gemini, and so it stresses that the necessity of good communication, having the facts and figures, avoiding double standards, and of not being cynical as a result of being too logical, is a critical issue for you – on both an external, material and professional level (Level One), and an internal, emotional and private level (Level Two).

The following people and matters are 'under Saturn' in the world at large, and could be what bring Saturn's influence to you:

Communications, primary and secondary education, local government and neighbourhood issues, siblings, Geminian types.

SATURN travelling through CANCER

This teacher stresses the necessity of nurturance, of maternal, domestic and family values, obeying the natural order, and not taking things personally.

So when during your Saturn Cycle you are experiencing a:

WAXING SQUARE – this would mean that your Natal Saturn is in Aries, and so it tests your practical ability to follow your more individual pursuits whilst also having to attend to your private, family and domestic duties and concerns, and not to let the one jeopardize or impinge on the other. Drawing the line between the two without getting too emotional is the goal to aim for.

WAXING TRINE – this would mean that your Natal Saturn is in Pisces, and so your imagination, sensitivity and spirituality gain support, reward and/or guidance from someone in the family or on the home front, or in terms of accommodation, or from your sympathies and emotional awareness.

OPPOSITION – this would mean that your Natal Saturn is in Capricorn, so it makes it apparent to you that a balance must be struck between the pressures of career and external authorities on the one hand, and your private, family and domestic duties on the other. You become aware of how much these two areas of your life are interdependent.

WANING TRINE – this would mean that your Natal Saturn is in Scorpio, and so your deepest emotional needs and involvements, and financial concerns, gain support, reward and/or guidance from someone in the family or on the home front, or in terms of accommodation, or from your sympathies and emotional awareness.

WANING SQUARE – this would mean that your Natal Saturn is in Libra, and so it tests your practical ability to maintain your social life and relationships whilst also having to attend to private, family and domestic affairs, and not to let the one jeopardize or impinge on the other. Establishing your emotional priorities is a key issue.

RETURN – this would mean that your Natal Saturn is also in Cancer, and so it stresses that the necessity of nurturance, of maternal, domestic and family values, obeying the natural order, and/or not taking things personally, is a critical issue for you, as too is your emotional state – on both an external, material and professional level (Level One), and an internal, emotional and private level (Level Two).

The following people and matters are 'under Saturn' in the world at large, and could be what bring Saturn's influence to you:

Nature, natural methods, nature conservation, mother(hood), family, national security, housing, carers, Cancerian types.

SATURN travelling through LEO

This teacher stresses the necessity of creativity, recognition, taking pride but not being conceited, and of not being a slave to your image or act.

So when during your Saturn Cycle you are experiencing a:

WAXING SQUARE – this would mean that your Natal Saturn is in Taurus, and so it tests your material position, earning power and sense of self-worth, in spite or because of how confident or important you think you ought to feel. Avoiding going beyond your means could be a key issue, as also would be letting pride compromise material or practical concerns, or allowing excessive caution to inhibit romantic or creative interests.

WAXING TRINE – this would mean that your Natal Saturn is in Aries, and so your independence, courage and forthrightness gain support, reward and/or guidance from someone or something in a position to confer honour or advantage upon you, or from your own sense of confidence, creativity and dignity.

OPPOSITION – this would mean that your Natal Saturn is in Aquarius, so it makes it apparent to you that a balance must be struck between considerations for society as a whole on the one hand, and your own creative and seemingly selfish interests on the other. You become aware of how much these two areas of your life are interdependent.

WANING TRINE – this would mean that your Natal Saturn is in Sagittarius, and so for your faith, enterprise and/or adventurousness you gain support, reward and/or guidance from someone or something in a position to confer honour or advantage upon you, or from your own sense of confidence, creativity and dignity.

WANING SQUARE – this would mean that your Natal Saturn is in Scorpio, and so it tests your ability to express your deepest emotional and sexual needs whilst at the same not losing face or appearing to be weak, and not to let the one jeopardize or impinge on the other. Establishing your emotional priorities is a key issue.

RETURN – this would mean that your Natal Saturn is also in Leo, and so it stresses that the necessity of creativity, recognition, taking pride but not being conceited, and of not being a slave to your image or act, is critical for you – on both an external, material and professional level (Level One), and an internal, emotional and private level (Level Two).

The following people and matters are 'under Saturn' in the world at large, and could be what bring Saturn's influence to you:

Rulers or those in charge, parents and children, enterprise, entertainment, creativity, royalty, privilege, Leonine types.

SATURN travelling through VIRGO

This teacher stresses the necessity of maintaining good health, rewarding employment, purity without fussiness, and of putting things in order.

So when during your Saturn Cycle you are experiencing a:

WAXING SQUARE – this would mean that your Natal Saturn is in Gemini, and so it tests your practical ability to communicate and make contacts, and manage day-to-day affairs whilst also allotting time for study, attending to health matters, and for being on your own, and not to let the one jeopardize or impinge on the other. It also tests your level of clear thinking, and that you are not missing the wood for the trees.

WAXING TRINE – this would mean that your Natal Saturn is in Taurus, and so your material and physical needs and concerns gain support, reward and/or guidance from someone in a service profession or helping capacity, or from your own sense of service and health.

OPPOSITION – this would mean that your Natal Saturn is in Pisces, so it makes it apparent to you that you have to relate to the emotional dimension of life, but also to pay attention to the details and practical considerations. You become aware of how much these two areas of your life are interdependent.

WANING TRINE – this would mean that your Natal Saturn is in Capricorn, and so your career, discipline, hard work and worldliness gain support, reward and/or guidance from someone in a service profession or helping capacity, or from your own sense of service and health.

WANING SQUARE – this would mean that your Natal Saturn is in Sagittarius, and so it tests your faith and adventurousness while you are also having to attend to work efficiently and be of service, and not to let the one jeopardize or impinge on the other. Assigning equal amounts of attention to both the general and the particular is a key issue.

RETURN – this would mean that your Natal Saturn is also in Virgo, and so it stresses that the necessity of maintaining good health, rewarding employment, purity without fussiness, and of putting things in order is a critical issue, but also that you should not be too hard on yourself or others – on both an external, material and professional level (Level One), and an internal, emotional and private level (Level Two).

The following people and matters are 'under Saturn' in the world at large, and could be what bring Saturn's influence to you:

All forms of service, bureaucracy, employment, health, personnel, animal care, Virgoan types.

SATURN travelling through LIBRA

This teacher stresses the necessity of adhering to principles of fair play and harmony in your dealings with others, especially with your partner.

So when during your Saturn Cycle you are experiencing a:

WAXING SQUARE – this would mean that your Natal Saturn is in Cancer, and so it tests your practical ability to manage your private, family and domestic duties and concerns whilst also having to attend to your social life generally, and love life in particular, and not to let the one jeopardize or impinge on the other. Combining being diplomatic and sympathetic, impartial and personal, could be a challenging issue.

WAXING TRINE – this would mean that your Natal Saturn is in Gemini, and so your communication skills and need to be in the know find support, reward and/or guidance from your partner or someone who is able to see matters in a fair and impartial light, which could just be yourself.

OPPOSITION – this would mean that your Natal Saturn is in Aries, and so it makes it apparent to you that a balance must be struck between your more individual pursuits and concerns on the one hand, and the pressures of relationship commitments and social considerations on the other. You become aware of how much these two areas of your life are interdependent.

WANING TRINE – this would mean that your Natal Saturn is in Aquarius, and so your social values, originality and pet concerns gain support, reward and/or guidance from your partner or someone who is able to see matters in a fair and impartial light, which could just be yourself.

WANING SQUARE – this would mean that your Natal Saturn is in Capricorn, and so it tests your practical ability to deal with career and external authorities and pressures whilst also having to attend to your social life generally, and love life in particular, and not to let the one jeopardize or impinge on the other. Establishing a balance here is a key issue.

RETURN – this would mean that your Natal Saturn is also in Libra, and so it stresses that the necessity of adhering to principles of fair play and harmony in your dealings with others, especially with your partner, is a critical issue for you – on both an external, material and professional level (Level One), and an internal, emotional and private level (Level Two).

The following people and matters are 'under Saturn' in the world at large, and could be what bring Saturn's influence to you:

Judiciary, arbitrators, peacemakers, diplomats, artists and fashion, Libran types.

SATURN travelling through SCORPIO

This teacher stresses the necessity of dealing with issues of power, intrigue, intimacy, sexuality, death, taxes, legacies and other people's resources.

So when during your Saturn Cycle you are experiencing a:

WAXING SQUARE – this would mean that your Natal Saturn is in Leo, and so it tests your level of outer confidence or how important you think you ought to feel, whilst also putting you under pressure in your intimate or business life, and not to let the one jeopardize or impinge on the other. Not being afraid to express your most genuine feelings with creative style is the challenge.

WAXING TRINE – this would mean that your Natal Saturn is in Cancer, and so your family and domestic concerns find support, reward and/or guidance from someone or something who is in a position of financial or psychological influence, or from your own sense of insight and emotional conviction.

OPPOSITION – this would mean that your Natal Saturn is in Taurus, so it makes it apparent to you that a balance must be struck between your own physical attractiveness and earning power on the one hand, and your emotional nature and desires, and dependency upon the resources of others, on the other. You become aware of how much what you have is dependent upon what you want, and that emotional factors are as important as material ones.

WANING TRINE – this would mean that your Natal Saturn is in Pisces, and so your imagination, sensitivity and spirituality gain support, reward and/or guidance from someone or something who is in a position of financial or psychological influence, or from your own sense of insight and emotional conviction.

WANING SQUARE – this would mean that your Natal Saturn is in Aquarius, and so your social values, originality and pet concerns have to be squared with your more private desires and possibly hidden side, and not to let the one jeopardize or impinge on the other. Hypocrisy and scandal are to be avoided at all costs.

RETURN – this would mean that your Natal Saturn is also in Scorpio, and so it stresses that the necessity of dealing with issues of power, intrigue, intimacy, sexuality, death, taxes, legacies and/or other people's resources, is a critical one for you – on both an external, material and professional level (Level One), and an internal, emotional and private level (Level Two).

The following people and matters are 'under Saturn' in the world at large, and could be what bring Saturn's influence to you:

Powermongers, tax officials, vice, police, miners, occultism, sexuality, secrets/espionage, intrigue, psychology, medics, Scorpionic types.

SATURN travelling through SAGITTARIUS

This teacher stresses the necessity of faith without dogmatism, and of thinking and acting in a philosophical and yet practical manner.

So when during your Saturn Cycle you are experiencing a:

WAXING SQUARE – this would mean that your Natal Saturn is in Virgo, and so it tests your practical ability to work efficiently, pay attention to details and health matters, and/or be of service whilst also taking into consideration the bigger picture and/or issues of law or morality, and not to let the one jeopardize or impinge on the other. You are learning to see the wood as well as the trees.

WAXING TRINE – this would mean that your Natal Saturn is in Leo, and so your style and creativity find support, reward and/or guidance from someone or something who is far-reaching or far-seeing, or from your own sense of faith and vision.

OPPOSITION – this would mean that your Natal Saturn is in Gemini, so it makes it apparent to you that there is a difference between the letter of the law and the spirit of it, between ideals and everyday affairs – and that you must create a balance between the two. You become aware of how much what you say to others must be backed up by believing in what you say, and seeing its place in the bigger picture.

WANING TRINE – this would mean that your Natal Saturn is in Aries, and so your forcefulness, initiative and independence gain support, reward and/or guidance from someone or something who is far-reaching or far-seeing, or from your own sense of faith and vision.

WANING SQUARE – this would mean that your Natal Saturn is in Pisces, and so your imagination, sensitivity and spirituality have to be squared with a more robust and philosophical outlook, without letting the one jeopardize or impinge on the other. Not allowing your sensitivity and philosophy to render you materially or emotionally ineffectual could become a issue.

RETURN – this would mean that your Natal Saturn is also in Sagittarius, and so it stresses that the necessity of faith without dogmatism, and of thinking and acting in a moral, philosophical and yet practical manner, while avoiding excessiveness, is a critical issue for you – on both an external, material and professional level (Level One), and an internal, emotional and private level (Level Two).

The following people and matters are 'under Saturn' in the world at large, and could be what bring Saturn's influence to you:

Travel, religion, higher education, ethics, clergy, law, Sagittarian types.

SATURN travelling through CAPRICORN

This teacher stresses the necessity of organization, being businesslike, objective and responsible, and of dealing with all forms of authority.

So when during your Saturn Cycle you are experiencing a:

WAXING SQUARE – this would mean that your Natal Saturn is in Libra, and so it tests your practical ability to deal with career pressures and external authorities whilst also having to attend to social and relationship needs and duties, and not to let the one jeopardize or impinge on the other. Learning or knowing where to draw the line is a key issue.

WAXING TRINE – this would mean that your Natal Saturn is in Virgo, and so good services rendered gain support, reward and/or guidance from those in authority, or from your own status and sense of responsibility.

OPPOSITION – this would mean that your Natal Saturn is in Cancer, so it makes it apparent to you that a balance must be struck between your private, family and domestic duties on the one hand, and those of career and external authorities and pressures on the other. You become aware of how much these two areas of your life are interdependent.

WANING TRINE – this would mean that your Natal Saturn is in Taurus, and so your worth and talent gain support, reward and/or guidance from those in authority, or from your own status and sense of responsibility.

WANING SQUARE – this would mean that your Natal Saturn is in Aries, and so it tests your practical ability to deal with your more individual pursuits and concerns whilst also having to attend to career pressures and external authorities, and not to let the one jeopardize or impinge on the other. Learning or exercising patience, yet also having the courage not to be anyone's puppet, is a key issue.

RETURN – this would mean that your Natal Saturn is also in Capricorn, and so it stresses that the necessity of organization, being businesslike, objective and responsible, and of dealing with all forms of authority, is a critical issue for you – on both an external, material and professional level (Level One), and an internal, emotional and private level (Level Two).

The following people and matters are 'under Saturn' in the world at large, and could be what bring Saturn's influence to you:

Bosses, organisations and corporations, government, structures and all who deal with them, builders, Capricornian types.

SATURN travelling through AQUARIUS

This teacher stresses the necessity of originality, inventiveness, sound social values and a sense of equality, impartiality and the unusual or unlikely.

So when during your Saturn Cycle you are experiencing a:

WAXING SQUARE – this would mean that your Natal Saturn is in Scorpio, and so it tests your practical ability to deal with issues relating to power, deep and hidden feelings, intimacy, sexuality, others resources, taxes or legacies whilst also maintaining your sense of freedom, being true to your own values and sense of progress, and not to let the one jeopardize or impinge on the other.

WAXING TRINE – this would mean that your Natal Saturn is in Libra, and so your artistic or social interests and partnership issues gain support, reward and/or guidance from the implementation of reforms, experimentation, impartiality and a forward-looking sense of social behaviour – be it from yourself or someone else.

OPPOSITION – this would mean that your Natal Saturn is in Leo, so it makes it apparent to you that a balance must be struck between your own creative and seemingly selfish interests and considerations on the one hand, and for society as a whole on the other. You become aware of how much these two areas of your life are interdependent.

WANING TRINE – this would mean that your Natal Saturn is in Gemini, and so your communication skills and your need to be in the know find support, reward and/or guidance from the implementation of reforms, experimentation, impartiality and a forward-looking sense of social behaviour – be it from yourself or someone else.

WANING SQUARE – this would mean that your Natal Saturn is in Taurus, and so it tests your material position and sense of self-worth on the one hand whilst maintaining your sense of freedom, being true to your own values and sense of progress on the other hand, and not to let the one jeopardize or impinge on the other. Learning to combine the traditional and the modern is a key issue.

RETURN – this would mean that your Natal Saturn is also in Aquarius, and so it emphasizes that the degree to which you are attending to your originality, inventiveness, sound social values and to a sense of equality and impartiality is a critical issue – on both an external, material and professional level (Level One), and an internal, emotional and private level (Level Two).

The following people and matters are 'under Saturn' in the world at large, and could be what bring Saturn's influence to you:

Alternative subjects and practitioners, science and technology, computers, unusual people and social scenes, inventors, Aquarian types.

♄

SATURN travelling through PISCES

This teacher stresses the necessity of sensitivity, imagination, spirituality and psychic awareness, and/or practical knowledge of psychic matters.

So when during your Saturn Cycle you are experiencing a:

WAXING SQUARE – this would mean that your Natal Saturn is in Sagittarius, and so it tests your ability to be expansive and travel around whilst simultaneously being sensitive to emotional and intangible elements – but to not let the one jeopardize or impinge on the other.

WAXING TRINE – this would mean that your Natal Saturn is in Scorpio, and so your deepest emotional needs and involvements gain support, reward and/or guidance from some sensitive or spiritual quarter or person, or from you yourself being emotionally or psychically attuned.

OPPOSITION – this would mean that your Natal Saturn is in Virgo, so it makes it apparent to you that you have not only to pay attention to the details and practical considerations but also to be attuned to the emotional dimension, and to the significance of sensitive or intangible elements. You become aware of how much these two areas of your life are interdependent.

WANING TRINE – this would mean that your Natal Saturn is in Cancer, and so your family and domestic concerns find support, reward and/or guidance from some sensitive or spiritual quarter or person, or from you yourself being emotionally or psychically attuned.

WANING SQUARE – this would mean that your Natal Saturn is in Gemini, and so it tests your practical ability to communicate and make contacts, and manage day-to-day affairs whilst simultaneously being sensitive to emotional and intangible elements, but not to let the one jeopardize or impinge on the other. Controlling any inclination to escape or dissemble is a key issue.

RETURN – this would mean that your Natal Saturn is also in Pisces, and so it stresses that the necessity of sensitivity, imagination, spirituality and psychic awareness, and/or practical knowledge of psychic matters is a critical issue for you – on both an external, material and professional level (Level One), and an internal, emotional and private level (Level Two).

The following people and matters are 'under Saturn' in the world at large, and could be what bring Saturn's influence to you:

Music and musicians, artists, psychics, anything or anyone involved with the sea, drugs, alcohol or chemicals, fascinations, addictions, Piscean types.

♄ ♄

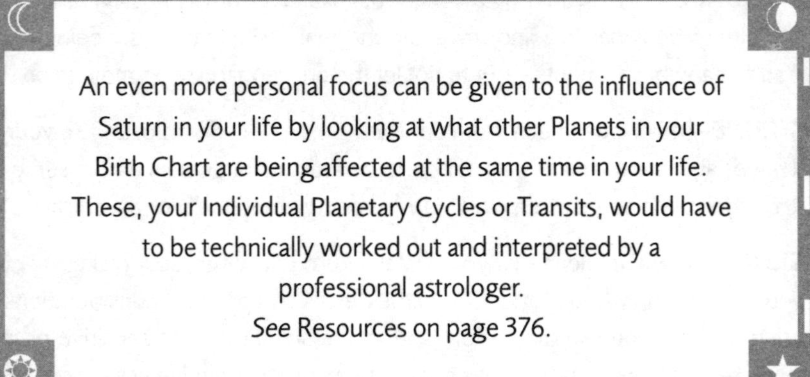

An even more personal focus can be given to the influence of Saturn in your life by looking at what other Planets in your Birth Chart are being affected at the same time in your life. These, your Individual Planetary Cycles or Transits, would have to be technically worked out and interpreted by a professional astrologer.

See Resources on page 376.

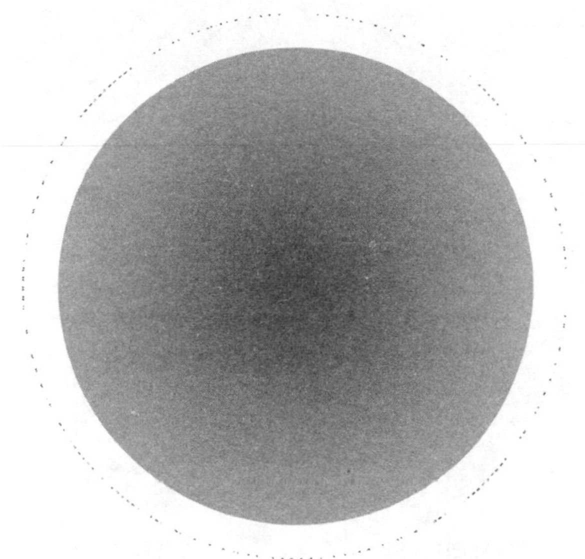

YOUR URANUS CYCLE

Change and Awakening

'If you hold to my teaching, you are really my disciples.
Then you will know the truth and the truth will set you free.'

John 8, 32

Key Influences

+	−
Awakening	Disrupting
Exciting	Shocking
Liberating	Destabilising
Reforming	Ruining

There is in all of us, to varying degrees, an urge to be free, whether from something or to do something. But as we have seen with the Moon and Saturn, there is also a need for emotional and material security. This compromises that urge to be free because on the face of it security is, of necessity, *bound* by familiarity, convention and a resistance to change. But all this begs the question as to what freedom really is. Essentially, freedom is having the urge, the space and the time to be or find out who you truly are, and to explore the unknown. And so who you truly are is the very thing that gets constrained by those needs for security, by staying with the 'Devil you know'. Consequently, the 'urge to be free to find out who you truly are' often has to find an original, surprising or alternative way of making itself felt in your life. Something has to change or shock us in order to free us – and because we all have that innate desire to be free, we therefore attract change and the unexpected into our lives.

Earlier on, in the section called 'The Story of your Life', I chose a Rocket Ship as a metaphor for Uranus because it symbolizes something which blasts one free from ordinary existence (that is, *whatever you feel trapped in or by*) as it launches one into space and the unknown (that is, *a new way of thinking and being*). In turn, this symbolizes what psychologist Carl Jung called the Individuation Process, through which one develops into a true individual – a true individual being someone 'in their own right', who rises above their emotional, social, cultural and political conditions and conditioning (the Moon and Saturn), and is not merely a product and prisoner of it. It is this conditioning which causes us to *react* to people and situations, rather than *choose* what to say or do, or not say and not do. This is what being free truly is – something called 'inner independence'. The 'Age of Aquarius', which is the 2160 year period into which the whole world is currently moving, is this freeing process happening on a mass, evolutionary scale.

And so Uranus – which initially is the freedom to find out who you truly are, and the urge to do so that ultimately propels you towards that inner independence – has to awaken us to this through anything which shocks or excites with what is unexpected or new. Whether you know it or not, there is in all of us an 'upgrading program' which, as we go through our life, bestows or thrusts upon us new and refreshing ideas, places, things and people. Or else it doesn't, but we wished it would, thus reflecting back at us that we have a desire for change that is either unrealistic or is being inhibited by a fear of losing our Lunar comfort and/or Saturnian stability. Ultimately, the first thing we have to change in order for something new to happen is our ideas or values – concerning life itself, ourselves, others or whatever it is we wish would change. To this end, Uranus will somehow introduce some radically new idea or way of looking at life, often rattling our cage in the process. Such may also involve the presence of some disruptive individual, relationship or event.

'A paradox is the truth standing on its head to attract attention'

from *The Cynics' Encyclopaedia*

The only constant is change itself. Everything must change sooner or later. But another trouble can be that, at one extreme, changes are often rung purely for the sake of it, such as, for example, a change in the educational or governmental system which does not work and actually sets a lot of people back in the process. Paradoxically, though, this eventually wakes us up to the folly of falling for such fads and spins. More superficially, there is what could be called 'novamania', an indiscriminate obsession for things new, as for instance when a new product replaces an old one which was perfectly all right. At the other extreme, change can be resisted because it is seen as threatening the security of the known. Such resistance can be very well-founded, but it can also just be fear of the unknown masquerading as caution.

There are many other superficial changes that we make in life, like having a new hairstyle or a 'make-over', or doing something that might shock a little, like wearing a ring through a nipple or dying one's hair bright pink. But when the influence of Uranus is strong in your life, more radical and more meaningful changes are called for. Uranus could be regarded as the Vanguard of Evolution for it is its impulse to experience what is new, unusual and exciting that eventually takes us away from the norm and what we are used to into, say, a different relationship, job, lifestyle, location or morality, and ultimately onwards to a new version of what it actually is to be human. Eventually this evolves collectively into the creation of a new society, then into a new world. But on a social level (as distinct from new inventions, products and art forms), the paradox is that it is not so much something *new* that we discover through all this change and experimentation and alternative lifestyle, but something that has always been: something that is *true*. This means that Uranus takes us closer and closer to our truth, individually and collectively. For example, homosexuality was once outlawed in the West, and still has the power to shock and be a focus for derision on the one hand, or for the assertion of the right to be free to be what you are, on the other. But generally speaking, it is now accepted as part and parcel of our society; it is simply a human truth. It is even vaunted as being 'cool' to be gay or to mix with gays; then again, the right to be gay can be used as carte blanche to leave individual shortcomings unattended to. Notwithstanding this, whether being gay is 'right' or 'wrong' is irrelevant because Uranus is impartiality itself and so is beyond judging; with laser-like precision it just lays bare the truth of the matter. It is as if the clannish familiarity of the Moon and the stuffy conventions of Saturn blind us to what we and the world and life and everything truly are.

On an esoteric note, Uranus is connected with the very nature of the Energy that is Life itself – electromagnetism. And seeing that Uranus rules the Unknown, it is interesting to note that it is still not quite clear what electricity actually is. By the same token, Uranus is the sexual impulse itself, in that it is the most basic function of any living creature's 'wiring' to 'interface' with the wiring of another creature – usually, but not always, of the same background, creed, race or even species (remember that Uranian impulse to shock, experiment and step beyond the norm). This is similar to the way in which electricity always seeks to go to earth; that is, we need another's body to absorb our 'charge' or to 'charge' us up. In the same sense, it is Uranus that sparks anything into action – like the 'electricity' you feel when encountering someone or something that is 'electrifying'; like the semi-conscious impulse to do or say something against one's 'better judgement'; or like the rebel who stirs up others to bring down, or again, paradoxically, resurrect old forms. Adolf Hitler, with Uranus his most prominent Planet, was a frightening example of this last type. (*See* page 359 for more regarding Hitler.)

Uranus also governs the science of anything, be it the circuitry of a computer, or the acupuncture meridians of the human body; the workings of any mechanism or the brain-mind. So the language of Uranus is symbolism, which includes subjects such as mathematics, algebra, computer-programming language and the very one you are reading now, astrology. In fact, astrology has been called the Algebra of Life.

From the above it should not be surprising to find that Uranian influences in one's life or personality can be rather emotionally detached and divorced from more safe or familiar feelings – or even from the so-called natural order of things. The security of the near-at-hand can be blown away by the icy blasts of Uranus. Such Uranian things as extra-marital affairs seem to split coldly asunder the family unit, while computers have perverted or even replaced normal emotional interaction. It has already been mooted that the harm computer games are doing to young males is going to have a heavy fall-out.

Uranus – Myth and Fact

Uranus was the firstborn of his mother, Gaia, who was Mother Earth herself. Effectively, Uranus was Heaven, and he then lay with his Mother, Earth. This parallels Genesis in the Bible – 'In the beginning God created the heavens and the earth'. From out of their union the created world itself came about. This is why Uranus has so much to do with sexuality, in the obvious sense and in terms of electricity and the body (Earth) creating the 'primordial soup' from which springs all life. However, one of his sons, Saturn – encouraged by his mother – saw his father as too powerful a being. And so he emasculated him, scything off his genitals and casting them into the sea. Saturn then took charge, giving rise to the 'Golden Age' which was an ordered world, but one which was patriarchal. We are still in that patriarchal state, but, of course, it is far from golden. This is because the primal male and female energies have become divorced, are at war with one another, recreating the impotence. And yet, from the foam of Uranus's severed genitals grew Venus (in Greek, Aphrodite, meaning *born of foam*). This symbolizes that through the Love, Harmony, Art and Grace of Venus we can again reunite as men and women, as fellow human beings. This is the goal and promise of the Uranus-ruled Age of Aquarius of which we are now on the cusp (*see* page 355).

As an actual planet, Uranus reflects its astrological significance as the Planet of Revolution and Freedom. It was discovered unexpectedly by William Herschel in 1781, the very same year that the American War of Independence ended at Yorktown, and eight years before the French Revolution began with the Storming of the Bastille (Saturn). The unique quality of Uranus is reflected in the fact that it is the only Planet to go around the Sun on its side, pointing one of its poles at it as it does so. The Planet of Heaven literally rolls around it all day!

But again the Uranian paradox pops up. Suddenly having emotional security threatened or taken away makes us very aware of how we truly feel, while computers are doing much to save the world by improving communication and education. A frightening amount of electricity is wasted by computers being left on all the time, but then doing this paradoxically prolongs the life of the Uranian components inside of them, thereby saving energy. Uranian technology enables us to wipe out rainforests at an alarming rate, and to clone species through genetic engineering. Such could spell disaster for us, or it might be a zigzag route to discovering the truths of our existence. Remember that the destructive technology of war has eventually given rise to many a life-saving device or substance. 'God (Truth) comes forth in the sign of the Arousing,' states Hexagram 51 'Shock' in that great Chinese book of wisdom, the *I Ching* (meaning *The Book of Changes*).

All this can be patently mind-boggling stuff, just like Uranus. But in the end, Uranus is not about logic and linear thought. It is actually Intuition itself. Intuition is not to be confused with instinct, which is the province of the Moon. Instinct is a sense of survival; intuition is a sense of the future and the truth, of where we are bound and what we truly are. However, trouble occurs with the 'mad scientist syndrome' where Uranian inventions are used merely for the sake of it, with no thought for our natural requirements and emotional sensibilities. For instance, weapons scientists distance themselves from feelings by classifying the crew of a tank as 'components'.

Ideally then, the Uranian influences in your life are best met with a balanced combination of instinct and intuition, of preserving what is natural or familiar (Moon) and practical or traditional (Saturn) with a willingness to change and experiment with the new. As ever, Uranus will both facilitate and accelerate whatever is going on, thus paradoxically making the process both easier and more difficult. However, finding this balance, hopefully, is what I shall be helping you to do through the interpretations of the Uranus Transits that begin on page 161.

Generally, and most importantly, observe how Uranian happenings are usually 'a means to an end' in that they trigger something off, rather than maintain it. It is the lightning bolt that can change everything in an instant; like the electrical charge that can start your car. However, if you and whatever or whoever you are interacting with are changing in the same direction and at the same rate, then again, paradoxically, such change is quite enduring – like the sparking of the sparkplugs that maintains the motion of your car. Two Uranian relationships come to mind here: gangsters Bonnie and Clyde met by chance and were both rebels with no other cause than that of stealing the money to buy 'freedom' and wantonly celebrating their outcast nature to the bloody end; Prince Charles and Camilla Parker-Bowles who, having carried on an illicit affair since before he even met Princess Diana, are probably made for one another and will last and last. The first instance illustrates our fascination with social outcasts making a killing (in both senses of the word) and our subsequently ambivalent feelings concerning their demise. The second case displays how our fairytale idea of true love is so often totally at odds with a love that really is true. Uranus, as the above definition of a paradox implies, turns everything on its head to get us to see it another way, to see its truth, and eventually live it another way.

The Path of Return

U ltimate
U ranian
U -turn

The ultimate experience of Uranus is when you go through what could be called your Personal Revolution. This is when it finally dawns on you that life is not what it appears to be, that everything is purely a reflection of your own nature and a creation of your own making. When this happens (which can be at any point in your life or lifetimes, but Uranus will probably be active then), you will already have got 'airborne' enough in your Rocket Ship of liberation to see that you have got as far as you can get living in the normal or conventional way. Such a way is living merely as a product of emotional, social, cultural and political conditioning. But now you begin to tread what is called the Path of Return which we touched upon when describing the 'Story of Your Life' on page 15. And so here is the Uranian 'super-paradox', for the point of *no* return marks the beginning of the Path *of* Return. Below is given a diagram of the Path and how it plots one's Soul's progress through three stages.

STAGE 1 – OUTWARD PATH

This Stage is characterized by having a coarse or insensitive physical disposition. One is driven by reaction, base desire and unconscious impulses. One is guided by how much one is fitting in with one's race, gang or whatever partisan group one identifies with. One's goals are those of physical gratification and getting whatever one can from life. Morality is of the 'Whatever I can get away with' variety.
This Stage or mode of living continues until it no longer satisfies. Soon thereafter one then arrives at Stage 2.

STAGE 3 – PATH OF RETURN

This Stage is characterized by a refined physical body, which is maintained with pure foods and non-impact or yogic exercise. One is driven by a sense of service to the greater whole, and guided through meditation and/or elevated thinking arrived at through truth-seeking. One's own wellbeing is identified with the wellbeing of all living things and the state of the environment.
One progresses along the Path through surrendering more and more the needs and wishes of the ego, and being guided.

STAGE 2 – POINT OF INTEGRATION AND RETURN

This Stage is characterized by having a well-balanced physical body that becomes increasingly sensitive to thoughts and feelings. One is driven by a desire to excel and guided by intellectual principles. All of this culminates in some form of relative worldly success. There then comes a critical point when difficulties arise as social and material ambitions cease to be entirely valid motivations, and/or the body starts to act up as a sign of this re-defining of priorities. One can feel caught 'between two worlds'. This is the point of no return as Stage 3, the Path of Return, becomes inevitable.

How Uranus's Influence Manifests in your Life

HERE ARE SOME OF THE MEANS THROUGH which Uranian change can make itself felt in your life, but by its very nature, Uranus does not easily lend itself to being fixed to some hard and fast definition.

♅ **Restlessness and/or Experimentation** – Whether you are consciously aware of it or not at the time, these are probably the first signs of the Uranian influence showing itself in your life – that is, in your own personality rather than coming from without. As such, it can give rise to any of the scenarios outlined below; it is the trip-switch that sets off a chain reaction of events. Restlessness, sometimes in the form of anxiety or rebelliousness, is probably the more unconscious of these two Uranian signs for it can show itself as taking risks you would not normally take, or as inadvertently looking for change in a negative form just so long as something different happens. Needless to say, you can get more than you bargained for this way. Experimentation can also be unconscious, but it is more likely to have a deliberate quality about it. Such could simply mean playing around with some kind of material, substance, device or programme to see what happens, or going somewhere you would not normally go, or involving yourself with types or relationships that are off your normal beat. Uranus is what makes us look for a kick or a buzz, an out-of-the-ordinary experience that, through contrast, awakens our minds, feelings or senses. More to the point, however, Uranus can lead you to experimenting with a new or alternative way of looking at life and those who live it. As one of these ways could be astrology itself, Uranus is possibly working with you right now! (*See* below for more concerning Alternatives.)

♅ **Disruption and Revolution** – If something is being disrupted, then the Uranian message is that it is therefore in need of being revolutionized, or at least reformed. And the more one resists change, then the more disruptive it gets. Uranus is rather like the explosive charge that is needed to break down or through something, and the more obstinate that something is, the bigger the explosive charge has to be. Naturally, any area of life can be disrupted and/or revolutionized in this way, but personal relationships are a 'popular' target for Uranian rays of change. The reason for this is that it's in our relationships that we are inclined to become complacent and get stuck in a rut – and to Uranus a rut is a red rag to a bull. Unfortunately, because we usually tend to resist change *within* a relationship, we may go and look for change *outside* of it. We then find ourselves suddenly involved in an affair, or the victim of one, because we had not looked to where a change was really needed. The trick is to be alive to what is wrong or insufficient in a relationship, one's partner or, most importantly, oneself, and raise the issue as soon as possible. This is 'rendering unto Uranus what is Uranus's'! This will usually create some kind of shock-wave, but at least the potential cause of future disruption will have been brought to the surface where it can be dealt with – hopefully as impartially and fairly as possible, which may require the help of a third party/counsellor. Through this means of exposing underlying feelings, there is a chance of a constructive change, both practically and in terms of how both parties view the relationship. In this way is pre-empted the need for a different sort of 'third party' to catalyse a more unstable and unpleasant kind of change.

But if all this fails, then you can be reasonably sure that a more radical change is called for – namely, divorce or separation. This is naturally a lot more difficult when children are involved. Unlike Uranus himself, who banished his children he didn't like, be frank and truthful with your children and you'll be surprised how this can smooth or clarify things. Whatever the case, be utterly truthful, but remember that the truth nearly always involves confronting something in yourself that you'd rather not admit to – what's called a home truth. The truth hurts – if it *is* the truth. If you do not feel that you have identified and eradicated some deep-seated negative habit or feeling, then you probably have not been truthful enough. It is no use saying that 'things will be different from now on' if nothing has changed in you.

Earlier it was pointed out that if whatever or whoever you are interacting with is changing in the same direction and at the same rate as yourself, then paradoxically such change is quite enduring. In this context this means that the 'other woman' or 'other man' could be the one for you if you are more than just passing catalysts of change to one another. But probably only time and circumstances will tell, because if it was begun purely for the thrill of it, such a relationship would not survive the more day-to-day, domestic aspects. Even then, the Uranian quality of such a relationship may create further changes within your new life together. Remember that if Uranian unpredictability and frequent adjustments are rife in your life and relationship, then complacency and security have little place there. Yet again though, the famous Uranian paradox is that an ability within a relationship to adapt to requirements and change with the times, *together*, is a sign and guarantee of durability.

Accidents and Coincidences are occurrences that are similar because they both depend upon one thing synchronizing with one or more other things in a way that creates 'a chance event'. But from a Uranian point of view, all accidents are 'looking to happen' and so-called coincidences are just more than usually noticeable examples of how all events are interlinked. In fact, when viewed this way, an accident *is* a coincidence. This phenomenon, which is called synchronicity, is actually the basis of astrology itself. A Planet being in a certain place in the sky is not 'making' something happen in your life down here on Earth, it is just that the two events *correspond* to one another in some way. The way in which everything is interlinked like this is currently beyond our total comprehension because mostly we see time in a linear, cause and effect, kind of way: 'A' happening makes 'B' happen which in turn makes 'C' happen. In so far as it goes, this is perfectly correct, but time is also radial – an event is something that is spreading out in all directions. And so, as psychologist Carl Jung defined synchronicity, certain events happen together because they *like* happening together. I gave a personal example of this back in the section 'Your Moon Cycle', where my car ran out of petrol on my way to see my doctor about being exhausted. Right at this moment this brings to mind another linked 'coincidence'. One day my car was hit by a Volvo driven by a doctor in a car park at 1.50 pm, almost a year after being hit by another Volvo driven by another doctor in the same car park at exactly the same time. I won't go into a thorough interpretation here – and remember the language of Uranus is symbolism – but I do know that I had a bit of an aversion towards Volvos and doctors!

The main point here is that accidents and coincidences *mean* something – they are messages from your unconscious mind. More precisely, they are messages from the uppermost part of your unconscious, where something is 'bubbling under' just waiting to happen and be revealed. Such messages also relate to health in that the part of the body causing problems symbolizes the psychological cause of the complaint. One example of this is that complaints of the pancreas represent the possibility that the person is too sweet or the sweetness has gone out of their life, the pancreas being responsible for controlling sugar levels in the body. If this area of Uranian symbolism interests you, then I refer you to the works of Louise L Hay.

Signs and Omens and all means of divination are, for the same reasons as accidents and coincidences, also the province of Uranus. As with dreams, we can take a sideways look at a 'chance event' or something that has been apparently randomly created (like a tarot spread or throwing coins using the *I Ching*) and interpret what it means and/or portends. Interpreting these 'chance

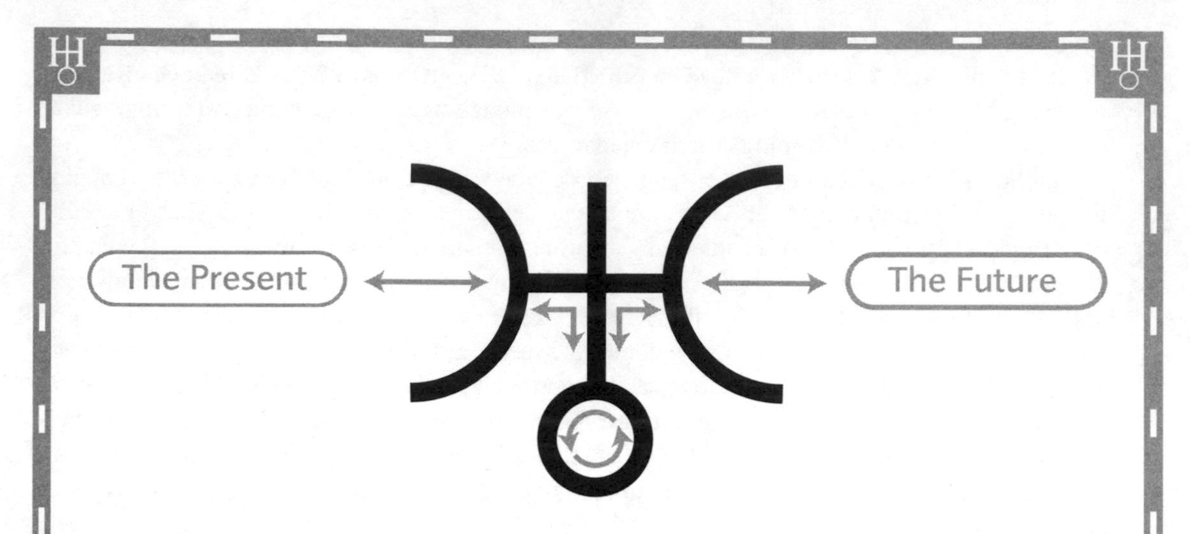

The glyph of Uranus, which usually appears somewhat like a television aerial (♅) is actually composed of the symbol of the semi-circle on the left, which symbolizes the Soul of the individual human being, and the semi-circle on the right, which symbolizes the Oversoul of this Solar System. The Oversoul receives and channels in Cosmic or Evolutionary Forces, which are effectively the Future in the making. Then through the medium of Materialization (symbolized by the Cross) and the Process of the Spirit (symbolized by the Circle), souls, both individually and collectively, are influenced in the Present to change, invent and reform. In so doing they create the Future as a material reality.

The Future could also be viewed in terms of the two other Outer Planets, Neptune and Pluto (called, along with Uranus, the Ambassadors of the Galaxy), while the Present is the Sun and all the Inner Planets as far as Saturn (the status quo). And so the Vision of Neptune and the Destiny of Pluto are collected by the right-hand 'dish' of Uranus, which then processes them, flings them from the left-hand dish to the Heart of the Sun, which then radiates them out to Earth and the other Planets as Creative Energy affecting all things, organic and inorganic. Saturn is the Ring-Pass-Not which contains and stabilizes all this. If this sounds far-fetched, that's because it is – but one day it will be found to be true. To quote Mark Chown from that very *scientific* magazine, the *New Scientist* on the subject of astrologers, 'Where they slipped up is not in being too way out, but in not being way out enough.'

events' in this way can be particularly helpful in emotionally relating. For example, if your partner appeared to be ignoring you in reality, how would you interpret that if it was in a dream? Most dream interpreters would say it was symbolic of a fear of rejection – and the same event in reality would be telling you that you are suffering from exactly the same thing. The trouble is we think we have more control over our 'real lives' that we do over our 'dream lives', but this an illusion that is evidenced by any life situation which we cannot change by force of will or reason, like being trapped by an adverse fate. And Uranus being Uranus, it also poses the necessity of being prepared to see and accept the truth. So, in the case of being ignored by one's partner, rather than sulking, fretting, pleading, persuading or pursuing him or her (and making matters worse), sort out the rejection problem by investigating your past experiences that comprise the part of your unconscious mind which is 'bubbling under'. This having been done, you would then no longer attract partners that rejected you, or feel rejected by a partner who seemed to be doing so. In other words, it is no coincidence that people who are afraid of being rejected, attract the very partners that do just that. And incidentally, Magic is also the province of Uranus because a true magician (as opposed to a conjuror) intuits what is bubbling under and appears to make it happen.

Alternative or Off-the-wall subjects and ideas are now what spring to mind. During the dawning of the Age of Aquarius (*see* the Glossary on page 355), more and more people have been interesting themselves in these 'things Uranian'. This is hardly surprising because Uranus is the Planetary Ruler of Aquarius – that is, the energies of the Sign and the Planet *correspond* to one another. The trouble here is that for every genuine Uranian or alternative item there are thousands of really wacky ones. But, as they say, life is a lottery, and very few come up with a winning combination. To put this another way, the very nature of experimenting with something new produces far more failures than successes, as every inventor or researcher knows. You just have to keep on experimenting until you are in sync with whatever it is that you are trying to find out or make happen. If you have ambitious hopes, then you had better be prepared for a long haul – or even an infinitely long haul. On the other hand, if you have no dream, aspiration or ideal to work towards, then you are never going to make one real. This brings us to another Uranian paradox.

Independence and Teamwork are seen to be inextricably linked, for the best team is one where all members know themselves and their role really well, and are free to perform it according to their proficiency and creativity. On the contrary, to regiment people and treat them as 'units' is doomed to disaster, as evidenced in the inefficiency of institutions like the British National Health Service or like the monumental example of what was the Soviet Union. The Uranian ideal is that we evolve from being a member of a small group like a family, football club or closed state, all of which tend not to see the bigger human picture, into being a member of the human race itself and serving *its* ends. The ultimate extension of this Uranian line of thinking is that of true anarchy: a situation where no government is needed because everyone is self-governing because they are self-aware. Notwithstanding the fact that this 'true anarchy' is one of those 'long hauls', Uranus can present you with the choice (freedom always implies choice) of being true to yourself rather than the closed group with which you are involved (which eventually brings you round to serving the greater whole), or sacrificing your ideals for the values of that group (and eventually finding out more about yourself by way of contrast). Again, it all depends whether you make a priority out of material stability, emotional security and rejecting the unknown, or if you are listening to 'a different drummer' (Thoreau) and are prepared to aim for your ideal or ultimate purpose for being, rather than just material ambition or emotional security.

♅ **Friends and Groups** – A very suitable and conscious way of responding to the influence of Uranus is through relationships and situations where there is not too much of an emotional agenda. Exploring or experimenting with new ways of experiencing and looking at life and yourself is made far easier when you do not feel pressured or compromised by the expectations and images that a partner or relative can have of you in the name of *their* security. Groups in the form of seminars and workshops can do wonders in revolutionizing your ideas and feelings, and with Uranus around 'a friend in need is a friend indeed'. Going to a workshop with a partner, or anyone else who *thinks* they know you, can be a difficult experience. But then, Uranus-wise, if you are open, honest and courageous enough to see the truth of each other, then you are on your way to a free and truly loving relationship – or out of a relationship that would have been anything but free and loving.

♅ **Science and Technology** – Involvement with gadgets, computers or anything to do with research or technology is an excellent means of getting into a faster lane or becoming more versed in the ways of the modern world – something which Uranus calls us to do, one way or the other. The down side to this is that such pursuits can detach some people so much from emotional interplay that it inhibits the development of emotional skills, or rather they can retreat to the computer to avoid such interplay.

♅ **Uranus Types** – These are people who are Aquarians, or who have an Aquarian/Eleventh House emphasis or strongly placed Uranus in their Birth Charts. In ordinary terms, such types would, on the positive side, be original, esoterically aware, scientific, impartial, open, cool, modern, avant garde, inventive or intuitive. On the negative side, they would be cranky, unstable, reactionary, cold, emotionless, erratic, impractical, hyperactive, unpredictable, irresponsible or reckless. Their professions could include such things as psychologist, astrologer, scientist, technician, mechanic, computer buff, inventor, researcher, flyer or any unusual or specialist occupation, or one that deals with freedom and rights. A Uranian type can be a fool or a genius, and go from being one to the other in a flash. Any of these types here mentioned could be the 'agency' or figure that brings Uranus into your life through possessing the qualities associated with this Planet and the Sign or House it rules. (*See* 'Planetary Effects, Figures and Responses' on page 259.)

Here follow the significant periods and meanings of your Uranus Cycle

★ Refer to the tables beginning on page 289 to determine what years are active for you personally (one to three years) within the age span given for any particular influence. Any influence will not be continuous throughout that period, but off and on, usually comprising three peaks. Occasionally, you may notice the 'Stretch Factor' when you experience an influence for somewhat longer than the period that has been given.

★ Please remember that there is also the Individual Cycle of Uranus to be considered (that is Uranus's Transits to other Planets in your Birth Chart that occur throughout your life), but these would have to be calculated by an astrologer or with an astrological computer program. If any of the foregoing has struck any chords with respect to certain life passages, then this is possibly because such Uranus Transits are what you are or were experiencing.

Waxing Square Between 17–23 years for 1–3 years

'BREAKING AWAY AND BREAKING THROUGH' – This is the time when most people leave their family home and start trying to make it on their own in the outside world. As Uranus is symbolic of flight, such a time could be compared to little birds learning to fly, or 'flying the nest'. Such 'breaking away' from the protections and limitations of the family home, and possibly one's background too, can be a natural and harmonious move or an unpleasant and rebellious one. A great deal depends upon your personal disposition, desire for adventure and urge to be free to do your own thing, and the attitude of your parents and family. The main thing to grasp with this significant turning point is that it is the individual right and natural inclination to find those 'wings'. Some may crash, return to base and relaunch, or others may choose or be obliged to stay put. As Uranus signifies what differs from the norm, then *not* 'flying the nest' could also be regarded as perfectly in order with this influence. Doing your own thing is doing *your own* thing.

However, if you are staying at home – or wherever it might be that you feel stuck in – for fear of discovering more about yourself and the way of the world, then this could be asking for problems later on. If this is the case, the chances are that your parent(s) or 'keeper(s)', for one reason or another, do not want to let go of you. Apart from checking out what is happening astrologically with them right now, it is highly important that you examine your motives or reasons for not leaping into the great unknown – which is what Uranus is prompting you to do. Staying in the same situation because your keeper(s) want it that way could mean that you are having to suppress a natural drive to become your own person and live your own truth. As mentioned in the introduction, Uranus has a definite bearing on one's sexuality, and any suppression of your 'breaking away' now can give rise to feelings of being sexually out of step later. This means that your peers will have gone, so to speak, through stages A to C, when you are still at stage A. Obviously, being 'out of sync' in this way can give rise to a variety of social and emotional problems. Then again, it may be your 'game plan' for your sexual development to pan out this way. Later on you may catch up as part of your unique way of discovering how you tick, sexually or otherwise. Still again, you may be someone who sits out the sexual dance altogether. All this applies even if sex isn't the issue, or the entire issue, but it just has to do with embarking upon something *new*. As ever with Uranus, it is not so much what you are doing that counts, but that it is your *own thing* that you are doing. Recognizing that you are a unique being that does not have to adhere to any conventions if you do not want to is the essence of Uranus and the discovery of your own truth. But if you are worried about not being 'one of the gang', then that would mean that you have to make the effort to be one of them or decide to be your own person at any cost. Whatever the situation, use this influence to break away and through to a place where you feel you are at least being true to yourself. After all, this is what this time in your life is primarily about – *being true to yourself.*

As part of this important process, there is always the possibility of falling somewhere between those two extremes of breaking away and staying put. Two words that could sum up such a state could be 'confusion' and/or 'rebellion'. Being confused would pose the need for an honest questioning and discovery of your own desires, and not allowing yourself to be influenced by anyone with a vested interest in your going or staying. Remember, 'Then you will know the truth, and the truth will set you free'. Rebelliousness, in the sense of rebelling against the people or situation with which you feel forced to stay is another expression of confusion, like making noises but not saying much. If you are unhappy with a situation, then just say so and why. You will then have a chance to discuss and negotiate, then decide whether to stay or move, in the knowledge of why you must do so. If you know you have to go but there is no agreement from whoever appears to be holding you back, then you must think again or fly the nest anyway.

All the various courses described above are to help you to devise your own code for living and loving that will guide you as an adult in the future, and, perhaps, ultimately develop you into that Uranian entity – a person in his or her own right. Whatever the case, try to discover or attain what you are after in manner that is as harmonious as possible, otherwise the inclination to detach yourself emotionally for the sake of doing your own thing can remove you from emotional interplay in more ways than you bargained for. (*See* 'Uranus – Myth and Fact' on page 153).

Flight, in the sense of 'finding your wings', is also a strong possibility at this time. By and large this would take the form of going to university, some other seat of learning or simply the 'school of life' where you can improve your 'power of flight'. The difference or novelty of such a life, with all the challenges and dramas it offers, is very much grist to the Uranian mill right now. Whether you are cautious or an 'Icarus' (someone who flies too close to the Sun), you have the opportunity in such a situation to get the measure and cut of your own personality.

Some, but relatively few, people really do 'take off' at this time, with respect to their own aims. The careers of film stars Greta Garbo and Marlon Brando both took off at this time, with Garbo mesmerizing audiences in *The Torrent* in 1926 at the age of 20, and with Brando overwhelming audiences on Broadway in *A Streetcar Named Desire* in 1947 at the age of 23. (*See* also page 117 for Brando's first Oscar.) This Uranian influence was also evident in Muhammed Ali's first big fight win in 1960 at the age of 22. A take-off that was to lead to even higher altitudes was that of the first man in space, Yuri Gagarin, who gained his first pilot's licence at 21 years of age (*see* also page 166). All of these examples contain a typical element of Uranian risk in that they all rose to fame but then later fell from grace in some way. Garbo died a recluse in obscurity (*see* page 179 for her Uranus Return); Brando suffered his son's indictment for murder, giving rise to the suicide of his daughter (*see* page 129) after much trial and tribulation; Ali fell and rose, and finally fell again suffering from Parkinson's disease (*see* Pluto Waxing Square on page 226, and Uranus Opposition on page 168); Yuri Gagarin died in an air crash just seven years after his historic space flight (Uranus opposing his natal Mars – an Individual Planetary Cycle not included in this book). Perhaps the fate of all these celebrities implies that the manner of one's ascent has a lot to do with the time and manner of one's descent. A meteoric rise can lead to burn-out – and Uranus invites the Unexpected. Being a success in the eyes of the public or one's peers is not necessarily being a success on one's own true and unique terms. And remember that Uranus is good at launching but not necessarily so good at maintaining altitude.

One more example of the Uranus Waxing Square demonstrates how its influence, like any Planetary effect, can be expressed through a child given birth to at that time. Queen Elizabeth II gave birth to Prince Charles during this period, and he has not conformed as a 'royal' is expected to. He has had his ongoing affair with Camilla Parker-Bowles, divorced his wife, the late Princess Diana, subscribes to alternative methods such as organic farming and talking to plants, is interested in the esoteric, supports minority groups, and is generally a bit of a misfit and emotionally misunderstood.

In Third World countries, the Uranus Waxing Square takes on the aspect of Evolution, that major dimension of Uranus's province. At this time of their lives, individuals now often leave the countryside and their parents' farms to become educated in the city, vowing never to return to those hard, agricultural roots. They are not only breaking away from their own pasts, but breaking through to a new future for their culture. This happened earlier – such as the Industrial Revolution in Europe in the 18th century – and is still happening – in other countries, giving rise to the 'city over country' culture that is now prevalent in the First World. So if you are wondering when or if to make the break, it might help you to be aware that you are actually part of a far greater process – which could just as well be a case of putting the country and nature before what could be regarded as the artificiality of city life.

Apart from all the above instances, other changes can occur or be enforced at this time. Whether it is to do with a personal relationship that ends or begins now, or a job or interest that shifts your gears, Uranus is saying it is time to move on.

On the opposite page you will see a flowchart of your Uranus Waxing Square, 'Breaking Away and Breaking Through', which gives you a graphic idea of what Uranus is bringing into your life, and what the various ways of responding to it can bring you.

JUPITER 'YEAR' for your Growth Mode

MOON 'PHASE' for Emotional Attunement

Having ascertained the actual years of your Uranus Waxing Square from the tables on page 289, refer to the Age Index on page 235 for the Jupiter 'Years' and Moon 'Phases' occurring during that period. If one of the Jupiter 'Years' is an 'Aquarius' one then this would further accentuate the influence of your Uranus Waxing Square in that year of your life.

Other Universal Planetary Cycles possibly active at this time

Check the tables on page 289 to see if any of these Planetary Cycles are active for you at the same time as your Uranus Waxing Square. Remember that any Trine influences may have to be actualized to make them that effective.

FIRST SATURN WANING TRINE (*see* page 113) This lends prudence and stability to a time when such could otherwise be in short supply. However, erring on the side of caution could find you safe and sound but bored – and lacking experience later. So weigh up the pros and cons, but first take on board the dynamic intent of the Uranus influence, because it is to the fore. Saturn might act as the brakes and steering, but Uranus is the engine itself!

FIRST SATURN WANING SQUARE (*see* page 114) This influence happening now is somewhat similar to that of the Saturn Waning Trine above, except that too much caution is more likely, but if not enough is exercised then there could well be some sort of mishap. On the other hand, such a 'brakes and accelerator' situation could leave you stalled. As this also includes a 'Capricorn' Jupiter Year, we can say that Saturn is to the fore, so if in doubt, wait – but not without pondering hard those pros and cons of taking on whatever it might be.

NEPTUNE SQUARING PLUTO (*see* page 193) Having both of these influences occurring simultaneously (or even consecutively) means that you should exercise considerable caution because a rocky ride could be on the cards. Or, as in the case cited above of Muhammed Ali, his success as a boxer began with his Olympic win in July 1960 , but it set a trend of over-confidence and seeming invulnerability that sadly culminated with Parkinson's disease brought on by head injury. One could say that there is the necessity of weighing one's strengths against one's weaknesses, while at the same time being experimental enough to find out which is which – which I suppose is what Ali did.

A MAJOR INFLUENCE occurs between the ages of 17 and 23*

URANUS WAXING SQUARE

Breaking Away and Breaking Through

The flowchart below shows how the Positive and Negative Uranus Traits *(left)*, that you may or may not have, encounter Uranus's influence *(arrow on right)*, which in turn transforms the Positive Traits into a Positive Outcome *(top)* and/or relegates the Negative Traits to a Negative Outcome *(bottom)*. Negative Outcomes can be made Positive through developing the Positive Traits.

POSITIVE OUTCOME

Discovery of one's individuality.
Sexually satisfying relationship(s).
Work experiences that are true to you.
Original expressions and creations.
Being equipped to follow your path.
(On your way to) being free.

POSITIVE TRAITS

Positively adventurous. Open to experimentation.
Determination to break free despite circumstances.
Prepared to make the most of a restrictive situation.
Flexible enough to go with any positive changes.
Firm enough not to be pushed from pillar to post.
Disciplined and sincere resolve to further yourself.

URANUS

New sexual attractions.
New choices and conflicts.
New job (opportunities).
New friends and colleagues.
New horizons beckoning.

Sudden chances and changes.
Urge to experiment.
Desire for freedom.
General disruption.
Restlessness.

NEGATIVE TRAITS

Fearful, overcautious outlook.
Feeling restricted and unable to get free.
Rebellious and not accepting limitations.
Stubborn refusal to try anything new.
Too easily led or dazzled by any prospect.
Feckless. Easily bored. Reckless.

NEGATIVE OUTCOME

Retarded experience of life.
Remaining bogged down in 'security'.
Running foul of authority.
Narrow-minded, unattractive personality.
Falling in with bad company.
Going nowhere. Accidents.

**see* tables on page 289 for when it is for you.

Waxing Trine Between 24–31 years for 1–3 years

'BECOMING YOUR OWN PERSON' – During this period you have the chance to launch yourself into a life that is far more conducive to realizing your unique capabilities. The secret is to be alive to this window of opportunity. So often we are too busy toeing the line and doing what others expect of us, rather than following our intuition and being true to some goal that goes beyond merely 'fitting in'. Friends, family and colleagues can have a vested interest in keeping you as you are, because it prevents them from feeling uncomfortable about the 'stuckness' of their own lives. On the other hand, it may well be a friend, family or a colleague that presents you with this chance to change. Group activities like workshops, clubs and seminars could also be what provide the opportunity to leap into the wide blue yonder.

So, keep an eye open for out-of-the-ordinary but significant openings, such as the chance to live in another location, to start a new job, or have an exciting and unusual person become an important part of your life. You may well be presented with the opportunity to embark on all of these!

On the opposite page you will see a flowchart showing 'The Uranus Slingshot Effect' which portrays graphically the nature of this opportunity. I have called it the 'Slingshot Effect' as this derives from the astronautical term for making use of the gravitational field of one planet in order to sling a space module millions of miles on to another planet. Simply follow the instructions given in the panel at the top. You will also see that the Uranus Trine occurs again between the ages of 51 to 59 years, but this would be the *Waning* Trine. The similarity between the two is that they both give you an opportunity to change your life for the better – to change your world. Come the Waning Trine, you will notice a remarkable connection with the Waxing Trine all those years before. The difference between them is that the Waxing Trine is more likely to manifest as some actual physical change or move, whereas the Waning Trine would have more to do with a change in outlook or inner attitude. But this isn't a hard and fast rule – things never are with Uranus around!

Two well-known people who made classic use of the Uranian Slingshot Effect were Yuri Gagarin and Howard Hughes. Yuri was the first man to be launched into space and go once round the Earth in *Vostok I* on 12 April 1961. His was a literal case of blasting free from the Earth in the Uranian Rocket Ship! Previously, the year after Yuri was born in fact, Howard Hughes set a new air speed record, flying his aeroplane at 351mph. Both of these feats were to do with flight, which is ruled by Uranus.

A third famous example demonstrates how a Trine influence may be used to plot one's way through a difficult patch. In June 1967, Muhammed Ali was convicted for refusing induction into the US Army, but he bore all the pressure that was obviously involved with this with conviction and dignity because Uranus kept him touch with his own principles and long-term goals.

JUPITER 'YEAR' for your Growth Mode

MOON 'PHASE' for Emotional Attunement

Having ascertained the actual years of your Uranus Waxing Trine from the Tables on page 289, then refer to the Age Index starting on page 235 for the Jupiter 'Years' and Moon 'Phases' occurring during that period.

AN OPPORTUNITY occurs between the ages of 24 to 31 and 51 to 59*

URANUS Waxing or Waning TRINE

The flowchart below shows how any of your Positive Uranus Attitudes and Actions *(bottom left)*, having attracted and made use of Uranus's influence *(Slingshot Effect within orb)*, allow you to attain certain Advantages *(top right)*.

REMEMBER – A trine only opens the way; it does not push or confront you. So you must deliberately recognize it and take advantage of it.

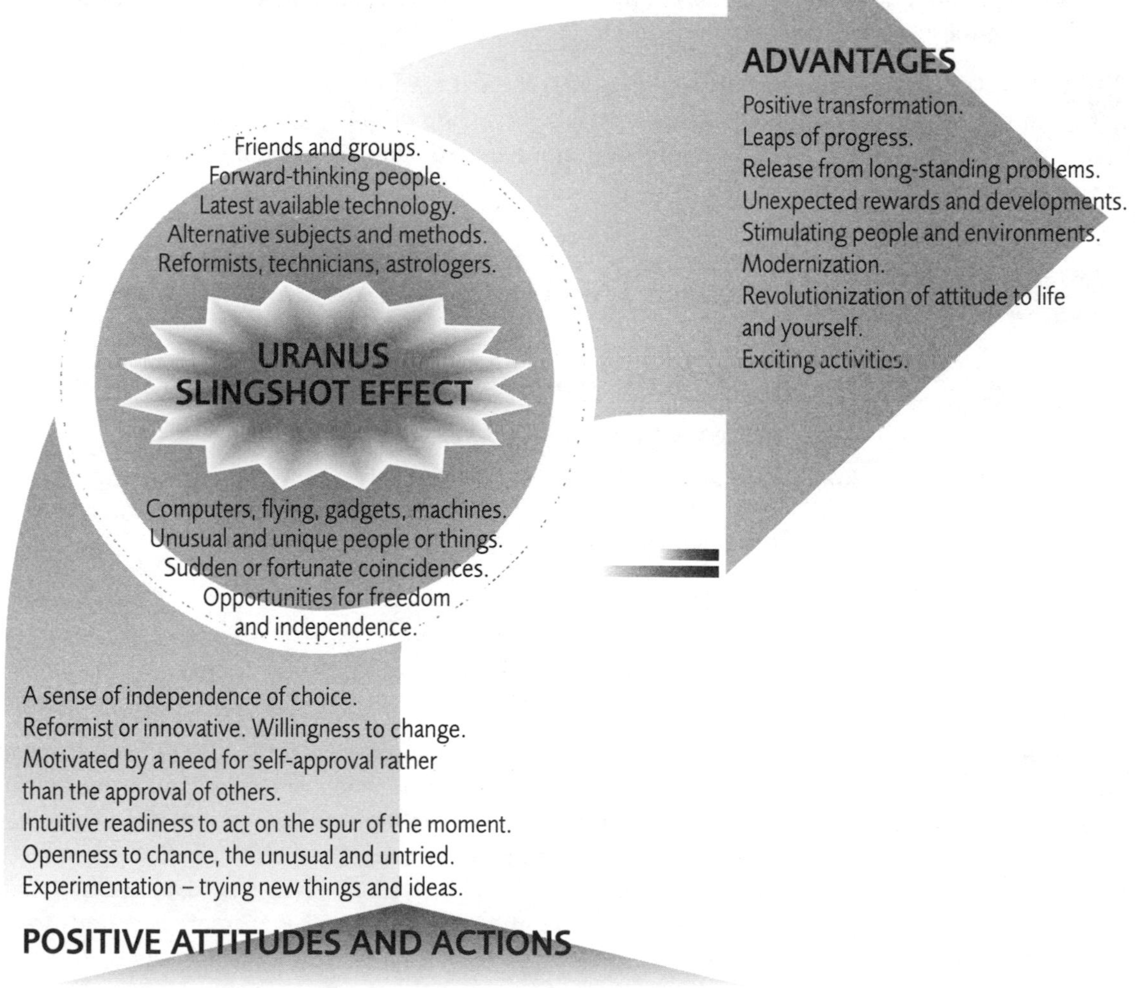

ADVANTAGES

Positive transformation.
Leaps of progress.
Release from long-standing problems.
Unexpected rewards and developments.
Stimulating people and environments.
Modernization.
Revolutionization of attitude to life and yourself.
Exciting activities.

Friends and groups.
Forward-thinking people.
Latest available technology.
Alternative subjects and methods.
Reformists, technicians, astrologers.

URANUS SLINGSHOT EFFECT

Computers, flying, gadgets, machines.
Unusual and unique people or things.
Sudden or fortunate coincidences.
Opportunities for freedom and independence.

A sense of independence of choice.
Reformist or innovative. Willingness to change.
Motivated by a need for self-approval rather than the approval of others.
Intuitive readiness to act on the spur of the moment.
Openness to chance, the unusual and untried.
Experimentation – trying new things and ideas.

POSITIVE ATTITUDES AND ACTIONS

**see* tables on page 289 for when it is for you.

Other Universal Planetary Cycles possibly active at this time

Check the tables on page 289 to see if any of these Planetary Cycles are active for you at the same time as your Uranus Waxing Trine.

SECOND EMERGING MOON PHASE (*see* page 51) This highly significant period happens between 27⅓ and 30¾ years of age, but is particularly significant during the first year of this period. Since it coincides with your Uranus Waxing Trine, then some event or change, especially on the home, family or emotional front, is very possible – a positive Course Correction, in fact. This was certainly true for Yuri Gagarin who went *around* our home, planet Earth, in space (*see* above and page 52).

FIRST SATURN RETURN (*see* page 115) This is one of the most important of all Planetary influences, occurring at 29 to 30 years of age. Coinciding with your Uranus Waxing Trine means that any changes you make now will be quite constructive and far-reaching, although you may not appreciate just how much so until years from now. In any event, circumstances will force you to realize that some sort of change is called for, thereby obliging you take advantage of the 'Slingshot Effect'. Howard Hughes achieved his record-breaking flight then, but in a way Saturn was indicating a weakness within strength (as it often does) for Hughes spent the last part of his life 'aloft' and aloof from the normality and sanity of the everyday world. He even died in flight.

PLUTO CONJOINING NEPTUNE (*see* page 221) This extraordinary influence is called the 'Illusion Buster' and the 'Inspiration Booster'. With it occurring during your Uranus Waxing Trine, it makes the likelihood of some major change very strong indeed, launching you into a world you never knew or only dreamt of.

Uranus Opposition | Between 37–45 years for 1–3 years

'A Memo from Heaven'.

You are now being reminded of what and who you really are –
and also of what your life on Earth is truly about.

Shake up! Break up! Wake up!
Yesterday is gone
Change is NOW!
Tomorrow is electric blue ...

'YOUR MID-LIFE CRISIS' – The message in this 'memo' is so monumental that it is difficult to avoid putting it across in the same way that it can get sent to you – by blowing your fuses. By 'fuses' I mean those basic limits we place upon the impressions we receive from the world around us, and upon what we express to that world in terms of our behaviour and personal style. These fuses are first put in place very early on – in the womb, at birth, or sometimes not until early childhood – and so, more often than not, people do not even know they are there. However, these fuses themselves are more to do the Moon, and not Uranus, because it is the Moon that protects us from shocks, the unfamiliar, chaos, and anything which is too sudden a change – the province of Uranus. Other fuses are installed later as we grow up, but they could be regarded as practicality, socialization or culturalization and are governed by

Saturn. With Uranus governing electricity, using fuses as a metaphor is very useful in conveying what this extraordinary Planet presently means in your life. And at some point it may become apparent that it is no longer just a metaphor – it really *is* electricity, *chi*, the life force, that we are dealing with and talking about where Uranus is concerned.

The trouble is that those 'fuses' are not just fuses – they are also blockages. Think of your body as an 'appliance' that is vitalized by the life force of electricity. Up until now you have got along, one way or the other, with your appliance being set up a certain way, with certain specifications. These specifications tell you what you like and dislike, what you feel allowed to do and not allowed to do. These specifications got installed early on, and were determined by the values, circumstances, customs and predilections of your parents, culture and the status quo. But actually, these were in themselves determined by what is called your fate – which has something to do with your soul and unconscious mind. Then along comes your Uranus Opposition which is like a power surge that courses through your body, and thereby your life. And if you are not ready for it, that surge quite simply blows one or more of those fuses.

Being 'not ready' means that certain parts of your life have become too attached to those specifications to which you have conformed in order to ensure a secure and stable life – whatever that might be in terms of those specifications. Remember that 'upgrading program' which was mentioned in the introduction – well, that is your Uranus Opposition kicking in.

So any aspects of your life – such as job, partnership, lifestyle, location, attitude, values and beliefs – that are in some way not conducive to your expressing your true identity (as distinct from the version that your fate initially determined for you), now begin to play up, get in your way, and generally seem to limit your freedom. The knee-jerk reaction would be to leave your job, spouse, home or whatever, and even become madly involved with something or someone highly unlikely. But this would be due to an urgent feeling that you had fallen into a rut of predictability, and hadn't really done what you once dreamt you'd do. Be on your guard against this desperate manner of relieving a feeling that life has passed you by, or trying to relive your adolescence in a fortyish-year-old body.

What this highly significant period of your life is actually asking you is 'What is unique about you, and how are you going to express that uniqueness?' This doesn't mean throwing the baby out with the bathwater by rejecting or ejecting anything or anyone that appears to represent what is stale and unstimulating and blocking your freedom. This would be like throwing away the mirror because you didn't like what you were seeing in it. For one thing, perhaps existing elements in your life could do with a shot in the arm. If the other people in your life are quite happy with the way things are, then you may need to start making some radical changes, which they can either accept or reject as they choose. But, as freedom is presently the name of the game, make sure that everyone has a chance to express themselves freely too. Bringing things out in the open is a very constructive way of dealing with this influence. Study very carefully any other Planetary influences that are occurring now, for they indicate in what ways and areas of your life changes may or must be made – and where you could be fooling yourself too. Overall, this is a great time to introduce something fresh and far-reaching into your life, particularly anything that furthers an ideal. So having a goal or aspiration is your truest guiding star right now. A current example of a celebrity doing nearly all these things during his Uranus Opposition (that is, saying and doing something against his better judgement, and bringing something into the open, and making a fool of himself, all at the same time) is ex-England football manager, Glenn Hoddle. His proclamation that disabled people are paying for sins committed in past lives didn't go down at all well and I dare say it will create waves for some time in quite subtle ways – even though he scored an own-goal first. He also had his Neptune Waxing Square (*see* page 198) going on at the same time, which astrologically explains the subject of the incident (reincarnation) and getting egg on his face (*see* also page 102).

To return to that 'electricity', that 'power surge', what I have said in the above two paragraphs is all very well as long as you are able to keep things under control mentally. This is a bit like fuses being blown when someone knows how to replace them – but not everyone does. In other words, being able to respond spontaneously in an appropriate fashion to a need for change, as and when it hits you, is quite rare and not so easy. So astrology comes into its own at, or preferably before, this time (Uranus also rules astrology, by the way). By knowing in advance that this is coming up for you or someone you know, changes can be made that facilitate or 'ride' that wave of increased power, that wind of change, when it arrives. So if you are stuck in a boring and unfulfilling job, start looking around or preparing yourself for something more 'you'. This may mean going it alone. If you are living in a situation where you do not feel at home or able to lead your life in the way that you want to, then plan that move. If you are in a relationship that is deadlocked and unhappy, then make some sort of temporary or permanent break from the relationship or make a radical change within it. The trick is to be alive to what is wanting to shift, and use Uranian 'chance means' to do so. For example, prior to the onset of my own Uranus Opposition, it became more and more obvious that where I was living and what I was doing was not taking me in the direction I wanted to go. Furthermore, the relationship I was in at the time reflected all of this. The trouble was that I did not know where to go. So I drew up a list of options and experimented with each one, eliminating them until there was only one left, having already agreed to myself that I would follow that last remaining option by the onset of my Uranus Opposition. On the face of it, this last option was the least attractive to my ego, but as things panned out it was the right one – and in an unexpected way. This is an important point, for Uranus is 'anti-ego' in that it is the force that really drives us as distinct from what we *think* is doing the driving. Again, think about the electricity (the force) and the appliance (the ego). A toaster, say, is totally dependent upon the electricity for its functioning, whereas the electricity exists in its own right. Or to put it another way, think of the transmission of television waves. *They* determine what pictures the TV can display, not the other way around.

So what other kinds of change can you expect during this time – or if you are in the middle of it, what signs of change should you be alive to, and subsequently act upon? On a physical level there is the possibility of the need for change manifesting as panic attacks, unfounded feelings of insecurity, hot flushes, the shakes, or any other kind of spasm or sensory distortion. Or then again, such anxieties can be projected on to the collective with over-concern for what is happening in the world at large as seen in the media. These are all signs that one or more of those fuses are blowing or are about to blow. Nervous complaints and conditions such as ME are also possible if the system is being overloaded. Each individual case will tell its own story, but generally speaking, any one of these indications is pointing to a urgent need to change the way you live or look at life. Using again the television analogy, it is as if a new channel is being beamed at the 'receiver' of your personality (which includes your physical body) but you are refusing to upgrade that receiver to allow it to channel or express it. This all poses the necessity of 'taking the back off' through having some form of therapy or consultation to discover and replace that 'dud component', that outmoded part of your personality and lifestyle.

Another important arena of Uranian focus is that of 'rights', which is the concern of social change, personally or collectively. At this time, if there is any situation that is severely abusing your rights or those of someone close to you, then you will be called to resolve it in some way. This could be on a strictly personal level, such as being in a unacceptably confining relationship, or more to do with some abuse of power by an external authority. The Uranus Opposition can be quite a time of reckoning in this respect. There is a 'Famous Five' here that dramatically demonstrate this point. At the time of their Uranus Oppositions, the following five champions of human rights were subject to crucial experiences (four of them fatal) as part and parcel of their respective crusades.

One who could be called an 'icon of freedom', Che Guevara, was killed by Bolivian troops at the age of 39. This was not such a surprising fate for such a radical and high profile activist for ending poverty in Latin America.

Similarly, Martin Luther King, champion of racial equality and harmony, was also a target of the hatred that opposes such truth and virtue. Although unlike Guevara, in that he espoused non-violence, he met a violent end, also at the age of 39. King was possibly more of a classic hero of freedom than Guevara; a few years before his assassination, while preaching in church when gunfire was heard outside, he declared, 'It may get me crucified. I may even die. But I want it said even if I die in the struggle, that "He died to make me free".'

Robert Kennedy, an outspoken champion of truth and justice, was also assassinated. He was 42 years of age.

John Lennon was assassinated at the age of 40, during his Uranus Opposition. Ironically, he was killed by the kind of working-class nonentity that he consciously strove to encourage.

The fifth of the 'Famous Five' is the happy exception to this rule of assassination – Nelson Mandela. He was arrested for allegedly conspiring to overthrow the then racist and oppressive South African Government when he was 44 years of age. But he lived to become his country's first black President 31 years later (*see* page 359 for more concerning Mandela).

Continuing with theme of freedom and rights, John F Kennedy, who was also assassinated, was experiencing his Uranus Opposition when he was elected President, very much on that kind of Uranian ticket.

These examples seem to indicate that the Uranus Opposition is a key point in one's 'freedom stakes'. Obviously, they are exceptional in that these men all lived and died (except Mandela) to show us how important a thing freedom is; even so (or is it all the more because?) they took quite a few liberties themselves!

One more example of a 'freedom-fighter' experiencing the ultimate during his Uranus Opposition was Billy Giles, ex-member of the Ulster Volunteer Force, a Protestant group in Northern Ireland. His story is remarkable in that following upon his assassinating a Catholic friend in 1982 as an act of retaliation for an IRA murder of a Catholic girl, he, in his own words, 'lost a part of myself that I'll never get back . . . Before I would have been classed as a decent young man. Then suddenly I turned into a killer. That's Northern Ireland.' Sixteen years later, a year after being released from prison, and during his Uranus Opposition, this inner truth of his being drove him to execute himself. Hopefully, this act of a man who was honest enough to admit that – apart from what he had done to his friend and his loved ones – he had abused and damaged his own humanity, will stand as a poignant example to others involved in the Peace Process.

Returning to more personal issues and the Uranus effect, any kind of situation or relationship that sets the proverbial cat amongst the pigeons is a 'favourite'. If something in your life needs stirring up, then Uranus will see to it that someone or something arrives on the scene to do the stirring, whether intentionally or not. When those fuses are blockages they will attract something to free it up. Such blockages can often be the result of social or religious indoctrination. Enormous conflicts can arise from feelings of 'forsaking a parent' or 'denying God' when a person is unavoidably drawn into something that challenges such allegiances. This is especially so when it is a case of one 'truth' opposing another. But the simple formula for truth as far as Uranus is concerned is that it 'will set you free' rather than make you feel guilty or judgemental. As Thomas Paine wrote in *The Rights of Man* '. . . and my religion is to do good'. Mind you, one had better be sure that what's seen as good *is* good. There is nothing worse than the 'Rights Blight' where people foist their idea of truth upon others when all they are really doing is trying to camouflage their own inadequacies. This too is something one may have to contend with, or be found guilty of, during the Uranus Opposition.

Earlier I pointed out that Uranus has a lot to do with sexuality, and so the Uranus Opposition will focus upon this in some way or other. In Uranian terms, sexual energy is like electricity, and an electric current will always seek to go to earth. And so it is as if, at a very primal level, we simply want to make sexual contact to 'earth' ourselves through the body of another. This sounds very unemotional and lacking in romance, but that is because Uranus is only concerned with the fundamental forces that drive us, not how we dress them up afterwards. None of this precludes sentiment and finer feelings, but unless you understand that this takes second place under the influence of Uranus, you are in for a shock – or rather another one. During my Uranus Opposition, I became involved with a white witch who initially was purely interested in me for my male sexual energy, which she 'earthed' very well – in fact, her 'craft name' meant 'Earth'. But later we fell out because the emotional side was lacking in our relationship. However, the point is that there was the mutual desire to earth or charge one another up in this way, and romance had to take a back seat. How, or if, your Uranus Opposition gets to you on a sexual level is, as ever with Uranus, down to your own unique personality and circumstances. But the point that I am trying to make here is that you will be primarily dealing with the energy that brings people closer together, often in spite of, or rather because of, their reservations. This Uranian energy is no respecter of your social standing, romantic illusions, sexual conventions or marital status. If there is a message 'bubbling under' that intends to blow the cobwebs away through some wild and off-the-wall affair that unsettles your settled life, then that is what it will happen! Uranus is very much the 'anything can happen' Planet.

On a more psychological or esoteric level, the sexual call is to become (more) aware of what is called your 'anima' if you are a male, or your 'animus' if you are a female. The anima is the feminine being inside a man, and the animus the male being inside a woman. The experience of these 'opposites within' can be particularly acute between two people at this time, or occasionally may manifest itself through experimenting with cross-dressing, homosexuality or group sex. But at a more intellectual or spiritual level, a man getting in touch with his anima (which means 'soul', that which earths the charge) is him becoming familiar and at one with his vulnerability, receptivity and passivity. Conversely, a woman getting in touch with her animus (meaning 'spirit', that which charges the earth) is her getting a handle on the more dynamic, objective and creative side of her being. However, the usual way that this is played out is through what is called 'anima or animus projection'. You can think you've fallen madly in love with someone when really you have been 'plugged in' to your own anima or animus through being respectively earthed or charged up by someone whose sexual energy resonates rather well with yours. If you are aware of this and take on board the fact that you are getting to know your other half through the sexual electricity between you and someone else, then you will become a more mature and whole person as a result. Furthermore, you can then build a lasting relationship upon such a foundation of self-completion – but not necessarily with that same person. On the other hand, if you insist on thinking you are 'in love' rather than becoming more sexually and psychologically aware, then don't be surprised if you suddenly become 'unplugged' as your opposite number gets a nasty attack of mismatched voltages or mistaken identity through feeling either suffocated, over-stimulated or emotionally neglected. Never forget that Uranus is there to wake you up, not make you feel cosy and romantic. If, as I have stressed in the introduction, your paths are destined to intertwine, then well and good – but it is Uranus that sets you off on that path. And talking of 'paths', it is your Uranus Opposition that is most likely to set you off on that Path of Return described on page 155.

To close on this crucially important influence, when all is said and done, the so-called 'Mid-life Crisis' *is* a crisis because we most of us are dead set upon 'normalizing' our lives and selves, when in truth, life and our true selves have something else in mind. So whatever happens comes as a shock – big or small. And, as I have warned previously, the bigger the resistance to Uranus's charge, then the bigger that explosive charge shall be. A tragic example of this was Paula Yates whose 'other plans' were shown, during her Uranus Opposition, to be sheer fantasies when her lover Michael Hutchence hung himself. And maybe John Lennon also needed to experience a shock, but in his case it was so great that it killed him. And, like Martin Luther King, his Uranian intuition foresaw it, and in a remarkably similar fashion – proclaming in one of his own songs that sooner or later they would 'crucify' him.

Naturally, do not expect such drastic and tragic events now – but expect the unexpected.

Overleaf you will see a flowchart of your Uranus Opposition, 'Your Mid-life Crisis', which gives you a graphic idea of what Uranus is bringing into your life, and what the various ways of responding to it can bring you.

JUPITER 'YEAR' for your Growth Mode

MOON 'PHASE' for Emotional Attunement

Having ascertained the actual years of your Uranus Opposition from the tables on page 289, then refer to the Age Index starting on page 235 for the Jupiter 'Years' and Moon 'Phases' occurring during that period.

Other Universal Planetary Cycles possibly active at this time

Check the tables on page 289 to see if any of these Planetary Cycles are active for you at the same time as your Uranus Opposition. Remember that any Trine influences may have to be actualized to make them that effective.

SECOND REALIZING MOON PHASE (*see* page 77) As this influence accentuates certain issues that you have been previously unaware of, the wake-up call of the Uranus Opposition can be particularly acute. How you experience all this has a great deal to do with whether you value an 'enhancement of your reality' or find it threatening and disturbing. Nelson Mandela, alluded to above, had these two powerful influences coinciding.

SECOND SATURN WAXING SQUARE (*see* page 119) Be prepared to be confronted with whatever needs renovating or refreshing in your life, and then take appropriate action. Letting the grass grow under your feet at this time could find you up to your neck in undesirable commitments later on. This combination of influences is particularly critical because caution is just as necessary as the changes.

SECOND SATURN WAXING TRINE (*see* page 121) Here you'd have a sobering or steadying influence that could possibly prevent you from overreacting to the Uranian wind of change. This could act as a welcome handrail.

SECOND SATURN OPPOSITION (*see* page 122) Being another Opposition, this stresses the 'writing on the wall' factor. Having to be practical and responsible and simultaneously move with times and ride the wave of Uranian excitement could prove very testing. The secret of success here would be to cut out non-essentials – once you have determined what those non-essentials are. Again, this is not easy, but it helps to know that these are your priorities.

NEPTUNE WAXING SQUARE (*see* page 198) Being willing to face the music is your best policy now. Failing to surrender to this need for absolute self-honesty could present you with a change

A MAJOR INFLUENCE occurs between the ages of 37 and 45*

THE URANUS OPPOSITION

Your Mid-life Crisis

The flowchart below shows how the Positive and Negative Uranus Traits *(left)*, that you may or may not have, encounter Uranus's influence *(arrow on right)*, which in turn transforms the Positive Traits into a Positive Outcome *(top)* and relegates the Negative Traits to a Negative Outcome *(bottom)*. Negative Outcomes can be made Positive through developing the Positive Traits.

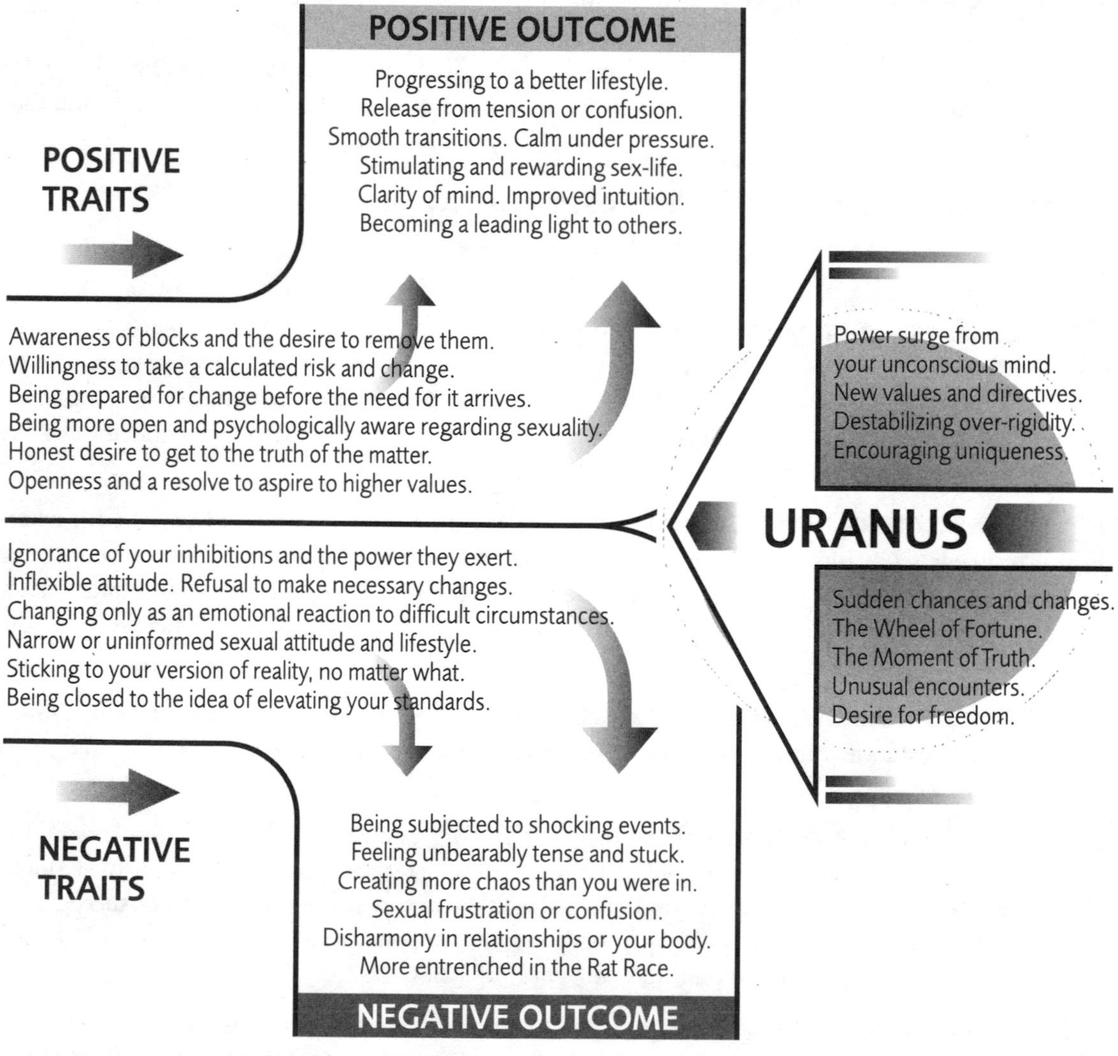

POSITIVE OUTCOME

Progressing to a better lifestyle.
Release from tension or confusion.
Smooth transitions. Calm under pressure.
Stimulating and rewarding sex-life.
Clarity of mind. Improved intuition.
Becoming a leading light to others.

POSITIVE TRAITS

Awareness of blocks and the desire to remove them.
Willingness to take a calculated risk and change.
Being prepared for change before the need for it arrives.
Being more open and psychologically aware regarding sexuality.
Honest desire to get to the truth of the matter.
Openness and a resolve to aspire to higher values.

URANUS

Power surge from your unconscious mind.
New values and directives.
Destabilizing over-rigidity.
Encouraging uniqueness.

Sudden chances and changes.
The Wheel of Fortune.
The Moment of Truth.
Unusual encounters.
Desire for freedom.

NEGATIVE TRAITS

Ignorance of your inhibitions and the power they exert.
Inflexible attitude. Refusal to make necessary changes.
Changing only as an emotional reaction to difficult circumstances.
Narrow or uninformed sexual attitude and lifestyle.
Sticking to your version of reality, no matter what.
Being closed to the idea of elevating your standards.

Being subjected to shocking events.
Feeling unbearably tense and stuck.
Creating more chaos than you were in.
Sexual frustration or confusion.
Disharmony in relationships or your body.
More entrenched in the Rat Race.

NEGATIVE OUTCOME

**see* tables on page 289 for when it is for you.

and/or loss that is painful indeed. Then again, such a rude awakening might by its very nature take you unawares – but the same rule of facing it still applies. As a result of such moral courage, you are then bound to take your whole sense of being alive to another, higher, level.

PLUTO WAXING SQUARE (*see* page 226) Taken together with your Uranus Opposition, this is not so much a wind of change as a typhoon! Rather like a beleaguered vessel at sea, stripping down to the bare essentials and jettisoning old attitudes and attachments is vital. Life is now being ruthless with you, so you too will have to be ruthless – with yourself primarily, and any others who persist in not getting the point. Coming through this will find you lighter and clearer about what matters in life, although maybe with an extra line or grey hair or two.

Waning Trine Between 51–59 years for 1–3 years

'A NEW LEASE OF LIFE' – Many astrological influences exist only in potential, and this is particularly the case with the Uranus Waning Trine. The reason for this is that what it presents – the opportunity to try new things, methods, relationships, pastimes or even a new lifestyle – can be strongly inhibited by the unwillingness to change that is often characteristic of the age group it happens to. But, astrologically, what this influence is actually saying is that this is precisely the time in your life that you need to adopt something new. You may have no trouble with this at all and be naturally young at heart and modern in mind, but it is all too possible that the things which really need a change or a 'face-lift' are the very things you are overlooking. They are also what are probably giving you the most trouble.

It seems that the nature of everything in one's life begins with one's attitude. For instance, if I think that going out for the evening to somewhere different is an unknown quantity with disappointment and inconvenience built into it, then that is probably what I will find – but because of this attitude I probably wouldn't go anyway. A case of nothing ventured, nothing gained. On the other hand, if I ventured forth with an attitude of 'There could be something just round this corner that could really help me or even change my life, a little or a lot, or it could just be a case of a change being as good as a rest; or I could be the one that does this for someone whom I meet around that corner.' This might not happen immediately, but would lead to an encounter that did so. You may not encounter anything or anyone of this nature at all the first time out. Giving up after one attempt would smack of the self-fulfilling prophecy that everything will remain just as it is – safe but boring. 'If you wish to change your life, then you must change your ways.'

Another thing that would create an improvement is the changing of a negative mental attitude. Assuming that so-and-so is never/always such-and-such, effectively keeps that person as what you have labelled them. In other words, give certain others the space and chance to change too. You will be happily surprised, and liberated from such a confining idea of them.

Often the *need* for change can accompany or precede this Transit of Uranus. For example, I had a client who was a professional illustrator who was getting less and less work because computer technology was fast taking over a large part of the area that skills such as his used to cover. He was convinced that the traditional way was the right way, and to a degree he had a point. Illustrations put together by amateurs with ready-made graphics or scanned originals could have a soulless quality about them. But this was mainly because those ready-made graphics or scanned originals were poorly done or chosen in the first place – by someone without his aesthetic awareness. And meanwhile he was getting into difficult financial straits. Eventually, and with great reluctance, he invested in a computer and software. Not only was he surprised at how sophisticated the technology had become since he first formed his prejudice against it, but also at how it gave him a type of artistic inspiration and faculty that he never knew existed. Not only did his business improve as well, but paradoxically he got more traditional illustration work through the new contacts he made.

However, 'all that is new is not gold'. Using 'agencies' of Uranus such as the right friends and contacts, the latest technology, and the 'face of youth and progress' can take you to a new level, but it won't necessarily keep you there. With Uranus, there can often be this hidden agenda of rapid ascent and equally rapid descent, as I have pointed out elsewhere in this section. A good example of this was Richard Nixon who became the 37th President of the United States under this influence, only to be undone by Watergate less than six years later. In fact, the seeds for his demise were sown long before (*see* page 199). The story of 'Tricky Dickie' also serves to show that with matters of Uranus it is vital to pursue and keep to the truth – at all times. So, if one is being economical with the truth during the lucky period that this Uranus influence bestows, then you are effectively abusing Lady Luck herself.

The real advantage of this influence is that it can introduce you to what the 'truth about truth' is. This has something to do with the fact that everything and everyone has a 'truth' to their very existence. This does not mean that they are necessarily honest people; it just means that there is an essential truth in the fact that they exist at all. There has to be something true for something to come into being – even if that something was 'evil'. As with all matters Uranian, this is quite an abstruse and esoteric point. For clarification, take as an example Adolf Hitler, who rose to power mainly because he intuited that the truth concerning German and other European peoples was a need for a sense of racial purity, heritage and superiority. Unfortunately, the 'lie' in him lost sight of the true meaning of this truth – that all such purity, heritage and superiority is entirely relative. It is true that the Aryan race has certain qualities that no other race has, and vice versa, but that does not mean that it should destroy or subjugate those other races. Hitler's 'lie' was his inability to see his own truth – that he was a hurt little boy getting his revenge – and the lie of many people on this Planet, that a sense of inadequacy looks for someone to blame or feel superior towards. Alternatively, maybe he was precisely aware of that second, collective, lie. If this was the case, then he was truly evil. And again, true to the negative Uranus type that he was, he engineered a rapid economic recovery before World War Two, but also gave rise to its rapid decline at the end of it.

And so, if you are seeking to find the real truth of the matter, in whatever context that may be, then now is the time to do so. To strike another Uranian esoteric note, bear in mind that it is the Truth that is presently looking for you! It may well be on its way to set you free from some burden or tie. In any event though, 'To thine own self be true' (Shakespeare).

On page 167 you will see a flowchart of 'The Uranus Slingshot Effect' which graphically portrays the nature of the opportunity this influence offers. I have called it the 'Slingshot Effect' as this derives from the astronautic term for making use of the gravitational field of one planet in order to sling a space module millions of miles towards another planet. Simply follow the instructions given in the panel at the top. You will also see that the Uranus Trine occurred before between the ages of 24 and 31, but this would be the *Waxing* Trine. The similarity between the two is that they both give you an opportunity to change your life for the better, to change your world, and come the Waning Trine you will notice a remarkable connection with this Waxing Trine all those years before. The difference between them is that this one, the earlier Waxing Trine, is more likely to manifest as some actual physical change or move without you necessarily thinking about it too much, whereas the Waning Trine has more to do with a change in outlook or inner attitude that has been consciously arrived at. But this isn't a hard and fast rule – things never are with Uranus around!

JUPITER 'YEAR' for your Growth Mode

MOON 'PHASE' for Emotional Attunement

Having ascertained the actual years of your Uranus Waning Trine from the tables on page 289, then refer to the Age Index beginning on page 235 for the Jupiter 'Years' and Moon 'Phases' occurring during that period.

Other Universal Planetary Cycles possibly active at this time

Check the tables on page 289 to see if any of these Planetary Cycles are active for you at the same time as your Uranus Waning Trine. Remember that any Trine influences may have to be actualized to make them that effective.

THIRD EMERGING MOON PHASE (*see* page 54) Positive changes can now occur on the home front. It is as if one era is being phased out while another is being phased in, and the influence of Uranus oils the wheels of this transition.

SECOND SATURN WANING SQUARE (*see* page 125) That you now wake up to the spiritual or alternative side of life is imperative. Stubbornly sticking with the 'Devil you know' would compound any problems. Give the unlikely a chance to prove its worth.

SECOND SATURN RETURN (*see* page 126) This highly important influence is called 'Your Second Life Progress Report' and so usually poses some kind of reckoning. As retirement or feeling stale and past one's sell-by date are possibilities at such a time, the Uranus Waning Trine's opportunity to refresh and renew one's attitude and way of living should be jumped at. This is rather like being presented with some lubricant at a time when one can seize up. The illustrator who got wise to computers is an example of this combination of influences at work.

NEPTUNE WAXING TRINE (*see* page 204) There are subtle changes going on now which may not be that noticeable to you in the midst of the more obvious events and situations in your life. Essentially though, this combination of detecting the truth of the matter and of the wisdom of allowing things to take their course can mean that you either transcend difficulties or they simply fade away, or that the one gives rise to the other. This is a good time for any spiritual pursuit.

NEPTUNE OPPOSING PLUTO (*see* page 207) The outstanding truth of the moment is that nothing and no-one lasts forever. Accepting this will see you through, as will looking for something new in life to replace what or who may have departed from it.

PLUTO WAXING TRINE (*see* page 231) The combined effect of these two influences is rather like power-assisted steering. If you have a difficult bend to negotiate or manoeuvre to perform, then this will give you the resources and foresight to do so smoothly. If all else is well, then you could be said to be cruising.

Waning Square | Between 59–65 years for 1–3 years

'MAKING A BREAK' – For most people this probably means retiring from a job, and possibly the lifestyle that went with it. Then again it could mean a change within your work situation. In any case, some form of change is in the air now, whether you have willed it or someone (or something) else has, despite the idea that at this age your life should have settled down. The Uranian paradox here is that far from settling down in the usual sense of the term, you are (supposed to be) making a radical move that should give you a new and different perspective upon your own life and life in general. The importance of some form of creative change in your life cannot be stressed enough. One of the saddest factors of the later stages of an individual's life can be that they no longer seem to have any role in life. This Uranian wind of change should therefore be welcomed with open arms (and mind) for 'Making a Break' would set you up for an active and meaningful autumn of life. To help you identify the signs and opportunities of change around at this time, it may help to consult the flowchart on page 165. Although this was designed for the Uranus *Waxing* Square for those aged between 17 and 23 years of age, by the very nature of Uranus's rejuvenating effect much of it will prove helpful. If nothing else it can only affect your outlook in a positive way, which is the basic quality of a Waning Planetary influence.

Here are two very famous examples of how the Uranus Waning Square can mean 'Making a Break'.

At this time in his life, Mao Tse-tung broke China away from Russia and initiated the 'Great Leap Forward'. This has been described as a 'crash-modernization programme, doomed to failure by poor planning'. Sounds pretty Uranian to me! Like Richard Nixon, cited in the Uranus Waning Trine section above, Mao had sown the seeds of his disaster earlier on with shaky ideals and visions. These were owing to 'bad management' of Neptune Transits in his early forties (*see* page 198). However, Mao did manage to 'Make a Break', even though it was ultimately more 'break' than 'make'. All the same, in the Uranian scheme of things it was all part of life's rich zigzag pattern.

Some years later, the leader of the country that Mao made a break from, Mikhail Gorbachev, resigned following the dissolution of the Soviet Union. He did this amidst typical Uranian turmoil and revolution – much of which he was the architect, which consequently lost him his popularity. This was all the more sudden considering that a year previously he was the recipient of the Nobel Peace Prize (*see* page 126).

The new horizons that can open up now are quite surprising, bearing testimony to the truth that it is never too late to start all over again.

JUPITER 'YEAR' for your Growth Mode

MOON 'PHASE' for Emotional Attunement

Having ascertained the actual years of your Uranus Waning Square from the tables on page 289, then refer to the Age Index starting on page 235 for the Jupiter 'Years' and Moon 'Phases' occurring during that period.

Other Universal Planetary Cycles possibly active at this time

Check the tables on page 289 to see if any of these Planetary Cycles are active for you at the same time as your Uranus Waning Square. Remember that any Trine influences may have to be actualized to make them that effective.

SECOND SATURN RETURN (*see* page 126) Having this milestone of an influence occurring alongside your Uranus Waning Square is very similar in effect to it happening at the same time as the Uranus Waning Trine, as described on page 175. The significant difference is that the need to change is far more imperative, also more likely to be resisted, and therefore liable to attract some enforced change that could be unpleasant. So consciously welcome any opportunity or ideas that are put your way that necessitate making some kind of alteration to your life-style.

THIRD SATURN WAXING SQUARE (*see* page 129) Changes that descend upon you now may be regarded as unwelcome to you at this time of your life. But this would be wrong thinking, because such would be forestalling your life becoming stale and meaningless.

NEPTUNE OPPOSING PLUTO (*see* page 207) The concurrence of this influence would greatly increase the possibility of some radical change. But whether it is positive or not greatly depends upon your outlook at this time. A gloomy and hopeless attitude would spiral you darkly downwards, whereas the realization that 'as one door closes another one opens' would have an amazingly refreshing effect upon you.

PLUTO WAXING TRINE (*see* page 231) At a deep level you hear a voice saying that change is exactly what you need – so listen and believe!

The Uranus Return — Between 82–84 years for 1–3 years

'ARRIVING AND DEPARTING BOTH' – This could have been more simply called 'Journey's End', but this would be too cosy or final a term for the completion of the cycle of unsentimental and paradoxical Uranus. So this is a quirky time for a quirky Planet, and as I have not yet got there myself (but I believe I will), it is probably best described in terms of examples that have already done so. Here they are in chronological order of the Uranus Return happening.

Sigmund Freud, the founding father of psychoanalysis, has to be granted the laurels of genius. Scorned or admired, he certainly made his mark on the very thing which he conceived – the subconscious. His ideas were totally radical and sometimes shocking, especially for the time he lived through. Uranus-style, he helped us become aware of what sex is really all about, along with what sexual repression does to people. This Uranian champion of freedom appropriately lived one whole cycle of this Planet of Freedom. Sadly, it was a Uranian end in the negative sense for his life was disrupted by the Nazis' persecution of the Jews, engineered by an opposing Uranian being, Adolf Hitler. He died on 23 September 1939 soon after his escape to freedom in London.

For Mao Tse-tung, one of the greatest revolutionaries of the 20th century, it is all too fitting that he should have lived one full cycle of radical Uranus. Actually, he died a little while before it was technically due, but then he always was, so to speak, a leap ahead. Mao died on 9 September 1976. (*See* page 359 for more about Mao.)

Here is one of my personal favourites. A man had been living all his life on the side of Mount St Helens in Washington State when it erupted in May 1980. Volcanoes are ruled by Uranus, and at that time the Sun was in the 'Fixed Earth' Sign of Taurus and Opposed by shocking Uranus. When rescue teams offered to escort him to safety (very un-Uranian), he refused. He said he had been there all his life and wasn't going to leave now. Uranus-style, he stayed and he went.

Greta Garbo was as much an oddity as an enigma, and died after years in seclusion, as was her Virgoan wont. One wonders why she should have 'chosen' to run the full-term of a cycle of Uranus. She died on 15 April 1990.

A happier example is that of a brother and sister, who had been separated at four years of age, and who finally found one another by chance 80 years later in late 1998 – a Happy Uranus Return.

Finally, and not so happily, except for those who are justifiably opposed to him, Augusto Pinochet who was being held in January 1999 in England while his fate as an enemy of human rights was being decided. Some might think it pointless to punish someone who is now an old man, but in Uranian terms this could be regarded as being beside the point because Pinochet is a symbol of what opposes Uranus – repression. Uranus rules freedom and rights, and so the issue has now come home to roost.

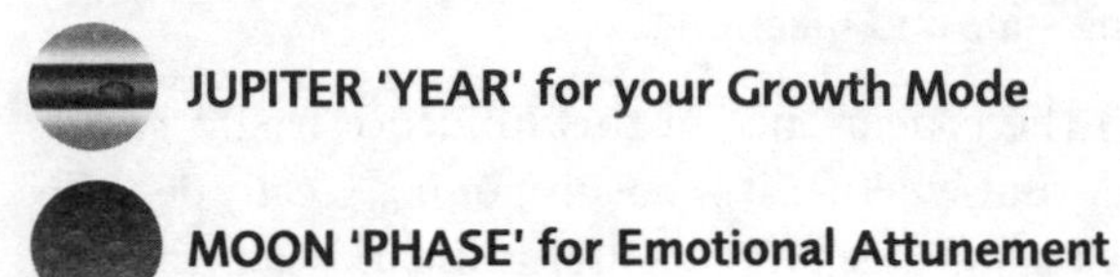

JUPITER 'YEAR' for your Growth Mode

MOON 'PHASE' for Emotional Attunement

Having ascertained the actual years of your Uranus Return from the tables on page 289, then refer to the Age Index starting on page 235 for the Jupiter 'Years' and Moon 'Phases' occurring during that period. If one of these Jupiter 'Years' is 'Aquarius' then this would further accentuate the influence of your Uranus Return in that year of your life.

Another Universal Planetary Cycles is active at this time

FOURTH EMERGING MOON PHASE (*see* page 56) The beautiful symmetry of astrology is demonstrated here as this, the Third Lunar Return as it is also called, coincides with the Uranus Return. Symbolically this is saying that the gamut of human emotion has been run three times – three being the number of completion – in the same time as the completion of an astrological lifetime, of a total voyage of discovery, as symbolized by a whole cycle of Uranus. This could be called a convergence of the cosmic (Uranus) and the commonplace (Moon). This is what this combination of Planetary influences means in theoretical terms, but personally I look forward to the practical experience of such a momentous time, and hope and intend to be as aware as possible of what life on Earth has been about, is about, and shall be about.

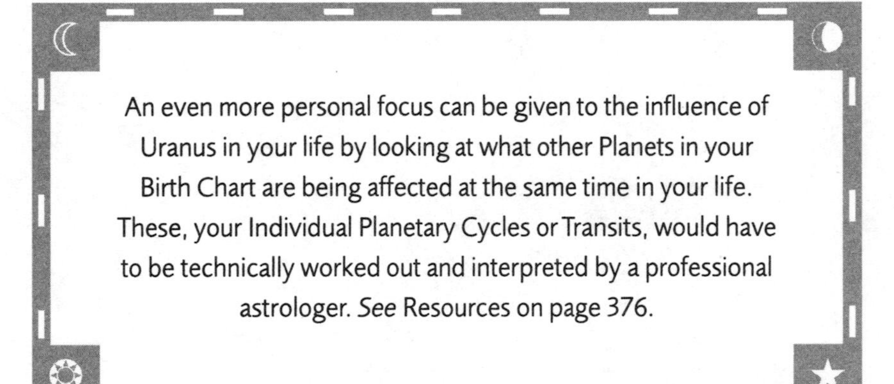

An even more personal focus can be given to the influence of Uranus in your life by looking at what other Planets in your Birth Chart are being affected at the same time in your life. These, your Individual Planetary Cycles or Transits, would have to be technically worked out and interpreted by a professional astrologer. *See* Resources on page 376.

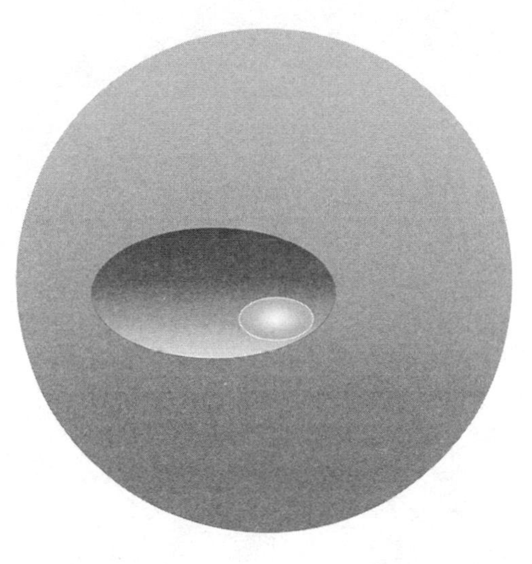

YOUR NEPTUNE CYCLE

Delusion and Enlightenment

So let our singing sun-heart be
Allowing all our eyes to see
Allowing all our love to free
The golden bird that's caged within
The illusions of our pain and sin

Key Influences

+	–
Enlightening	Deluding
Sensitizing	Weakening
Inspiring	Confusing
Relieving	Undermining

One does not have to be that honest with oneself to admit that at times we can all get the wrong end of the stick or be fooled by something or someone. At the time, however, we are not aware that this is happening, and consequently we might then do or say something which takes our misconceptions a stage or two further – often in the misguided attempt to evade what we erroneously think is harmful to us. Think about this a little more and you will realize that there are probably many of these misapprehensions, great and small, and that they build up into a sizeable cloud of delusion. Such a cloud is hovering over everyone all of the time to one degree or another. What is more, all these individual clouds progressively merge into an enormous one that affects the whole of humanity, obscuring our views both individually and collectively. An important point concerning this cloud is that usually people either draw a dubious benefit from it, suffer because of it, or both. An historical example of this would be slavery, an illusion that it was okay to treat fellow human beings as beasts of burden in order to profit from it and lighten one's own load. Many, if not all, politicians could be cited as examples of such mass delusion because we consistently fall for their promises, only to be let down or be further flannelled by them. Like my good friend said, 'A politician is usually as strong as the people are weak.' However, there is a more subtle but far-reaching delusion that underlies nearly all other delusions. This is the Illusion of Separation which has people believe that they and everyone and everything else are not interconnected. This, the Grand Illusion, extends from giving rise to such global matters as war, owing to a lack of international co-operation and understanding, and to personal ones where it is still not generally recognized how finely and utterly our feelings, thoughts and lifestyle are connected to one another, not to mention our bodies, and therefore our health. At last, though, this cloud, the Grand Illusion, is beginning to disperse with the dawning of the Aquarian Age, which is briefly described in the Glossary on page 355.

In talking about this concept of clouds, be they personal or collective, we have to consider the subject of reincarnation. It is a fact that more people in the world base their beliefs on this than those who do not. I think it is also true to say that more people than ever now believe in some form of reincarnation. Neptune governs reincarnation primarily because it has to do with anything that lacks a distinct or constant shape (like a cloud or water) and because it recognizes no boundary between one thing and another, or between one time and another. Neptune declares that everything is One and interconnected. And so it follows that the person that I am now is somehow connected to whatever came before me. On page 215, in the Pluto section, I explain all this in a more scientific way, but Neptune is almost anti-scientific in that it relies mostly on feeling and intuition, and on metaphor rather than specific definition. The feeling that 'I have been here before' is almost too common to mention, but personally I believe that so-called past lives are best regarded as metaphors for something that happened previously and is therefore affecting us now. And so your 'cloud' can have a long history, and therefore be extremely complex. It can also be mysterious and sublime.

On a personal level, this cloud – as a negative expression of Neptune – can be experienced in an infinite number of ways. Some prime ones are: falling in love with someone only to find out later that it was merely lust or wishful thinking on one or both of your parts; seeking sensation or advantage through means that can eventually undermine one's position – like the abuse of drugs or alcohol, for example; avoiding something which has negative connotations, and in so doing never experiencing a positive dimension concealed within it – like, for instance, not involving yourself for fear of failure and thereby ending up all alone or without something you need.

Such fascinations and aversions are endless, as are the effects they can produce. But the highly significant factor is that they must have all originated from some experience that happened previously and somehow gave us a false impression, which grew and grew like 'Chinese whispers'. For this reason, Neptune is classed as a 'karmic Planet' for it represents all those happenings in the past that have created those misconceptions, fascinations and aversions, and the addictions and phobias they can give rise to.

A positive expression of Neptune is anything that dispels such clouds, that pierces that veil of illusion, and relieves, enlightens or entertains in the process. Probably the very point of the arrow that does this piercing is compassion. It was compassion that saw through the horror of slavery and enlightened the masses as to how wrong it was. A more pragmatic view would cite changes in world trade and the balance of power as also having an influence upon the abolition of slavery. This is true up to a point, but does not account for the hounding and persistence of intransigent slave owners or traders who obviously still found it profitable. Such material factors are still an expression of the Neptunian knock-on effect of universality, that ultimately the good of the whole depends upon the parts of the whole, and vice versa. All the same, racial or any other kind of discrimination is a delusion, or the product of one, and it is still very much with us.

On a more personal and everyday level, the suffering that has been caused by some delusion or other is likewise initially relieved through compassion, or simply a good 'bedside manner'. A tragically common example of this is the child who does not receive the love, care and attention it needs because its parents or teachers were similarly afflicted. The delusion or cloud here is the one that prevents a child from being seen for what they are – a vulnerable and highly-sensitive being. It is only when that child at last encounters someone who has the compassion to recognize and minister to its condition that a healing can occur. This may not occur until they are an adult – or it may not occur at all, and so the tragedy perpetuates itself through history and lifetimes.

Perhaps an even more fundamental keyword for Neptune is sensitivity. Sensitivity is the creator of both compassion and of suffering. And sensitivity itself is vulnerable to delusion. When sensitivity, which is the essential quality of being human, is not recognized, it can turn in on itself and become over-sensitivity which in turn can devolve into feelings of weakness. Such weakness is then deluded into feeling that it has to suppress or escape from itself. Apart from sheer fantasy, we can see how those Neptunian substances, alcohol, tobacco and sedating drugs, are used in this way for they lower our sensitivity at the expense of physical and psychological wellbeing. Stimulant drugs do the opposite in that they stimulate the senses, blur the boundaries between 'real' and 'unreal', or 'open the Doors of Perception'. Still other drugs stimulate the body and nervous system by giving one more energy by 'overdrawing on one's energy account', an overdraft that should be paid back before it builds up to an impossible debt. But illusion being illusion, such drugs merely give the impression of relief, enlightenment or confidence. This is not to say that small doses are not useful, but the trouble is they can create addiction. Very few people can merely have a drink, cigarette or joint only when they really need one, or for occasional enjoyment. The subject of drugs is a vast one, and the social problems they give rise to are legion, but the Neptunian fact is that the human being has, or wants to have, access to other realms of perception and existence. However, our so-called civilized societies have no official recognition of this, and so the natural desire to seek contact with those realms goes underground, is outlawed, and still receives no guidance other than the purely technical or patently crass, not to mention the political. It is not surprising that Neptune (as drugs) just comes to mean an escape for most people, rather than illumination, and that one of Neptune's biggest and densest 'clouds' is comprised of the illusions concerning drugs. The inclination to blur the edges of our reality, with drugs, alcohol, or just plain evasiveness, is common, and a testament to our ignorance of the meaning of sensitivity in our lives.

(*See* also page 193, Neptune Squaring Pluto, for more concerning drugs.) We even blur the edge of the reality between drugs and alcohol – and between one drug and another. Furthermore, there is a major difference between naturally occurring drugs and those that are synthetically produced . Nature put the former here for some reason – either medicinal, practical or spiritual – and also included in them certain non-biochemical agents which balance out the substances which contain the psychoactive properties. The latter, however, have very dubious goals and unwelcome side-effects. Something else that pervades out culture like a drug is advertising, for it panders to our fantasies and thereby institutionalizes our delusions – especially those concerning what we are as men and women, as sexual beings. Neptune's universality, by it very nature, touches virtually everything.

All of this begs the question as to what sensitivity actually is, and as to how Neptune symbolizes something natural and good. To begin to answer this we have to enter that other dimension of Neptune's realm, that of mysticism. Mysticism is concerned with the invisible, subtle and spiritual aspects of existence, and the main theme running through most mystical teachings is that in the beginning we were pure spirit without any material form, or that we were not (entirely) of this Earth. So from a mystical, rather than scientific, point of view, the story of how everything was created is told in the words of myth – which some schoolboy aptly described as 'something which is true on the inside if not on the outside'. Most creation myths involve some kind of Fall, that is, a loss of connection with our original and pristine spiritual identity. It is also written that at certain times, a messenger comes from the spirit realm to teach us or remind us of our spiritual origins. One of these 'avatars' was Jesus Christ and, true to things Neptunian, his life and teachings are enlightening, inspiring and relieving, but also, owing to the fallen state of our interpretations of them, confusing and deluding.

Essentially, Christ was, and still is, saying that we are all one, and that we are all God's children. In mystical or quantum terms this is saying that we are all aspects or creations of the same energy source – the Sun. Indeed, it is a scientific fact that we are all literally 'stardust', comprised of solar material like everything else in the Solar System. The Maya of Southern Mexico were (and are) also aware of this, as evidenced in, among many other things, their greeting *In lak'ech* which means 'I am another (like) yourself', and their word for the human body *winclil* which literally means 'vibrating vessel or root'. In other words, a human being is seen as a receiver/transmitter that was originally attuned to that one central source of love, power and intelligence (the Sun), and that each individual human being has a unique way of expressing this.

Back to Neptune and sensitivity – a human being is *sensitivity incarnate* but we have long since fallen from this divine state of awareness. Such a Fall, it must be pointed out, was probably just the unavoidable result of going from a spiritual state into a physical one – or from energy into matter, from a wave to a particle. At root then, every Neptunian influence you experience is aimed at bringing you back to this state of spiritual awareness, that we are all one and that we are here on the material plane for a spiritual reason. However, there are a great many levels of expression regarding this. At one extreme, there is the alcoholic or drug addict who is eroding the very bounds of their personality in a deeply unconscious attempt to lose their sense of separateness, while at the same time trying to escape from, and eventually destroy, the very sensitivity which is the testament to their 'in-touchness' with everything and everyone, thereby descending into profound antisocial behaviour. This is why it is now becoming increasingly obvious that addicts must have a spiritual rehabilitation if they are to be cured at all. A Neptunian influence for them could push them further down the slippery slope or put them in touch with some road to recovery – and discovery.

At the other extreme of the Neptunian spectrum, we have the person who is consciously seeking and following some kind of spiritual reason for being, and involves themself with meditation and the study and practice of spiritual ways – all of which are means of dissolving the boundary between oneself and everything else. For them, Neptunian influences will carry them further along their path in a relatively straightforward way, or show them in what ways they are still subtly missing the point even though, or possibly because, they might deem themselves to be quite spiritually aware. Such 'falling from grace' ultimately happens to anyone who professes to be serving the greater whole while all the time they are simply serving themselves. Neptune has a decidedly humbling or humiliating effect, whatever your position or path in life. In fact, the more spiritually evolved or socially elevated one becomes, the worse it is if you stray — 'Lilies that fester smell far worse than weeds' (Shakespeare).

Generally speaking, the most common effect Neptune seems to have on us is first to confuse our senses and feelings or appeal to our finer ones. It then softens our feelings towards ourselves and others, or further refines our sensibilities and thereby the expression of them.

How Neptune's Influence Manifests in Your Life

Of the two extremes, enlightenment and delusion, the vast majority of us lie or waver somewhere between the two, and it is through those commonplace weaknesses or illusions that Neptune makes felt to us the current state of our sensitivity. This could manifest in many ways and, Neptune being Neptune, they tend to dissolve into one another, but here are some key ones.

♆ **Allergies** to food and other substances can occur and conditions arise that reflect a problem in maintaining one's immune system, which in Neptunian terms is that vital filter between our own physical estate and everything/everyone else's. This is the oneness of Neptune incorrectly or inadequately expressed. You (as 'receiver') could be allowing in too much or too little, being too accepting or too critical. Positively, Neptune would help lead you to the realization of this and an understanding of how to heal such complaints. Alternative medicines come under the rule of Neptune for they utilize, rather than suppress, our sensitivity – hence the efficacy of therapies such as homeopathy, flower remedies and Reiki. Viruses and bacteria are also ruled by Neptune, so under its heavier influence you can be more prone to infection, and again, be cured or relieved by such alternative or Neptunian methods as these, and through visualization as described below under Creative Imagination. Finally, the immune system is also dependent upon your Jupiter – Saturn Balance of Power (*see* page 265) and Protection along the Way (*see* page 271).

♆ **Psychic Experiences** of one kind or another are Neptunian. Psychism is simply the ability to 'pierce the veil', but this time in the sense of seeing (clairvoyance), hearing (clairaudience), sensing (psychometry) or communicating with (channelling/telepathy) what lies beyond the perception of usual five senses. Again this is Neptune's absence of boundaries at work, and the range of what can be reached through these means is also infinite. Seeing ghosts, communicating with the dead, strange so-called coincidences, recalling past lives, spiritual healing, UFOs, inspirational writing – these are some of the more likely areas of encounter with Neptune's realm. Drugs, because they artificially induce altered states, would also come under this heading.

♆ **Emotional Confusion and/or Enlightenment** – To a great extent, the way in which we manage our emotional security and wellbeing is determined by setting, unconsciously or otherwise, limits (Saturn) on our involvement with other people and things. Prejudices such as towards other races, classes, cultures and the opposite sex, are all born of this. But through keeping our distance, we may not allow ourselves to feel that closely or intensely, and Neptune, wanting to include more and more in our sphere of acceptance and relationship, will see to it that something occurs

The glyph of Neptune, which usually appears as his trident ♆, is actually composed of the Cross which symbolizes Matter or the World, and the Semi-circle which symbolizes the Soul. So the glyph symbolizes the essence of Neptune as the Soul wounded by Matter or by being in the World. However, by being in the World, the Soul spiritualizes and saves the World. See also how the glyph is pictographic of Christ upon the Cross.

(like falling in love) that forces us to do just that (when we sooner or later realize that romance is merely 'Love's calling card' and not Love itself). Finding ourselves involved with people and situations of a different background is also a Neptunian 'ruse' to get us to accept more. Consequently we would then have to track down and heal those aversions, illusions and phobias, born from experiences in our past, near or distant. The learning of tolerance would become highly important, along with the attenuation of critical or overly sceptical thinking. The cultivation of unconditional love is the ultimate goal here, but one has to be careful not to make that a condition in itself. Conversely, letting in too much emotionally may be the trouble, and Neptune will make such excessive openness unacceptable, possibly with a view to realizing the necessity of caring for yourself as much as for others, and exercising some necessary discrimination (*see* Psychic Protection on page 271). Not caring for oneself enough is a form of Victimhood (*see* below). It is this process of emotional, and, consequently, mental, refinement that brings about enlightenment, for we learn to accept others and see them as part of ourselves, and thereby to accept ourselves as well. By the same token, it is worth noting that people who are emotionally 'shut down' still have a pronounced emotional effect upon others. It is as if such people present the biggest blips of all on our Neptunian radar screens, resisting unity because they do not believe in it being possible – possibly as the result of experiencing closeness in a negative or wounding way. Neptune also offers us one solution to emotional confusion in sexual relationships, by making the relationship platonic, either permanently or just long enough to 'let the dust settle'. The trouble is that negatively

Neptune-driven prevailing social values disincline us to take this step because we feel that not being sexual somehow suggests being inadequate. Often, however, the alternative to this is trying to artificially induce 'sexiness' through such Neptunian means as drugs and fantasy. These means may work, but tend to create other, more complex or dangerous, illusion-born problems in the process. Above and beyond all this there is the avenue of 'spiritualizing' sexual interaction through the likes of Taoism or Tantra.

♆ **Creative Imagination** – as Neptune has to do with universality, the universal language of music and the arts is also an important expression of this Planet. So involvement, as a practitioner or spectator, is likely to come more to the fore under Neptune's influence. Inspiration and spiritual elevation are more likely to occur as one becomes more alive to the highs and significance of all art forms. There is another use of creative imagination that is close to meditation, one of the most pure and applied expressions of Neptune. This is visualization, the means by which you can heal ills, protect loved ones, receive higher guidance for help with problems, and influence others in a magical and positive way. This works by simply 'allowing' a person (including yourself) or thing into your mind and then seeing them as being healed, protected, discovering a solution (in time), or letting a person be what they are without any interference. (*See also* Protection along the Way on page 271.) Hypnosis, hypnotherapy or any kind of dream or trance work also comes under this heading. Negative effects can be created by not giving creative expression to Neptune's influence in these ways that I have suggested and described. Such effects would take the form of absent-mindedness, neurotic worrying, or at worst mental derangement. Being naïvely open to suggestion could also apply here, as in letting one's imagination get the better of you.

♆ **Via Negativa** – means the 'way that is not' and is often an unavoidable road that Neptune, or rather our fascinations and illusions, tempt us down. We sometimes have to find out what is right for us by experiencing to the limit that which is not! But in Neptune's view *Via Negativa* is right for us too for how else would we learn? All roads lead to Rome. To put it another way, what this amounts to is the experience of disillusion, which simply means the dispelling of illusion. Allowing others to do this is also sometimes highly necessary. Another, possibly more subtle, version of *Via Negativa* is evasion. We may be tempted to avoid things which we do not want to experience, because of those aversions, but we wind up getting nowhere. Sooner or later the music has to be faced.

♆ **Victimhood or Selflessness** – are two faces of the same Neptunian coin. Victimhood is the face worn by those for whom suffering, or accepting peace at any price, has reached a point where the world then treats them in the very way to which they have become accustomed. A victim is someone who is denying their own effectiveness as an individual, possibly as a result of having being made to feel all 'receiver' and no 'transmitter'; that is, life seems to have only happened to them and never the other way around – sometimes to the extreme of being abused. This is possibly owing to taking too far Christ's admonition to 'turn the other cheek', and winding up getting one slap in the face after another. As one wit parodied: 'Blessed are the meek for they shall inherit the Earth – if it's all right by everyone else'. In truth, though, it suited the political and ecclesiastical powers to misuse the scriptures to foster the illusion of the masses being unable to govern themselves or of being 'hopeless sinners' or 'evil' (hence, witch-hunting), in order to manipulate them to co-operate towards their selfish ends. To counter all of this, the victim must take responsibility for their own being (and the spiritual or karmic reason for it) and begin to 'transmit' or express themselves for what they are.

The other face, selflessness, is as bright as victimhood is dull, but as the crowning glory atop Neptune's tree, it takes some reaching. All the same, it is selflessness to which any spiritual seeker aspires for it is in such a state that they experience themselves as the channel or facet of the One, which was always Neptune's intention. So, for the victim, Neptune's influence can make that victimhood all the more uncomfortable, until eventually it becomes unacceptable. But beware masochism, which is victimhood taken to its most confusing and deluded end. For the one seeking selflessness, Neptune offers many an opportunity to make a sacrifice of whatever the ego is attached to in order to relieve and enlighten oneself and others, be they near or far. Meditation and other spiritual or mystical pursuits are also a sure way of embracing and experiencing the Neptunian Oneness and Beautiful Mystery of all things. Ultimately, enlightenment is not only 'seeing the light' in a Road-to-Damascus-like fashion, but also no longer having to carry the load of guilt from the past through having cleared one's karma and redeemed oneself.

♆ **Do-gooders or Doing Good** – if you wish to save the world, first make sure you are not in need of saving yourself! Unfortunately, Neptune is commonly expressed through a kind of false idealism whereby an individual by-passes their own weaknesses and blind spots through avidly tending to what they perceive as the needs and deficiencies of others 'less fortunate'. Or they might espouse some metaphysical philosophy that puts them beyond the obligations of an ordinary mortal. Politicians and charismatic preachers often fall into this negative category too. The frequent scandals that bedevil such people are a testament to Neptune's wonderful ability to force an individual to bare their soul one way or the other, sooner or later.

Doing good, on the other hand, is closely allied to the sacrifice of the ego and its attachments as described above under Selflessness. True 'good works' are not so much rare, but by their very nature they usually cut a pretty low profile. Apart from being some aid worker out in the field, in the more everyday world Neptune's influence can quite often put one in a situation where the Neptunian manner of handling it is to adopt a modest and humble position, allowing things to work themselves out without putting one's oar in or trying to force the issue. In such predicaments, going with the flow or steering around obstacles can be the only way to avoid conflicts that probably create more fall-out than the original problem. All in all, identifying and living up to your ideals and principles is an essential of this type of Neptunian influence.

♆ **Neptune Types** – are people who are Pisceans, or have a Piscean/Twelfth House emphasis or strongly placed Neptune in their Birth Charts. In ordinary terms, such types would be creative, sensitive, visionary, psychically aware, refined and subtle, on the positive side, or evasive, weak, deceptive, addictive and victimlike, on the negative side. Such a person might be a musician, artist, dancer, entertainer, healer/doctor, chemist, sailor, advertising executive, psychic, drug dealer or wine merchant, or have any strange or water-related occupation. As such they would be the 'agency' or figure that brought Neptune into your life through possessing the qualities associated with this Planet and the Sign that it.rules. (*See* more about Neptunian agencies or figures and how to deal with them in Planetary Effects, Figures and Responses on page 259.)

Here follow the significant periods and meanings of your Neptune Cycle

★ Refer to the tables on page 289 to determine what ages are active for you personally (one to three) years) within the age span given for any particular influence. For 'Neptune Opposing Pluto', *see* the special table on page 352. Any influence will not be continuous throughout that period, but off and on, usually comprising three peaks. Occasionally you may notice the 'stretch factor' when you experience an influence for somewhat longer than the period that has been technically determined.

★ Pluto's eccentric orbit – owing to Pluto's highly irregular cycle, note that for pre-1930 births the age span of Neptune Squaring Pluto can be somewhat greater than the one given in the heading for the interpretation and for the flowchart of this influence.

★ Please remember that there is also the individual Cycle of Neptune to be considered (that is, Neptune's Transits to other Planets in your Birth Chart that occur throughout your life), but these would have to be calculated by an astrologer or with an astrological computer program. If any of the foregoing has struck a chord with respect to certain life passages, then this is possibly because such Neptune Transits are what you are or were experiencing.

Neptune Squaring Pluto | Between 10–23 years for 1–3 years

'ADOLESCENCE STRIKES' – In astrological terms, these adolescence or teenage years are seen as the rising sap of a young person's destiny and desire (Pluto) being met with their own fascinations and aversions, but more particularly, those of the world in which they are growing up (all symbolized by Neptune). Viewed in this way, we can appreciate why these years set one's emotional and social patterns for years to come, and why they are so often beset with confusions and confrontations. But an important and interesting point to note here is that from scanning the tables (*see* page 289), we can see that Neptune Squaring Pluto occurred to individuals earlier and earlier in their lives as the 20th century progressed. This means that since about 1930 this powerful Planetary influence has been prevailing upon people from as young as ten to their very early twenties. It has been said that young people grow up more quickly nowadays and are more rebellious, and from an astrological standpoint this would appear to be true, but as an evolutionary development that is peculiar to the late 20th century, rather than just due to socio-cultural circumstances. As we go back through the earlier part of the 20th century this influence occurred later and later in life. In other words, by the latter part of the 20th century very young people were having to deal with this 'society's illusions *versus* my destiny' type of influence. As this book is more likely to be about, or read by people born since 1930, I have mainly addressed the following text to such individuals, who have had to deal with intensely evolving forces at a younger and more tender age than did their forebears.

We are all born with a blueprint or destiny that, apart from in exceptional cases, only begins to show itself when you become aware of something or someone opposing or encouraging it. Seeing that Nature has decreed for the part of history mentioned above that this confrontation initially occurs during your youth, it is most likely to manifest merely as a set of powerful compulsions – either to immerse yourself in something, or to avoid or resist it. However, on top of this you are also subject, and highly sensitive to, parental and peer pressure which would traditionally be at odds with one another. These pressures would include, on the one hand, authority figures such as teachers and the current political climate, and, on the other hand, prevailing fashions and trends. Parents are particularly noted, notorious even, for imposing their values, ideals and unrealized ambitions upon their children – something which ultimately leads to acute dissatisfaction and resentment. Young girls shifting their preference from playing with dolls to going out with boys is often missed, denied or resisted. Consequently, young women can feel especially victimized at this time. Taken to the extreme, we have the anorexic, who needs so much to feel in control of her own body and fate that she can starve herself to death – the reason for this neurotic need for control being the result of parental domination or neglect, or some previous trauma where she was made to feel without any control at all.

All this gives the impression of our 'Rocket Ship of selfhood' (*see* 'The Story of Your Life' on page 15 and 'Your Uranus Cycle' on page 149) trying to achieve escape velocity – that is, enough personal thrust, sense of identity and power to attain an idea or feeling of your own reason for being, rather than simply being a product of your environment. But at the same time, that environment – (the above-mentioned pressures) is something you are dependent upon to give you a sense of identity, and an idea of what surroundings you do or do not want to find yourself in. Such an environment can, Neptune-style, benefit you at first but undo you later. The beginning of Elvis Presley's Neptune Squaring Pluto at 18 years of age was when he first impressed a studio manager with his voice and style, leading directly to fame and fortune, but also to his ultimate fate. At the same age, Muhammed Ali won his first Olympic gold medal, setting him on the road to both fame and ill-fortune. Incidentally, both Presley and Ali were

Capricorn, a Sign that is often more focused upon achievement itself, rather than what is actually being achieved.

Ultimately, though, everything depends upon what your fate actually is. If yours is to be a conventional path through life, then you will have relatively conventional teenage experiences, co-operating with the status quo enough to prepare yourself for a career, or to become, say, a wife and mother. One recent survey in the UK discovered that young people now want these conventional things more than in the last 30 years or so. Maybe they are beginning to understand Neptune Squaring Pluto, although the shadow of AIDS and self-destructive drug-taking is still with young people (*see* below). However, such conforming will not necessarily continue indefinitely, as future Planetary Transits might reveal. At the other extreme, if your fate decrees, at this time you may find something occurring that affects you strongly for years to come – possibly for the whole of your life. However, you will need to be alive to such an intimation from fate, for it may come in the form of a disappointment or seem insignificant at the time. In my own case, at 15 I got ridiculed in front of the whole class for writing a short story that I thought was really good, but the teacher was highly critical of it. It took me many years to regain the confidence to write, which is now how I mainly earn a living.

Apart from being of an unconventional or conventional bent, there is also the *strength* of your fate to consider. If you are an individual who has a particularly strong and conscious idea of what you want to be in life, then those outside pressures of authority and fashion will only influence you enough to contribute to your development, rather than pervert, confuse or suppress it. Conversely, if your sense of destiny is vague, subtle or weak, then external pressures and influences will respectively confuse you, or quicken your idea of what your destiny is, or simply submerge you in the mainstream of society, happily or not, employed or unemployed. But all this is in danger of giving an impression that Neptune Squaring Pluto is really in aid of dispelling,that you have a limited number of choices. The trouble usually, however, is that many people of that age do not know what choices there are other than those set before them. For this reason, if no other, you may be driven to experiment in order to find out what else is on offer apart from what is in that official 'shop window'. Such experimentation will also attract you if you have a strong sense of destiny, and whether you are conventional or unconventional.

This experimentation – what has been traditionally labelled rather patronizingly as 'rebellious youth' – is a highly necessary part of growing up and discovering yourself. It is also essential to the evolution of the human race itself; if a cave-dweller's son or daughter had not behaved differently to their parents we would still be there! Yet again, the era in which you are growing up determines what's on offer. However, one thing has always been 'on offer' because it is biological rather than merely social – sex. And another, more recent, area of experimentation is drugs.

Sex is the 'force that through the green-fuse drives' virtually everything (Dylan Thomas). No matter how civilized and human we like to think we are, we are still motivated by the desire to achieve sexual success or satisfaction, whether that is in the form of having an orgasm, being seen to be an attractive member of society, creating children, or conversely, avoiding the challenge, embarrassment or responsibility that such entails. It is during Neptune Squaring Pluto that you encounter such delights or dilemmas – and particularly so if it should coincide with your pubescence.

In the light of what Neptune Squaring Pluto means as a Planetary energy, at this time the strength and nature of your sexuality comes into contact with whatever fashions, trends, myths or dangers are prevailing concerning sexual practices generally. Depending on what kind of person you are, your experiences can range from being immersed in whatever most others are regarding as sexually acceptable and desirable, to having to follow a more idiosyncratic route. The former route is conventional and, by its very nature, encounters highs and lows that one is able to talk about and share, and gain knowledge and maturity in the process. Being a 'stud', 'easy lay', 'teenage bride/mother' or 'average Joe/

Joanna' are some of the categories one could get placed in (or whatever would be the equivalent for your generation). As such, Neptune Squaring Pluto can pass without you thinking anything remarkable has happened. But this is because of that all-consuming human need to 'fit in' and be 'normal', or rather appear to be so, especially on the sexual front, while suppressing any feelings that make you feel 'different'.

But the conventionally perceived sexuality of adolescence is probably as reliable as one of those sexual surveys. Where human sexuality is concerned everyone knows (apart from the sexual surveyors, it would seem) that there is a strong inclination to lie or exaggerate. The truth is probably a lot more varied and peculiar, but because of this 'conventional image' many are confused by their sexuality when it first makes itself felt. A good example of this would be the shy or sensitive person who feels pressured into doing something that they are not yet ready for, or that they form an idea of what is required of them which they later feel bound to live up to. Consequently, they eventually become neurotic and self-conscious, or worse, about their sexuality. Any latent feelings of rejection are exacerbated, any sensitivity is made to feel a weakness, and worst of all, they can be forced into some relationship that negatively colours many or all subsequent relationships. Or they can get funnelled into a longstanding lifestyle that does not suit them or that they are not ready for – all because prevailing fashions and concepts of sexual behaviour were based upon an illusion.

In so-called primitive societies there exist such things as vision quests and puberty rites, or other rites of passage. In so-called civilized society we are expected to make the transition from childhood to adulthood with guidelines that we can hardly respect, without a culturally or psychological inspired ritual that would give us a sense of being a growing creature in a society that recognizes the deep significance of such a transition.

So, whether it is you who is going through this Transit, or someone close to you, like a son or daughter, the best you can do in the circumstances is to identify, honour and affirm whatever your or their deepest feelings and inclinations are concerning sex – or for that matter, any emotional issue. This would also include the recognition that one *is* making a transition, and that it has long-lasting effects, but does not last that long in itself. Being understood a little means that your 'Rocket Ship of selfhood' does not have to fight so hard against the 'gravity' of prevailing standards, and so reduces the possibility of destructive rebelliousness, which otherwise would invite more repression and misunderstanding, which in turn would attract more rebelliousness, ad nauseam. In other words, you should be – or be encouraged to be – true to yourself, but avoid the obvious dangers of disease and unwanted pregnancy.

Drugs can be all too hard to resist as an avenue of experimentation. What is needed is education as to what is what, but unfortunately, Neptune's myth again descends – this time upon the nature of such drug education. It is relatively straightforward merely saying what does what and how dangerous it is. But what is not usually pointed out by authorities or parents (or even ex-addict teachers) is what is the real and basic reason behind the impulse to take a drug in the first place. Peer-pressure, curiosity, rebelliousness and boredom are some of the relatively superficial reasons given, but from an astrological/esoteric viewpoint it is down to the natural urge to alter one's state of consciousness – or ironically, to escape from an overly materialistic society that does not recognize this urge to explore 'unseen realms'. One could say that adolescence or pubescence (via hormones, drugs in themselves) does precisely that, but there is often a feeling that there is something other than the everyday reality that we are apparently stuck with. The more bleak that reality is, the more that feeling grows, until escape then appears to be the only option – but often into a hell far worse than the reality from which escape was sought. So, short of writing a book on the esoteric whys and wherefores of drugs, suffice to say that drug experimentation has the search for 'another reality' behind it, not just an escape from this one.

Hopefully, a system of education will develop that produces guides to those 'other realities', much like the shamans or medicine men of those so-called primitive societies. (*See* also the introduction of this Neptune section for more regarding drugs.)

Neptune Squaring Pluto is about becoming more aware of all your deepest urges at a time when they are burgeoning, and when the environment in which you are living is going to impose itself upon you by the bucketful. Getting in touch with your deepest values and desires is the key, and then being able to 'pick and mix' in a way that suits you and not what some friend, advertisement or authority figure claims is right for you. It may be so, but only you will know. Experience is the great teacher in life, and whatever your experiences teach you now will colour many years of your life to come. The vital fact is that amidst all these experiences will be a big clue as to what you are supposed to be doing in this world. But that clue could be obscured by a fear of being seen to be different. So if you feel odd or a misfit, don't bury that feeling – it's really your destiny calling from the shadows.

On the opposite page you will see a flowchart of Neptune Squaring your Pluto, 'Adolescence Strikes', which graphically portrays the challenges and opportunities that this influence offers.

JUPITER 'YEAR' for your Growth Mode

MOON 'PHASE' for Emotional Attunement

Having ascertained the actual years of your Neptune Squaring Pluto from the tables on page 289, then refer to the Age Index on page 235 for the Jupiter 'Years' and Moon 'Phases' occurring during that period. If one of these Jupiter 'Years' is a 'Piscean' one or a 'Scorpio' one, then this would further accentuate the influence of Neptune Squaring Pluto in that year of you life.

Other Universal Planetary Cycles possibly active at this time

Check the tables on page 289 to see if any of these Planetary Cycles are active for you at the same time as Neptune Squaring Pluto. Remember that any Trine influences may have to be actualized to make them that effective.

FIRST REALIZING MOON PHASE (*see* page 75) The contrast between 'you' and 'them' is all the more acute under this influence. This can either give you a stronger sense of what you and your peers are about by virtue of comparison with others, or can find you feeling alienated if you do not have a particularly close-knit group of friends. In the second case, a sympathetic ear would be a life-line, but if this is not forthcoming then it has to be said that you are making some deeply personal realization that may not be fully appreciated until some years later.

FIRST SATURN WAXING TRINE (*see* page 111) Supportive older people would be available with this influence prevailing, making the potentially rough ride of Neptune Squaring Pluto into a more manageable one, offering guidance and advice that you would be wise to accept.

FIRST SATURN OPPOSITION (*see* page 111) This happening as well now is somewhat similar to the 'First Realizing Moon Phase' described above. This, however, would stress the possibility of alienation or hardship, and that some karmic factor was being brought into play. This means that events at this time would be very fated, and therefore unavoidable. Yet they would be serving a deep and longterm purpose.

A MAJOR INFLUENCE occurs between the ages of 10 and 23*

NEPTUNE SQUARING PLUTO

Adolescence Strikes

The flowchart below shows how the Positive and Negative Pluto Traits *(left)*, that you may or may not have, encounter Neptune's influence *(arrow on right)*, which in turn transforms the Positive Traits into a Positive Outcome *(top)* and relegates the Negative Traits to a Negative Outcome *(bottom)*. Negative Outcomes can be made Positive through developing the Positive Traits.

POSITIVE OUTCOME

Healthy career path or basis for one.
Sound relationships born of full-blown emotional and physical experiences.
Realizing yourself as the unique creative being that you potentially are.
Being happy in your own company.
Being sure of yourself through being true to yourself.

Strong sense of having your own destiny.
Healthy urge to express/satisfy your desires.
Conscious and/or guided intentions.
Unconscious but necessary urge to be alone.
Initially scrutinizing what you are tempted by.
Deep, maybe partly unconscious, sense of what is right for you.

NEPTUNE

Fashion, cultural and political influences.
Parents' ideals or illusions.
Temptation. Delusion.
Inspiration. Illumination.

Undermining influences.
Sensitizing influences.
Irritating people/situations.
Artistic creative people.
Drugs, sex and music.

Weak sense of having your own destiny.
Blind or wanton desire or compulsion.
Destructive or self-destuctive instincts.
Neurotic aloneness or untreated problems.
Paranoid reluctance to try anything unknown.
Shallow or limited idea of yourself and life.

NEGATIVE PLUTO TRAITS

Losing virginity too soon.
Unwanted pregnancy/parenthood.
Receiving a false impression of love and life.
Misspent youth. Poor prospects. Misfit.
Missing out on experiences that are vital to a healthy and satisfying adult life.
Damage to physical and mental health.

NEGATIVE OUTCOME

**see* tables on page 289 for when it is for you. Can occur later for pre-1930 births.

FIRST SATURN WANING TRINE (*see* page 113) As with First Saturn Waxing Trine, supportive older people would be available with this influence prevailing, making the potentially rough ride of Neptune Squaring Pluto into a more manageable one, offering guidance and advice that you would be wise to accept.

FIRST SATURN WANING SQUARE (*see* page 114) Somehow or other, with both these influences prevailing, you have to find some way of combining discipline with exploring the wilder side of life, or of keeping them apart without letting one cancel out the other.

URANUS WAXING SQUARE (*see* page 162) As this is the 'Breaking Away and Breaking Through' influence, the rebellious aspect of Neptune Squaring Pluto is emphasized. On the negative side it can lead to some serious scrapes, while on the positive, you certainly won't, or shouldn't, allow yourself to be dragged down or overlooked.

Neptune Waxing Square | Between 38–43 years for 1–3 years

'FASCINATIONS AND ILLUSIONS' – The road of your life now is beset with both curses and charms, but it is often not recognised until later, probably not until some time after this transit is over, *what was curse and what was charm*. Events that happen now seem to conspire to catch you off your guard, but this is not merely to make a fool out of you. The ultimate aim of Neptune is to put us in touch with a dimension of ourselves and life that is beyond reason or being clearly defined – a dimension that is ultimately spiritual, or at least, more soft and compliant. But in so doing we are made susceptible to thinking, feeling and doing things that would not be regarded as 'practical' in the usual sense of the word. So this is a time when you can really be a led a dance, or lead others a dance.

Actually, it is you yourself that is leading the dance, but it is a side of yourself that you may not be that aware of – the blind side. The more in touch you are with your illusions and fantasies, the more you observe and express your heartfelt sentiments and ideals, and the more you employ and investigate the powers of the imagination, then the less trouble this Transit will cause you. Where you continue to be blind to your blind spots, you must expect to experience spectacular flops and confusion within confusion. Then again, bear in mind that disillusionment does rid you of illusions you no longer need – painful though this may be. If you prefer your illusions to reality, which is quite possible under Neptune's spell, then that is another matter which we will explore later.

If you cast your mind back to the beginning of this Neptune section, you'll remember we talked about all those misunderstandings and misconceptions, wounds and whims, be they little or large, that we all accumulate throughout life – and maybe inherit from past lives. Such illusions cast a spell over us that means that we are then susceptible to being taken in or frightened off by whatever harks back to that initial false impression. Such an illusion can be anything you like – or don't like – and it could have got 'installed' in your upbringing through the influence of your mother's or father's own illusions, or those of a teacher, a first love encounter, a traumatic hospital experience, or anything that strongly influenced how you now think the world works. Say, for example, you first fell in love with someone who then rejected you, you might then labour under the illusion that the only people you fall in love with are ones that sooner or later dump you, thus making it a self-fulfilling prophecy. Another example could be your first job interview being an embarrassing failure, making all subsequent interviews become strongly charged with that possibility. Yet another, a doctor remarks on your big tummy at a school medical inspection, leaving you with a complex about the curve of your abdomen for years to come. Then again, quite the reverse could be the case – a first love being deceptively easy, getting your first job without a hitch, or, more subtly, someone making a sarcastic or excessively flattering compliment about how you look and you taking it literally, thereby giving you a false confidence.

Now along comes your Neptune Waxing Square and one or more of these illusions comes home to roost. And so, following our examples, you get dumped or deceived by your partner; you lose or fail to get a really important position; you go for cosmetic surgery that has a negative outcome. But, yet again, you won't see these events occurring if you're not looking. Failing this you could compound any such illusion and thicken that cloud of unreality that envelops your relationships, work, self-image or anything made vulnerable by your illusions. So this is a time when self-honesty is a real must! If you don't face the music then you'll never know the score – which brings us back to the possibility of preferring to remain with your own escapist version of reality, despite the warning signs.

The prime feature of being a human being is that of free will. No matter what you might believe, you are always free to think what you want. We can always define our reality to get it to fit our preconceptions of what life ought to be like. This does not necessarily mean that what it 'ought to be like' is to our liking. After all, those previous Neptunian delusory experiences can just as easily persuade us to 'want' a negative reality. For example, I refer you back to the case of anorexics on page 193.

So at this time, reality may be impinging upon you to such a degree that you have to bend the truth quite considerably in order to maintain your idea of life as it 'ought to be' – that is, a life where you haven't got to confront your illusions. World-class examples of this Neptunian process of sowing or reaping delusion and undoing were Mao Tse-tung, Richard Nixon and Malcolm X. During his Neptune Waxing Square at 41 years of age, Mao embarked on the Long March, literally leading his followers towards the realization of his vision. Yet most of us in the Free World would agree that what he created was far from ideal. He even outlawed love! Interestingly, he died when Neptune came round to its Opposition (a time of reckoning, not included in this book) at 82, just short of his Uranus Return (*see* page 179). Richard Nixon similarly first found influence and leadership when he became US Vice-President at 39, only to go down the drain of Watergate 22 years later. Finally, black activist Malcolm X was assassinated by fellow black Muslims at 40 years of age. Whether he died in vain, was a martyr to his cause, or was just someone else who had built their castle on sand, is open to question.

Another, more everyday or personal, example of Neptune's Waxing Square could be that one is in a relationship that (yet again) seems to invade your delicate inner space so much that you behave in a manner that effectively (yet again) destroys that relationship. But this time you are around 40 years of age and the odds are getting longer on forming a decent relationship from scratch. But this prospect may also tempt you to rationalize it as 'Who needs 'em?' or 'I'm happy in my own world'. In effect, there is nothing that anyone can say, especially your erstwhile partner, that will persuade you otherwise – that you are consigning yourself to a lonely and peculiar old age, that you are not developing emotionally or spiritually, that such self-containment will also block off feelings and inspirations that are essential to your wellbeing in other areas of your life. Then again, another relationship could come along, which *really* invades your inner space! But Neptune is a slow-moving Planet, which means to say that if you want to find where you went wrong when it's too late, it's okay by Neptune, there are lifetimes to get it right. Neptune, it must be noted, specializes in grabbing you by your Achilles heel.

Alternatively, you can decipher the Neptunian messages that life is bringing you in the form of what hurts but is true. My personal experience of Neptune's Waxing Square involved someone who flattered my ego so much that I embarked on a relationship with her even though I had little real desire for her. I was simply fascinated – like a snake's prey. The upshot was a three-and-a-half year relationship in which I learnt much about the mystical and trance-worlds of Neptune (which added to my overall understanding and ability) but the crowning realization was that she only cared for her image of me, not the vulnerable and unsure me on the inside. And this was my longstanding Neptunian pattern:

I had always gone for that kind of partner because I had grown up in a school and family environment where one did not show one's feelings, just presented a strong, stiff upper-lip image. Now Neptune was making that lip decidedly wobbly! But to give an idea of the stubbornness of such a pattern (Neptune rules addiction, remember), I jammed another relationship in at the end of my Neptune Waxing Square which had the same formula. Three years and Pluto's Waxing Square later (*see* page 226), and I finally let go of this pattern, and soon after attracted my present partner who is definitely in touch with the real emotional me.

A famous example of Neptune's propensity for dissolving our emotional defences and softening our edges is the singer Madonna. During her Neptune Waxing Square she noticeably softened, becoming more 'female' as a result of being a mother and practising yoga. *Ray of Light*, her album of that time (1998), was also patently Neptunian, being ethereal, confessorial, spiritual, mantric, with frequent references to the ocean, swimming, the other shore, and other Neptunian imagery.

Now there is a meta-message in all this. Not only does Neptune tell us how foolish it is to keep getting sucked in by our delusions, and how we can benefit from changing our negative patterns, but it can also inspire us with a greater sense of what life is all about. Perhaps this can be expressed in terms of seeing your life as a river, as has been done elsewhere in this book. We begin at the source with our birth, and progress through various types of terrain or experience until we at last reach the open sea. However, at certain points we can get bogged down or misdirected, become stagnant, go underground, or even dry up altogether. It is Neptune that governs the inexorable flow to the sea of reunion and togetherness, and which brings us the vision, beauty and promise of such a goal. As such it will always try to free us up, get us back into the flow when yet again we get stuck. Of course, it is easy to refute or ridicule this concept with scientific rationale or rank cynicism – but despite this, a feeling, longing or sorrow persists.

Going back to music and poetry, these and any other art forms are avenues of expression that this Neptune influence enriches, and in turn they give positive form to the boost to your imagination that Neptune is currently providing. This would include a greater involvement with art generally in the sense of its appreciation or promotion.

Another good way of handling Neptune is to 'render unto Neptune what is Neptune's', meaning that as Neptune is ultimately about getting in touch with the mystical side of existence, a conscious involvement with some metaphysical discipline would be the best and most profitable way of channelling its influence. Because Neptune makes us more sensitive to all impressions, it is therefore a good idea to make use of a technique or body of knowledge that studies expressly the world of psychic phenomena and capability. Psychism is simply sensitivity taken to such a point that it is able to see beyond what the usual five senses are aware of. However, you may be 'getting psychic' under Neptune's influence, but either not be aware of it or else not know how to handle it. You could then start over-reacting to what appear to others to be fairly ordinary situations. Put quite simply, over-sensitivity is usually unschooled psychic sensitivity. So if you have not already, embark upon some metaphysical course or journey at this time. This could include such things as yoga, meditation, dream- or trance-work, dance-work (like that of Gabrielle Roth), or anything that constructively uses and connects you with your imagination or higher mental faculties (opens your third eye). This would also include forms of healing that treat the 'subtle body' like homeopathy or Ayurvedic medicine. If you find any such Neptunian avenue not to your liking because you are of a more rational or scientific frame of mind, then you would either find yourself attracted to someone who is involved in such areas, or Neptune would subject you to its influence from another angle. More likely than not, such an approach would be more confusing as your dominant left brain (reason) is increasingly hard-pressed by your right brain (intuition), which Neptune is currently in the process of stimulating. At its extreme, such a need to find a rational explanation, as distinct from the truth, can lead to mental problems like absent-mindedness or feeling paranoid. As this happens, at some point you will own up to the fact that the mystical and mysterious is very much a part of life in general, and of your life in particular. This means that you recognize that we are affected by subtle or invisible elements, and that through relating to them on their own level, they not only become less troublesome but actually fascinating, uplifting, healing and enlightening.

Failing this, though, you will either just muddle through, blaming whatever confusion you are experiencing on the opposite sex, the political climate, family, or anything else that gets under your skin, or you will resort to some way of resisting it. Such resistance could take the form of one of those Neptunian 'substances', namely, drugs and alcohol. All these substances either increase our sensitivity or suppress it. Whatever is your particular poison, if you have not followed one of the positive expressions described above, then Neptune's Waxing Square could force you to 'up the dosage'. This means that you'd try harder to escape or block out whatever music it is that you are trying not to face. However, you may not even know that this is what is going on. The monumentally famous example of coming to this negative Neptunian nadir was the King himself, Elvis Presley. He died of a gigantic drug overdose during his Neptune Waxing Square, after years of refusing to recognize the true significance of his own vulnerability and of suppressing it with barbiturate abuse. Incidentally, note how artists/musicians and drugs/alcohol so often go together – Neptune is as Neptune does! Another world-class example of this was Al Capone who during Neptune's Waxing Square was paroled from prison for good behaviour, only to enjoy his freedom suffering from semi-paralysis and syphilis, dying seven years later at 48.

Here again we can see how, just as in life 'one thing leads to another', a Planetary influence occurring at one time in your life will in some way affect another happening later. The 'trajectory of inclinations' that was set during your previous Neptune Transit, Neptune Squaring Pluto (*see* page 193), now reaches a point when your strengths and weaknesses, developed at that stage of your 'Rocket Ship of selfhood', go critical (*see* 'The Story of Your Life' on page 15 and 'Your Uranus Cycle' on page 149). In Presley's case, if you read about Neptune Squaring Pluto, you will see that at this time he made his first important impression as a musician-performer. But true to Neptune and Pluto's hidden agendas, this was also the beginning of his undoing.

Earlier we referred to how experiences in 'past lives' could be affecting our thoughts and feelings in this life. Under Neptune's current influence such effects could become more 'active' or troublesome. So this could be a time when you become more aware of the possibility and significance of past lives. For example, you may find yourself going in for some 'regression therapy', where you are hypnotically induced and made to recall previous incarnations, along with how they bear upon your current state. This, or some other kind of encounter with reincarnation, is all the more consequential because it can tip you into that Neptunian ocean of possibilities that can be confusing in itself. Hopefully, around this age you have got at least one foot on firm ground, but there is still the possibility of losing sight of dry land.

Fascinating as involvement with past lives or any other kind of psychic experience might be, it can leave you open to all manner of crackpot theories or mystical cults. Possibly the only way to weed out such phoneys is to keep reminding yourself that there are no quick fixes or shortcuts to enlightenment – and keep a large pinch of salt handy at all times! Be that as it may, one of those witches or wizards, mystics or mirages, fakes or fairies, could lead you down that *Via Negativa*, that 'way that is not', I described in the introduction to this section. And as I said, this would all be par for the Neptunian course. However, there are certain individuals and sects around that can do lasting damage, so remember 'There is nothing you can get soon that is anything like Neptune'.

Neptune is very seductive. Feelings of euphoria can alight upon you out of the blue, possibly following some subtle remark or facial expression. But then, like the proverbial bubble, it bursts, either quite unaccountably or because some other little incident broke the spell. Just as likely will be occasional moods of lethargy or apathy, brought on by your (possibly unconscious) need to use your energy and time to pursue some goal more worthwhile than merely earning and spending. Nevertheless, occasionally feeling 'at sea' is more or less unavoidable. By the same token, you could attract time-wasters and drifters at this point.

Such openness and susceptibility can find many expressions. A friend of mine was having the time of his life in Rio de Janeiro during his Neptune Waxing Square, when one day on the beach he trod on the spine of a poisonous fish. This almost killed him, but made him question the somewhat decadent lifestyle he was leading. Also on the physical level, be wary of infections at this time because you are more susceptible generally.

Not least of all, Neptunian susceptibility is most prone to making you fall in love, as mentioned before. The famous pair of rose-tinted spectacles can get very easily jammed on your nose right now, and temptation abounds. Short of keeping in touch with common sense and the so-called real world, Neptune's agenda is so strongly unconscious that it is often impossible to see what's really coming. I have called romance 'Love's calling card'. In other words, if you knew the real reasons for which Neptune wants to get you involved with an irresistible someone, you'd probably run a mile. But you do get involved, caught in the famous Tender Trap, and like as not, that Neptunian reason will be to force you to face some cloud of yours, some aversion or addiction, or very likely, the inclination to be a 'doormat' or victim in a relationship. If you are an earthy and logical type then there is the possibility of being attracted to someone vague and sensitive, imaginative and mystical. Ultimately, you are learning compassion and tolerance – for others *and* yourself. Then and only then are you able to make a conscious surrender or a sacrifice, which is the positive side of the coin to victimhood.

Finally, it must be re-emphasized that the best way of dealing with this Neptunian time in your life is to pursue something patently spiritual, healing or creative. Apart from those already suggested, any practice that serves the whole, gets you more in touch with your inner being, or utilizes your imagination, is recommended. Even so, assuming that in this way you will encounter no reefs would be asking for trouble. It's the reefs and clouds that you're looking for, as it is they that are blocking the way.

On the opposite page you will see a flowchart of your Neptune Waxing Square, 'Fascinations and Illusions', which graphically portrays the challenges and opportunities that this influence offers.

JUPITER 'YEAR' for your Growth Mode

MOON 'PHASE' for Emotional Attunement

Having ascertained the actual years of your Neptune Waxing Square from the tables on page 289, then refer to the Age Index on page 235 for the Jupiter 'Years' and Moon 'Phases' occurring during that period.

Other Universal Planetary Cycles possibly active at this time

Check the tables on page 289 to see if any of these Planetary Cycles are active for you at the same time as your Neptune Waxing Square. Remember that any Trine influences may have to be actualized to make them that effective.

SECOND REALIZING MOON PHASE (*see* page 77) This Phase is inclined to spell out certain contrasts and differences that exist within your life in particular, and life in general. Seeing that your Neptune Waxing Square wants to fuzz the line between one thing or person and another, you could be in for some 'interesting times'. A probable experience is that of becoming involved with some-

A MAJOR INFLUENCE occurs between the ages of 38 and 43*

NEPTUNE WAXING SQUARE

Fascinations and Illusions

The flowchart below shows how the Positive and Negative Neptune Traits *(left)*, that you may or may not have, encounter Neptune's influence *(arrow on right)*, which in turn transforms the Positive Traits into a Positive Outcome *(top)* and relegates the Negative Traits to a Negative Outcome *(bottom)*. Negative Outcomes can be made Positive through developing the Positive Traits.

POSITIVE OUTCOME

Profound emotional enlightenment.
Inspiring and uplifting creations.
A life enriched through helping others.
Subtle ways of seeing and dealing with life.
Development of psychic abilities.
Opening your heart other than through breaking it. Releasing angst.

POSITIVE TRAITS

Expressing yourself creatively.
Admitting to selfishness. Making sacrifices.
Being prepared to go with the flow.
Practical spiritual/mystical pursuits like yoga/meditation.
The self-honesty and humility to look at your blind spots and self-undoing patterns.

NEPTUNE

Increased sensitivity.
Tempters and tricksters.
Undermining influences.
False or genuine gurus.
Creative inspiration.

Dissolving boundaries.
Conventionally unsolvable problems.
Past events calling for resolution and healing.

NEGATIVE TRAITS

Addictive or weak personality. Carelessness with health.
Escapist, overly romantic, gullible. Seducible. Martyr.
Rationalizing your way around your blind spots or simply refusing to admit you have any.
Stubborn, childish egocentricity. Willfulness.
Blind faith/spirital opportunism.

NEGATIVE OUTCOME

Throwing yourself away.
Infections. Psychosomatic complaints.
Confusion. Scandal and embarrassment.
Horribly compromising situations.
Robbing Peter to pay Paul.
Erosion of a more stable lifestyle.
Dissipation. Deception. Derangement.
Being trapped in a negative relationship.

*see tables on page 289 for when it is for you.

one who is quite different to you but who fascinates you. This would be saying that they embody that side of yourself of which you are not that aware, and which you need to get to know and accept. Sooner or later this may prove confusing and painful, but you stand to learn something very important about the way you behave and emotionally tick.

SECOND SATURN WAXING TRINE (*see* page 121) Having this Saturn influence prevailing now will definitely help you to keep your feet on the ground – or someone you respect will be at hand to make sure that you do.

SECOND SATURN OPPOSITION (*see* page 122) As Saturn does not take kindly to you kidding yourself in any way, coming down to earth with a rude bump is highly likely if you are or have been looking at life or others in a fanciful or deluded manner.

URANUS OPPOSITION (*see* page 168) This is 'Your Mid-Life Crisis', and if it happens alongside your Neptune Waxing Square it indicates some highly important point of reckoning for you. This is bound to mean having a few myths exploded, while at the same time exploring some new ones. It would be advisable to study well the texts for both of these influences, because the way in which you manage them is going to affect the remainder of your life.

PLUTO WAXING SQUARE (*see* page 226) Combined with your Neptune Waxing Square, this smacks of having to extricate yourself from some habit or attachment once and for all. At first, you will probably see such a wrench as impossible or unwelcome. But this Planetary twosome is saying either kick whatever it is into touch, or it will kick *you* into touch.

Neptune Waxing Trine | Between 52–58 years for 1–3 years

'SEEING THE LIGHT' – Two major qualities of Neptune manifest at this time in your life. The one is that Neptune, as the 'top of the Planetary tree', ultimately represents the highest and best of human nature: to have a sense of compassion, beauty and the sublime; to live and let live; to be at one with all things and all people. The other quality is that of mildness, passivity and non-action – all of which mean that ironically it is all too easy to miss out on the first major quality, especially because this is a Trine aspect which is a mild influence in itself. Again I am made to think of that cruel but true and funny distortion of one of the Beatitudes: 'Blessed are the meek for they shall inherit the Earth – if it's all right by everyone else!'

Essentially, the Neptune Waxing Trine offers you the line of least resistance, but in our culture, which is confrontational and self-actuating, such a path is often regarded as either ineffectual or malingering. But in positive Neptunian terms, following the line of least resistance is being able to detect the natural path to a goal or objective, or through difficult circumstances, rather like the river finds its way down to the sea. This simile is doubly appropriate because the critical issue with Neptune is that its goals and objectives are, like the sea, mysterious in that they concern something that is beyond our understanding and control in the usual sense of the words. Neptunian intentions and motivations are for the good of all, and so they tend to run counter to the selfish intentions and motivations that are conditioned into us. The fact that we dump all manner of poisonous waste into our oceans as part of our 'take, make and throw away' culture, illustrates how little awareness we have of everything the sea represents.

At this time in your life, you stand to discover that 'way down to the sea' through sensing where things want to go, as distinct from where your more linear and self-orientated self thinks they want to go. Where 'the river' wants to go (down to the sea) is best for you, simply because it is best for everyone – and you can't really work that one out in your head! All you can do is allow yourself to be led by reading the signs, following your heart, maintaining emotional balance, and attuning yourself to a higher level of consciousness. This is what Neptune's Waxing Trine offers you – and if you do not accept it then you probably will not notice much difference in your life.

Let us look at a method of 'going with the flow', of following the river down to the sea. I use this, not necessarily to entreat you to employ such a method, but in order to demonstrate further the subtle and spiritual way of Neptune – because, as I say, it is hard to grasp. There is a popular spiritual discipline called *Nichiren Shoshu* Buddhism which has as the central part of its practice a chant or mantra. This mantra goes *Nam-myoho-renge-kyo* (pronounce the 'myo' and 'kyo' as one syllable as in 'To-kyo') and is repeated over and over again many times. Its literal meaning is 'Mystic Law of the Lotus Sutra', but without going into how or why, essentially it is saying 'I wholly submit to what the truth of my being has in store for me'. Such a chant is rather like a self-hypnosis that trains every cell of one's being to follow the path destined for that being, much as a human embryo eventually grows into an adult human being which in turn forms a part of that whole called the human race. Going back to our original simile, it is like the river calling to the sea.

I'm swimming in the River swimming in me
I'm swimming in the River down to the Sea
I'm swimming in the River
With the River
Swimming in me

And so during this time in your life, providing that you have developed at least a measure of acceptance of the way you are and the ways of others, you find it far easier to steer round obstacles rather than meet them head on with all the hassle and energy wastage that such would involve. Being able to pick up on the best in people is also something that comes more easily now. If you are pursuing any creative, artistic activity then the flow of inspiration should be fuller and clearer during this time. The Muses are now very much with you – at least, in proportion to how much you are open to them, or how hard you are working at it. And as indicated above, some spiritual path that is particularly suitable to you can magically appear at this time. Yet this is very much a case of 'When the pupil is ready, the master will appear'.

So, all in all, you are now in line for feeling more a part of the cosmic scheme of things. Failing this, you should at least begin to detect that there is an easier way of living life, that there is, in fact, a 'way'. The beauty of knowing you have the Neptune Waxing Trine is that you are more likely to be alive to these subtle possibilities than you would otherwise be. It is as if a still small voice is calling, 'This way, it is far easier and infinitely better' – but you may need to ignore the roar of the traffic of your ego and other inferior voices in order to hear it. Concentrate, and that 'still small voice' will develop into your spiritual guide through life, taking you to undreamed of heights.

Overleaf you will see a flowchart of your Neptune Waxing Trine or 'Neptune's Line of Least Resistance' which graphically portrays the nature of the opportunity that this influence offers.

AN OPPORTUNITY occurs between the ages of 52 and 58*

NEPTUNE WAXING TRINE

The flowchart below shows how any of your Positive Neptune Attitudes and Actions *(bottom left)*, having attracted and made use of Neptune's influence *(Slingshot Effect within orb)*, allow you to attain certain Advantages *(top right)*.

REMEMBER – A trine only opens the way; it does not push or confront you, so you must deliberately recognize it and take advantage of it.

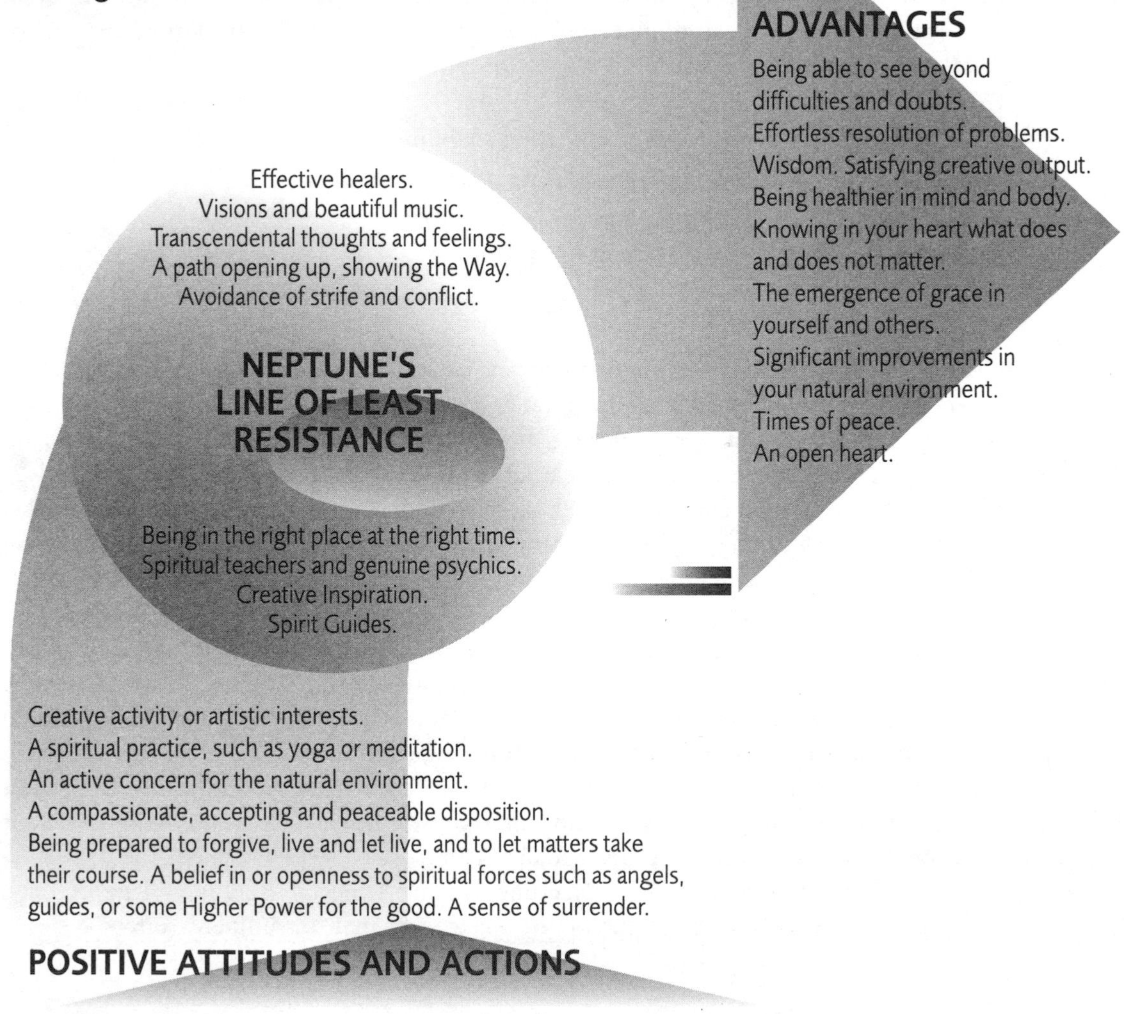

ADVANTAGES

Being able to see beyond difficulties and doubts.
Effortless resolution of problems.
Wisdom. Satisfying creative output.
Being healthier in mind and body.
Knowing in your heart what does and does not matter.
The emergence of grace in yourself and others.
Significant improvements in your natural environment.
Times of peace.
An open heart.

Effective healers.
Visions and beautiful music.
Transcendental thoughts and feelings.
A path opening up, showing the Way.
Avoidance of strife and conflict.

NEPTUNE'S LINE OF LEAST RESISTANCE

Being in the right place at the right time.
Spiritual teachers and genuine psychics.
Creative Inspiration.
Spirit Guides.

Creative activity or artistic interests.
A spiritual practice, such as yoga or meditation.
An active concern for the natural environment.
A compassionate, accepting and peaceable disposition.
Being prepared to forgive, live and let live, and to let matters take their course. A belief in or openness to spiritual forces such as angels, guides, or some Higher Power for the good. A sense of surrender.

POSITIVE ATTITUDES AND ACTIONS

*see tables on page 289 for when it is for you.

JUPITER 'YEAR' for your Growth Mode

MOON 'PHASE' for Emotional Attunement

Having ascertained the actual years of your Neptune Waxing Trine from the tables on page 289, then refer to the Age Index on page 235 for the Jupiter 'Years' and Moon 'Phases' occurring during that period.

Other Universal Planetary Cycles possibly active at this time

Check the tables on page 289 to see if any of these Planetary Cycles are active for you at the same time as your Neptune Waxing Trine. Remember that any Trine influences may have to be actualized to make them that effective.

THIRD EMERGING MOON PHASE (*see* page 54) At best this could amount to a kind of spiritual rebirth for you as the gamut of emotional experience that you have run now invites you to surrender into a new way of being and relating, mainly facilitated by a developed sense of 'live and let live'. Short of this, a more sensitive awareness now serves to ease life's pressures, particularly on the home and family front and, irrespective of your sex, with your dealings with females in general.

SECOND SATURN WANING SQUARE (*see* page 125) This makes it more imperative that you find that line of least resistance. If not, the difficulties you find yourself in can become quite debilitating and frustrating as you vainly wrestle with them, rather than finding a way around them.

SECOND SATURN RETURN (*see* page 126) Any finer or spiritual aspects of your life must now 'walk their talk' in that you may well be called upon to prove that your ideals can be lived up to. Alternatively, events could force you to develop a more evolved sense of spirituality in your life.

URANUS WANING TRINE (*see* page 175) Getting into a decidedly more spiritual and esoterically informed awareness of life is the theme here. Some, and possibly you yourself, would simply see this as a mellowing of attitude and behaviour. You now have it in you to allow things to breeze along more, to take it easier and not try so hard.

PLUTO WAXING TRINE (*see* page 231) If you have been 'on the case' in the sense of improving yourself and developing a more spiritual outlook on life, then this combination will elevate you to a higher level of seeing and being. People who have your deepest needs and interests at heart come to your aid.

Neptune Opposing Pluto | Between 51–76 years for 1–3 years

* The time that this influence occurs for you personally is quite variable. *See* the special table on page 352.

'INTIMATIONS OF MORTALITY' – During the early part, and most of the latter part of the 20th century, this influence occurs for everyone at the same time as or soon after experiencing Neptune Waxing Trine, 'Seeing the Light'. Astrologically, this means that the spiritual insights and development of the Neptune's Waxing Trine have been 'put there' (in the Cosmic scheme of things) to help you with these 'Intimations of Mortality'. This is because such 'Intimations' can present you with a feeling or fear of inevitable or imminent death. This could be for yourself, someone you know, or a more generalized sense of foreboding. Then again, if anyone close to you should die at this time, or should you see death close up, this would also give you 'Intimations of Mortality'. How any of this affects you personally would depend greatly upon your idea of death for yourself or someone else. Whatever the case, however, this influence and the (feelings of) endings that it can bring could be more positively interpreted as something in your life and the way you live having to 'die' – or more practically, as attending to the state of your physical health, in a curative or preventative way, rather than just worrying about it.

You can, and probably do, sense that there is something about yourself that is 'past its sell-by date'. This could be a way you have of not letting people getting too close to you, a certain superficiality that protects you, or possibly the way you habitually put off the inevitable. Whatever the case – and hopefully you are old enough and honest enough to admit to it – the time has come to trust someone to know you better, to find a new and better way of protecting yourself (*see* Protection along the Way on page 271), or to take the plunge into the unknown or whatever it is that entails the exorcism of your innermost fears or desires.

Failing this 'dying', your life could begin to feel quite stale and directionless. Rationalizing such a state of stagnancy as just being part of growing older would merely be your way of not letting go of the Devil you know. Perhaps you ought to tell yourself that the opposite to the Devil you know is not the Devil you don't know, but the *God* you don't know. This means to say that possibly for quite some time you have been sticking to some belief, attitude, aversion or habit – and the people, places and lifestyle that go with it – because you assumed that the alternatives were only degrees of 'worse' rather than 'better'. The founding father of modern psychology, Sigmund Freud, had a term for this condition: anal retention. So it could be time let go of your 'shit', that is, negative ideas of yourself, life and others that were formed in your childhood, and which you still cling to as a security blanket.

If at this time your experience of death is quite literally one of grief, then indeed you are having to let a part of you die so that it might be transformed. As has so often, and so rightly, been said – death is not an ending but a beginning, not a cul-de-sac but a gateway to another realm. Neptune Opposing Pluto is trying to put you on more intimate terms with this spiritual fact. You now have the opportunity to regard whomever you have lost – either now or at any previous time in your life – as still being very much in existence, but not on the physical plane (that is, not unless they have already reincarnated). As such, they can still be reached by you if let go of the part of you that misses them so on the physical level. Unfortunately, such physical attachment precludes a clear psychic connection with them. This is rather like grabbing the arm of someone jumping into a lift just as the doors are closing. While you hang on to their arm, the lift cannot move and neither can you have more than just a frustrating connection with them. Let go and they can go to the next level where they are free to communicate with you from their level to yours, and vice versa – as opposed to being 'lost and in between'.

JUPITER 'YEAR' for your Growth Mode

MOON 'PHASE' for Emotional Attunement

Having ascertained the actual years of your Neptune Opposing Pluto from the table on page 289, refer to the Age Index on page 235 for the Jupiter 'Years' and Moon 'Phases' occurring during that period. If one of these Jupiter 'Years' is 'Pisces' then this would further accentuate the influence of Neptune Opposing Pluto in that year of you life.

Other Universal Planetary Cycles possibly active at this time

Check the tables on page 289 to see if any of these Planetary Cycles are active for you at the same time as Neptune Opposing Pluto. Remember that any Trine influences may have to be actualized to make them that effective.

SECOND SATURN RETURN (*see* page 126) As this is a time when one can 'feel one's age', the above description of Neptune Opposing Pluto is all too applicable. Mortality and the question of what has (my) life been all about tend to preoccupy your thoughts. But again, what you make out of this can vary from seeing it in a depressed and hopeless way as a result of not having a positive philosophy, to being inspired by the inevitable and by the wisdom that can be in evidence during the last chapter or so of your life, gathered through your past experiences.

URANUS WANING TRINE (*see* page 175) If you already have a positive handle on Neptune Opposing Pluto, then this influence could bring the knowledge or people that will enable you to develop it.

URANUS WANING SQUARE (*see* page 177) This influence occurring now makes it more imperative that you wake up to the idea of there being more to life than meets the eye, that you realize that all of it is in the eye of the beholder. Without such a transcendent or youthful viewpoint, you could be in for a rocky ride as changes and disruptions attempt to jolt you into this awareness.

NEPTUNE WAXING TRINE (*see* page 204) Having this happening now should give you an almost poetic awareness of the pattern and beauty of life on earth. Not to see this must mean that cynicism has become deeply ingrained. This may have some justification, but the challenge here is still to see past what apparent misfortune means in the conventional sense.

PLUTO WAXING TRINE (*see* page 231) Your sense of mortality should be positively uplifting with this influence happening alongside Neptune Opposing Pluto. Nevertheless, there is always the possibility of past experiences having bred in you a sense of there being 'nothing left'. To this I can only entreat you to delve deeper, for your sense of depth and profundity now stands to be made all the greater.

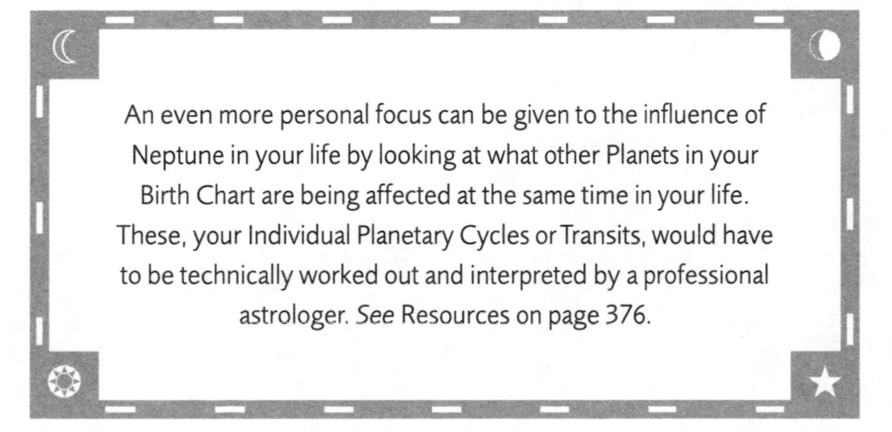

An even more personal focus can be given to the influence of Neptune in your life by looking at what other Planets in your Birth Chart are being affected at the same time in your life. These, your Individual Planetary Cycles or Transits, would have to be technically worked out and interpreted by a professional astrologer. *See* Resources on page 376.

The Wave, the Wave the Dolphin rides –
Of Sirens' sad, alluring lays
We lost Sailors dream
In our deep womb of green –
The Dolphin sighs, and plies his way

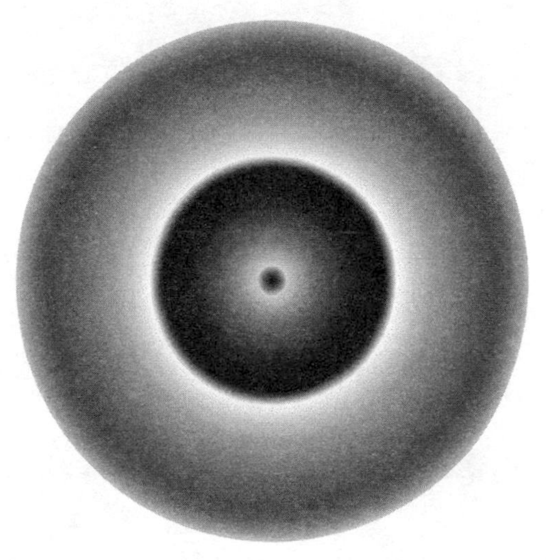

YOUR PLUTO CYCLE

Endings and Beginnings

Delve in my moist and shady den
And you will find true wealth within.
Feel your darkness, feel your pain,
Fall to me and rise again –
Immortals steeped in innocence.

Key Influences

+	-
Regenerating	Degenerating
Influencing	Manipulating
Deepening	Degrading
Eliminating	Annihilating

Pluto is the Alpha and Omega influence in our lives, the Death and Rebirth Cycle of inevitable endings and new beginnings. It is the Seed of Potential in all of us – our Fate or Destiny, what we could be, what we are supposed to become. But like the Underworld which is Pluto's domain, for most people their destiny is hidden from them, and is something for which they have to delve or be thrust into the depths of their being if they wish to discover it – or even if they don't wish to. Only the famous or infamous, the powerful, are usually regarded as being 'men or women of destiny' who influence the thoughts and feelings of a whole culture, transforming it as they do so.

And so the Pluto Cycle is geared and timed to give us clues and experiences that intimate what our individual fates are about. The trouble is that most of us are frightened, probably unconsciously, of living out such a profound thing as a 'destiny'. In the 20th century (that of the planet Pluto's discovery), the most popular book has been J R R Tolkien's *The Lord of the Rings*, in which a band of hobbits, little creatures living little lives in a little place called the Shire where very little happens, are suddenly thrust into a great drama of Good versus Evil. Most of the other hobbits live their lives gossiping, blaming and concealing themselves. This perfectly reflects our unconscious longing for a life more moving and profound, but which is something that we can, it seems, only live out vicariously through pieces of writing, or more recently on film, television (especially 'soaps') or the 'wannabe' pop culture.

However, none of this means to say we should all become 'men or women of destiny'! It is really a case of Pluto desiring that we connect with our own core sense of power and purpose, as opposed to hiding ourselves away in some hobbit-hole of security at the expense of limiting our potential. *In all of us there is there is the power to influence the world to a small but significant degree*. The problem, though, is that in order to do so, we have to confront whatever it is in ourselves that obstructs our connecting with that powerful 'inner truth', which is probably a better term than 'destiny'. The Plutonian metaphor that suggests itself here is one of mining for gold. In order to tap into our personal reservoir of power, we have to engage with the rock and rubble of our past, and penetrate it. In depth psychology (also Pluto's realm) this is called our Shadow. Our Shadow is, in the negative sense, our personal store of 'things that been did and things that been hid'. It is also our connection with the collective shadow of mankind's murky past, the Sodom and Gomorrah of our unconscious mind.

All this brings to mind the Three Monkeys' 'Hear no evil, see no evil, speak no evil'. Meanwhile the power mongers of this world carry on plundering and polluting it, perpetuating the 'take, make and throw away' culture that is destroying us. But Pluto is not merely some programme to enlist your help in restoring the Earth in the usual ecological sense. It is far more fundamental than that, for it says that the first and foremost thing you must purge and regenerate is yourself. The gigantic waste, refuse and recycling problems of our world only reflect our individual ones. This could be what is meant in the Bible when it says we are all 'born of sin' – we are all born with something in us that needs eliminating, often through purging, in order that we might discover what has the power to change the world for the better. Pluto's edict is to ensure that we should all endeavour to do this, and in so doing create a critical mass of transformed and transformative human beings that can make a positive difference, initially to ourselves, and ultimately to the world around us.

So, whether we like it or not, there is this seed in all of us that is, to varying degrees, wanting to germinate and develop – like an acorn into an oak tree. And like a journey down into the Underworld, Pluto as a Planetary influence will sink us into situations and relationships that deepen and intensify our sense of who and what we are, eliminating what we thought we were in the process. But if we still persist in a hobbit-like way of resisting it, Pluto has to find a more circuitous and seemingly devious route to call our attention to the fact that there is this power and destiny to life in which each of us is included, not merely invited to. And so as with other Planetary influences, the ways in which we are made to do this are many, and on first impression, may not seem that connected.

How Pluto's Influence Manifests in Your Life

Here are some of Pluto's favourite areas of interest, where it can most easily make its powerful and invisible decrees felt – where one door closes while another one opens, even though sometimes we get left alone in the corridor for an uncomfortable while.

♀ **Intimacy** – Becoming closely involved with someone, or something for that matter, is probably Pluto's prime predilection as far as making its influence felt. In the vast majority of cases this means sexual intimacy, for as soon as we are on sexually intimate terms with someone we have effectively interpenetrated each other on more than just a physical level, but also on a psychological one, where Pluto operates. Anyone who thinks that penetrative sex makes little or no difference to a relationship will eventually feel Pluto's grip on their collar in some other, heavier, fashion. Sexual involvement changes us, which is to say that staying sexually involved with one person for any length of time has the potential to cause a part of us to die. That part is what it is in each of us that resists a merging with another through not offering up certain fixed emotions and thoughts which get in the way of union. These thoughts and emotions are the 'rock and rubble' referred to above, those negative or outworn attitudes which must be intensified, brought to the surface, and got rid of. This is a painful process, but not identifying and relinquishing these blocks will ultimately spell the death of the relationship itself, or else make it into a living death. Progressively passing the purging tests of Pluto means that we progressively create a deeper and deeper bond with our sexual partner. It may well be the case that it is the actual manner of sexual expression that must change, or the sexual side of the relationship will die. That crude injunction 'What he/she needs is a good fuck!' is in essence often quite true, for it is sexual involvement that shifts personal rubbish and narrow attitudes like nothing else, although of course such a statement is not usually uttered with this psychological sentiment in mind.

Any discourse on this dimension of Pluto's realm would not be complete without mentioning its prime biological purpose – the procreation of the species. Suffice to say that this most basic product of sex, children and parenthood, is or should be the most life-transforming event of all in anybody's life. So in any event, Pluto uses sexual intimacy and desire to entice us into a deeper intimacy with ourselves and our inner truth, that place where the nucleus of our being dwells, the very root and power source of our being. But getting there can, as I say, be too much for many people, for their fear of intimacy (whether with another or with their own inner being) born of some psychological maladjustment to which we are all more or less subject, aborts Pluto's mission. That is, until the next time, or when being desperately alone becomes unacceptable.

♀ **Obsessions or Convictions** – An obsession is an unconscious and therefore somewhat out-of-control conviction. A conviction is a deeply-seated feeling or belief that acts as a highly valuable motivation and guidance system. Pluto's influence here is aimed at transforming an obsession into a conviction, or eliminating any conviction that is now inappropriate and consequently misguiding. As usual, Pluto will firstly intensify whatever it wishes to transform, take it to critical

The glyph of Pluto ♇ is composed of the symbol of the Cross, which symbolizes Matter or the World, below, the Circle of Spirit above, and the Semi-circle which symbolizes the Soul caught between the two. So the glyph symbolizes the process whereby Spirit or Energy is converted into Matter, that is, how an idea or intention is forced to manifest itself as something real, inevitably, through the agency of desire. Biologically, this is the sperm penetrating the egg, and bringing forth physical life, through the medium of desire – with all that such a process involves. In reverse, that Matter or physical being eventually dies and returns as pure Soul to the Spirit to be Energy once more, and in terms of the memories and ideas that have been left behind. The Cycle then begins again. The Second Law of Physics, the Law of the Conservation of Energy, echoes Pluto in stating that energy neither is born nor dies, but manifests continually in different forms. Pluto and this Law simply assert that everything, be it spiritual, physical, mental or emotional, is destined to be transformed into another state, through the medium of sheer existence.

mass – that is, a state which cannot be endured or maintained – and then either achieve its objective or abort mission. A positive outcome depends absolutely upon becoming aware of what the obsession or preoccupation is actually telling you and acting upon it, or in the case of a conviction, recognizing that it is past its sell-by date and letting go of it.

Obsessions can be over anything, such as another person, health, body image, or any subject that calls intense attention to itself, or rather to what lies behind it. Some obsessions, for example a consuming interest in study, will eventually lead to some kind of accomplishment and will burn-out having served their purpose, simply because they had a purpose in the first place. At the other extreme, conditions which could be loosely or clinically described as obsessive compulsive disorders

– anorexia, bulimia, or some such other serious complaint – would obviously require therapy, being indicative of some deeply-seated psychological disturbance. Obsessions in the middle area between those two extremes, which are naturally the most common, need addressing, but in a way that can either be done by oneself, or with a little help from a friend, loved one or counsellor. If, for example, you have an obsession with another person, or another person is obsessed with you, the secret is to track down the actual feeling or state that the object of the obsession represents, much as one would interpret a dream figure. For example, being preoccupied with someone who was unavailable could reflect your own reluctance to commit, for whatever reason.

Someone being preoccupied with you could indicate that they were calling your attention to something that you were refusing to look at and accept in your own personality. An extreme example of this is 'stalking', which would mean that whatever it was in yourself that needed looking at required urgent attention. With something as grave as harassment or rape, the issue of how much of a part the victim is playing is a very charged one. Obviously the perpetrator has a problem, but from an astrological point of view, a victim is also seen to have some Planetary influence in play that simply suggests that they had, for some reason or other, strayed into Pluto's realm. There is nothing new in this; the Greco-Roman myth, the Rape of Persephone, is about an innocent whom the Lord of the Underworld, Pluto, took by force into his realm. You can read up on this myth in order to form your own idea of what it means, but the classic meaning is that there is a depth, power and darkness to the female that at some point she must be immersed in, one way or another, and thereby be transformed into a stronger person. Indeed, this is often the case with rape victims. This interpretation is based upon the 'mythical fact' that Persephone's mother, Demeter (the Earth Goddess), was so powerful as to have dominion over not just the surface of the Earth, but the bowels of it too. So Persephone represents that deep and hidden part of femininity that Pluto made conscious through laying claim to her. He then made Persephone eat the seed of a pomegranate, and thus established an indissoluble bond with her. This underlines the earlier point concerning the anguish and depths that one has to go to in order to find true union with another being; Pluto and Persephone had one of the most stable marriages in all mythology. Interestingly, Persephone's name before her enforced transformation was Kore, meaning 'maiden' or 'unmarried girl' in Greek, which can be seen in English as a pun about Mother Earth – Earth's *core*. None of this is to say that, as far as rape is concerned, a female (or a male or a child, for that matter) is 'asking for it' in the way that is crassly put forward by some people, but to emphasize the astrological fact that there is always a reason for anything happening to anybody, especially when Pluto (that is, fate) is involved. Or it can be an event that points to the fact that our society has no control over our collective Shadow, and that therefore we cannot protect our innocents from the Shadow figures that are abroad to haunt, molest or kill them. Neither should one expect to be stalked, raped or harassed when Pluto is abroad, but at the same time, it would best to avoid dubious situations and people, and be aware of your Shadow (*see* Shadowlands below).

With regard to convictions that are no longer appropriate, they are a menace because by their very nature they often have a history of being reliable and essential to one's existence, and so one is strongly attached to them. If such is the case, you have to have the sense to see that if a conviction or value is creating more trouble than you and yours can bear, it is probably time to let go of it. In time, you will then find that the very reason for having such a conviction is no longer necessary. I had a white client whose daughter was going out with a black man, and he was strongly of the opinion that blacks going with whites was a bad idea, while not actually being prejudiced against blacks themselves. With difficulty, I managed to help him let go of this 'conviction' and bless his daughter's partnership. He is now the proud grandfather of two coffee-coloured kids!

♀ **Shadowlands** – Pluto's domain, the Underworld, was also called the World of Shadows.

My realm wherein all rots and dies,
The beauty above ground belies.
The blooms I feed, your world to grace,
But you dare not behold my face,
And lose yourselves in vain pretence.

We all have a dark side to our nature (the 'Id') which, for one reason or another, we keep hidden. Socially, this is largely desirable and necessary, but the trouble is that when such thoughts and feelings are denied and driven underground they have a habit of surfacing when we least want them to, and possibly in a guise that we do not recognize as having anything to do with us. The vital Plutonian point is that these thoughts and feelings, our Shadow, have great potency, and without them we can eventually become weak, unstable, and lacking in authenticity or the power to survive; they also give life and vibrancy to our outer expression. Sexually, for example, the desire for a 'bit of rough' or the caveman/tart type is the Shadow making itself felt.

More particularly, however, when Pluto is abroad our Shadow can manifest physically. Someone I know who was very angry at the world, but for all the world appeared to be a soft and easygoing guy, was cut up by a van at a road junction while out cycling. Incensed, he beat upon the side of the van, whereupon a group of thugs leapt out the back and chased him. Fortunately, being on a bike, he was able to lose them through narrow alleys. The Shadow can also manifest itself through some kind of involvement with the 'underworld' of vice, crime or sleaze. But our Shadow can play itself out in as many ways as there are personality expressions, and most not as obvious or scary as this example. Whatever the case, if Pluto presents you with something or someone that frightens, offends or uncomfortably attracts you, then be aware that your own shadow is being projected on to the outside world and that you best become aware that it is definitely something to do with you. But our social conditioning to be 'nice' or 'civilized' is so strong that we find this hard to do. But it is doubly advisable to endeavour to do so, because on the one hand we are then no longer manipulated by or in the power of someone or something else, and on the other, we stand to reaffirm that part of ourself that is not neurotically concerned with what the world thinks of us. The Pluto part of ourself exists in its own right, which accounts for our feeling so alone in there and so possibly compromised by a need for company and approval. Indeed, feeling or being alone can be very much a part of Pluto's influence in our lives, with a view to becoming at ease with our aloneness because we are at ease with our Shadows. At another extreme though, is the possibility of caring too little what others think, and so Pluto would then make such insularity too lonesome, thereby demanding that one come out of one's cavern and be more open and less antisocial.

Possibly the most disturbing thing about all the Shadows that are denied is that they build up into a 'Collective Shadow' which in turn seeks to play itself out through some unstable individual of low vibration. This is the psychopath, of whom sometimes people are later heard to say, 'he was such a nice/quiet man, but a bit of a loner'.

♀ **Psychology and the Occult** – Much of the above might appear strange or unlikely when viewed from a conventional or superficial standpoint. This is because the 'ordinary world' is more or less oblivious to the reality of the Underworld, that there is fundamental power to life on Earth that underlies and influences everything we say, do and feel – 'One ring to rule them all . . . and in the darkness bind them' (*The Lord of the Rings*). Nothing and no-one is independent of this power, much in the same way that the state of a subterranean water table will affect all the surface water in that region. Any brand of psychology that lives up to its name (it means the 'study of the soul' not 'of the mind' as it is lately so often misinterpreted and applied) is aware of these deep and dark forces. The Occult, a word which simply means 'hidden', is misunderstood by the general public even more – as is the fact that the two subjects are just different aspects of the same thing. And the Occult in itself is neither just black magic nor white magic (any more than a language is good or bad) but the area of practice and study that appertains to these forces that bind us all – such as fear and desire. The occultist learns how to contact and utilize these forces, but it is entirely up to him or her as to whether they do this towards selfish and negative ends or spiritually positive ones. Likewise, if Pluto's influence leads you into involvement with psychology or the Occult on any level, it is the condition of your desires and fears that will determine what knowledge and experience of it you attract, and how you use that knowledge and experience. With regard to the Occult, beware that if you choose the negative or dark path you will invoke evil, something which is akin to the sump of all humanity's fears and misdeeds, and it *is* characterized by demons and poltergeists. Follow the positive and you will empower yourself to heal and beneficially influence others; and this realm *is* inhabited by angels and other helpful forces.

♇ **Death and Rebirth** – Everything and everyone must die. This is the most inevitable fact of life. Although this is so, it does not mean to say that every Plutonian influence is going to bring about the death of a person or animal – in fact, most of them do not, for if they did we would have all died out long ago! But the chances are that something, in the sense of an inanimate object, like a machine or device, or a belief or relationship, a job or home, or simply anything of an abstract nature, will come to an end in some way. Or a *part* of it could die and so render it in need of repair. All this naturally applies to your own personality too, for as has been mentioned above, the death and transformation of certain aspects of your life and being are central to Pluto's purpose, as is the coming into being of something fresh or refreshed. With regard to death, literal and physical, the same Plutonian law applies in that death is not merely an end, but a beginning as well. Reincarnation, when viewed in this light and especially the meaning of Pluto's glyph on page 215, can be seen to be a fact of physics, for in quantum terms everything is energy, and a human being is a package of energy. Looked at positively, it is often a death or end of something that releases us into a new chapter of life. To a smaller or greater degree, this might come as some form of disaster – indeed, the word means 'star of destiny', which sounds like Pluto itself. A lesser form of disaster is a crisis, something else which is essential to the regenerative process of Pluto, and which I have referred to as 'critical mass' elsewhere.

♇ **The Powers that Be** – Government agencies and big corporations, and the people that run them (which includes police and armed services), can figure in your life, directly or indirectly, during Pluto Transits. It may simply be that your lifestyle is made better or worse by the activities of these powers that be, or then again, you might find yourself more closely involved with them – after all, some can be described as Plutocrats. Money itself comes under this heading, and your financial situation can be boosted or become in need of an overhaul during such times. On a more esoteric level, the influence or interference in your life of the state or a corporation can be viewed as commensurate to the amount you recognize them as actually *having* the power. Insofar as we tend to shy away from our destiny or inner truth, we are also denying the independence that our own personal power would allow us, and so we become more dependent upon these so-called powers that be.

♀ **Intrigues and Conundrums** – Complex and convoluted relationships or situations can arise because the issues that created them have to be rooted out and sorted out. There is no easy or superficial answer. Again this entails looking more deeply into the motivations that one has for being involved, and to one's role in whatever that case might be. Hypocrisy and secrecy can abound, as can lying to conceal one's truth from oneself, along with the damage that these things do. The answer is to be as straight and truthful as possible with everyone, starting with oneself.

♀ **Plumbing** – Pluto governs any kind of hidden watercourse, and so improvements or problems can arise as Pluto attempts to draw attention to your inner emotional workings in a more roundabout way. 'Plumbing' can also apply to your own inner piping, such as intestines and reproductive organs, and the same symbolism would appertain. Digestive problems reflect difficulty in letting go of something or someone, possibly from the past. Pluto governs any kind of waste. Complaints of the reproductive system can mean a number of things, but creativity or sexual attitude are the main ones. Menstrual problems could well pose a combination of these, but they are all the more complex because they can reflect our culture's misunderstanding of what being female truly means; like for example, the Latin term for it, *cursum* (flow or course), somehow got perverted into 'curse'.

♀ **Pluto Types** – include people who are Scorpios, or have a Scorpionic/Eighth House emphasis or strongly placed Pluto in their Birth Charts. In ordinary terms, such types would be genuinely powerful, deep, profound, psychologically and/or sexually aware and compelling, on the positive side, or manipulative, inaccessible, power-hungry, disempowered, cruel and tyrannical, on the negative side. Their professions could include such things as psychologist, police officer, criminal, occultist, diver, miner or plumber, or any occupation dealing with death, sex, birth, big business, secrets, destruction or waste. As such, they would be the 'agency' or figure that brought Pluto into your life through possessing the qualities associated with this Planet and the Sign or House it rules. (There is more about Plutonian agencies or figures and how to deal with them in Planetary Effects, Figures and Responses on page 259).

In the end, destiny is like that the river surrounding the Underworld, the Styx, coursing through our unconscious minds, and occasionally surfacing to submerge or wash away the world we knew in order that it might make way for a new or renewed way of being, and so get us somewhat nearer to discovering not just our particular reason for being on this planet, but eventually the profound reason for the existence of the Earth herself. Also, when this river surfaces it also gives us an access to our inner being and that all-important inner truth – although this surfacing occurs in stages, until one is ready to really look at oneself. After all, Pluto is the Underworld, that is, everything this planet is – or you are – apart from what it is seen to be on the surface. And this conceals the traces of who and what came before us, and therefore where we came from, in the sense of Earth mysteries and ancient civilizations such as the Egyptians, Mayans and ones even less well-known or not known of at all – as yet. Somehow our common roots and our individual roots are bound up together, and in delving into the one we ultimately discover the other.

Here follow the significant periods and meanings of your Pluto Cycle

★ Refer to the tables on page 289 to determine what ages are active for you personally (one to three years) within the age span given for any particular influence. Any influence will not be continuous throughout that period, but off and on, usually comprising three peaks. Occasionally you may notice the 'Stretch Factor' when you experience an influence for somewhat longer than the period that has been technically determined.

★ Pluto's Eccentric Orbit – owing to Pluto's highly irregular cycle, note from the Age Index and the tables that Pluto's active Cycles can occur earlier for pre-1919 births, or later for pre-1937 or post-1984 births, than the ages given in the headings for the interpretations and flowcharts that are set out below.

★ Remember that there is also the Individual Cycle of Pluto to be considered (that is, Pluto's Transits to other planets in your birth chart that occur throughout your life), but these would have to be calculated by an astrologer or with an astrological computer program. If some of the foregoing has struck any chords with respect to certain life passages, then this is possibly because such Pluto Transits are what you are or were experiencing.

Pluto Conjoining Neptune | Between 22–23 years for 1–3 years

'THE INSPIRATION BOOSTER AND THE ILLUSION BUSTER' – The one to three years during which you experience this influence are vitally important to your development as a creative and evolving individual. This is like the second stage of your 'Rocket Ship' that fought to achieve 'escape velocity' from the negative aspects of your cultural and family background during your Neptune Squaring Pluto and/or First Uranus Waxing Square back in your teens or early twenties (*see* pages 193 and 162).

This time, the 'thrust' that you created then in terms of your mental and emotional development and direction gets another chance to enhance itself, an opportunity to boost itself. It also gives you the opportunity, or rather poses the necessity, of unloading some baggage (in the form of past attachments, addictions or regrets) – or the baggage that you should have unloaded back then. This may also involve having a 'second adolescence' as you do, or finish off doing, what you didn't do or complete at that time. This quite possibly entails sexual experimentation or sowing some wild oats. If you are committed to a partner at this time, this would either compromise such a need, or render it unnecessary because you had already 'done the rounds' and had settled down. But another alternative is that the relationship you are in no longer seems to satisfy you or mean enough to you. If this is the case, then the opportunity to embark upon a new one, or some other kind of venture, is very likely to present itself. It may even present itself when you think your existing relationship is okay, which means you'll have a crisis of confrontation on your hands. If this should happen, it is Pluto telling your Neptune that your present relationship has been living a lie in some way. Usually that lie is born of staying with the 'Devil you know' because one or both of you do not want to look at the real reasons for being with the other – reasons that are suspect in some way, hence the reluctance to look at them. By 'suspect' I mean habit, security only, fear of being alone or not finding anyone else, or some such negative reason for remaining in a relationship. Alternatively again, it could be your partner that wants out, but this would amount to the same thing – it is time for you to move on.

On the other hand, you may decide, or be forced to decide, to regenerate an existing relationship by clearing out those cobwebs of negative reasons, and rediscovering why you got together in the first place. You would have to be brutally honest in doing this, otherwise you'll go round in circles, aimlessly drift apart and back together again, and generally create an awful lot of heartache – and possibly not just for yourselves. All of this would come under the heading of the 'baggage' referred to above, for there could be childhood reasons for your being in that existing relationship, like for example, having a partner as a mother or father substitute. If such is the case, you'd have to discover if you are more to one another than that, or come to terms with the need for such a substitute. In any event, you'd need to track down the illusion that is infecting your relationship.

Alternatively, at this time, not being in any relationship at all could be felt to be unacceptable – as a result of there being some change on your social scene, like a friend moving away, for example. This would also pose some childhood issue that needed looking into. Yet again, a brand new relationship could offer itself to you, and you'd have to 'clear your decks' in the sense of clearing out your 'single' ways.

In any event, what is happening now is like a power surge from your unconscious mind, which means to say that you are receiving an important one of those Course Corrections referred to earlier, which more likely than not involves quite dramatic changes in a number of areas of your life. It is all in aid of getting you more attuned to who you are, where you've come from and where you're supposed to be going. In the process of this happening therefore, you have to dig a lot more deeply in order to understand yourself, through such pursuits as psychology, astrology, creative expression, or anything that gets you more in touch with the real you. Of necessity, this involves the gradual, yet sometimes quite drastic, elimination of the old you, with its relatively superficial values and pastimes. At the same time, you'll go through an intensification of your ideals and yearnings, and a keener suffering because of your illusions. However, you probably won't notice just how much you have changed over this period until some years hence. Indeed you will look back then, and regard this as a major turning point in your life; so it is advisable to live through this time with a sense of destiny being at work, and not to treat it lightly.

The actual experiences that you have now can vary enormously, but they are probably deeply moving, with moments of great significance and intimacy. (Reading or reviewing the areas of Pluto's interest given above in the introduction will be of help here.) There are also times of acute pain as you strive towards whatever you are trying to express or make contact with – or as you wrestle with what you're attempting to evade! Be mindful that it is these very moments of poignant pain and pleasure that are a testament to the fact that you are a lot closer than usual to the heart and soul of the matter. And to differing degrees, this roller-coaster ride will be shared by other people of your own age, for this is a generational influence. But you experience it very much in your own special way, or at least you should, for this influence is in aid of putting you in touch with what is essentially 'you', and certainly not about going along with the crowd, although at times paradoxically you find yourself being wonderfully carried along by a tide of collective emotion. So mixing with people of a different age group than your own is a good idea now, for the contrast will give you a sharper idea of where you're at. You are now setting a trend for the rest of your life, that will profoundly influence your future work and relationships: so enjoy, endure and envision.

On the subject of work, at this time the issue of ideals and the current state of your emotional sensitivity can come to the fore. This means that whatever work you are doing will need to have something about it that satisfies your longings and spiritual goals. There are several notable examples here of people who made a high profile out of their principles and beliefs at this time of their lives. At the age of 29, Muhammed Ali's conviction, for refusing to be drafted into the US Army on religious grounds was reversed, thereby vindicating him and showing him to be a man that could stand his ground outside as well as inside the boxing ring (*see* page 359 for more about Ali). At 27 years of age, Yasser Arafat co-founded the Al Fatah movement, a terrorist group formed to assert his people's right to a State of their own (Palestine). However, demonstrating the delusory but ultimately peaceful nature of Neptune, he later renounced all connections with terrorism (*see* page 126). Along similar lines of idealism, at 26 years of age Fidel Castro took part in attacks on two army barracks in Santiago de Cuba, for which he received a prison sentence, as the beginnings of his ultimate overthrow of the decadent Batista regime six years later. At 27 years of age, Martin Luther King led the famous 381-day boycott of Montgomery city buses in the name of black civil rights (*see* page 52). Black activist Malcolm Little joined the Nation of Islam at 27 years, taking the name of Malcolm X (*see* also page 199). John Lennon returned his MBE in peace protest at 29 years of age (*see* page 171).

At a more ordinary level of existence, idealism will manifest in a more mundane way. So if you are in a dead-end job, then you could well feel extremely discontented with it and go for something closer to your heart than just your wallet. This may involve the arts, music or some form of healing or mystical practice. But be wary of the injunction 'Don't give up your day job' because there is the possibility of your possessing a lot more aspiration than ability. So consider any such intentions carefully, perhaps embarking upon some suitable course while you're still in secure employment. Then again, you may not be in work at all, and this Transit will force you to focus upon why that is, possibly you are daydreaming, unmotivated, lazy or apathetic. You might just have to face reality and bite the bullet, especially if this period coincides with your Saturn Return (*see* below). Then again, such a predicament could lead you to some 'service' such as voluntary or charity work.

Yet another alternative is that you already are involved with a Neptunian pursuit such as the performing or visual arts, writing, healing/helping, or some patently spiritual interest. If this is the case, there is going to be a shift in how you go about expressing this. This is the 'Inspiration Boost' and it can take you to greater heights and/or new forms of self-expression. There are several examples of significant points being reached in the lives of famous performers at this time. Greta Garbo began drama study at 17 years of age – the Pluto Neptune Conjunction occurring somewhat earlier for her generation. Marilyn Monroe's first big film success *Gentlemen Prefer Blondes* had its premiere when she was 27, a few months after she admitted to being the nude in that famous calendar spread. (*See* page 359 to be directed to more concerning both Garbo and Monroe.) Elizabeth Taylor won her first Oscar for *Butterfield Eight* at 29, while at the start of her Pluto Conjoining Neptune she changed her religion to Judaism. Marlon Brando appeared in his first film, *The Men*, at 26 years of age. Madonna released her first album at the start of Pluto Conjoining her Neptune. Finally, here is an example of the transformative power (Pluto) in artistic terms (Neptune) that this influence can bring. At this time in his life (1969), Bob Dylan's surreal hard-edged lyrical and musical style changed to a romantic sentimental one with a country backing. Incidentally, this followed upon his near-death, neck-breaking, motorcycle accident which occurred under the Individual Planetary Cycle (not included in this book) of Neptune Opposing his Moon in Taurus, the Sign which rules the neck.

By and large then, this is a potentially very rich time in terms of connecting with your true life track, but you have, as I say, to deal with your illusory goals and ideas in the process – or sometimes before you can proceed at all. In my own life I was aspiring to be a singer/songwriter during this Transit, but I got into astrology at the same time. Through quite a severe bout of disillusionment I had to admit that I didn't quite have what it took to be a 'star', and got into another kind of star-dom instead! But it was only after my musical ambitions had thoroughly receded that I was able to see and follow my true star.

Pluto Conjoining your Neptune can also bring mystical or peak experiences. These occurrences happen quite spontaneously, as the result of some kind of trauma or drug, or occasioned through some metaphysical practice. Whatever the case, they have a life-long effect upon one's idea of life and self. As these are usually quite personal it is hard to cite examples, but I know of such things as UFO sightings, visions and important dreams, through my own or friends' experiences. Such events may also involve the sea in some way.

Another kind of peak experience was enjoyed by Queen Elizabeth II when she ascended the British throne in 1952, being crowned on television the following year. Yuri Gagarin, at 27, hit a literally high point by being the first man in space (*see* also page 359.)

Finally, and fascinatingly, Pluto Conjoining Neptune could manifest in your life as a baby. Any major Transit, or Universal or Individual Planetary Cycle, which is actually happening to you as a parent (far more so if you are the mother), can be felt and expressed in this way. This means that the nature of a baby born and/or conceived during this time will take on the characteristics of Pluto and/or Neptune. Depending on other factors, such as the personalities and situations of the mother and father, such a baby will progressively display the strengths and/or weakness of such Planetary qualities. On the positive side, the baby can be felt to be special in some way and grow up to have great psychological insight and power (Pluto) and/or unusual sensitivity and creative gifts (Neptune). On the negative side, there could be an insecurity and 'cut-offness' present which alienates the child (Pluto) and/or a weakness or even an affliction of some kind (Neptune). Needless to say, there can be a combination of both the positive and the negative. Although such a child is a major proportion of the symbolism of Pluto

Conjoining Neptune, you as an individual or as a parent can also experience it in your own right, in the ways described above.

On the opposite page you will see a flowchart of Pluto Conjoining your Neptune, which graphically portrays the challenges and opportunities that this influence offers.

JUPITER 'YEAR' for your Growth Mode

MOON 'PHASE' for Emotional Attunement

Having ascertained the actual years of Pluto Conjoining your Neptune from the tables on page 289, refer to the Age Index on page 235 for the Jupiter 'Years' and Moon 'Phases' occurring during that period.

Other Universal Planetary Cycles possibly active at this time

Check the tables on page 289 to see if any of these Planetary Cycles are active for you at the same time as Pluto Conjoining your Neptune. Remember that any Trine influences may have to be actualized to make them that effective.

SECOND EMERGING MOON PHASE (*see* page 51) This, your 'First Emotional Round-Up', begins at 27 years of age. Should this happen during Pluto Conjoining your Neptune, then it should be quite an emotional time, with your family, domestic and background issues strongly affecting your ability to make the most of this momentous and inspiring period. The values that you inherited from your roots, past and upbringing need to be consciously reviewed with an eye to sorting out which ones still suit you and which ones do not. If your family helps and supports you in your present situation, then well and good, and count yourself lucky. If not, you will have to make adjustments in your attitude to them, or even a wrench from them, if you are to be true to your own destiny.

FIRST SATURN WANING SQUARE (*see* page 114) With this Saturn influence concurring, Pluto Conjoining your Neptune poses, now or at a later date, either rapid advancement on emotional and/or creative levels, if you take some calculated risks – or stagnation if you merely play safe.

FIRST SATURN RETURN (*see* page 115) This, one of the most important Planetary influences in your life happens at 29 to 30 years of age. It cannot be emphasized enough how significant it is to have this synchronize with Pluto Conjoining your Neptune because it means you have two major life Course Corrections happening in concert. This means that a powerful message is coming to you from your 'Mission Control' or unconscious mind. The highs and lows, ease and difficulty, of Pluto Conjoining Neptune will therefore be greatly stressed. Consequently, you can, if you read and obey the writing on the wall, find yourself being given a boost that will last well into the future. However, not seeing your own blind spot(s) now could mean that some time from now you become unstuck – a bit like betting on the wrong horse in a rather long race. It all comes down to being true to your highest values and ideals – for they are your guiding star right now.

URANUS WAXING TRINE (*see* page 166) This useful and refreshing influence happens for a few years between the ages of 24 and 31. If it does coincide with Pluto Conjoining your Neptune then the chances are that you will get the best out of it through being given and taking lucky chances to further invent yourself and realize your creative potentials.

A MAJOR INFLUENCE occurs between the ages of 22 and 34*

PLUTO CONJOINING NEPTUNE

Inspiration Booster and Illusion Buster

The flowchart below shows how the Positive and Negative Neptune Traits *(left)*, that you may or may not have, encounter Pluto's influence *(arrow on right)*, which in turn transforms the Positive Traits into a Positive Outcome *(top)* and degenerates the Negative Traits into a Negative Outcome *(bottom)*. Negative Outcomes can be made Positive through developing the Positive Traits.

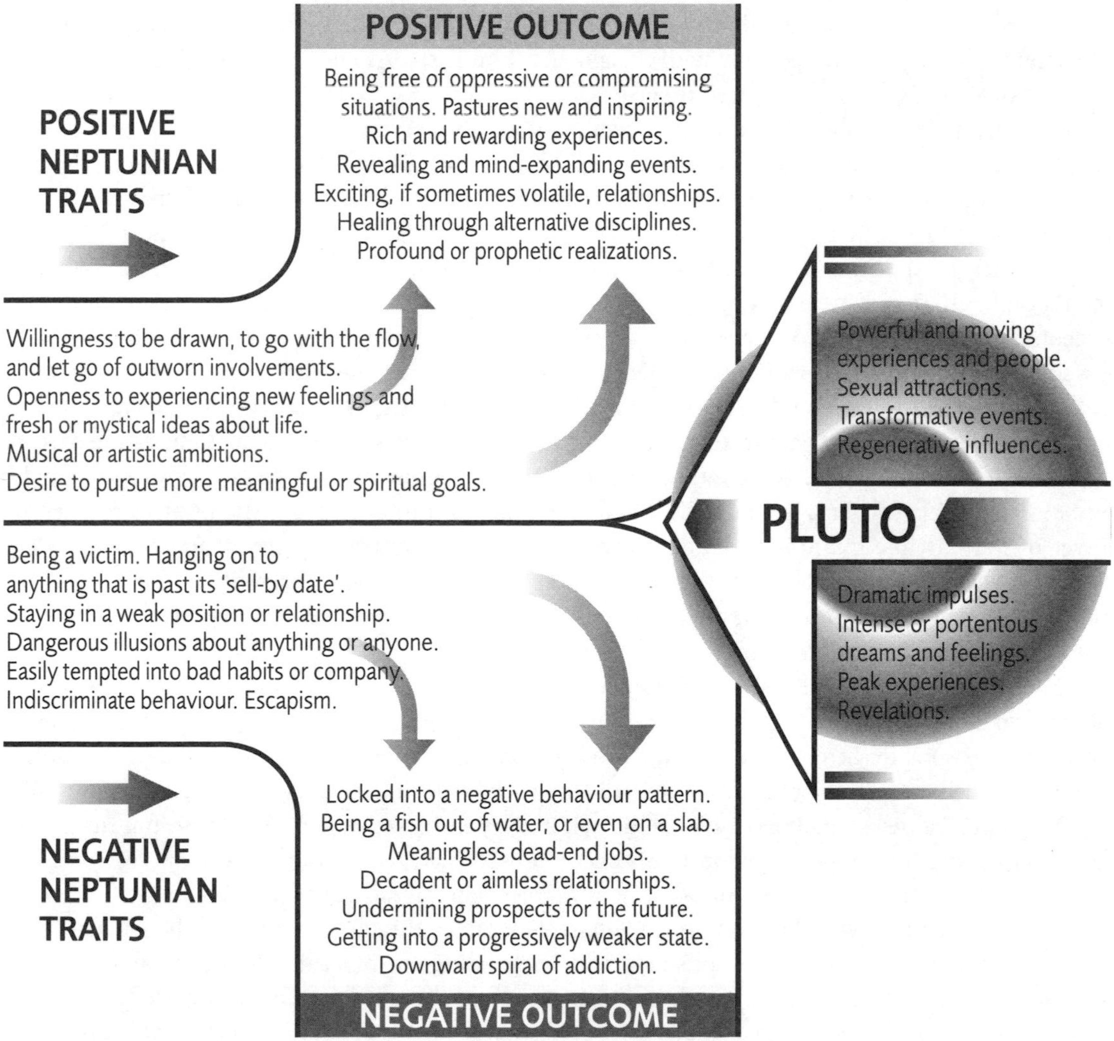

POSITIVE OUTCOME

Being free of oppressive or compromising situations. Pastures new and inspiring.
Rich and rewarding experiences.
Revealing and mind-expanding events.
Exciting, if sometimes volatile, relationships.
Healing through alternative disciplines.
Profound or prophetic realizations.

POSITIVE NEPTUNIAN TRAITS

Willingness to be drawn, to go with the flow, and let go of outworn involvements.
Openness to experiencing new feelings and fresh or mystical ideas about life.
Musical or artistic ambitions.
Desire to pursue more meaningful or spiritual goals.

PLUTO

Powerful and moving experiences and people.
Sexual attractions.
Transformative events.
Regenerative influences.

Dramatic impulses.
Intense or portentous dreams and feelings.
Peak experiences.
Revelations.

NEGATIVE NEPTUNIAN TRAITS

Being a victim. Hanging on to anything that is past its 'sell-by date'.
Staying in a weak position or relationship.
Dangerous illusions about anything or anyone.
Easily tempted into bad habits or company.
Indiscriminate behaviour. Escapism.

NEGATIVE OUTCOME

Locked into a negative behaviour pattern.
Being a fish out of water, or even on a slab.
Meaningless dead-end jobs.
Decadent or aimless relationships.
Undermining prospects for the future.
Getting into a progressively weaker state.
Downward spiral of addiction.

**see* tables on page 289 for when it is for you. Can occur earlier for pre-1919 births.

Pluto Waxing Square | Between 34–47 years for 1–3 years

'HAVING TO LET GO' – Metaphorically, this is like a sweeping curve in the road of your life that inexorably leaves certain things behind as you travel along it. Attempting to hang on to whatever these things might be will effectively hinder and confuse what the future has in store for you.

Looking through my client files, it would seem that the most common occurrence for people during this time is divorce or the break-up of a relationship – often longstanding ones. This does not have to be the case if it is merely an element in your relationship or manner of relating that needs letting go of. However, whether or not this is so will necessitate some hard and honest investigation. Such investigation firstly will mean looking at why you are in the relationship, and secondly, if such a reason is still valid or appropriate.

If the main reason for your staying together has been your children, and they are now ready to leave the nest, or they are independent of you in some way, then the absence of such a reason's validity should become evident. You might find yourself asking, 'What am I doing here with you?' There may be an answer to this question other than 'Not a lot!', and you'll need to discover it with some soul-searching, together and apart, and possibly with the aid of a third party. Amidst all of this there could also be the issue of letting go of your children themselves.

Then again, there could be another outworn reason for staying together. Material or emotional support that is no longer needed or present could be one. With regard to emotional support, the great difficulty could be that one of you no longer needs it but the other still does, or rather *thinks* they do. Such a situation can unfortunately manifest itself as one of you being interested in someone else, with all the heartache that this involves. But if you are the victim here, the message is still the same: you must let go of that person, and in order to do so you must discover what it is in you that clings to that someone you cannot have – or that you do not want in your life any more. Carrying on clinging would be asking for more pain as, with Pluto around, victims tend to become even more victimized. Short of seeking out some professional help here, may I suggest that for the time being you endeavour to concentrate on the part of you that can stand on its own two feet until you are able to attend to the part that can't, and to avoid obsessing and fretting over that part for the time being. To do the reverse would be very destructive, particularly of yourself. It is highly important to accentuate the positive regenerative power of Pluto, while minimizing its degenerative power, simply because this Planetary energy only knows 'up' or 'down', with nothing in the middle. To put it another way, presently Pluto is the part of you that has your future interests at heart, and it knows that you must leave something behind. 'They that drink of the old wine have no place for the new.'

Having been alone for quite some time could also be a feeling that Pluto now intensifies. The reasons for this can be numerous, so I refer you to my book *Do it Yourself Relationship Astrology* where are explored the various reasons for being alone according to Planetary indications in your Birth Chart. One notable example of such intense aloneness was Howard Hughes who during his Pluto Waxing Square – at 61, it occurring later for his generation – isolated himself in the penthouse suite of the Desert Inn Hotel in Las Vegas, remaining there more or less until his death in 1976 (*see* pages 114 and 166).

Alternatively (and hopefully not additionally), the material side of your life – that is, work and money – could be the focus of Pluto Waxing Square. Essentially though, a need for a change here would still have the same fundamental reasons behind it – depending on something that is no longer dependable. Such could be a failing business, an unhealthy market situation (maybe leading to redundancy), an ambition that has proved unrealizable, or simply an attitude to work and money that is unsuitable.

Again, Pluto demands that you look deep inside yourself for the reason for such a predicament, and by way of reward it will then show you a resource that you didn't know you had.

Money and possessions can also become a problem because of settlements and dividing up assets. One way of looking at this that affords a useful insight into such a problem, is to see shared property as symbolic of what you have invested in the relationship. The Plutonian secret here is still one of letting go. The more we hang on emotionally, the more we find that our opposite number hangs on too – but to material things, or worse still, children. Through truly letting go emotionally, with grace, we find that an amicable settlement becomes possible. 'If a runaway horse is truly yours it shall return of its own volition'. Failing this, litigation can become a particularly expensive way, both emotionally and materially, of experiencing Pluto's pressure.

Generally speaking, the letting go of negative emotions can be the most essential response to this influence, with the hardest and most destructive feelings being those of resentment and jealousy. Forgiveness might be the only antidote. Powerlessness could also be a feeling to let go of.

Another area that involves both the emotional and material is home life. A necessity to move away from a home and an area you have grown attached to can arise now. If this proves to be a downward move, then meticulously analyse what elements in your life brought about such an eventuality, and deal with them. However, it is just as likely that such a move will find you in a neighbourhood that is more stimulating, and therefore regenerative for you and yours – given time.

Whether the situation in your life that calls for change is emotional or material, or a combination of the two, Pluto is not doing it to you just for the hell of it, even though at times that is how it could feel. Pluto is always saying that there is a straight and open road around that bend that leads into a better future – but you do have to negotiate that bend.

Another rather stark expression of Pluto's degenerative influence is on the physical level, especially if you are experiencing it at the older end of the age span given above. Quite simply, one's body does become less resilient as one grows older, and it also loses its elasticity and attractive appearance. Yet again, Pluto forces us to look at what we have within in terms of mental, emotional and spiritual worth, rather than depending on our physical assets. That giant of a man, Muhammed Ali, whom I have cited many times as an example in this book, was diagnosed as having Parkinson's disease at 42 years of age, during his Pluto Waxing Square. Regardless of the cause of this, it was still forcing him to turn away from a purely physical or egoistic way of expressing himself; as someone said recently, his eyes now appear to be looking inwards rather than outwards. As far as men are concerned generally, sexual potency can become an issue now, which could put a strain on relationship. Such could cause a man to embark upon a relationship, possibly with a younger woman, for reasons of proving himself. This may work out, or at least boost his ego for a bit – but he may have to pay a greater price at a later date, like being himself traded in for a younger model. There is always Viagra, said to transform one's sex life, but as far as I know Pluto does not really come in a synthetic form – after all it rules what is genuine! With women, the ageing process is more usually an issue of physical looks. So much here rests upon how much a woman has depended on them in the past, and whether she has cultivated her inner being. Needless to say, the same would apply to men.

Continuing with the physical aspect of Pluto's effect, health issues can arise at this time if there has previously been a degenerative style of living. Although this includes the obvious example of an unhealthy and indulgent lifestyle coming to some sort of crisis, we must also consider the more subtle and psychological dimension which is so much a part of Pluto's domain. If there is some imbalance or tension in one's way of living, then it could manifest physically at this time. Once more, Pluto-wise, one would have to delve inside to detect what was creating that imbalance or tension, and take steps to eliminate it. This could stem from some event or trauma that occurred a long time ago without being resolved, and is now surfacing as an ailment. Resisting the idea that one's psychological state affects or even determines one's physical condition could result in one suffering longer and harder than necessary. Pluto chooses whatever way it can in order to draw attention to the interior state of your being.

Sooner or later, that most taboo-ridden aspect of Pluto, death itself, has to be considered. This is not to say that a death is bound to occur at this time – astrology should never be used to make such a prediction – but death does happen. After all, death is the ultimate 'letting go' – whether it is of one's own body and life, or of someone close to us.

Being around death can also manifest as working in a hospice or the like. Communicating with the dead through a medium or some such means can also be an experience that Pluto attracts.

For my own part, my father died during my Pluto Waxing Square, and because I was well prepared for it, the letting go was relatively easy. A death that moved the whole world, however, was that of Princess Diana (born 1 July 1961), at 36 years of age, and at the start of her Pluto Waxing Square. It was reported that shortly before this tragic event, she had decided to 'let go' of the negative aspects of her life – but it would seem her fate had even more than just that in store. In any event, from an astrological point of view, her death was definitely fated. Pluto, as I say, has its own deep and profound agenda. Even after her death, and while her Pluto Waxing Square is still active (at the time of writing), Diana continues to have a lot of dirt dished on her, for Pluto governs all types of rubbish and the need to clear it. On the same theme, film star Tom Cruise and his wife, Nicole Kidman, were recently been accused by a newspaper of living a lie, of them both being gay and pretending to have a healthy marriage for the sake of their professional image. Tom sued the newspaper for libel, and won. This has all been happening during his Pluto Waxing Square, a time when you may have to prove that you (and yours) are genuine and free from anything shady or dubious.

Another notable, or rather notorious, example of the Pluto Waxing Square, is that of Michael Stone who was experiencing it when he murdered a mother, her daughter and their dog, and seriously wounding the second daughter (the Kent murders, 1996). I mention this horrific event, not, of course, to imply that your own experience of Pluto will be anything like this. It does, however, demonstrate how negative the expression of Pluto can be when manifested through an individual of an already deeply degenerate nature. I also refer you back to the concept of the Collective Shadow, under Shadowlands at the start of this Pluto Section. On this same issue, it is worth noting that there was some doubt cast over the validity of evidence against him. If it does turn out that he is innocent, it then throws suspicion upon our need to have a Collective Shadow figure to lock up and thereby attain a false sense of security – because the real killer is still 'out there'.

Whatever form the Pluto Waxing Square takes, it will always take something away, and put something else there to replace it. Knowing and accepting what it is that needs to be let go of is the key, the importance of which I cannot emphasize enough. A positive example of someone who let go of one life and embarked upon another is Brigitte Bardot who gave up acting at 49 to dedicate herself to animal welfare, something of which since then she has proved herself to be a devoted and powerful champion.

To end, as I began, with a metaphor – when you have to abandon ship, don't go back to look for your purse, for you might lose something far more valuable.

On the opposite page you will see a flowchart of your Pluto Waxing Square, 'Having to Let Go', which graphically portrays the challenges and opportunities that this influence offers.

A MAJOR INFLUENCE occurs between the ages of 34 and 47*

PLUTO WAXING SQUARE

Having to Let Go

The flowchart below shows how the Positive and Negative Pluto Traits *(left)*, that you may or may not have, encounter Pluto's influence *(arrow on right)*, which in turn transforms the Positive Traits into a Positive Outcome *(top)* and degenerates the Negative Traits into a Negative Outcome *(bottom)*. Negative Outcomes can be made Positive through developing the Positive Traits.

POSITIVE OUTCOME

Being free (eventually) of burdensome or manipulative involvements. Clearing you and yours of salacious rumours and poisonous opinions. Being rid of anything negative. Sexually and emotionally rewarding relationship. Psychological healing.

POSITIVE TRAITS

Being committed to resolving issues. Being prepared to cut your losses and let go when it is seen to be necessary or unavoidable. An intense desire to get to the bottom of things and root out the truth. Psychological awareness. A willingness to take the plunge into intimacy.

PLUTO

Inevitable endings. Powerful and moving experiences and people. Sexual attractions. Transformative events.

Relationship crises. Dilemmas demanding deep commitment and/or investigation. Emotional and/or material deadlock.

NEGATIVE TRAITS

Fear of or refusing to let go of outworn attachments or attitudes. Self-destructive or victimlike behaviour. Staying in a relationship for negative reasons. Trying to have your cake and eat it. Being/becoming involved with dubious people. Greedy or dishonourable motivations.

NEGATIVE OUTCOME

Separation or divorce. Material losses/difficulties. Bad feeling. Getting your fingers burnt – or worse. Falling between two stools. Being manipulated by others. Blame and recrimination from others. Long-term relationship problems. Way ahead blocked for some time.

**see* tables on page 289 for when it is for you. Can occur later for pre-1937 and post-1992 births.

JUPITER 'YEAR' for your Growth Mode

MOON 'PHASE' for Emotional Attunement

Having ascertained the actual years of your Pluto Waxing Square from the tables on page 289, then refer to the Age Index on page 235 for the Jupiter 'Years' and Moon 'Phases' occurring during that period. If one of these Jupiter 'Years' is a 'Scorpio' one then this would further accentuate the influence of Pluto's Waxing Square in that year of you life.

Other Universal Planetary Cycles possibly active at this time

Check the tables on page 289 to see if any of these Planetary Cycles are active for you at the same time as your Pluto Waxing Square. Remember that any Trine influences may have to be actualized to make them that effective.

SECOND REALIZING MOON PHASE (*see* page 77) This begins at around 41 years of age, lasting three and a half years, and emphasizes any emotional difficulty that Pluto brings, but eventually it does make it clear to you what you do and do not want – and more to the point, what you do and do not *need*. This is because the emotional lie of the land becomes so much more distinct now, enabling you to see both the path ahead and the one you have been travelling along.

SECOND SATURN WAXING SQUARE (*see* page 119) This lends weight to the effect of your Pluto Waxing Square in that it becomes only too clear what has to go and what has to change. This can be quite hard on you if you are stubborn.

SECOND SATURN WAXING TRINE (*see* page 121) If this slots into your Pluto Waxing Square, then it will enable you to keep things on an even keel. Count yourself lucky if it does because most other transits that could happen at this time are challenging, all being part of the mid-life experience. Your sense of order, duty and sobriety serve you well here.

SECOND SATURN OPPOSITION (*see* page 122) This influence, which happens between 43 and 44 years of age, really does force you to read the writing on the wall. So if it should happen during your Pluto Waxing Square, then reality will be speaking quite loudly to you, making the right way and the wrong way quite evident. As such, it will make things both harder, because the truth will be presented quite harshly to you, and easier, because there will be no mistaking what you have to do.

URANUS OPPOSITION (*see* page 168) This is your 'Mid-life Crisis' which lasts for two years or so between 38 and 45 years of age. If your Pluto Waxing Square does happen at the same time as this, then you can be sure that it's going to pack a very definite punch – very much so in the case of Muhammed Ali, for he hung up his gloves during this influence and then was diagnosed as having Parkinson's disease. The feelings that Uranus provokes of wanting to make off and do something entirely different can make the Plutonian pressure to let go a lot easier. Or conversely, not being able to let go of heavy attachments can make you feel fit to burst. Try to tread a middle path by letting your feelings be known to whoever matters or will listen, and then you'll be able to see some kind resolution. Bottling it all up in any way would be a bad idea considering the revolutionary energies that are abounding right now. Whatever the case, a radical move is called for – but just make sure that it is conscious and constructive rather than a knee-jerk reaction where you burn your boats and come to regret it later.

NEPTUNE WAXING SQUARE (*see* page 198) This is a tricky time, occurring for a few years between 39 and 43 years of age, and I advise you to study carefully what it means. If it does coincide with your Pluto Waxing Square then confusion and feelings of helplessness are more or less unavoidable – at least, to some degree. I also advise you to heartily admit to any such feelings. It is vital that you think not only twice, but three or four times, before making any decision that's going to change your life and that of those close to you. When Neptune's around, it is best to follow the line of least resistance, while at the same time keeping a weather eye open for a way out or through. Try to let things develop of their own accord, while only fixing what obviously can be fixed. More often than not, though, this passage of your life is going to be that *Via Negativa* described on page 190. With Pluto involved, it is most probably going to be the bitter end of something that was just not meant to happen or work out. Cutting your losses and telling yourself that there are plenty more fish in the sea could well be your best policy now.

Pluto Waxing Trine **Between 48–63 years for 1–3 years**

'OVERDRIVE' – By now your life has most probably evolved to a level where at least some parts of it are up and running, and are to a degree self-regenerating. If not, this influence could help you to get into a higher gear at last. In any event, you do have a wealth of experience to draw upon, and the fact that you have got this far is a testament to the fact that at the very least you can survive.

Whatever gear you are in right now, there is still the opportunity with Pluto's Waxing Trine to rid yourself of any hindering habits or debilitating attitudes that you are hanging on to. This would include the people or involvements that correspond to such habits and attitudes. All you need do is just drop them, for by now you should know what does and what doesn't matter as a result of having some sense of your own power and forward motion. Indeed, you might now take an active interest in some kind of psychological discipline. How you and others operate at a deep level can be something that you become (more) interested in, and might even become adept at. Such may even take you towards a more active involvement in matters psychological or occult, where you realize how you can influence the world around you with the power of inner truth. Such power is comprised of an intimate but unemotionally charged awareness of what is to come, with which you are able to influence others – invisibly and wordlessly. In this, or some similar way, you can be a source of wisdom for others.

Alternatively, Pluto could express itself now on a purely external or material level. Inheriting money, attaining outright ownership or reaping interest are a likelihood now, considering the age of relatives and the duration of investments or regular payments. Apart from this, money matters more or less take care of themselves – that is, unless you have been particularly careless or hard up. Preparing yourself, materially or emotionally, for some inevitable development is also well-starred now.

It as if the waters of life have been tumbling for so long now over the terrain of your own life-track that you know in your bones what's what. Experience has simply taught you what you can and cannot do, what you can influence and what you have to let go of, and you know enough about the difference between the genuine and the false not to be too fazed by much, and not to waste your energy on involvements that do not merit it.

Overleaf you will see a flowchart of your Pluto Waxing Trine, which graphically portrays the challenges and opportunities that this influence offers.

AN OPPORTUNITY occurs between the ages of 48 and 63*

PLUTO WAXING TRINE

The flowchart below shows how any of your Positive Pluto Attitudes and Actions *(bottom left)*, having attracted and made use of Pluto's influence *(Slingshot Effect within orb)*, allow you to attain certain Advantages *(top right)*.

REMEMBER – A trine only opens the way; it does not push or confront you. So you must deliberately recognize it and take advantage of it.

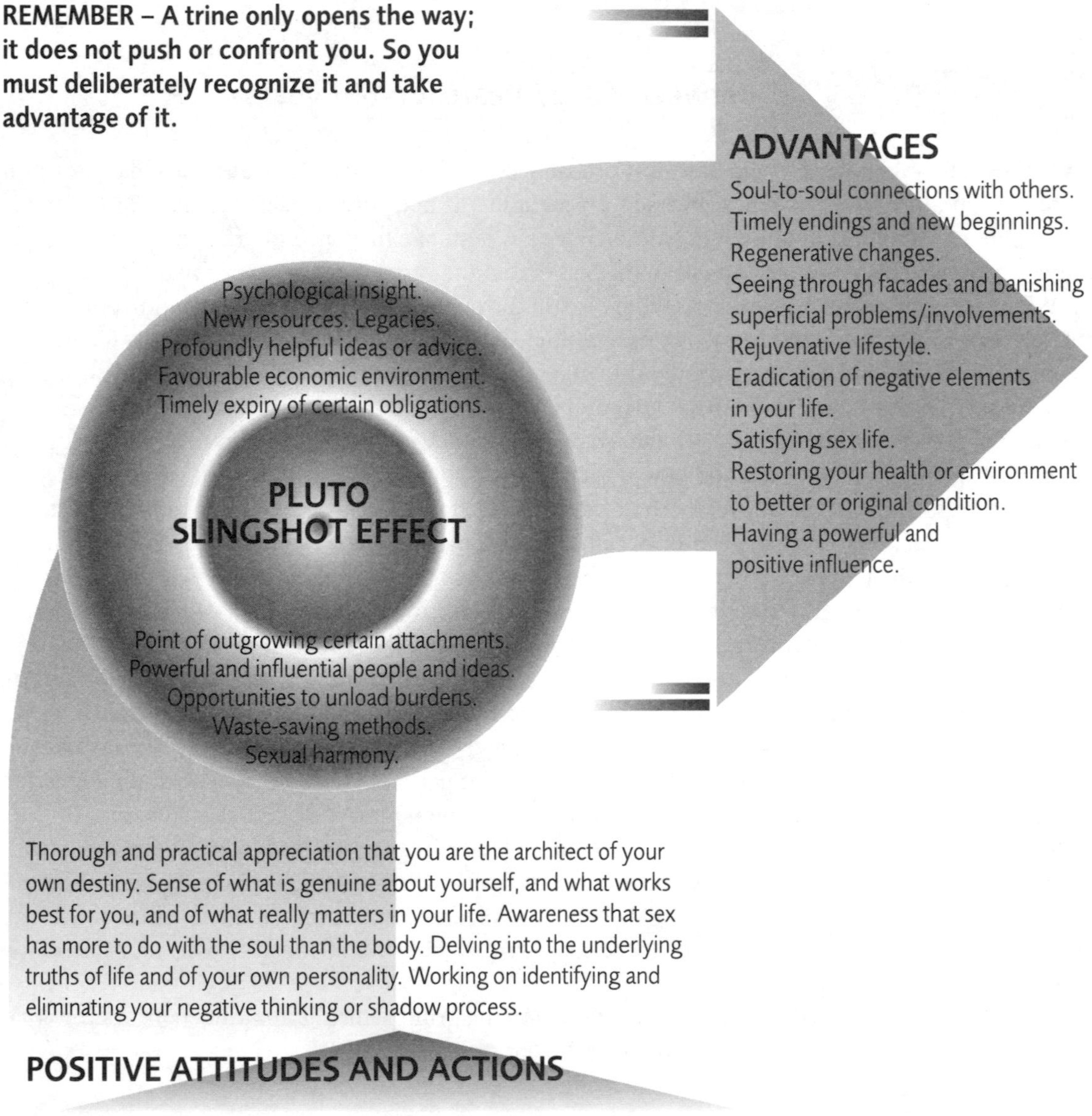

ADVANTAGES

Soul-to-soul connections with others.
Timely endings and new beginnings.
Regenerative changes.
Seeing through facades and banishing superficial problems/involvements.
Rejuvenative lifestyle.
Eradication of negative elements in your life.
Satisfying sex life.
Restoring your health or environment to better or original condition.
Having a powerful and positive influence.

Psychological insight.
New resources. Legacies.
Profoundly helpful ideas or advice.
Favourable economic environment.
Timely expiry of certain obligations.

PLUTO SLINGSHOT EFFECT

Point of outgrowing certain attachments.
Powerful and influential people and ideas.
Opportunities to unload burdens.
Waste-saving methods.
Sexual harmony.

Thorough and practical appreciation that you are the architect of your own destiny. Sense of what is genuine about yourself, and what works best for you, and of what really matters in your life. Awareness that sex has more to do with the soul than the body. Delving into the underlying truths of life and of your own personality. Working on identifying and eliminating your negative thinking or shadow process.

POSITIVE ATTITUDES AND ACTIONS

*see tables on page 289 for when it is for you. Can occur later for pre-1928 and post-1984 births.

JUPITER 'YEAR' for your Growth Mode

MOON 'PHASE' for Emotional Attunement

Having ascertained the actual years of your Pluto Waxing Trine from the tables on page 289, then refer to the Age Index on page 235 for the Jupiter 'Years' and Moon 'Phases' occurring during that period.

Other Universal Planetary Cycles possibly active at this time

Check the tables on page 289 to see if any of these Planetary Cycles are active for you at the same time as your Pluto Waxing Trine. Remember that any Trine influences may have to be actualized to make them that effective.

THIRD EMERGING MOON PHASE (*see* page 54) This highly important point in your life, your "Second Emotional Round-Up", begins at $54\frac{2}{3}$ years of age, and if it is concurrent with your Pluto Waxing Trine then a major focus for the latter's regenerative energy is that of mother and family, home and roots. You will see that an important aspect of your emotional life at this time is how you feel towards those who brought you into the world in the first place, along with others who shared your early life, such as brothers and sisters. Whether a parent is dead or alive, if your relationship with them is still not that healthy, then now is your chance to rectify this. If they are no longer in this world, you will, if you sincerely wish to, find ways and means of simply communicating with them in order to make peace or amends. If they are alive, then successfully conveying your most fundamental feelings to them can happen now. This may be embarrassing or painful, but nothing compared to the negative feelings that could arise when they are no longer physically available. Generally speaking, with Pluto Waxing Trine happening alongside this, your Second Lunar Return, you are now able to put all family and domestic issues on a firmer footing or into a clearer perspective.

SECOND SATURN WANING TRINE (*see* page 124) Smooth and sweeping changes of a very positive variety can now be implemented; they may even happen without you lifting a finger – but don't bank on it.

SECOND SATURN WANING SQUARE (*see* page 125) While this is happening, making good use of the advantages of your Pluto Waxing Trine is possibly mandatory rather than a matter of choice. This is because certain circumstances could arise that force you to make a definite stand over some issue, and either *be* the authority or bow down to who or whatever *is* the authority.

SECOND SATURN RETURN (*see* page 126) Having this 'Your Second Life Progress Report' also occurring at this time suggests that the 'report' should be favourable.

URANUS WANING TRINE (*see* page 175) This provides luck and a good sense of direction to the 'Overdrive' of your Pluto Waxing Trine. So, wherever you are headed, you can now get there – swiftly and safely, most likely.

URANUS WANING SQUARE (*see* page 177) Something should occur that means that you have to make the most of what 'Overdrive' offers you. A sudden change in circumstances, for example, would oblige you to muster and use your knowledge and experience in earnest, and forge it into wisdom.

NEPTUNE WAXING TRINE (*see* page 204) Spiritual and creative forces also come to your aid now, giving your Pluto Waxing Trine an aspect of divine inspiration, if only in the sense of becoming a lot surer of what (your) life is about.

NEPTUNE OPPOSING PLUTO (*see* page 207) This influence will lend a note of poignancy to the power of your Pluto Waxing Trine. The seas may get choppy or even stormy – but you have a reliable vessel to carry you on.

An even more personal focus can be given to the influence of Pluto in your life by looking at what other Planets in your birth chart are being affected at the same time in your life. These, your Individual Planetary Cycles or Transits, would have to be technically worked out and interpreted by a professional astrologer. *See* Resources on page 376.

Chapter Three

THE AGE INDEX

Your Whole Life Journey at a Glance

The Age Index gives an overall view of all the Universal Planetary Cycles that will occur during anyone's life. After familiarizing yourself with this, simply use the Age Index to determine the times when these Planetary Cycles were active for you personally, and then go to the previous chapter, 'The Planetary Cycles of Your Life', for descriptions of those influences of your immediate interest. Alternatively, you can enter all the Active Planetary Cycles during your life on to a 'Personal Astro-Life Plan' chart, explained below, after you have discovered how to use the Age Index in general.

To use the Age Index, note that . . .

- The various Planetary influences are listed as columns under the main heading ACTIVE PLANETARY CYCLES. The name of any actual influence is given in the shaded box that intersects a Planet column and the row corresponding to your age (given in first column), and the density of the shade indicates whether it is a Major, Medium or Mild influence.

  For Jupiter, the year 'Sign' or 'Growth Mode' is given along with the page number where that Year/Mode is described in this book. No shading is used for Jupiter as its influence varies considerably, although I have designated them as Medium in the Jupiter section. Before reading the description of any Jupiter influence, first digest the Jupiter introduction on page 23.

  For the Moon, the various 'Phases' are depicted as shaded strips of graduated density. The densest shade is at the beginning, indicating when that 'Phase' is at its most powerful, while the actual name of the 'Phase' is given along with the page number where it is described.

  For Saturn, Uranus, Neptune and Pluto, the page on which the Active Cycle is actually described is given in the last column of the row where the actual name of the influence is written.

  So, for example, at age 29 you would have a 'Virgo' Jupiter Year or Growth Mode (described on page 32), be mid-way through your Second Emerging Moon 'Phase' (described on page 51), your First Return of Saturn (described on page 115), and possibly (according to your month and year of birth – *see* tables beginning on page 289) your Uranus Waxing Trine (page 166) and Pluto Conjoining your Neptune (page 221).
- The types of Planetary influence are as follows:

  RETURN = Beginning of new cycle, end of an old one. For example, 1st RETURN = end of 1st Cycle, start of 2nd one). Note that the first year of an 'Emerging Moon Phase' is also its Return.
  Square = Challenging or hard parts of a cycle. Forces progress.
  Trine = Flowing or easy parts of a cycle. Releases opportunity.
  Waxing = Sowing or developing half of a cycle.
  Waning = Reaping or reflective half of a cycle.
  Opposition = Realizational or confrontational part of cycle.

  So for example, a Waxing Square would be the challenging or hard part of a sowing or developmental phase, with respect to the Planet Cycle in effect.
- When you see two arrows like this  this means that this Planetary Transit will be occurring over a one- to three-year period within this time frame. When this occurs will vary according to the date you were born, and the tables on page 289 will show when it is for you.

- Ages where no apparent Universal Planetary Cycle is active may be regarded as transitional years, but an Individual Planetary Cycle may well be active at that time. There is always a Jupiter 'Year' or a Moon 'Phase' active.
- There are three *Individual* Planetary Cycles (*see* page 6) included in this book, which are Neptune (in the sky) Squaring Pluto (in your chart), Pluto (in the sky) Conjoining Neptune (in your chart), and finally Neptune (in the sky) Opposing Pluto (in your chart). These are made possible owing to these two Planets travelling unusually 'in tandem' for the most part of the 20th century. The page number of the description of these is again given in the last column, alongside the word 'Conjoining' (Neptune) in the Pluto column, and 'Squaring' (Pluto) and 'Opposing' (Pluto) in the Neptune column. Again, consult the tables beginning on page 289 to discover exactly when the first two Transits occur in your life. With Neptune Opposing Pluto, consult the individual table on page 352.
- Following the Age Index itself, you will find 'The Stages of Your Life' and 'The Three Ages of Man' which give you further overviews of your whole life in astrological terms.
- The next stage is to plot in the times in your life when Planetary cycles were active. You will see that the Jupiter, Moon and Saturn influences occur at set times for everybody, but for Uranus, Neptune and Pluto, consult the tables on page 289 for the actual age that their influences occur or occurred for you personally. Then you can enter those ages in the relevant places on a Personal Astro-Life Plan chart which you'll find beginning on page 361 at the back of the book. (I recommend that you photocopy this as you'll find a set of separate sheets easier to use, and you'll also have spare charts for other people). A good way of entering these periods of active Planetary Cycles is simply to draw in a box with a highlighter pen (including the ages when occurring), as in the example Astro-Life Plan (fully completed with life events entered) on page 248. In this example, you will find the very first of these Outer Planet influences, Neptune Squaring Pluto, in the Neptune column on page 249 occurring between the ages of 15 and 17. In the Age Index itself you will see that Neptune Square Pluto can occur within quite a large period of time, but you can discover from the tables when this influence occurred for you personally, then draw in a box on the Personal Astro-Life Plan chart in the Neptune column, in the rows corresponding to those ages.
- Bear in mind that any influence might have to get underway before you feel its effect, or alternatively it might begin a bit earlier – so don't expect it to start precisely on your birthday!

The Three Ages of Man

All of the Cycles listed in this book can also be seen as being divided into three basic Cycles – the Ages of Man. Each 'Age of Man' is measured as 28 years, that being a third of the Cycle of Uranus, and also the average of the Cycles of the Moon (27 years) and Saturn (29 years), or what is called the Cycle of the Self Point. The first Cycle is for the development of the personality, the second for the development of oneself as a being who creatively contributes to society, and the third is one of spiritual development and teaching others, preparing oneself for the next world. The numerology of these three 28-year Cycles runs like this:

Personality	from	0 – 28	=	2 + 8	=	10	=	1 + 0	=	**1st Age**
Social Being	from	29 – 56	=	5 + 6	=	11	=	1 + 1	=	**2nd Age**
Spirituality	from	57 – 84	=	8 + 4	=	12	=	1 + 2	=	**3rd Age**

ACTIVE PLANETARY CYCLES

INFLUENCE> MAJOR | MEDIUM | MILD

AGE	JUPITER	MOON	SATURN	URANUS*	NEPTUNE*	PLUTO*	PAGE
BIRTH 0	'Aries' (27)	1st Emerging (49)					9
1	'Taurus' (28)						
2	'Gemini' (29)						
3	'Cancer' (30)	1st Striving (58)					
4	'Leo' (31)						
5	'Virgo' (32)						
6	'Libra' (33)		1st Waxing Square				110
7	'Scorpio' (34)	1st Deciding (63)					
8	'Sagittarius' (35)						
9	'Capricorn' (36)		1st Waxing Trine				111
10	'Aquarius' (37)	1st Adjusting (69)			Squaring Pluto		
11	'Pisces' (38)						
12	'Aries' (27)						
13	'Taurus' (28)						
14	'Gemini' (29)	1st Realizing (75)	1st Opposition				111
15	'Cancer' (30)						193
16	'Leo' (31)						
17	'Virgo' (32)	1st Sharing (81)		Waxing Square			
18	'Libra' (33)						
19	'Scorpio' (34)		1st Waning Trine				113
20	'Sagittarius' (35)	Understanding (87)					162
21	'Capricorn' (36)		1st Waning Square				114
22	'Aquarius' (37)				(Can occur later for pre-1930 births)	Conjoining Neptune (Can occur earlier for pre-1919 births)	
23	'Pisces' (38)						
24	'Aries' (27)	1st Releasing (93)		Waxing Trine			
25	'Taurus' (28)						
26	'Gemini' (29)						221
27	'Cancer' (30)	2nd Emerging (1st Return) (51)					
28	'Leo' (31)						166
29	'Virgo' (32)		1st RETURN				115
30	'Libra' (33)						
31	'Scorpio' (34)	2nd Striving (59)					

*See tables beginning on page 289 to discover the exact times of the one to three years in your life that Uranus, Neptune and Pluto Cycles are active.

ACTIVE PLANETARY CYCLES

INFLUENCE> MAJOR | MEDIUM | MILD

AGE	JUPITER	MOON	SATURN	URANUS*	NEPTUNE*	PLUTO*	PAGE
32	'Sagittarius' (35)	Striving (cont) (59)				Conjoining Neptune (cont)	221
33	'Capricorn' (36)						
34	'Aquarius' (37)	2nd Deciding (64)					
35	'Pisces' (38)						
36	'Aries' (27)		2nd Waxing Square				119
37	'Taurus' (28)						
38	'Gemini' (29)	2nd Adjusting (71)	2nd Waxing Trine			Waxing Square	121
39	'Cancer' (30)						226
40	'Leo' (31)				Waxing Square		198
41	'Virgo' (32)	2nd Realizing (77)		Opposition			168
42	'Libra' (33)						
43	'Scorpio' (34)		2nd Opposition				122
44	'Sagittarius' (35)	2nd Sharing (83)					
45	'Capricorn' (36)					(Can occur later for pre-1937 and post-1992 births)	
46	'Aquarius' (37)						
47	'Pisces' (38)						
48	'Aries' (27)	Understanding (89)	2nd Waning Trine				124
49	'Taurus' (28)						
50	'Gemini' (29)						
51	'Cancer' (30)	2nd Releasing (95)	2nd Waning Square				125
52	'Leo' (31)						
53	'Virgo' (32)						
54	'Libra' (33)					Waxing Trine	231
55	'Scorpio' (34)	3rd Emerging (2nd Return) (54)			Waxing Trine		204
56	'Sagittarius' (35)			Waning Trine			175
57	'Capricorn' (36)						
58	'Aquarius' (37)	3rd Striving (60)	2nd RETURN				126
59	'Pisces' (38)						
60	'Aries' (27)				(51 yrs) Opp. Pluto (76 yrs) (Variable – *see* Table page 289)		207
61	'Taurus' (28)					(Can occur later for pre-1928 and post-1984 births)	
62	'Gemini' (29)	3rd Deciding (66)		Waning Square			177
63	'Cancer' (30)						

**See* tables beginning on page 289 to discover the exact times of the one to three years in your life that Uranus, Neptune and Pluto Cycles are active.

ACTIVE PLANETARY CYCLES

INFLUENCE> MAJOR | MEDIUM | MILD

AGE	JUPITER	MOON	SATURN	URANUS*	NEPTUNE*	PLUTO*	PAGE
64	'Leo' (31)			Waning Square ↓			177
65	'Virgo' (32)	3rd Adjusting (73)	3rd Waxing Square				129
66	'Libra' (33)						
67	'Scorpio' (34)						
68	'Sagittarius' (35)	3rd Realizing (78)	3rd Waxing Trine				130
69	'Capricorn' (36)						
70	'Aquarius' (37)						
71	'Pisces' (38)						
72	'Aries' (27)	3rd Sharing (84)					
73	'Taurus' (28)		3rd Opposition				130
74	'Gemini' (29)						
75	'Cancer' (30)	Understanding (90)					
76	'Leo' (31)						
77	'Virgo' (32)						
78	'Libra' (33)	3rd Releasing (97)	3rd Waning Trine				131
79	'Scorpio' (34)						
80	'Sagittarius' (35)		3rd Waning Square				131
81	'Capricorn' (36)						
82	'Aquarius' (37)	4th Emerging (3rd Return) (56)		↑ RETURN ↓			
83	'Pisces' (38)						179
84	'Aries' (27)						

*See tables beginning on page 289 to discover the exact times of the one to three years in your life that Uranus, Neptune and Pluto Cycles are active.

Make sure you have read 'How to Use the Age Index' on page 237.

The Stages Of Your Life

Here are all the major active Planetary Cycles as I have described them in the last chapter, given in chronological order

Age*	Phase	Cycle or Transit	Page
0	Birth	The Start of All Cycles	9
0–3	'Your Birthing'	First Emerging Moon Phase	49
6–7	'The Die is Cast'	First Saturn Waxing Square	110
6–10	'The Die is Cast'	First Deciding Moon Phase	63
9–10	'Releasing Your Creativity'	First Saturn Waxing Trine	111
10–23	'Adolescence Strikes'	Neptune Squaring Pluto	193
13–17	'Emotional Discovery'	First Realising Moon Phase	75
14–15	'Me Versus the Rest'	First Saturn Oppostion	111
17–23	'Breaking Away and Breaking Through'	Uranus Waxing Square	162
19–20	'Finding Your Niche'	First Saturn Waning Trine	113
21–22	'Coming of Age'	First Saturn Waning Square	114
22–34	'The Inspiration Booster and The Illusion Buster'	Pluto Conjoining Neptune	221
24–31	'Becoming Your Own Person'	Uranus Waxing Trine	166
27–30	'Your First Emotional Round–Up'	Second Emerging Moon Phase	51
29–30	'Your First Life Progress Report'	First Saturn Return	115
34–47	'Having to Let Go'	Pluto Waxing Square	226
36–37	'Having Your Wheels Tapped'	Second Saturn Waxing Square	119
37–45	'Your Mid–Life Crisis'	Uranus Opposition	168
38–39	'Following Your Path'	Second Saturn Waxing Trine	121
38–43	'Fascinations and Illusions'	Neptune Waxing Square	198
41–44	'Emotional Awareness'	Second Realising Moon Phase	77
43–44	'The Old Versus The New'	Second Saturn Opposition	122
48–49	'Finding Your Truth'	Second Saturn Waning Trine	124
48–63	'Overdrive'	Pluto Waxing Trine	231
51–52	'Spiritual Maturity'	Second Saturn Waning Square	125
51–59	'A New Lease of Life'	Uranus Waning Trine	175
51–76	'Intimations of Mortality'	Neptune Opposing Pluto	207
52–58	'Seeing the Light'	Neptune Waxing Trine	204
54–58	'Your Second Emotional Round–Up'	Third Emerging Moon Phase	54
58–59	'Your Second Life Progress Report'	Second Saturn Return	126
59–65	'Making A Break'	Uranus Waning Square	177
65–66	'Reforging or Shunted to the Sidings'	Third Saturn Waxing Square	129
68–69	'Trusting Your Fate'	Third Saturn Waxing Trine	130
68–71	'Emotional Wisdom'	Third Realising Moon Phase	78
73–74	'The Old Versus The New – Two'	Third Saturn Opposition	130
78–79	'Knowing Your Heart'	Third Saturn Waning Trine	131
80–81	'Final Acceptance'	Third Saturn Waning Square	131
82–85	'Your Last Emotional Round–Up'	Fourth Emerging Moon Phase	56
82–84	'Arriving & Departing Both'	Uranus Return	179

* During age periods given as more than two years, influences occur for one to three years. *See* tables on page 289 to discover exactly when they occur for you. Also note from the Age Index itself and Tables that some Pluto and Neptune influences can occur earlier or later than the ages given here for some births before 1937 or after 1984.

Chapter Four

YOUR PERSONAL ASTRO-LIFE PLAN

You can now draw up your personal Astro-Life Plan in order to see and feel the nature and course of your life so far. Quite simply, your Personal Astro-Life Plan is a record of your personal history set against the significance of your Universal Planetary Cycles. Immediately after creating it, and also as you do so, you will experience a sensation of reassurance and accomplishment – not just for having drawn up your Plan, but also because you will automatically feel that your life has a story and purpose all of its own. Consequently, reading or re-reading the various interpretations for active Planetary Cycles for any particular part of your life will then take on extra meaning and significance. Feeling the shape and energy of your life in this way can amount to a sort of therapy, creating a direct effect on your life as you live it now, including your dream-life. Here is how to draw up Your Personal Astro-Life Plan.

1 Photocopy the set of Personal Astro-Life Plan chart blanks that you will find at the back of the book on page 361. You can work directly in the book if you wish, but it is easier to use a separate photocopied set – and you'll also have a master set for other people's Plans.
2 In the second column, headed 'AGE', you will see the ages of your life from birth to 84 years, row by row, as also given in the Age Index. However, on this chart there is more space so that you can fill in your personal history. Now in the first column, headed 'YEAR', enter all the calendar years that correspond to these ages of your life, making sure to insert a little line where each calendar year begins, approximately in relation to your actual age. So, in the example chart starting on page 248, which is my Personal Astro-Life Plan, I put the year of my birth, 1944, in the 'YEAR' column, then I insert a little line approximately where 1944 ended and 1945 began (a third of the way down to the left of the first year of the 'AGE' column as I was born in early September). I then insert all the years, and the little lines separating them, all the way through my life to date and beyond. All this sounds a bit fiddly but it is really quite simple, and it is well worth doing because, as all the influences are indexed to your age, inserting the relative calendar year in this way greatly assists you in remembering when past events actually occurred.
3 Now enter the years (by ages) of all the Uranus, Neptune and Pluto influences that you noted from the tables on page 289. Do all the Uranus influences first, then all Neptune's, and finally all Pluto's. On my example on page 250, the first one I enter is 'Uranus Waxing Square' which occurred for me between the ages of 20 and 21. This is best done by using a highlighter pen, creating a thick border for the relevant period.Then you can later write the events of your life history over it (Step 4). Also write in the name of the influence itself. Finally, if you wish, enter in the far right column the page number of this book where the influence's interpretation is given, according to the Age Index. You will see that all the Jupiter Years, Moon 'Phases' and Saturn influences have already been entered because they are the same for everyone. The very last influence I enter will be my Pluto Waxing Trine on page 253.
4 Now enter any personally significant events for all the years of your life so far – and to do this I suggest you use a pencil with rubber attached. Do this quite freely, for nobody is going to see it unless you want them to. Start off by entering anything significant about your birth and time in the womb, as I have done. How you actually write all this in is up to you, but the method I have used is quite easy and effective (bubbles and arrows). Having entered your life history to date (or as you do so if you are familiar enough with what each influence means), you can now go about seeing how your life events correspond to those Active Planetary Cycles that you entered in Step 3. In order to do this, you will

need either to have remembered what those influences mean from having read them earlier, or to look them up as you go – or a combination of the two. Personally, I recommend that you always read the interpretations again after entering events because it means that you are more likely to make vital connections and also jog your memory further. In this way a sort of chain reaction of events and astrological correspondences is set in motion, sometimes with quite extraordinary results as things fall into place.

5. This is optional, but it is a good idea to jot down (probably on a separate sheet of paper) what all of these events, taken together with the Planetary influences, mean to you personally. This would be like a commentary upon your life, including the inferences and conclusions that you draw from it all. Also, if you wish, you can arrange *future* events with your Astro-Life Plan.
6. You have now completed your Personal Astro-Life Plan. It is important that you do all this in a way that feels comfortable to you, so do not feel that you have to do it exactly in the way or order that I have described here.

Tips for compiling and using your Personal Astro-Life Plan

- The significance of your Jupiter Years can often be very noticeable and rewarding.
- On completing your Personal Astro-Life Plan of your life so far, you will then find that present circumstances and Planetary influences are seen more clearly and in a new light.
- Never forget that certain events will not be accountable to Universal Planetary Cycles (the subject of this book), but rather will be signified by Individual Planetary Cycles, which can be obtained from an astrologer or astrological service (*see* Resources, page 376).
- Do not feel obliged to keep strictly to the ages/years given for any particular Planetary influence. Owing to a multitude of astrological reasons, any influence (*apart* from Jupiter Years and Moon 'Phases') can be operational for anything up to a year before *or* after (but not both) the period given. So if an event and a Planetary influence synchronize but are up to a year out, then believe the synchronicity and ignore the technicality.
- Realize and appreciate the patterns of your life that emerge from constructing your Plan. The more you connect the course of your life with the Planetary Cycles and their meanings, the more on-course and meaningful you will feel your life to be. This is astrology at its most fundamental.
- It may be a good idea to draw up your Personal Astro-Life Plan again more clearly (in ink), for this brings a clarity and definiteness to your life itself. You can also add in further memories as you do so.
- There follows a table showing what could be generally regarded as the most important of all ages/influences of your life, so pay special attention to any events at these times for they hold keys to the meaning and direction of your life as a whole and into the future.

Age	Influence/Cycle Active	Significance
0	Birth/Start of all cycles	This is a microcosm of your life
6 – 7	Saturn Waxing Square and First Deciding Moon Phase begins	'The Die is Cast' – events now set a major pattern for life
14 – 15	First Saturn Opposition	'Me Versus Them' – sets pattern for relationships
10 – 23*	Neptune Squaring Pluto	'Adolescence Strikes'
17 – 23*	Uranus Waxing Square	'Breaking Away and Breaking Through'
22 – 34*	Pluto Conjoining Neptune	'The Inspiration Booster and the Illusion Buster'
27 – 28	First Lunar Return (Second Emerging Moon Phase begins)	'Your First Emotional Round-Up'
29 – 30	First Saturn Return	'Your First Life Progress Report'
34 – 47*	Pluto Waxing Square	'Having to Let Go'
38 – 43*	Neptune Waxing Square	'Fascinations and Illusions'
37 – 45*	Uranus Opposition	'Your Mid-life Crisis'
51 – 76*	Neptune Opposing Pluto	'Intimations of Mortality'
54 – 55	Second Lunar Return (Third Emerging Moon Phase begins)	'Your Second Emotional Round-Up'
58 – 59	Second Saturn Return	'Your Second Life Progress Report'

* for one to three years between these ages (*see* Tables on page 289)

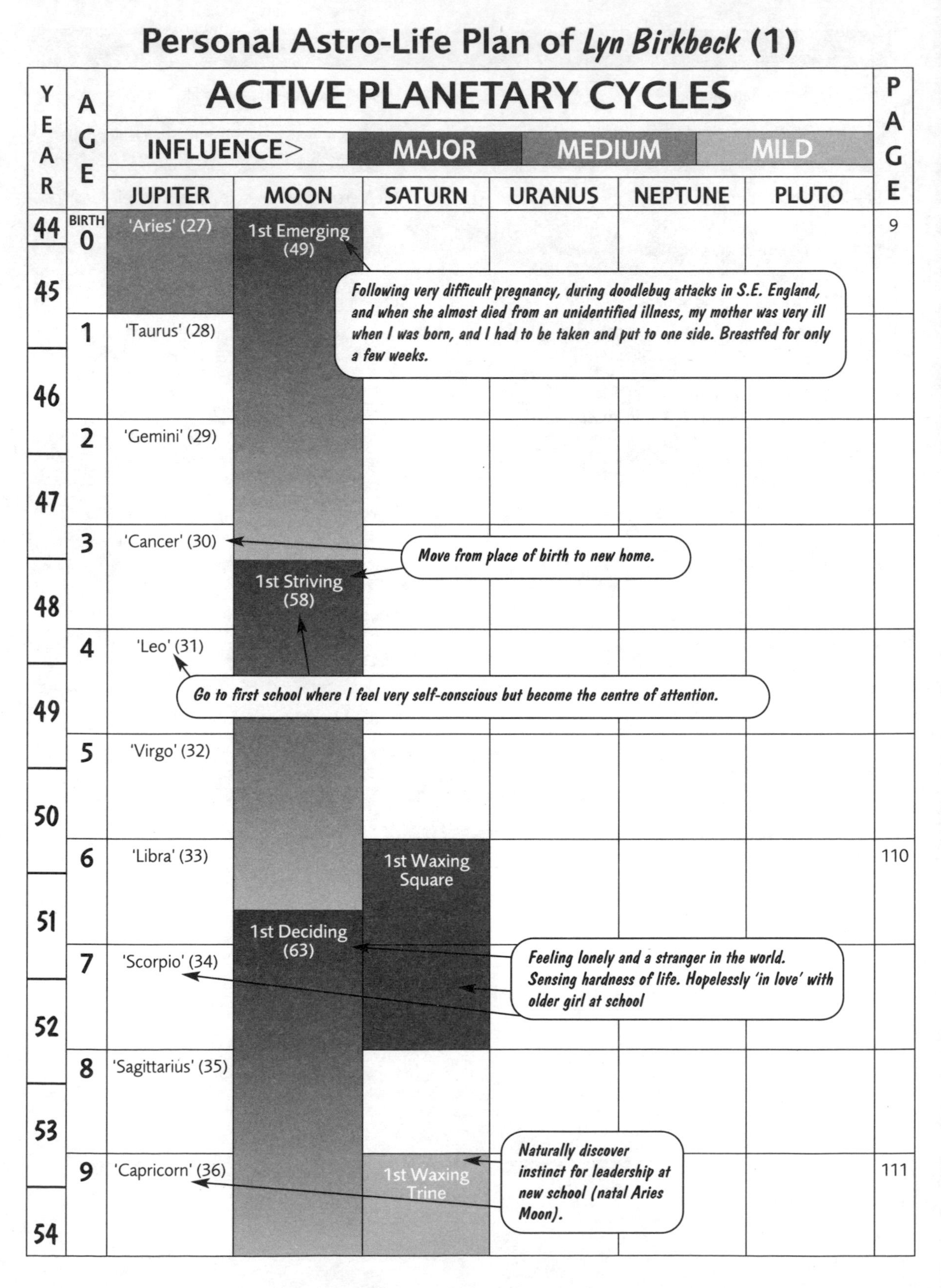
Personal Astro-Life Plan of Lyn Birkbeck (1)
YEAR
AGE
ACTIVE PLANETARY CYCLES
INFLUENCE>
MAJOR
MEDIUM
MILD
PAGE
JUPITER
MOON
SATURN
URANUS
NEPTUNE
PLUTO
44
BIRTH
0
'Aries' (27)
1st Emerging (49)
9
45
Following very difficult pregnancy, during doodlebug attacks in S.E. England, and when she almost died from an unidentified illness, my mother was very ill when I was born, and I had to be taken and put to one side. Breastfed for only a few weeks.
1
'Taurus' (28)
46
2
'Gemini' (29)
47
3
'Cancer' (30)
Move from place of birth to new home.
1st Striving (58)
48
4
'Leo' (31)
Go to first school where I feel very self-conscious but become the centre of attention.
49
5
'Virgo' (32)
50
6
'Libra' (33)
1st Waxing Square
110
51
1st Deciding (63)
7
'Scorpio' (34)
Feeling lonely and a stranger in the world. Sensing hardness of life. Hopelessly 'in love' with older girl at school
52
8
'Sagittarius' (35)
53
9
'Capricorn' (36)
1st Waxing Trine
Naturally discover instinct for leadership at new school (natal Aries Moon).
111
54

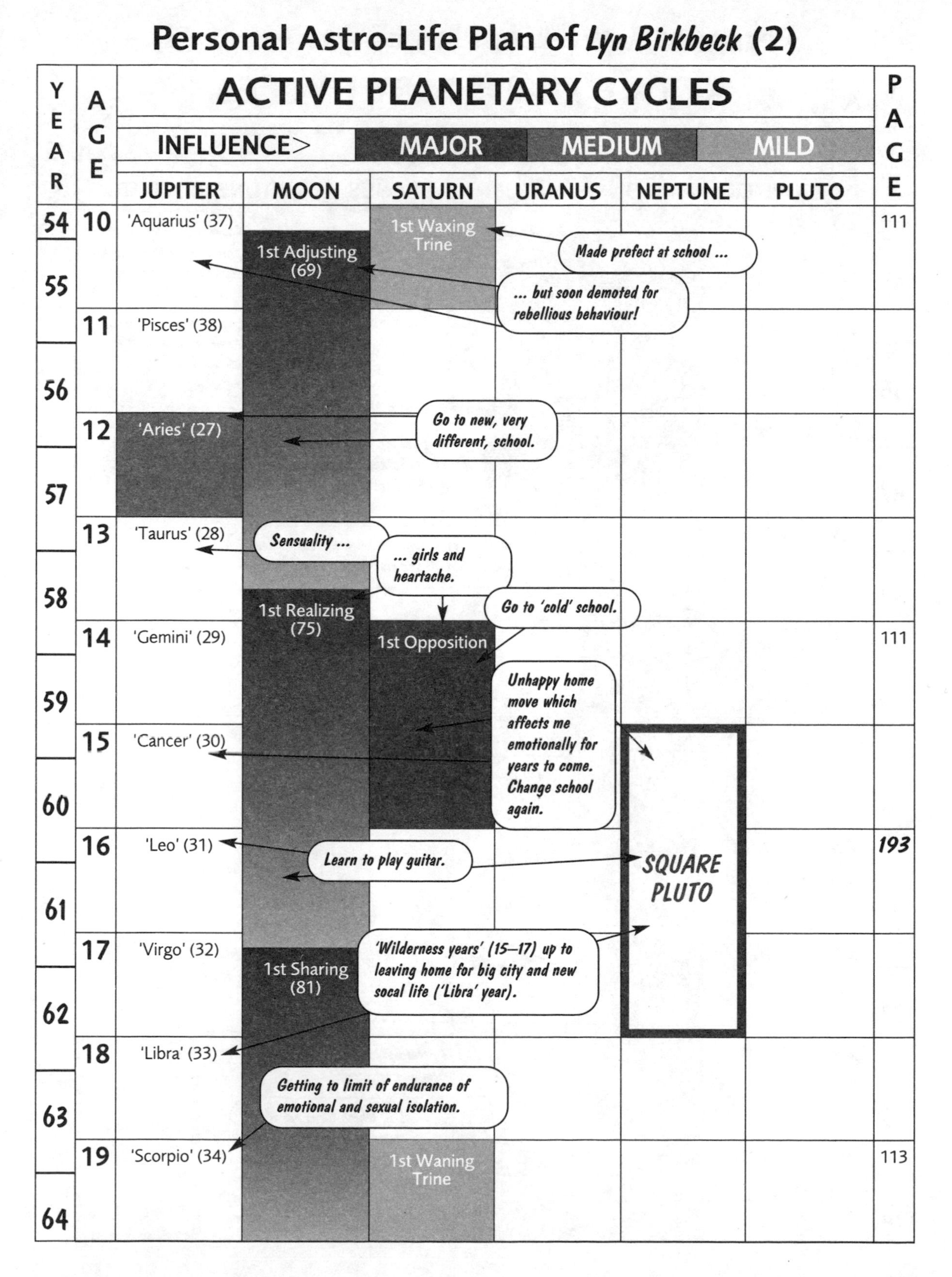

Personal Astro-Life Plan of Lyn Birkbeck (2)
YEAR
AGE
ACTIVE PLANETARY CYCLES
INFLUENCE>
MAJOR
MEDIUM
MILD
JUPITER
MOON
SATURN
URANUS
NEPTUNE
PLUTO
PAGE
54
55
56
57
58
59
60
61
62
63
64
10
11
12
13
14
15
16
17
18
19
'Aquarius' (37)
'Pisces' (38)
'Aries' (27)
'Taurus' (28)
'Gemini' (29)
'Cancer' (30)
'Leo' (31)
'Virgo' (32)
'Libra' (33)
'Scorpio' (34)
1st Adjusting (69)
1st Realizing (75)
1st Sharing (81)
1st Waxing Trine
1st Opposition
1st Waning Trine
SQUARE PLUTO
111
111
193
113
Made prefect at school ...
... but soon demoted for rebellious behaviour!
Go to new, very different, school.
Sensuality ...
... girls and heartache.
Go to 'cold' school.
Unhappy home move which affects me emotionally for years to come. Change school again.
Learn to play guitar.
'Wilderness years' (15–17) up to leaving home for big city and new socal life ('Libra' year).
Getting to limit of endurance of emotional and sexual isolation.

Personal Astro-Life Plan of *Lyn Birkbeck* (3)

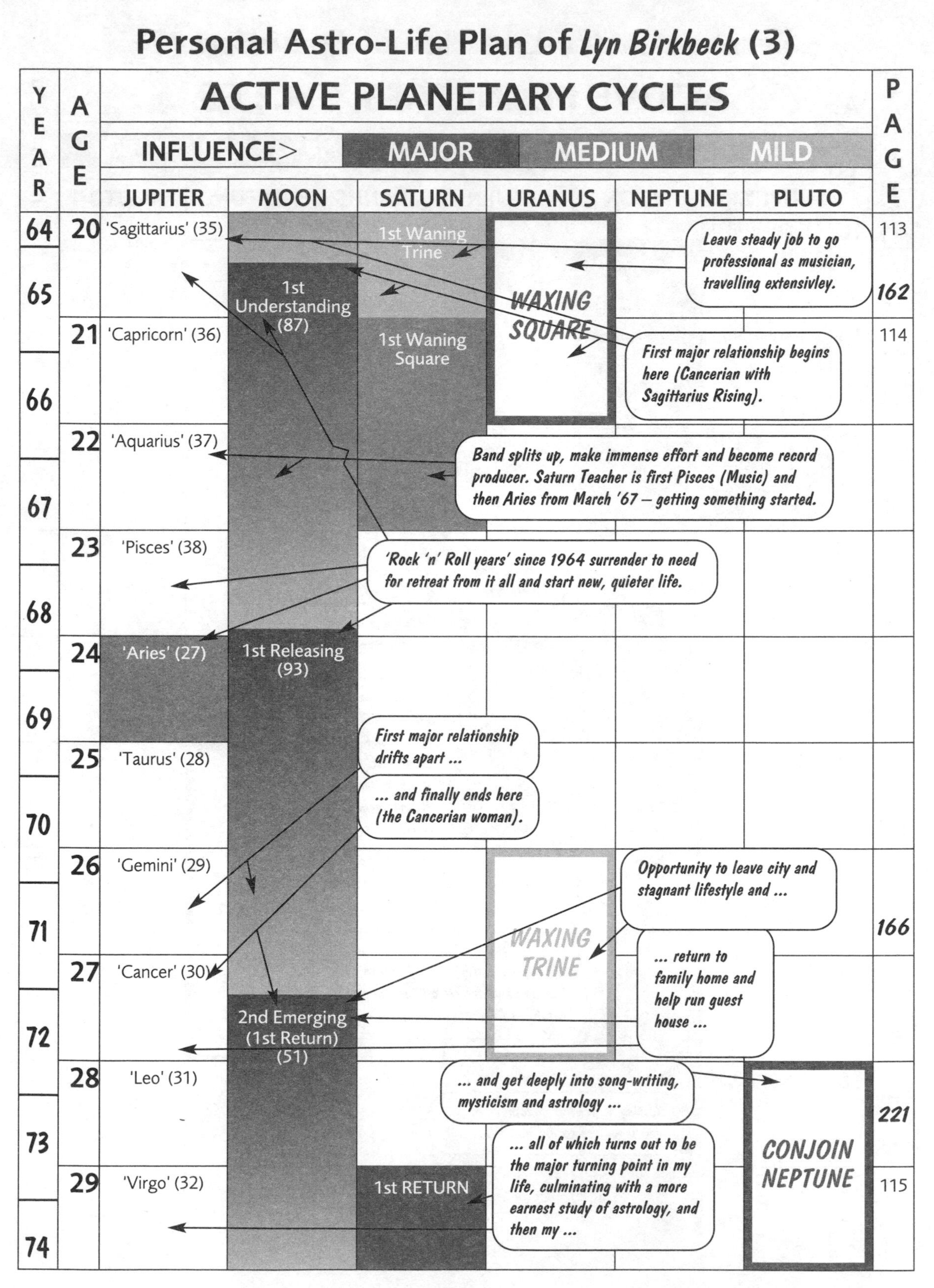

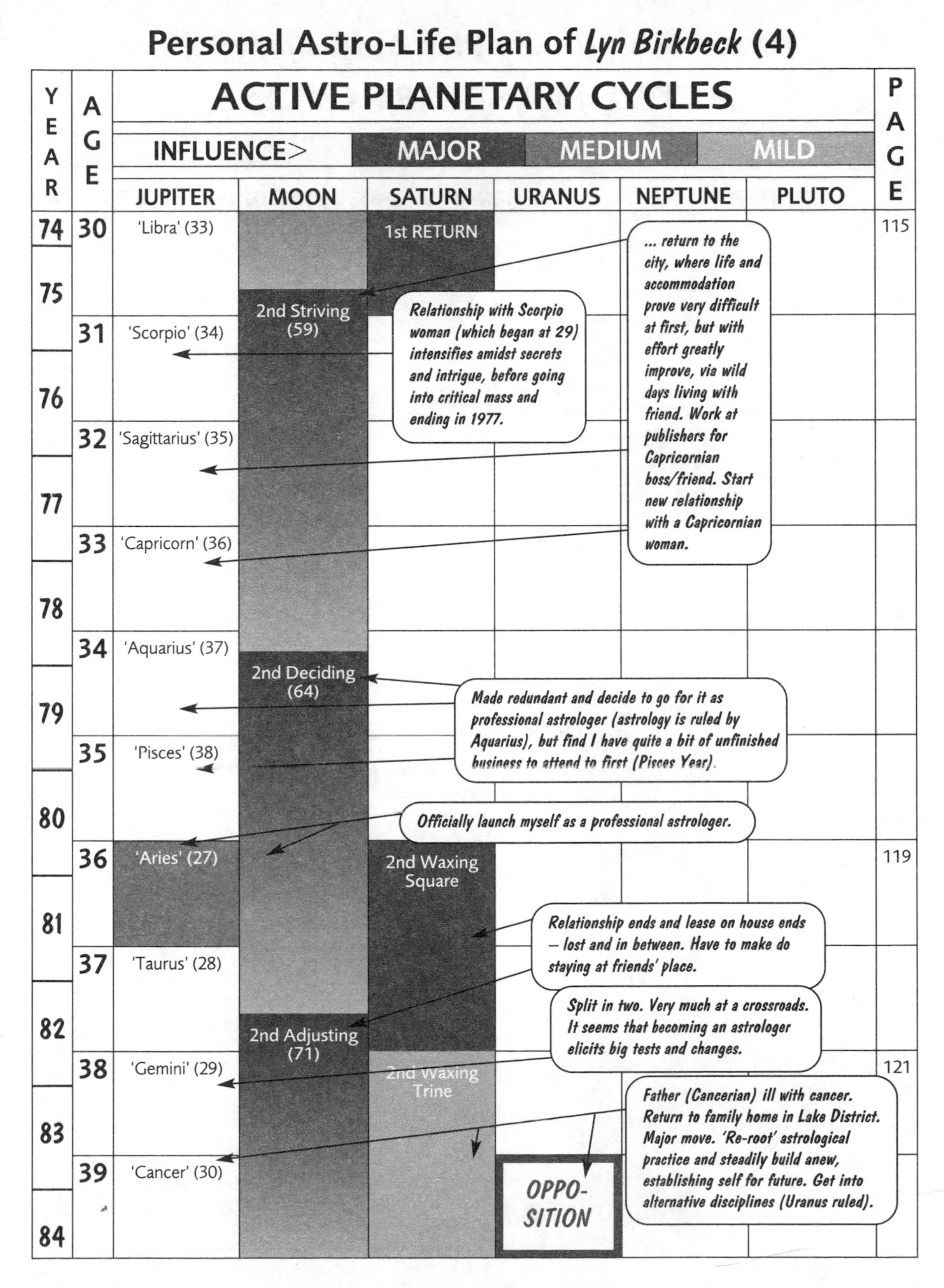
Personal Astro-Life Plan of Lyn Birkbeck (4)
YEAR
AGE
ACTIVE PLANETARY CYCLES
INFLUENCE>
MAJOR
MEDIUM
MILD
PAGE
JUPITER
MOON
SATURN
URANUS
NEPTUNE
PLUTO
74
75
76
77
78
79
80
81
82
83
84
30
31
32
33
34
35
36
37
38
39
'Libra' (33)
'Scorpio' (34)
'Sagittarius' (35)
'Capricorn' (36)
'Aquarius' (37)
'Pisces' (38)
'Aries' (27)
'Taurus' (28)
'Gemini' (29)
'Cancer' (30)
2nd Striving (59)
2nd Deciding (64)
2nd Adjusting (71)
1st RETURN
2nd Waxing Square
2nd Waxing Trine
OPPO-SITION
115
119
121
... return to the city, where life and accommodation prove very difficult at first, but with effort greatly improve, via wild days living with friend. Work at publishers for Capricornian boss/friend. Start new relationship with a Capricornian woman.
Relationship with Scorpio woman (which began at 29) intensifies amidst secrets and intrigue, before going into critical mass and ending in 1977.
Made redundant and decide to go for it as professional astrologer (astrology is ruled by Aquarius), but find I have quite a bit of unfinished business to attend to first (Pisces Year).
Officially launch myself as a professional astrologer.
Relationship ends and lease on house ends – lost and in between. Have to make do staying at friends' place.
Split in two. Very much at a crossroads. It seems that becoming an astrologer elicits big tests and changes.
Father (Cancerian) ill with cancer. Return to family home in Lake District. Major move. 'Re-root' astrological practice and steadily build anew, establishing self for future. Get into alternative disciplines (Uranus ruled).

Personal Astro-Life Plan of *Lyn Birkbeck* (5)

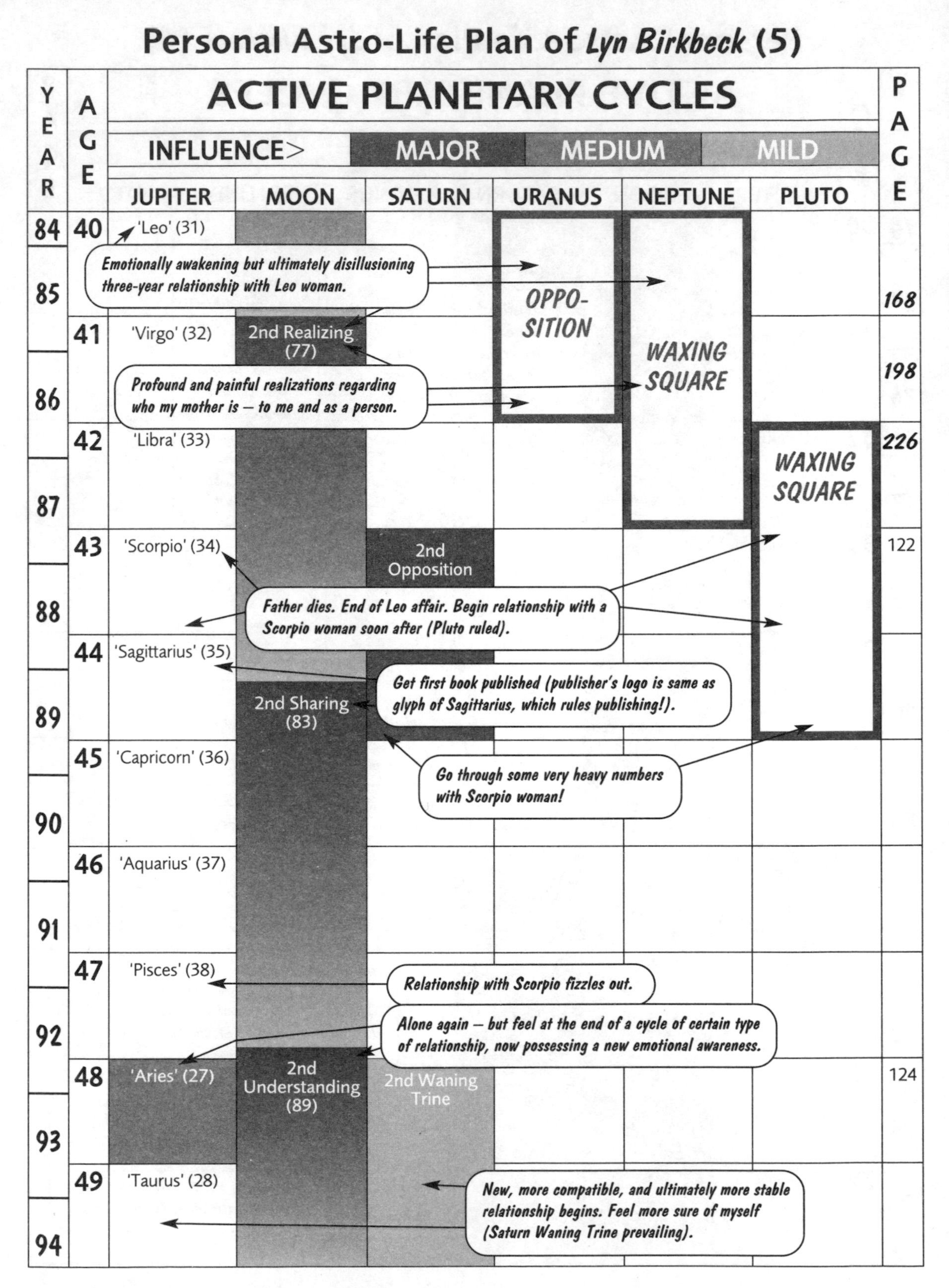

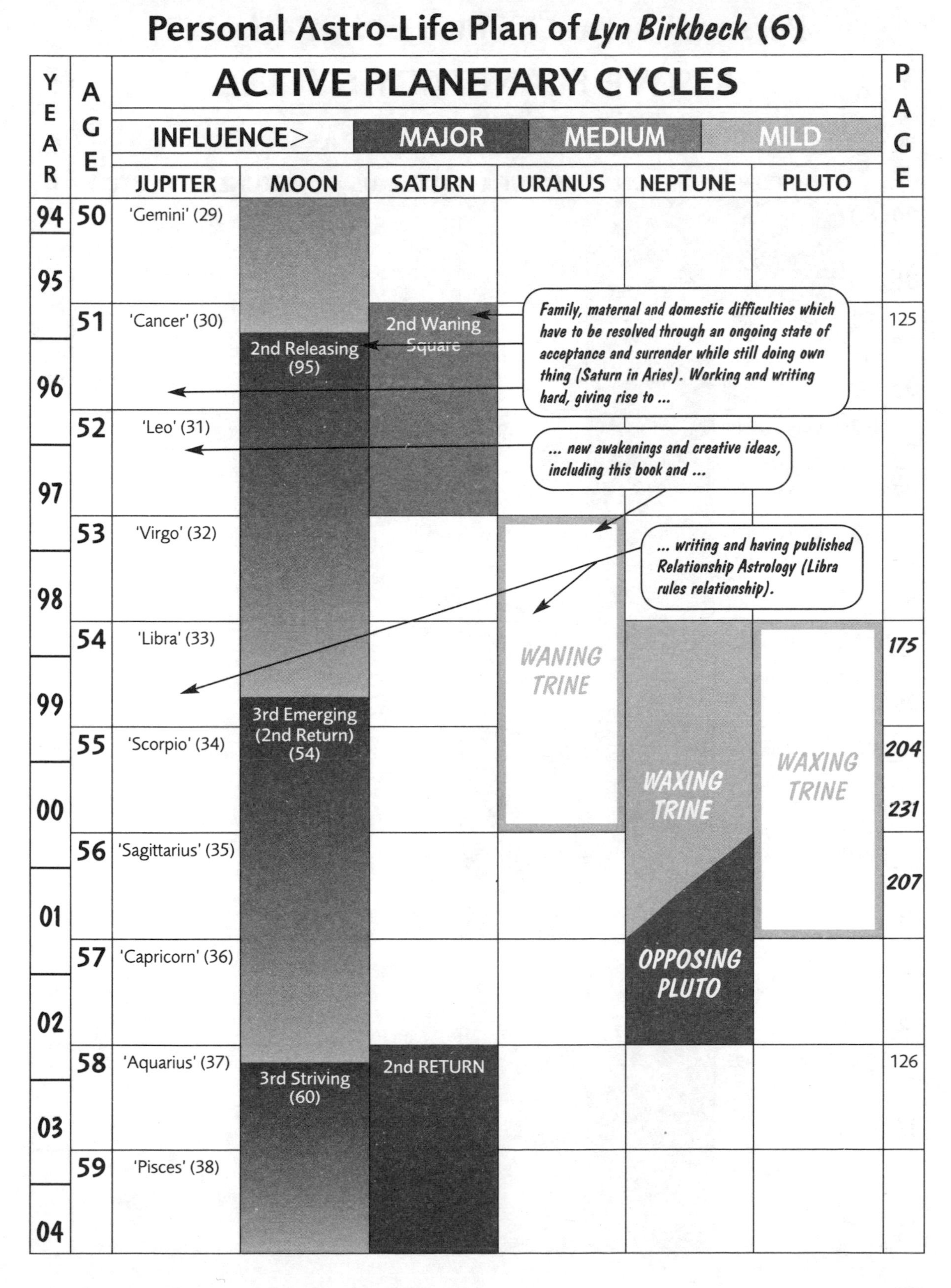
Personal Astro-Life Plan of *Lyn Birkbeck* (6)
YEAR
AGE
ACTIVE PLANETARY CYCLES
INFLUENCE>
MAJOR
MEDIUM
MILD
JUPITER
MOON
SATURN
URANUS
NEPTUNE
PLUTO
PAGE
94
95
96
97
98
99
00
01
02
03
04
50
51
52
53
54
55
56
57
58
59
'Gemini' (29)
'Cancer' (30)
'Leo' (31)
'Virgo' (32)
'Libra' (33)
'Scorpio' (34)
'Sagittarius' (35)
'Capricorn' (36)
'Aquarius' (37)
'Pisces' (38)
2nd Releasing (95)
3rd Emerging (2nd Return) (54)
3rd Striving (60)
2nd Waning Square
2nd RETURN
WANING TRINE
WAXING TRINE
OPPOSING PLUTO
WAXING TRINE
125
175
204
231
207
126
Family, maternal and domestic difficulties which have to be resolved through an ongoing state of acceptance and surrender while still doing own thing (Saturn in Aries). Working and writing hard, giving rise to ...
... new awakenings and creative ideas, including this book and ...
... writing and having published Relationship Astrology (Libra rules relationship).

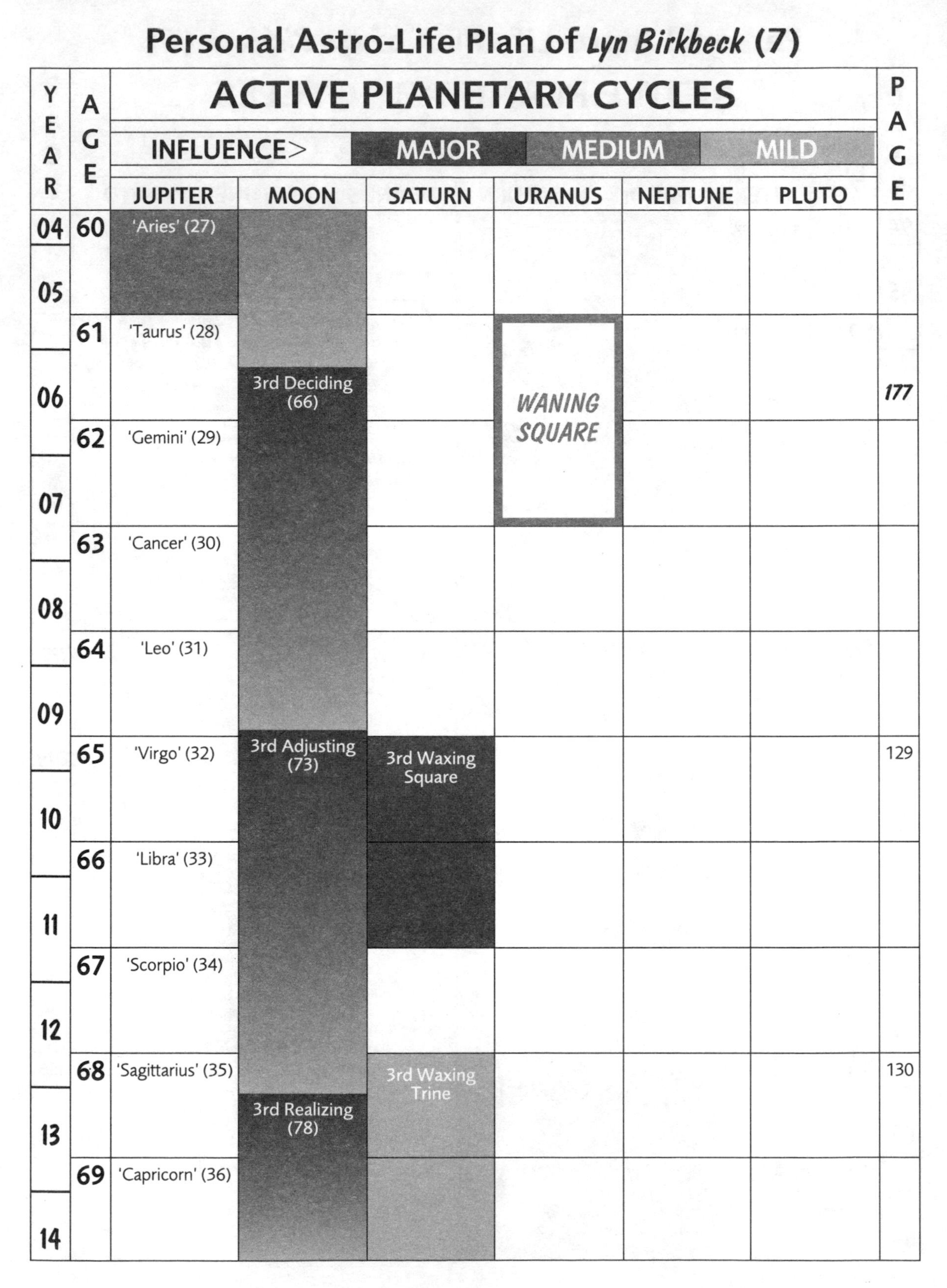
Personal Astro-Life Plan of *Lyn Birkbeck* (7)
ACTIVE PLANETARY CYCLES
YEAR
AGE
PAGE
INFLUENCE>
MAJOR
MEDIUM
MILD
JUPITER
MOON
SATURN
URANUS
NEPTUNE
PLUTO
04
05
06
07
08
09
10
11
12
13
14
60
61
62
63
64
65
66
67
68
69
'Aries' (27)
'Taurus' (28)
'Gemini' (29)
'Cancer' (30)
'Leo' (31)
'Virgo' (32)
'Libra' (33)
'Scorpio' (34)
'Sagittarius' (35)
'Capricorn' (36)
3rd Deciding (66)
3rd Adjusting (73)
3rd Realizing (78)
3rd Waxing Square
3rd Waxing Trine
WANING SQUARE
177
129
130

Personal Astro-Life Plan of *Lyn Birkbeck* (8)

ACTIVE PLANETARY CYCLES

INFLUENCE> MAJOR | MEDIUM | MILD

YEAR	AGE	JUPITER	MOON	SATURN	URANUS	NEPTUNE	PLUTO	PAGE
14, 15	70	'Aquarius' (37)						
16	71	'Pisces' (38)						
17	72	'Aries' (27)	3rd Sharing (84)					
18	73	'Taurus' (28)		3rd Opposition				130
19	74	'Gemini' (29)						
20	75	'Cancer' (30)	3rd Understanding (90)					
21	76	'Leo' (31)						
22	77	'Virgo' (32)						
23	78	'Libra' (33)	3rd Releasing (97)	3rd Waning Trine				131
24	79	'Scorpio' (34)						

Personal Astro-Life Plan of *Lyn Birkbeck* (9)

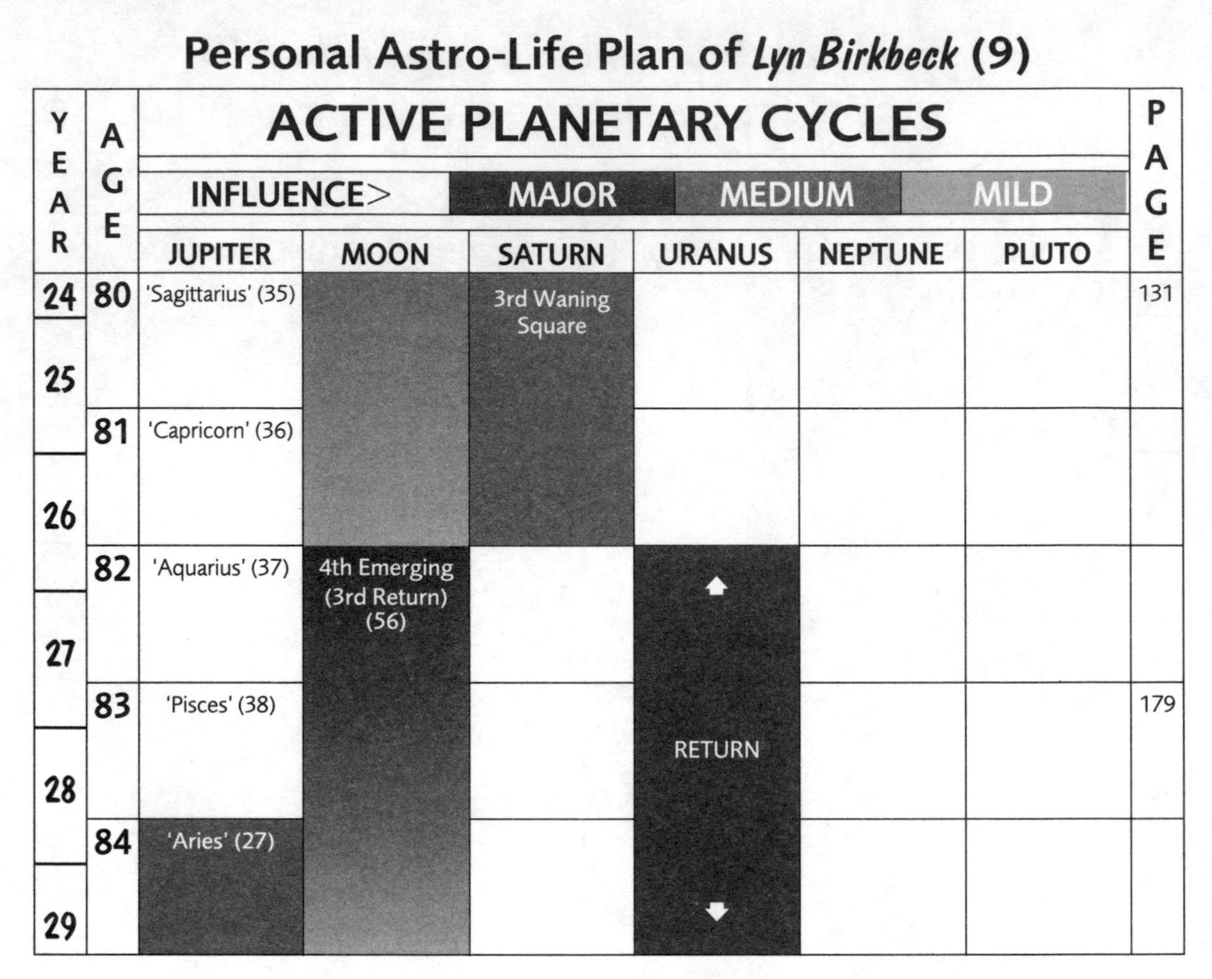

ACTIVE PLANETARY CYCLES

INFLUENCE> MAJOR | MEDIUM | MILD

YEAR	AGE	JUPITER	MOON	SATURN	URANUS	NEPTUNE	PLUTO	PAGE
24 25	80	'Sagittarius' (35)		3rd Waning Square				131
26	81	'Capricorn' (36)						
27	82	'Aquarius' (37)	4th Emerging (3rd Return) (56)		↑			
28	83	'Pisces' (38)			RETURN			179
29	84	'Aries' (27)			↓			

Chapter Five

MANAGING YOUR PLANETARY CYCLES AND YOUR LIFE

Planetary effects, figures and responses

IN ORDER TO HELP YOU MANAGE ANY problematic issues that life presents you with during any particular Saturn, Uranus, Neptune or Pluto influence or set of influences, the following pages offer key-words for all these Planetary Effects, along with the Figures (types of people, things, events or situations) that they will possibly manifest as, and with recommended Positive Responses to adopt and Negative Responses to avoid. These lists do not pretend to be complete or all-purpose, but they can at the very least provide useful clues, suggestions and guidance. To use them, follow the procedure below.

1 Turn to the page of the Planet the Effects of which you are having to deal with (*see* below for Effects).
2 Find a word in the 'EFFECTS' column that corresponds as nearly as possible to the issue you are addressing and enter that Effect in Box A of the form for that Planet given below the Effects, Figures and Responses lists, or on the same form at the back of the book (photocopy some first!). Enter more than one Effect if you need to.
3 Find the word or words in the 'FIGURES' column that corresponds as nearly as possible to the person, thing or state you are involved with, and enter that Figure or Figures in Box B. This may be more than one type, or more than one of the same type. (Note that the Figures are not aligned with the Effects, but are simply presented in alphabetical order.)
4 Find the word or words in the 'POSITIVE RESPONSES' column that seems appropriate to the Effect and Figure, and enter that response or those responses in Box C.
5 Find the word or words in the 'NEGATIVE RESPONSES' column that seems appropriate to avoid, and enter that response or those responses in Box D.
6 You now have a simple Statement and a Solution regarding your issue or question.

Here is an example of an issue arising during a Saturn influence.

Saturn Management: Statement/Solution

When SATURN is creating an EFFECT or ISSUE of ***Increased responsibilities***	**Box A**
Imposed/brought on by ***Elderly person***	**Box B**
then POSITIVE RESPONSE(S) towards this could be to ***Be patient and unassuming and/or do whatever your duty is***	**Box C**
while NEGATIVE RESPONSE(S) would be ***Feeling compromised or acting immaturely***	**Box D**

You may find that the Effects or Figures that you are experiencing correspond to a Planet that you do not appear to be technically under the influence of. This is probably because you are experiencing an Individual Planetary Cycle regarding that Planet (*see* page 6). If this is the case, simply use the suggested management of that Planet as appropriate. Similarly, if you are under the influence of more than one Planet, you might find that the Positive Response of the one Planet suits well the Effects/ Figures of the other Planet.

SATURN

Effects

- Bones, teeth, skin, hair, knees
- Consolidating matters
- Constriction/dissatisfaction
- Delay/lateness/slowness
- Deprivation/distress
- Doubt/inadequacy
- Drudgery/hardness
- Failure/dysfunction
- Fear/anxiety
- Feeling your age
- Heaviness/oppression
- Increased responsibilities
- Increased/onerous pressure
- Limitation/imposing boundaries
- Material/financial difficulties
- Order (enforced)
- Physical condition/deterioration
- Physical instability/weakness
- Reality taking precedence over feelings
- Separation/differences
- Slowing things down
- Stress
- Structural considerations
- Test/examination
- Time and timing

Figures

- Ageing
- Authority (figure)
- Bitter person
- Builder/construction worker
- Cold, unfeeling person
- Controlling person
- Definers of boundaries
- Depressed/ive person
- Disciplinarian
- Elderly person
- Important person/relationship
- Material predicament/issues
- Mean/dry person
- Official
- Older person
- Patriarch
- Posture (bad)
- Professional person/relationship
- Restrictive person/relationship
- Rigid attitude
- Saturnine person
- Serious relationship
- Teacher
- The Devil
- The world at large
- Wet blanket

Positive Responses

- Be patient and unassuming
- Be sober and serious-minded
- Do whatever your duty is
- Exercise discipline/your body
- Face reality/do what you must
- Focus attention/be objective
- Get organized/be economical
- Look and learn and prepare/study
- Lose weight/tighten your belt
- Make more effort/measure up
- Seek professional advice
- Use step-by-step procedure
- Utilize your time/be punctual

Negative Responses

- Acting immaturely/prematurely
- Arrogance/defiance
- Being too cautious
- Being too conventional
- Depression/pessimism
- Excessiveness/imprudence
- Fatigue/feebleness
- Feeling compromised/hurried
- Feeling cornered/thwarted
- Feeling inadequate/hopeless
- Health abuse

Saturn Management: Statement/Solution*

When SATURN is creating an EFFECT or ISSUE of	**Box A**
Imposed/brought on by	**Box B**
then POSITIVE RESPONSE(S) towards this could be to	**Box C**
while NEGATIVE RESPONSE(S) would be	**Box D**

* For further Saturn Management, *see* 'The Jupiter–Saturn Balance of Power' on page 265

URANUS

Effects	Figures	Positive Responses
Accelerated pace	Activist	Be cool, objective, to the point
Accident/shock	Astrology/astrologer	Be intuitive and stick to the truth
Alienation	Eccentric	Be open and non-judgemental
Alternative/esoteric subjects	Electrician	Befriend the situation/person
Birth	Engineer	Determine what situation means
Change/turnaround	Flyer	Make necessary changes only
Circulation	Fool	Move with the times
Coincidence	Genius	Obtain sufficient information
Disruption/upset	Homosexual	Retreat, live to fight another day
Divorce/relationship crisis	Internet/computer	Streamline/simplify
Electricity	Intuitive type	Bear with them kindly
Freedom/rights	Leader/catalyst	Take a calculated risk
Future development	Loose cannon	Take with a pinch of salt
Modernization	Machine	**Negative Responses**
Principles	Media (person)	
Reforming matters	Oddball	Acting out of fear or anxiety
Remoteness/detachedness	Pathfinder	Acting or behaving unnaturally
Revolutionizing matters	Psychologist	Going along with the crowd
Signs (Omens)	Radical/interloper	Hedging your bets
Spasm/hyperactivity	Rebel	Loosing sight of feelings
Surgery/emergency	Reckless person	Losing your cool/panicking
Strike/unrest	Researcher	Not being true to yourself
Technological matters	Robot	Not questioning things
Truth (moment of)	Scientist/technician	Pretending not to care
Unprecedented situation	Socialist/communist	Rationalizing to miss the point
Unpredictability/surprise	Specialist	Remaining uninformed
Unusualness/unconventionality	Trade union(ist)	Resisting any change at all
	Unusual type	Violent rebellion or reaction

Uranus Management: Statement/Solution

When URANUS is creating an EFFECT or ISSUE of	Box A
Imposed/brought on by	Box B
then POSITIVE RESPONSE(S) towards this could be to	Box C
while NEGATIVE RESPONSE(S) would be	Box D

NEPTUNE

Effects

- Absentmindedness
- Addiction/alcoholism
- Anaesthesia/fainting/trance
- Astral phenomena/ghosts
- Charismatic appeal/glamour
- Chemicals/oils/gas
- Confusion/strangeness
- Conscience
- Deception/suspicion/lies
- Delusion/disillusion-ing/-ment
- Enigma/vagueness/mystery
- Frustration
- Hallucination/hysteria/insanity
- Healing (bogus)
- Hospitalization
- Hygiene (obsessive)
- Illogicality
- Infatuation/gullibility
- Infection/poisoning
- Lack or loss of energy
- Metaphysics/spiritualism
- Pining/longing
- Scandal
- Temptation/weakness
- Vulnerability/oversensitivity
- Water (= emotional) problems

Figures

- Addict/alcoholic
- Chemical industry
- Chronically sick (person)
- Confidence trickster
- Doctor/quack
- Drink, drug or tobacco (industry)
- Drug dealer
- Emotional type
- Guru/visionary (suspect/false)
- Haunted/spooky places
- Hypnotist (suspect)
- Illusionist
- Imagination
- Institution
- Medium/clairvoyant/psychic
- Music(ian)/dance(r)/artist
- Neurotic person
- Pharmaceutical industry
- Poisonous creature/person
- Seducible/wanton person
- Seductive/teasing person
- Unreliable person
- Unstable person
- Victim/pathetic, helpless person
- Waterworks/plumber
- Witch/wizard

Positive Responses

- Accept your fate
- Be gentle on yourself; then other
- Check out/deal with victimhood
- Be compassionate but tough
- Consider karmic reasons
- Get the facts/check credentials
- Let things be/live and let live
- Look deep within for the answer
- Practise yoga/deep breathing
- Research their background
- Seek to understand it/them
- Trust good track record only
- Try to see the bigger picture

Negative Responses

- Being afraid to face the music
- Disregarding spiritual aspect
- Going in blindfold/half-cocked
- Imagining it will go away
- Mistaking (your) weakness for compassion
- Not being emotionally honest
- Not getting it in writing
- Not questioning matters at all
- Not thinking things through
- Thinking/hoping for the best

Neptune Management: Statement/Solution

When NEPTUNE is creating an EFFECT or ISSUE of	**Box A**
Imposed/brought on by	**Box B**
then POSITIVE RESPONSE(S) towards this could be to	**Box C**
while NEGATIVE RESPONSE(S) would be	**Box D**

PLUTO

Effects

Abuse/rape
Compulsive feelings
Constipation/toxicity
Corruption/underhandedness
Criminal activity/behaviour
Decadence/degradation
Disempowerment
End/death of something/one
Evil/horror/nightmare
Feeling convinced but unhappy
Feeling very alone
Guilt/remorse
Inaccessibility
Intense uncomfortable feelings
Intrigue/undercurrents
Manipulation/coercion
Matters relating to death
Occult (trouble with the)
Powerlessness
Psychological/emotional pain
Sexual disease/lust
Smokescreens/secretiveness
Stalking/spying
The other side
Unsatisfied desire/loss
Violence/disaster

Figures

Criminal/hoodlum/mugger
Cruel/treacherous person
Dark person/black magician
Destructive person/thing
Fanatical person/thing
Forces (armed or otherwise)
Hard to reach person/thing
Insurance agent
Intimate/blackmailer
Jealous/desperate person
Kidnapper/cuckolder
Monster/devil/immoral person
Natural processes of decay
Nihilistic person
Obsessive/intense person
Overpowering person/thing
Pest/nuisance/menace
Poisonous creature
Police officer
Pornography/pervert
Powerful person/thing
Prostitute
Secretive person/thing
Taboo breaker/keeper
Trash/revolting person/thing
Tyrant/dictatorial person

Positive Responses

Accept/own your own dark side
Be strong and admit weaknesses
Confess your own wrongdoing
Delve into yourself, non-judging
Discover and surrender to the Light
Discover the root cause
Forgive yourself and/or other
Hand it over to Higher Power
Psychically protect yourself (*see* page 271)
Pull out and cut your losses
Purify/detoxify your mind/body
Raise feelings higher up body
Seek psychological/spiritual help

Negative Responses

Being controlled by something
Being hateful and unforgiving
Denying own part in it/blaming
Getting in deeper/too deep
Giving in to base desires/feelings
Going it alone (without the Light)
Guiltily hiding yourself or something/someone you don't respect
Thinking two wrongs make a right
Thinking worst of yourself/others
Torturing/tormenting self/other

Pluto Management: Statement/Solution

When PLUTO is creating an EFFECT or ISSUE of	Box A
Imposed/brought on by	Box B
then POSITIVE RESPONSE(S) towards this could be to	Box C
while NEGATIVE RESPONSE(S) would be	Box D

The Jupiter–Saturn Balance of Power

'And with joy we'll persevere'

MANAGING TO LIVE HEALTHILY AND SUCCESSFULLY in the midst of the experiences that the Planetary influences symbolize can be greatly assisted by understanding how to balance the effects of these, the two biggest Planets in our Solar System.

You Are a Physical Being

Although this might seem obvious, we do tend to forget what being physical actually means. One of the most important aspects of physical existence is pressure. We can appreciate this by thinking of an automobile tyre. The tyre itself has an internal air pressure to counteract the external air pressure of our atmosphere. When the internal pressure is too low or too high in relation to the external pressure, then the vehicle becomes unstable while travelling.

Relating all this to being human is quite simple. We travel around, or live, with an internal sense of life and ourselves. This 'internal pressure' is what we need in order to respond to and contend with the pressures of living in the world of people and things – the 'external pressure'. Astrologically, Jupiter represents the internal pressure, while Saturn stands for the external pressure; and you cannot, and should not, consider the one without the other. When our internal pressure or faith in ourself is too low then the everyday pressures of our daily lives become too hard to bear – eventually it can even crush us or put us 'off the road'. Alternatively, when our internal pressure is too high or we have an inflated idea of our capabilities and prospects, then through ignoring the limitations placed on us by the external pressure (which includes your physical body) we can burn out. Alternatively, external pressure has to compensate in some way – for instance, causing us to fall foul of someone in authority, or someone who has an emotional hold on us.

The tyre metaphor again applies here in that lower tyre pressure is more secure as there is more grip on the road surface (like for wet weather driving), but it uses up more energy and so limits how far we can go. With higher tyre pressure there is less grip and less energy used – but driving is more precarious. In both cases there is far less control over the vehicle (you, that is!), especially if an emergency or the unexpected should happen.

The principle of pressure can also be seen in terms of something highly applicable to our own physical health and existence: blood pressure. With low blood pressure we feel weak and faint, so, psychologically speaking, we need 'pumping up' with a stronger sense of ourselves in order to cope with life. This can also mean feigning humility (being a victim) in order to avoid asserting oneself. With high blood pressure we are over-compensating for what we feel the pressures of the outside world to be – and so need to calm down, or be let down. This could also mean feigning being full of oneself as a defence mechanism.

All of the above can actually be seen in the symbols or glyphs for Jupiter and Saturn. Both are comprised of two other symbols – the Crescent for the Moon and the Cross for Matter. The Moon represents your emotional being – that is, your personal sense of existing. So, Jupiter, on the left, the Crescent on

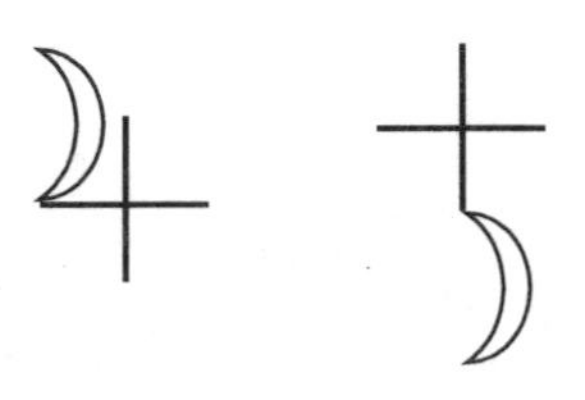

top of the Cross, is the personal, emotional sense taking precedence over the material world and its pressures and considerations. Saturn, on the other hand, is those material pressures and considerations taking precedence over your emotional or personal state of being.

The following tables show the various states that can be experienced through having either too much Saturn and not enough Jupiter, or the reverse, along with suggestions for appropriate remedies.

Both Jupiter and Saturn effects can result from over-compensation, that is, too much Jupiterian behaviour can eventually attract more Saturn influences in to your life, or conversely, too much Saturnian behaviour will attract more Jupiter influences. For example, high blood pressure (too much Jupiter) could be a condition compensating for feeling inhibited (too much Saturn).

♄ ♃

Too much SATURN and not enough JUPITER

- *Feeling sluggish and/or depressed*
- *Feeling intimidated by life and others*
- *A sense of hopelessness* • *Lacking in confidence*
- *Fearfulness* • *Worry and anxiety*
- *Low blood pressure*

REMEDY (adopt one or more that appeals to you)

Cultivate optimism; look on the bright side; accentuate the positive, attenuate the negative. Laugh – a sense of humour is a sense of proportion.

Fill your life with a faith in someone or something, but preferably yourself. Follow your dreams.

Be more assertive and ebullient. Be theatrical, outgoing and larger than life.

Attract joy into your life by being true to your responsibilities, exercising impartiality, and protecting the sanctity of your inner being (Moon).

Do yogic breathing exercises. This literally 'pumps' you up, filling you with the spirit of enthusiasm – *pneuma*, as in pneumatic, is the Greek for 'spirit' or 'breath of life'.

Acquire a philosophy of life than enables you to see the bigger picture, thereby cutting your perceived problems down to size or putting them into a more meaningful perspective.

Be more adventurous; feel the fear and do it anyway. Get out, do something different. Let experience be your teacher rather than worry be your tormentor.

Be more open and spontaneous; have fewer agendas and rigid expectations.

Take vigorous physical exercise, let off steam, have a good scream and shout.

Make the most of your Jupiter 'Year' or Growth Mode.

♃ ♄

Too much JUPITER and not enough SATURN

- *Over-optimism resulting in deflation or not being taken seriously by others*
- *Over-commitment; promising more than you can deliver*
- *Excessiveness and indulgence, giving rise to health or weight problems, lack of control and wasted time and energy*
- *False confidence created by alcohol, drugs, delusions of grandeur, etc.*
- *Lack of discipline*
- *High blood pressure*

REMEDY (adopt one or more that appeals to you)

Give yourself readily achievable goals, rather than pie-in-the-sky projects.

Ask yourself why you are trying to please or impress others more than yourself. Make the distinction between what you expect of yourself and what you have been led to believe others expect of you (but probably do not).

Remember that actions speak louder than words. Do it, don't just think about it. Walk your talk.

That which is not written down does not exist.

Practise the art of humility. True greatness veils itself.

By your works shall they know you – not by your intentions.

Identify the 'gap' in your life that you are trying to fill, and then you will know how to successfully fill it. This would be 'creatively indulging' yourself.

Channel excessiveness into a productive pursuit to benefit yourself *and* others.

Use your time constructively – draw up a plan and schedule and keep to it.

Become aware of how you overreact to/overcompensate for what you feel to be inadequacies, rather than being clear what those actual shortcomings are.

Take up a programme of exercise, study or relaxation – preferably one involving others that make sure you stay the course.

Give actuality to words of wisdom.

Protection Along the Way

IT IS BECOMING MORE AND MORE ACCEPTED that we have an 'aura' – a non-physical or quasi-physical envelope that surrounds our physical bodies rather like the Earth herself has an atmosphere that protects her (or rather her inhabitants) from unwelcome outside influences. The aura is egg-like in shape, and extends from beneath the surface of the skin for a distance that can vary between several inches to several feet, depending on the individual's health, energy state and spiritual development.

The aura, again like an atmosphere or the skin itself, is multi-layered. There is some debate as to the nature and names of those parts or layers, simply because such descriptions are dependent upon various clairvoyants' perceptions of them. Science does not at present recognize the aura because it has not got instruments sensitive enough to detect them. There are in fact methods that do this to a degree, like Kirlian photography, but presently the mainstream scientific fraternity chooses largely to ignore them. Such ignorance is reflected in the fact that this fraternity was oblivious enough to the importance and nature of our planet's protective atmosphere to allow a hole in the ozone layer to be created by pollution – a 'by-product' of science.

Likewise, we too as individual beings can get holes in our 'ozone layers', caused by our own polluting thoughts, feelings or habits, which can leave us vulnerable to those of others. Being made vulnerable, our aura can then get more holes punched in it from the outside. Although we cannot do much about how others manage, abuse or pollute their own auras – with how they think, feel and live, we can protect our own. Furthermore, it is not just those 'holes' that can make one vulnerable, but also whatever is *in* one's aura. If one is going around with a negative thought or feeling, this will be present in the aura much like an oil slick in the sea or a poisonous cloud of gas in the air. And, so to speak, oil is drawn to oil, and gas to gas, in that one can thus draw negative people and experiences to oneself. In turn, all of this can affect one's actual physical health, because the aura interpenetrates the physical body – especially the nervous and endocrine (hormone-secreting) systems. Again, using the analogy of planet Earth, this is on a par with greenhouse gases causing global warming, which in turn causes changes in climate and sea levels.

Now, inasmuch as the Planets affect our state of being, we can also use them as a means of classifying the types of 'climate' we are having to live through and with, and thereby choose an appropriate method to deal with that 'climate'. What follows is a number of means of protecting yourself 'along the way' according to whatever Planetary influence you are under. However – and it's a big however – the Planetary influence may not be so simply determined by what this book alone can tell you. This is because the Planetary influence in question could be other than one of the active Universal Planetary Cycles which are the subject of this book. It could be owing to an Individual Planetary Cycle being active. Furthermore, it may be the result of an ongoing Planetary influence from a Natal Planet – that is, a Planetary influence you were born with as part of your personality. In this respect, it should never be forgotten that you yourself are a Planetary influence. So when reading the various methods of protection that are given below, choose one (or more) that appeals to you intuitively or according to the description of what they are for. It is often another, compensatory Planetary energy that does the trick; for example, if you are 'under Uranus' (alienated), it is possibly a Moon method of protection

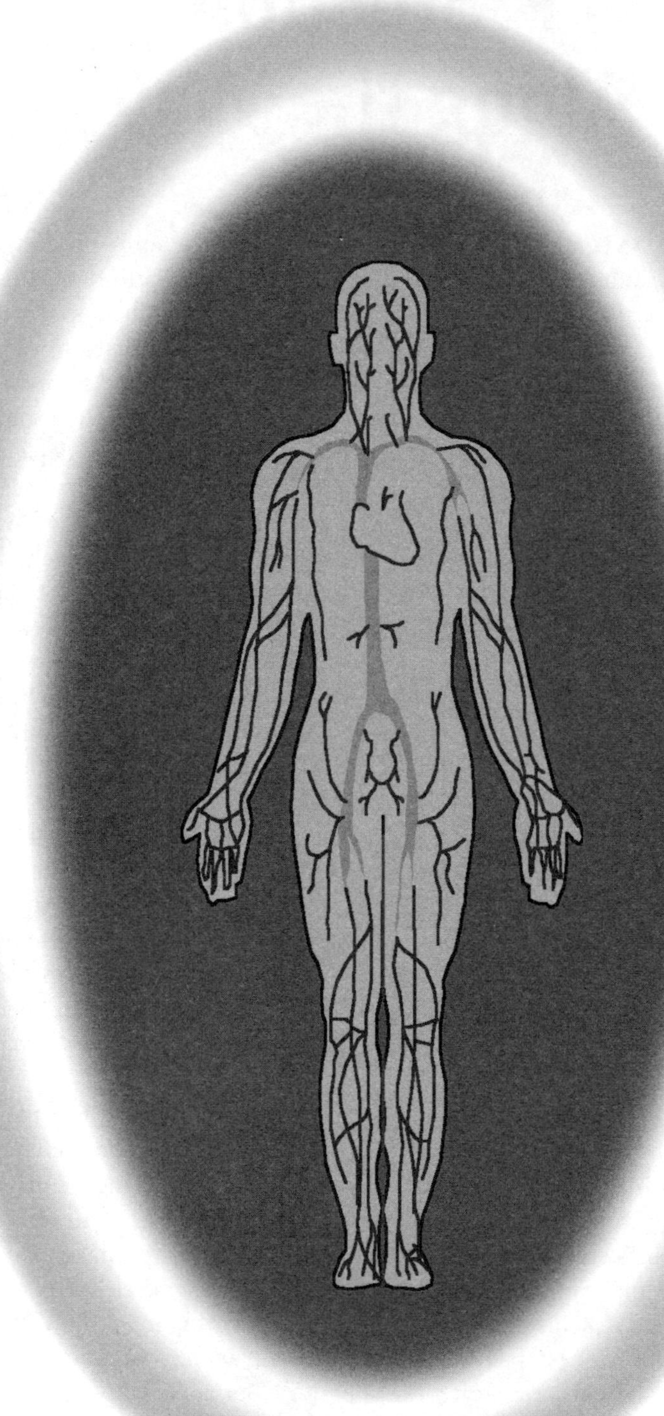

(comforting) you require; if you are 'under Saturn' (depressed) it may be a Jupiter method (inflating) that you would benefit from. Although the Planetary category may concur with a particular Universal Planetary Cycle that is now active for you, these categories are more a way of presenting the whole subject of psychic or auric protection to you in an 'astro-smart' fashion.

Jupiter

Breathing

Breathing is not only a means of protection in the context of Jupiterian influences, but is also what you first need to master to properly utilize most of the methods outlined under the other Planet headings. In line with its pneumatic function as described in the last section 'The Jupiter–Saturn Balance of Power', it that most basic of life forces – the air we breathe. Most significantly, this means of protection has more to do with the *way* you breathe. Most people breathe inefficiently and shallowly, using only a relatively small part of their lungs, and breathing from the chest rather than the diaphragm and abdomen. This is a bit like a blacksmith trying to get a decent fire going by blowing at it through a straw rather than using bellows – and it is your protective aura that is the 'fire' here.

There are many books (mostly on yoga) about how to breathe properly and in ways that vitalize and calm the system, so I shall not go into detail here regarding the various exercises. Suffice to say here that, if you have not done so already, you should learn to breathe more *deeply* to give you more energy, more *evenly* and *slowly* to relax you, and by using your diaphragm and abdomen to pull the air through your nostrils and down into your lungs rather than thinking that it is your nose, mouth or chest that does the breathing. Initially, at least, throughout the day one needs to cultivate an awareness of how one is breathing. You may be surprised to find how quickly or shallowly you are breathing at any given time, not to mention the times you find you are actually not breathing at all! Incidentally, there are also some breathing exercises that entail rapid breathing (to energize) or alternate nostril breathing (to calm).

Breathing is the only human life function that is both conscious and unconscious. All the rest, including our heart and digestive organs, proceed automatically, and there is little most people can do to override them. But we can do so with breathing. Someone once said that 'breathing can make giants of us all' in that through certain exercises and proper breathing throughout the day, we can become more centred, relaxed and in command of ourselves. We can pump ourselves full of life energy.

The Bubble

This method works directly upon your aura. Breathe in the way described above and imagine that you are filling up its egg-like shape until it is well inflated. Make sure that it is covering all of you, as in the illustration opposite. Now gain a sense of how this egg-shaped bubble or balloon is protecting you both through cushioning you against the outer environment as well as keeping bad vibrations out with its 'skin'. Now if you wish to, and in the manner of your Jupiter 'Growth Mode' described earlier, you can feel yourself pumped up, resilient, bouncy and ready to make the most of life.

Saturn

Earthing

Like the Breathing of Jupiter, the Earthing of Saturn is basic to the practice of most means of protection. You are, after all, an Earthling – so you need Earthing to feel safe and at home here. Remembering that your aura is like your personal atmosphere, the next step is to imagine it merging with the aura or atmosphere of the Earth herself. This you can do simply with visualization (while breathing properly) or by more direct means such as walking or standing barefoot on the ground itself – preferably in a natural setting, but anywhere will do. Standing against a tree or actually hugging one is something I personally find very Earthing and reassuring. You might choose to 'adopt' your own special tree, to which you may give a name of your choosing. If you are good at visualization, you can simply imagine that you are a tree with your roots sinking deep into the rich and ancient earth. Finally, a useful and practical way of Earthing yourself is to get down to some physical and earthy task, like energetic housework, making something out of wood, stone or clay, or, best of all, gardening.

Knowing Where You Stand

This is simply a case of knowing where you are geographically. Apart from having a strictly practical application in that it helps to know the names of streets and landmarks when in a strange environment, it is also more basic in that your mind–body system has almost primitive requirements in this respect that must be met. So wherever you are, be aware of in what direction lie the four cardinal points (North, South, East and West), and therefore where the Sun will rise (East) and set (West) and be at noon (due South in the Northern Hemisphere and due North in the Southern Hemisphere). Your system is also reassured by knowing where the nearest large body of water is, where the nearest hills or mountains are, what kind of soil you are living on, in what direction your friends or anyone important to you lives, and any other co-ordinate which is important to you.

Knowing Where You Are Going

This has to do with what I call your 'Slipstream of Purpose'. Discovering your purpose in life is important for many reasons. One that is less obvious is that it gives you a sense of forward-going motion which, like any moving object, has a slipstream. As such, this slipstream causes your aura to be 'massaged' and consequently made more vital and resilient. The first thing to affirm to yourself is that you do have a purpose, even if you have not yet discovered what it is. If this is the case, just tell yourself that your purpose is that of *discovering* your purpose! This is not a cop-out; far from it. If you are in a 'seeking mode', then your aura will respond accordingly. Just think how some unemployed or inactive people can become listless and vulnerable to negative influences. Also, if you know that you are after something, it gives you a stronger sense of what your priorities are and what your time is for, which means that you do not allow yourself to be or feel put upon by others.

The Moon

Protection Itself

As the Moon is the astrological symbol of protection generally, it could be said that all of the methods given here are related to the Moon. This connects with the themes of mother, home and family, as discussed in the Moon section beginning on page 39, and means that the emotional attunement that is facilitated by living with awareness of your Moon Cycle is protecting in itself.

Listening to the World

This is a passive, lunar version of 'Knowing Where You Stand' as described for Saturn. In this case, though, you sit somewhere you are not going to be disturbed (preferably outside) and listen in an intent but relaxed way to whatever sounds are going on around you. Do not try to analyse or identify the noises, just aurally acknowledge them. After a short while, you will hear sounds that you were not even aware of, until eventually you have a sense of 'listening to the world'. This is quite integrating, and automatically instils a trust in the great 'out there'. It can also be quite mystical.

Protecting Your Home

As your home is the place where you should feel most protected of all, the method for protecting it is quite elaborate, but well worth doing. However, if your home is not detached or semi-detached, it is not that practical, and I suggest that you simply use the Bubble method given above, enclosing your home in a bubble, rather than, or as well as, yourself. So assuming that your home is detached or semi-detached, and preferably has some garden, this is what you do. Buy enough salt to lay a trail around the perimeter of your abode, and obtain a small branch from a tree that forks in three directions, meaning that there are four arms in all. Trim this branch to about hand-size and then cut each of the four arms off, leaving a fifth piece which was the junction of the four arms. Now mark the cut on each of the four arms with N (north), S (south), E (east) and W (west), and the corresponding cuts on the fifth piece also. Then, using a compass, locate the four cardinal points on the perimeter of your abode, as viewed from as near the centre of your home as possible, and place each arm somewhere at the appropriate point. Keep the fifth piece by for the time being. Now, starting at the East point, lay a trail of salt in a clockwise direction if you live in the Northern Hemisphere, or in an anti-clockwise direction if you live in the Southern Hemisphere, until you come back to where you started. Use a salt container with a spout, and make sure that the trail is as unbroken as possible, and connects with all four arms of wood. Bury or hide the pieces of wood in their respective quarters when you have done so. In some cases it will be impossible to make the trail continuous, for example, with a dividing wall or in a semi-detached house. This does not matter too much because if you are not able to lay down the salt, the chances are that any intruder won't be able to cross the area either. As you are laying this trail, it helps to chant a mantra, if you know one – if not, use the one given below under Neptune (*see* page 278). Ritualize the whole process further if you wish to or know how to. Finally take the remaining salt and the fifth piece of wood and tie them up in a cloth. Then hide the bundle (together with any other significant objects if you wish) in some secret place inside your home, as near the centre of it as possible.

Here's a little anecdote that bears some testament to the efficacy of this method. Not long after protecting my own home in this way, I was looking out of an upstairs window one day when a car came roaring into our drive. Inside I could see a bunch of undesirables drinking and shouting. Before I had any chance to react one way or the other, they suddenly seemed to panic. The driver reversed the car so quickly as to lose control and get stuck on an old tree-stump. This made him panic so much that he ripped his exhaust off trying to get free of the stump. Eventually they were able to make their escape – much to my relief and amazement!

The Divine Umbilical

This is protecting in the sense of securing and emotionally supporting yourself. Begin by sitting, and then relaxing and centering yourself. Then begin humming as deeply as you can, while focusing the vibration of the humming at a point two or three inches above your navel. When you have attained a strong sense of that part of your body, form a ring with the thumb and forefinger of your left hand and place the opening of that ring over this spot so that the outside of your thumb and the thumb-side of your forefinger touch the skin. Now raise your right arm up in the air, bringing all the fingers and thumb of that hand together so that the tips are side by side rather than meeting, thereby forming a nozzle-like shape. Imagine that your right hand and arm is an umbilical cord coming from a divine source (like the Milky Way, the Goddess, or whoever or whatever you can regard as a source of nurturance), then bring it down to meet and insert into your ringed left hand. Do this slowly and imagine divine sustenance entering your solar plexus area. The operation resembles an airplane refuelling in flight from a tanker (mother-ship). You can do this for as long or as many times as you feel you need to. Remember that this is very much a ritual, and so the lead-up to it (breathing, centring, humming, etc) is just as important as the 'refuelling' itself. Sit quietly for a while afterwards to 'digest' what you have been 'fed'.

Belonging

These two methods protect against feelings of 'not-belonging', which can leave one open to negative outside influences. The first method can be done alone or with a partner or close friend. This simply involves gently massaging one another or yourself, which promotes closeness, familiarity and a feeling of being in touch. The second method involves making the following affirmation to yourself at least once daily, in a relaxed state: 'I feel welcome on Planet Earth, wherever I am, whoever I am with, and whatever I am doing. I feel welcome on Planet Earth.' Repeat this three or more times each session, either out loud or to yourself. Focus the vibration or energy of your voice over your solar plexus. It also helps if you lie on your back with the floorboards running the length of your body. Again follow the breathing, relaxing, and centring process. This affirmation will also greatly help to decrease any sudden, even violent, mood swings or panic attacks that you may be experiencing.

Uranus

Tie-cutting

This technique is the opposite to the Divine Umbilical (given above under the Moon), as it removes from your aura the influences of anyone who is 'under your skin' to the point of debilitating or preoccupying you in an unpleasant way. In effect, such a person is feeding off something that you have within you – something to which you initially gave them access, whether consciously or unconsciously, because that part needed to be contacted in some way. In effect then, they have a 'line' into you – a line like an umbilical cord that you must sever. Sit and centre yourself and imagine the other person opposite you. Focus upon their connection with you and visualize it as this line or cord, observing where it connects with them and with you. As you do this, you should feel what it is that connects the two of you – possibly despite what you *think* connects you. In any event, at some point you have to start ruthlessly cutting that cord. Use an imaginary cutting device of some sort (such as a knife, scalpel, guillotine or laser) and keep severing until it is cut and does not grow back together. If it does try to grow back – which is common – focus more closely upon what it is in you that makes the other person so attached to you, or what it is about them that you are attached to. As their attention is unwelcome, the chances are that it is something in yourself which you do not want to look at and come to terms with. The truth of the matter is that until you do so, the cord will keep growing back. Such negative ties can be very persistent – simply because one can persist in not looking at what is causing that tie to be there in the first place.

Sacred Symbols

This is simply the well-known practice of wearing or visualizing certain sacred symbols. These symbols could be one of your own personal predilection, like the glyph for your Sun or Moon Sign (*see* page 285) or one of the following well-known ones.

The Cross

The Star of David

The Celtic Cross

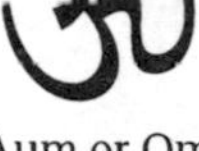
Aum or Om

The Star of Venus

If you are visualizing a symbol to protect yourself, I suggest that you place them over the appropriate chakra as given on page 279. Chakras are symbolic energy centres aligned anatomically to the spinal cord; each relates to an attribute as described in the picture.

Neptune

Clothing in Light

This method is good because you can – true to the selflessness that is Neptune – protect someone else with it. Imagine your own body or that of another to be clothed in light. First imagine absolutely every part of the surface of the body to be covered in a serene blue light to the depth of about three to six inches. (The shade of blue with which the robes of the Virgin Mary are traditionally depicted is best.) Then line the blue coating on the inside, next to your/their body, with a thinner layer of gold. The effectiveness of this method rests mostly upon concentrating on every little spot of the surface of the body – not forgetting armpits and behind the ears, for example!

Love Thine Enemy

This is probably one of the most effective methods of protection because it is the whole concept of 'enemy' that gets us into trouble in the first place. Not only that, but there has to be something in one's own personality and aura that is attracting such an 'enemy'. The secret is to recognize or imagine your enemy as someone whose perceived negativity has its roots in some misunderstanding or mishandling, or some far-distant karmic deed. To see them as vulnerable and originally pure is the key, but it is a difficult thing to do. The reason for this is that you are searching your own heart for a goodness that 'blesses' them or forgives them. In other words, you are neutralizing their darkness with your light – or the Light from a Higher Power. You cannot pretend to do this; if you are pretending it will simply have no effect. If you find that you cannot genuinely do this, then in order to prepare yourself better, I recommend you read about the concept of the Mirror in my book *Do it Yourself Relationship Astrology*.

Mantras

Using mantras, or the reciting or chanting of certain words and phrases, is a highly effective means of not only bringing protection but of attuning oneself to higher levels of consciousness. There are many mantras, but here is one which is central to a popular spiritual discipline called Nichiren Shoshu Buddhism. This mantra goes *Nam-myoho-renge-kyo* (pronounce the 'myo' and 'kyo' as one syllable, as in 'To-kyo') and is repeated over and over again many times. Its literal meaning is 'Mystic Law of the Lotus Sutra', but without going into how or why, essentially it is saying 'I wholly submit to what the truth of my being has in store for me'. This works rather like a self-hypnosis that trains every cell of one's being to follow the path destined for that being, much as a human embryo eventually grows into a adult human being which in turn forms a part of that whole called the human race. This is somewhat similar to the 'Knowing Where You Are Going' protection (*see* page 274), because it protects you from ill-fortune – ill-fortune being something that either does not come to you because you are on your true path, or else it is something that you have to deal with along your path and are therefore capable of dealing with.

Chakras as Flowers

Here we touch upon one of the most basic tenets of esoteric anatomy and thought – the Chakras. The Chakras are the seven focal points of energy within your aura – rather like the main junctions in a complex system of roads or power lines. As with the aura itself, there is some difference in ideas about the nature and number of the Chakras, but here I give the most widely accepted version. These are depicted as flowers in the illustration on page xx. Although the word *chakra* actually means 'wheel', and they have been psychically perceived as wheels or whorls, I have set them out as flowers because this particular imagery works well for our purpose of protection. According to what your issue or problem is (regarding what each chakra is associated with, as given in the picture) visualize the appropriate chakra/flower closing down petal by petal. Concentrate very carefully as you do this, being very aware of how some petals might be harder to close. So if, for example, you find someone emotionally disturbs you, close down the petals of your Solar Plexus Chakra. If you are concerned with whether that Chakra will open again, it will – just like real flowers do. However, some Chakras can be semi-permanently and/or partially closed. There is probably a good reason for this, and one should not go trying to force the petals open. Again, like a real flower, this would damage it as a whole. The whole subject of Chakras is extremely complex and working with them beyond a simple exercise such as this can be hazardous. Even using this safe exercise, you may find that by closing, say, your Solar Plexus Chakra you then feel a sensation of tightness in or around your heart, (this would be telling you that the better remedy for being emotionally upset by someone would be to learn how to love them by opening your heart (chakra) more with a method such as 'Love Thine Enemy' given above. This is because the Chakras are all interlinked in a very subtle way – a more complete knowledge of which is way beyond the scope of this book.

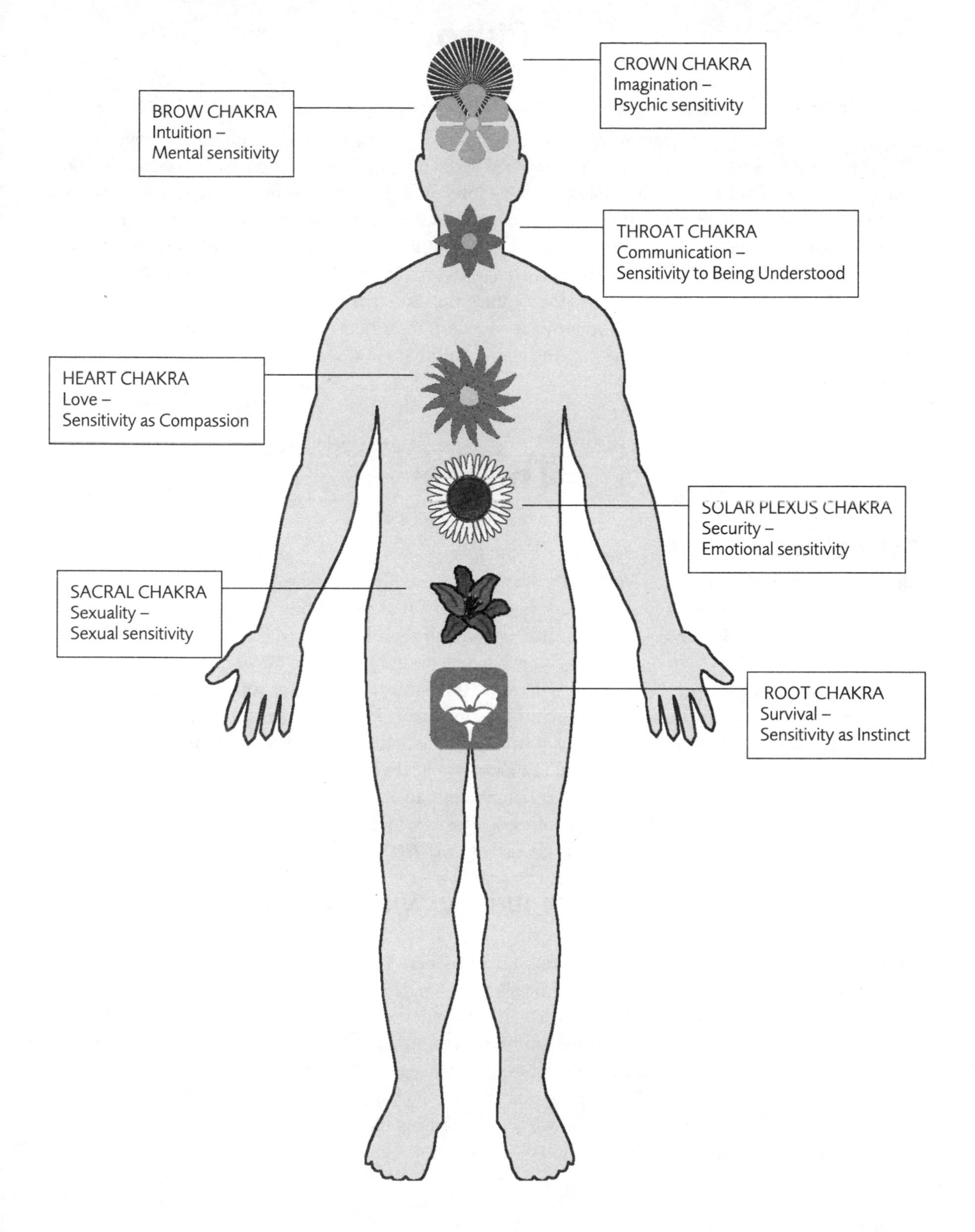
CROWN CHAKRA
Imagination –
Psychic sensitivity
BROW CHAKRA
Intuition –
Mental sensitivity
THROAT CHAKRA
Communication –
Sensitivity to Being Understood
HEART CHAKRA
Love –
Sensitivity as Compassion
SOLAR PLEXUS CHAKRA
Security –
Emotional sensitivity
SACRAL CHAKRA
Sexuality –
Sexual sensitivity
ROOT CHAKRA
Survival –
Sensitivity as Instinct

The Chakras

Pluto

Genuineness

One of the most protective qualities of all is being yourself – which is what genuine astrology aims to help you to do. For instance, when you have worked with your Shadow, as described in the Pluto section, you are no longer so afraid or vulnerable to the Shadows of others. For example, I might feel threatened by violent types of people or situations, but to a considerable degree this would be down to repressed violence in my own personality. John Lennon was a great promoter of peace in the world, but he died a violent death – like a lot of peacemakers (*see* page 168 under the Uranus Opposition). This is not to say that their works are not highly positive – they are. But it is their more deeply-seated personal reason for promoting peace that is the provocative question. When I met John in 1967 while making 'Magical Mystery Tour', I was alarmed at the aggression that exuded from him – even though he was being friendly.

The Sun

Solar Chant

This chant from a Mayan Solar Meditation was given to me by a Mayan Daykeeper or shaman. The Mayans of Southern Mexico were and are highly aware that everything in life is dependent upon the Sun – it is, after all, the Solar System that we and all the Planets exist in. The Sun is seen as the Father who protects his children, and this chant is based around the sacred Mayan name for the Sun. So in performing this chant you are basically entreating the Father to look after you, His Child.

Simply chant the Mayan word for the Sun for 3 sets of 7, 13 or 20 times for each set – the choice being yours. However, please note that this is to be practised only during daylight hours.

The word is *K'IN*, which is pronounced with a hard, clicking *K'* as if you were saying 'kick' without sounding the 'i'. Physically, the back of your tongue slaps against your soft palate (the back of the roof of your mouth). This *K'* is followed, as smoothly as possible, by the *I* which sounds like *EEEEE*. Finally the *N* is sounded in the normal way, but is elongated: *NNNNN*.

K ' EEEEEEENNNNNNNN

I find it helps to sound this word with a pout, as if you were kissing the Sun – the sound then travels from the back of the mouth to the front with ease. One single chant takes one whole exhalation, allotting as much time and breath to the *N* as the *I*. This is important when you consider the onomatopoeic way the word is made up: *K'* represents the ignition of the Sun; *I* represents its light and heat travelling; *N* represents it settling upon whatever it reaches. Inhale deeply before each chant/exhalation.

As stated above, there are three sets or rounds of the chant, each one comprising of 7, 13 or 20 *K'IN*s. Centre and be still within between each set. Performing this chant outside, facing the Sun, is best – or if indoors, simply facing in the direction of the Sun as a sign of respect.

First Set

Chant *K'IN* to the Pleiades, the Seven Sisters, or 'seven stars in the sky' – or 'the Seven Solar Systems' as the Mayans call them. This is in aid of your higher levels of tuning and vibration, the Pleiades being a star cluster of the highest spiritual significance. It is suitable therefore to chant at as high a pitch as you can, which will be in falsetto if you are male. Centre, be still and sense the greater pattern of existence.

Second Set

Chant *K'IN* to the Sun itself (in actuality if possible, or to a candle or solar icon). This is pitched at one octave below the high note of the first set. This chant is for 'all your relations', which means anyone or anything to whom you wish solar energy to be sent. This is unconditional love, so avoid any thought or feeling other than a compassionate one and the name and image of who/what you are chanting/sending to. Alternatively, simply focus upon your heart, imagining it as the Sun radiating or unfolding like a flower as it greets the Sun in the morning. Again, centre, be still and experience oneness.

Third Set

This round is for yourself. This time you chant an octave below the last one, so that it should be pretty deep and resonant. This is so you feel the chant (around your solar plexus or even lower down) more than you may hear it. Also, in order to internalize it further, close your eyes if you wish, and in the middle of the chant and at the very end, close your mouth, while still sounding. So in effect it will sound like this:

K ' EEEEEEEmmmmmmmEEEEEEENNNNNNNmmmmmmmm

While sounding the *NNNNNNN* make sure that you keep your tongue connected to the front of the roof of your mouth, even when your mouth is closed (*mmmmmmmm*) at the end. This is important because your tongue is like a switch which completes a circuit of energy that goes up the spine, down through the mouth, and down the front of the body to start again at the groin. While performing this round of *K'IN*, make sure that you think of and nurture only yourself. Centre, be still and think I AM THAT I AM.

This completes the essentials of this Solar Meditation.

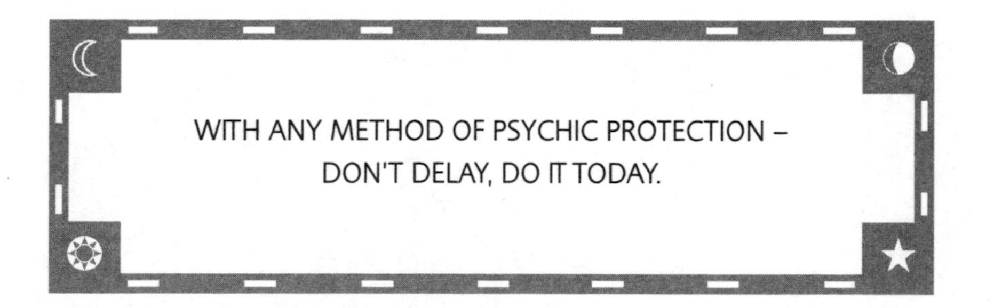

Chapter Six

THE ZODIACAL LIFE-STREAM

'Then the angel showed me the river of the water of life,
as clear as crystal, flowing down from the throne of God and of the Lamb
down the middle of the great street of the city. On each side of the river
stood the tree of life, bearing twelve crops of fruit, yielding its
fruit every month. And the leaves of the tree
are for the healing of the nations.'

Revelation 22, 1–2

Popular and well-known as Sun Signs are, they are not usually appreciated for making up a cycle in themselves, even though everyone is aware of that cycle's name – the Zodiac.

The Zodiac is traditionally seen as a cycle that parallels the unfolding of the four seasons, embracing Nature's budding, blooming, fruiting, dying, sleeping, rebirth process. However, this seasonal analogy, beautiful as it is, does not quite work when viewed from one's own position in the human drama of thinking, speaking, feeling and doing. Moreover, it does not apply to the seasons of the Southern Hemisphere.

To this end, a different metaphor for the Zodiac could be that of a river flowing from source to ocean and back to source. This analogy, I hope you will find, contains the emotional flow that is necessary for a healthy life, a sense of going on a meaningful and eternal journey, and the understanding that in being part of some whole there lies a good reason for why you are as you are.

Above all, there is a spiritual and even romantic quality to this river analogy, for it creatively places us all in that great production where 'all the world's a stage'.

I call this the Zodiacal Life-Stream. In the explanation that follows, the Sign glyphs are as follows:

♈ = Aries	♋ = Cancer	♎ = Libra	♑ = Capricorn
♉ = Taurus	♌ = Leo	♏ = Scorpio	♒ = Aquarius
♊ = Gemini	♍ = Virgo	♐ = Sagittarius	♓ = Pisces

Jupiter Growth Modes can also be further understood by following the Zodiacal Life Stream. For example, for a 'Virgo' year use the Virgo part of the Stream to deepen your understanding of that year of your life, or simply as an affirmation.

The Zodiacal

♒ At last the Stream is liberated on to the wetlands and the delta, bringing irrigation and refreshment to all flora and fauna. And so I bear the precious water of knowledge and life, distributing it evenly and impartially – never resisting or shrinking from the truth, for I know it spreads inexorably. And as all beings are thus nourished and informed with the truths of life, eventually a great awakening occurs amongst the people as the accumulated knowledge that the Stream brings frees us from fear . . .

♓ And finally all the many individual streams of the one Stream empty into the ocean. We realize that our individual longings are ultimately a longing for the same thing – the mysterious sea of peace and acceptance. And so I feel all the life-streams within me as the one Life-stream, and I am faithfully led forever on to inspire and relieve – or I merely crave some non-existent shortcut. For as vapour from the sea rises up into the sky, and falls as rain, or sleet or snow, so too shall we go, and return to the Source . . .

♈ I am the first, straight from the Source, energetically emerging from the hidden depths of the Earth and Primordial Waters, fresh and sparkling. I tumble onward – impetuously and enthusiastically. When strong in flow, I feel independent, and lead others, showing them the way to push on. When the flow is weak, I admit to it, tarry awhile, and allow more energy to accumulate until I can once more exert and assert myself . . .

♉ As I encounter the richness and resistance of the Earth, I savour and ponder Her nature and worth, and in so doing realize my own talents and abilities. When strong in flow, I feel abundant and fertile, and bring the pleasure and reassurance of Nature's goodness to myself and others. When the flow is weak, in order to avoid becoming bogged down, I allow chance and change into my life, freeing and refreshing me . . .

♑ The Stream, having become full, is now tested, as I endure its course through the narrow divides of the canyon. Hemmed in by the steep cliffs of Earthly conditions, I use my natural senses of resourcefulness and ambition to amount to something through negotiating the cataracts of adversity that I now encounter. My power to progress is not in question, for it is inescapable; but my ultimate achievement is to exercise only the control necessary to remain upright and aware of my emotional depths . . .

We must now remember that we all are Sons and Daughters of the Sun. As Mother Earth travels the year around Father Sun, She and we Her Children follow the Course of the River of Life, the Zodiac, beginning in Aries ♈, through Taurus, on anti-clockwise and around to Pisces, and then to begin again. According to our date of birth, each one of us embodies and plays out a particular stretch of the Life-Stream.

♊ The Stream now divides and criss-crosses with other streams. I bubble and froth with the diversity of Life and the amusement and knowledge that it brings. When the flow is strong, I tirelessly make more and more connections, informing and delighting myself and others. When the flow is weak I allow myself to slow down a little, that I might appreciate my multi-faceted expression of the Life-Stream's possibilities . . .

Life-Stream

♐ The crisis of merging now passed, the Stream is freed on to the open plain, where It runs wide and majestic. Upon both banks may be seen sweeping vistas of opportunity and adventure which I am called to take up and explore. So as the Stream proceeds It gathers many experiences that allow me to learn and teach, teach and learn. And lest I flood the banks with excesses or arrogant enthusiasm, I always remember my small and trickling origins – and again the Stream is good . . .

♏ With the merging of two Streams it now becomes imperative that my former self dies. Should I resist this merging or attempt to coerce the Other solely for my own ends, then inevitably I shall be cast out, alone, drowned in my own undercurrent. Yet when I sacrifice to the true intimacy of truly joining, the power of the Life-Stream is doubled. With such combined Flow the strength of my inner convictions can move mountains . . .

♎ The Life-Stream now meets another Stream of equal strength and importance – I sense the natural give and take that must now occur at this place of confluence. Both have equal right of way, and turbulence arises as alternately we each separate and unite, surrender and hold on to our individual identities, yet knowing that merging is inevitable. And so, naturally, I balance as the Stream swings between separateness and unity . . .

Through simply being true to our Sun Sign, our part in the Life-Stream, the Essence of Life itself shall guide and support us, vitalize and protect us. The Stream flows inexorably on, around and around, year after year, through the Earthly Seasons that correspond to them. And so we too, like the River, through keeping to our part of the Course, will inevitably find our true Path.

♍ Yet now the Stream is checked by difficult terrain, and it becomes critical as to how I proceed. And should I lose a sense of the Flow altogether, I simply remain true to myself and before long I am once more aware of the Stream coursing through me and within me, guiding me and buoying me up. Such experiences create in me an intricate knowledge of the nature of things, enabling me to help others through difficult times . . .

♋ I now settle into a pool of safety and security where I may prosper from Nature's surrounding bounty, and dream in my silent depths. When clear and therefore maintained and fed by the Stream's input and output, I am able and willing to nourish myself and others. When clogged by the silt of stale memories and unnecessary emotions, I become aware of those fresh thoughts and feelings that are readily offered, allowing myself to receive them and be restored by them . . .

♌ Now the Life-Stream's Force has risen and accumulated as much as It can, and I overflow and burst forth to show myself to the World – full and ebullient. Being so strong in flow I readily and easily express how it feels to have the Force coursing through and around me, showing and reminding others of Its greatness. When, inevitably, this tide of enthusiasm ebbs, I gracefully and happily languish until once more the Force rushes through . . .

THE TABLES

For when Outer Planetary Cycles are active in your life

for all months of birth between 1900 and 2012

HOW TO USE THE TABLES

THESE TABLES WILL SHOW YOU THE AGES in your life when the Cycles of Uranus, Neptune and Pluto are active, according to the month and year in which you were born. These Cycles, along with the pages in the book where their meanings are described, are as follows.

Active Planetary Cycle	Page No	Active Planetary Cycle	Page No
Uranus Waxing Square	162	Neptune Squaring Pluto	193
Uranus Waxing Trine	166	Neptune Waxing Square	198
Uranus Opposition	168	Neptune Waxing Trine*	204
Uranus Waning Trine	175	Pluto Conjoining Neptune	221
Uranus Waning Square	177	Pluto Waxing Square	226
Uranus Return	179	Pluto Waxing Trine^Δ	231

* Neptune Opposing Pluto (*see* page 207) happens at around the same time as the Neptune Waxing Trine. The table on page 352 shows you exactly when.

Δ Non-applicable is a term used in the tables regarding the Pluto Waxing Trine when the nature of that influence does not practically apply to the advanced age (over 84) at which it is occurring. The reason for this is that Pluto swings way out into space at a certain point and therefore takes longer to travel through the Zodiac.

An example of how to use the tables is given below.

Example Birth Date: 3 FEBRUARY 1969

1 Go to the year in question. 1969 is on page 329.

2 Find the month in the far-left column. This is always given twice, once for all the active Uranus Cycles given in the upper half of the 1969 section, and once for the active Neptune and Pluto Cycles in the lower half of the 1969 section. Looking along the row to the right of each of these we find that the ages these Outer Planetary Cycles are active for this February birth date are as follows.

Active Planetary Cycle	Ages	Active Planetary Cycle	Ages
Uranus Waxing Square	19–20	Neptune Squaring Pluto	11–13
Uranus Waxing Trine	27–28	Neptune Waxing Square	41–42
Uranus Opposition	41–43	Neptune Waxing Trine*	55–56
Uranus Waning Trine	56–58	Pluto Conjoining Neptune	24–26
Uranus Waning Square	63–65	Pluto Waxing Square	36–38
Uranus Return	83–84	Pluto Waxing Trine^Δ	51–52

3 Now refer to the table on page 352 to find out the ages when Neptune Opposing Pluto was Active.

4 Go to the page in question (as given above) to discover the meaning of any of these Active Planetary Cycles. However, as your read them, bear the following points in mind.

Duration

Usually, these Outer Planetary Cycles are given as active for a period or age span of two to three years, or very occasionally it may be given as four years. However, these cycles are not necessarily going to last that complete term. For instance, in our example, Pluto Conjoining Neptune is given as occurring from 24 to 26 years of age, but the individual concerned is probably not going to feel this influence suddenly cut in on the day he or she turns 24, and then feel it cut out as soon as they turn 27. In reality, they will experience it on and off for a proportion of that period. This can be better explained by using a graphic representation of the given age span of Pluto Conjoining Neptune for our example. The arrow represents a birthday.

➧ 24 years of age (25th year of life)	➧ 25 years of age (26th year of life)	➧ 26 years of age (27th year of life)

This means that they could experience the maximum time in that period, which is just under three years.

➧ 24 years of age	➧ 25 years of age	➧ 26 years of age

Or they could experience the minimum time in that period, which is just over one year.

➧ 24 years of age	➧ 25 years of age	➧ 26 years of age

Or they could experience any variation with that period, like this for instance:

➧ 24 years of age	➧ 25 years of age	➧ 26 years of age

And if the given age span was for just two years, say 24 to 25, then for similar reasons its duration could be less than a year:

Do not be concerned if the age that you personally experience any particular Outer Planetary Cycle is at variance by up to a year either way with the given age span in the tables – this is simply an unavoidable anomaly of the month-by-month nature of these tables. You also need to consider the Stretch Factor, which is also mentioned at the beginning of the descriptions for the influences of all of the Planets. The Stretch Factor means that one can sometimes experience a Planetary influence for considerably longer that is given in the tables, usually owing to other Planetary influences coming to bear that are not covered by this book.

Effect

Planetary influences can be likened to winds. Within the duration of any active Planetary cycle you could occasionally experience it as a strong gust when its effect is unmistakable and undeniable. At the other extreme, there could be periods of calm when you feel that it has gone away – but do anticipate this (either in hope or dread) until you are well clear of the end of the active period. And then in between these two extremes there are the variations that wind itself can produce, such as steady but not too strong, a light breeze, eddies or draughts.

Long-lasting Effect

It is highly important to bear in mind that any Planetary influence can create an effect that lasts for a long time after it is technically over – even for the rest of one's life. Continuing with the wind metaphor, a hurricane can devastate an area, or a timely change of wind can bring life-giving rain. In real terms, a Planetary influence can, for example, manifest as an encounter with someone that leads to a life-long and life-changing relationship. The more powerful Planetary influences, like those listed on page 291, are likely to manifest as something that has such a long-lasting effect – either through one's conscious creation, as an act of fate, or as a combination of the two.

Collective Experience

At certain times there are events which obviously affect a lot of people at once. People often ask questions such as 'how can everyone in an air-crash be experiencing similar Planetary influences.' The answer is that they are not, they are all experiencing some kind of effect that is entirely personal to them, but the common event that creates that effect is the air-crash. Indeed, it could be postulated that having so many people at varying critical astrological stages in their personal lives, in the same place and at the same time, is what actually provokes the accident. Then there is something called Collective Karma to consider. If one is caught up in the fates and fortunes of the collective to which one belongs which could be anything from a football club to a whole country, or even the entire human race, rather than being true to one's individual conscience and destiny, then one is more prone to experience the highs and lows of that collective above and beyond one's own. Going to war or being involved in a war because of the dictates of a nation's policies is a significant example of this. It must also be pointed out that a collective experience such as this is what can spark off a more individual stance – like refusing to go to war. Talking of historical events like wars, there are also situations where a number of people, who were born during the same period, experience such an historical event as giving form to a Planetary influence that they are all undergoing at that time. For example, nearly everyone born during the first half of the year 1900 would have experienced their Uranus Opposition (at ages 44–45) at the end of World War Two. Now this could have been experienced as anything from a time of disruption and displacement – mostly for the losing side – or liberation and jubilation, mostly for the winning side. There are many subtleties and variations concerning this issue of collective experience, and as well as Universal Planetary Cycles, one also has to take into account Individual Planetary Cycles, which are not included in this book.

1900

URANUS CYCLE 0–84 YEARS						
m.o.b.	WAXING SQUARE	WAXING TRINE	OPPOSITION	WANING TRINE	WANING SQUARE	RETURN
JAN	22–23	29–31	44–45	57–58	63–65	83–84
FEB	22–23	29–31	44–45	57–59	64–65	83–84
MAR	22–23	30–31	44–45	57–59	64–65	83–84
APR	21–22	29–30	44–45	57–58	64–65	83–84
MAY	21–22	29–30	44–45	57–58	64–65	83–84
JUN	20–22	28–29	43–44	57–58	63–64	82–84
JUL	20–22	28–29	43–44	57–58	63–64	82–84
AUG	20–22	28–29	42–44	56–57	63–64	82–84
SEP	20–22	28–29	42–44	56–57	63–64	82–84
OCT	20–22	28–29	43–44	56–57	63–64	82–84
NOV	21–22	28–30	43–44	56–58	63–64	83–84
DEC	21–23	29–30	43–45	57–58	63–64	83–84

NEPTUNE CYCLE				PLUTO CYCLE		
m.o.b.	SQUARING PLUTO	WAXING SQUARE	WAXING TRINE	CONJ.NEPTUNE	WAXING SQUARE	WAXING TRINE
JAN	34–36	39–41	53–55	7–10	63–66	76–78
FEB	34–36	38–41	52–54	6–10	63–65	76–78
MAR	34–36	38–41	52–54	6–10	63–65	76–78
APR	34–36	39–41	52–54	6–10	63–66	76–78
MAY	35–37	39–41	52–55	7–11	64–66	77–79
JUN	35–37	40–42	53–55	8–11	64–66	77–79
JUL	35–37	40–42	54–56	9–12	64–67	77–79
AUG	35–37	40–42	54–56	10–13	65–66	77–79
SEP	35–37	41–42	54–56	10–13	65–66	78–79
OCT	35–37	41–42	54–56	10–14	64–66	77–78
NOV	34–36	40–42	54–55	9–13	64–66	77–78
DEC	34–36	39–41	53–55	9–12	63–65	76–78

1901

URANUS CYCLE 0–84 YEARS						
m.o.b.	WAXING SQUARE	WAXING TRINE	OPPOSITION	WANING TRINE	WANING SQUARE	RETURN
JAN	22–23	29–31	44–45	57–58	63–65	83–84
FEB	22–23	29–31	44–45	57–59	64–65	83–84
MAR	22–23	30–31	44–45	57–59	64–65	83–84
APR	21–22	29–30	44–45	57–58	64–65	83–84
MAY	21–22	29–30	44–45	57–58	64–65	83–84
JUN	20–22	28–30	43–44	57–58	63–64	82–84
JUL	20–22	28–29	43–44	57–58	63–64	82–84
AUG	20–22	28–29	42–44	56–57	63–64	82–84
SEP	20–22	28–29	42–44	56–57	63–64	82–84
OCT	20–22	28–29	43–44	56–57	63–64	82–84
NOV	21–22	28–30	43–44	56–58	63–64	83–84
DEC	21–23	29–30	43–45	57–58	63–64	83–84

NEPTUNE CYCLE				PLUTO CYCLE		
m.o.b.	SQUARING PLUTO	WAXING SQUARE	WAXING TRINE	CONJ.NEPTUNE	WAXING SQUARE	WAXING TRINE
JAN	34–36	39–41	53–55	8–12	63–65	76–78
FEB	34–36	38–41	52–54	7–11	63–65	76–78
MAR	34–36	38–41	52–54	7–11	63–65	76–78
APR	34–36	39–41	52–54	7–11	63–66	76–78
MAY	34–36	39–41	52–55	8–12	63–66	76–78
JUN	34–36	40–42	53–55	9–13	63–66	76–78
JUL	34–36	40–42	54–56	10–13	64–66	76–78
AUG	34–36	40–42	54–56	11–13	64–66	76–78
SEP	35–36	41–42	54–56	11–13	64–65	77–78
OCT	34–36	41–42	54–56	11–13	64–65	76–77
NOV	34–36	40–42	54–55	11–14	64–65	76–77
DEC	34–36	39–41	53–55	11–14	63–65	75–77

1902

	URANUS CYCLE 0–84 YEARS					
m.o.b.	WAXING SQUARE	WAXING TRINE	OPPOSITION	WANING TRINE	WANING SQUARE	RETURN
JAN	22–23	29–31	44–45	57–58	63–65	83–84
FEB	22–23	29–31	44–45	57–59	64–65	83–84
MAR	22–23	30–31	44–45	57–59	64–65	83–84
APR	21–22	29–30	44–45	57–58	64–65	83–84
MAY	21–22	29–30	44–45	57–58	64–65	83–84
JUN	20–22	28–30	43–44	57–58	63–64	83–84
JUL	20–22	28–29	43–44	57–58	63–64	83–84
AUG	20–22	28–29	42–44	56–57	63–64	82–84
SEP	20–22	28–29	42–44	56–57	62–63	82–84
OCT	20–22	28–29	43–44	56–57	62–63	82–84
NOV	21–22	28–30	43–44	56–58	63–64	83–84
DEC	21–23	29–30	43–45	56–58	63–64	83–84

	NEPTUNE CYCLE			PLUTO CYCLE		
m.o.b.	SQUARING PLUTO	WAXING SQUARE	WAXING TRINE	CONJ.NEPTUNE	WAXING SQUARE	WAXING TRINE
JAN	34–36	39–41	53–55	10–13	63–65	75–77
FEB	34–36	38–41	52–54	9–13	63–64	75–77
MAR	33–35	38–41	52–54	8–12	62–64	75–77
APR	33–35	39–41	52–54	8–12	62–65	75–77
MAY	34–36	39–41	52–55	9–12	63–65	76–78
JUN	34–36	40–42	53–55	10–13	63–65	76–78
JUL	34–36	40–42	54–56	10–13	63–66	76–78
AUG	34–36	40–42	54–56	11–14	64–65	76–78
SEP	34–36	41–42	54–56	12–15	64–65	76–78
OCT	34–36	41–42	54–56	12–15	63–65	75–77
NOV	33–35	40–42	54–55	11–15	63–65	75–77
DEC	33–35	39–41	53–55	11–15	62–64	75–77

1903

	URANUS CYCLE 0–84 YEARS					
m.o.b.	WAXING SQUARE	WAXING TRINE	OPPOSITION	WANING TRINE	WANING SQUARE	RETURN
JAN	22–23	29–31	44–45	57–58	63–65	83–84
FEB	22–23	29–31	44–45	57–59	64–65	83–84
MAR	22–23	30–31	44–45	57–59	64–65	83–84
APR	21–22	29–30	44–45	57–58	64–65	83–84
MAY	21–22	29–30	44–45	57–58	64–65	83–84
JUN	21–22	29–30	43–44	57–58	63–64	83–84
JUL	20–22	28–29	43–44	57–58	63–64	83–84
AUG	20–22	28–29	42–44	56–57	63–64	82–84
SEP	20–22	28–29	42–44	56–57	62–63	82–83
OCT	20–22	28–29	43–44	56–57	62–63	82–84
NOV	21–22	28–30	43–44	56–58	63–64	83–84
DEC	21–23	29–30	43–45	56–58	63–64	83–84

	NEPTUNE CYCLE			PLUTO CYCLE		
m.o.b.	SQUARING PLUTO	WAXING SQUARE	WAXING TRINE	CONJ.NEPTUNE	WAXING SQUARE	WAXING TRINE
JAN	33–35	39–41	53–55	11–14	62–64	75–77
FEB	33–35	39–41	52–54	10–14	62–64	75–77
MAR	32–35	39–41	52–54	9–13	62–64	75–76
APR	32–35	39–41	52–54	9–13	62–65	75–77
MAY	33–35	39–41	52–55	10–13	62–65	75–77
JUN	33–35	39–42	53–55	11–14	62–65	75–77
JUL	33–35	39–42	54–56	11–14	63–66	75–77
AUG	33–35	40–42	54–56	12–15	63–64	75–77
SEP	34–35	41–42	55–56	13–16	63–64	76–77
OCT	34–35	41–42	54–56	13–16	63–64	75–77
NOV	33–35	40–42	54–55	12–16	62–64	74–76
DEC	32–35	39–41	53–55	12–16	62–64	74–76

1904

URANUS CYCLE 0–84 YEARS						
m.o.b.	WAXING SQUARE	WAXING TRINE	OPPOSITION	WANING TRINE	WANING SQUARE	RETURN
JAN	22–23	29–31	44–45	57–58	63–65	83–84
FEB	22–23	29–31	44–45	57–59	63–65	83–84
MAR	22–23	30–31	44–45	57–58	63–65	83–84
APR	21–22	29–30	44–45	57–58	64–65	83–84
MAY	21–22	29–30	44–45	57–58	64–65	83–84
JUN	21–22	29–30	43–44	57–58	63–64	83–84
JUL	20–22	28–29	43–44	57–58	63–64	83–84
AUG	20–22	28–29	42–44	56–57	63–64	82–84
SEP	20–22	28–29	42–44	56–57	62–63	82–83
OCT	20–22	28–29	43–44	56–57	62–63	82–84
NOV	21–22	28–30	43–44	56–57	62–63	83–84
DEC	21–23	29–30	43–44	56–57	63–64	83–84

NEPTUNE CYCLE				PLUTO CYCLE		
m.o.b.	SQUARING PLUTO	WAXING SQUARE	WAXING TRINE	CONJ.NEPTUNE	WAXING SQUARE	WAXING TRINE
JAN	32–35	39–41	53–55	12–15	62–64	74–76
FEB	32–35	39–41	52–54	11–15	62–63	74–76
MAR	32–34	39–41	52–54	10–14	61–63	74–76
APR	32–35	39–41	52–54	10–14	61–64	74–76
MAY	33–35	39–41	52–55	11–14	62–64	74–76
JUN	33–35	39–42	53–55	12–15	62–64	75–76
JUL	33–35	39–42	54–56	12–15	62–64	75–77
AUG	33–35	40–42	54–56	13–16	62–63	75–77
SEP	33–35	41–42	55–56	14–17	63–64	76–77
OCT	33–35	41–42	54–56	14–17	62–64	75–77
NOV	33–35	40–42	54–55	13–17	62–64	74–76
DEC	32–34	39–41	53–55	13–17	61–63	74–76

1905

URANUS CYCLE 0–84 YEARS						
m.o.b.	WAXING SQUARE	WAXING TRINE	OPPOSITION	WANING TRINE	WANING SQUARE	RETURN
JAN	22–23	29–31	44–45	57–58	63–65	83–84
FEB	22–23	29–31	44–45	57–58	64–65	83–84
MAR	22–23	30–31	44–45	57–58	64–65	83–84
APR	21–22	29–30	44–45	57–58	64–65	83–84
MAY	21–22	29–30	44–45	57–58	64–65	83–84
JUN	21–22	29–30	43–44	57–58	63–64	83–84
JUL	20–22	28–29	43–44	57–58	63–64	83–84
AUG	20–22	28–29	42–44	56–57	63–64	82–84
SEP	20–22	28–29	42–44	56–57	62–63	82–83
OCT	20–22	28–29	43–44	56–57	62–63	82–84
NOV	21–22	28–30	43–44	56–57	62–63	83–84
DEC	21–23	29–30	43–44	56–57	62–64	83–84

NEPTUNE CYCLE				PLUTO CYCLE		
m.o.b.	SQUARING PLUTO	WAXING SQUARE	WAXING TRINE	CONJ.NEPTUNE	WAXING SQUARE	WAXING TRINE
JAN	32–34	39–41	53–55	13–16	61–63	74–76
FEB	32–34	39–41	52–55	12–16	61–63	74–76
MAR	31–34	39–41	52–55	11–15	61–63	73–75
APR	32–34	39–41	52–55	11–15	61–63	73–75
MAY	32–34	39–41	52–55	12–16	61–63	74–76
JUN	32–34	39–42	53–55	13–16	61–63	75–76
JUL	32–34	39–42	54–56	13–16	62–63	74–76
AUG	33–34	40–42	54–56	14–17	62–63	74–76
SEP	33–34	41–42	55–57	15–18	62–63	74–76
OCT	32–34	41–42	54–56	15–18	62–63	74–76
NOV	31–34	40–42	54–55	14–18	61–63	73–75
DEC	31–33	39–41	53–55	14–18	61–63	73–75

1906

URANUS CYCLE 0–84 YEARS						
m.o.b.	WAXING SQUARE	WAXING TRINE	OPPOSITION	WANING TRINE	WANING SQUARE	RETURN
JAN	22–23	29–31	44–45	57–58	63–64	83–84
FEB	22–23	29–31	44–45	57–58	63–64	83–84
MAR	22–23	30–31	44–45	57–58	63–64	83–84
APR	21–22	29–30	44–45	57–58	63–64	83–84
MAY	21–23	29–30	44–45	57–58	63–64	83–84
JUN	21–23	29–30	44–45	57–58	63–64	83–84
JUL	20–22	28–29	43–44	57–58	63–64	83–84
AUG	20–22	28–29	42–44	56–57	63–64	82–84
SEP	20–22	28–29	42–44	56–57	62–63	82–83
OCT	20–22	28–29	43–44	56–57	62–63	82–84
NOV	21–22	28–30	43–44	56–57	62–63	83–84
DEC	21–23	29–30	43–44	56–57	62–63	83–84

NEPTUNE CYCLE				PLUTO CYCLE		
m.o.b.	SQUARING PLUTO	WAXING SQUARE	WAXING TRINE	CONJ.NEPTUNE	WAXING SQUARE	WAXING TRINE
JAN	31–33	39–41	53–55	14–16	61–63	73–75
FEB	31–33	39–41	52–55	13–17	61–63	73–75
MAR	31–33	39–41	52–55	12–16	60–62	73–75
APR	31–33	39–41	52–55	12–16	60–62	73–75
MAY	32–34	39–41	52–55	14–17	61–63	73–75
JUN	32–34	39–42	53–55	14–17	61–63	73–75
JUL	32–34	39–42	54–56	14–17	61–63	73–75
AUG	32–34	40–42	54–56	15–18	61–63	73–75
SEP	32–34	41–42	55–57	16–19	61–63	74–75
OCT	32–34	41–42	54–56	16–19	61–63	73–75
NOV	31–34	40–42	54–55	15–19	60–62	73–75
DEC	31–33	39–41	53–55	15–19	60–62	73–74

1907

URANUS CYCLE 0–84 YEARS						
m.o.b.	WAXING SQUARE	WAXING TRINE	OPPOSITION	WANING TRINE	WANING SQUARE	RETURN
JAN	22–23	29–31	44–45	57–58	63–64	83–84
FEB	22–23	29–31	44–45	57–58	63–64	83–84
MAR	22–23	30–31	44–45	57–58	63–64	83–84
APR	21–22	29–30	44–45	57–58	63–64	83–84
MAY	21–23	29–30	44–45	57–58	63–64	83–84
JUN	21–23	29–30	44–45	57–58	63–64	83–84
JUL	20–22	28–29	43–44	57–58	63–64	83–84
AUG	20–22	28–29	42–44	56–57	63–64	82–84
SEP	21–22	28–29	42–43	56–57	62–63	82–83
OCT	21–22	28–29	43–44	56–57	62–63	82–84
NOV	21–22	28–30	43–44	56–57	62–63	83–84
DEC	21–23	29–30	43–44	56–57	62–63	83–84

NEPTUNE CYCLE				PLUTO CYCLE		
m.o.b.	SQUARING PLUTO	WAXING SQUARE	WAXING TRINE	CONJ.NEPTUNE	WAXING SQUARE	WAXING TRINE
JAN	31–33	39–41	53–55	15–17	60–62	73–74
FEB	31–33	39–41	52–55	14–18	60–62	72–74
MAR	30–33	39–41	52–55	13–17	60–62	72–74
APR	30–33	39–41	52–55	13–17	60–62	73–75
MAY	31–33	39–41	52–55	15–18	60–62	73–75
JUN	31–33	39–42	53–55	15–18	60–62	73–75
JUL	31–33	39–42	54–56	15–18	60–62	73–75
AUG	31–33	40–42	54–56	16–19	60–62	73–75
SEP	31–33	41–42	55–57	17–20	61–62	73–75
OCT	31–33	41–42	54–56	17–20	61–62	73–75
NOV	30–33	40–42	54–55	16–20	60–62	72–74
DEC	30–32	39–41	54–55	16–20	60–62	72–74

1908

URANUS CYCLE 0–84 YEARS						
m.o.b.	WAXING SQUARE	WAXING TRINE	OPPOSITION	WANING TRINE	WANING SQUARE	RETURN
JAN	22–23	29–31	44–45	56–57	63–64	83–84
FEB	22–23	29–31	44–45	57–58	63–64	83–84
MAR	22–23	30–31	44–45	57–58	63–64	83–84
APR	22–23	29–30	44–45	57–58	63–64	83–84
MAY	22–23	29–30	44–45	56–58	63–64	83–84
JUN	22–23	29–30	44–45	56–58	63–64	83–84
JUL	21–22	28–29	43–44	56–58	63–64	83–84
AUG	21–22	28–29	42–44	56–57	63–64	82–84
SEP	21–22	28–29	42–43	55–56	62–63	82–83
OCT	21–22	28–29	43–44	55–56	62–63	82–84
NOV	21–22	28–30	43–44	55–57	62–63	83–84
DEC	21–23	29–30	43–44	55–57	62–63	83–84

NEPTUNE CYCLE				PLUTO CYCLE		
m.o.b.	SQUARING PLUTO	WAXING SQUARE	WAXING TRINE	CONJ.NEPTUNE	WAXING SQUARE	WAXING TRINE
JAN	30–32	39–41	53–55	16–18	60–62	72–74
FEB	30–32	39–41	52–55	15–19	59–62	72–74
MAR	30–32	39–41	52–55	14–18	59–61	72–73
APR	30–32	39–41	52–55	14–18	59–61	72–74
MAY	30–33	39–41	52–55	16–19	60–62	72–74
JUN	30–33	39–42	53–55	16–19	60–62	72–74
JUL	31–33	39–42	54–56	16–19	60–62	72–74
AUG	31–33	40–42	54–56	17–20	60–62	73–74
SEP	31–33	41–42	55–57	18–21	60–62	73–74
OCT	31–33	41–42	54–56	18–21	60–62	72–74
NOV	30–33	40–42	54–56	17–21	59–61	71–73
DEC	30–32	39–41	54–56	17–20	59–61	71–73

1909

URANUS CYCLE 0–84 YEARS						
m.o.b.	WAXING SQUARE	WAXING TRINE	OPPOSITION	WANING TRINE	WANING SQUARE	RETURN
JAN	22–23	29–31	44–45	56–57	63–64	83–84
FEB	22–23	29–31	44–45	57–58	63–64	83–84
MAR	22–23	30–31	44–45	57–58	63–64	83–84
APR	22–23	29–30	44–45	57–58	63–64	83–84
MAY	22–23	29–30	44–45	56–58	63–64	83–84
JUN	22–23	29–30	44–45	56–58	63–64	83–84
JUL	21–22	28–29	43–44	56–58	63–64	83–84
AUG	21–22	28–29	42–44	56–57	63–64	82–84
SEP	21–22	28–30	42–43	55–56	62–63	82–83
OCT	21–22	28–30	43–44	55–56	62–63	82–84
NOV	21–22	28–30	43–44	55–57	62–63	83–84
DEC	21–23	29–30	43–44	55–57	62–63	83–84

NEPTUNE CYCLE				PLUTO CYCLE		
m.o.b.	SQUARING PLUTO	WAXING SQUARE	WAXING TRINE	CONJ.NEPTUNE	WAXING SQUARE	WAXING TRINE
JAN	30–32	39–41	53–55	17–19	59–61	71–73
FEB	30–32	39–41	52–55	16–20	59–61	71–73
MAR	29–31	39–41	52–55	15–19	58–61	71–73
APR	29–31	39–41	52–55	15–19	58–61	71–73
MAY	30–32	39–41	52–55	17–20	59–61	71–73
JUN	30–32	39–42	53–55	17–20	59–61	71–73
JUL	30–32	39–42	54–56	17–20	59–61	71–73
AUG	31–32	40–42	54–56	18–21	59–61	71–73
SEP	31–32	41–42	55–57	19–21	60–61	72–74
OCT	30–32	41–42	54–56	19–21	60–61	72–74
NOV	30–32	40–42	54–56	18–21	59–61	71–73
DEC	29–31	39–41	54–56	18–21	59–60	71–73

1910

URANUS CYCLE 0–84 YEARS						
m.o.b.	WAXING SQUARE	WAXING TRINE	OPPOSITION	WANING TRINE	WANING SQUARE	RETURN
JAN	22–23	29–30	43–44	55–57	62–63	83–84
FEB	22–23	29–31	43–44	56–58	62–63	83–84
MAR	22–23	30–31	44–45	56–57	63–64	83–84
APR	22–23	30–31	44–45	56–58	63–64	83–84
MAY	22–23	30–31	44–45	56–58	63–64	83–84
JUN	22–23	29–30	43–45	56–57	63–64	83–84
JUL	21–23	29–30	43–44	56–57	62–63	83–84
AUG	21–22	28–30	42–43	56–57	62–63	82–84
SEP	21–22	28–30	42–43	55–56	62–63	82–83
OCT	21–22	28–30	42–43	55–56	62–63	82–83
NOV	21–22	28–30	42–43	55–56	62–63	82–83
DEC	21–23	29–30	42–44	55–56	62–63	82–83

NEPTUNE CYCLE				PLUTO CYCLE		
m.o.b.	SQUARING PLUTO	WAXING SQUARE	WAXING TRINE	CONJ.NEPTUNE	WAXING SQUARE	WAXING TRINE
JAN	29–31	39–41	53–55	17–21	58–60	70–72
FEB	29–31	39–41	53–55	17–21	58–60	70–72
MAR	29–31	39–41	52–55	16–20	58–60	70–72
APR	29–31	39–41	52–55	16–20	58–60	71–73
MAY	29–31	39–41	53–55	17–20	58–60	71–73
JUN	29–32	39–42	53–55	17–20	59–61	71–73
JUL	30–32	40–42	54–56	17–20	59–61	71–73
AUG	30–32	40–42	54–56	19–22	59–61	71–73
SEP	30–32	41–43	55–57	19–22	59–61	71–73
OCT	29–31	41–42	55–57	19–22	59–61	71–72
NOV	29–31	41–42	55–56	19–22	58–60	70–72
DEC	29–31	40–42	54–56	19–22	58–60	70–72

1911

URANUS CYCLE 0–84 YEARS						
m.o.b.	WAXING SQUARE	WAXING TRINE	OPPOSITION	WANING TRINE	WANING SQUARE	RETURN
JAN	22–23	29–30	43–44	55–57	62–63	83–84
FEB	22–23	29–31	43–44	56–58	62–63	83–84
MAR	22–23	30–31	43–45	56–57	62–64	83–84
APR	22–23	30–31	44–45	56–58	63–64	83–84
MAY	22–23	30–31	44–45	56–58	63–64	83–84
JUN	22–23	29–30	43–45	56–57	62–64	83–84
JUL	21–23	29–30	43–44	56–57	62–63	83–84
AUG	21–22	28–30	42–43	56–57	62–63	82–84
SEP	21–22	28–30	42–43	55–56	62–63	82–84
OCT	21–22	28–30	42–43	55–56	62–63	82–83
NOV	21–22	28–30	42–43	55–56	62–63	82–83
DEC	21–23	29–30	42–44	55–56	62–63	83–84

NEPTUNE CYCLE				PLUTO CYCLE		
m.o.b.	SQUARING PLUTO	WAXING SQUARE	WAXING TRINE	CONJ.NEPTUNE	WAXING SQUARE	WAXING TRINE
JAN	29–31	39–41	53–55	18–21	58–60	70–72
FEB	28–30	39–41	53–55	17–21	57–60	70–72
MAR	28–30	39–41	52–55	17–20	57–60	70–72
APR	28–31	39–41	52–55	17–20	58–60	70–72
MAY	29–31	39–41	53–55	17–20	58–60	70–72
JUN	29–31	39–42	53–55	18–21	58–60	70–72
JUL	29–31	40–42	54–56	19–21	58–60	71–72
AUG	29–31	40–42	54–56	20–22	59–60	71–73
SEP	29–31	41–43	55–57	20–23	59–60	71–72
OCT	29–31	41–42	55–57	20–23	59–60	71–72
NOV	29–31	41–42	55–56	20–23	58–60	71–72
DEC	28–30	40–42	54–56	20–23	57–59	70–71

1912

URANUS CYCLE 0–84 YEARS						
m.o.b.	WAXING SQUARE	WAXING TRINE	OPPOSITION	WANING TRINE	WANING SQUARE	RETURN
JAN	22–23	29–30	43–44	55–57	62–63	83–84
FEB	22–23	29–31	43–44	56–58	62–63	83–84
MAR	22–23	30–31	43–45	56–57	62–64	83–84
APR	22–23	30–31	44–45	56–58	63–64	83–84
MAY	22–23	30–31	44–45	56–58	63–64	83–84
JUN	22–23	29–30	43–44	56–57	62–64	83–84
JUL	21–23	29–30	43–44	56–57	62–63	83–84
AUG	21–22	28–30	42–43	56–57	62–63	82–84
SEP	21–22	28–30	42–43	55–56	62–63	82–84
OCT	21–22	28–30	42–43	55–56	62–63	82–83
NOV	21–22	28–30	42–43	55–56	61–63	82–83
DEC	21–23	28–30	42–43	55–56	61–62	82–83

NEPTUNE CYCLE				PLUTO CYCLE		
m.o.b.	SQUARING PLUTO	WAXING SQUARE	WAXING TRINE	CONJ.NEPTUNE	WAXING SQUARE	WAXING TRINE
JAN	28–30	39–41	53–55	19–22	57–59	70–71
FEB	28–30	39–41	53–55	18–21	57–59	70–71
MAR	28–30	39–41	53–55	18–21	57–59	69–71
APR	28–30	39–41	53–55	18–21	57–59	70–72
MAY	28–30	39–41	53–55	19–22	57–59	70–72
JUN	28–30	39–42	53–55	19–22	57–59	70–72
JUL	29–31	40–42	54–56	19–21	57–59	70–72
AUG	29–31	40–42	54–56	20–22	57–59	70–72
SEP	29–31	41–43	55–57	20–23	58–60	70–72
OCT	28–30	41–42	55–57	20–23	58–60	71–72
NOV	28–30	41–42	55–56	20–23	58–60	71–72
DEC	28–30	40–42	54–56	20–23	57–59	69–71

1913

URANUS CYCLE 0–84 YEARS						
m.o.b.	WAXING SQUARE	WAXING TRINE	OPPOSITION	WANING TRINE	WANING SQUARE	RETURN
JAN	22–23	29–30	43–44	55–57	62–63	83–84
FEB	22–23	29–31	43–44	56–58	62–63	83–84
MAR	22–23	30–31	43–44	56–57	62–63	83–84
APR	22–23	30–31	43–44	56–58	63–64	83–84
MAY	22–23	30–31	43–44	56–58	63–64	83–84
JUN	22–23	29–30	43–44	56–57	62–64	83–84
JUL	21–23	29–30	43–44	56–57	62–63	83–84
AUG	21–22	28–30	42–43	56–57	62–63	82–84
SEP	21–22	28–30	42–43	55–56	62–63	82–84
OCT	21–22	28–30	42–43	55–56	62–63	82–83
NOV	21–22	28–30	42–43	55–56	61–62	82–83
DEC	21–23	28–30	42–43	55–56	61–62	82–83

NEPTUNE CYCLE				PLUTO CYCLE		
m.o.b.	SQUARING PLUTO	WAXING SQUARE	WAXING TRINE	CONJ.NEPTUNE	WAXING SQUARE	WAXING TRINE
JAN	28–30	39–41	53–55	20–23	57–59	69–71
FEB	28–30	39–41	53–55	20–23	57–59	69–71
MAR	27–29	39–41	53–55	19–22	56–58	69–70
APR	27–29	39–41	53–55	19–22	56–58	69–70
MAY	28–30	39–41	53–55	19–22	57–59	69–71
JUN	28–30	39–42	53–55	20–22	57–59	69–71
JUL	28–30	40–42	54–56	20–22	57–59	69–71
AUG	29–30	40–42	54–56	21–23	57–59	70–71
SEP	29–30	41–43	55–57	21–24	58–59	70–71
OCT	28–30	41–42	55–57	21–24	57–59	70–71
NOV	27–29	41–42	55–56	21–24	57–59	69–71
DEC	27–29	40–42	54–56	21–24	56–58	69–70

1914

URANUS CYCLE 0–84 YEARS						
m.o.b.	WAXING SQUARE	WAXING TRINE	OPPOSITION	WANING TRINE	WANING SQUARE	RETURN
JAN	22–23	29–30	43–44	55–57	62–63	83–84
FEB	22–23	29–31	43–44	56–58	62–63	83–84
MAR	22–23	30–31	43–44	56–57	62–63	83–84
APR	22–23	30–31	43–44	56–58	62–63	83–84
MAY	22–23	30–31	43–44	56–58	62–63	83–84
JUN	22–23	29–30	43–44	56–57	62–63	83–84
JUL	21–23	29–30	43–44	56–57	62–63	83–84
AUG	21–22	28–30	42–43	56–57	62–63	82–84
SEP	21–22	28–30	42–43	55–56	61–62	82–84
OCT	21–22	28–30	42–43	55–56	61–62	82–83
NOV	21–22	28–30	42–43	55–56	61–62	82–83
DEC	21–23	28–30	42–43	55–56	61–62	82–83

NEPTUNE CYCLE				PLUTO CYCLE		
m.o.b.	SQUARING PLUTO	WAXING SQUARE	WAXING TRINE	CONJ.NEPTUNE	WAXING SQUARE	WAXING TRINE
JAN	27–29	39–41	53–55	21–24	56–58	69–70
FEB	27–29	39–41	53–55	20–22	56–58	69–70
MAR	27–29	39–41	53–55	19–23	56–58	68–70
APR	27–29	39–41	53–55	19–23	56–58	68–70
MAY	27–29	39–41	53–55	20–23	56–58	68–70
JUN	27–29	39–42	53–55	20–23	56–58	68–70
JUL	27–29	40–42	54–56	20–23	56–58	68–70
AUG	28–30	40–42	54–56	21–24	56–58	68–70
SEP	28–30	41–43	55–57	22–25	57–58	69–71
OCT	28–30	41–42	55–57	22–25	57–58	69–71
NOV	27–29	41–42	55–56	22–25	56–58	68–70
DEC	27–29	40–42	54–56	22–25	56–58	68–70

1915

URANUS CYCLE 0–84 YEARS						
m.o.b.	WAXING SQUARE	WAXING TRINE	OPPOSITION	WANING TRINE	WANING SQUARE	RETURN
JAN	22–23	29–30	43–44	55–57	62–63	83–84
FEB	22–23	29–31	43–44	56–58	62–63	83–84
MAR	22–23	29–31	43–44	55–57	62–63	83–84
APR	22–23	30–31	43–44	56–58	62–63	83–84
MAY	22–23	30–31	43–44	56–58	62–63	83–84
JUN	22–23	29–30	43–44	56–57	62–63	83–84
JUL	21–23	29–30	43–44	56–57	62–63	83–84
AUG	21–22	28–30	42–43	56–57	62–63	82–84
SEP	21–22	28–29	42–43	55–56	61–62	82–84
OCT	21–22	28–29	42–43	55–56	61–62	82–83
NOV	21–22	28–30	42–43	55–56	61–62	82–83
DEC	21–23	28–30	42–43	54–55	61–62	82–83

NEPTUNE CYCLE				PLUTO CYCLE		
m.o.b.	SQUARING PLUTO	WAXING SQUARE	WAXING TRINE	CONJ.NEPTUNE	WAXING SQUARE	WAXING TRINE
JAN	27–29	39–41	53–55	22–25	56–58	68–70
FEB	26–28	39–41	53–55	21–24	56–58	68–70
MAR	26–28	39–41	53–55	20–23	55–57	67–69
APR	26–28	39–41	53–55	20–23	55–57	67–69
MAY	27–29	39–41	53–55	20–23	56–58	68–70
JUN	27–29	39–41	53–55	21–24	56–58	68–70
JUL	27–29	40–42	54–56	21–24	56–58	68–70
AUG	28–30	40–42	54–56	22–25	56–58	68–70
SEP	28–39	41–42	55–57	23–25	57–58	69–70
OCT	27–39	41–42	55–57	23–25	57–58	68–70
NOV	27–29	41–42	55–56	22–25	55–57	67–69
DEC	26–28	40–42	54–56	22–25	55–57	67–69

1916

URANUS CYCLE 0–84 YEARS						
m.o.b.	WAXING SQUARE	WAXING TRINE	OPPOSITION	WANING TRINE	WANING SQUARE	RETURN
JAN	22–23	29–30	43–44	55–57	61–63	83–84
FEB	22–23	29–31	43–44	56–57	62–63	83–84
MAR	22–23	29–31	43–44	55–56	62–63	83–84
APR	22–23	30–31	43–44	56–57	62–63	83–84
MAY	22–23	30–31	43–44	56–58	62–63	83–84
JUN	22–23	29–30	43–44	56–57	62–63	83–84
JUL	21–23	29–30	43–44	56–57	62–63	83–84
AUG	21–22	28–30	42–43	56–57	62–63	82–84
SEP	21–22	28–29	42–43	55–56	61–62	82–84
OCT	21–22	28–29	42–43	55–56	61–62	82–83
NOV	21–22	28–29	42–43	55–56	61–62	82–83
DEC	21–22	28–30	41–43	54–55	61–62	82–83

	NEPTUNE CYCLE			PLUTO CYCLE		
m.o.b.	SQUARING PLUTO	WAXING SQUARE	WAXING TRINE	CONJ.NEPTUNE	WAXING SQUARE	WAXING TRINE
JAN	26–28	39–41	53–55	22–25	55–57	67–69
FEB	26–28	39–41	53–55	21–24	55–57	67–69
MAR	26–28	39–41	53–55	21–24	55–57	67–69
APR	26–28	39–41	53–55	21–24	55–57	67–69
MAY	26–28	39–41	53–55	21–24	55–57	67–69
JUN	26–28	39–41	53–55	21–24	55–57	67–69
JUL	27–29	40–42	54–56	21–24	55–57	68–70
AUG	27–39	40–42	54–56	22–25	55–57	68–70
SEP	27–29	41–42	55–57	23–26	56–57	69–70
OCT	27–39	41–42	55–57	23–26	56–57	68–70
NOV	26–28	41–42	55–56	23–26	55–57	67–69
DEC	26–28	40–42	54–56	23–26	55–57	67–69

1917

URANUS CYCLE 0–84 YEARS						
m.o.b.	WAXING SQUARE	WAXING TRINE	OPPOSITION	WANING TRINE	WANING SQUARE	RETURN
JAN	22–23	29–30	42–43	55–57	61–63	83–84
FEB	22–23	29–31	42–44	56–57	61–63	83–84
MAR	22–23	29–31	42–44	55–56	61–63	83–84
APR	22–23	30–31	43–44	56–57	62–63	83–84
MAY	22–23	30–31	43–44	56–57	62–63	83–84
JUN	22–23	29–30	43–44	55–57	62–63	83–84
JUL	21–23	29–30	43–44	56–57	62–63	83–84
AUG	21–22	28–30	42–43	56–57	62–63	82–84
SEP	21–22	28–29	42–43	55–56	61–62	82–84
OCT	21–22	28–29	42–43	55–56	61–62	82–83
NOV	21–22	28–29	42–43	55–56	61–62	82–83
DEC	21–22	28–29	41–43	54–55	61–62	82–83

	NEPTUNE CYCLE			PLUTO CYCLE		
m.o.b.	SQUARING PLUTO	WAXING SQUARE	WAXING TRINE	CONJ.NEPTUNE	WAXING SQUARE	WAXING TRINE
JAN	26–28	39–41	53–55	22–25	55–57	67–69
FEB	25–27	39–41	53–55	21–24	55–57	67–69
MAR	25–27	39–41	53–55	21–24	54–56	66–68
APR	25–27	39–41	53–55	21–24	54–56	66–68
MAY	26–28	39–41	53–55	21–24	55–57	67–69
JUN	26–28	39–41	53–55	21–24	55–57	67–69
JUL	27–29	40–42	54–56	22–25	55–57	67–69
AUG	27–39	40–42	54–56	23–25	55–57	67–69
SEP	27–29	41–42	55–57	24–26	56–57	67–69
OCT	27–39	41–42	55–57	23–26	56–57	67–69
NOV	26–28	41–42	55–56	23–26	54–56	66–68
DEC	25–27	40–42	54–56	23–26	54–56	66–68

1918

URANUS CYCLE 0–84 YEARS						
m.o.b.	WAXING SQUARE	WAXING TRINE	OPPOSITION	WANING TRINE	WANING SQUARE	RETURN
JAN	22–23	29–30	42–43	55–57	61–63	83–84
FEB	22–23	29–30	42–44	56–57	61–63	83–84
MAR	22–23	29–30	42–44	55–56	61–63	83–84
APR	22–23	30–31	43–44	56–57	62–63	83–84
MAY	22–23	30–31	43–44	56–57	62–63	83–84
JUN	22–23	29–30	43–44	55–57	62–63	83–84
JUL	21–23	29–30	43–44	56–57	62–63	83–84
AUG	21–22	28–30	42–43	56–57	62–63	82–84
SEP	21–22	28–29	41–43	55–56	61–62	82–84
OCT	21–22	28–29	41–43	55–56	61–62	82–83
NOV	21–22	28–29	41–43	55–56	61–62	82–83
DEC	21–22	28–29	41–42	54–55	60–61	82–83

NEPTUNE CYCLE				PLUTO CYCLE		
m.o.b.	SQUARING PLUTO	WAXING SQUARE	WAXING TRINE	CONJ.NEPTUNE	WAXING SQUARE	WAXING TRINE
JAN	25–27	39–41	53–55	22–25	54–56	66–68
FEB	25–27	39–41	53–55	22–25	54–56	66–68
MAR	25–27	39–41	53–55	22–25	54–56	66–67
APR	25–27	39–41	53–55	22–25	54–56	66–67
MAY	25–27	39–41	53–55	22–25	54–56	66–68
JUN	25–27	39–41	53–55	22–25	54–56	66–68
JUL	25–27	40–42	54–56	22–25	54–56	66–68
AUG	26–28	40–42	54–56	23–25	54–56	67–68
SEP	26–28	41–42	54–56	24–26	55–56	67–68
OCT	26–28	41–42	54–56	24–26	55–56	67–68
NOV	25–27	41–42	55–56	24–26	54–56	66–67
DEC	25–27	40–42	54–56	24–27	54–56	66–67

1919

URANUS CYCLE 0–84 YEARS						
m.o.b.	WAXING SQUARE	WAXING TRINE	OPPOSITION	WANING TRINE	WANING SQUARE	RETURN
JAN	22–23	29–30	42–43	55–57	60–62	83–84
FEB	22–23	29–30	42–43	56–57	61–62	83–84
MAR	22–23	29–30	42–43	55–56	61–62	83–84
APR	22–23	29–30	43–44	56–57	61–63	83–84
MAY	22–23	29–30	43–44	56–57	62–63	83–84
JUN	22–23	29–30	43–44	55–56	62–63	83–84
JUL	21–23	29–30	43–44	56–57	62–63	83–84
AUG	21–22	28–30	42–43	56–57	62–63	82–84
SEP	21–22	28–29	41–43	55–56	61–62	82–84
OCT	21–22	28–29	41–43	55–56	61–62	82–83
NOV	21–22	28–29	41–43	55–56	61–62	82–83
DEC	21–22	28–29	41–42	54–55	60–61	82–83

NEPTUNE CYCLE				PLUTO CYCLE		
m.o.b.	SQUARING PLUTO	WAXING SQUARE	WAXING TRINE	CONJ.NEPTUNE	WAXING SQUARE	WAXING TRINE
JAN	25–27	39–41	53–55	24–27	54–56	66–67
FEB	25–27	39–41	53–55	23–26	54–56	65–67
MAR	24–26	39–41	53–55	23–25	53–55	65–67
APR	24–26	39–41	53–55	23–25	53–55	65–67
MAY	25–27	39–41	53–55	23–25	54–56	66–67
JUN	25–27	39–41	53–55	23–25	54–56	66–67
JUL	25–27	40–42	54–56	23–25	54–56	66–67
AUG	26–28	40–42	54–56	23–26	54–56	66–68
SEP	26–28	41–42	54–56	24–27	54–56	66–68
OCT	25–27	41–42	54–56	24–27	54–56	66–68
NOV	25–27	41–42	55–56	24–27	53–55	66–67
DEC	24–26	40–42	54–56	24–27	53–55	66–67

1920

URANUS CYCLE 0–84 YEARS						
m.o.b.	WAXING SQUARE	WAXING TRINE	OPPOSITION	WANING TRINE	WANING SQUARE	RETURN
JAN	21–23	28–29	41–42	54–55	61–62	82–84
FEB	21–23	28–30	41–43	54–55	61–62	83–84
MAR	22–23	29–30	42–43	54–56	61–62	83–84
APR	22–23	29–30	42–43	55–56	61–62	83–84
MAY	22–23	29–30	42–44	55–56	61–62	83–84
JUN	22–23	29–30	42–44	55–56	62–63	83–84
JUL	22–23	29–30	42–43	55–56	61–63	83–84
AUG	21–23	28–29	42–43	55–56	61–63	83–84
SEP	21–22	28–29	41–42	54–55	61–62	82–84
OCT	21–22	28–29	41–42	54–55	61–62	82–84
NOV	21–22	28–29	41–42	54–55	60–61	82–84
DEC	21–22	28–29	41–42	54–55	60–61	82–84

NEPTUNE CYCLE				PLUTO CYCLE		
m.o.b.	SQUARING PLUTO	WAXING SQUARE	WAXING TRINE	CONJ.NEPTUNE	WAXING SQUARE	WAXING TRINE
JAN	24–26	40–42	54–56	24–27	53–55	65–67
FEB	24–26	40–42	53–55	23–26	53–55	65–66
MAR	24–26	39–41	53–55	23–26	53–55	64–66
APR	24–26	39–41	53–55	23–26	53–55	65–66
MAY	24–26	39–41	53–55	23–26	53–55	65–67
JUN	24–26	39–41	53–55	23–26	53–55	65–67
JUL	25–27	40–42	53–56	24–26	53–55	65–67
AUG	25–27	40–42	54–56	25–27	54–55	66–67
SEP	25–27	41–43	54–57	25–27	54–55	66–67
OCT	25–27	41–43	55–57	25–28	54–55	66–67
NOV	24–26	41–43	55–57	25–28	53–55	65–66
DEC	24–26	41–42	55–56	25–28	53–55	65–66

1921

URANUS CYCLE 0–84 YEARS						
m.o.b.	WAXING SQUARE	WAXING TRINE	OPPOSITION	WANING TRINE	WANING SQUARE	RETURN
JAN	21–23	28–29	41–42	54–55	60–62	82–84
FEB	21–23	28–30	41–43	54–55	61–62	83–84
MAR	22–23	29–30	42–43	54–56	61–62	83–84
APR	22–23	29–30	42–43	55–56	61–62	83–84
MAY	22–23	29–30	42–43	55–56	61–62	83–84
JUN	22–23	29–30	42–43	55–56	61–63	83–84
JUL	22–23	29–30	42–43	55–56	61–63	83–84
AUG	21–23	28–29	42–43	55–56	61–63	83–84
SEP	21–22	28–29	41–42	54–55	61–62	82–84
OCT	21–22	28–29	41–42	54–55	61–62	82–84
NOV	21–22	28–29	41–42	54–55	60–61	82–84
DEC	21–22	28–29	41–42	53–54	60–61	82–84

NEPTUNE CYCLE				PLUTO CYCLE		
m.o.b.	SQUARING PLUTO	WAXING SQUARE	WAXING TRINE	CONJ.NEPTUNE	WAXING SQUARE	WAXING TRINE
JAN	24–26	40–42	54–56	24–27	53–55	65–67
FEB	24–26	40–42	53–55	24–26	53–55	65–67
MAR	23–25	39–41	53–55	24–26	52–55	65–67
APR	23–25	39–41	53–55	24–26	52–55	66–67
MAY	24–26	39–41	53–55	24–26	53–55	64–66
JUN	24–26	39–41	53–55	25–27	53–55	64–66
JUL	25–27	40–42	53–56	25–27	53–55	64–66
AUG	25–27	40–42	54–56	26–28	53–55	65–67
SEP	25–26	41–43	54–57	26–28	53–55	65–67
OCT	25–26	41–43	55–57	26–28	53–55	65–67
NOV	24–26	41–43	55–57	25–28	52–54	64–66
DEC	24–26	41–42	55–56	25–28	52–54	64–66

1922

URANUS CYCLE 0–84 YEARS						
m.o.b.	WAXING SQUARE	WAXING TRINE	OPPOSITION	WANING TRINE	WANING SQUARE	RETURN
JAN	21–23	28–29	41–42	53–54	60–62	82–84
FEB	21–23	28–30	41–43	54–55	60–62	83–84
MAR	22–23	28–30	41–43	54–55	61–62	83–84
APR	22–23	29–30	42–43	55–56	61–62	83–84
MAY	22–23	29–30	42–43	55–56	61–62	83–84
JUN	22–23	29–30	42–43	55–56	61–63	83–84
JUL	22–23	29–30	42–43	55–56	61–63	83–84
AUG	21–23	28–29	41–42	55–56	61–63	83–84
SEP	21–22	28–29	41–42	54–55	61–62	82–84
OCT	21–22	28–29	41–42	54–55	61–62	82–84
NOV	21–22	28–29	41–42	54–55	60–61	82–84
DEC	21–22	27–29	40–42	53–54	60–61	82–84

	NEPTUNE CYCLE			PLUTO CYCLE		
m.o.b.	SQUARING PLUTO	WAXING SQUARE	WAXING TRINE	CONJ.NEPTUNE	WAXING SQUARE	WAXING TRINE
JAN	24–26	40–42	54–56	24–27	52–54	65–67
FEB	24–26	40–42	53–55	24–27	52–54	63–65
MAR	23–25	39–41	53–55	24–27	52–54	63–65
APR	23–25	39–41	53–55	24–27	52–54	63–65
MAY	23–25	39–41	53–55	24–27	52–54	64–66
JUN	23–25	39–41	53–55	23–27	52–54	64–66
JUL	23–25	40–42	53–56	24–27	52–54	64–66
AUG	23–25	40–42	54–56	25–27	52–54	65–66
SEP	24–26	41–43	54–57	26–28	53–55	65–66
OCT	24–26	41–43	55–57	26–28	53–55	65–67
NOV	23–25	41–43	55–57	26–28	52–54	64–67
DEC	23–25	41–42	55–56	26–28	52–54	64–67

1923

URANUS CYCLE 0–84 YEARS						
m.o.b.	WAXING SQUARE	WAXING TRINE	OPPOSITION	WANING TRINE	WANING SQUARE	RETURN
JAN	21–23	28–29	41–42	53–54	60–62	82–84
FEB	21–23	28–29	41–42	54–55	60–62	83–84
MAR	22–23	28–30	41–42	54–55	60–62	83–84
APR	22–23	29–30	42–43	55–56	61–62	83–84
MAY	22–23	29–30	42–43	55–56	61–62	83–84
JUN	22–23	29–30	42–43	55–56	61–63	83–84
JUL	22–23	29–30	42–43	55–56	61–63	83–84
AUG	21–23	28–29	41–42	55–56	61–63	83–84
SEP	21–22	28–29	41–42	54–55	61–62	82–84
OCT	21–22	28–29	41–42	54–55	61–62	82–84
NOV	21–22	28–29	41–42	54–55	60–61	82–84
DEC	20–22	27–29	40–41	53–54	60–61	82–84

	NEPTUNE CYCLE			PLUTO CYCLE		
m.o.b.	SQUARING PLUTO	WAXING SQUARE	WAXING TRINE	CONJ.NEPTUNE	WAXING SQUARE	WAXING TRINE
JAN	23–25	40–42	54–56	24–27	52–54	64–66
FEB	23–25	40–42	53–55	24–27	52–54	63–65
MAR	22–24	39–41	53–55	24–27	51–53	63–65
APR	22–24	39–41	53–55	24–27	51–53	63–65
MAY	22–24	39–41	53–55	25–27	51–53	63–65
JUN	23–25	39–41	53–55	25–27	52–54	63–65
JUL	23–25	40–42	53–56	25–27	52–54	63–65
AUG	23–25	40–42	54–56	25–27	52–54	64–66
SEP	24–26	41–43	54–57	26–28	52–54	64–66
OCT	24–26	41–43	55–57	26–28	52–54	64–66
NOV	23–25	41–43	55–57	26–29	51–53	63–65
DEC	23–25	41–42	55–56	26–29	51–53	63–65

1924

URANUS CYCLE 0–84 YEARS						
m.o.b.	WAXING SQUARE	WAXING TRINE	OPPOSITION	WANING TRINE	WANING SQUARE	RETURN
JAN	20–22	28–29	41–42	53–54	60–62	82–84
FEB	21–23	28–29	41–42	54–55	60–62	83–84
MAR	21–23	28–29	41–42	54–55	60–62	83–84
APR	22–23	29–30	41–42	55–56	61–62	83–84
MAY	22–23	29–30	42–43	55–56	61–62	83–84
JUN	22–23	29–30	42–43	54–56	61–62	83–84
JUL	22–23	29–30	42–43	54–56	61–63	83–84
AUG	21–23	28–29	41–42	54–55	61–63	83–84
SEP	21–22	28–29	41–42	54–55	61–62	82–84
OCT	21–22	28–29	41–42	54–55	61–62	82–84
NOV	21–22	28–29	41–42	54–55	60–61	82–84
DEC	20–22	27–28	40–41	53–54	60–61	82–83

NEPTUNE CYCLE				PLUTO CYCLE		
m.o.b.	SQUARING PLUTO	WAXING SQUARE	WAXING TRINE	CONJ.NEPTUNE	WAXING SQUARE	WAXING TRINE
JAN	23–25	40–42	54–56	26–28	51–53	63–65
FEB	23–25	40–42	53–55	26–28	51–53	63–65
MAR	22–24	39–41	53–55	25–27	51–52	62–64
APR	22–24	39–41	53–55	25–27	51–52	62–64
MAY	22–24	39–41	53–55	25–27	51–53	63–65
JUN	22–25	39–41	53–55	25–27	51–53	63–65
JUL	23–25	40–42	53–56	25–27	51–53	63–65
AUG	23–25	40–42	54–56	26–28	52–53	63–65
SEP	23–25	41–43	54–57	27–28	52–53	63–65
OCT	23–25	41–43	55–57	27–28	52–53	63–65
NOV	22–24	41–43	55–57	26–29	51–52	62–64
DEC	22–24	41–42	55–56	26–29	51–52	62–64

1925

URANUS CYCLE 0–84 YEARS						
m.o.b.	WAXING SQUARE	WAXING TRINE	OPPOSITION	WANING TRINE	WANING SQUARE	RETURN
JAN	20–22	28–29	40–41	53–54	60–62	82–83
FEB	21–23	28–29	41–42	53–55	60–62	83–84
MAR	21–23	28–29	41–42	53–55	60–62	83–84
APR	22–23	29–30	41–42	53–55	61–62	83–84
MAY	22–23	29–30	41–42	54–56	61–62	83–84
JUN	22–23	29–30	41–43	54–56	61–62	83–84
JUL	22–23	29–30	41–43	54–56	61–63	83–84
AUG	21–23	28–29	41–42	54–55	61–63	83–84
SEP	21–22	28–29	41–42	54–55	61–62	82–84
OCT	21–22	28–29	41–42	54–55	61–62	82–84
NOV	21–22	28–29	41–42	54–55	60–61	82–84
DEC	20–22	27–28	40–41	53–54	60–61	82–83

NEPTUNE CYCLE				PLUTO CYCLE		
m.o.b.	SQUARING PLUTO	WAXING SQUARE	WAXING TRINE	CONJ.NEPTUNE	WAXING SQUARE	WAXING TRINE
JAN	22–24	40–42	54–56	26–28	51–52	62–64
FEB	22–24	40–42	53–55	26–28	51–52	62–64
MAR	21–23	39–41	53–55	25–28	50–52	62–64
APR	21–23	39–41	53–55	25–28	50–52	62–64
MAY	22–24	39–41	53–55	25–28	50–52	62–64
JUN	22–24	39–41	53–55	25–28	50–52	62–64
JUL	22–24	40–42	53–56	25–28	50–52	62–64
AUG	23–25	40–42	54–56	26–28	51–53	63–65
SEP	23–24	41–43	54–57	27–29	51–53	63–65
OCT	23–24	41–43	55–57	27–29	51–53	63–65
NOV	22–24	41–43	55–57	27–29	50–52	62–64
DEC	22–24	41–42	55–56	27–29	50–52	62–64

1926

URANUS CYCLE 0–84 YEARS						
m.o.b.	WAXING SQUARE	WAXING TRINE	OPPOSITION	WANING TRINE	WANING SQUARE	RETURN
JAN	20–22	28–29	40–41	53–54	60–62	82–83
FEB	21–22	28–29	40–42	53–55	60–62	83–84
MAR	21–22	28–29	40–42	53–55	60–61	83–84
APR	22–23	29–30	41–42	53–55	61–62	83–84
MAY	22–23	29–30	41–43	54–55	61–62	83–84
JUN	22–23	29–30	41–43	54–55	61–62	83–84
JUL	22–23	29–30	41–43	54–55	61–62	83–84
AUG	21–23	28–29	41–42	54–55	61–62	83–84
SEP	21–22	28–29	41–42	54–55	60–62	83–84
OCT	21–22	28–29	41–42	54–55	61–62	83–84
NOV	21–22	28–29	41–42	54–55	60–61	82–84
DEC	20–22	27–28	40–41	53–54	60–61	82–83

	NEPTUNE CYCLE			PLUTO CYCLE		
m.o.b.	SQUARING PLUTO	WAXING SQUARE	WAXING TRINE	CONJ.NEPTUNE	WAXING SQUARE	WAXING TRINE
JAN	22–24	40–42	54–56	26–28	50–52	62–64
FEB	22–24	40–42	53–55	26–28	49–51	61–63
MAR	21–23	39–41	53–55	26–28	49–51	61–63
APR	21–23	39–41	53–55	25–28	49–51	61–63
MAY	21–23	39–41	53–55	25–28	50–52	62–63
JUN	21–24	39–41	53–55	25–28	50–52	62–63
JUL	22–24	40–42	53–56	25–28	50–52	62–63
AUG	22–24	40–42	54–56	26–28	51–52	62–64
SEP	22–24	41–43	54–57	27–29	51–52	62–64
OCT	21–23	41–43	55–57	27–29	51–52	62–64
NOV	21–23	41–43	55–57	27–29	50–51	61–63
DEC	21–23	41–42	55–56	27–29	50–51	61–63

1927

URANUS CYCLE 0–84 YEARS						
m.o.b.	WAXING SQUARE	WAXING TRINE	OPPOSITION	WANING TRINE	WANING SQUARE	RETURN
JAN	20–22	28–29	40–41	53–54	60–62	82–83
FEB	21–22	28–29	40–42	53–54	60–61	83–84
MAR	21–22	28–29	40–42	53–54	60–61	83–84
APR	22–23	28–29	41–42	53–55	61–62	83–84
MAY	22–23	28–30	41–42	54–55	61–62	83–84
JUN	22–23	28–30	41–42	54–55	61–62	83–84
JUL	22–23	28–29	41–42	54–55	61–62	83–84
AUG	21–23	28–29	41–42	54–55	61–62	83–84
SEP	21–22	27–28	41–42	54–55	60–62	83–84
OCT	21–22	27–28	41–42	54–55	61–62	83–84
NOV	21–22	27–28	40–42	53–54	60–61	82–84
DEC	20–21	27–28	39–41	53–54	60–61	82–83

	NEPTUNE CYCLE			PLUTO CYCLE		
m.o.b.	SQUARING PLUTO	WAXING SQUARE	WAXING TRINE	CONJ.NEPTUNE	WAXING SQUARE	WAXING TRINE
JAN	21–23	40–42	54–56	26–28	50–51	61–63
FEB	21–23	40–42	53–55	26–28	49–51	61–63
MAR	20–23	39–41	53–55	26–28	49–51	61–62
APR	21–23	39–41	53–55	26–28	49–51	61–63
MAY	21–23	39–41	53–55	26–28	49–51	61–63
JUN	21–23	39–41	53–55	26–28	49–51	61–63
JUL	21–23	40–42	53–56	26–28	49–51	61–63
AUG	22–24	40–42	54–56	26–28	50–52	62–64
SEP	22–24	41–43	54–56	27–29	50–52	62–64
OCT	21–23	41–43	55–57	27–29	50–52	62–64
NOV	21–23	41–43	55–57	27–29	50–52	61–63
DEC	21–23	41–42	55–56	27–29	50–52	61–62

1928

URANUS CYCLE 0–84 YEARS						
m.o.b.	WAXING SQUARE	WAXING TRINE	OPPOSITION	WANING TRINE	WANING SQUARE	RETURN
JAN	20–22	27–28	39–41	53–54	60–61	82–83
FEB	21–22	27–28	40–41	53–54	60–61	83–84
MAR	21–22	27–29	40–41	53–54	60–61	83–84
APR	22–23	28–29	41–42	53–55	61–62	83–84
MAY	22–23	28–30	41–42	54–55	61–62	83–84
JUN	22–23	28–30	41–42	54–55	61–62	83–84
JUL	22–23	28–29	41–42	54–55	61–62	83–84
AUG	21–23	28–29	41–42	54–55	61–62	83–84
SEP	21–22	27–28	41–42	54–55	60–62	83–84
OCT	21–22	27–28	40–41	53–54	61–62	83–84
NOV	20–21	27–28	40–41	53–54	60–61	82–84
DEC	20–21	27–28	39–40	52–53	60–61	82–83

NEPTUNE CYCLE				PLUTO CYCLE		
m.o.b.	SQUARING PLUTO	WAXING SQUARE	WAXING TRINE	CONJ.NEPTUNE	WAXING SQUARE	WAXING TRINE
JAN	21–23	40–42	54–56	26–28	49–51	61–62
FEB	21–23	40–42	53–55	26–28	49–51	60–62
MAR	20–22	39–41	53–55	26–29	48–50	60–62
APR	20–22	39–41	53–55	26–29	48–50	60–62
MAY	20–22	39–41	53–55	26–28	49–51	60–62
JUN	20–22	39–41	53–55	26–28	49–51	60–62
JUL	20–22	40–42	53–56	27–29	49–51	60–62
AUG	21–23	40–42	54–56	27–29	50–51	61–63
SEP	21–23	41–43	54–56	28–29	50–51	61–63
OCT	21–23	41–43	55–57	28–29	50–51	61–63
NOV	21–22	41–43	55–57	28–29	49–50	61–63
DEC	21–22	41–42	55–56	28–29	49–50	61–62

1929

URANUS CYCLE 0–84 YEARS						
m.o.b.	WAXING SQUARE	WAXING TRINE	OPPOSITION	WANING TRINE	WANING SQUARE	RETURN
JAN	20–22	27–28	39–40	53–54	60–61	82–83
FEB	21–22	27–28	40–41	53–54	60–61	83–84
MAR	21–22	27–29	40–41	53–54	60–61	83–84
APR	22–23	28–29	41–42	53–55	61–62	83–84
MAY	22–23	28–30	41–42	54–55	61–62	83–84
JUN	22–23	28–30	41–42	54–55	61–62	83–84
JUL	22–23	28–29	41–42	54–55	61–62	83–84
AUG	21–23	28–29	41–42	54–55	61–62	83–84
SEP	21–22	27–28	41–42	53–55	60–62	83–84
OCT	21–22	27–28	40–41	53–54	61–62	83–84
NOV	20–21	26–28	40–41	53–54	60–61	82–84
DEC	20–21	26–28	39–40	52–53	60–61	82–83

NEPTUNE CYCLE				PLUTO CYCLE		
m.o.b.	SQUARING PLUTO	WAXING SQUARE	WAXING TRINE	CONJ.NEPTUNE	WAXING SQUARE	WAXING TRINE
JAN	21–23	40–42	54–56	26–28	49–50	61–62
FEB	21–23	40–42	53–55	26–28	48–50	60–62
MAR	20–22	39–41	53–55	26–29	48–50	60–61
APR	20–22	39–41	53–55	26–29	48–50	60–61
MAY	20–22	39–41	53–55	26–28	48–50	60–62
JUN	20–22	39–41	53–55	26–29	48–50	60–62
JUL	20–22	40–42	53–56	27–29	49–51	60–62
AUG	21–23	40–42	54–56	27–29	49–51	61–62
SEP	21–23	41–43	54–56	28–29	49–51	61–62
OCT	21–23	41–43	55–57	28–29	48–50	61–62
NOV	20–22	41–43	55–57	28–29	48–50	60–61
DEC	20–22	41–42	55–56	28–29	48–50	60–61

1930

URANUS CYCLE 0–84 YEARS						
m.o.b.	WAXING SQUARE	WAXING TRINE	OPPOSITION	WANING TRINE	WANING SQUARE	RETURN
JAN	20–21	26–28	39–40	52–53	60–61	82–84
FEB	20–22	27–28	39–41	53–54	60–61	82–84
MAR	20–22	27–28	40–41	53–54	60–61	83–84
APR	21–22	27–29	40–41	53–54	60–61	83–84
MAY	21–22	28–29	40–42	53–55	60–62	83–84
JUN	21–22	28–29	41–42	54–55	61–62	83–84
JUL	21–22	28–29	41–42	54–55	61–62	83–84
AUG	21–22	28–29	41–42	54–55	61–62	83–84
SEP	21–22	27–28	40–42	53–55	60–62	83–84
OCT	20–21	27–28	40–41	53–54	60–62	82–84
NOV	20–21	26–28	39–40	53–54	60–61	82–84
DEC	20–21	26–28	39–40	52–53	60–61	82–83

NEPTUNE CYCLE				PLUTO CYCLE		
m.o.b.	SQUARING PLUTO	WAXING SQUARE	WAXING TRINE	CONJ.NEPTUNE	WAXING SQUARE	WAXING TRINE
JAN	20–22	41–42	55–56	27–30	48–50	59–61
FEB	19–21	40–42	54–56	27–29	48–49	59–61
MAR	19–21	39–42	53–55	27–29	47–49	59–61
APR	19–21	39–41	53–55	26–29	47–49	59–61
MAY	19–21	39–41	53–55	26–29	48–49	59–61
JUN	20–22	39–41	53–55	26–29	48–50	59–61
JUL	20–22	39–42	53–55	27–29	48–50	59–61
AUG	20–22	40–42	54–56	27–29	48–50	59–61
SEP	21–23	40–43	54–56	27–29	49–50	60–62
OCT	21–23	41–43	55–57	28–30	49–50	60–62
NOV	20–22	41–43	55–57	28–30	48–50	60–62
DEC	20–22	41–43	55–57	28–30	48–50	59–61

1931

URANUS CYCLE 0–84 YEARS						
m.o.b.	WAXING SQUARE	WAXING TRINE	OPPOSITION	WANING TRINE	WANING SQUARE	RETURN
JAN	20–21	26–28	39–40	52–53	60–61	82–84
FEB	20–22	27–28	39–40	52–54	60–61	82–84
MAR	20–22	27–28	39–41	52–54	60–61	82–84
APR	21–22	27–29	40–42	53–55	60–61	82–84
MAY	21–22	28–29	40–42	53–55	60–62	82–84
JUN	21–22	28–29	40–42	53–55	60–62	82–84
JUL	21–22	28–29	41–42	53–55	60–62	83–84
AUG	21–22	28–29	41–42	53–55	60–62	83–84
SEP	21–22	27–28	40–42	53–55	60–62	83–84
OCT	20–22	27–28	40–41	53–54	60–62	82–84
NOV	20–22	26–28	39–40	53–54	60–61	82–84
DEC	20–22	26–27	39–40	52–53	60–61	82–84

NEPTUNE CYCLE				PLUTO CYCLE		
m.o.b.	SQUARING PLUTO	WAXING SQUARE	WAXING TRINE	CONJ.NEPTUNE	WAXING SQUARE	WAXING TRINE
JAN	20–22	41–42	55–56	28–30	48–49	59–61
FEB	19–21	40–42	54–56	27–29	48–49	58–60
MAR	19–21	39–42	53–55	27–29	47–49	58–60
APR	19–21	39–41	53–55	26–29	47–49	58–60
MAY	19–21	39–41	53–55	26–29	47–49	58–60
JUN	19–21	39–41	53–55	26–29	47–49	59–61
JUL	19–21	39–42	53–55	27–29	47–49	59–61
AUG	20–22	40–42	54–56	27–29	47–49	59–61
SEP	20–22	40–43	54–56	28–30	48–50	60–62
OCT	20–22	41–43	55–57	28–30	48–50	60–62
NOV	19–21	41–43	55–57	28–30	47–49	60–62
DEC	19–21	41–43	55–57	28–30	47–49	59–61

1932

URANUS CYCLE 0–84 YEARS						
m.o.b.	WAXING SQUARE	WAXING TRINE	OPPOSITION	WANING TRINE	WANING SQUARE	RETURN
JAN	20–21	26–28	39–40	52–53	60–61	82–84
FEB	20–21	27–28	39–40	52–54	60–61	82–84
MAR	20–21	27–28	39–40	52–54	60–61	83–84
APR	20–21	27–29	40–41	53–55	60–61	83–84
MAY	21–22	28–29	40–41	53–55	60–62	83–84
JUN	21–22	28–29	40–41	53–55	60–62	83–84
JUL	21–22	28–29	40–41	53–55	60–62	83–84
AUG	21–22	28–29	40–41	53–55	60–62	83–84
SEP	20–21	27–28	40–41	53–55	60–62	83–84
OCT	20–21	27–28	40–41	53–54	60–62	82–84
NOV	20–21	26–28	39–40	53–54	60–61	82–84
DEC	19–21	26–27	39–40	52–53	60–61	82–83

NEPTUNE CYCLE				PLUTO CYCLE		
m.o.b.	SQUARING PLUTO	WAXING SQUARE	WAXING TRINE	CONJ.NEPTUNE	WAXING SQUARE	WAXING TRINE
JAN	19–21	41–42	55–56	28–30	47–49	59–61
FEB	19–21	40–42	54–56	27–29	47–49	58–60
MAR	18–20	39–42	53–55	27–29	46–48	58–60
APR	18–20	39–41	53–55	27–29	46–48	58–60
MAY	19–21	39–41	53–55	27–29	47–49	58–60
JUN	19–21	39–41	53–55	27–29	47–49	59–61
JUL	19–21	39–42	53–55	27–29	47–49	59–61
AUG	20–22	40–42	54–56	27–29	47–49	59–61
SEP	20–22	40–43	54–56	28–30	48–50	59–61
OCT	20–22	41–43	55–57	28–30	48–50	59–61
NOV	19–21	41–43	55–57	28–30	47–49	59–61
DEC	19–21	41–43	55–57	28–30	47–49	59–61

1933

URANUS CYCLE 0–84 YEARS						
m.o.b.	WAXING SQUARE	WAXING TRINE	OPPOSITION	WANING TRINE	WANING SQUARE	RETURN
JAN	19–21	26–28	39–40	52–53	60–61	82–84
FEB	20–21	26–28	39–40	52–54	59–61	82–84
MAR	20–21	26–28	39–40	52–54	59–61	83–84
APR	20–21	26–28	40–41	53–55	60–61	83–84
MAY	21–22	27–29	40–41	53–55	60–62	83–84
JUN	21–22	27–29	40–41	53–55	60–62	83–84
JUL	21–22	27–29	40–41	53–55	60–62	83–84
AUG	21–22	27–29	40–41	53–55	60–62	83–84
SEP	20–21	27–28	40–41	53–55	60–62	83–84
OCT	20–21	27–28	40–41	53–54	60–62	82–84
NOV	20–21	26–28	39–40	53–54	60–61	82–84
DEC	19–21	26–27	39–40	52–53	60–61	82–84

NEPTUNE CYCLE				PLUTO CYCLE		
m.o.b.	SQUARING PLUTO	WAXING SQUARE	WAXING TRINE	CONJ.NEPTUNE	WAXING SQUARE	WAXING TRINE
JAN	19–21	41–42	55–56	28–30	46–48	58–60
FEB	18–20	40–42	54–56	27–29	46–48	58–60
MAR	18–20	39–42	53–55	27–29	46–48	57–59
APR	18–20	39–41	53–55	27–29	46–48	57–59
MAY	18–20	39–41	53–55	27–29	46–48	58–60
JUN	18–20	39–41	53–55	27–29	46–48	58–60
JUL	18–20	39–42	53–55	27–29	47–49	58–60
AUG	19–21	40–42	54–56	27–29	47–49	59–61
SEP	19–21	40–43	54–56	28–30	46–48	59–61
OCT	19–21	41–43	55–57	28–30	46–48	59–61
NOV	19–21	41–43	55–57	28–30	46–48	58–60
DEC	19–21	41–43	55–57	28–30	46–48	58–60

1934

URANUS CYCLE 0–84 YEARS						
m.o.b.	WAXING SQUARE	WAXING TRINE	OPPOSITION	WANING TRINE	WANING SQUARE	RETURN
JAN	20–21	26–28	39–40	52–53	60–61	82–84
FEB	20–21	26–28	39–40	52–53	59–61	82–84
MAR	20–21	26–28	39–40	52–53	59–61	83–84
APR	20–21	26–28	40–41	53–54	60–61	83–84
MAY	21–22	27–29	40–41	53–55	60–62	83–84
JUN	21–22	27–29	40–41	53–55	60–62	83–84
JUL	21–22	27–29	40–41	53–55	60–62	83–84
AUG	21–22	27–29	40–41	53–55	60–62	83–84
SEP	20–21	27–28	40–41	53–55	60–62	83–84
OCT	20–21	27–28	40–41	53–54	60–62	82–84
NOV	20–21	26–28	39–40	53–54	60–61	82–84
DEC	19–21	26–27	39–40	52–53	60–61	82–84

	NEPTUNE CYCLE			PLUTO CYCLE		
m.o.b.	SQUARING PLUTO	WAXING SQUARE	WAXING TRINE	CONJ.NEPTUNE	WAXING SQUARE	WAXING TRINE
JAN	19–21	41–42	55–56	28–30	46–48	58–60
FEB	18–20	40–42	54–56	27–29	45–47	58–60
MAR	17–20	40–42	53–55	27–30	45–47	57–59
APR	17–20	39–41	53–55	27–30	45–47	57–59
MAY	18–20	39–41	53–55	27–29	46–48	57–59
JUN	18–20	39–41	53–55	27–29	46–48	57–59
JUL	18–20	39–42	53–55	27–29	47–49	57–59
AUG	19–21	40–42	54–56	27–29	47–49	58–60
SEP	19–21	40–43	54–56	28–30	47–49	58–60
OCT	19–21	41–43	55–57	28–30	47–49	58–60
NOV	18–20	41–43	55–57	28–30	46–48	58–60
DEC	18–20	41–43	55–57	28–30	46–48	58–60

1935

URANUS CYCLE 0–84 YEARS						
m.o.b.	WAXING SQUARE	WAXING TRINE	OPPOSITION	WANING TRINE	WANING SQUARE	RETURN
JAN	19–21	26–27	39–40	52–53	60–61	82–84
FEB	20–21	26–27	39–40	52–53	59–61	82–84
MAR	20–21	26–27	39–40	52–53	59–60	83–84
APR	20–21	26–28	40–41	53–54	60–61	83–84
MAY	21–22	26–28	40–41	53–55	60–62	83–84
JUN	21–22	27–28	40–41	53–55	60–62	83–84
JUL	21–22	27–28	40–41	53–55	60–62	83–84
AUG	21–22	27–28	40–41	53–55	60–62	83–84
SEP	20–21	27–28	40–41	53–55	60–62	83–84
OCT	20–21	27–28	40–41	53–54	60–62	82–84
NOV	20–21	26–27	39–40	53–54	60–61	82–84
DEC	19–20	26–27	38–39	52–53	60–61	82–84

	NEPTUNE CYCLE			PLUTO CYCLE		
m.o.b.	SQUARING PLUTO	WAXING SQUARE	WAXING TRINE	CONJ.NEPTUNE	WAXING SQUARE	WAXING TRINE
JAN	18–20	41–42	55–56	28–30	45–47	57–59
FEB	18–20	40–42	54–56	27–29	45–47	57–59
MAR	17–19	40–42	53–55	27–30	45–47	56–58
APR	17–20	39–41	53–55	27–30	45–47	56–58
MAY	17–20	39–41	53–55	27–29	45–47	57–59
JUN	17–20	39–41	53–55	27–29	45–47	57–59
JUL	18–20	39–42	53–55	27–29	45–47	57–59
AUG	18–20	40–42	54–56	27–29	45–47	58–60
SEP	18–20	40–43	54–56	28–30	46–48	58–60
OCT	18–20	41–43	55–57	28–30	46–48	58–60
NOV	18–20	41–43	55–57	28–30	45–47	57–59
DEC	18–20	41–43	55–57	28–30	45–47	57–59

1936

URANUS CYCLE 0–84 YEARS						
m.o.b.	WAXING SQUARE	WAXING TRINE	OPPOSITION	WANING TRINE	WANING SQUARE	RETURN
JAN	19–21	26–27	38–39	52–53	60–61	82–84
FEB	19–21	26–27	38–39	52–53	59–61	82–84
MAR	19–21	26–27	39–40	52–53	59–60	82–84
APR	20–21	26–28	40–41	53–54	60–61	82–84
MAY	21–22	26–28	40–41	53–54	60–62	83–84
JUN	21–22	27–28	40–41	53–54	60–62	83–84
JUL	21–22	27–28	40–41	53–55	60–62	83–84
AUG	20–21	27–28	40–41	53–55	60–62	83–84
SEP	20–21	27–28	40–41	53–55	60–62	83–84
OCT	20–21	27–28	40–41	53–54	60–62	82–84
NOV	19–20	26–27	39–40	53–54	60–61	82–84
DEC	19–20	25–27	38–39	52–53	60–61	82–84

NEPTUNE CYCLE				PLUTO CYCLE		
m.o.b.	SQUARING PLUTO	WAXING SQUARE	WAXING TRINE	CONJ.NEPTUNE	WAXING SQUARE	WAXING TRINE
JAN	18–20	41–42	55–56	28–30	45–47	57–59
FEB	18–20	40–42	54–56	27–29	44–46	56–58
MAR	17–19	40–42	53–55	27–30	44–46	56–58
APR	17–19	39–41	53–55	27–30	44–46	56–58
MAY	17–19	39–41	53–55	27–29	44–46	56–58
JUN	17–19	39–41	53–55	27–29	45–47	56–58
JUL	17–19	39–42	53–55	27–29	45–47	56–58
AUG	17–19	40–42	54–56	27–29	45–47	56–58
SEP	18–20	40–43	54–56	28–30	46–48	57–59
OCT	18–20	41–43	55–57	28–30	46–48	57–59
NOV	17–19	41–43	55–57	28–30	45–47	57–59
DEC	17–19	41–43	55–57	28–30	45–47	57–59

1937

URANUS CYCLE 0–84 YEARS						
m.o.b.	WAXING SQUARE	WAXING TRINE	OPPOSITION	WANING TRINE	WANING SQUARE	RETURN
JAN	19–21	25–27	38–39	52–53	60–61	82–84
FEB	19–21	25–27	38–39	52–53	59–61	82–84
MAR	19–21	26–27	39–40	52–53	59–60	82–84
APR	20–21	26–28	39–40	53–54	60–61	82–84
MAY	20–21	26–28	39–40	53–54	60–62	83–84
JUN	20–21	27–28	39–40	53–54	60–62	83–84
JUL	20–21	27–28	40–41	53–55	60–62	83–84
AUG	19–20	27–28	40–41	53–55	60–62	83–84
SEP	19–20	27–28	40–41	53–55	60–62	83–84
OCT	19–20	27–28	40–41	53–54	60–62	82–84
NOV	19–20	26–27	39–40	53–54	60–61	82–84
DEC	19–20	25–27	38–39	52–53	60–61	82–84

NEPTUNE CYCLE				PLUTO CYCLE		
m.o.b.	SQUARING PLUTO	WAXING SQUARE	WAXING TRINE	CONJ.NEPTUNE	WAXING SQUARE	WAXING TRINE
JAN	17–19	41–42	55–56	28–30	45–47	57–59
FEB	17–19	40–42	54–56	28–30	44–46	56–58
MAR	16–18	40–42	54–56	28–30	44–46	56–58
APR	16–18	39–41	54–56	28–30	44–46	56–58
MAY	16–18	39–41	53–55	27–29	44–46	56–58
JUN	17–19	39–41	53–55	27–29	44–46	56–58
JUL	17–19	39–42	53–55	27–29	44–46	56–58
AUG	17–19	40–42	54–56	27–29	44–46	56–58
SEP	18–20	40–42	54–56	28–30	45–47	57–59
OCT	18–20	41–43	55–57	28–30	45–47	57–59
NOV	17–19	41–43	55–57	28–30	44–46	57–59
DEC	17–19	41–43	55–57	28–30	44–46	57–59

1938

	URANUS CYCLE 0–84 YEARS					
m.o.b.	WAXING SQUARE	WAXING TRINE	OPPOSITION	WANING TRINE	WANING SQUARE	RETURN
JAN	19–21	25–27	38–39	52–53	60–61	82–84
FEB	19–21	25–27	38–39	52–53	59–61	82–84
MAR	19–20	26–27	38–39	52–53	59–60	82–84
APR	19–20	26–28	38–40	53–54	60–61	82–84
MAY	20–21	26–28	39–40	53–54	60–62	83–84
JUN	20–21	27–28	39–40	53–54	60–62	83–84
JUL	20–21	27–28	40–41	53–55	60–62	83–84
AUG	20–21	27–28	40–41	53–55	60–62	83–84
SEP	20–21	27–28	40–41	53–55	60–62	83–84
OCT	20–21	27–28	40–41	53–54	60–62	82–84
NOV	19–20	26–27	39–40	53–54	60–61	82–84
DEC	19–20	25–26	38–39	52–53	60–61	82–84

	NEPTUNE CYCLE			PLUTO CYCLE		
m.o.b.	SQUARING PLUTO	WAXING SQUARE	WAXING TRINE	CONJ.NEPTUNE	WAXING SQUARE	WAXING TRINE
JAN	17–19	41–42	55–56	28–30	45–47	56–58
FEB	17–19	40–42	54–56	28–30	44–46	56–58
MAR	16–18	40–42	54–56	28–30	44–46	55–57
APR	16–18	39–41	54–56	28–30	44–46	55–57
MAY	16–18	39–41	53–55	27–29	44–46	55–57
JUN	16–18	39–41	53–55	27–29	44–46	55–57
JUL	16–18	39–42	53–55	27–29	44–46	55–57
AUG	16–18	40–42	54–56	27–29	44–46	55–57
SEP	17–19	40–42	54–56	28–30	45–47	56–58
OCT	17–19	41–43	55–57	28–30	45–47	56–58
NOV	17–19	41–43	55–57	28–30	44–46	56–58
DEC	17–19	41–43	55–57	28–30	44–46	56–58

1939

	URANUS CYCLE 0–84 YEARS					
m.o.b.	WAXING SQUARE	WAXING TRINE	OPPOSITION	WANING TRINE	WANING SQUARE	RETURN
JAN	19–21	25–27	38–39	52–53	60–61	82–84
FEB	19–21	25–27	38–39	52–53	59–61	82–84
MAR	19–20	26–27	38–39	52–53	59–60	82–84
APR	19–20	26–28	38–40	53–54	60–61	82–84
MAY	20–21	26–28	39–40	53–54	60–61	83–84
JUN	20–21	27–28	39–40	53–54	60–61	83–84
JUL	20–21	27–28	40–41	53–54	60–62	83–84
AUG	20–21	27–28	40–41	53–54	60–62	83–84
SEP	20–21	27–28	40–41	53–54	60–62	83–84
OCT	20–21	27–28	40–41	53–54	60–62	82–84
NOV	19–20	26–27	39–40	53–54	60–61	82–84
DEC	19–20	25–26	38–39	52–53	60–61	82–84

	NEPTUNE CYCLE			PLUTO CYCLE		
m.o.b.	SQUARING PLUTO	WAXING SQUARE	WAXING TRINE	CONJ.NEPTUNE	WAXING SQUARE	WAXING TRINE
JAN	17–19	41–42	55–56	28–30	44–46	56–58
FEB	16–18	40–42	54–56	28–30	43–45	56–58
MAR	16–18	40–42	54–56	28–30	43–45	55–57
APR	16–18	39–41	54–56	28–30	43–45	55–57
MAY	16–18	39–41	53–55	27–29	43–45	55–57
JUN	16–18	39–41	53–55	27–29	43–45	55–57
JUL	16–18	39–42	53–55	27–29	43–45	55–57
AUG	16–18	40–42	54–56	27–29	44–46	55–57
SEP	17–19	40–42	54–56	28–30	44–46	56–58
OCT	17–19	41–43	55–57	28–30	43–45	56–58
NOV	16–18	41–43	55–57	28–30	43–45	56–58
DEC	16–18	41–43	55–57	28–30	43–45	56–58

1940

URANUS CYCLE 0–84 YEARS						
m.o.b.	WAXING SQUARE	WAXING TRINE	OPPOSITION	WANING TRINE	WANING SQUARE	RETURN
JAN	19–20	26–26	38–39	52–54	60–61	82–84
FEB	19–20	25–26	38–39	52–54	60–61	82–84
MAR	19–20	25–26	38–39	52–53	59–60	82–84
APR	20–21	25–26	38–39	53–54	59–60	82–84
MAY	20–21	26–27	39–40	53–54	60–61	82–84
JUN	20–21	26–27	39–40	53–54	60–61	83–84
JUL	20–21	26–27	39–40	53–55	60–61	83–84
AUG	20–21	27–28	39–41	53–55	61–62	83–84
SEP	20–21	27–28	39–41	53–55	61–62	83–84
OCT	19–20	26–27	39–41	53–55	60–61	82–84
NOV	19–20	26–27	38–39	53–55	60–61	82–84
DEC	19–20	25–26	38–39	52–54	60–61	82–84

NEPTUNE CYCLE				PLUTO CYCLE		
m.o.b.	SQUARING PLUTO	WAXING SQUARE	WAXING TRINE	CONJ.NEPTUNE	WAXING SQUARE	WAXING TRINE
JAN	16–18	41–43	55–57	28–30	42 44	55–57
FEB	15–17	40–42	54–56	28–30	43–44	54–56
MAR	15–17	40–42	54–56	28–30	43–44	54–56
APR	15–17	40–42	54–56	28–30	43–45	54–56
MAY	16–18	39–41	53–55	27–29	43–45	54–56
JUN	16–18	39–41	53 55	27–29	43–45	54–56
JUL	16–18	39–41	53–55	27–29	43–45	54–56
AUG	17–19	39–41	54–56	28–30	44–46	55–57
SEP	17–19	40–42	54–56	28–30	44–46	55–57
OCT	17–19	40–42	54–56	28–30	44–46	55–57
NOV	16–18	40–42	55–57	28–30	43–45	55–57
DEC	16 18	41–43	55–57	28–30	43–45	55–57

1941

URANUS CYCLE 0–84 YEARS						
m.o.b.	WAXING SQUARE	WAXING TRINE	OPPOSITION	WANING TRINE	WANING SQUARE	RETURN
JAN	19–20	25–26	38–39	52–54	60–61	82–84
FEB	19–20	25–26	38–39	52–54	60–61	82–84
MAR	19–20	25–26	38–39	52–53	60–61	82–84
APR	20–21	25–26	38–39	53–54	60–61	82–84
MAY	20–21	26–27	39–40	53–54	60–61	82–84
JUN	20–21	26–27	39–40	53–54	60–61	83–84
JUL	20–21	26–27	39–40	53–55	60–61	83–84
AUG	20–21	27–28	39–41	53–55	61–62	83–84
SEP	20–21	27–28	39–41	53–55	61–62	83–84
OCT	19–20	26–27	39–41	53–55	61–62	82–84
NOV	19–20	26–27	38–39	53–55	60–61	82–84
DEC	19–20	25–26	38–39	53–54	60–61	82–84

NEPTUNE CYCLE				PLUTO CYCLE		
m.o.b.	SQUARING PLUTO	WAXING SQUARE	WAXING TRINE	CONJ.NEPTUNE	WAXING SQUARE	WAXING TRINE
JAN	16–18	41–43	55–57	28–30	43–45	55–57
FEB	15–17	40–42	54–56	28–30	42–44	54–56
MAR	15–17	40–42	54–56	28–30	42–44	54–56
APR	15–17	40–42	54–56	28–30	42–44	54–56
MAY	15–17	39–41	53–55	27–29	42–44	54–56
JUN	15–17	39–41	53–55	27–29	42–44	54–56
JUL	15–17	40–42	53–55	27–29	42–44	54–56
AUG	15–17	40–42	54–56	28–30	43–45	55–57
SEP	16–18	40–42	54–56	28–30	43–45	55–57
OCT	16–18	40–42	54–56	28–30	43–45	55–57
NOV	16–18	41–43	55–57	28–30	43–44	55–57
DEC	16–18	41–43	55–57	28–30	43–44	55–57

1942

URANUS CYCLE 0–84 YEARS						
m.o.b.	WAXING SQUARE	WAXING TRINE	OPPOSITION	WANING TRINE	WANING SQUARE	RETURN
JAN	19–20	26–26	38–39	52–54	60–61	82–84
FEB	19–20	25–26	38–39	52–54	60–61	82–84
MAR	19–20	25–26	38–39	52–53	60–61	82–84
APR	20–21	25–26	38–39	53–54	60–61	82–84
MAY	20–21	26–27	39–40	53–54	60–61	82–84
JUN	20–21	26–27	39–40	53–54	60–61	83–84
JUL	20–21	26–27	39–40	53–55	60–61	83–84
AUG	20–21	26–27	39–41	53–55	61–62	83–84
SEP	20–21	26–27	39–41	53–55	61–62	83–84
OCT	19–20	25–26	39–41	53–55	61–62	82–84
NOV	19–20	25–26	38–39	53–55	60–61	82–84
DEC	19–20	25–26	38–39	52–54	60–61	82–84

	NEPTUNE CYCLE			PLUTO CYCLE		
m.o.b.	SQUARING PLUTO	WAXING SQUARE	WAXING TRINE	CONJ.NEPTUNE	WAXING SQUARE	WAXING TRINE
JAN	16–18	41–43	55–57	28–30	42–44	55–57
FEB	15–17	40–42	54–56	28–30	42–44	54–56
MAR	15–17	40–42	54–56	28–30	42–44	53–55
APR	15–17	40–42	54–56	28–30	42–44	53–55
MAY	15–17	39–41	53–55	27–29	42–44	53–55
JUN	15–17	39–41	53–55	27–29	42–44	54–55
JUL	15–17	40–42	53–55	27–29	42–44	54–56
AUG	16–18	40–42	54–56	27–29	42–44	54–56
SEP	16–18	40–42	54–56	28–29	42–44	55–56
OCT	16–18	40–42	54–56	28–29	42–44	55–56
NOV	16–18	41–43	55–57	28–29	42–44	55–56
DEC	16–18	41–43	55–57	28–29	41–43	55–56

1943

URANUS CYCLE 0–84 YEARS						
m.o.b.	WAXING SQUARE	WAXING TRINE	OPPOSITION	WANING TRINE	WANING SQUARE	RETURN
JAN	18–20	26–26	38–39	52–54	60–61	82–84
FEB	18–20	25–26	38–39	52–54	60–61	82–84
MAR	18–20	25–26	38–39	52–53	60–61	82–84
APR	19–21	25–26	38–39	53–54	60–61	82–84
MAY	19–21	26–27	39–40	53–54	60–61	82–84
JUN	19–21	26–27	39–40	53–54	60–61	83–84
JUL	19–21	26–27	39–40	53–55	60–61	83–84
AUG	19–21	26–27	39–41	53–55	61–62	83–84
SEP	20–21	26–27	39–41	53–55	61–62	83–84
OCT	20–21	25–26	39–41	53–55	61–62	82–84
NOV	19–20	25–26	38–39	53–55	60–61	82–84
DEC	19–20	25–26	38–39	53–54	60–61	82–84

	NEPTUNE CYCLE			PLUTO CYCLE		
m.o.b.	SQUARING PLUTO	WAXING SQUARE	WAXING TRINE	CONJ.NEPTUNE	WAXING SQUARE	WAXING TRINE
JAN	15–17	41–43	55–57	27–29	41–43	54–56
FEB	15–17	40–42	54–56	27–29	41–43	54–56
MAR	14–16	40–42	54–56	28–29	41–43	53–55
APR	14–16	40–42	54–56	27–29	41–43	53–55
MAY	14–17	39–41	53–55	27–29	41–43	53–55
JUN	14–17	39–41	53–55	27–29	41–43	53–55
JUL	14–17	40–42	53–55	28–30	41–43	53–55
AUG	15–18	40–42	54–56	28–30	41–43	53–55
SEP	16–18	40–42	54–56	28–30	42–44	53–55
OCT	15–18	40–42	54–56	28–30	42–44	53–55
NOV	15–18	41–43	55–57	28–30	42–44	53–55
DEC	15–17	41–43	55–57	29–30	42–43	54–55

1944

URANUS CYCLE 0–84 YEARS						
m.o.b.	WAXING SQUARE	WAXING TRINE	OPPOSITION	WANING TRINE	WANING SQUARE	RETURN
JAN	18–20	25–26	38–39	52–53	60–61	82–84
FEB	18–19	25–26	37–39	52–53	60–61	82–84
MAR	18–20	25–26	37–39	52–53	60–61	82–84
APR	19–20	25–26	38–40	53–54	60–61	82–84
MAY	19–20	25–26	38–40	53–54	60–61	82–84
JUN	19–21	26–27	39–40	53–54	60–62	83–84
JUL	20–21	26–27	39–40	53–55	60–62	83–84
AUG	20–21	26–27	39–41	53–55	61–62	83–84
SEP	20–21	26–27	39–41	53–55	61–62	83–84
OCT	19–20	26–27	39–41	53–55	61–62	82–84
NOV	19–20	26–27	39–41	53–55	60–62	82–84
DEC	18–20	26–27	39–40	53–54	60–62	82–84

NEPTUNE CYCLE				PLUTO CYCLE		
m.o.b.	SQUARING PLUTO	WAXING SQUARE	WAXING TRINE	CONJ.NEPTUNE	WAXING SQUARE	WAXING TRINE
JAN	15–17	41–43	55–57	28–30	41–43	54–55
FEB	14–16	40–42	54–56	28–30	41–43	54–55
MAR	14–16	40–42	54–56	28–30	41–43	53–54
APR	14–16	40–42	54–56	28–30	41–43	53–55
MAY	14–16	39–41	53–55	27–29	41–43	53–55
JUN	14–16	39–41	53–55	27–29	41–43	53–55
JUL	14–16	40–42	53–55	27–29	41–43	53–55
AUG	15–17	40–42	54–56	28–29	42–44	54–56
SEP	15–17	40–42	54–56	28–29	42–44	54–56
OCT	15–17	40–42	54–56	28–30	42–44	54–56
NOV	15–17	41–43	55–57	28–30	42–44	54–56
DEC	15–17	41–43	55–57	28–30	42–43	54–55

1945

URANUS CYCLE 0–84 YEARS						
m.o.b.	WAXING SQUARE	WAXING TRINE	OPPOSITION	WANING TRINE	WANING SQUARE	RETURN
JAN	18–20	25–26	38–40	52–54	60–61	82–84
FEB	18–20	24–26	38–40	52–54	60–61	82–84
MAR	18–19	24–26	37–39	51–53	60–61	82–84
APR	18–19	24–26	37–39	51–53	60–62	82–84
MAY	19–20	25–27	38–40	53–54	60–62	82–84
JUN	19–20	25–27	38–40	53–54	60–62	83–84
JUL	19–20	25–27	38–40	53–54	60–62	83–84
AUG	20–21	25–27	39–41	53–55	60–62	83–84
SEP	20–21	26–27	39–41	53–55	61–62	83–84
OCT	20–21	26–27	39–41	53–55	61–62	82–84
NOV	19–20	25–26	39–40	53–54	60–62	82–84
DEC	18–20	25–26	39–40	53–54	60–62	82–84

NEPTUNE CYCLE				PLUTO CYCLE		
m.o.b.	SQUARING PLUTO	WAXING SQUARE	WAXING TRINE	CONJ.NEPTUNE	WAXING SQUARE	WAXING TRINE
JAN	15–17	41–43	55–57	28–30	41–43	53–55
FEB	14–16	40–42	54–56	28–30	41–43	53–55
MAR	14–16	40–42	54–56	27–29	40–42	52–54
APR	14–16	40–42	54–56	27–29	40–42	52–54
MAY	14–16	39–41	53–55	27–29	41–42	52–54
JUN	14–16	39–41	53–55	27–29	41–42	52–54
JUL	14–16	39–41	53–55	27–29	42–43	52–54
AUG	15–17	40–42	54–56	28–29	42–43	52–54
SEP	15–17	40–42	54–56	28–29	42–43	53–55
OCT	15–17	40–42	54–56	28–29	42–43	53–55
NOV	15–17	41–43	55–57	28–30	42–43	53–55
DEC	15–17	41–43	55–57	28–30	42–43	54–55

1946

URANUS CYCLE 0–84 YEARS						
m.o.b.	WAXING SQUARE	WAXING TRINE	OPPOSITION	WANING TRINE	WANING SQUARE	RETURN
JAN	18–20	24–26	39–40	53–54	60–62	82–84
FEB	18–19	24–26	38–40	52–53	60–62	82–84
MAR	18–19	24–26	38–39	52–53	60–61	82–84
APR	18–19	24–26	38–40	52–53	60–62	82–84
MAY	18–19	25–27	38–40	53–54	60–62	82–84
JUN	19–20	25–27	38–40	53–54	60–62	83–84
JUL	19–20	25–27	38–40	53–54	60–62	83–84
AUG	20–21	25–27	39–41	53–55	61–63	83–84
SEP	20–21	26–27	39–41	53–55	61–63	83–84
OCT	19–20	26–27	39–41	53–55	61–63	83–84
NOV	19–20	25–27	39–41	53–55	60–62	83–84
DEC	18–20	25–26	39–40	53–54	60–62	83–84

NEPTUNE CYCLE				PLUTO CYCLE		
m.o.b.	SQUARING PLUTO	WAXING SQUARE	WAXING TRINE	CONJ.NEPTUNE	WAXING SQUARE	WAXING TRINE
JAN	14–16	41–43	55–57	28–30	42–43	54–55
FEB	13–16	40–42	54–56	28–30	42–43	53–55
MAR	13–16	40–42	54–56	27–29	40–42	52–54
APR	13–16	40–42	54–56	27–29	40–42	52–54
MAY	13–16	39–41	53–55	27–29	40–42	52–54
JUN	13–16	39–41	53–55	27–29	40–42	52–54
JUL	14–16	40–42	53–55	27–29	40–42	52–54
AUG	14–16	40–42	54–56	27–29	40–42	53–55
SEP	15–17	40–42	54–56	27–29	41–43	53–55
OCT	15–17	40–42	54–56	27–29	41–43	53–55
NOV	15–17	41–43	55–57	28–29	41–43	53–55
DEC	15–17	41–43	55–57	28–29	41–43	53–55

1947

URANUS CYCLE 0–84 YEARS						
m.o.b.	WAXING SQUARE	WAXING TRINE	OPPOSITION	WANING TRINE	WANING SQUARE	RETURN
JAN	18–20	25–26	39–40	53–54	60–62	82–84
FEB	18–20	24–25	38–40	52–53	60–62	82–84
MAR	18–19	24–25	37–39	52–53	60–61	82–84
APR	18–20	24–25	38–40	52–53	60–62	82–84
MAY	18–20	25–26	38–40	53–54	60–62	82–84
JUN	19–20	25–26	38–40	53–54	60–62	83–84
JUL	19–20	25–26	38–40	53–54	60–62	83–84
AUG	19–20	26–27	39–41	53–55	61–63	83–84
SEP	20–21	26–27	39–41	54–55	61–63	83–84
OCT	20–21	26–27	39–41	54–55	61–63	83–84
NOV	19–20	25–26	39–41	53–55	61–62	83–84
DEC	19–20	25–26	39–40	53–54	61–62	83–84

NEPTUNE CYCLE				PLUTO CYCLE		
m.o.b.	SQUARING PLUTO	WAXING SQUARE	WAXING TRINE	CONJ.NEPTUNE	WAXING SQUARE	WAXING TRINE
JAN	14–16	41–43	55–57	28–29	40–42	52–54
FEB	14–16	40–42	54–56	27–29	40–42	51–53
MAR	13–15	40–42	54–56	27–29	40–41	51–53
APR	13–15	40–42	54–56	27–29	40–42	52–54
MAY	13–15	39–41	53–55	27–29	40–42	52–54
JUN	13–15	39–41	53–55	27–29	40–42	52–54
JUL	13–15	40–42	53–55	27–29	40–42	53–55
AUG	14–16	40–42	54–56	27–29	40–42	53–55
SEP	14–16	40–42	54–56	27–29	41–42	53–55
OCT	14–16	40–42	54–56	27–29	41–42	53–55
NOV	14–16	41–43	55–57	27–29	41–42	53–54
DEC	14–16	41–43	55–57	28–29	41–42	53–54

1948

URANUS CYCLE 0–84 YEARS						
m.o.b.	WAXING SQUARE	WAXING TRINE	OPPOSITION	WANING TRINE	WANING SQUARE	RETURN
JAN	19–20	24–25	39–40	53–54	60–62	82–84
FEB	18–20	24–25	38–40	52–53	60–62	82–84
MAR	18–19	24–25	37–39	52–53	60–61	82–84
APR	18–20	25–26	38–40	52–53	60–62	82–84
MAY	18–20	25–26	38–40	53–54	60–62	82–84
JUN	19–20	25–26	38–40	53–54	60–62	83–84
JUL	19–20	25–26	38–40	53–54	60–62	83–84
AUG	20–21	26–27	39–41	53–55	61–63	83–84
SEP	20–21	26–27	39–41	54–55	61–63	83–84
OCT	19–20	26–27	39–41	53–55	61–63	83–84
NOV	19–20	25–26	39–41	53–55	61–62	83–84
DEC	19–20	25–26	39–40	53–55	61–62	83–84

NEPTUNE CYCLE				PLUTO CYCLE		
m.o.b.	SQUARING PLUTO	WAXING SQUARE	WAXING TRINE	CONJ.NEPTUNE	WAXING SQUARE	WAXING TRINE
JAN	14–16	41–43	55–57	28–29	40–42	52–54
FEB	13–15	40–42	54–56	27–29	40–42	52–54
MAR	13–15	40–42	54–56	27–29	39–41	51–53
APR	13–15	40–42	54–56	27–29	39–41	51–53
MAY	12–15	39–41	53–55	26–28	39–41	51–53
JUN	12–15	39–41	53–55	26–28	39–41	51–53
JUL	13–15	40–42	53–55	26–28	39–41	51–53
AUG	13–15	40–42	53–56	27–29	40–42	51–53
SEP	14–16	40–42	53–56	27–29	40–42	51–53
OCT	14–16	40–42	54–56	27–29	40–42	52–54
NOV	14–16	41–43	55–57	28–29	40–42	52–54
DEC	14–16	41–43	55–57	28–29	40–42	52–54

1949

URANUS CYCLE 0–84 YEARS						
m.o.b.	WAXING SQUARE	WAXING TRINE	OPPOSITION	WANING TRINE	WANING SQUARE	RETURN
JAN	18–20	24–25	39–40	53–54	60–62	82–84
FEB	18–20	24–25	38–40	52–53	60–62	82–84
MAR	18–19	24–25	38–39	52–53	60–61	82–84
APR	18–20	25–26	38–40	52–53	60–62	82–84
MAY	18–20	25–26	38–40	53–54	60–62	82–84
JUN	19–20	25–26	38–40	53–54	60–62	83–84
JUL	19–20	25–26	38–40	53–54	60–62	83–84
AUG	19–20	26–27	39–41	53–55	61–63	83–84
SEP	19–20	26–27	39–41	54–55	61–63	83–84
OCT	18–20	26–27	39–41	53–55	61–63	83–84
NOV	18–20	25–26	39–41	53–55	61–62	83–84
DEC	18–19	25–26	39–40	53–55	61–62	83–84

NEPTUNE CYCLE				PLUTO CYCLE		
m.o.b.	SQUARING PLUTO	WAXING SQUARE	WAXING TRINE	CONJ.NEPTUNE	WAXING SQUARE	WAXING TRINE
JAN	14–16	41–43	55–57	28–29	40–42	52–54
FEB	13–15	40–42	54–56	27–29	39–41	51–53
MAR	13–15	40–42	54–56	27–29	39–41	51–53
APR	13–15	40–42	54–56	27–29	39–41	51–53
MAY	13–15	39–41	53–55	26–28	39–41	51–53
JUN	13–15	39–41	53–55	26–28	39–41	51–53
JUL	13–15	40–42	53–55	26–28	39–41	51–53
AUG	14–16	40–42	53–56	27–29	39–41	52–54
SEP	14–16	40–42	53–56	27–29	40–42	52–54
OCT	14–16	40–42	54–56	27–29	40–42	52–54
NOV	14–16	41–43	55–57	28–29	40–42	52–54
DEC	14–16	41–43	55–57	28–29	40–42	52–54

1950

	URANUS CYCLE 0–84 YEARS					
m.o.b.	WAXING SQUARE	WAXING TRINE	OPPOSITION	WANING TRINE	WANING SQUARE	RETURN
JAN	18–19	25–26	39–40	54–55	61–62	83–84
FEB	18–19	24–25	38–40	53–54	60–61	82–83
MAR	18–19	24–25	37–39	53–54	60–61	82–83
APR	18–20	24–25	37–39	53–54	60–62	82–83
MAY	18–20	25–26	37–39	53–54	60–62	83–84
JUN	18–20	25–26	38–39	53–54	60–62	83–84
JUL	18–20	25–26	38–39	53–54	60–62	83–84
AUG	18–20	26–27	39–41	54–55	60–62	83–84
SEP	19–20	26–27	39–41	54–55	61–63	83–84
OCT	19–20	26–27	39–41	54–55	61–63	83–84
NOV	19–20	25–26	39–40	54–55	61–63	83–84
DEC	18–19	25–26	39–40	54–55	61–63	83–84

	NEPTUNE CYCLE			PLUTO CYCLE		
m.o.b.	SQUARING PLUTO	WAXING SQUARE	WAXING TRINE	CONJ.NEPTUNE	WAXING SQUARE	WAXING TRINE
JAN	13–15	41–42	55–56	27–29	39–41	51–53
FEB	13–15	41–42	55–56	27–29	39–41	51–53
MAR	12–14	40–42	54–56	27–29	38–40	50–52
APR	12–14	40–42	53–55	27–29	38–40	50–52
MAY	12–14	39–41	53–55	26–28	38–40	50–52
JUN	12–15	39–41	53–55	26–28	39–40	50–52
JUL	13–15	39–41	53–55	27–28	39–41	51–53
AUG	13–15	39–42	53–55	27–28	39–41	51–53
SEP	14–16	40–42	53–56	27–29	40–41	52–54
OCT	14–16	40–42	54–56	27–29	40–42	52–54
NOV	14–15	41–43	54–57	27–29	40–41	52–54
DEC	14–15	41–43	54–57	27–29	40–41	52–54

1951

	URANUS CYCLE 0–84 YEARS					
m.o.b.	WAXING SQUARE	WAXING TRINE	OPPOSITION	WANING TRINE	WANING SQUARE	RETURN
JAN	18–19	25–26	38–40	54–55	61–63	83–84
FEB	18–19	24–25	38–40	53–54	61–63	82–83
MAR	18–19	24–25	38–39	53–54	60–61	82–83
APR	18–20	24–25	38–39	53–54	60–62	82–83
MAY	18–20	25–26	38–39	53–54	60–62	83–84
JUN	18–20	25–26	38–39	53–54	60–62	83–84
JUL	18–20	25–26	38–39	53–54	60–62	83–84
AUG	18–20	26–27	39–41	54–55	60–62	83–84
SEP	19–20	26–27	39–41	54–55	60–63	83–84
OCT	19–20	26–27	39–41	54–55	60–63	83–84
NOV	19–20	25–26	39–40	54–55	61–63	83–84
DEC	19–20	25–26	39–40	54–55	61–63	83–84

	NEPTUNE CYCLE			PLUTO CYCLE		
m.o.b.	SQUARING PLUTO	WAXING SQUARE	WAXING TRINE	CONJ.NEPTUNE	WAXING SQUARE	WAXING TRINE
JAN	13–15	41–42	55–56	27–29	39–41	51–53
FEB	13–15	41–42	55–56	27–29	39–41	51–53
MAR	12–14	40–42	54–56	27–29	38–40	50–52
APR	12–14	40–42	53–55	27–29	38–40	50–52
MAY	12–14	39–41	53–55	26–28	38–40	50–52
JUN	12–15	39–41	53–55	26–28	39–40	50–52
JUL	13–15	39–41	53–55	27–28	39–41	51–53
AUG	13–15	39–42	53–55	27–28	39–41	51–53
SEP	14–15	40–42	53–56	27–29	39–41	51–53
OCT	14–15	40–42	54–56	27–29	39–41	51–53
NOV	14–15	41–43	54–57	27–29	39–41	51–53
DEC	14–15	41–43	55–57	27–29	39–41	52–53

1952

URANUS CYCLE 0–84 YEARS						
m.o.b.	WAXING SQUARE	WAXING TRINE	OPPOSITION	WANING TRINE	WANING SQUARE	RETURN
JAN	18–19	25–26	38–40	54–55	61–63	83–84
FEB	18–19	24–25	38–40	53–54	61–63	82–83
MAR	18–19	24–25	38–39	53–54	61–62	82–83
APR	18–19	24–25	38–39	53–54	61–62	82–83
MAY	18–19	25–26	38–39	53–54	61–62	83–84
JUN	18–19	25–26	38–39	53–54	61–62	83–84
JUL	18–20	25–26	38–39	53–54	62–63	83–84
AUG	18–20	26–27	39–41	54–55	62–63	83–84
SEP	19–20	26–27	39–41	54–56	62–63	83–84
OCT	19–20	26–27	39–41	54–56	61–63	83–84
NOV	19–20	25–26	39–40	54–56	61–63	83–84
DEC	18–19	25–26	39–40	54–55	61–63	83–84

NEPTUNE CYCLE				PLUTO CYCLE		
m.o.b.	SQUARING PLUTO	WAXING SQUARE	WAXING TRINE	CONJ.NEPTUNE	WAXING SQUARE	WAXING TRINE
JAN	13–15	41–42	55–56	27–29	39–41	51–53
FEB	13–15	41–42	55–56	27–29	39–41	50–52
MAR	12–14	40–42	54–56	27–29	38–40	50–52
APR	12–14	40–42	53–55	27–29	38–40	50–52
MAY	12–14	39–41	53–55	26–28	38–40	50–52
JUN	12–15	39–41	53–55	26–28	39–40	50–52
JUL	13–15	39–41	53–55	27–28	39–41	51–53
AUG	13–15	39–42	53–55	27–28	39–41	51–53
SEP	13–15	40–42	53–55	27–28	39–41	51–53
OCT	13–15	40–42	53–55	27–29	39–41	51–53
NOV	13–15	41–43	54–57	27–29	39–41	51–53
DEC	13–15	41–43	55–57	27–29	39–40	51–53

1953

URANUS CYCLE 0–84 YEARS						
m.o.b.	WAXING SQUARE	WAXING TRINE	OPPOSITION	WANING TRINE	WANING SQUARE	RETURN
JAN	18–19	25–26	38–40	54–55	61–63	83–84
FEB	18–19	24–25	38–40	53–54	61–63	82–83
MAR	18–19	24–25	38–39	53–54	61–63	82–83
APR	18–19	24–25	38–39	53–54	61–63	82–83
MAY	18–19	24–26	38–39	53–54	61–62	83–84
JUN	18–19	24–26	38–40	53–54	61–62	83–84
JUL	18–20	25–26	38–40	53–54	62–63	83–84
AUG	18–20	26–27	39–41	54–55	62–63	83–84
SEP	19–20	26–27	39–41	54–56	62–63	83–84
OCT	19–20	26–27	39–41	54–56	62–63	83–84
NOV	19–20	25–26	39–40	54–56	62–63	83–84
DEC	18–19	25–26	39–40	54–56	62–63	83–84

NEPTUNE CYCLE				PLUTO CYCLE		
m.o.b.	SQUARING PLUTO	WAXING SQUARE	WAXING TRINE	CONJ.NEPTUNE	WAXING SQUARE	WAXING TRINE
JAN	13–15	41–42	55–56	27–29	38–40	51–53
FEB	13–15	41–42	55–56	27–29	38–40	50–52
MAR	12–14	40–42	54–56	27–29	37–39	50–52
APR	12–14	40–42	53–55	27–29	37–39	50–52
MAY	12–14	39–41	53–55	26–28	37–39	50–52
JUN	12–15	39–41	53–55	26–28	37–39	50–52
JUL	13–15	39–41	53–55	26–28	37–39	51–53
AUG	13–15	39–42	53–55	26–28	37–39	51–53
SEP	13–15	40–42	53–55	26–28	38–40	51–53
OCT	13–15	40–42	53–55	26–28	38–40	51–53
NOV	13–15	41–43	54–57	26–28	38–40	51–53
DEC	13–15	41–43	55–57	26–28	39–40	51–53

1954

URANUS CYCLE 0–84 YEARS						
m.o.b.	WAXING SQUARE	WAXING TRINE	OPPOSITION	WANING TRINE	WANING SQUARE	RETURN
JAN	17–19	25–26	38–40	54–55	61–63	83–84
FEB	17–19	24–25	38–40	53–54	61–63	82–83
MAR	17–19	24–25	38–39	53–54	61–62	82–83
APR	17–19	24–25	38–39	53–54	61–62	82–83
MAY	17–19	24–26	38–39	53–54	61–62	83–84
JUN	18–19	24–26	38–40	53–54	61–62	83–84
JUL	18–20	25–26	38–40	53–54	62–63	83–84
AUG	18–20	26–27	39–41	54–55	62–63	83–84
SEP	19–20	26–27	39–41	54–56	62–63	83–84
OCT	19–20	26–27	39–41	54–56	62–63	83–84
NOV	19–20	25–26	39–41	54–56	62–63	83–84
DEC	18–19	25–26	39–41	54–56	62–63	83–84

NEPTUNE CYCLE				PLUTO CYCLE		
m.o.b.	SQUARING PLUTO	WAXING SQUARE	WAXING TRINE	CONJ.NEPTUNE	WAXING SQUARE	WAXING TRINE
JAN	12–15	41–42	55–56	27–28	38–40	51–53
FEB	12–14	41–42	55–56	27–28	38–40	50–52
MAR	11–14	40–42	54–56	27–28	37–39	50–52
APR	11–14	40–42	53–55	26–28	37–39	50–52
MAY	11–14	39–41	53–55	26–28	37–39	49–51
JUN	11–14	39–41	53–55	26–28	37–39	49–51
JUL	12–14	39–41	53–55	26–28	37–39	50–52
AUG	12–15	39–42	53–55	26–28	37–39	50–52
SEP	13–15	40–42	54–55	26–28	38–40	51–53
OCT	13–15	40–42	53–55	26–28	38–40	51–53
NOV	13–15	41–43	54–57	26–28	38–40	51–53
DEC	13–15	41–43	55–57	27–28	38–39	51–53

1955

URANUS CYCLE 0–84 YEARS						
m.o.b.	WAXING SQUARE	WAXING TRINE	OPPOSITION	WANING TRINE	WANING SQUARE	RETURN
JAN	17–19	25–26	38–40	54–55	61–63	83–84
FEB	17–19	24–25	38–40	53–54	61–63	82–84
MAR	17–19	24–25	38–39	53–54	61–62	82–84
APR	17–19	24–25	38–39	53–54	61–62	82–84
MAY	17–19	24–26	38–39	53–55	61–62	83–84
JUN	18–19	24–26	38–40	53–55	61–62	83–84
JUL	18–20	25–26	38–40	53–55	62–63	83–84
AUG	18–20	25–27	39–41	54–55	62–63	83–84
SEP	19–20	25–27	39–41	54–56	62–63	83–84
OCT	19–20	25–27	39–41	54–56	61–63	83–84
NOV	19–20	25–27	39–41	54–56	62–64	83–84
DEC	18–19	25–26	40–41	55–56	62–64	83–84

NEPTUNE CYCLE				PLUTO CYCLE		
m.o.b.	SQUARING PLUTO	WAXING SQUARE	WAXING TRINE	CONJ.NEPTUNE	WAXING SQUARE	WAXING TRINE
JAN	12–15	41–42	55–56	27–28	38–40	50–52
FEB	12–14	41–42	55–56	27–28	38–40	50–52
MAR	11–13	40–42	55–56	27–28	37–39	49–51
APR	11–13	40–42	54–56	26–28	37–39	49–51
MAY	11–13	39–41	53–55	26–28	37–39	49–51
JUN	11–13	39–41	53–55	26–27	37–39	49–51
JUL	12–14	39–41	53–55	26–28	37–39	49–51
AUG	12–15	39–42	53–55	26–28	37–39	49–51
SEP	13–15	40–42	53–55	26–28	38–40	50–52
OCT	13–15	40–42	53–55	26–28	38–40	50–52
NOV	13–14	41–43	54–57	26–28	38–40	51–53
DEC	13–14	41–43	55–57	27–28	38–39	51–53

1956

URANUS CYCLE 0–84 YEARS						
m.o.b.	WAXING SQUARE	WAXING TRINE	OPPOSITION	WANING TRINE	WANING SQUARE	RETURN
JAN	17–19	25–26	39–41	54–55	61–63	83–84
FEB	17–19	24–25	38–40	54–55	61–63	82–83
MAR	17–19	24–25	38–39	54–55	61–62	82–83
APR	17–19	24–25	38–39	53–55	61–62	82–83
MAY	17–19	24–26	38–39	53–55	61–62	83–84
JUN	18–19	24–26	38–40	53–55	61–62	83–84
JUL	18–20	24–26	38–40	53–55	62–63	83–84
AUG	18–20	25–27	39–41	54–56	62–63	83–84
SEP	19–20	25–27	39–41	54–56	62–63	83–84
OCT	19–20	25–27	39–41	54–56	61–63	83–84
NOV	19–20	26–27	39–41	54–56	62–64	83–84
DEC	18–19	26–27	40–41	55–56	62–64	83–84

NEPTUNE CYCLE				PLUTO CYCLE		
m.o.b.	SQUARING PLUTO	WAXING SQUARE	WAXING TRINE	CONJ.NEPTUNE	WAXING SQUARE	WAXING TRINE
JAN	12–14	41–42	55–56	27–28	38–39	50–52
FEB	12–14	41–42	55–56	26–28	37–38	50–52
MAR	11–13	40–42	55–56	26–28	37–38	49–51
APR	11–13	40–42	54–56	26–28	37–38	49–51
MAY	11–13	39–41	53–55	25–27	36–38	49–51
JUN	11–13	39–41	53–55	25–27	36–38	49–51
JUL	12–14	39–41	53–55	25–27	36–38	49–51
AUG	12–14	39–42	53–55	25–27	37–39	49–51
SEP	12–14	40–42	53–55	26–27	38–39	50–52
OCT	12–14	40–42	53–55	26–28	38–39	50–52
NOV	13–14	41–43	54–57	26–28	38–39	51–53
DEC	13–14	41–43	55–57	26–28	38–40	51–53

1957

URANUS CYCLE 0–84 YEARS						
m.o.b.	WAXING SQUARE	WAXING TRINE	OPPOSITION	WANING TRINE	WANING SQUARE	RETURN
JAN	17–19	25–26	39–41	54–55	61–63	83–84
FEB	17–19	24–25	38–40	53–55	61–63	82–84
MAR	17–19	24–25	38–39	53–55	61–62	82–84
APR	17–19	24–25	38–39	53–55	61–62	82–84
MAY	17–19	24–26	38–39	53–55	61–62	83–84
JUN	18–19	24–26	38–40	53–55	61–62	83–84
JUL	18–20	24–26	38–40	53–55	62–63	83–84
AUG	18–20	25–27	39–41	54–56	62–63	83–84
SEP	19–20	25–27	40–41	55–56	62–64	83–84
OCT	19–20	25–27	40–41	55–56	62–64	83–84
NOV	19–20	26–27	40–41	55–56	62–64	83–84
DEC	18–19	26–27	40–41	55–56	62–64	83–84

NEPTUNE CYCLE				PLUTO CYCLE		
m.o.b.	SQUARING PLUTO	WAXING SQUARE	WAXING TRINE	CONJ.NEPTUNE	WAXING SQUARE	WAXING TRINE
JAN	12–14	41–42	55–56	26–28	37–39	50–52
FEB	12–14	41–42	55–56	26–28	37–39	50–52
MAR	11–13	40–42	55–56	26–28	36–38	49–51
APR	11–13	40–42	54–56	26–28	36–38	49–51
MAY	11–13	39–41	53–55	25–27	36–38	49–51
JUN	11–13	39–41	53–55	25–27	36–38	49–51
JUL	12–14	39–41	53–55	25–27	36–38	49–51
AUG	12–14	39–42	53–55	25–27	36–38	49–51
SEP	12–14	39–42	53–55	26–27	37–39	50–52
OCT	12–14	39–42	53–55	26–28	37–39	50–52
NOV	13–14	40–43	54–57	26–28	37–39	51–53
DEC	13–14	41–43	54–57	26–28	38–39	51–53

1958

m.o.b.	URANUS CYCLE 0–84 YEARS					
	WAXING SQUARE	WAXING TRINE	OPPOSITION	WANING TRINE	WANING SQUARE	RETURN
JAN	17–19	25–26	40–41	54–55	62–63	83–84
FEB	17–19	24–25	40–41	54–55	62–63	83–84
MAR	17–19	24–25	39–40	54–55	62–63	83–84
APR	17–19	24–25	38–39	54–55	61–62	83–84
MAY	17–19	24–26	38–39	54–55	61–62	83–84
JUN	18–19	24–26	38–40	54–55	61–62	83–84
JUL	18–20	24–26	38–40	54–55	62–63	83–84
AUG	18–20	25–27	39–41	54–56	62–63	83–84
SEP	19–20	25–27	40–41	55–56	62–64	83–84
OCT	19–20	25–27	40–41	55–56	62–64	83–84
NOV	19–20	26–27	40–41	55–57	62–64	83–84
DEC	19–20	26–27	40–41	55–57	63–64	83–84

m.o.b.	NEPTUNE CYCLE			PLUTO CYCLE		
	SQUARING PLUTO	WAXING SQUARE	WAXING TRINE	CONJ.NEPTUNE	WAXING SQUARE	WAXING TRINE
JAN	12–14	41–42	55–56	26–28	37–39	50–52
FEB	12–14	41–42	55–56	26–28	37–39	50–52
MAR	11–13	40–42	55–56	26–28	36–38	49–51
APR	11–13	40–42	54–56	26–28	36–38	49–51
MAY	11–13	39–41	53–55	25–27	36–38	49–51
JUN	11–13	39–41	53–55	25–27	36–38	49–51
JUL	12–14	39–41	53–55	25–27	36–38	49–51
AUG	12–14	39–42	53–55	25–27	36–38	49–51
SEP	12–14	39–42	53–55	25–27	37–39	50–52
OCT	12–14	39–42	53–55	25–27	37–39	50–52
NOV	13–14	40–43	54–57	26–28	37–39	50–52
DEC	13–14	41–43	54–57	26–28	37–39	50–52

1959

m.o.b.	URANUS CYCLE 0–84 YEARS					
	WAXING SQUARE	WAXING TRINE	OPPOSITION	WANING TRINE	WANING SQUARE	RETURN
JAN	19–20	25–26	39–41	54–56	62–63	83–84
FEB	18–20	24–25	39–41	54–56	62–63	82–84
MAR	17–19	24–25	39–40	54–55	62–63	82–84
APR	17–19	24–25	38–40	54–55	62–63	82–84
MAY	17–19	24–26	38–40	54–55	62–63	82–84
JUN	18–19	24–26	38–40	54–55	62–63	82–84
JUL	18–20	24–26	38–40	54–55	62–63	82–84
AUG	18–20	25–27	39–41	54–56	62–63	83–84
SEP	19–20	25–27	40–41	55–56	62–64	83–84
OCT	19–20	25–27	40–41	55–56	62–64	83–84
NOV	19–20	26–27	40–42	55–57	62–64	83–84
DEC	19–20	26–27	40–42	55–57	63–64	83–84

m.o.b.	NEPTUNE CYCLE			PLUTO CYCLE		
	SQUARING PLUTO	WAXING SQUARE	WAXING TRINE	CONJ.NEPTUNE	WAXING SQUARE	WAXING TRINE
JAN	12–14	41–42	55–56	26–28	37–39	50–52
FEB	12–14	41–42	55–56	26–28	37–39	50–52
MAR	11–13	40–42	55–56	26–28	36–38	49–51
APR	11–13	40–42	54–56	26–28	36–38	49–51
MAY	11–13	39–41	53–55	25–27	36–38	48–51
JUN	11–13	39–41	53–55	25–27	36–38	48–51
JUL	12–14	39–41	53–55	25–27	36–38	49–51
AUG	12–14	39–42	53–55	25–27	36–38	49–51
SEP	12–14	39–42	53–55	25–27	37–39	50–52
OCT	12–14	39–42	53–55	25–27	37–39	50–52
NOV	12–14	40–43	54–57	25–27	37–39	50–52
DEC	12–14	41–43	54–57	26–27	37–39	50–52

1960

URANUS CYCLE 0–84 YEARS						
m.o.b.	WAXING SQUARE	WAXING TRINE	OPPOSITION	WANING TRINE	WANING SQUARE	RETURN
JAN	18–19	25–26	40–41	55–57	63–64	82–84
FEB	18–19	25–26	40–41	55–56	62–64	82–84
MAR	18–19	24–26	39–40	55–56	62–63	82–84
APR	18–19	24–25	39–40	54–55	62–63	82–84
MAY	18–19	24–25	39–40	54–55	62–63	82–84
JUN	18–19	24–26	39–40	54–55	62–63	82–84
JUL	18–19	25–26	39–40	54–56	62–63	82–84
AUG	18–20	25–26	39–41	54–56	62–63	82–84
SEP	19–20	25–27	40–41	55–56	63–64	83–84
OCT	19–20	26–27	40–42	55–57	63–64	83–84
NOV	19–20	26–27	40–42	56–57	63–64	83–84
DEC	19–20	26–27	40–42	56–57	63–64	83–84

NEPTUNE CYCLE				PLUTO CYCLE		
m.o.b.	SQUARING PLUTO	WAXING SQUARE	WAXING TRINE	CONJ.NEPTUNE	WAXING SQUARE	WAXING TRINE
JAN	12–13	41–43	55–56	26–28	37–38	50–51
FEB	11–13	41–42	55–56	26–28	36–38	49–51
MAR	11–13	41–42	55–56	26–28	36–38	49–51
APR	10–13	40–42	54–55	25–27	35–37	48–50
MAY	10–12	39–42	53–55	25–27	35–37	48–50
JUN	10–13	39–41	53–55	25–27	35–37	48–50
JUL	11–13	39–41	52–55	25–27	36–38	49–51
AUG	11–13	39–41	53–55	25–27	36–38	49–51
SEP	12–14	39–42	53–55	25–27	36–38	50–52
OCT	12–14	40–42	53–56	25–27	37–39	50–52
NOV	12–14	40–42	54–56	25–27	37–38	50–52
DEC	12–14	41–43	54–56	26–27	37–38	50–52

1961

URANUS CYCLE 0–84 YEARS						
m.o.b.	WAXING SQUARE	WAXING TRINE	OPPOSITION	WANING TRINE	WANING SQUARE	RETURN
JAN	18–19	25–26	40–41	55–57	63–64	83–84
FEB	18–19	25–26	40–41	55–56	62–64	82–84
MAR	18–19	25–26	39–40	55–56	62–64	82–84
APR	18–19	24–25	39–40	54–55	62–64	82–84
MAY	18–19	24–25	39–40	54–55	62–64	82–84
JUN	18–19	24–26	39–40	55–56	62–64	82–84
JUL	18–19	25–26	39–40	55–57	62–64	82–84
AUG	18–20	25–27	39–41	56–57	62–64	82–84
SEP	19–20	26–28	40–41	56–57	62–64	83–84
OCT	19–20	26–28	40–42	56–57	62–64	83–84
NOV	19–20	26–27	40–42	56–57	62–64	83–84
DEC	19–20	26–27	40–42	56–57	63–64	83–84

NEPTUNE CYCLE				PLUTO CYCLE		
m.o.b.	SQUARING PLUTO	WAXING SQUARE	WAXING TRINE	CONJ.NEPTUNE	WAXING SQUARE	WAXING TRINE
JAN	12–13	41–43	55–56	26–27	36–38	50–52
FEB	11–13	41–42	55–56	26–27	36–38	49–51
MAR	11–13	41–42	55–56	26–27	35–37	49–51
APR	10–13	40–42	54–55	25–27	35–37	49–51
MAY	10–12	39–42	53–55	25–27	35–37	49–51
JUN	10–13	39–41	53–55	25–27	35–37	49–51
JUL	11–13	39–41	52–55	25–27	36–38	49–51
AUG	11–13	39–41	53–55	25–27	36–38	49–52
SEP	11–14	39–41	53–55	25–27	36–38	49–52
OCT	11–14	40–42	53–56	25–27	37–39	50–52
NOV	11–14	40–42	54–56	25–27	37–38	50–52
DEC	12–14	41–43	54–56	26–27	37–38	50–52

1962

URANUS CYCLE 0–84 YEARS						
m.o.b.	WAXING SQUARE	WAXING TRINE	OPPOSITION	WANING TRINE	WANING SQUARE	RETURN
JAN	18–19	25–26	40–41	55–57	63–64	83–84
FEB	18–19	25–26	40–41	55–56	62–64	82–84
MAR	18–19	25–26	39–40	55–56	62–64	82–84
APR	18–19	24–25	39–40	54–55	62–64	82–84
MAY	18–19	24–25	39–40	54–55	62–64	82–84
JUN	18–19	24–26	39–40	54–55	62–63	82–84
JUL	18–19	25–26	39–40	54–56	62–64	82–84
AUG	18–20	25–27	39–41	54–56	62–64	82–84
SEP	19–20	26–27	40–42	55–56	62–64	83–84
OCT	19–20	26–27	40–42	55–57	62–64	83–84
NOV	19–20	26–27	40–42	56–57	63–65	83–84
DEC	19–20	26–27	41–42	56–57	63–65	83–84

NEPTUNE CYCLE				PLUTO CYCLE		
m.o.b.	SQUARING PLUTO	WAXING SQUARE	WAXING TRINE	CONJ.NEPTUNE	WAXING SQUARE	WAXING TRINE
JAN	12–13	41–43	55–56	25–27	36–38	50–52
FEB	11–13	41–42	55–56	25–27	36–38	49–51
MAR	11–13	41–42	55–56	25–27	35–37	49–51
APR	10–13	40–42	54–55	25–27	35–37	49–51
MAY	10–12	39–42	53–55	25–26	35–37	48–50
JUN	10–13	39–41	53–55	25–26	35–37	48–50
JUL	11–13	39–41	52–55	25–26	36–38	49–51
AUG	11–13	39–41	53–55	25–26	36–38	49–52
SEP	11–14	39–41	53–55	25–27	36–38	49–52
OCT	11–14	40–42	53–56	25–27	37–39	50–52
NOV	11–14	40–42	54–56	25–27	37–38	50–52
DEC	12–14	41–43	54–56	26–27	37–38	50–52

1963

URANUS CYCLE 0–84 YEARS						
m.o.b.	WAXING SQUARE	WAXING TRINE	OPPOSITION	WANING TRINE	WANING SQUARE	RETURN
JAN	19–20	25–26	40–41	55–57	63–65	83–84
FEB	18–20	25–26	40–41	55–56	62–64	82–84
MAR	18–19	25–26	40–41	55–56	62–64	82–84
APR	18–19	24–25	39–41	54–56	62–64	82–84
MAY	18–19	24–25	39–40	54–56	62–64	82–84
JUN	18–19	24–26	39–40	55–56	62–63	82–84
JUL	18–19	25–26	39–40	54–56	62–64	82–84
AUG	18–20	25–27	39–41	54–56	62–64	82–84
SEP	19–20	26–27	40–41	55–57	62–64	83–84
OCT	19–20	26–27	40–42	55–57	62–64	83–84
NOV	19–20	26–27	40–42	56–58	63–65	83–84
DEC	19–20	26–27	41–43	56–58	63–65	83–84

NEPTUNE CYCLE				PLUTO CYCLE		
m.o.b.	SQUARING PLUTO	WAXING SQUARE	WAXING TRINE	CONJ.NEPTUNE	WAXING SQUARE	WAXING TRINE
JAN	11–13	41–43	55–56	26–27	36–38	50–52
FEB	11–13	41–42	55–56	25–27	36–38	49–51
MAR	10–13	41–42	55–56	25–26	35–37	49–51
APR	10–12	40–42	54–55	24–26	35–37	49–51
MAY	10–12	39–42	53–55	24–26	35–37	48–50
JUN	10–12	39–41	53–55	24–26	35–37	48–50
JUL	10–12	39–41	52–55	24–26	36–38	49–51
AUG	11–13	39–41	53–55	24–26	36–38	49–52
SEP	11–13	39–41	53–55	25–26	36–38	49–52
OCT	12–13	40–42	53–56	25–27	37–39	50–52
NOV	12–14	40–42	54–56	25–27	37–38	50–52
DEC	12–14	41–43	54–56	25–27	37–38	50–52

1964

URANUS CYCLE 0–84 YEARS						
m.o.b.	WAXING SQUARE	WAXING TRINE	OPPOSITION	WANING TRINE	WANING SQUARE	RETURN
JAN	19–20	26–27	40–42	55–57	63–65	83–84
FEB	18–20	26–27	40–42	55–57	63–64	82–84
MAR	18–19	25–26	40–41	55–57	63–65	82–84
APR	18–19	24–25	39–41	55–57	62–64	82–84
MAY	18–19	24–25	39–41	54–56	62–63	82–84
JUN	18–19	25–26	39–41	55–56	62–63	82–84
JUL	18–19	25–26	39–41	54–56	62–63	82–84
AUG	18–20	25–27	39–41	54–56	62–64	82–84
SEP	19–20	26–27	40–42	55–57	63–64	83–84
OCT	19–20	26–27	40–42	55–57	63–64	83–84
NOV	19–20	26–27	40–42	56–58	63–65	83–84
DEC	19–20	27–28	41–43	56–58	63–65	83–84

NEPTUNE CYCLE				PLUTO CYCLE		
m.o.b.	SQUARING PLUTO	WAXING SQUARE	WAXING TRINE	CONJ.NEPTUNE	WAXING SQUARE	WAXING TRINE
JAN	11–13	41–43	55–56	25–27	36–38	50–52
FEB	11–13	41–42	55–56	25–27	36–38	49–51
MAR	10–13	41–42	55–56	25–27	35–37	49–51
APR	10–12	40–42	54–55	24–26	35–37	49–51
MAY	10–12	39–42	53–55	24–26	35–37	48–50
JUN	10–12	39–41	53–55	24–26	35–37	48–50
JUL	10–12	39–41	52–55	24–26	36–38	49–51
AUG	11–13	39–41	53–55	24–26	36–38	49–52
SEP	11–13	39–41	53–55	24–26	36–38	49–52
OCT	12–13	40–42	53–56	25–27	37–39	50–52
NOV	12–14	40–42	54–56	25–27	37–38	50–52
DEC	12–14	41–43	54–56	25–27	37–38	50–52

1965

URANUS CYCLE 0–84 YEARS						
m.o.b.	WAXING SQUARE	WAXING TRINE	OPPOSITION	WANING TRINE	WANING SQUARE	RETURN
JAN	19–20	26–27	40–42	55–57	63–65	83–84
FEB	18–20	26–27	40–42	55–57	63–64	83–84
MAR	18–19	25–26	40–41	56–57	63–65	83–84
APR	18–19	25–26	39–41	55–57	62–64	82–84
MAY	18–19	25–26	39–41	55–56	62–63	82–84
JUN	18–19	25–26	39–41	55–56	62–63	82–84
JUL	18–19	25–26	39–41	55–56	62–63	82–84
AUG	18–20	25–27	39–41	55–56	62–64	82–84
SEP	19–20	26–27	40–42	55–57	63–64	82–84
OCT	19–20	26–28	40–42	55–57	63–64	82–84
NOV	19–20	26–28	40–42	56–58	63–65	82–84
DEC	20–21	27–28	41–43	56–58	63–65	83–84

NEPTUNE CYCLE				PLUTO CYCLE		
m.o.b.	SQUARING PLUTO	WAXING SQUARE	WAXING TRINE	CONJ.NEPTUNE	WAXING SQUARE	WAXING TRINE
JAN	11–13	41–43	55–56	25–27	36–38	50–52
FEB	11–13	41–42	55–56	25–27	36–38	49–51
MAR	10–13	41–42	55–56	25–27	35–37	49–51
APR	10–12	40–42	54–55	24–26	35–37	49–51
MAY	10–12	39–42	53–55	24–26	35–37	48–50
JUN	10–12	39–41	53–55	24–26	34–36	48–50
JUL	10–12	39–41	52–55	24–26	34–36	49–51
AUG	11–13	39–41	53–55	24–26	35–37	49–52
SEP	11–13	39–41	53–55	24–26	36–38	49–52
OCT	12–13	40–42	53–56	24–26	36–38	50–52
NOV	12–14	40–42	54–56	24–26	36–38	50–52
DEC	12–14	41–43	54–56	25–26	36–38	50–52

1966

URANUS CYCLE 0–84 YEARS						
m.o.b.	WAXING SQUARE	WAXING TRINE	OPPOSITION	WANING TRINE	WANING SQUARE	RETURN
JAN	19–20	26–27	40–42	55–57	63–65	83–84
FEB	18–20	26–27	40–42	55–57	63–64	83–84
MAR	18–19	25–27	41–42	56–57	63–64	83–84
APR	18–19	25–27	41–42	56–57	63–64	82–84
MAY	18–19	25–27	40–41	55–57	63–64	82–84
JUN	18–19	25–27	40–41	55–56	63–64	82–84
JUL	18–19	25–27	40–42	55–56	63–64	82–84
AUG	18–20	25–27	40–42	55–56	63–64	82–84
SEP	19–20	26–27	41–42	56–57	63–64	83–84
OCT	19–20	26–28	41–42	56–57	63–64	83–84
NOV	19–20	26–28	42–43	56–58	64–65	83–84
DEC	20–21	27–28	42–43	57–58	64–65	83–84

	NEPTUNE CYCLE			PLUTO CYCLE		
m.o.b.	SQUARING PLUTO	WAXING SQUARE	WAXING TRINE	CONJ.NEPTUNE	WAXING SQUARE	WAXING TRINE
JAN	11–13	41–43	55–56	25–27	36–38	50–52
FEB	11–13	41–42	55–56	25–27	36–38	49–51
MAR	10–13	41–42	55–56	25–27	35–37	49–51
APR	10–12	40–42	54–55	24–26	35–37	49–51
MAY	10–12	39–42	53–55	24–26	35–37	49–51
JUN	10–12	39–41	53–55	24–26	34–36	49–51
JUL	10–12	39–41	52–55	24–26	34–36	49–51
AUG	11–13	39–41	53–55	24–26	35–37	49–52
SEP	11–13	39–41	53–55	24–26	35–37	49–52
OCT	12–13	40–42	53–56	24–26	35–37	50–52
NOV	12–14	40–42	54–56	24–26	36–38	50–52
DEC	12–14	40–43	54–56	25–26	36–38	51–53

1967

URANUS CYCLE 0–84 YEARS						
m.o.b.	WAXING SQUARE	WAXING TRINE	OPPOSITION	WANING TRINE	WANING SQUARE	RETURN
JAN	19–20	27–28	41–43	57–58	63–65	83–84
FEB	18–20	26–27	41–43	56–58	63–64	83–84
MAR	18–20	25–26	41–42	56–58	63–64	83–84
APR	18–20	25–26	41–42	56–57	63–64	82–84
MAY	18–19	25–26	40–41	55–57	63–64	82–84
JUN	18–19	25–26	40–41	55–56	63–64	82–84
JUL	18–19	25–26	40–42	55–56	63–64	82–84
AUG	18–20	25–27	40–42	55–56	63–64	82–84
SEP	19–20	26–27	41–42	56–57	63–64	83–84
OCT	19–20	26–28	41–42	56–57	63–64	83–84
NOV	19–20	26–28	42–43	56–58	64–65	83–84
DEC	20–21	27–28	42–43	57–58	64–65	83–84

	NEPTUNE CYCLE			PLUTO CYCLE		
m.o.b.	SQUARING PLUTO	WAXING SQUARE	WAXING TRINE	CONJ.NEPTUNE	WAXING SQUARE	WAXING TRINE
JAN	11–13	41–43	55–56	25–26	36–38	50–52
FEB	11–13	41–42	55–56	25–26	36–38	49–51
MAR	10–13	41–42	55–56	25–26	35–37	49–51
APR	10–12	40–42	54–55	24–26	35–37	49–51
MAY	10–12	39–42	53–55	24–26	35–37	49–51
JUN	10–12	39–41	53–55	24–26	34–36	49–51
JUL	10–12	39–41	52–55	24–26	34–36	49–51
AUG	11–13	39–41	53–55	24–26	35–37	49–52
SEP	11–13	39–41	53–55	24–26	35–37	50–52
OCT	12–13	40–42	53–56	24–26	35–37	50–52
NOV	12–14	40–42	54–56	24–26	36–38	50–52
DEC	12–14	40–43	54–56	25–26	36–38	51–53

1968

URANUS CYCLE 0–84 YEARS						
m.o.b.	WAXING SQUARE	WAXING TRINE	OPPOSITION	WANING TRINE	WANING SQUARE	RETURN
JAN	19–20	27–28	41–43	57–58	63–65	83–84
FEB	18–20	26–27	41–43	56–58	63–65	83–84
MAR	18–20	25–26	41–42	56–58	63–65	83–84
APR	18–20	25–26	41–42	56–57	63–64	82–84
MAY	18–19	25–26	40–41	55–57	63–64	82–84
JUN	18–19	25–26	40–41	55–56	63–64	82–84
JUL	18–19	25–26	40–42	55–56	63–64	82–84
AUG	18–20	25–27	40–42	55–56	63–64	82–84
SEP	19–20	26–27	41–42	56–57	63–64	83–84
OCT	19–20	26–28	41–42	56–57	63–64	83–84
NOV	19–20	26–28	42–43	56–58	64–65	83–84
DEC	20–21	27–28	42–44	57–58	64–65	83–84

	NEPTUNE CYCLE			PLUTO CYCLE		
m.o.b.	SQUARING PLUTO	WAXING SQUARE	WAXING TRINE	CONJ.NEPTUNE	WAXING SQUARE	WAXING TRINE
JAN	11–13	41–43	55–56	25–26	36–38	50–52
FEB	11–13	41–42	55–56	24–26	36–38	49–51
MAR	10–13	41–42	55–56	24–26	35–37	49–51
APR	10–12	40–42	54–55	24–26	35–37	48–51
MAY	10–12	39–42	53–55	24–26	35–37	48–51
JUN	10–12	39–41	53–55	24–26	34–36	48–51
JUL	10–12	39–41	52–55	24–26	34–36	49–51
AUG	11–13	39–41	53–55	24–26	35–37	49–52
SEP	11–13	39–41	53–55	24–25	35–37	50–52
OCT	12–13	40–42	53–56	24–26	35–37	50–52
NOV	12–14	40–42	54–56	24–26	36–38	50–52
DEC	12–14	40–42	54–56	25–26	36–38	51–53

1969

URANUS CYCLE 0–84 YEARS						
m.o.b.	WAXING SQUARE	WAXING TRINE	OPPOSITION	WANING TRINE	WANING SQUARE	RETURN
JAN	20–21	27–28	41–43	57–58	63–65	83–84
FEB	19–20	27–28	41–43	56–58	63–65	83–84
MAR	19–20	26–27	42–43	56–58	63–65	83–84
APR	18–20	25–27	41–42	56–57	63–64	82–84
MAY	18–19	25–27	40–42	55–57	63–64	82–84
JUN	18–19	25–27	40–42	55–57	63–64	82–83
JUL	18–19	25–27	40–42	55–57	63–64	82–83
AUG	18–20	25–27	40–42	55–57	63–64	82–84
SEP	19–20	26–27	41–43	56–57	63–64	83–84
OCT	19–20	26–28	41–43	56–57	63–64	83–84
NOV	19–20	26–28	41–43	56–58	64–65	83–84
DEC	20–21	27–28	41–43	57–58	64–65	83–84

	NEPTUNE CYCLE			PLUTO CYCLE		
m.o.b.	SQUARING PLUTO	WAXING SQUARE	WAXING TRINE	CONJ.NEPTUNE	WAXING SQUARE	WAXING TRINE
JAN	12–13	41–43	55–56	25–26	36–38	51–53
FEB	11–13	41–42	55–56	24–26	36–38	51–52
MAR	11–13	41–42	54–56	24–26	35–37	50–52
APR	10–12	40–42	54–55	24–26	35–37	50–52
MAY	10–12	39–42	53–55	24–26	35–37	49–51
JUN	10–12	39–41	53–55	24–25	34–36	49–51
JUL	10–12	39–41	52–55	23–25	34–36	49–51
AUG	11–13	39–41	53–55	23–25	35–37	49–52
SEP	11–13	39–41	53–55	23–25	35–37	50–52
OCT	12–13	40–42	53–56	24–26	35–37	50–52
NOV	12–14	40–42	54–56	24–26	36–38	50–52
DEC	12–14	40–42	54–56	24–26	36–38	51–53

1970

URANUS CYCLE 0–84 YEARS						
m.o.b.	WAXING SQUARE	WAXING TRINE	OPPOSITION	WANING TRINE	WANING SQUARE	RETURN
JAN	20–21	27–28	42–44	57–58	64–65	83–84
FEB	19–20	27–28	42–43	57–58	64–65	83–84
MAR	19–20	26–27	42–43	57–58	64–65	83–84
APR	18–20	26–27	41–42	56–57	63–64	82–84
MAY	18–20	25–27	41–42	56–57	63–64	82–84
JUN	18–20	25–27	40–42	56–57	63–64	82–84
JUL	18–20	25–27	40–42	56–57	63–64	82–84
AUG	19–20	26–27	41–42	56–57	63–64	83–84
SEP	19–20	26–28	41–43	56–57	63–64	83–84
OCT	19–21	27–28	42–43	56–58	63–65	83–84
NOV	19–21	27–28	42–44	57–58	64–65	83–84
DEC	20–21	27–28	42–44	57–58	64–65	83–84

	NEPTUNE CYCLE			PLUTO CYCLE		
m.o.b.	SQUARING PLUTO	WAXING SQUARE	WAXING TRINE	CONJ.NEPTUNE	WAXING SQUARE	WAXING TRINE
JAN	12–13	41–42	54–56	25–26	36–37	51–53
FEB	11–13	41–42	54–56	25–26	35–37	51–52
MAR	11–13	41–42	54–56	24–26	35–37	50–52
APR	10–12	40–42	54–55	24–26	34–37	49–52
MAY	10–12	40–42	53–55	24–26	34–36	49–51
JUN	10–12	39–41	53–55	23–25	34–36	49–51
JUL	10–12	39–41	52–54	23–25	34–36	49–51
AUG	11–13	39–41	52–54	23–25	35–37	49–52
SEP	11–13	39–41	52–55	23–25	35–37	50–52
OCT	12–14	39–41	53–55	24–25	36–37	51–53
NOV	12–14	40–42	53–55	24–25	36–37	51–54
DEC	12–14	40–42	54–56	24–25	36–38	52–54

1971

URANUS CYCLE 0–84 YEARS						
m.o.b.	WAXING SQUARE	WAXING TRINE	OPPOSITION	WANING TRINE	WANING SQUARE	RETURN
JAN	20–21	27–28	42–44	57–58	64–65	83–84
FEB	19–20	27–28	42–43	57–58	64–65	83–84
MAR	19–20	27–28	42–43	57–58	64–65	83–84
APR	18–20	26–27	41–42	56–57	63–64	82–84
MAY	18–20	25–27	41–42	56–57	63–64	82–84
JUN	18–20	25–27	40–42	56–57	63–64	82–84
JUL	18–20	25–27	40–42	56–57	63–64	82–84
AUG	19–20	26–27	41–42	56–57	63–64	83–84
SEP	19–20	26–28	41–43	56–57	63–64	83–84
OCT	19–21	27–28	42–43	56–58	63–65	83–84
NOV	19–21	27–28	42–44	57–58	64–65	83–84
DEC	20–21	27–28	43–44	57–58	64–65	83–84

	NEPTUNE CYCLE			PLUTO CYCLE		
m.o.b.	SQUARING PLUTO	WAXING SQUARE	WAXING TRINE	CONJ.NEPTUNE	WAXING SQUARE	WAXING TRINE
JAN	12–13	41–42	54–56	25–26	36–37	51–53
FEB	11–13	41–42	54–56	25–26	35–37	51–52
MAR	12–13	41–42	54–56	24–26	35–37	50–52
APR	11–13	40–42	54–55	24–26	34–37	49–52
MAY	10–12	40–42	53–55	24–26	34–36	49–52
JUN	10–12	39–41	53–55	23–25	34–36	49–52
JUL	10–12	39–41	52–54	23–25	34–36	49–52
AUG	11–13	39–41	52–54	23–25	35–37	49–52
SEP	11–13	39–41	52–54	23–25	35–37	50–52
OCT	12–14	39–41	53–55	24–25	36–37	51–53
NOV	12–14	40–42	53–55	24–25	36–37	51–54
DEC	12–14	40–42	54–56	24–25	36–38	52–54

1972

URANUS CYCLE 0–84 YEARS						
m.o.b.	WAXING SQUARE	WAXING TRINE	OPPOSITION	WANING TRINE	WANING SQUARE	RETURN
JAN	20–21	27–28	43–44	57–58	64–65	83–84
FEB	19–20	27–28	43–44	57–58	64–65	83–84
MAR	19–20	27–28	42–43	57–58	64–65	83–84
APR	18–20	26–27	42–43	56–57	63–64	82–84
MAY	18–20	26–27	41–42	56–57	63–64	82–84
JUN	18–20	26–27	41–42	56–57	63–64	82–84
JUL	18–20	25–27	40–42	56–57	63–64	82–84
AUG	19–20	26–27	41–42	56–57	63–64	83–84
SEP	19–20	26–28	41–43	56–57	63–64	83–84
OCT	19–21	27–28	42–43	56–58	63–65	83–84
NOV	19–21	27–29	42–44	57–58	64–65	83–84
DEC	20–21	28–29	43–44	57–59	64–65	83–84

NEPTUNE CYCLE				PLUTO CYCLE		
m.o.b.	SQUARING PLUTO	WAXING SQUARE	WAXING TRINE	CONJ.NEPTUNE	WAXING SQUARE	WAXING TRINE
JAN	12–13	41–42	54–56	25–26	36–37	51–53
FEB	11–13	41–42	54–56	25–26	35–37	51–53
MAR	11–13	41–42	54–56	24–26	35–37	50–53
APR	11–13	40–42	54–55	24–26	34–37	49–52
MAY	10–12	40–42	53–55	24–26	34–36	49–52
JUN	10–12	39–41	53–55	23–25	34–36	49–52
JUL	10–12	39–41	52–54	23–25	34–36	49–52
AUG	11–13	39–41	52–54	23–25	35–37	49–52
SEP	11–13	39–41	52–54	23–25	35–37	50–52
OCT	12–14	39–41	53–55	24–25	36–37	51–53
NOV	12–14	40–42	53–55	24–25	36–37	51–54
DEC	12–14	40–42	54–56	24–25	36–38	52–54

1973

URANUS CYCLE 0–84 YEARS						
m.o.b.	WAXING SQUARE	WAXING TRINE	OPPOSITION	WANING TRINE	WANING SQUARE	RETURN
JAN	20–21	28–29	43–44	57–59	64–65	83–84
FEB	20–21	27–28	43–44	57–59	64–65	83–84
MAR	19–21	27–28	43–44	57–58	64–65	83–84
APR	18–20	26–27	42–43	56–57	63–64	82–84
MAY	18–20	26–27	41–42	56–57	63–64	82–84
JUN	18–20	26–27	41–42	56–57	63–64	82–84
JUL	18–20	25–27	40–42	56–57	63–64	82–84
AUG	19–20	26–27	41–42	56–57	63–64	83–84
SEP	19–21	26–28	42–43	56–57	63–64	83–84
OCT	19–21	27–28	42–43	56–58	63–65	83–84
NOV	19–21	27–29	42–44	57–58	64–65	83–84
DEC	20–21	28–29	43–44	57–59	64–65	83–84

NEPTUNE CYCLE				PLUTO CYCLE		
m.o.b.	SQUARING PLUTO	WAXING SQUARE	WAXING TRINE	CONJ.NEPTUNE	WAXING SQUARE	WAXING TRINE
JAN	12–13	41–42	54–56	25–26	36–38	52–54
FEB	11–13	41–42	54–56	25–26	35–37	51–53
MAR	11–13	41–42	54–56	24–26	35–37	51–53
APR	11–13	40–42	54–55	24–26	34–37	50–53
MAY	10–12	40–42	53–55	24–26	34–36	49–52
JUN	10–12	39–41	53–55	23–25	34–36	50–52
JUL	10–12	39–41	52–54	23–25	34–36	50–52
AUG	11–13	39–41	52–54	23–25	35–37	50–53
SEP	11–13	39–41	52–54	23–25	35–37	51–54
OCT	12–14	39–41	53–55	24–25	36–37	51–54
NOV	12–14	40–42	53–55	24–25	36–37	52–55
DEC	12–14	40–42	54–56	24–25	37–38	53–55

1974

URANUS CYCLE 0–84 YEARS						
m.o.b.	WAXING SQUARE	WAXING TRINE	OPPOSITION	WANING TRINE	WANING SQUARE	RETURN
JAN	20–21	28–29	43–44	57–59	64–65	83–84
FEB	20–21	28–29	43–44	57–59	64–65	83–84
MAR	20–21	27–28	43–44	57–59	64–65	83–84
APR	19–20	26–27	42–43	56–57	63–64	82–84
MAY	19–20	26–27	41–43	56–57	63–64	82–84
JUN	19–20	26–27	41–43	56–57	63–64	82–84
JUL	19–20	26–27	41–43	56–57	63–64	82–84
AUG	19–20	26–27	41–42	56–57	63–64	83–84
SEP	19–21	27–28	42–43	56–57	63–64	83–84
OCT	19–21	27–28	42–43	56–58	63–65	83–84
NOV	19–21	28–29	42–44	57–59	64–65	83–84
DEC	20–21	28–29	43–44	57–59	64–65	83–84

	NEPTUNE CYCLE			PLUTO CYCLE		
m.o.b.	SQUARING PLUTO	WAXING SQUARE	WAXING TRINE	CONJ.NEPTUNE	WAXING SQUARE	WAXING TRINE
JAN	12–13	41–42	54–56	25–26	36–38	52–54
FEB	11–13	41–42	54–56	25–26	35–37	52–54
MAR	11–13	41–42	54–56	24–26	35–37	52–54
APR	11–13	40–42	54–55	24–26	34–37	51–53
MAY	10–12	40–42	53–55	24–26	34–37	50–53
JUN	10–12	39–41	53–55	23–25	34–37	50–53
JUL	10–12	39–41	52–54	23–25	34–37	50–52
AUG	11–13	39–41	52–54	23–25	35–38	50–53
SEP	11–13	39–41	52–54	23–25	35–38	51–54
OCT	12–14	39–41	53–55	24–25	36–38	51–54
NOV	12–14	40–42	53–55	24–25	37–38	52–55
DEC	13–14	40–42	53–56	24–25	37–38	53–55

1975

URANUS CYCLE 0–84 YEARS						
m.o.b.	WAXING SQUARE	WAXING TRINE	OPPOSITION	WANING TRINE	WANING SQUARE	RETURN
JAN	20–21	28–29	43–44	57–59	64–65	83–84
FEB	20–21	28–29	43–44	57–59	64–65	83–84
MAR	20–21	27–28	43–44	57–59	64–65	83–84
APR	19–21	27–28	42–43	56–57	63–64	82–84
MAY	19–20	27–28	41–43	56–57	63–64	82–84
JUN	19–20	26–28	41–43	56–57	63–64	82–84
JUL	19–20	26–28	41–43	56–57	63–64	82–84
AUG	19–21	26–28	41–42	56–57	63–64	83–84
SEP	19–21	27–28	42–43	56–57	63–64	83–84
OCT	19–21	27–28	42–43	56–58	63–65	83–84
NOV	19–21	28–29	42–44	57–59	64–65	83–84
DEC	20–21	28–29	43–44	57–59	64–65	83–84

	NEPTUNE CYCLE			PLUTO CYCLE		
m.o.b.	SQUARING PLUTO	WAXING SQUARE	WAXING TRINE	CONJ.NEPTUNE	WAXING SQUARE	WAXING TRINE
JAN	12–13	41–42	54–56	25–26	36–38	53–55
FEB	11–13	41–42	54–56	25–26	36–38	52–54
MAR	11–13	41–42	54–56	24–26	36–38	52–54
APR	11–13	40–42	54–55	24–26	35–38	51–53
MAY	10–13	40–42	53–55	24–26	35–37	50–53
JUN	10–13	39–41	53–55	23–25	35–37	50–53
JUL	11–13	39–41	52–54	23–25	35–38	50–52
AUG	11–13	39–41	52–54	23–25	35–38	50–53
SEP	11–13	39–41	52–54	23–25	35–38	51–54
OCT	12–14	39–41	53–55	24–25	36–38	51–54
NOV	12–14	40–42	53–55	24–25	37–38	52–55
DEC	13–14	40–42	53–56	24–25	37–39	53–56

1976

URANUS CYCLE 0–84 YEARS						
m.o.b.	WAXING SQUARE	WAXING TRINE	OPPOSITION	WANING TRINE	WANING SQUARE	RETURN
JAN	20–21	28–29	43–44	57–59	64–65	83–84
FEB	20–21	28–29	43–44	57–59	64–65	83–84
MAR	20–21	28–29	43–44	57–59	64–65	83–84
APR	19–21	27–28	42–43	56–57	63–64	82–84
MAY	19–21	27–28	42–43	56–57	63–64	82–84
JUN	19–20	26–28	41–43	56–57	63–64	82–84
JUL	19–20	26–28	41–43	56–57	63–64	82–84
AUG	19–21	26–28	42–43	56–57	63–64	83–84
SEP	19–21	27–28	42–43	56–57	63–64	83–84
OCT	19–21	27–28	42–43	56–58	63–65	83–84
NOV	19–21	28–29	42–44	57–59	64–65	83–84
DEC	20–21	28–29	43–45	57–59	64–65	83–84

	NEPTUNE CYCLE			PLUTO CYCLE		
m.o.b.	SQUARING PLUTO	WAXING SQUARE	WAXING TRINE	CONJ.NEPTUNE	WAXING SQUARE	WAXING TRINE
JAN	12–13	41–42	54–56	25–26	36–38	53–56
FEB	11–13	41–42	54–56	25–26	36–38	53–56
MAR	11–13	41–42	54–56	24–26	36–38	53–55
APR	11–13	40–42	54–55	24–26	35–38	52–54
MAY	10–13	40–42	53–55	24–26	35–37	51–54
JUN	10–13	39–41	53–55	23–25	35–37	51–54
JUL	11–13	39–41	52–54	23–25	35–38	51–54
AUG	11–13	39–41	52–54	23–25	35–38	51–54
SEP	11–14	39–41	52–54	22–24	36–38	52–55
OCT	12–14	39–41	53–55	23–25	36–38	52–55
NOV	12–14	40–42	53–55	23–25	37–38	53–56
DEC	13–14	40–42	53–55	23–25	37–39	54–57

URANUS CYCLE 0–84 YEARS						
m.o.b.	WAXING SQUARE	WAXING TRINE	OPPOSITION	WANING TRINE	WANING SQUARE	RETURN
JAN	20–21	28–29	43–45	57–59	64–65	83–84
FEB	20–21	28–29	43–44	57–59	64–65	83–84
MAR	20–21	28–29	43–44	57–59	64–65	83–84
APR	20–21	27–28	42–43	57–58	63–64	82–84
MAY	19–21	27–28	42–43	56–58	63–64	82–84
JUN	19–21	27–28	42–43	57–58	63–64	82–84
JUL	19–21	26–28	42–43	56–58	63–64	82–84
AUG	19–21	26–28	42–43	56–57	63–64	82–84
SEP	19–21	27–28	42–43	56–57	63–64	82–84
OCT	19–21	27–28	42–43	56–58	63–65	83–84
NOV	19–21	28–29	42–44	57–59	64–65	83–84
DEC	20–21	28–30	43–45	57–59	64–65	83–84

	NEPTUNE CYCLE			PLUTO CYCLE		
m.o.b.	SQUARING PLUTO	WAXING SQUARE	WAXING TRINE	CONJ.NEPTUNE	WAXING SQUARE	WAXING TRINE
JAN	12–13	41–42	54–56	23–25	36–38	53–56
FEB	12–13	41–42	54–56	23–25	36–38	53–56
MAR	12–14	41–42	54–56	23–25	36–38	53–55
APR	12–14	40–42	54–55	23–25	35–38	52–54
MAY	11–13	40–42	53–55	23–25	35–37	51–54
JUN	11–13	39–41	53–55	23–25	35–37	51–54
JUL	11–13	39–41	52–54	23–25	35–38	51–54
AUG	11–14	39–41	52–54	23–25	35–38	52–55
SEP	11–14	39–41	52–54	22–24	36–38	53–55
OCT	12–14	39–41	53–55	23–25	36–38	54–56
NOV	12–14	40–42	53–55	23–25	37–38	54–57
DEC	13–14	40–42	53–55	23–25	37–39	55–57

1978

URANUS CYCLE 0–84 YEARS						
m.o.b.	WAXING SQUARE	WAXING TRINE	OPPOSITION	WANING TRINE	WANING SQUARE	RETURN
JAN	21–22	28–30	43–45	57–59	64–65	83–84
FEB	21–22	28–29	43–45	57–59	64–65	83–84
MAR	21–22	28–29	44–45	58–59	64–65	83–84
APR	20–21	27–28	43–44	57–58	63–64	82–84
MAY	19–21	27–28	42–43	57–58	63–64	82–84
JUN	19–21	27–28	42–43	57–58	63–64	82–84
JUL	19–21	27–28	42–43	56–58	63–64	82–84
AUG	19–21	27–28	42–43	56–57	63–64	82–84
SEP	19–21	27–28	42–43	56–57	63–64	82–84
OCT	20–21	27–28	42–43	56–58	63–65	83–84
NOV	20–21	28–29	42–44	57–58	64–65	83–84
DEC	21–22	28–30	43–45	57–58	64–65	83–84

NEPTUNE CYCLE				PLUTO CYCLE		
m.o.b.	SQUARING PLUTO	WAXING SQUARE	WAXING TRINE	CONJ.NEPTUNE	WAXING SQUARE	WAXING TRINE
JAN	12–13	41–42	54–56	23–25	37–39	54–56
FEB	12–13	41–42	54–56	23–25	36–39	54–56
MAR	12–14	41–42	54–56	23–25	36–39	54–56
APR	12–14	40–42	54–55	23–25	35–38	53–55
MAY	12–14	40–42	53–55	23–25	35–37	52–55
JUN	11–13	39–41	53–55	23–25	35–37	52–55
JUL	11–13	39–41	52–54	23–25	35–38	52–55
AUG	12–14	39–41	52–54	23–25	35–38	53–55
SEP	12–14	39–41	52–54	22–24	36–38	53–56
OCT	12–14	39–41	53–55	23–25	36–38	54–56
NOV	12–14	40–42	53–55	23–25	37–39	54–57
DEC	13–14	40–42	53–55	23–24	38–40	55–58

1979

URANUS CYCLE 0–84 YEARS						
m.o.b.	WAXING SQUARE	WAXING TRINE	OPPOSITION	WANING TRINE	WANING SQUARE	RETURN
JAN	21–22	28–30	43–45	57–59	64–65	83–84
FEB	21–22	28–29	43–45	57–59	64–65	83–84
MAR	21–22	28–29	44–45	58–59	64–65	82–84
APR	20–21	27–28	43–44	57–58	63–64	82–84
MAY	19–21	27–28	42–43	57–58	63–64	82–84
JUN	19–21	27–28	42–43	57–58	63–64	82–84
JUL	19–21	27–28	42–43	56–58	63–64	82–84
AUG	19–21	27–28	42–43	56–57	63–64	82–84
SEP	20–21	27–29	42–43	56–57	63–64	82–84
OCT	20–21	28–29	42–43	56–58	63–65	82–84
NOV	20–21	28–29	42–44	57–58	64–65	83–84
DEC	21–22	28–30	43–45	57–58	64–65	83–84

NEPTUNE CYCLE				PLUTO CYCLE		
m.o.b.	SQUARING PLUTO	WAXING SQUARE	WAXING TRINE	CONJ.NEPTUNE	WAXING SQUARE	WAXING TRINE
JAN	12–13	41–42	54–56	23–25	37–39	54–57
FEB	12–13	41–42	54–56	23–25	37–39	54–57
MAR	12–14	41–42	54–56	23–25	37–39	55–57
APR	12–14	40–42	54–55	23–25	36–38	54–57
MAY	12–14	40–42	53–55	23–25	35–38	54–57
JUN	12–14	39–41	53–55	23–25	35–38	54–57
JUL	12–14	39–41	52–54	23–25	35–38	53–56
AUG	12–14	39–41	52–54	23–25	35–38	53–55
SEP	12–14	39–41	52–54	22–24	36–39	53–55
OCT	12–14	39–41	53–55	23–25	36–39	54–57
NOV	12–14	40–42	53–55	23–25	37–39	55–58
DEC	13–14	40–42	53–55	23–24	38–40	56–59

1980

URANUS CYCLE 0–84 YEARS						
m.o.b.	WAXING SQUARE	WAXING TRINE	OPPOSITION	WANING TRINE	WANING SQUARE	RETURN
JAN	21–22	29–30	44–45	57–59	64–65	83–84
FEB	21–22	29–30	44–45	58–59	64–65	83–84
MAR	21–22	29–30	44–45	58–59	64–65	83–84
APR	21–22	28–29	44–45	57–59	64–65	83–84
MAY	20–21	28–29	43–44	57–58	64–65	82–84
JUN	20–21	27–29	43–44	57–58	63–65	82–84
JUL	19–21	27–28	42–43	56–57	63–64	82–84
AUG	19–21	27–28	42–43	56–57	63–64	82–84
SEP	20–21	27–29	42–44	56–57	63–64	82–84
OCT	20–21	28–29	42–44	56–58	63–64	82–84
NOV	20–22	28–30	43–44	57–58	63–64	83–84
DEC	21–22	28–30	43–45	57–58	63–65	83–84

NEPTUNE CYCLE				PLUTO CYCLE		
m.o.b.	SQUARING PLUTO	WAXING SQUARE	WAXING TRINE	CONJ.NEPTUNE	WAXING SQUARE	WAXING TRINE
JAN	13–15	40–42	54–56	23–25	38–40	56–58
FEB	13–14	41–42	54–56	23–25	38–40	56–58
MAR	12–14	41–42	54–56	23–25	37–39	55–58
APR	12–14	41–42	54–55	23–25	36–39	54–57
MAY	11–13	40–42	53–55	23–25	36–38	54–57
JUN	11–13	39–41	53–55	23–25	36–38	53–56
JUL	11–13	39–41	52–54	22–24	36–38	53–56
AUG	11–14	39–41	52–54	22–24	36–38	54–57
SEP	12–14	39–41	52–54	22–24	37–39	54–57
OCT	12–14	39–41	52–54	22–24	37–40	55–58
NOV	13–15	39–41	53–55	23–24	38–40	56–59
DEC	13–15	40–42	53–55	23–24	38–40	57–59

1981

URANUS CYCLE 0–84 YEARS						
m.o.b.	WAXING SQUARE	WAXING TRINE	OPPOSITION	WANING TRINE	WANING SQUARE	RETURN
JAN	21–22	29–30	44–45	57–59	64–65	83–84
FEB	21–22	29–30	44–45	58–59	64–65	83–84
MAR	21–22	29–30	44–45	58–59	64–65	83–84
APR	21–22	28–29	44–45	57–59	64–65	83–84
MAY	20–21	28–29	43–44	57–58	64–65	82–84
JUN	20–21	27–29	43–44	57–58	63–65	82–84
JUL	19–21	27–28	42–43	56–57	63–64	82–84
AUG	19–21	27–28	42–43	56–57	63–64	82–84
SEP	20–21	27–29	42–44	56–57	63–64	82–84
OCT	20–21	28–29	42–44	56–58	63–64	82–84
NOV	20–22	28–30	43–44	57–58	63–64	83–84
DEC	21–22	29–30	43–45	57–58	63–65	83–84

NEPTUNE CYCLE				PLUTO CYCLE		
m.o.b.	SQUARING PLUTO	WAXING SQUARE	WAXING TRINE	CONJ.NEPTUNE	WAXING SQUARE	WAXING TRINE
JAN	13–15	40–42	54–56	23–25	38–40	56–58
FEB	13–15	41–42	54–56	23–25	38–40	56–58
MAR	12–14	41–42	54–56	23–25	38–40	56–58
APR	12–14	41–42	54–55	23–25	37–39	55–57
MAY	11–13	40–42	53–55	23–25	36–38	54–57
JUN	11–13	39–41	53–55	23–25	36–39	54–57
JUL	11–13	39–41	52–54	22–24	36–38	54–57
AUG	11–14	39–41	52–54	22–24	36–38	55–58
SEP	12–14	39–41	52–54	22–24	37–39	55–58
OCT	12–14	39–41	52–54	22–24	37–40	56–59
NOV	13–15	39–41	53–55	23–24	38–40	57–60
DEC	13–15	40–42	53–55	23–24	39–41	57–60

1982

URANUS CYCLE 0–84 YEARS						
m.o.b.	WAXING SQUARE	WAXING TRINE	OPPOSITION	WANING TRINE	WANING SQUARE	RETURN
JAN	21–22	29–30	44–45	57–59	64–65	83–84
FEB	21–22	29–30	44–45	58–59	64–65	83–84
MAR	21–22	29–30	44–45	58–59	64–65	83–84
APR	21–22	28–29	44–45	57–59	64–65	83–84
MAY	20–21	28–29	43–44	57–58	64–65	82–84
JUN	20–21	28–29	43–44	57–58	63–65	82–84
JUL	19–21	27–29	42–43	56–57	63–64	82–84
AUG	19–21	27–29	42–43	56–57	63–64	82–84
SEP	20–21	27–29	42–44	56–57	63–64	82–84
OCT	20–21	28–29	42–44	56–58	63–64	82–84
NOV	20–22	28–30	43–44	57–58	63–64	83–84
DEC	21–22	29–30	43–45	57–58	63–64	83–84

NEPTUNE CYCLE				PLUTO CYCLE		
m.o.b.	SQUARING PLUTO	WAXING SQUARE	WAXING TRINE	CONJ.NEPTUNE	WAXING SQUARE	WAXING TRINE
JAN	13–15	40–42	54–56	23–25	39–41	57–59
FEB	13–15	41–42	54–56	23–25	38–40	57–59
MAR	13–15	41–42	54–56	23–25	38–40	57–59
APR	12–14	41–42	54–55	23–25	37–39	56–58
MAY	11–14	40–42	53–55	23–25	37–39	55–58
JUN	11–14	39–41	53–55	23–25	37–39	55–58
JUL	11–14	39–41	52–54	22–24	37–39	55–58
AUG	12–14	39–41	52–54	22–24	37–39	56–59
SEP	12–14	39–41	52–54	22–24	37–40	56–59
OCT	12–14	39–41	52–54	22–24	37–40	57–60
NOV	13–15	39–41	53–55	23–24	38–40	58–61
DEC	14–15	40–42	53–55	23–24	39–41	58–61

1983

URANUS CYCLE 0–84 YEARS						
m.o.b.	WAXING SQUARE	WAXING TRINE	OPPOSITION	WANING TRINE	WANING SQUARE	RETURN
JAN	21–22	29–30	44–45	57–59	64–65	83–84
FEB	21–22	29–30	44–45	58–59	64–65	83–84
MAR	21–22	29–30	44–45	58–59	64–65	83–84
APR	21–22	28–29	44–45	57–59	64–65	83–84
MAY	20–22	28–29	43–44	57–58	64–65	82–84
JUN	20–22	28–29	43–44	57–58	63–65	82–84
JUL	20–21	27–29	42–43	56–57	63–64	82–84
AUG	20–21	27–29	42–43	56–57	63–64	82–84
SEP	20–21	27–29	42–44	56–57	63–64	82–84
OCT	20–21	28–29	42–44	56–58	63–64	82–84
NOV	20–22	28–30	43–44	57–58	63–64	83–84
DEC	21–22	29–30	43–45	57–58	63–64	83–84

NEPTUNE CYCLE				PLUTO CYCLE		
m.o.b.	SQUARING PLUTO	WAXING SQUARE	WAXING TRINE	CONJ.NEPTUNE	WAXING SQUARE	WAXING TRINE
JAN	13–15	40–42	54–56	23–25	39–41	58–60
FEB	13–15	41–42	54–56	23–25	39–41	58–60
MAR	13–15	41–42	54–56	23–25	39–41	58–60
APR	12–14	41–42	54–55	23–25	38–40	57–59
MAY	12–14	40–42	53–55	23–25	37–39	56–59
JUN	12–14	39–41	53–55	23–25	37–39	56–59
JUL	12–14	39–41	52–54	22–24	37–39	56–59
AUG	12–14	39–41	52–54	22–24	37–39	56–59
SEP	12–14	39–41	52–54	22–24	38–40	56–59
OCT	12–14	39–41	52–54	22–24	38–40	57–60
NOV	13–15	39–41	53–55	23–24	39–41	58–61
DEC	14–16	40–42	53–55	23–24	40–42	59–62

1984

URANUS CYCLE 0–84 YEARS						
m.o.b.	WAXING SQUARE	WAXING TRINE	OPPOSITION	WANING TRINE	WANING SQUARE	RETURN
JAN	21–22	29–30	44–45	57–59	63–65	83–84
FEB	21–23	29–30	44–45	57–59	64–65	83–84
MAR	22–23	30–31	44–45	57–59	64–65	83–84
APR	21–22	29–31	44–45	57–59	64–65	83–84
MAY	20–22	28–29	43–44	57–58	64–65	82–84
JUN	20–22	28–29	43–44	57–58	63–65	82–84
JUL	20–22	28–29	42–43	56–57	63–64	82–84
AUG	20–22	28–29	42–43	56–57	63–64	82–84
SEP	20–22	28–29	42–44	56–57	63–64	82–84
OCT	20–22	28–29	42–44	56–58	63–64	82–84
NOV	20–22	28–30	43–44	57–58	63–64	83–84
DEC	21–23	29–30	43–45	57–58	63–64	83–84

NEPTUNE CYCLE				PLUTO CYCLE		
m.o.b.	SQUARING PLUTO	WAXING SQUARE	WAXING TRINE	CONJ.NEPTUNE	WAXING SQUARE	WAXING TRINE
JAN	13–15	40–42	54–56	23–25	40–42	59–61
FEB	13–15	41–42	54–56	23–25	39–41	59–61
MAR	13–15	41–42	54–56	23–25	39–41	59–61
APR	12–14	41–42	54–55	23–25	38–40	58–60
MAY	12–14	40–42	53–55	23–25	37–39	57–60
JUN	12–14	39–41	53–55	23–25	37–40	56–60
JUL	12–14	39–41	52–54	22–24	37–40	56–60
AUG	12–14	39–41	52–54	22–24	38–40	57–60
SEP	12–14	39–41	52–54	22–24	38–41	57–60
OCT	12–14	39–41	52–54	22–24	38–41	58–61
NOV	13–15	39–41	53–55	23–24	39–41	59–62
DEC	14–16	39–42	53–55	23–24	40–42	60–63

1985

URANUS CYCLE 0–84 YEARS						
m.o.b.	WAXING SQUARE	WAXING TRINE	OPPOSITION	WANING TRINE	WANING SQUARE	RETURN
JAN	21–22	29–30	44–45	57–59	64–65	83–84
FEB	21–23	29–30	44–45	57–59	64–65	83–84
MAR	22–23	30–31	44–45	57–59	64–65	83–84
APR	21–22	29–31	44–45	57–59	64–65	83–84
MAY	20–22	28–30	43–44	57–58	64–65	82–84
JUN	20–22	28–30	43–44	57–58	63–64	82–84
JUL	20–22	28–30	42–43	56–57	63–64	82–84
AUG	20–22	28–29	42–43	56–57	62–64	82–84
SEP	20–22	28–29	42–44	56–57	62–63	82–84
OCT	20–22	28–29	42–44	56–58	62–64	82–84
NOV	20–22	28–30	43–44	57–58	63–64	83–84
DEC	21–23	29–30	43–45	57–58	63–64	83–84

NEPTUNE CYCLE				PLUTO CYCLE		
m.o.b.	SQUARING PLUTO	WAXING SQUARE	WAXING TRINE	CONJ.NEPTUNE	WAXING SQUARE	WAXING TRINE
JAN	14–16	40–42	54–56	23–25	40–42	60–62
FEB	13–15	41–42	54–56	23–25	40–42	60–62
MAR	13–15	41–42	54–56	23–25	40–42	60–62
APR	12–14	41–42	54–55	23–25	39–41	59–61
MAY	12–14	40–42	53–55	23–25	38–41	58–61
JUN	12–14	39–41	53–55	23–25	38–41	57–60
JUL	12–14	39–41	52–54	22–24	38–41	57–61
AUG	12–14	38–41	52–54	22–24	38–41	58–61
SEP	12–15	38–41	52–54	22–24	38–41	58–61
OCT	12–15	39–41	52–54	22–24	38–41	59–62
NOV	13–15	39–41	53–55	23–24	39–41	60–63
DEC	14–16	39–42	53–55	23–24	40–42	61–64

1986

URANUS CYCLE 0–84 YEARS						
m.o.b.	WAXING SQUARE	WAXING TRINE	OPPOSITION	WANING TRINE	WANING SQUARE	RETURN
JAN	21–22	29–30	44–45	57–59	63–65	83–84
FEB	21–23	29–30	44–45	57–59	64–65	83–84
MAR	22–23	30–31	44–45	57–59	64–65	83–84
APR	21–22	29–31	44–45	57–59	64–65	83–84
MAY	21–22	28–30	43–44	57–58	64–65	83–84
JUN	21–22	28–30	43–44	57–58	63–64	83–84
JUL	20–22	28–30	42–43	56–57	63–64	82–84
AUG	20–22	28–29	42–43	56–57	62–64	82–84
SEP	20–22	28–29	42–44	56–57	62–63	82–84
OCT	20–22	28–29	42–44	56–58	62–64	82–84
NOV	20–22	28–30	43–44	57–58	63–64	83–84
DEC	21–23	29–30	43–45	57–58	63–64	83–84

NEPTUNE CYCLE				PLUTO CYCLE		
m.o.b.	SQUARING PLUTO	WAXING SQUARE	WAXING TRINE	CONJ.NEPTUNE	WAXING SQUARE	WAXING TRINE
JAN	14–16	40–42	54–56	23–25	40–42	61–63
FEB	14–16	41–42	54–56	23–25	40–42	61–63
MAR	13–15	41–42	54–56	23–25	40–42	61–63
APR	12–15	41–42	54–55	23–25	39–41	60–62
MAY	12–15	40–42	53–55	23–25	38–41	59–62
JUN	12–14	39–41	53–55	23–25	38–41	58–61
JUL	12–14	39–41	52–54	22–24	38–41	58–62
AUG	12–14	38–41	52–54	22–24	38–41	59–62
SEP	13–15	38–41	52–54	22–24	39–42	59–62
OCT	13–15	39–41	52–54	22–24	39–42	60–63
NOV	13–15	39–41	53–55	23–24	40–42	61–64
DEC	14–16	39–41	53–55	23–24	41–43	61–65

1987

URANUS CYCLE 0–84 YEARS						
m.o.b.	WAXING SQUARE	WAXING TRINE	OPPOSITION	WANING TRINE	WANING SQUARE	RETURN
JAN	21–22	29–30	44–45	57–58	63–65	83–84
FEB	21–23	29–30	44–45	57–59	64–65	83–84
MAR	22–23	30–31	44–45	57–59	64–65	83–84
APR	21–22	29–31	44–45	57–58	64–65	83–84
MAY	21–22	28–30	43–44	57–58	64–65	83–84
JUN	21–22	28–30	43–44	57–58	63–64	83–84
JUL	20–22	28–30	42–43	56–57	63–64	82–84
AUG	20–22	28–29	42–43	56–57	62–64	82–84
SEP	20–22	28–29	42–44	56–57	62–63	82–83
OCT	20–22	28–29	42–44	56–57	62–64	82–84
NOV	20–22	28–30	43–44	56–58	63–64	83–84
DEC	21–23	29–30	43–45	57–58	63–64	83–84

NEPTUNE CYCLE				PLUTO CYCLE		
m.o.b.	SQUARING PLUTO	WAXING SQUARE	WAXING TRINE	CONJ.NEPTUNE	WAXING SQUARE	WAXING TRINE
JAN	14–16	40–42	54–56	23–25	41–43	62–64
FEB	14–16	41–42	54–56	23–25	41–43	62–64
MAR	14–16	41–42	54–56	23–25	41–43	62–64
APR	13–15	41–42	54–55	23–25	40–42	61–63
MAY	12–15	40–42	53–55	23–25	39–42	60–63
JUN	12–15	39–41	53–55	23–25	39–42	59–62
JUL	12–15	39–41	52–54	22–24	39–42	59–63
AUG	13–15	38–41	52–54	22–24	39–42	60–63
SEP	13–15	38–41	52–54	22–24	39–42	60–63
OCT	13–15	39–41	52–54	22–24	40–43	61–64
NOV	14–16	39–41	53–55	23–24	41–43	62–65
DEC	14–16	39–41	53–55	23–24	42–44	62–66

1988

URANUS CYCLE 0–84 YEARS						
m.o.b.	WAXING SQUARE	WAXING TRINE	OPPOSITION	WANING TRINE	WANING SQUARE	RETURN
JAN	21–22	29–30	44–45	57–58	63–64	83–84
FEB	21–23	29–30	44–45	57–58	63–65	83–84
MAR	22–23	30–31	44–45	57–58	63–65	83–84
APR	21–22	29–31	44–45	57–58	63–65	83–84
MAY	21–22	28–30	44–45	57–58	63–65	83–84
JUN	21–22	28–30	43–44	57–58	63–64	83–84
JUL	20–22	28–30	42–43	56–57	63–64	82–84
AUG	20–22	28–29	42–43	56–57	62–64	82–84
SEP	20–22	28–29	42–44	56–57	62–63	82–83
OCT	20–22	28–29	42–44	56–57	62–64	82–84
NOV	20–22	28–30	43–44	56–57	62–64	83–84
DEC	21–23	29–30	43–45	56–57	62–64	83–84

NEPTUNE CYCLE				PLUTO CYCLE		
m.o.b.	SQUARING PLUTO	WAXING SQUARE	WAXING TRINE	CONJ.NEPTUNE	WAXING SQUARE	WAXING TRINE
JAN	14–16	40–42	54–56	23–25	42–44	63–65
FEB	14–16	41–42	54–56	23–26	42–44	63–65
MAR	14–16	41–42	54–56	23–26	42–44	63–65
APR	13–15	41–42	54–55	23–26	41–43	62–64
MAY	13–15	40–42	53–55	23–25	40–43	61–64
JUN	13–15	39–41	53–55	23–25	40–42	60–64
JUL	13–15	39–41	52–54	22–24	40–42	60–64
AUG	13–15	38–41	52–54	22–24	40–42	61–64
SEP	13–15	38–41	52–54	22–24	40–43	61–64
OCT	14–16	39–41	52–54	22–24	40–43	62–65
NOV	14–16	39–41	53–55	23–24	41–43	63–66
DEC	14–16	39–41	53–55	23–25	42–44	63–67

1989

URANUS CYCLE 0–84 YEARS						
m.o.b.	WAXING SQUARE	WAXING TRINE	OPPOSITION	WANING TRINE	WANING SQUARE	RETURN
JAN	21–22	29–30	44–45	57–58	63–64	83–84
FEB	21–23	29–30	44–45	57–58	63–65	83–84
MAR	22–23	30–31	44–45	57–58	63–65	83–84
APR	21–22	29–31	44–45	57–58	63–65	83–84
MAY	21–22	28–30	44–45	57–58	63–65	83–84
JUN	21–22	28–30	43–44	57–58	63–64	83–84
JUL	20–22	28–30	42–43	56–57	63–64	82–84
AUG	20–22	28–29	42–43	56–57	62–64	82–84
SEP	20–22	28–29	42–44	56–57	62–63	82–83
OCT	20–22	28–29	42–44	56–57	62–64	82–84
NOV	20–22	28–30	43–44	56–57	62–64	83–84
DEC	21–23	29–30	43–45	56–57	62–64	83–84

NEPTUNE CYCLE				PLUTO CYCLE		
m.o.b.	SQUARING PLUTO	WAXING SQUARE	WAXING TRINE	CONJ.NEPTUNE	WAXING SQUARE	WAXING TRINE
JAN	14–16	40–42	54–56	23–25	42–44	64–66
FEB	14–16	41–42	54–56	23–26	42–44	64–66
MAR	14–16	41–42	54–56	24–26	42–45	64–66
APR	13–15	41–42	54–55	23–26	41–44	63–65
MAY	13–15	40–42	53–55	23–25	41–44	62–65
JUN	13–15	39–41	53–55	23–25	40–43	61–65
JUL	13–15	39–41	52–54	22–24	40–43	61–65
AUG	13–15	38–41	52–54	22–24	41–43	62–66
SEP	13–15	38–41	52–54	22–24	41–43	62–65
OCT	13–15	39–41	52–54	22–24	41–44	63–66
NOV	14–16	39–41	53–55	23–24	42–44	64–67
DEC	15–17	39–41	53–55	23–25	43–45	64–68

1990

URANUS CYCLE 0–84 YEARS						
m.o.b.	WAXING SQUARE	WAXING TRINE	OPPOSITION	WANING TRINE	WANING SQUARE	RETURN
JAN	22–23	29–31	43–45	56–58	63–64	83–84
FEB	22–23	30–31	44–45	57–58	63–64	83–84
MAR	22–23	30–31	44–45	57–58	63–64	83–84
APR	22–23	30–31	44–45	57–58	63–64	83–84
MAY	22–23	29–30	44–45	57–58	63–64	83–84
JUN	21–23	29–30	44–45	57–58	63–64	83–84
JUL	21–22	29–30	43–44	56–57	62–63	83–84
AUG	21–22	28–30	43–44	56–57	62–63	82–84
SEP	21–22	28–29	42–44	56–57	62–63	82–83
OCT	21–22	28–30	42–44	55–56	62–63	82–83
NOV	21–22	28–30	43–44	56–57	62–63	82–83
DEC	21–23	29–30	43–44	56–57	62–63	82–83

NEPTUNE CYCLE				PLUTO CYCLE		
m.o.b.	SQUARING PLUTO	WAXING SQUARE	WAXING TRINE	CONJ.NEPTUNE	WAXING SQUARE	WAXING TRINE
JAN	15–17	40–42	53–55	23–25	43–45	66–67
FEB	15–16	40–42	54–56	23–25	43–45	66–67
MAR	15–16	40–42	54–56	24–26	43–45	66–67
APR	14–16	40–42	54–56	24–26	42–45	65–66
MAY	13–16	40–42	54–55	23–26	41–44	64–66
JUN	13–15	39–41	53–55	23–25	41–44	63–65
JUL	13–15	39–41	52–54	23–25	41–44	63–65
AUG	13–15	39–41	52–54	22–25	41–44	63–65
SEP	13–15	38–40	52–54	22–24	41–44	63–65
OCT	14–16	38–41	52–54	22–24	42–45	64–66
NOV	14–16	39–41	52–54	22–25	43–45	65–67
DEC	15–17	39–41	52–55	23–25	43–46	66–68

1991

URANUS CYCLE 0–84 YEARS						
m.o.b.	WAXING SQUARE	WAXING TRINE	OPPOSITION	WANING TRINE	WANING SQUARE	RETURN
JAN	22–23	29–31	43–45	56–58	63–64	83–84
FEB	22–23	30–31	44–45	57–58	63–64	83–84
MAR	22–23	30–31	44–45	57–58	63–64	83–84
APR	22–23	30–31	44–45	57–58	63–64	83–84
MAY	22–23	29–30	44–45	57–58	63–64	83–84
JUN	21–23	29–30	44–45	57–58	63–64	83–84
JUL	21–22	29–30	43–44	56–57	62–63	83–84
AUG	21–22	28–30	43–44	56–57	62–63	82–84
SEP	21–22	28–29	42–43	56–57	62–63	82–83
OCT	21–22	28–30	42–44	55–56	62–63	82–83
NOV	21–22	28–30	43–44	56–57	62–63	82–83
DEC	21–23	29–30	43–44	56–57	62–63	82–83

NEPTUNE CYCLE				PLUTO CYCLE		
m.o.b.	SQUARING PLUTO	WAXING SQUARE	WAXING TRINE	CONJ.NEPTUNE	WAXING SQUARE	WAXING TRINE
JAN	15–17	40–42	53–55	23–25	44–46	67–68
FEB	15–16	40–42	54–56	23–25	44–46	67–68
MAR	15–16	40–42	54–56	24–26	44–46	67–68
APR	14–16	40–42	54–56	24–26	43–46	66–67
MAY	13–16	40–42	54–55	23–26	42–45	65–67
JUN	13–15	39–41	53–55	23–26	42–45	64–66
JUL	13–15	39–41	52–54	23–25	42–45	64–66
AUG	13–15	39–41	52–54	22–25	42–45	64–66
SEP	13–15	38–40	52–54	22–25	42–45	64–66
OCT	14–16	38–41	52–54	22–24	43–46	65–67
NOV	14–16	39–41	52–54	22–25	44–46	66–68
DEC	15–17	39–41	52–55	23–25	44–47	67–69

1992

URANUS CYCLE 0–84 YEARS						
m.o.b.	WAXING SQUARE	WAXING TRINE	OPPOSITION	WANING TRINE	WANING SQUARE	RETURN
JAN	22–23	29–31	43–45	56–58	63–64	83–84
FEB	22–23	30–31	44–45	57–58	63–64	83–84
MAR	22–23	30–31	44–45	57–58	63–64	83–84
APR	22–23	30–31	44–45	57–58	63–64	83–84
MAY	22–23	29–30	44–45	56–58	63–64	83–84
JUN	21–23	29–30	44–45	56–58	63–64	83–84
JUL	21–22	29–30	43–44	56–57	62–63	83–84
AUG	21–22	28–30	43–44	56–57	62–63	82–84
SEP	21–22	28–29	42–43	56–57	62–63	82–83
OCT	21–22	28–30	42–44	55–56	62–63	82–83
NOV	21–22	28–30	42–44	55–57	62–63	82–83
DEC	21–23	29–30	43–44	55–57	62–63	82–83

NEPTUNE CYCLE				PLUTO CYCLE		
m.o.b.	SQUARING PLUTO	WAXING SQUARE	WAXING TRINE	CONJ.NEPTUNE	WAXING SQUARE	WAXING TRINE
JAN	15–17	40–42	53–55	23–25	45–47	68–69
FEB	15–17	40–42	54–56	23–25	45–47	68–69
MAR	15–17	40–42	54–56	24–26	45–47	68–69
APR	14–16	40–42	54–56	24–26	44–47	67–68
MAY	13–16	40–42	54–55	24–26	43–46	66–67
JUN	13–16	39–41	53–55	24–26	43–45	66–67
JUL	13–16	39–41	52–54	23–25	43–45	66–67
AUG	13–16	39–41	52–54	22–25	42–45	66–67
SEP	13–16	38–40	52–54	22–25	42–45	66–67
OCT	14–16	38–41	52–54	22–24	43–46	67–68
NOV	14–16	39–41	52–54	22–25	44–47	67–69
DEC	15–17	39–41	52–54	23–25	45–48	68–70

1993

URANUS CYCLE 0–84 YEARS						
m.o.b.	WAXING SQUARE	WAXING TRINE	OPPOSITION	WANING TRINE	WANING SQUARE	RETURN
JAN	22–23	29–31	43–45	55–57	63–64	83–84
FEB	22–23	30–31	44–45	56–58	63–64	83–84
MAR	22–23	30–31	44–45	56–58	63–64	83–84
APR	22–23	30–31	44–45	56–58	63–64	83–84
MAY	22–23	29–30	44–45	56–58	63–64	83–84
JUN	22–23	29–30	44–45	56–58	63–64	83–84
JUL	21–22	29–30	43–44	56–57	62–63	83–84
AUG	21–22	28–30	43–44	56–57	62–63	82–84
SEP	21–22	28–29	42–43	56–57	62–63	82–83
OCT	21–22	28–30	42–44	55–56	62–63	82–83
NOV	21–22	28–30	42–44	55–57	62–63	82–83
DEC	21–23	29–30	43–44	55–57	62–63	82–83

NEPTUNE CYCLE				PLUTO CYCLE		
m.o.b.	SQUARING PLUTO	WAXING SQUARE	WAXING TRINE	CONJ.NEPTUNE	WAXING SQUARE	WAXING TRINE
JAN	15–17	40–42	53–55	23–25	46–48	69–70
FEB	15–17	40–42	54–56	23–25	46–48	69–70
MAR	15–16	40–42	54–56	24–26	46–48	69–70
APR	14–16	40–42	54–56	24–26	45–47	68–69
MAY	14–16	40–42	54–55	24–26	44–46	67–68
JUN	14–16	39–41	53–55	24–26	43–46	67–68
JUL	14–16	39–41	52–54	23–25	43–46	67–68
AUG	14–16	39–41	52–54	22–25	43–46	67–68
SEP	14–16	38–40	52–54	22–25	43–46	67–68
OCT	14–16	38–41	52–54	22–24	43–46	68–69
NOV	14–16	39–41	52–54	22–25	44–47	68–70
DEC	15–17	39–41	52–54	23–25	45–48	69–71

1994

URANUS CYCLE 0–84 YEARS						
m.o.b.	WAXING SQUARE	WAXING TRINE	OPPOSITION	WANING TRINE	WANING SQUARE	RETURN
JAN	22–23	29–31	43–45	55–57	63–64	83–84
FEB	22–23	30–31	43–45	56–57	63–64	83–84
MAR	22–23	30–31	44–45	56–58	63–64	83–84
APR	22–23	30–31	44–45	56–58	63–64	83–84
MAY	22–23	29–30	44–45	56–58	63–64	83–84
JUN	22–23	29–30	44–45	56–58	63–64	83–84
JUL	21–22	29–30	43–44	56–57	62–63	83–84
AUG	21–22	28–30	43–44	56–57	62–63	82–84
SEP	21–22	28–29	42–43	56–57	62–63	82–83
OCT	21–22	28–30	42–44	55–56	62–63	82–83
NOV	21–22	28–30	42–44	55–56	62–63	82–83
DEC	21–23	29–30	42–44	55–56	62–63	82–83

NEPTUNE CYCLE				PLUTO CYCLE		
m.o.b.	SQUARING PLUTO	WAXING SQUARE	WAXING TRINE	CONJ.NEPTUNE	WAXING SQUARE	WAXING TRINE
JAN	15–17	40–42	53–55	23–25	46–48	70–72
FEB	15–17	40–42	54–56	23–25	46–48	70–72
MAR	15–17	40–42	54–56	24–26	46–48	70–72
APR	14–16	40–42	54–56	24–26	45–47	69–70
MAY	14–16	40–42	54–55	24–26	45–47	68–70
JUN	14–16	39–41	53–55	24–26	44–47	68–70
JUL	14–16	39–41	52–54	23–25	44–47	68–70
AUG	14–16	39–41	52–54	22–25	44–47	68–70
SEP	14–16	38–40	52–54	22–25	44–47	68–70
OCT	14–16	38–41	52–54	23–25	44–47	69–70
NOV	14–16	39–41	52–54	23–25	45–48	70–71
DEC	15–17	39–41	52–54	23–25	46–49	71–72

1995

URANUS CYCLE 0–84 YEARS						
m.o.b.	WAXING SQUARE	WAXING TRINE	OPPOSITION	WANING TRINE	WANING SQUARE	RETURN
JAN	22–23	29–31	43–45	55–57	62–63	83–84
FEB	22–23	30–31	43–45	56–57	62–63	83–84
MAR	22–23	30–31	43–45	56–57	62–63	83–84
APR	22–23	30–31	43–45	56–57	62–64	83–84
MAY	22–23	29–30	44–45	56–57	62–64	83–84
JUN	22–23	29–30	44–45	56–57	62–64	83–84
JUL	21–22	29–30	43–44	56–57	62–63	83–84
AUG	21–22	28–30	43–44	56–57	62–63	82–84
SEP	21–22	28–29	42–43	56–57	62–63	82–83
OCT	21–22	28–30	42–43	55–56	62–63	82–83
NOV	21–22	28–30	42–44	55–56	62–63	82–83
DEC	21–23	29–30	42–44	55–56	62–63	82–83

NEPTUNE CYCLE				PLUTO CYCLE		
m.o.b.	SQUARING PLUTO	WAXING SQUARE	WAXING TRINE	CONJ.NEPTUNE	WAXING SQUARE	WAXING TRINE
JAN	15–17	40–42	53–55	23–25	46–49	72–73
FEB	15–17	40–42	54–56	23–25	46–49	72–73
MAR	15–17	40–42	54–56	24–26	46–49	72–73
APR	14–16	40–42	54–56	24–26	45–48	70–71
MAY	14–16	40–42	54–55	24–26	45–48	69–71
JUN	14–16	39–41	53–55	24–26	45–48	69–71
JUL	14–16	39–41	52–54	23–25	45–48	69–71
AUG	14–16	39–41	52–54	22–25	45–48	69–71
SEP	14–16	38–40	52–54	22–25	45–48	69–71
OCT	14–16	38–41	52–54	23–25	45–48	70–72
NOV	14–16	39–41	52–54	23–25	46–49	71–73
DEC	15–17	39–41	52–54	23–25	47–50	72–74

1996

URANUS CYCLE 0–84 YEARS						
m.o.b.	WAXING SQUARE	WAXING TRINE	OPPOSITION	WANING TRINE	WANING SQUARE	RETURN
JAN	22–23	29–31	43–44	55–57	62–63	83–84
FEB	22–23	30–31	43–44	56–57	62–63	83–84
MAR	22–23	30–31	43–45	56–57	62–64	83–84
APR	22–23	30–31	43–45	56–57	62–64	83–84
MAY	22–23	29–30	43–44	56–57	62–64	83–84
JUN	22–23	29–30	43–44	56–57	62–64	83–84
JUL	21–22	29–30	43–44	56–57	62–63	83–84
AUG	21–22	28–30	43–44	56–57	62–63	82–84
SEP	21–22	28–29	42–43	55–56	62–63	82–83
OCT	21–22	28–30	42–43	55–56	62–63	82–83
NOV	21–22	28–30	42–43	55–56	62–63	82–83
DEC	21–23	28–30	42–43	55–56	61–62	82–83

NEPTUNE CYCLE				PLUTO CYCLE		
m.o.b.	SQUARING PLUTO	WAXING SQUARE	WAXING TRINE	CONJ.NEPTUNE	WAXING SQUARE	WAXING TRINE
JAN	15–17	40–42	53–55	23–25	47–49	73–74
FEB	15–17	40–42	53–55	24–26	48–50	73–74
MAR	16–17	40–42	54–56	24–27	48–50	73–74
APR	15–16	40–42	54–56	24–27	46–49	72–73
MAY	14–16	40–42	54–55	24–27	46–49	71–73
JUN	14–16	39–41	53–55	24–27	46–49	70–72
JUL	14–16	39–41	52–54	23–26	45–48	70–72
AUG	14–16	39–41	52–54	23–26	45–48	70–72
SEP	14–16	38–40	52–54	23–25	45–48	70–72
OCT	14–16	38–41	52–54	23–25	46–49	71–73
NOV	14–16	39–41	52–54	23–25	47–50	72–74
DEC	15–17	39–41	52–54	24–26	48–51	73–75

1997

URANUS CYCLE 0–84 YEARS						
m.o.b.	WAXING SQUARE	WAXING TRINE	OPPOSITION	WANING TRINE	WANING SQUARE	RETURN
JAN	22–23	29–31	43–44	55–57	62–63	83–84
FEB	22–23	30–31	43–44	56–57	62–63	83–84
MAR	22–23	30–31	43–44	56–57	62–63	83–84
APR	22–23	30–31	43–44	56–57	62–64	83–84
MAY	22–23	29–30	43–44	56–57	62–64	83–84
JUN	22–23	29–30	43–44	56–57	62–64	83–84
JUL	21–22	29–30	43–44	56–57	62–63	83–84
AUG	21–22	28–30	43–44	56–57	62–63	82–84
SEP	21–22	28–29	42–43	55–56	62–63	82–83
OCT	21–22	28–30	42–43	55–56	62–63	82–83
NOV	21–22	28–30	42–43	55–56	62–63	82–83
DEC	21–23	28–30	42–43	55–56	61–62	82–83

NEPTUNE CYCLE				PLUTO CYCLE		
m.o.b.	SQUARING PLUTO	WAXING SQUARE	WAXING TRINE	CONJ.NEPTUNE	WAXING SQUARE	WAXING TRINE
JAN	15–17	40–42	53–55	24–26	49–51	74–75
FEB	15–17	40–42	53–55	24–26	49–51	74–75
MAR	16–17	40–42	53–55	24–27	49–51	74–75
APR	15–17	40–42	54–56	24–27	48–51	73–76
MAY	14–17	40–42	54–55	24–27	47–50	72–75
JUN	14–16	39–41	53–55	24–27	46–50	72–73
JUL	14–16	39–41	52–54	23–26	46–50	72–73
AUG	14–16	39–41	52–54	23–26	46–49	72–73
SEP	14–16	38–40	52–54	23–26	46–49	72–73
OCT	14–16	38–41	52–54	23–26	46–49	72–74
NOV	14–16	39–41	52–54	24–26	47–50	73–75
DEC	15–17	39–41	52–54	24–26	48–51	74–76

1998

URANUS CYCLE 0–84 YEARS						
m.o.b.	WAXING SQUARE	WAXING TRINE	OPPOSITION	WANING TRINE	WANING SQUARE	RETURN
JAN	22–23	29–31	43–44	55–57	62–63	83–84
FEB	22–23	30–31	43–44	55–57	62–63	83–84
MAR	22–23	30–31	43–44	56–57	62–63	83–84
APR	22–23	30–31	43–44	56–57	62–63	83–84
MAY	22–23	29–30	43–44	56–57	62–63	83–84
JUN	22–23	29–30	43–44	56–57	62–63	83–84
JUL	21–22	29–30	43–44	56–57	62–63	83–84
AUG	21–22	28–30	43–44	56–57	62–63	82–84
SEP	21–22	28–29	42–43	55–56	61–62	82–83
OCT	21–22	28–30	42–43	55–56	61–62	82–83
NOV	21–22	28–30	42–43	55–56	61–62	82–83
DEC	21–23	28–30	42–43	55–56	61–62	82–83

	NEPTUNE CYCLE			PLUTO CYCLE		
m.o.b.	SQUARING PLUTO	WAXING SQUARE	WAXING TRINE	CONJ.NEPTUNE	WAXING SQUARE	WAXING TRINE
JAN	15–17	40–42	53–55	24–26	50–52	75–77
FEB	15–17	40–42	53–55	24–26	50–52	76–77
MAR	16–17	40–42	53–55	24–27	50–52	76–77
APR	15–17	40–42	54–56	24–27	49–52	75–76
MAY	14–17	40–42	54–55	24–27	48–51	74–76
JUN	14–17	39–41	53–55	25–27	47–51	73–75
JUL	14–16	39–41	52–54	24–27	47–51	73–75
AUG	14–16	39–41	52–54	23–26	47–50	72–74
SEP	14–16	38–40	52–54	23–26	47–50	72–74
OCT	14–16	38–41	52–54	23–26	47–50	73–75
NOV	14–16	39–41	52–54	23–26	48–51	74–76
DEC	15–17	39–41	52–54	23–26	49–52	75–77

1999

URANUS CYCLE 0–84 YEARS						
m.o.b.	WAXING SQUARE	WAXING TRINE	OPPOSITION	WANING TRINE	WANING SQUARE	RETURN
JAN	22–23	29–31	43–44	55–57	62–63	83–84
FEB	22–23	30–31	43–44	55–57	62–63	83–84
MAR	22–23	30–31	43–44	55–57	62–63	83–84
APR	22–23	30–31	43–44	56–57	62–63	83–84
MAY	22–23	29–30	43–44	56–57	62–63	83–84
JUN	22–23	29–30	43–44	56–57	62–63	83–84
JUL	21–22	29–30	43–44	56–57	62–63	83–84
AUG	21–22	28–30	43–44	56–57	62–63	82–84
SEP	21–22	28–29	42–43	55–56	61–62	82–83
OCT	21–22	28–29	42–43	55–56	61–62	82–83
NOV	21–22	28–30	42–43	55–56	61–62	82–83
DEC	21–23	28–30	42–43	54–55	61–62	82–83

	NEPTUNE CYCLE			PLUTO CYCLE		
m.o.b.	SQUARING PLUTO	WAXING SQUARE	WAXING TRINE	CONJ.NEPTUNE	WAXING SQUARE	WAXING TRINE
JAN	15–17	40–42	53–55	24–26	51–53	76–78
FEB	15–17	40–42	53–55	24–26	51–53	77–78
MAR	16–18	40–42	53–55	25–27	51–53	77–78
APR	15–17	40–42	54–56	25–28	50–52	76–77
MAY	15–17	40–42	54–55	25–28	49–51	75–77
JUN	15–17	39–41	53–55	25–28	48–51	74–76
JUL	14–16	39–41	52–54	24–27	48–51	74–76
AUG	14–16	39–41	52–54	24–26	48–51	73–75
SEP	14–16	38–40	52–54	24–26	48–51	73–75
OCT	14–16	38–41	52–54	24–26	48–51	74–76
NOV	15–17	39–41	52–54	24–26	49–52	75–77
DEC	16–18	39–41	52–54	24–26	50–53	76–78

2000

URANUS CYCLE 0–84 YEARS						
m.o.b.	WAXING SQUARE	WAXING TRINE	OPPOSITION	WANING TRINE	WANING SQUARE	RETURN
JAN	21–23	29–30	42–43	55–56	61–62	83–84
FEB	22–23	29–30	42–44	55–56	61–63	83–84
MAR	22–23	29–31	43–44	55–56	62–63	83–84
APR	22–23	29–31	43–44	55–57	62–63	83–84
MAY	22–23	29–31	43–44	56–57	62–63	83–84
JUN	22–23	29–30	43–44	56–57	62–63	83–84
JUL	22–23	29–30	43–44	55–57	62–63	83–84
AUG	21–23	29–30	42–43	55–56	62–63	82–84
SEP	21–22	28–29	42–43	55–56	61–62	82–84
OCT	21–22	28–29	41–42	54–55	61–62	82–84
NOV	21–22	28–29	41–43	54–55	61–62	82–84
DEC	21–22	28–30	41–43	54–55	61–62	82–83

NEPTUNE CYCLE				PLUTO CYCLE		
m.o.b.	SQUARING PLUTO	WAXING SQUARE	WAXING TRINE	CONJ.NEPTUNE	WAXING SQUARE	WAXING TRINE
JAN	16–18	39–41	52–55	25–27	51–54	77–79
FEB	16–18	40–42	53–55	25–27	51–54	78–79
MAR	16–18	40–42	53–55	25–28	51–54	78–79
APR	16–18	40–42	53–55	25–28	51–54	77–78
MAY	15–17	40–42	53–55	25–28	50–53	76–78
JUN	15–17	40–41	53–55	25–28	49–52	75–77
JUL	14–16	39–41	52–54	25–28	49–52	75–77
AUG	14–16	39–41	52–54	25–28	48–52	74–76
SEP	14–16	38–40	52–54	24–27	48–52	74–76
OCT	15–17	38–40	52–54	24–27	49–52	75–77
NOV	15–17	38–40	52–54	24–27	50–53	76–78
DEC	16–18	39–41	52–54	24–27	51–54	77–79

2001

URANUS CYCLE 0–84 YEARS						
m.o.b.	WAXING SQUARE	WAXING TRINE	OPPOSITION	WANING TRINE	WANING SQUARE	RETURN
JAN	21–23	29–30	42–43	55–56	61–62	83–84
FEB	22–23	29–30	42–44	55–56	61–63	83–84
MAR	22–23	29–31	42–44	55–57	62–63	83–84
APR	22–23	29–31	43–44	55–57	62–63	83–84
MAY	22–23	29–31	43–44	56–57	62–63	83–84
JUN	22–23	29–30	43–44	56–57	62–63	83–84
JUL	22–23	29–30	43–44	55–57	62–63	83–84
AUG	21–23	29–30	42–43	55–56	62–63	82–84
SEP	21–22	28–30	42–43	55–56	61–62	82–84
OCT	21–22	28–30	42–43	54–55	61–62	82–84
NOV	21–22	28–29	41–43	54–55	61–62	82–84
DEC	21–22	28–29	41–43	54–55	61–62	82–83

NEPTUNE CYCLE				PLUTO CYCLE		
m.o.b.	SQUARING PLUTO	WAXING SQUARE	WAXING TRINE	CONJ.NEPTUNE	WAXING SQUARE	WAXING TRINE
JAN	16–18	39–41	52–55	25–27	52–55	78–80
FEB	16–18	40–42	53–55	25–27	52–55	79–80
MAR	16–18	40–42	53–55	26–28	52–55	79–80
APR	16–18	40–42	53–55	26–28	52–55	78–79
MAY	15–17	40–42	53–55	26–28	51–54	77–79
JUN	15–17	40–41	53–55	26–28	51–54	76–78
JUL	14–16	39–41	52–54	26–28	51–54	76–78
AUG	14–16	39–41	52–54	26–28	50–53	75–77
SEP	14–16	38–40	52–54	24–27	50–53	75–77
OCT	15–17	38–40	52–54	24–27	50–53	76–78
NOV	15–17	38–40	52–54	24–27	51–54	77–79
DEC	16–18	39–41	52–54	25–27	51–55	78–80

2002

URANUS CYCLE 0–84 YEARS						
m.o.b.	WAXING SQUARE	WAXING TRINE	OPPOSITION	WANING TRINE	WANING SQUARE	RETURN
JAN	21–23	28–29	42–43	55–56	61–62	83–84
FEB	22–23	29–30	42–44	55–56	61–63	83–84
MAR	22–23	29–30	42–44	55–57	61–63	83–84
APR	22–23	29–31	43–44	55–57	62–63	83–84
MAY	22–23	29–31	43–44	56–57	62–63	83–84
JUN	22–23	29–30	43–44	56–57	62–63	83–84
JUL	22–23	29–30	43–44	55–57	62–63	83–84
AUG	21–23	29–30	42–43	55–56	62–63	82–84
SEP	21–22	28–29	41–42	55–56	61–62	82–84
OCT	21–22	28–29	41–42	54–55	61–62	82–84
NOV	21–22	28–29	41–42	54–55	61–62	82–84
DEC	21–22	28–29	41–42	54–55	60–61	82–83

	NEPTUNE CYCLE			PLUTO CYCLE		
m.o.b.	SQUARING PLUTO	WAXING SQUARE	WAXING TRINE	CONJ.NEPTUNE	WAXING SQUARE	WAXING TRINE
JAN	16–18	39–41	52–55	25–27	52–55	79–81
FEB	16–18	40–42	53–55	25–27	52–55	80–81
MAR	16–18	40–42	53–55	26–28	53–56	80–81
APR	16–18	40–42	53–55	26–28	53–56	79–80
MAY	15–17	40–42	53–55	26–28	52–55	78–80
JUN	15–17	40–41	53–55	26–29	51–54	77–79
JUL	14–16	39–41	52–54	26–29	51–54	77–79
AUG	14–16	39–41	52–54	26–28	50–54	76–78
SEP	14–17	38–40	52–54	25–27	50–55	76–78
OCT	15–17	38–40	52–54	25–27	50–55	77–79
NOV	15–17	38–40	52–54	25–27	51–55	78–80
DEC	16–18	38–41	52–54	25–27	52–56	79–81

2003

URANUS CYCLE 0–84 YEARS						
m.o.b.	WAXING SQUARE	WAXING TRINE	OPPOSITION	WANING TRINE	WANING SQUARE	RETURN
JAN	21–23	28–29	42–43	55–56	61–62	83–84
FEB	22–23	29–30	42–43	55–56	61–62	83–84
MAR	22–23	29–30	42–43	55–56	61–62	83–84
APR	22–23	29–31	43–44	55–56	62–63	83–84
MAY	22–23	29–31	43–44	55–56	62–63	83–84
JUN	22–23	29–30	43–44	55–56	62–63	83–84
JUL	22–23	29–30	43–44	55–56	62–63	83–84
AUG	21–23	29–30	42–43	55–56	62–63	82–84
SEP	21–22	28–29	42–43	55–56	61–62	82–84
OCT	21–22	28–29	41–42	54–55	61–62	82–84
NOV	21–22	28–29	41–42	54–55	61–62	82–84
DEC	21–22	28–29	41–42	54–55	60–61	82–83

	NEPTUNE CYCLE			PLUTO CYCLE		
m.o.b.	SQUARING PLUTO	WAXING SQUARE	WAXING TRINE	CONJ.NEPTUNE	WAXING SQUARE	WAXING TRINE
JAN	16–18	39–41	52–55	25–27	53–57	80–81
FEB	16–18	40–42	53–55	26–28	53–57	81–82
MAR	16–18	40–42	53–55	27–29	54–57	82–83
APR	16–18	40–42	53–55	27–29	54–57	81–82
MAY	15–17	40–42	53–55	27–29	53–56	80–82
JUN	15–17	40–41	53–55	27–29	52–55	79–81
JUL	14–16	39–41	52–54	26–29	52–55	79–81
AUG	14–16	39–41	52–54	26–29	52–55	78–80
SEP	14–17	38–40	52–54	25–28	51–54	78–80
OCT	15–17	38–40	52–54	25–28	51–54	79–81
NOV	15–17	38–40	52–54	25–27	52–55	80–82
DEC	16–18	38–41	52–54	25–27	53–56	81–83

2004

URANUS CYCLE 0–84 YEARS						
m.o.b.	WAXING SQUARE	WAXING TRINE	OPPOSITION	WANING TRINE	WANING SQUARE	RETURN
JAN	21–23	28–29	42–43	54–55	61–62	83–84
FEB	22–23	29–30	42–43	54–56	61–62	83–84
MAR	22–23	29–30	42–43	54–56	61–62	83–84
APR	22–23	29–31	42–43	55–56	62–63	83–84
MAY	22–23	29–31	42–43	55–56	62–63	83–84
JUN	22–23	29–30	42–43	55–56	62–63	83–84
JUL	22–23	29–30	42–43	55–56	62–63	83–84
AUG	21–23	29–30	42–43	55–56	62–63	82–84
SEP	21–22	28–29	41–42	54–55	61–62	82–84
OCT	21–22	28–29	41–42	54–55	61–62	82–84
NOV	21–22	28–29	41–42	54–55	61–62	82–84
DEC	21–22	28–29	41–42	54–55	60–61	82–83

NEPTUNE CYCLE				PLUTO CYCLE		
m.o.b.	SQUARING PLUTO	WAXING SQUARE	WAXING TRINE	CONJ.NEPTUNE	WAXING SQUARE	WAXING TRINE
JAN	16–18	39–41	52–55	25–27	54–57	82–83
FEB	16–18	40–42	53–55	26–28	54–57	83–84
MAR	16–18	40–42	53–55	27–29	55–57	83–84
APR	16–18	40–42	53–55	27–29	55–57	83–84
MAY	15–17	40–42	53–55	27–29	54–57	82–83
JUN	15–17	40–41	53–55	27–30	53–56	81–83
JUL	14–16	39–41	52–54	27–30	53–56	81–83
AUG	14–16	39–41	52–54	26–29	53–56	81–83
SEP	14–17	38–40	52–54	25–28	52–55	81–83
OCT	15–17	38–40	52–54	25–28	52–55	82–84
NOV	15–17	38–40	52–54	25–28	53–56	82–84
DEC	16–18	38–41	52–54	26–28	54–57	84–85

2005

URANUS CYCLE 0–84 YEARS						
m.o.b.	WAXING SQUARE	WAXING TRINE	OPPOSITION	WANING TRINE	WANING SQUARE	RETURN
JAN	21–23	28–29	42–43	54–55	61–62	83–84
FEB	22–23	29–30	42–43	54–55	61–62	83–84
MAR	22–23	29–30	42–43	54–55	61–62	83–84
APR	22–23	29–31	42–43	55–56	62–63	83–84
MAY	22–23	29–31	42–43	55–56	62–63	83–84
JUN	22–23	29–30	42–43	55–56	61–63	83–84
JUL	22–23	29–30	42–43	55–56	61–63	83–84
AUG	21–23	29–30	42–43	55–56	61–62	82–84
SEP	21–22	28–29	41–42	54–55	61–62	82–84
OCT	21–22	28–29	41–42	54–55	61–62	82–84
NOV	21–22	28–29	41–42	54–55	61–62	82–84
DEC	21–22	28–29	40–42	53–54	60–61	82–83

NEPTUNE CYCLE				PLUTO CYCLE		
m.o.b.	SQUARING PLUTO	WAXING SQUARE	WAXING TRINE	CONJ.NEPTUNE	WAXING SQUARE	WAXING TRINE
JAN	16–18	39–41	52–55	26–28	55–58	non-applicable
FEB	16–18	40–42	53–55	27–29	55–58	non-applicable
MAR	16–18	40–42	53–55	27–30	56–58	non-applicable
APR	16–18	40–42	53–55	27–30	56–58	non-applicable
MAY	15–17	40–42	53–55	27–30	55–57	non-applicable
JUN	15–17	40–41	53–55	27–30	54–57	83–84
JUL	14–16	39–41	52–54	27–30	54–57	83–84
AUG	14–16	39–41	52–54	26–29	54–57	83–84
SEP	14–17	38–40	52–54	26–29	53–56	83–84
OCT	15–17	38–40	52–54	26–29	53–56	non-applicable
NOV	15–17	38–40	52–54	26–28	54–57	non-applicable
DEC	16–18	38–40	52–54	26–28	55–58	non-applicable

2006

URANUS CYCLE 0–84 YEARS						
m.o.b.	WAXING SQUARE	WAXING TRINE	OPPOSITION	WANING TRINE	WANING SQUARE	RETURN
JAN	21–23	28–29	41–43	54–55	61–62	83–84
FEB	22–23	28–30	41–43	54–55	61–62	83–84
MAR	22–23	28–30	41–43	54–55	61–62	83–84
APR	22–23	29–31	42–43	55–56	61–62	83–84
MAY	22–23	29–31	42–43	55–56	61–63	83–84
JUN	22–23	29–30	42–43	55–56	61–63	83–84
JUL	22–23	29–30	42–43	55–56	61–63	83–84
AUG	21–23	29–30	42–43	55–56	61–62	82–84
SEP	21–22	28–29	41–42	54–55	61–62	82–84
OCT	21–22	28–29	41–42	54–55	61–62	82–84
NOV	21–22	28–29	41–42	54–55	61–62	82–84
DEC	21–22	27–29	40–42	53–54	60–61	82–83

	NEPTUNE CYCLE			PLUTO CYCLE		
m.o.b.	SQUARING PLUTO	WAXING SQUARE	WAXING TRINE	CONJ.NEPTUNE	WAXING SQUARE	WAXING TRINE
JAN	16–18	39–41	52–55	26–28	56–58	non-applicable
FEB	16–18	40–42	53–55	27–29	56–58	non-applicable
MAR	16–18	40–42	53–55	28–30	57–59	non-applicable
APR	16–18	40–42	53–55	28–30	57–59	non-applicable
MAY	15–17	40–42	53–55	28–30	56–58	non-applicable
JUN	15–17	40–41	53–55	28–31	55–58	non-applicable
JUL	14–16	39–41	52–54	28–31	55–58	non-applicable
AUG	14–16	39–41	52–54	27–30	55–58	non-applicable
SEP	14–17	38–40	52–54	26–29	54–57	non-applicable
OCT	15–17	38–40	52–54	26–29	54–57	non-applicable
NOV	15–17	38–40	52–54	26–29	55–58	non-applicable
DEC	15–18	38–40	52–54	26–29	55–59	non-applicable

2007

URANUS CYCLE 0–84 YEARS						
m.o.b.	WAXING SQUARE	WAXING TRINE	OPPOSITION	WANING TRINE	WANING SQUARE	RETURN
JAN	21–23	28–29	41–42	54–55	60–61	83–84
FEB	22–23	28–30	41–42	54–55	60–62	83–84
MAR	22–23	28–30	41–42	54–55	60–62	83–84
APR	22–23	29–31	42–43	55–56	61–62	83–84
MAY	22–23	29–31	42–43	55–56	61–63	83–84
JUN	22–23	29–30	42–43	55–56	61–63	83–84
JUL	22–23	29–30	42–43	55–56	61–63	83–84
AUG	21–23	29–30	42–43	55–56	61–62	82–84
SEP	21–22	28–29	41–42	54–55	61–62	82–84
OCT	21–22	28–29	41–42	54–55	61–62	82–84
NOV	21–22	28–29	41–42	54–55	61–62	82–84
DEC	21–22	27–29	40–41	53–54	60–61	82–83

	NEPTUNE CYCLE			PLUTO CYCLE		
m.o.b.	SQUARING PLUTO	WAXING SQUARE	WAXING TRINE	CONJ.NEPTUNE	WAXING SQUARE	WAXING TRINE
JAN	16–18	39–41	52–55	26–29	56–59	non-applicable
FEB	16–18	40–42	53–55	27–29	57–60	non-applicable
MAR	16–18	40–42	53–55	28–30	58–60	non-applicable
APR	16–18	40–42	53–55	28–30	58–60	non-applicable
MAY	15–17	40–42	53–55	28–30	57–59	non-applicable
JUN	15–17	40–41	53–55	28–31	56–59	non-applicable
JUL	14–16	39–41	52–54	28–31	56–59	non-applicable
AUG	14–16	39–41	52–54	27–30	55–58	non-applicable
SEP	14–17	38–40	52–54	27–30	54–58	non-applicable
OCT	15–17	38–40	52–54	27–30	54–58	non-applicable
NOV	15–17	38–40	52–54	27–30	55–59	non-applicable
DEC	15–18	38–40	52–54	27–30	56–60	non-applicable

2008

URANUS CYCLE 0–84 YEARS						
m.o.b.	WAXING SQUARE	WAXING TRINE	OPPOSITION	WANING TRINE	WANING SQUARE	RETURN
JAN	21–23	28–29	41–42	54–55	60–61	83–84
FEB	21–23	28–29	41–42	54–55	60–62	83–84
MAR	21–23	28–29	41–42	54–56	60–62	83–84
APR	22–23	29–30	42–43	55–56	61–62	83–84
MAY	22–23	29–30	42–43	55–56	61–62	83–84
JUN	22–23	29–30	42–43	55–56	61–62	83–84
JUL	22–23	29–30	42–43	55–56	61–62	83–84
AUG	21–23	29–30	42–43	54–56	61–62	82–84
SEP	21–22	28–29	41–42	54–55	61–62	82–84
OCT	21–22	28–29	41–42	54–55	61–62	82–84
NOV	21–22	28–29	41–42	54–55	61–62	82–84
DEC	20–22	27–28	40–41	53–54	60–61	82–83

NEPTUNE CYCLE				PLUTO CYCLE		
m.o.b.	SQUARING PLUTO	WAXING SQUARE	WAXING TRINE	CONJ.NEPTUNE	WAXING SQUARE	WAXING TRINE
JAN	16–18	39–41	52–55	27–30	57–60	non applicable
FEB	16–18	40–42	53–55	28–31	58–61	non-applicable
MAR	16–18	40–42	53–55	29–31	59–61	non-applicable
APR	16–18	40–42	53–55	29–31	59–61	non-applicable
MAY	15–17	40–42	53–55	29–31	58–60	non-applicable
JUN	15–17	40–41	53–55	29–32	57–60	non-applicable
JUL	14–16	39–41	52–54	28–31	57–60	non-applicable
AUG	14–16	39–41	52–54	27–30	56–59	non-applicable
SEP	14–17	38–40	52–54	27–30	55–59	non-applicable
OCT	15–17	38–40	52–54	27–30	55–59	non-applicable
NOV	15–17	38–40	52–54	27–30	56–60	non-applicable
DEC	15–17	38–40	52–54	27–30	57–61	non-applicable

2009

URANUS CYCLE 0–84 YEARS						
m.o.b.	WAXING SQUARE	WAXING TRINE	OPPOSITION	WANING TRINE	WANING SQUARE	RETURN
JAN	21–23	28–29	41–42	54–55	60–61	83–84
FEB	21–23	28–29	41–42	54–55	60–62	83–84
MAR	21–23	28–29	41–42	54–56	60–62	83–84
APR	22–23	29–30	41–43	55–56	61–62	83–84
MAY	22–23	29–30	41–43	55–56	61–62	83–84
JUN	22–23	29–30	41–43	55–56	61–62	83–84
JUL	22–23	29–30	42–43	55–56	61–62	83–84
AUG	21–23	29–30	42–43	54–56	61–62	82–84
SEP	21–22	28–29	41–42	54–55	61–62	82–84
OCT	21–22	28–29	41–42	54–55	61–62	82–84
NOV	21–22	28–29	41–42	54–55	61–62	82–84
DEC	20–22	27–28	40–41	53–54	60–61	82–83

NEPTUNE CYCLE				PLUTO CYCLE		
m.o.b.	SQUARING PLUTO	WAXING SQUARE	WAXING TRINE	CONJ.NEPTUNE	WAXING SQUARE	WAXING TRINE
JAN	16–18	39–41	52–55	27–30	57–61	non-applicable
FEB	16–18	40–42	53–55	28–31	58–62	non-applicable
MAR	16–18	40–42	53–55	29–31	59–62	non-applicable
APR	16–18	40–42	53–55	29–31	59–62	non-applicable
MAY	15–17	40–42	53–55	29–31	58–62	non-applicable
JUN	15–17	40–41	53–55	29–32	58–61	non-applicable
JUL	14–16	39–41	52–54	28–31	58–61	non-applicable
AUG	14–16	39–41	52–54	28–31	57–60	non-applicable
SEP	14–17	38–40	52–54	28–31	56–60	non-applicable
OCT	15–17	38–40	52–54	28–31	56–60	non-applicable
NOV	15–17	38–40	52–54	28–31	57–61	non-applicable
DEC	15–17	38–40	52–54	28–31	58–62	non-applicable

2010

URANUS CYCLE 0–84 YEARS						
m.o.b.	WAXING SQUARE	WAXING TRINE	OPPOSITION	WANING TRINE	WANING SQUARE	RETURN
JAN	20–22	27–28	40–41	53–54	60–61	82–84
FEB	21–22	27–29	40–41	53–54	60–61	82–84
MAR	21–22	28–29	40–42	53–55	60–61	83–84
APR	21–23	28–29	41–42	54–55	60–62	83–84
MAY	22–23	28–29	41–42	54–55	61–62	83–84
JUN	22–23	29–30	41–43	54–55	61–62	83–84
JUL	21–22	28–29	41–43	54–55	61–62	83–84
AUG	21–22	28–29	41–42	54–55	61–62	83–84
SEP	21–22	28–29	41–42	54–55	60–62	83–84
OCT	20–22	27–28	40–41	53–54	60–62	83–84
NOV	20–22	27–28	40–41	53–54	60–61	82–83
DEC	20–22	27–28	40–41	53–54	60–61	82–83

NEPTUNE CYCLE				PLUTO CYCLE		
m.o.b.	SQUARING PLUTO	WAXING SQUARE	WAXING TRINE	CONJ.NEPTUNE	WAXING SQUARE	WAXING TRINE
JAN	16–18	39–41	52–54	28–31	59–62	non-applicable
FEB	16–18	39–41	52–55	29–31	60–63	non-applicable
MAR	16–18	40–42	53–55	30–32	60–63	non-applicable
APR	16–18	40–42	53–55	30–33	60–63	non-applicable
MAY	15–17	40–42	54–55	30–33	60–63	non-applicable
JUN	15–17	40–41	54–55	30–33	59–62	non-applicable
JUL	15–17	40–41	53–54	30–33	58–61	non-applicable
AUG	14–17	39–41	52–54	29–32	57–61	non-applicable
SEP	14–16	38–40	52–54	28–31	57–61	non-applicable
OCT	14–17	38–40	51–54	28–31	57–61	non-applicable
NOV	15–17	38–40	51–53	28–31	58–62	non-applicable
DEC	15–17	38–40	51–54	28–31	59–62	non-applicable

2011

URANUS CYCLE 0–84 YEARS						
m.o.b.	WAXING SQUARE	WAXING TRINE	OPPOSITION	WANING TRINE	WANING SQUARE	RETURN
JAN	20–22	27–28	40–41	53–54	60–61	82–84
FEB	21–22	27–29	40–41	53–54	60–61	82–84
MAR	21–22	28–29	40–42	53–54	60–61	83–84
APR	21–23	28–29	41–42	54–55	60–62	83–84
MAY	22–23	28–29	41–42	54–55	61–62	83–84
JUN	22–23	28–30	41–42	54–55	61–62	83–84
JUL	21–22	28–29	41–43	54–55	61–62	83–84
AUG	21–22	28–29	41–42	54–55	61–62	83–84
SEP	21–22	28–29	41–42	54–55	60–62	83–84
OCT	20–22	27–28	40–41	53–54	60–62	83–84
NOV	20–22	27–28	40–41	53–54	60–61	82–83
DEC	20–21	27–28	39–41	53–54	60–61	82–83

NEPTUNE CYCLE				PLUTO CYCLE		
m.o.b.	SQUARING PLUTO	WAXING SQUARE	WAXING TRINE	CONJ.NEPTUNE	WAXING SQUARE	WAXING TRINE
JAN	16–18	39–41	52–54	28–31	59–62	non-applicable
FEB	16–18	39–41	52–55	29–31	61–63	non-applicable
MAR	16–18	39–41	53–55	30–32	61–63	non-applicable
APR	16–18	40–42	53–55	30–33	61–63	non-applicable
MAY	15–17	40–42	54–55	30–33	61–63	non-applicable
JUN	15–17	40–41	54–55	31–34	60–63	non-applicable
JUL	15–17	40–41	53–54	30–33	59–62	non-applicable
AUG	14–17	39–41	52–54	30–33	58–62	non-applicable
SEP	14–16	38–40	52–54	29–32	58–62	non-applicable
OCT	14–17	38–40	51–54	29–32	58–62	non-applicable
NOV	15–17	38–40	51–53	29–32	59–63	non-applicable
DEC	15–17	38–40	51–54	29–32	60–63	non-applicable

URANUS CYCLE 0–84 YEARS						
m.o.b.	WAXING SQUARE	WAXING TRINE	OPPOSITION	WANING TRINE	WANING SQUARE	RETURN
JAN	20–21	27–28	39–41	53–54	60–61	82–84
FEB	21–22	27–29	40–41	53–54	60–61	82–84
MAR	21–22	27–29	40–41	53–54	60–61	83–84
APR	21–23	28–29	41–42	54–55	60–62	83–84
MAY	22–23	28–29	41–42	54–55	61–62	83–84
JUN	22–23	28–29	41–42	54–55	61–62	83–84
JUL	21–22	28–29	41–43	54–55	61–62	83–84
AUG	21–22	28–29	41–42	54–55	61–62	83–84
SEP	21–22	27–28	41–42	54–55	60–62	83–84
OCT	20–22	27–28	40–41	53–54	60–62	83–84
NOV	20–22	27–28	40–41	53–54	60–61	82–83
DEC	20–21	27–28	39–40	53–54	60–61	82–83

NEPTUNE CYCLE				PLUTO CYCLE		
m.o.b.	SQUARING PLUTO	WAXING SQUARE	WAXING TRINE	CONJ.NEPTUNE	WAXING SQUARE	WAXING TRINE
JAN	16–18	39–41	52–54	29–32	60–63	non-applicable
FEB	16–18	39–41	52–55	30–33	61–63	non-applicable
MAR	16–17	39–41	53–55	31–33	62–64	non-applicable
APR	15–17	40–42	53–55	31–33	62–64	non-applicable
MAY	15–17	40–42	54–55	31–34	62–64	non-applicable
JUN	15–17	40–41	54–55	31–34	61–64	non-applicable
JUL	15–17	40–41	53–54	30–33	60–63	non-applicable
AUG	14–17	39–41	52–54	30–33	59–63	non-applicable
SEP	14–16	38–40	52–54	30–33	59–63	non-applicable
OCT	14–17	38–40	51–54	29–32	59–63	non-applicable
NOV	15–17	38–40	51–53	29–32	60–63	non-applicable
DEC	15–16	38–40	51–54	29–32	61–64	non-applicable

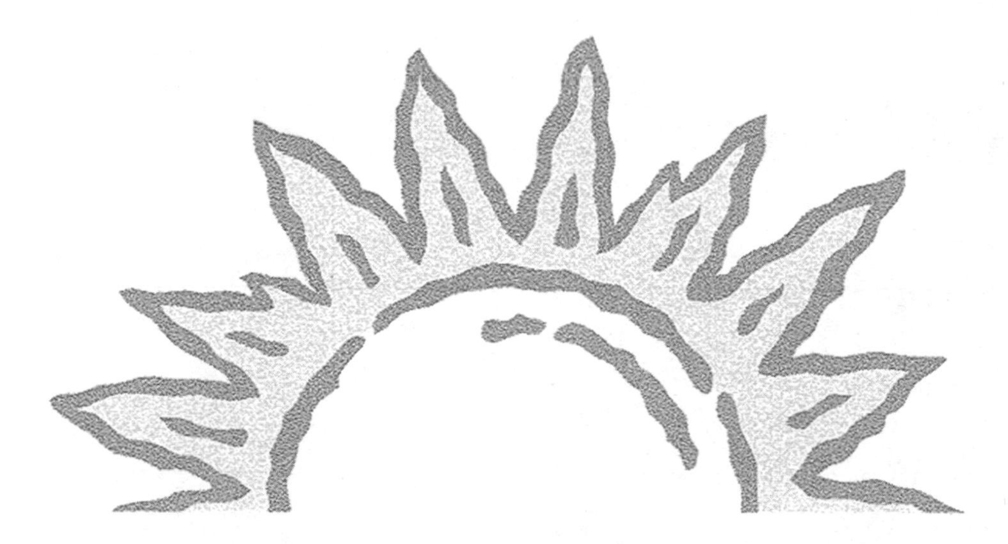

These Tables finish with the end of the Mayan Great Year, a 26,000 year cycle, current at the time of writing. The New Great Year begins at the Winter Solstice of 2012, when a Galactic Frequency Shift is predicted, heralding in the New Time of a New Humanity. This coincides with the Dawning of the Age of Aquarius (*see* page 355).

TIMES OF NEPTUNE OPPOSING PLUTO

FOR MOST BIRTHS DURING MOST OF THE 20th century and the beginning of the 21st, this Planetary influence occurs close to the Neptune Waxing Trine. For births from 1950 to 1982 it actually occurred during or before the Neptune Waxing Trine. In the table below, look in the shaded columns for the period wherein falls your year of birth. To the right of this in the next column you'll see a figure giving the number of years *after* the age that your Neptune Waxing Trine started (*see* tables on previous pages). In some cases there is a minus number of years, meaning that Neptune Opposing Pluto begins that many years *before* the start of your Neptune Waxing Trine; a zero figure indicates that the two influences occur simultaneously. The duration of Neptune Opposing Pluto is approximately as long as your Neptune Waxing Trine (two to three years).

Using the same example birth date as we used for the main tables, 3 February 1969, we find that this Cycle is active two years before the start of the subject's Neptune Waxing Trine. Referring back to the main tables, we can see that this occurs at 55 to 56 years of age, so their Neptune Opposing Pluto would occur at 53 to 55 years. Here is another example: I was born in September 1944, so my Neptune Waxing Trine occured between the ages of 54 and 56 (*see* page 317). Looking at the table below, we see that for 1944 births, Neptune Opposing Pluto started two years after the starting age of 54, at 56, just as my Neptune Waxing Trine was ending.

NEPTUNE OPPOSING PLUTO

Year of Birth	Years from Start of Neptune Waxing Trine	Year of Birth	Years from Start of Neptune Waxing Trine	Year of Birth	Years from Start of Neptune Waxing Trine
1900–1902	23	1919	13	1939–1942	3
1903	22	1920–1921	12	1943–1945	2
1904	21	1922–1923	11	1946–1949	1
1905–1906	20	1924	10	1950–1956	0
1907–1908	19	1925–1927	9	1957–1963	–1
1909	18	1928–1930	8	1964–1971	–2
1910–1912	17	1931–1932	7	1972–1978	–1
1913–1914	16	1933	6	1979–1982	0
1915	15	1934–1936	5	1983–1987	1
1916–1918	14	1937–1938	4	1988–2012	2

APPENDIX

GLOSSARY

THIS INCLUDES ASTROLOGICAL AND PSYCHOLOGICAL TERMS AND CONCEPTS, AS WELL AS SOME ORIGINAL ONES (DENOTED BY AN ASTERISK) THAT APPLY ESPECIALLY TO THE CONTEXT OF THIS BOOK. CROSS REFERENCES APPEAR IN *CAPITAL ITALICS*. REFERRING TO MY PREVIOUS BOOKS *DO IT YOURSELF ASTROLOGY* AND *DO IT YOURSELF RELATIONSHIP ASTROLOGY* (BOTH ELEMENT BOOKS) WILL INFORM YOU OF MORE ASTROLOGICAL AND PSYCHOLOGICAL CONCEPTS.

ACTIVE PLANETARY CYCLES* – *PLANETARY CYCLES* when they are influential in your life.

AGE INDEX* – A chart which shows you an overall view of all the *UNIVERSAL PLANETARY CYCLES* occurring through your life.

AGE OF AQUARIUS – The *ASTROLOGICAL AGE* we are just entering which is concerned with an awakening to the truth, especially regarding humanity and human nature, and which is necessarily disruptive. During this time we will increasingly free ourselves from the fears and illusions that have dogged humanity for so long (especially during the preceding Piscean Age), eventually giving rise to a peak in civilization.

ASCENDANT – The Eastern point or horizon on a *BIRTH CHART* where the *PLANETS* and *SIGNS* are seen to rise. The Sign on the Ascendant at any given time is called the Rising Sign.

ASPECTS – Certain angular relationships between one *PLANET* and another that indicate the quality of influence that the individual experiences during the *SIGNIFICANT POINTS* of a Planetary Cycle. (ASPECTS also occur between the *PLANETS* in an individual's *BIRTH CHART* and between one Birth Chart and another). The ASPECTS used in this book are: *CONJUNCTION* (0° between one Planet and another); *SQUARE* (90°); *TRINE* (120°); *OPPOSITION* (180°).

ASTROLOGICAL AGE – A 2160 year period of time which is governed by a certain *SIGN*, the quality of which determines the character of this evolutionary era. *See also AGE OF AQUARIUS.*

BIRTH CHART – A map of where the *PLANETS, SIGNS* and Houses are positioned at the time of birth.

COMPENSATION – What astrologer Liz Greene called the commonest human psychological trait, it describes how we respond to characteristics in ourselves or others by going to the opposite extreme. For example, one might feel shy or under pressure and compensate by being very bold and extrovert. This is explored more in the *JUPITER–SATURN BALANCE OF POWER.*

CONJOINING – An *ASPECT* where one Planet is in the same place in the sky or *ZODIAC* as another Planet. It intensifies any relevant issues at that time, calling for some new form of expression or point of departure. Also called the *CONJUNCTION*, and a *RETURN* when occurring during a *UNIVERSAL PLANETARY CYCLE*. For example, when Saturn CONJOINS itself after completing its 29.5 year orbit around the Sun, it is also said to be making its *RETURN.*

CONJUNCTION – *See CONJOINING.*

COURSE CORRECTIONS* – Critical *SIGNIFICANT POINTS* which, as *ASPECTS*, are *CONJOINING, SQUARE* or *OPPOSING*.

EGO – In the context of this book I have mainly used the word ego to mean that vain, childish, over-sensitive and self-important part of oneself that is inherently inept at responding to reality in an objective or balanced way. As such, it is a false or incomplete sense of oneself. However, it is important to state that ego, in the sense of having a strong feeling of one's self-significance, is essential – especially when growing up or when confronted by someone with an ego of the negative variety here described.

ESOTERIC – Teachings and information about life that are reserved for people who have reached a certain level of awareness. The word literally means 'inside the temple'. The esoteric word is that more people than ever have now reached that 'certain level of awareness'.
GOOD RUNS* – *SIGNIFICANT POINTS* that are harmonious and supportive. As *ASPECTS*, these are usually *TRINE* influences, but sometimes can be *CONJOINING*.
GO-ZONE* – Fruitful area of appreciation and understanding during a *JUPITER YEAR*.
GROWTH MODE* – The nature and quality of growth and development that can be taken advantage of during any particular *JUPITER YEAR*.
INDIVIDUAL PLANETARY CYCLE – A *PLANETARY CYCLE* that is seen in terms of its relationship to other *PLANETARY CYCLES*. For example, Saturn SQUARING Moon would be a *SIGNIFICANT POINT* in an INDIVIDUAL PLANETARY CYCLE, but these are not covered by this book because they take complex calculation for each individual.
JUPITER–SATURN BALANCE OF POWER* – The critical equilibrium between expansion and contraction, faith and reality.
JUPITER YEAR – The particular *SIGN* qualities of any year of your life that encourage growth and development.
MEMO FROM MERCURY* – Passages in the book which have the sole purpose of guiding you through it, and keeping you mindful of important points. They are signified by a watermark of a caduceus.
MOON PHASES – Eight three-year-and-five-month periods which go to make up one whole Moon Cycle of 27⅓ years. They reveal your emotional bias and inclinations during those times. 'PHASES' is put in inverted commas because they are not the same as the actual Moon Phases we see during the lunar month, but are symbolic of them.
OPPOSING – An *ASPECT* where one *PLANET* is in the opposite place in the sky or *ZODIAC* to another *PLANET*. It grants or forces awareness concerning any relevant matter in your life at that time. Also called the *OPPOSITION*.
OPPOSITION – *See OPPOSING*.
OUTER PLANETARY CYCLES – *PLANETARY CYCLES* of Uranus, Neptune and Pluto.
PATH OF RETURN – *ESOTERIC* concept that sees human life as having three critical stages: Outward Path; Point of Integration and Return; Path of Return (*see* Your Uranus Cycle on page 149).
PERSONAL ASTRO-LIFE PLAN* – A personalised version of the *AGE INDEX*.
PERSONAL UNCONSCIOUS – *See THE UNCONSCIOUS*.
PHASE – *See MOON 'PHASES'*.
PLANETARY CYCLE – The period of time taken by any *PLANET* to go around the Sun; the interaction it has with other PLANETARY CYCLES, which are called *SIGNIFICANT POINTS*. There are two distinct types: *UNIVERSAL PLANETARY CYCLES* and *INDIVIDUAL PLANETARY CYCLES*.
PLANETARY MANAGEMENT* – Recommended means of responding to and dealing with various planetary influences (*see* Planetary Effects, Figures and Responses on page 259).
PLANETS – The Sun, Moon and all the PLANETS of our Solar System which symbolize, along with the *SIGNS*, the various energies or influences operating within and upon a *BIRTH CHART*.
PROGRESSIONS – Symbolic positions of the *PLANETS* and their relationship to the *PLANETS* in one's *BIRTH CHART*; the ongoing influences of your Moon Cycle and some of your *INDIVIDUAL PLANETARY CYCLES*.
PROJECTION – The psychological phenomenon of seeing in other people and the world at large what is actually or also a trait of one's own. This is done because that trait is regarded as either too bad or too good to belong to oneself. This is an unconscious function – until of course you become aware of it. This

book is, to quite a degree, aimed at making you aware of it, simply because *your* Planetary influences are happening *to you*, and therefore whatever happens is ultimately entirely of your own making or doing, consciously or unconsciously.

RETURN – *See CONJOINING.*

SIGN – One of the SIGNS of the *ZODIAC*. SIGNS lend certain qualities to the *PLANETS* operating through them at any given time.

SIGNIFICANT POINTS* – Particular periods during *PLANETARY CYCLES* and their influence upon your life at those times. There are two types: *COURSE CORRECTIONS* and *GOOD RUNS.*

STAGES OF YOUR LIFE* – The major *ACTIVE PLANETARY CYCLES* of your life.

STORY OF YOUR LIFE* – A conceptual version of an individual human life, with the *PLANETS* symbolizing its various parts and stages.

SQUARE – An *ASPECT* or *COURSE CORRECTION*, which is challenging in some way, and demands action, decision and resolution. There are two types: *WAXING TRINE* and *WANING TRINE*. Sometimes expressed as *SQUARING*, as in one *PLANET* Squaring another.

SQUARING – *See SQUARE.*

STRETCH FACTOR* – How the influence of any *TRANSIT* can last longer than has been technically determined, that is, given in the *TABLES*.

TABLES – Listings starting on page 289 that give the ages in your life when the *OUTER PLANETARY CYCLES* are affecting you.

TRANSITS – The *PLANETS* in the sky and their relationship to the *PLANETS* in one's *BIRTH CHART*; the ongoing influences of both *UNIVERSAL PLANETARY CYCLES* and *INDIVIDUAL PLANETARY CYCLES*.

THE UNCONSCIOUS – If human beings are seen as pieces in some gigantic cosmic game, then the Unconscious is what is actually playing that game. Your *PERSONAL UNCONSCIOUS* is your particular connection with the Unconscious, and therefore partly determines, along with your conscious will and senses, your fate. Astrology is a map and mapper of the Unconscious, which is why it is able to predict or provide an insight into what is determining your fate in terms of *TRANSITS* and *PROGRESSIONS*. One can also view the Unconscious as a great sea upon which the little boat of your personality (a mixture of consciousness and unconsciousness) is floating or travelling, as it negotiates and navigates the currents of *PLANETARY CYCLES*.

TRINE – A harmonious and supportive *ASPECT* or *GOOD RUN*. There are two types: *WAXING TRINE* and *WANING TRINE*. Sometimes expressed as *TRINING* as in one *PLANET* Trining another.

TRINING – *See TRINE.*

UNCONSCIOUS – Describing anything that lies in *THE UNCONSCIOUS.*

UNIVERSAL PLANETARY CYCLES* – Technically called *Generic* Planetary Cycles, a *PLANETARY CYCLE* that is seen purely in relation to the *PLANET* concerned, as distinct from *INDIVIDUAL PLANETARY CYCLES*. For example, Saturn *OPPOSING* Saturn is a *SIGNIFICANT POINT* occurring during the Universal Cycle of Saturn. UNIVERSAL PLANETARY CYCLES are the main subject matter of this book.

WANING SQUARE – The *SQUARE* that occurs *after* the *OPPOSING ASPECT* has happened, and is usually more conscious and introspective in effect.

WANING TRINE – The *TRINE* that occurs *after* the *OPPOSING ASPECT* has happened, and is usually more conscious and introspective in effect.

WAXING SQUARE – The *SQUARE* that occurs *before* the *OPPOSING ASPECT* happens, and is usually more spontaneous and proactive in effect.

WAXING TRINE – The *TRINE* that occurs *before* the *OPPOSING ASPECT* happens, and is usually more spontaneous and proactive in effect.

ZODIAC – An imaginary band that encircles the Earth which acts as (1) a grid upon which to fix the position of any *PLANET* or point in the sky, and as (2) a set of 12 influences, the *SIGNS*. The ZODIAC should not be confused with the Constellations of Signs, which are the actual stars in the skies, which once upon a time did coincide with this, the ZODIAC of Signs.
ZODIACAL LIFE-STREAM* – A graphic concept that illustrates the essential meaning of your Sun Sign, that is, your life.

Birth Data of Famous People

NAME	DATE	TIME*	PLACE	CITED ON PAGES
Adolf Hitler	23 April 1889	18.00	Braunau, Austria	55,122,152,176,179
Al Capone	17 January 1899		Brooklyn, NYC	119,201
Andy Warhol	29 October 1930	18.00	Pittsburgh, Pennsylvania	55
Augusto Pinochet	25 November 1915		Santiago, Chile	179
Benito Mussolini	29 July 1883		Predappio, Italy	55,66
Bob Dylan	24 May 1942	21.05	Duluth, Minnesota	223
Boris Yeltsin	1 February 1931	16.45	Yekaterinburg, Russia	55,117,126
Brian Jones	28 February 1942		Cheltenham, England	52
Brigitte Bardot	28 September 1934		Paris, France	114,228
Elizabeth Taylor	27 February 1932	19.56	London, England	53,70,117,223
Elvis Presley	8 January 1935	04.35	Tupelo, Mississipi	193,201
Emmeline Pankhurst	14 July 1858	21.30	Manchester, England	79
Fidel Castro	13 August 1927		Biran, Cuba	59,64,120,222
Glenn Hoddle	27 October 1957		Hayes, England	102,169–70
Grace Kelly	12 November 1929		Philadelphia, Pennsylvania	114
Greta Garbo	18 September 1905	19.00	Stockholm, Sweden	114,163,180,223
Howard Hughes	24 December 1905		Houston, Texas	114,117,119,166,226
Humphrey Bogart	25 December 1899		New York City	55
Indira Gandhi	19 November 1917	23.11	Allahabad, India	126
Jackie Onassis	28 July 1929		Southampton, NY	114
Janis Joplin	19 January 1943	09.45	Port Arthur, Texas	52
Jim Morrison	8 December 1943	11.55	Melbourne, Florida	52
Jimi Hendrix	27 November 1942	10.15	Seattle, Washington	52
John Kennedy	29 May 1917	15.00	Brookline, Massachusetts	122,171
John Lennon	9 October 1940	18.30	Liverpool, England	52,117,171,173,222,280
Madonna	16 August 1958	(19.00)	Rochester, Michigan	200,223
Mahatma Gandhi	2 October 1869	07.12	Porbandar, India	114
Malcolm X	19 May 1925		Lansing, Michigan	199,222
Mao Tse-tung	26 December 1893		Shaoshan, China	178–9,199
Margaret Thatcher	13 October 1925	9.00	Grantham, England	55,73,114,129
Marie Curie	7 November 1867	13.30	Warsaw, Poland	117,120
Marilyn Monroe	1 June 1926	9.30	Los Angeles, California	53,65,119,120,223
Marlon Brando	3 April 1924	23.00	Omaha, Nebraska	117,129,163,223
Martin Luther King	15 January 1929	(12.00)	Atlanta, Georgia	52,120,171,173,222
Mikhail Gorbachev	2 March 1931	7.33	Privolnoye, Russia	55,114,126,178
Muhammed Ali	17 January 1942	(17.00)	Louisville, Kentucky	59,94,117,163,166,193,222,227
Nelson Mandela	18 July 1918		Umtata, South Africa	52,78,79,90,123,171
Paula Yates	24 April 1959		Colwyn Bay, Wales	173
Peter Sellers	8 September 1925	05.00	Portsmouth, England	54
Princess Diana	1 July 1961	19.45	Sandringham, England	120,154,163,228
Queen Elizabeth II	21 April 1926	01.40	London, England	53,73,114,163,223
Richard Nixon	9 January 1913	21.44	Yorba Linda, California	55,66,176,178,199
Rod Stewart	10 January 1945	02.00	London, England	95
Robert Kennedy	20 November 1925		Brookline, Massachusetts	171
Sigmund Freud	6 May 1856	18.30	Freiburg, Moravia	179,208
Yasser Arafat	24 August 1929		Jerusalem, Israel	52,126,222
Yuri Gagarin	9 March 1934		Gagarin, Russia	52,114,163,166,223

* Birth time is given only when known, and in parentheses when unconfirmed or approximated.
NOTE: If an individual is included in the book but not in the above list, it is because only their age is known and not their actual birth details.

CHART BLANKS

Personal Astro-Life Plan of

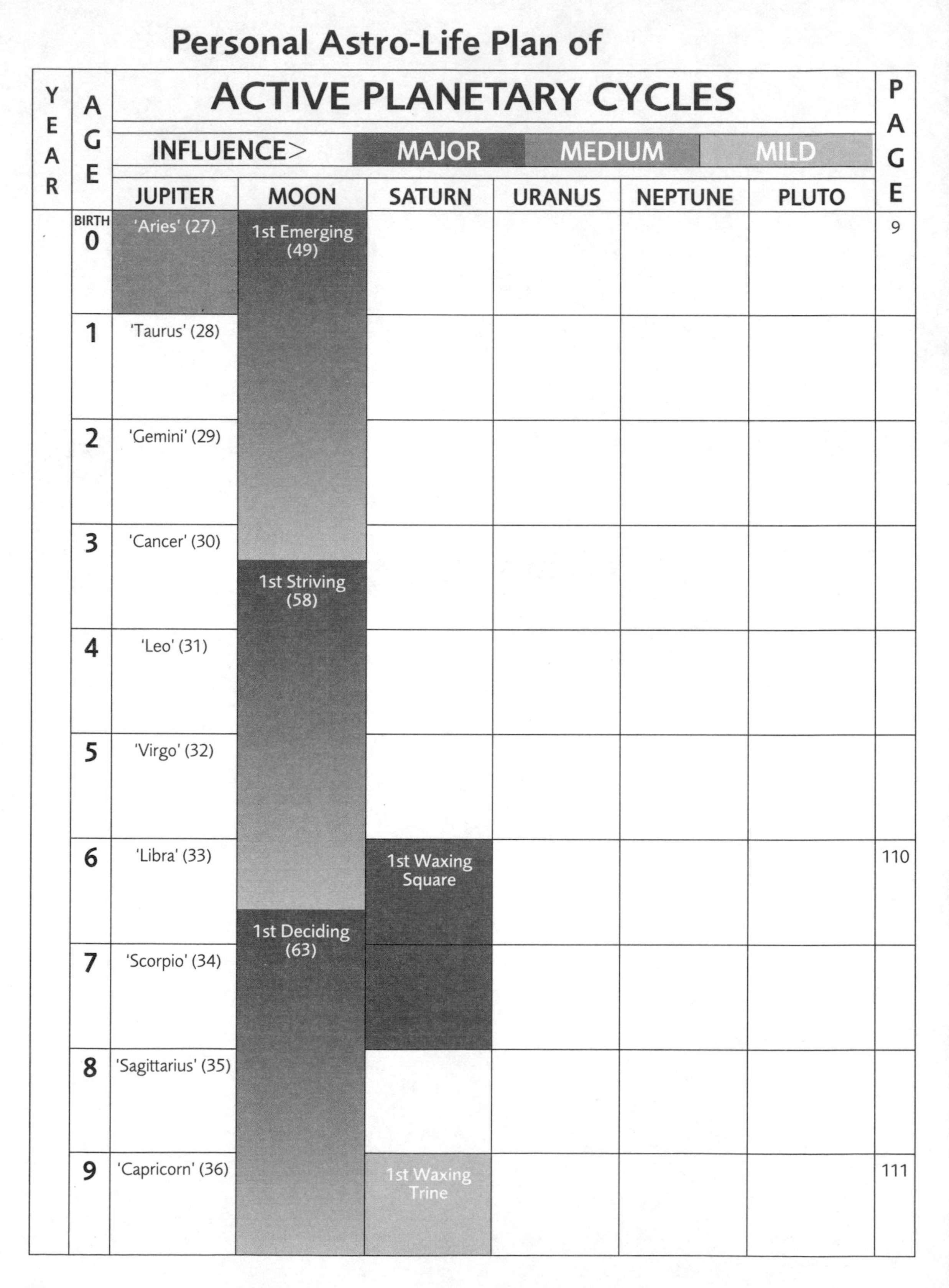

YEAR	AGE	JUPITER	MOON	SATURN	URANUS	NEPTUNE	PLUTO	PAGE
	BIRTH 0	'Aries' (27)	1st Emerging (49)					9
	1	'Taurus' (28)						
	2	'Gemini' (29)						
	3	'Cancer' (30)	1st Striving (58)					
	4	'Leo' (31)						
	5	'Virgo' (32)						
	6	'Libra' (33)	1st Deciding (63)	1st Waxing Square				110
	7	'Scorpio' (34)						
	8	'Sagittarius' (35)						
	9	'Capricorn' (36)		1st Waxing Trine				111

ACTIVE PLANETARY CYCLES

INFLUENCE> MAJOR MEDIUM MILD

ACTIVE PLANETARY CYCLES

INFLUENCE> MAJOR | MEDIUM | MILD

YEAR	AGE	JUPITER	MOON	SATURN	URANUS	NEPTUNE	PLUTO	PAGE
	10	'Aquarius' (37)	1st Adjusting (69)	1st Waxing Trine				111
	11	'Pisces' (38)						
	12	'Aries' (27)						
	13	'Taurus' (28)	1st Realizing (75)					
	14	'Gemini' (29)		1st Opposition				111
	15	'Cancer' (30)						
	16	'Leo' (31)						
	17	'Virgo' (32)	1st Sharing (81)					
	18	'Libra' (33)						
	19	'Scorpio' (34)		1st Waning Trine				113

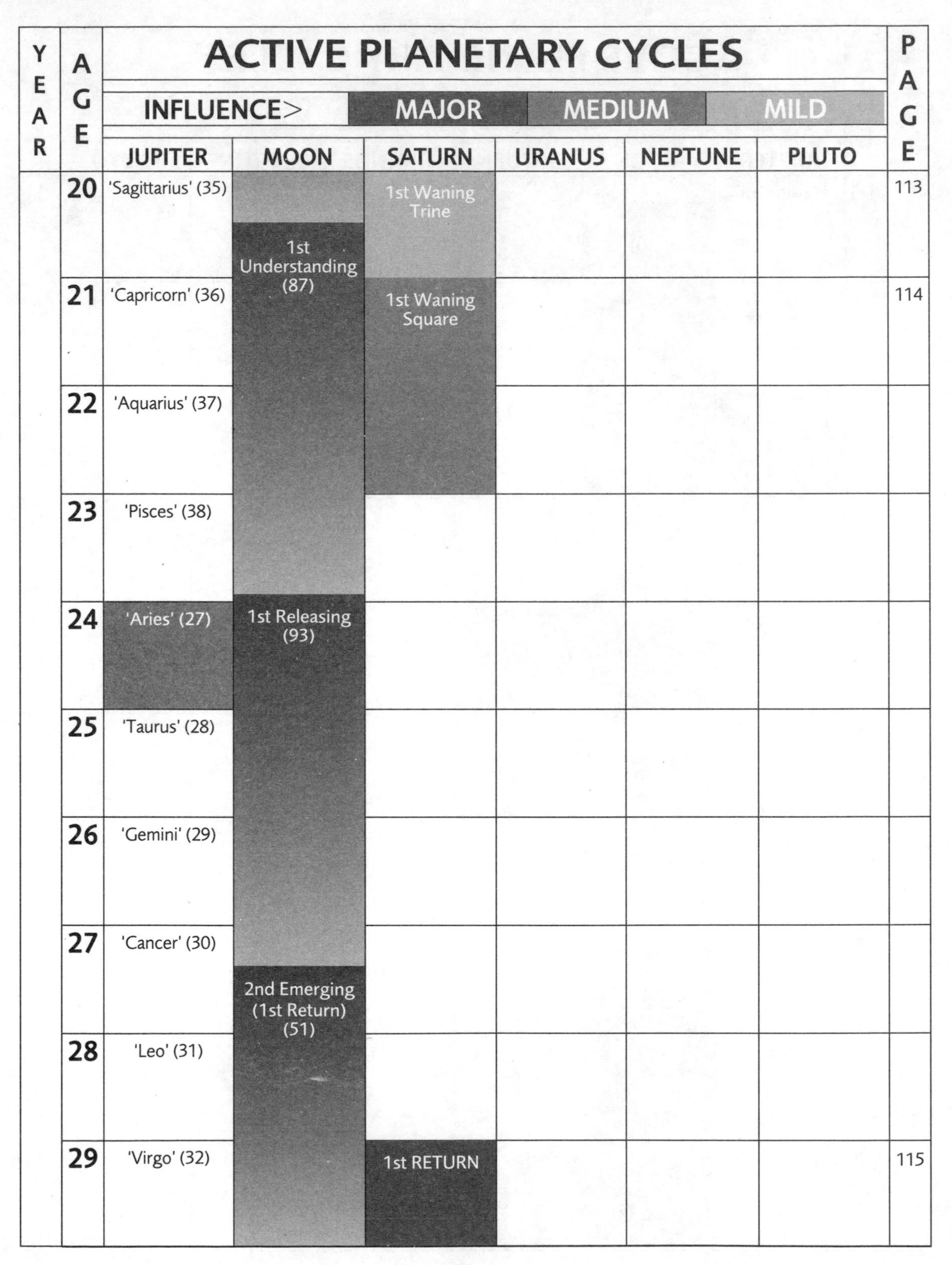

ACTIVE PLANETARY CYCLES

INFLUENCE> MAJOR | MEDIUM | MILD

YEAR	AGE	JUPITER	MOON	SATURN	URANUS	NEPTUNE	PLUTO	PAGE
	20	'Sagittarius' (35)	1st Understanding (87)	1st Waning Trine				113
	21	'Capricorn' (36)		1st Waning Square				114
	22	'Aquarius' (37)						
	23	'Pisces' (38)						
	24	'Aries' (27)	1st Releasing (93)					
	25	'Taurus' (28)						
	26	'Gemini' (29)						
	27	'Cancer' (30)	2nd Emerging (1st Return) (51)					
	28	'Leo' (31)						
	29	'Virgo' (32)		1st RETURN				115

ACTIVE PLANETARY CYCLES

INFLUENCE> MAJOR | MEDIUM | MILD

YEAR	AGE	JUPITER	MOON	SATURN	URANUS	NEPTUNE	PLUTO	PAGE
	30	'Libra' (33)		1st RETURN				115
	31	'Scorpio' (34)	2nd Striving (59)					
	32	'Sagittarius' (35)						
	33	'Capricorn' (36)						
	34	'Aquarius' (37)	2nd Deciding (64)					
	35	'Pisces' (38)						
	36	'Aries' (27)		2nd Waxing Square				119
	37	'Taurus' (28)						
	38	'Gemini' (29)	2nd Adjusting (71)	2nd Waxing Trine				121
	39	'Cancer' (30)						

ACTIVE PLANETARY CYCLES

INFLUENCE> MAJOR | MEDIUM | MILD

YEAR	AGE	JUPITER	MOON	SATURN	URANUS	NEPTUNE	PLUTO	PAGE
	40	'Leo' (31)						
	41	'Virgo' (32)	2nd Realizing (77)					
	42	'Libra' (33)						
	43	'Scorpio' (34)		2nd Opposition				122
	44	'Sagittarius' (35)	2nd Sharing (83)					
	45	'Capricorn' (36)						
	46	'Aquarius' (37)						
	47	'Pisces' (38)						
	48	'Aries' (27)	2nd Understanding (89)	2nd Waning Trine				124
	49	'Taurus' (28)						

ACTIVE PLANETARY CYCLES

INFLUENCE>	MAJOR	MEDIUM	MILD

YEAR	AGE	JUPITER	MOON	SATURN	URANUS	NEPTUNE	PLUTO	PAGE
	50	'Gemini' (29)						
	51	'Cancer' (30)	2nd Releasing (95)	2nd Waning Square				125
	52	'Leo' (31)						
	53	'Virgo' (32)						
	54	'Libra' (33)	3rd Emerging (2nd Return) (54)					
	55	'Scorpio' (34)						
	56	'Sagittarius' (35)						
	57	'Capricorn' (36)						
	58	'Aquarius' (37)	3rd Striving (60)	2nd RETURN				126
	59	'Pisces' (38)						

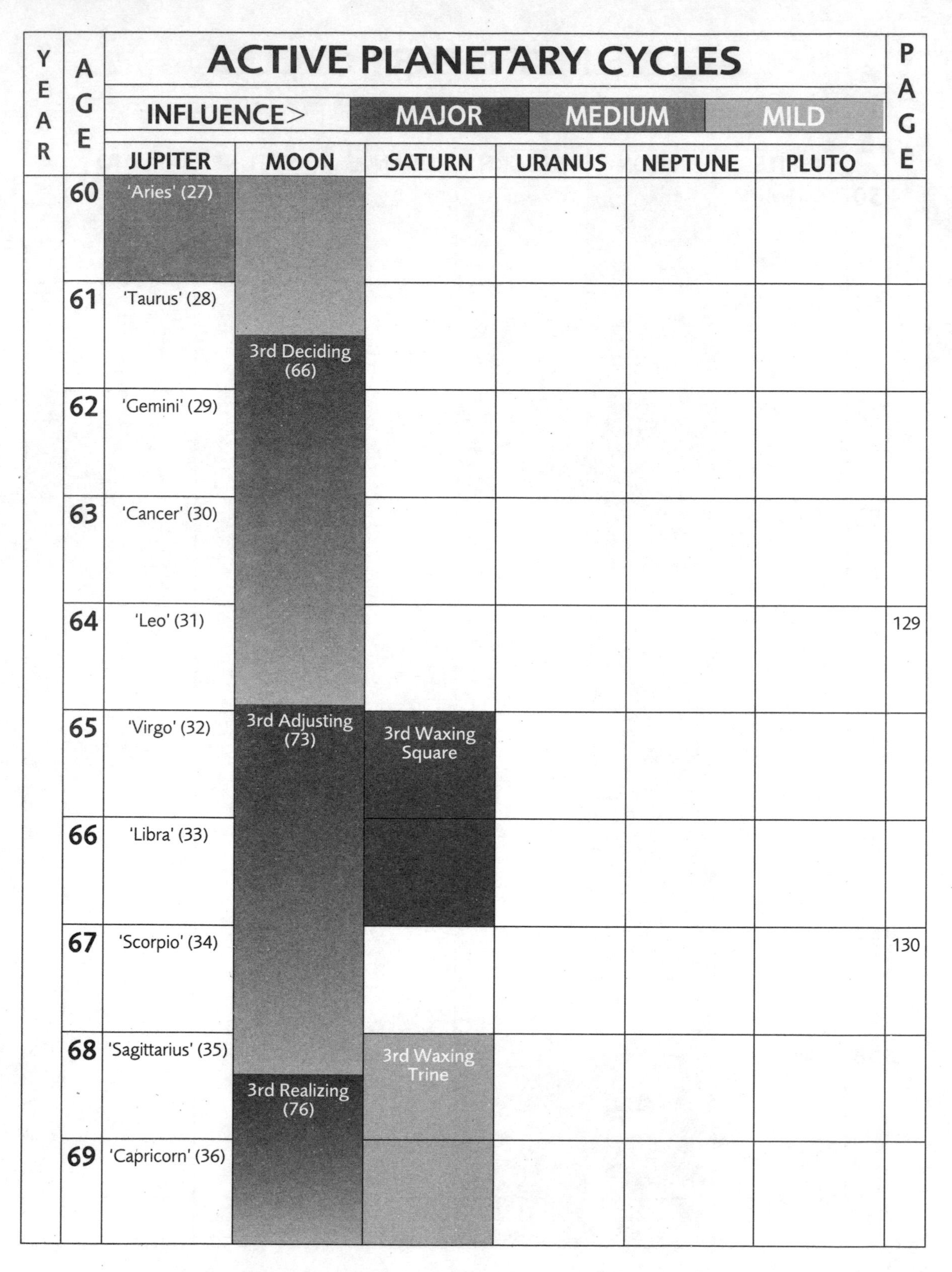

YEAR	AGE	JUPITER	MOON	SATURN	URANUS	NEPTUNE	PLUTO	PAGE
		ACTIVE PLANETARY CYCLES						
		INFLUENCE>	MAJOR	MEDIUM	MILD			
	60	'Aries' (27)						
	61	'Taurus' (28)	3rd Deciding (66)					
	62	'Gemini' (29)						
	63	'Cancer' (30)						
	64	'Leo' (31)						129
	65	'Virgo' (32)	3rd Adjusting (73)	3rd Waxing Square				
	66	'Libra' (33)						
	67	'Scorpio' (34)						130
	68	'Sagittarius' (35)	3rd Realizing (76)	3rd Waxing Trine				
	69	'Capricorn' (36)						

ACTIVE PLANETARY CYCLES

INFLUENCE> | MAJOR | MEDIUM | MILD

YEAR	AGE	JUPITER	MOON	SATURN	URANUS	NEPTUNE	PLUTO	PAGE
	70	'Aquarius' (37)						
	71	'Pisces' (38)	3rd Sharing (84)					
	72	'Aries' (27)						
	73	'Taurus' (28)		3rd Opposition				130
	74	'Gemini' (29)						
	75	'Cancer' (30)	3rd Understanding (90)					
	76	'Leo' (31)						
	77	'Virgo' (32)						
	78	'Libra' (33)	3rd Releasing (97)	3rd Waning Trine				131
	79	'Scorpio' (34)						

YEAR	AGE	JUPITER	MOON	SATURN	URANUS	NEPTUNE	PLUTO	PAGE
		ACTIVE PLANETARY CYCLES						
		INFLUENCE>	MAJOR	MEDIUM	MILD			
	80	'Sagittarius' (35)		3rd Waning Square				131
	81	'Capricorn' (36)						
	82	'Aquarius' (37)	4th Emerging (3rd Return) (56)					
	83	'Pisces' (38)			RETURN			179
	84	'Aries' (27)						

SATURN

Effects	Figures	Positive Responses
Bones, teeth, skin, hair, knees	Ageing	Be patient and unassuming
Consolidating matters	Authority (figure)	Be sober and serious-minded
Constriction/dissatisfaction	Bitter person	Do whatever your duty is
Delay/lateness/slowness	Builder/construction worker	Exercise discipline/your body
Deprivation/distress	Cold, unfeeling person	Face reality/do what you must
Doubt/inadequacy	Controlling person	Focus attention/be objective
Drudgery/hardness	Definers of boundaries	Get organized/be economical
Failure/dysfunction	Depressed/ive person	Look and learn and prepare/study
Fear/anxiety	Disciplinarian	Lose weight/tighten your belt
Feeling your age	Elderly person	Make more effort/measure up
Heaviness/oppression	Important person/relationship	Seek professional advice
Increased responsibilities	Material predicament/issues	Use step-by-step procedure
Increased/onerous pressure	Mean/dry person	Utilize your time/be punctual
Limitation/imposing boundaries	Official	**Negative Responses**
Material/financial difficulties	Older person	
Order (enforced)	Patriarch	Acting immaturely/prematurely
Physical condition/deterioration	Posture (bad)	Arrogance/defiance
Physical instability/weakness	Professional person/relationship	Being too cautious
Reality taking precedence over feelings	Restrictive person/relationship	Being too conventional
	Rigid attitude	Depression/pessimism
Separation/differences	Saturnine person	Excessiveness/imprudence
Slowing things down	Serious relationship	Fatigue/feebleness
Stress	Teacher	Feeling compromised/hurried
Structural considerations	The Devil	Feeling cornered/thwarted
Test/examination	The world at large	Feeling inadequate/hopeless
Time and timing	Wet blanket	Health abuse

Saturn Management: Statement/Solution*

When SATURN is creating an EFFECT or ISSUE of	**Box A**
Imposed/brought on by	**Box B**
then POSITIVE RESPONSE(S) towards this could be to	**Box C**
while NEGATIVE RESPONSE(S) would be	**Box D**

* For further Saturn Management, *see* 'The Jupiter–Saturn Balance of Power' on p265.

URANUS

Effects

Accelerated pace
Accident/shock
Alienation
Alternative/esoteric subjects
Birth
Change/turnaround
Circulation
Coincidence
Disruption/upset
Divorce/relationship crisis
Electricity
Freedom/rights
Future development
Modernization
Principles
Reforming matters
Remoteness/detachedness
Revolutionizing matters
Signs (omens)
Spasm/hyperactivity
Surgery/emergency
Strike/unrest
Technological matters
Truth (moment of)
Unprecedented situation
Unpredictability/surprise
Unusualness/unconventionality

Figures

Activist
Astrology/astrologer
Eccentric
Electrician
Engineer
Flyer
Fool
Genius
Homosexual
Internet/computer
Intuitive type
Leader/catalyst
Loose cannon
Machine
Media (person)
Oddball
Pathfinder
Psychologist
Radical/interloper
Rebel
Reckless person
Researcher
Robot
Scientist/technician
Socialist/communist
Specialist
Trade union(ist)
Unusual type

Positive Responses

Be cool, objective, to the point
Be intuitive and stick to the truth
Be open and non-judgemental
Befriend the situation/person
Determine what situation means
Make necessary changes only
Move with the times
Obtain sufficient information
Retreat, live to fight another day
Streamline/simplify
Bear with them kindly
Take a calculated risk
Take with a pinch of salt

Negative Responses

Acting out of fear or anxiety
Acting or behaving unnaturally
Going along with the crowd
Hedging your bets
Loosing sight of feelings
Losing your cool/Panicking
Not being true to yourself
Not questioning things
Pretending not to care
Rationalizing to miss the point
Remaining uninformed
Resisting any change at all
Violent rebellion or reaction

Uranus Management: Statement/Solution

When URANUS is creating an EFFECT or ISSUE of	Box A
Imposed/brought on by	Box B
then POSITIVE RESPONSE(S) towards this could be to	Box C
while NEGATIVE RESPONSE(S) would be	Box D

NEPTUNE

Effects	Figures	Positive Responses
Absentmindedness	Addict/alcoholic	Accept your fate
Addiction/alcoholism	Chemical industry	Be gentle on yourself; then other
Anaesthesia/fainting/trance	Chronically sick (person)	Check out/deal with victimhood
Astral phenomena/ghosts	Confidence trickster	Be compassionate but tough
Charismatic appeal/glamour	Doctor/quack	Consider karmic reasons
Chemicals/oils/gas	Drink, drug or tobacco (industry)	Get the facts/check credentials
Confusion/strangeness	Drug dealer	Let things be/live and let live
Conscience	Emotional type	Look deep within for the answer
Deception/suspicion/lies	Guru/visionary (suspect/false)	Practise yoga/deep breathing
Delusion/disillusion-ing/-ment	Haunted/spooky places	Research their background
Enigma/vagueness/mystery	Hypnotist (suspect)	Seek to understand it/them
Frustration	Illusionist	Trust good track record only
Hallucination/hysteria/insanity	Imagination	Try to see the bigger picture
Healing (bogus)	Institution	**Negative Responses**
Hospitalization	Medium/clairvoyant/psychic	Being afraid to face the music
Hygiene (obsessive)	Music(ian)/dance(r)/artist	Disregarding spiritual aspect
Illogicality	Neurotic person	Going in blindfold/half-cocked
Infatuation/gullibility	Pharmaceutical industry	Imagining it will go away
Infection/poisoning	Poisonous creature/person	Mistaking (your) weakness for compassion
Lack or loss of energy	Seducible/wanton person	Not being emotionally honest
Metaphysics/spiritualism	Seductive/teasing person	Not getting it in writing
Pining/longing	Unreliable person	Not questioning matters at all
Scandal	Unstable person	Not thinking things through
Temptation/weakness	Victim/pathetic, helpless person	Thinking/hoping for the best
Vulnerability/oversensitivity	Waterworks/plumber	
Water (= emotional) problems	Witch/wizard	

Neptune Management: Statement/Solution

When NEPTUNE is creating an EFFECT or ISSUE of	Box A
Imposed/brought on by	Box B
then POSITIVE RESPONSE(S) towards this could be to	Box C
while NEGATIVE RESPONSE(S) would be	Box D

PLUTO

Effects

- Abuse/rape
- Compulsive feelings
- Constipation/toxicity
- Corruption/underhandedness
- Criminal activity/behaviour
- Decadence/degradation
- Disempowerment
- End/death of something/one
- Evil/horror/nightmare
- Feeling convinced but unhappy
- Feeling very alone
- Guilt/remorse
- Inaccessibility
- Intense uncomfortable feelings
- Intrigue/undercurrents
- Manipulation/coercion
- Matters relating to death
- Occult (trouble with the)
- Powerlessness
- Psychological/emotional pain
- Sexual disease/lust
- Smokescreens/secretiveness
- Stalking/spying
- The other side
- Unsatisfied desire/loss
- Violence/disaster

Figures

- Criminal/hoodlum/mugger
- Cruel/treacherous person
- Dark person/black magician
- Destructive person/thing
- Fanatical person/thing
- Forces (armed or otherwise)
- Hard to reach person/thing
- Insurance agent
- Intimate/blackmailer
- Jealous/desperate person
- Kidnapper/cuckolder
- Monster/devil/immoral person
- Natural processes of decay
- Nihilistic person
- Obsessive/intense person
- Overpowering person/thing
- Pest/nuisance/menace
- Poisonous creature
- Police officer
- Pornography/pervert
- Powerful person/thing
- Prostitute
- Secretive person/thing
- Taboo breaker/keeper
- Trash/revolting person/thing
- Tyrant/dictatorial person

Positive Responses

- Accept/own your own dark side
- Be strong and admit weaknesses
- Confess your own wrongdoing
- Delve into yourself, non-judging
- Discover and surrender to the Light
- Discover the root cause
- Forgive yourself and/or other
- Hand it over to Higher Power
- Psychically protect yourself (*see* page 281)
- Pull out and cut your losses
- Purify/detoxify your mind/body
- Raise feelings higher up body
- Seek psychological/spiritual help

Negative Responses

- Being controlled by something
- Being hateful and unforgiving
- Denying own part in it/blaming
- Getting in deeper/too deep
- Giving in to base desires/feelings
- Going it alone (without the Light)
- Guiltily hiding yourself or something/someone you don't respect
- Thinking two wrongs make a right
- Thinking worst of yourself/others
- Torturing/tormenting self/other

Pluto Management: Statement/Solution

When PLUTO is creating an EFFECT or ISSUE of	Box A
Imposed/brought on by	Box B
then POSITIVE RESPONSE(S) towards this could be to	Box C
while NEGATIVE RESPONSE(S) would be	Box D

Further Reading

Anthony, Carol K, *A Guide to the I Ching,* Anthony Press, Stowe, Mass, 1988

Arroyo, Stephen, *Astrology, Karma and Transformation,* CRCS Publications, California, 1978

Baker, Dr Douglas, [all books]

Baynes, C F, Richard Wilhelm and Vary F Baynes, *The I Ching,* Princeton University Press, Princeton, 1967

Birkbeck, Lyn, *Do It Yourself Astrology,* Element Books, Shaftesbury, 1996

Birkbeck, Lyn, *Do It Yourself Relationship Astrology,* Element Books, Shaftesbury, 1999

Bloom, William, *Psychic Protection,* Piatkus, London, 1997

Castenada, Carlos, [all books]

Causton, Richard, *Nichiren Shoshu Buddhism,* Rider, London, 1998

Clow, Barbara Hand, *The Liquid Light of Sex,* Bear & Co, Santa Fe, NM, 1996

Goldsmith, Joel, *The Art of Meditation,* Harper, SanFransisco, 1990

Graves, Robert (intro), *The New Larousse Encyclopaedia of Mythology,* Hamlyn, London, 1959

Greene, Liz, *Relating: An Astrological Guide to Living With Others on a Small Planet,* Coventure, London, 1978

— *Saturn: A New Look at an Old Devil,* Weiser, New York, 1976

Hay, Louise L, *Heal Your Body,* Hay House, Carlsbad, CA, 1994

Icke, David, *The Robot's Rebellion: The Story of the Spiritual Renaissance,* Gateway Books, Bath, 1994

Karcher, Stephen, *How to Use the I Ching,* Element Books, Shaftesbury, 1997

Playfair, Guy Lyon, and Scot Hill, *The Cycles of Heaven,* Pan, London, 1979

Ruperti, Alexander, *Cycles of Becoming,* CRCS Publications, Davis, California, 1978

Seymour, Percy, *The Scientific Basis of Astrology,* Foulsham, Slough, 1997

Sky, Michael, *Breathing,* Bear & Co, Santa Fe, NM, 1997

RESOURCES

If you wish to take your interest in astrology further, I recommend that you contact:

AUSTRALIA
Federation of Australian Astrologers
Lynda Hill
20 Harley Road
Avalon NSW 2107
Tel: + 61-2-918-9539

CANADA
Association Canadiennes des Astrologues Francophones
Denise Chrzanwska
CP 1715
Succ 'B'
Montreal HSB 3LB
Tel: + 1-514-831-4153
Fax: + 1-514-521-1502

Astrolinguistics Institute
Anne Black
2182 Cubbon Drive
VICTORIA
British Columbia V8R 1R5
Tel: + 1-604-370-1874
Fax: + 1-604-370-1891
e-mail: ablack@islandnet.com

NEW ZEALAND
Astrological Society of New Zealand
Joy Dowler
5266 Wellesley Street
AUCKLAND 1003

Astrological Foundation Inc.
Hamish Saunders
41 New North Road
Eden Terrace
AUCKLAND 1003
Tel/Fax: +64-9-373-5304

SOUTH AFRICA
Astrological Society of South Africa
Cynthia Thorburn
PO BOX 2968
RIVONIA 2128
Tel +27-11-864-1436

UNITED KINGDOM
The Astrological Association
396 Caledonian Road
London N1 1DN
+ 171-700-6479

USA
American Federation of Astrologers
Robert Cooper
PO Box 22040
Tempe
AZ 85285-2040
Tel: 602-838-1751
Fax: 602-838-8293

Association for Astrological Networking
8306 Wilshire Blvd
Suite 537
Beverley Hills
CA 90211

Any other queries or correspondence for Lyn Birkbeck to be sent (enclose SAE if from UK) to:

Lyn Birkbeck (ALP)
c/o Element Books Ltd, The Old Schoolhouse, The Courtyard,
Bell Street, SHAFTESBURY, Dorset SP7 8BP, England

or e-mail Lyn at: lynbirkbeck@talk21.com
or at: www.msp–online.com